Nanotechnology in Biology and Medicine

Methods, Devices, and Applications

Second Edition

Nanotechnology in Biology and Medicine

Methods, Devices, and Applications

Second Edition

Edited by
Tuan Vo-Dinh

CRC Press
Taylor & Francis Group
Boca Raton London New York

CRC Press is an imprint of the
Taylor & Francis Group, an **informa** business

CRC Press
Taylor & Francis Group
6000 Broken Sound Parkway NW, Suite 300
Boca Raton, FL 33487-2742

First issued in paperback 2019

© 2018 by Taylor & Francis Group, LLC
CRC Press is an imprint of Taylor & Francis Group, an Informa business

No claim to original U.S. Government works

ISBN-13: 978-1-4398-9378-4 (hbk)
ISBN-13: 978-0-367-86690-7 (pbk)

Library of Congress Cataloging–in–Publication Data

Names: Vo-Dinh, Tuan, editor.
Title: Nanotechnology in biology and medicine : methods, devices, and applications / [edited by] Tuan Vo-Dinh.
Description: Second edition. | Boca Raton : Taylor & Francis, 2017. | Includes bibliographical references and index.
Identifiers: LCCN 2017020719| ISBN 9781439893784 (hardback : alk. paper) | ISBN 9781315374581 (ebook)
Subjects: | MESH: Nanotechnology | Biomedical Engineering--methods
Classification: LCC R857.N34 | NLM QT 36.5 | DDC 610.28--dc23
LC record available at https://lccn.loc.gov/2017020719

Visit the Taylor & Francis Web site at
http://www.taylorandfrancis.com

and the CRC Press Web site at
http://www.crcpress.com

This book is dedicated to

the pioneers whose visions have

sailed to the outer edges of the universe,

pierced into the inner world of the atom, and

unlocked the mysteries of the human cell.

Contents

Section II Applications in Biology and Medicine

Preface

The second edition of *Nanotechnology in Biology and Medicine* is intended to serve as an authoritative reference source for a broad audience involved in the research, teaching, learning, and practice of nanotechnology in life sciences. Nanotechnology, which involves research on and the development of materials and species at length scales between 1 and 100 nm, has been revolutionizing many important scientific fields, ranging from biology to medicine. This technology, which is on the scale of molecules, has enabled the development of devices smaller and more efficient than anything currently available. To understand complex biological nanosystems at the cellular level, we urgently need to develop a next-generation nanotechnology tool kit. It is believed that the new advances in genetic engineering, genomics, proteomics, medicine, and biotechnology will depend on our mastering of nanotechnology in the coming decades. The integration of nanotechnology, material sciences, molecular biology, and medicine opens the possibility of detecting and manipulating atoms and molecules using nanodevices, which have the potential for a wide variety of biological research topics and medical uses at the cellular level.

Today, the amount of research in the biomedical sciences and engineering at the molecular level is growing exponentially because of the availability of new investigative nanotools for molecular biology and medicine. These tools are capable of detecting biomarkers for early disease detection as well as probing the nanometer world and will make it possible to characterize the chemical and mechanical properties of cells; discover novel phenomena and processes; and provide science with a wide range of tools, materials, devices, and systems with unique characteristics.

The combination of molecular biology, medicine, and nanotechnology has already led to a new generation of devices for probing the cell machinery and elucidating molecular-level life processes heretofore invisible to human inquiry. Tracking biochemical processes within intracellular environments can now be performed *in vivo* with the use of fluorescent molecular probes and nanosensors. With powerful microscopic tools using near-field optics, scientists are now able to explore the biochemical processes and submicroscopic structures of living cells at unprecedented resolutions. It is now possible to develop nanocarriers for targeted delivery of drugs that have their shells conjugated with antibodies for targeting antigens and fluorescent chromophores for *in vivo* tracking.

The second edition of this monograph presents the most recent scientific and technological advances of nanotechnology for use in biology and medicine. Each chapter provides introductory material with an overview of the topic of interest; a description of methods, protocols, instrumentation, and applications; and a collection of published data with an extensive list of references for further details.

The goal of this book is to provide a comprehensive overview of the most recent advances in instrumentation, methods, and applications in areas of nanobiotechnology, integrating interdisciplinary research and development of interest to scientists, engineers, manufacturers, teachers, and students. It is our hope that this handbook will stimulate a greater appreciation of the usefulness, efficiency, and potential of nanotechnology in biology and medicine.

Tuan Vo-Dinh
Duke University
Durham, North Carolina

MATLAB® is a registered trademark of The MathWorks, Inc. For product information, please contact:

The MathWorks, Inc.
3 Apple Hill Drive
Natick, MA 01760-2098 USA
Tel: 508 647 7000
Fax: 508-647-7001
E-mail: info@mathworks.com
Web: www.mathworks.com

Acknowledgments

The completion of this work has been made possible with the assistance of many friends and colleagues. It is a great pleasure for me to acknowledge, with deep gratitude, the contribution of the contributors of the chapters in the second edition of this book. I wish to thank many colleagues in academia, federal laboratories, and industry for their kind help in reading and commenting on various chapters of the manuscript. My gratitude is extended to all my present and past students, postdoctoral associates, and colleagues, who have been collaborating and traveling with me on this exciting journey of scientific discovery to explore and bring nanotechnology and medicine to the service of society.

I gratefully acknowledge the support of the National Institutes of Health, the Department of Energy, the Defense Advanced Research Projects Agency, the Department of the Army, the Defense Advanced Research Projects Agency, the Army Medical Research and Material Command, the Department of Justice, the Federal Bureau of Investigation, the Office of Naval Research, the Environmental Protection Agency, the Fitzpatrick Foundation, the R. Eugene and Susie E. Goodson Endowment Fund, and the Wallace Coulter Foundation.

The completion of this work has been made possible with the encouragement, love, and inspiration of my wife, Kim-Chi, and my daughter, Jade.

Editor

Tuan Vo-Dinh is R. Eugene and Susie E. Goodson distinguished professor of biomedical engineering, professor of chemistry, and director of the Fitzpatrick Institute for Photonics at Duke University. A native of Vietnam and a naturalized U.S. citizen, Dr. Vo-Dinh completed high school education in Saigon (now Ho Chi Minh City). He continued his studies in Europe where he received a BS in physics in 1970 from EPFL (Ecole Polytechnique Federal de Lausanne) in Lausanne, Switzerland, and a PhD in physical chemistry in 1975 from ETH (Swiss Federal Institute of Technology) in Zurich, Switzerland. Before joining Duke University in 2006, Dr. Vo-Dinh was director of the Center for Advanced Biomedical Photonics, group leader of Advanced Biomedical Science and Technology Group, and a corporate fellow, one of the highest honors for distinguished scientists at Oak Ridge National Laboratory (ORNL). His research has focused on the development of advanced technologies for the protection of the environment and the improvement of human health. His research activities involve nano-biophotonics, nanosensors, laser spectroscopy, molecular imaging, medical diagnostics, cancer detection, chemical sensors, biosensors, and biochips.

Dr. Vo-Dinh has authored over 400 publications in peer-reviewed scientific journals. He is the author of a textbook on spectroscopy and editor of six books. He holds over 37 U.S. and international patents, 5 of which been licensed to private companies for commercial development. Dr. Vo-Dinh has presented over 200 invited lectures at international meetings in universities and research institutions. He has chaired over 20 international conferences in his field of research and served on various national and international scientific committees. He also serves the scientific community through his participation in a wide range of governmental and industrial boards and advisory committees.

Dr. Vo-Dinh has received seven R&D 100 Awards for Most Technologically Significant Advance in Research and Development for his pioneering research and inventions of innovative technologies. He has received the Gold Medal Award, Society for Applied Spectroscopy (1988); the Languedoc-Roussillon Award (France) (1989); the Scientist of the Year Award, ORNL (1992); the Thomas Jefferson Award, Martin Marietta Corporation (1992); two Awards for Excellence in Technology Transfer, Federal Laboratory Consortium (1995, 1986); the Inventor of the Year Award, Tennessee Inventors Association (1996); and the Lockheed Martin Technology Commercialization Award (1998), the Distinguished Inventors Award, UT-Battelle (2003), and the Distinguished Scientist of the Year Award, ORNL (2003). In 1997, Dr. Vo-Dinh was presented the Exceptional Services Award for distinguished contribution to a healthy citizenry from the U.S. Department of Energy. In 2011, Dr. Vo-Dinh received the Award for Spectrochemical Analysis from the American Chemical Society (ACS) Division of Analytical Chemistry.

Contributors

Olaoluwa Adeniba
Department of Mechanical Science and
    Engineering
University of Illinois
Urbana, Illinois

S.-K. Ahn
Oak Ridge National Laboratory
Oak Ridge, Tennessee

Nina G. Argibay
Department of Biological Sciences
Nova Southeastern University
Fort Lauderdale, Florida

Lane A. Baker
Departments of Chemistry and
    Anesthesiology
University of Florida
Gainesville, Florida

Natalia Barkalina
Nuffield Department of Obstetrics and
    Gynaecology
University of Oxford
Oxford, United Kingdom

M. D. Barnes
Department of Chemistry
University of Massachusetts
Amherst, Massachusetts

Rashid Bashir
Department of Bioengineering
and
Department of Electrical and Computer
    Engineering
University of Illinois
Urbana, Illinois

Rachel L. Beingessner
Department of Chemistry
Massachusetts Institute of Technology
Cambridge, Massachusetts

Sean Brahim
Center for Bioelectronics, Biosensors, and
    Biochips (C3B)
Virginia Commonwealth University
Richmond, Virginia

J.-M.Y. Carrillo
Computational Sciences and Engineering
    Division
and
Center for Nanophase Materials Sciences
Oak Ridge National Laboratory
Oak Ridge, Tennessee

Andrzej Chałupniak
Catalan Institute of Nanoscience and
    Nanotechnology (ICN2)
and
CSIC
and
The Barcelona Institute of Science and
    Technology
Barcelona, Spain

Kui Chen
Oak Ridge National Laboratory
Oak Ridge, Tennessee

Jadwiga Chroboczek
Therex, TIMC-IMAG, CNRS UMR 5525
UJF, Domaine de la Merci
La Tronche, France

and

Institute of Biochemistry and Biophysics
Polish Academy of Sciences
Warsaw, Poland

Jarrod Clark
Kaplan Clinical Research Laboratory
City of Hope Medical Center
Duarte, California

Elise A. Corbin
Department of Bioengineering
University of Illinois
Urbana, Illinois

and

Department of Cardiology
University of Pennsylvania Philadelphia
Pennsylvania

Maxime Couture
Département de Chimie
Université de Montréal
Montreal, Québec, Canada

Kevin Coward
Nuffield Department of Obstetrics and
    Gynaecology
University of Oxford
Oxford, United Kingdom

Travis J.A. Craddock
Institute for Neuroimmune Medicine
and
Department of Neuroscience
and
Department of Computer Science
and
Department of Clinical Immunology
Nova Southeastern University
Fort Lauderdale, Florida

Bridget M. Crawford
Department of Biomedical Engineering
and
Fitzpatrick Institute for Photonics
Duke University
Durham, North Carolina

Lien Davidson
Nuffield Department of Obstetrics and
    Gynaecology
University of Oxford
Oxford, United Kingdom

Tejal A. Desai
Department of Physiology
University of California
San Francisco, California

Mitchel J. Doktycz
Oak Ridge National Laboratory
Oak Ridge, Tennessee

Mostafa A. El-Sayed
School of Chemistry and Biochemistry
Georgia Institute of Technology
Atlanta, Georgia

M. Nance Ericson
Oak Ridge National Laboratory
Oak Ridge, Tennessee

Andrew M. Fales
Fitzpatrick Institute for Photonics
and
Department of Biomedical Engineering
Duke University
Durham, North Carolina

Hicham Fenniri
Departments of Chemical Engineering
    Bioengineering and Chemistry
Northeastern University
Boston, Massachusetts

Emmanuel Fort
Institut Langevin
ESPCI Paris
Paris, France

Sarah Francis
Nuffield Department of Obstetrics and
    Gynaecology
University of Oxford
Oxford, United Kingdom

Naveen Gandra
Fitzpatrick Institute for Photonics
and
Department of Biomedical Engineering
Duke University
Durham, North Carolina

Samuel Grésillon
Institut Langevin
Université Pierre et Marie Curie
Paris, France

Guy D. Griffin
Oak Ridge National Laboratory
Oak Ridge, Tennessee

Michael A. Guillorn
Oak Ridge National Laboratory
Oak Ridge, Tennessee

Anthony Guiseppi-Elie
Center for Bioelectronics, Biosensors, and
    Biochips (C3B)
Virginia Commonwealth University
Richmond, Virginia

Amit Gupta
Royole Corporation
Fremont, California

Amanda J. Haes
Department of Chemistry
Northwestern University
Evanston, Illinois

R. J. Harrison
Computational Sciences and Engineering
    Division
Oak Ridge National Laboratory
Oak Ridge, Tennessee

H.P. Ho
Department of Electronic Engineering
The Chinese University of Hong Kong
Hong Kong, China

Xiaohua Huang
Department of Chemistry
The University of Memphis
Memphis, Tennessee

Celine Jones
Nuffield Department of Obstetrics and
    Gynaecology
University of Oxford
Oxford, United Kingdom

Paul M. Kasili
Departments of Biomedical Engineering
    and Chemistry
Duke University
Durham, North Carolina

and

Oak Ridge National Laboratory
Oak Ridge, Tennessee

and

Science and Engineering Department
Bunker Hill Community College
Boston, Massachusetts

Bruce Klitzman
Department of Surgery
Duke University Medical Center
Durham, North Carolina

S.K. Kong
School of Life Sciences
The Chinese University of Hong Kong
Hong Kong, China

Leo Kretzner
Kaplan Clinical Research Laboratory
City of Hope Medical Center
Duarte, California

Katarzyna Lamparska-Kupsik
Kaplan Clinical Research Laboratory
City of Hope Medical Center
Duarte, California

Yang Liu
Department of Biomedical Engineering
and
Department of Chemistry
and
Fitzpatrick Institute of Photonics
Duke University
Durham, North Carolina

F.C. Loo
Department of Electronic Engineering
and
School of Life Sciences
The Chinese University of Hong Kong
Hong Kong, China

Charles R. Martin
Departments of Chemistry and
    Anesthesiology
University of Florida
Gainesville, Florida

Jean-Francois Masson
Département de Chimie
Université de Montréal
Montreal, Québec, Canada

Timothy E. McKnight
Oak Ridge National Laboratory
Oak Ridge, Tennessee

Anatoli V. Melechko
Oak Ridge National Laboratory
Oak Ridge, Tennessee

Arben Merkoçi
Catalan Institute of Nanoscience and
    Nanotechnology
and
Consejo Superior de Investigaciones
    Científicas
and
Barcelona Institute of Science and
    Technology
and
Institució Catalana de Recerca i Estudis
    Avançats
Barcelona, Spain

Vladimir I. Merkulov
Oak Ridge National Laboratory
Oak Ridge, Tennessee

Kristofer Munson
Kaplan Clinical Research Laboratory
City of Hope Medical Center
Duarte, California

Hoan T. Ngo
Fitzpatrick Institute for Photonics
and
Department of Biomedical Engineering
Duke University
Durham, North Carolina

D. W. Noid
Computational Sciences and Engineering
    Division
Oak Ridge National Laboratory
Oak Ridge, Tennessee

Anjali Pal
Department of Civil Engineering
Indian Institute of Technology
Kharagpur, India

Tarasankar Pal
Department of Chemistry
Indian Institute of Technology
Kharagpur, India

Gregory M. Palmer
Department of Radiation Oncology
Duke University Medical Center
Durham, North Carolina

Sudipa Panigrahi
Department of Chemistry
Indian Institute of Technology
Kharagpur, India

Kidong Park
Division of Electrical and Computer
    Engineering
Louisiana State University
Baton Rouge, Louisiana

Hugo-Pierre Poirier-Richard
Département de Chimie
Université de Montréal
Montreal, Québec, Canada

Ketul C. Popat
Department of Physiology
University of California
San Francisco, California

Janna K. Register
Fitzpatrick Institute for Photonics
and
Department of Biomedical Engineering
Duke University
Durham, North Carolina

Ajit Sadana
Chemical Engineering Department
University of Mississippi
Oxford, Mississippi

Neeti Sadana
Department of Anesthesiology
The University of Oklahoma Health
    sciences Center
Oklahoma City, Oklahoma

Sadhana Sharma
Department of Physiology and Biophysics
University of Illinois
Chicago, Illinois

W. A. Shelton
Computational Sciences and Engineering
    Division
Oak Ridge National Laboratory
Oak Ridge, Tennessee

Olga Shimoni
School of Mathematical and Physical
    Sciences
University of Technology Sydney
New South Wales, Australia

Nikhil K. Shukla
Center for Bioelectronics, Biosensors, and
    Biochips (C3B)
Virginia Commonwealth University
Richmond, Virginia

Michael L. Simpson
Oak Ridge National Laboratory
Oak Ridge, Tennessee
and
University of Tennessee
Knoxville, Tennessee

Elizabeth Singer
Kaplan Clinical Research Laboratory
City of Hope Medical Center
Duarte, California

Baljit Singh
Faculty of Veterinary Medicine
University of Calgary
Calgary, Canada

Robert P. Smith
Department of Biological Sciences
Nova Southeastern University
Fort Lauderdale, Florida

Steven S. Smith
Kaplan Clinical Research Laboratory
City of Hope Medical Center
Duarte, California

Douglas A. Stuart
Department of Chemistry
Northwestern University
Evanston, Illinois

B. G. Sumpter
Computational Sciences and Engineering
    Division
and
Center for Nanophase Materials
    Sciences
Oak Ridge National Laboratory
Oak Ridge, Tennessee

Mark T. Swihart
Department of Chemical and Biological
    Engineering
The University at Buffalo
The State University of New York
Buffalo, New York

Inga Szurgot
Institute of Biochemistry and Biophysics
Polish Academy of Sciences
Warsaw, Poland

Steve M. Taylor
Fitzpatrick Institute for Photonics
and
Department of Medicine and Duke Global
    Health Institute
Duke University
Durham, North Carolina

Stella M. Valenzuela
School of Life Sciences
University of Technology Sydney
New South Wales, Australia

Richard P. Van Duyne
Department of Chemistry
Northwestern University
Evanston, Illinois

Eric M. Vazquez
Department of Biological Sciences
Florida International University
Miami, Florida

Tuan Vo-Dinh
Department of Biomedical Engineering
and
Fitzpatrick Institute for Photonics
and
Department of Chemistry
Duke University
Durham, North Carolina

and

Oak Ridge National Laboratory
Oak Ridge, Tennessee

and

Molecular Discovery Research
Platform Technology & Science
GlaxoSmithKline
Collegeville, Pennsylvania

Musundi B. Wabuyele
Molecular Discovery Research
Platform Technology & Science
GlaxoSmithKline
Collegeville, Pennsylvania

Hsin-Neng Wang
Department of Biomedical Engineering
and
Fitzpatrick Institute for Photonics
Duke University
Durham, North Carolina

Thomas J. Webster
Departments of Chemical Engineering
Northeastern University
Boston, Massachusetts

Cortney E. Wilson
Department of Biological Sciences
Nova Southeastern University
Fort Lauderdale, Florida

S.Y. Wu
Department of Electronic Engineering
The Chinese University of Hong Kong
Hong Kong, China

Ashkan YekrangSafakar
Division of Electrical and Computer
    Engineering
Louisiana State University
Baton Rouge, Louisiana

Hsiangkuo Yuan
Department of Biomedical Engineering
and
Fitzpatrick Institute of Photonics
Duke University
Durham, North Carolina

1

Nanotechnology at the Frontier of Biology and Medicine

Tuan Vo-Dinh

CONTENTS

1.1 Introduction

In order to obtain a perspective on the dimension at the nanoscale level, it is useful to look up in the dimension scale, up to the outer edges of our "local universe," the Milky Way, a galaxy of 100–400 billion stars. The universe revealed to us has the dimension of 50,000 light-years from the outer edges to its center. A light-year is the distance that light travels in one year at the speed of ~300 million (300,000,000) meters per second, which corresponds to ~10,000,000,000,000,000 (16 zeros) or 10^{16} m. The meter, a dimension unit closest to everyday human experience, is often considered as the basic dimension of reference for human beings. Therefore, the distance from the center to the outer edge of the Milky Way is 500,000,000,000,000,000,000 or 5×10^{20} m. Let us now look down in the other direction of the dimensional scale, down to a nanometer, which is a billion (1,000,000,000) times smaller than a meter (i.e., 10^{-9} m). The word *nano* is derived from the Greek word meaning "dwarf." In dimensional scaling, *nano* refers to 10^{-9}—that is, one billionth of a unit. A human hair has a diameter of ~10 μ, which is 10,000 nm. Diameters of atoms are in the order of tenths of (10^{-1}) nanometers, whereas the diameter of a DNA strand is about a few nanometers. Thus, *nanotechnology* is a general term that refers to the techniques and methods for studying, designing, and fabricating things at the level of atoms and molecules. The initial concept of investigating materials and biological systems at the nanoscale dates to more than 40 years ago, when Richard Feynman presented a lecture in 1959 at the annual meeting of the American Physical Society at the California Institute of Technology. This lecture, entitled "There's Plenty of Room at the Bottom," is generally considered to be the first look into the world of materials, species, and structures at nanoscale levels. Thinking small, however, is not a new idea. Thousands of years ago, the Greek philosophers Leucippus and Democritus had suggested that all matter was made from tiny particles like atoms. It is only now with the advent of nanotechnology that has led to the development of a new generation of instruments capable of revealing the structure of these tiny particles conceived since the Hellenic Ages.

It is now generally accepted that nanotechnology involves research and development on materials and species at length scales from 1 to 100 nm. Nanotechnology is very important to biology since many biological species have molecular structures at the nanoscale levels. These species comprise a wide variety of basic structures such as proteins, polymers, carbohydrates (sugars), and lipids, which have a great variety of chemical, physical, and functional properties. Individual molecules, when organized into controlled and defined nanosystems, have new structures and exhibit new properties. This structural variety and the versatility of these biological nanomaterials and systems have important implications for the design and development of new and artificial assemblies that are critical to biological and medical applications. The development of a next-generation nanotechnology tool kit is critical to understand the inner world of complex biological nanosystems at the cellular level. Since nanotechnology involves technology on the scale of molecules it has the potential of developing devices smaller and more efficient than anything currently available. Traditionally defined disciplines, such as chemistry, biology, and materials science, also deal with atoms and molecules, which are of nanometer sizes. But nanotechnology differs from traditional disciplines in a very fundamental aspect. For example, whereas chemistry (or biology and materials science) deals with atoms and molecules at the bulk level (we do not see the molecules in chemical solutions), nanotechnology seeks to actually "manipulate" individual atoms and molecules in very specific ways, thus creating new materials having new properties and new functions. It is this "bottom-up" capability that makes nanotechnology a unique new field of research of undreamed possibilities and potential. Our mastering of nanotechnology could unleash breakthroughs in genetic engineering, genomics, proteomics, and medicine in the coming decades. If we can assemble biological systems and devices at the atomic and molecular levels, we will achieve versatility in design, a precision in construction, and a control in operation heretofore hardly imagined.

1.2 Cellular Nanomachines and the Building Blocks of Life

Nanotechnology is of great importance to molecular biology and medicine because life processes are maintained by the action of a series of biological molecular nanomachines in the cell machinery. By evolutionary modification over trillions of generations, living organisms have perfected an armory of molecular machines, structures, and processes. The living cell, with its myriad of biological components, may be considered the ultimate "nano factory." Figure 1.1 shows a schematic diagram of the cell with its various components. Some typical sizes of nucleic acids, proteins, and biological species are shown in Table 1.1. Nucleic acids and proteins are important cellular components, which play a critical role in maintaining the operation of the cell. DNA is a polymeric chain made up of subunits called nucleotides. The polymer is referred to as a "polynucleotide." Each nucleotide is made up of a sugar, a phosphate, and a base. There are four different types of nucleotides found in DNA, differing only in the nitrogenous base: adenine (A), guanine (G), cytosine (C), and thymine (T). The basic structure of the DNA molecule is helical, with the bases being stacked on top of each other. DNA normally has a double-stranded conformation, with two polynucleotide chains held together by weak thermodynamic forces. Two DNA strands form a helical spiral, winding around a helix axis in a right-handed spiral. The two polynucleotide chains run in opposite directions. The sugar-phosphate backbones of

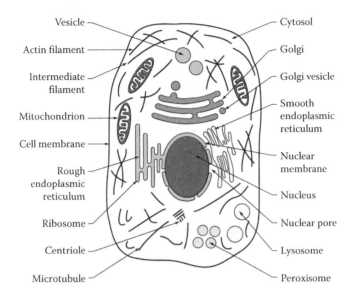

FIGURE 1.1
Schematic diagram of a cell and its components.

the two DNA strands wind around the helix axis like the railing of a spiral staircase. The bases of the individual nucleotides are inside the helix, stacked on top of each other like the steps of a spiral staircase.

Genes and proteins are intimately connected. The genetic code encrypted in the DNA is transcribed into a corresponding sequence of RNA, which is then read by the ribosome to construct a sequence of amino acids, which is the backbone of a specific protein. The amino acid chain folds up into a three-dimensional shape and becomes a specific protein, which is designed to perform a particular function. Ribosomes are important molecular nanomachines that build proteins essential to the functioning of the cell. Although the size of a typical ribosome is only 8000 nm³, this nanomachine is capable of manufacturing almost any protein by stringing together amino acids in a precise linear sequence following instructions from a messenger RNA (mRNA) copied from the host DNA. To perform its molecular manufacturing task, the ribosome takes hold of a specific transfer RNA (tRNA), which in turn is chemically bonded by a specific enzyme to a specific amino acid. It has the means to grasp the growing polypeptide and to cause the specific amino acid to react with, and be added to, the end of the polypeptide. In other words, DNA can be considered to be the biological software of the cellular machinery, whereas ribosomes are large-scale molecular constructors, and enzymes are functional molecular-sized assemblers.

TABLE 1.1

Typical Nanosizes of Cellular Species

Biological Species	Example	Typical Size	Typical Mol. Weight
Small assemblies	Ribosome	20-nm sphere	10^5–10^7
Large assemblies	Viruses	100-nm sphere	10^7–10^{12}
Nucleic acids	tRNA	10-nm rod	10^4–10^5
Small proteins	Chymotrypsin	4-nm sphere	10^4–10^5
Large proteins	Aspartate transcarbamoylase	7-nm sphere	10^5–10^7

Proteins are nanoscale components that are essential in biology and medicine [1]. They consist of long chains of polymeric molecules assembled from a large number of amino acids like beads on a necklace. There are 20 basic amino acids. The sequence of the amino acids in the polymer backbone, determined by the genetic code, is the primary structure of any given protein. Typical polypeptide chains contain about 100–600 amino acid molecules and have a molecular weight of about 15,000–70,000 Da. Since amino acids have hydrophilic, hydrophobic, and amphiphilic groups, in the aqueous environment of the cell they tend to fold to form a locally ordered, three-dimensional structure, called the secondary structure, which is characterized by a low-energy configuration with the hydrophilic groups outside and the hydrophobic groups inside. In general, simple proteins have a natural configuration referred to as the α-helix configuration. Another natural secondary configuration is a β-sheet. These two secondary configurations (α-helix and β-sheet) are the building blocks that assemble to form the final tertiary structure, which is held together by extensive secondary interactions, such as van der Waals bonding. The tertiary structure is the complete three-dimensional structure of one indivisible protein unit, that is, one single covalent species. Sometimes, several proteins are bound together to form supramolecular aggregates, which make up a quaternary structure. The quaternary structure, which is the highest level of structure, is formed by the noncovalent association of independent tertiary structure units.

Determination of the three-dimensional structure of proteins, which requires analytical tools capable of measurement precision at the nanoscale level, is essential in understanding their functions. Knowledge of the primary structure provides little information about the function of proteins. To carry out their function, proteins must take on a specific conformation, often referred to as an active form, by folding themselves. The three-dimensional structure of bovine serum albumin, illustrated in Figure 1.2, shows that the

FIGURE 1.2
Three-dimensional nanostructure of a biological molecule, bovine serum albumin.

molecule exhibits a folded conformation. The folded conformation of some proteins, such as egg albumin, can be unfolded by heating. Heating produces an irreversible folding conformation change of albumin which turns white. Albumin is said to be denatured in this form. Denatured albumin cannot be reversed into its natural state. However, some proteins can be denatured and renatured repeatedly—that is, they can be unfolded and refolded to their natural configuration. Diseases such as Alzheimer's, cystic fibrosis, "mad cow" disease, an inherited form of emphysema, and even many cancers are believed to result from protein misfolding.

There are a wide variety of proteins, which are "nanomachines" capable of performing a number of specific tasks. Enzymes are important proteins providing the driving force for biochemical reactions. Antibodies are another type of proteins that are designed to recognize invading elements and allow the immune system to neutralize and eliminate unwanted invaders. Since diseases, therapy, and drugs can alter protein profiles, a determination of protein profiles can provide useful information for understanding disease and designing therapy. Therefore, understanding the structure, metabolism, and function of cellular components such as proteins at the nanoscale (molecular) level is essential to our understanding of biological processes and monitoring the health status of a living organism in order to effectively diagnose and ultimately prevent disease. Molecular machines in the simplest cells involve nanoscale manipulators for building molecule-sized objects. They are used to build proteins and other molecules atom by atom according to defined instructions encrypted in the DNA. The cellular machinery uses rotating bearings that are found in many forms: for example, some protein systems found in the simplest bacteria serve as clamps that encircle DNA and slide along its length. Human cells contain a rotary motor that is used to generate energy. Various types of molecule-selective pumps are used by cells to transfer and carry ions, amino acids, sugars, vitamins, and nutrients needed for the normal functioning of the cell. Cells also use molecular sensors, which can detect the concentration of surrounding molecules and compute the proper functional outcome. The movement of another well-known molecular motor, myosin, along double-helical filaments of a protein called actin ($\sim$10 nm across) produces the contraction of muscle cells during each heartbeat.

1.3 New Generation of Nanotools

Nanotechnology has triggered a revolution in many important areas in molecular biology and medicine, especially in the detection and manipulation of biological species at the molecular and cellular level. The convergence of nanotechnology, molecular biology, and medicine will open new possibilities in detecting and manipulating atoms and molecules using nanodevices, with the potential for a wide variety of medical uses at the cellular level. Today, the amount of research in biomedical science and engineering at the molecular level is growing exponentially due to the availability of new analytical tools based on nanotechnology. Novel microscopic devices using near-field optics allow scientists to explore the biochemical processes and nanoscale structures of living cells at unprecedented resolutions. The optical detection sensitivity and the high resolution of near-field scanning optical microscopy (NSOM) were used to detect the cellular localization and activity of ATP-binding cassette (ABC) proteins associated with multidrug resistance (MDR) [2]. Drug resistance can be associated with several cellular mechanisms ranging from reduced drug uptake to reduction of drug sensitivity due to genetic alterations. MDR is therefore a phenomenon that indicates a variety of strategies that cancer cells are able to develop in order to resist the cytotoxic effects of anticancer drugs. Figure 1.3 shows images of single Chinese hamster ovary (CHO) cells incubated with drugs using nanoimaging tools, such as confocal microscopy and NSOM, which are now readily available to biomedical researchers. These new analytical tools are capable of probing the nanometer world and will make it possible to characterize the chemical and mechanical properties of cells, discover novel phenomena and processes, and provide science with a wide range of tools, materials, devices, and systems with unique characteristics.

The combination of nanotechnology and molecular biology has produced a new generation of devices capable of probing the cell machinery and elucidating molecular-level life processes heretofore invisible to human inquiry. Nanocarriers having antibodies for recognizing target species and spectroscopic labels (fluorescence, Raman) for *in vivo* tracking have been developed for seamless diagnostic and therapeutic operations. Tracking biochemical processes within intracellular environments is possible with molecular nanoprobes and nanosensors. Optical nanosensors have been designed to detect individual biochemical species in subcellular locations throughout a living cell [3]. The nanosensors

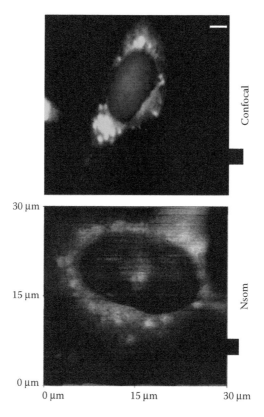

FIGURE 1.3
Nanoimaging of a CHO cell incubated with cancer drugs, doxorubicin and verapamil: (top) Confocal microscopy: fluorescence image obtained using a mercury arc lamp with following filter sets: λ_{ex}, 470–490 nm, λ_{em} 520–560 m for verapamil and λ_{ex}, 510–560 nm, λ_{em} 580 nm band-pass filter sets for doxorubicin; (bottom) Near-field Scanning Microscopy image obtained using a near-field scanning optical microscope. A 488-nm laser source was used for the excitation of verapamil and 532-nm laser was used to excite doxorubicin.

were fabricated with optical fibers pulled down to tips with distal ends having nanoscale sizes (30–40 nm). Laser light is launched into the fiber and the resulting evanescent field at the tip of the fiber is used to excite target molecules bound to the antibody molecules. A photodetector is used to detect the fluorescence originated from the analyte molecules. Dynamic information of signaling processes inside living cells is important to fundamental biological understanding of cellular processes. Many traditional microscopy techniques involve incubation of cells with fluorescent dyes or nanoparticles (NPs) and examining the interaction of these dyes with compounds of interest. However, when a dye or NP is delivered into a cell, it is transported to certain intracellular sites that may or may not be where it is most likely to stay and not to areas where the investigator would like to monitor. The fluorescence signals which are supposed to reflect the interaction of the dyes with chemicals of interest are generally directly related to the dye concentration as opposed to the analyte concentration. Only with optical nanosensors can excitation light be delivered to specific locations inside cells. Figure 1.4 shows a fiberoptic nanobiosensor developed for monitoring biomarkers of DNA damage [4] or an apoptotic signaling pathway in a single cell [5]. An important advantage of the optical sensing modality is its capability to measure biological parameters in a noninvasive or minimally invasive manner due to the very small size of the nanoprobe. The capability to detect important biological molecules at ultratrace concentrations *in vivo* is central to many advanced diagnostic techniques. Early detection of diseases will be made possible by tracking down trace amounts of biomarkers in tissue. Due to their very small sizes, nanosensors are an important technology that can be used to measure biotargets in a living cell without significantly affecting cell viability. Following measurements using the nanobiosensor, cells have been shown to survive and undergo mitosis. Biomedical nanosensors, which have been used to investigate the effect of cancer drugs in cells [5], will play an important role in the future of medicine. Combined with the exquisite molecular recognition of bioreceptor probes, nanosensors could serve as powerful tools capable for exploring biomolecular processes in subcompartments of living cells. They have a great potential to provide the necessary tools to investigate multiprotein molecular machines of complex living systems and the complex network that controls the assembly and operation of these machines in a living cell. Future developments would

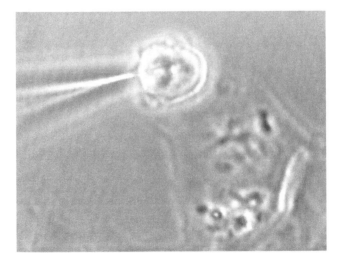

FIGURE 1.4
Fiberoptics nanosensor for single-cell analysis.

lead to the development of nanosensors equipped with nanotool sets that enable tracking, assembly, and disassembly of multiprotein molecular machines and their individual components. These nanosensors would have multifunctional probes that could measure the structure of biological components in single cells. With traditional analytical tools, scientists are handicapped in investigating the workings of individual genes and proteins by breaking the cell apart and studying its individual components *in vitro*. The advent of nanosensors will hopefully permit research on entire networks of genes and proteins in an entire living cell *in vivo* in a systems biology approach.

The goal of understanding the structure and function of proteins as integrated processes in cells, often referred to as "system biology," presents a formidable challenge, much more difficult than that associated with the determination of the human genome. Therefore, proteomics, which involves determination of the structure and function of proteins in cells, could be a research area that requires the use of nanotechnology-based techniques. Proteomics research directions can be categorized as structural and functional. Structural proteomics, or protein expression, measures the number and types of proteins present in normal and diseased cells. This approach is useful in defining the structure of proteins in a cell. However, the role of a protein in a disease is not defined simply by knowledge of its structure. An important function of proteins is in the transmission of signals through intricate protein pathways. Proteins interact with each other and with other organic molecules to form pathways. Functional proteomics involves the identification of protein interactions and signaling pathways within cells and their relationship to disease processes. Elucidating the role proteins play in signaling pathways allows a better understanding of their function in cellular behavior and permits diagnosis of disease and, ultimately, identification of potential drug targets for preventive treatment.

A wide variety of nanoprobes (NPs, dendrimers, quantum dots, etc.) have been developed for cellular diagnostics. The development of metallic nanoprobes that can produce a surface-enhancement effect for ultrasensitive biochemical analysis is another area of active nanoscale research. *Plasmonics* refers to the research area that deals with enhanced electromagnetic properties of metallic nanostructures. The term is derived from *plasmons*, which are the quanta associated with longitudinal waves propagating in matter through the collective motion of large numbers of electrons. Incident light irradiating these surfaces excites conduction electrons in the metal and induces excitation of surface plasmons, which in turn leads to enormous electromagnetic enhancement for ultrasensitive detection of spectral signatures through surface-enhanced Raman scattering (SERS) [6]. Metallic nanostructures such as nanowave (i.e., metallic film coated on NP arrays) have been developed for gene detection [7] and cellular imaging using SERS.

NP-mediated thermal therapy has recently demonstrated the potential to combine the advantages of precise cancer cell ablation. As one of the first effective systemic cancer treatments, hyperthermia (HT) aims to increase tumor temperature above the normal value ($\sim36°C$) to trigger local and systemic antitumor effects and/or ablate cancer cells. While HT at high temperature ($>55°C$) can actually induce immediate thermal death (ablation) to targeted tumors, HT at mild fever-range can be used to improve drug delivery to tumors, improve cancer cell sensitivity to other therapies, and trigger potent systemic anticancer immune responses [8–11]. Traditional HT modalities such as microwaves, radiofrequency, and ultrasound can control macroscopic heating around the tumor region, but cannot precisely target or ablate cancer cells in a timely manner. Cancer treatment using photothermal therapy (PTT), which exploits high temperature transduced from photon energy is a promising method offering high efficiency and specificity because cancer cells are more sensitive to elevated temperature ($>42°C$) than normal cells. Nanomedicine has attracted increasing attention in recent years since it offers great promise to provide personalized cancer therapy with improved treatment efficiency and specificity. NPs have a natural propensity to extravasate from the tumor vascular network and accumulate in and around cancer cells due to the enhanced permeability and retention (EPR) effect. The choice of the exogenous photothermal agents is made on the basis of their strong absorption cross sections and efficient light-to-heat conversion. This feature greatly minimizes the amount of laser energy needed to induce local damage of the diseased cells, making the therapy method less invasive. A problem associated with the use of dye molecules is their photobleaching under laser irradiation. Therefore, NPs such as gold NPs have recently been used. The use of plasmonics-enhanced photothermal properties of metal NPs including nanoshells, nanorods, and nanostars for PTT has been reported [12–18]. Gold nanoshells have been used for targeted bimodal or trimodal cancer therapy because they can be tuned to absorb NIR light, which can penetrate tissue and can be designed to be specifically targeted and delivered to cancer cells. The promising role of nanoshells in PTT of tumors has been demonstrated [12,13]. Among the various types of NPs, gold nanostars (GNS) whose sharp branches create a "lightning rod" effect that enhances the local electromagnetic (EM) field dramatically, are the most effective in converting light into heat for PTT [16,17]. The unique tip-enhanced plasmonics property of GNS can be optimally tuned in the near-infrared (NIR) "therapeutic" optical window, where photons can travel further in healthy tissue to be "captured" and converted into heat by GNS taken up in cancer cells [19]. GNS probe have been developed for multimodality theranostics including SERS detection, x-ray computed tomography (CT), two-photon luminescence (TPL) imaging, and PTT [20]. *In vivo* PTT with near-infrared (NIR) laser under maximum permissible exposure (MPE) led to ablation of aggressive tumors containing GNS, but has no effect in the absence of GNS. These multifunctional GNS nanoprobes have potential to be used for *in vivo* biosensing, preoperative imaging with CT, intraoperative detection with optical methods (SERS and TPL), as well as image-guided PTT.

Figure 1.5 (left) shows the Transmission Electron Microscopy (TEM) images of gold nanostars and the theoretical simulation of electromagnetic $|E|$ in the vicinity of the nanostars in response to a z-polarized plane wave incident E-field of unit amplitude, propagating in the y-direction, and with a wavelength of 800 nm, showing strong electromagnetic field at the tips ("lightning rod" effect) [17]. Also, nanostars exhibit a strong two-photon action cross section (10^6 Goeppert-Mayer units—a value higher than quantum dots or organic fluorophore), allowing the possibility of real-time particle imaging by two-photon absorption microscopy. Figure 1.5 (middle) shows the TPL microscopy images of photothermal-triggered tumor blood–brain barrier permeation examined through a cranial window: tumor vessels prior to (a) and 48 h after (c) laser irradiation [15,16]. The ability to safely target single tumor cells with a high level of efficacy and specificity can be achieved with GNS that have multiple sharp branches acting like "lightning rods" by efficiently converting light into heat. This will allow for a significant reduction of the laser energy needed to precisely destroy the targeted cancer cells in which GNS preferentially accumulate due to the enhanced permeation and retention effect. In a study, PEG-coated AuNS were injected in mice and accumulated in tumors. Figure 1.5 (Right) shows images of mice before and after PTT with and without injected GNS. Note, the dramatic tumor size reduction due to PTT in the mouse injected with AuNS [19].

A new photoimmunotherapy is based on the Synergistic Immuno Photothermal Nanotherapy (SYMPHONY) concept, which combines anti-PD-L1 immunotherapy with GNS-mediated PTT for a two-pronged treatment modality aimed at achieving three main

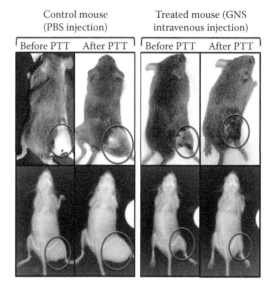

FIGURE 1.5
Photographs (top) and x-ray images (bottom) of mice before and after photothermal therapy (PTT) with and without injected GNS. Note, the dramatic tumor size reduction due to PTT in the mouse injected with GNS. (Adapted from Liu, Y et al. 2015. *Theranostics* 5, 946–960. doi:10.7150/thno.11974.)

goals: (1) GNS-mediated photothermal heating and ablation of the primary tumor, (2) induction of a strong immunogenic cell lethality, and (3) reversal of factors contributing to immune suppression. One therapeutic arm uses laser light to irradiate the primary tumor area where GNS have accumulated, resulting in generation of heat, which kills the primary tumor cells (Figure 1.6) [21]. Not only is there an immediate killing effect at the site treated with light, but this treatment also results in a general activation of the immune system, as evidenced by the fact that distant tumors that are not treated with light also show cancer cell killing. The second therapeutic arm involves administration of PD-L1 immune checkpoint blockade to disable cancer resistance. Many cancers exploit immune checkpoints to evade the anticancer immune response. Immune checkpoint inhibition is a promising immunotherapy that aims to reverse signals from immunosuppressive tumor microenvironments. Programmed death-ligand 1 (PD-L1), a protein overexpressed by many cancers, contributes to the suppression of the immune system and cancer immune evasion. PD-L1 binds to its receptor, PD-1 found on activated T cells, and inhibits cytotoxic T-cell function, thus escaping the immune response. To reverse tumor-mediated immunosuppression, therapeutic anti-PD-1/PD-L1 antibodies have been designed to block the PD-L1/PD-1 interaction. By suppressing this tumor defense, the tumor cells are now vulnerable to the killing action of immune system cells that have been primed against the tumor by the NP phototherapy [21].

A novel type of nanoprobe could be used in assays that require rapid, ultrahigh throughput identification of genomic material (unique genomes or single nucleotide variations) and multiplex detection techniques of small molecules for drug discovery. This nanoprobe, referred to as "molecular sentinel" (MS), illustrated in Figure 1.6, involves a nanoprobe having a Raman label at one end, which is immobilized onto a metallic NP via a thiol group attached on the other end to form a SERS nanoprobe [22]. The metal NP is used as a signal-enhancing platform for the SERS signal associated with the label. Therefore,

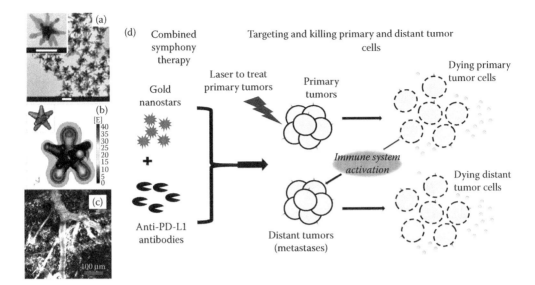

FIGURE 1.6
(a) Transmission Electron Microscopy (TEM) image of GNS. Scale bar, 20 nm; (b) Theoretical 3D polarization-averaged theoretical model, the plasmon peak position and intensity correlate with the branch aspect ratio and branch length/number. (Adapted from Yuan, H et al. 2012. Gold nanostars: Surfactant-free synthesis, 3D modelling, and two-photon photoluminescence imaging. *Nanotechnology* 23(7), 075102. doi:10.1088/0957-4484/23/7/075102.); (c). Two-photon photoluminescence (TPL) imaging of GNS nanoparticles in tumor under window chamber; due to their unique plasmonic properties, GNS emit strong TPL allowing direct particle visualization under multiphoton microscopy. The GNS are shown as white color under TPL imaging. The scale bar is 100 μm. (Adapted from Yuan, H et al. 2012. TAT Peptide-functionalized gold nanostars: Enhanced intracellular delivery and efficient NIR photothermal therapy using ultralow irradiance. *J. Am. Chem. Soc.* 134, 11358–11361. doi:10.1021/ja304180y; Yuan, H et al. 2012. Gold nanostars: Surfactant-free synthesis, 3D modelling, and two-photon photoluminescence imaging. *Nanotechnology* 23(7), 075102. doi:10.1088/0957-4484/23/7/075102.); (d) Principle of Synergistic Immuno Photo Nanotherapy (SYMPHONY) Modality: by disabling the tumor immune resistance using anti-PD-L1 antibodies and simultaneously ablating individual cancer cells using gold nanostars-enabled photothermal therapy, SYMPHONY can trigger a powerful thermally enhanced systemic immune activation to rapidly eradicate locally aggressive as well as distant metastatic cancer. (Adapted from Liu, Y et al. 2016. Synergistic immuno photothermal nanotherapy (SYMPHONY) to treat unresectable and metastatic cancers.)

in designing the SERS nanoprobe, the hairpin configuration has the Raman label in contact or close proximity (<1 nm) to the NPs, thus providing a SERS signal (Figure 1.5). Hybridization with the target DNA opens the hairpin and physically separates the Raman label from the NPs, thus decreasing the SERS effect and quenching the SERS signal upon excitation. The application of the SERS MS nanoprobes in real-time PCR could greatly improve molecular genotyping due to many advantages, such as spectral selectivity due to the sharp, narrow, and molecular-specific vibrational band from Raman labels and the use of a single-laser excitation for multiple labels, which will offer higher multiplexing capabilities over conventional optical detection methodologies. A novel "turn-on" plasmonics-based nanobiosensor, referred to as "inverse Molecular Sentinel (iMS)" nanoprobe was developed for the detection of DNA sequences of interest with an "OFF-to-ON" signal switch [23]. As shown in Figure 1.7, the "stem-loop" DNA probe, having a Raman label at one end, is immobilized onto a nanostar via a metal–thiol bond. A single-stranded linear DNA molecule serving as a "placeholder" strand is partially hybridized to the stem-loop probe via a placeholder-binding region, keeping the Raman dye away from the nanostar

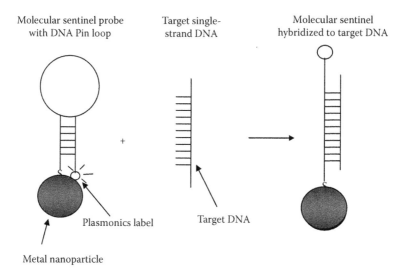

Molecular sentinel probe with DNA Pin loop

Target single-strand DNA

Molecular sentinel hybridized to target DNA

+

Plasmonics label

Target DNA

Metal nanoparticle

FIGURE 1.7
SERS molecular sentinels nanoprobes: SERS signal is observed when the MS probe is in the hairpin conformation (closed-state), whereas in the open state (hybridization to target DNA) the SERS signal is diminished. (Taken from Wust, P. et al. 2002. *Lancet Oncol.* 3, 487–497. doi:10.1016/s1470-2045(02)00818-5).

surface. In this configuration (i.e., in the absence of the target), the probe is "open" with low SERS intensity ("Off" status) as the plasmon field enhancement decreases significantly with increasing distance from the surface. Upon exposure to a target sequence, the placeholder strand leaves the nanostar surface following a nonenzymatic strand-displacement process: the target first binds to the toehold region (i.e., an overhang region of the probe-placeholder conjugate) (Intermediate I) and begins displacing the DNA probe from the placeholder via branch migration (Intermediate II), and finally releases the placeholder from the NP system. This allows the stem-loop structure to "close" and moves the Raman label onto the plasmonics-active surface of the nanostar yielding a strong SERS signal ("On" status). SERS MS technologies could provide useful tools for molecular genotyping with many advantages, such as spectral selectivity due to the sharp, narrow specific vibrational band from Raman labels. Exploitation of the information from the human genome project will make nanomedicine an exciting reality. Current research has indicated that many diseases such as cancer occur as the result of the gradual buildup of genetic changes in single cells. Nanotools based on MS nanoprobes are capable of identifying specific subsets of genes encoded within the human genome that can cause the development of cancer. These techniques can detect infectious diseases such as malaria and dengue [24] and identify early cancer biomarkers, such as microRNAs [25], or the molecular alterations that distinguish a cancer cell from a normal cell at the earliest stages, that is, at the gene level long before manifestation of the disease could be normally detected.

Nanotechnology-based devices and techniques have provided important tools to measure fundamental parameters of biological species at the molecular level. Optical tweezer techniques can trap small particles via radiation pressure in the focal volume of a high-intensity, focused beam of light. This technique, also called optical trapping, could move small cells or subcellular organelles by the use of a guided, focused beam [26]. Ingenious optical trapping systems have also been used to measure the force exerted by individual motor proteins [27]. The optical tweezer method uses the momentum of focused laser beams to hold and stretch single collagen molecules bound to polystyrene beads. The collagen molecules are stretched through the beads using the optical laser tweezer system, and the deformation of the bound collagen molecules is measured as the relative displacement of the microbeads, which are examined by optical microscopy. Optical trapping at the nanoscale allows controlled manipulation and assembly of individual and multiple nanostructures, force measurement with femtonewton resolution, and development of biosensors [28].

Optical technologies using near-field optics or subdiffraction optical resolution allow scientists to analyze nanoscale structures and monitor biomolecular processes of living cells at unprecedented resolutions. Near-field microscopy allows the investigation of the complex electromagnetic fields that surround nanophotonic structures. At nanoscale dimensions, light–matter interactions are defined by an object's geometry and not just by the optical properties of its constituent materials, thus permitting near-field mapping and imaging of nanosystems, leading to a better understanding of the underlying biological processes and studying many processes such as extraordinary optical transmission, light generation, and propagation through photonic crystal waveguides and fibers [29], and the optical response of nanoantennas [30]. NSOM has permitted the investigation of the cellular localization and activity of ABC proteins associated with MDR [31]. Overcoming the diffraction-limited resolution of confocal microscopes, super-resolution microscopy tools such as photo-activated localization microscopy (PALM), stochastic optical reconstruction microscopy (STORM), and stimulated emission depletion (STED) microscopy provide new tools for ultrahigh-resolution imaging and investigation of materials and biological systems at the nanometer scale level [32–35]. Nanopore arrays, where DNA molecules are transported through a small hole in a metal film, have been developed for DNA sequencing. In these devices, the read out of the base-pair sequence is achieved with the use of fluorescent markers that are selectively bound to different base pairs. Signals from multiple nanopores can be detected simultaneously, allowing parallelization with large nanopore arrays [36].

1.4 Conclusion

Research in nanotechnology is experiencing an explosive growth. The technologies illustrated above are just a few examples of a new generation of nanotools developed in this and other laboratories. As illustrated in the various chapters, nanotechnology-related research in laboratories around the world holds the promise of providing the critical tools for a wide variety of biological and biomedical applications. These new analytical tools are capable of probing the nanometer world and will make it possible to characterize the chemical and mechanical properties of cells, probe the working of molecular protein machines, discover novel phenomena and processes, and provide science with a wide range of tools, materials, devices, and systems with unique capabilities. They could ultimately lead to the development of new modalities for early diagnostics and medical treatment and prevention beyond the cellular level to that of individual organelles. Medical applications of nanomaterials could revolutionize biology and healthcare in much the same way that materials science changed medicine decades ago with the introduction of synthetic heart valves, nylon arteries, and artificial joints. The futuristic vision of nanorobots patrolling inside our body armed with antibody-based nanoprobes and nanolaser beams that recognize one cell at a time and kill diseased cells might someday become a practical reality.

References

1. Vo-Dinh, T., Ed. 2005. *Protein Nanotechnology*. Humana Press, Totowa, New Jersey.
2. Wabuyele, M. B., Culha, M., Griffin, G. D., Viallet, P. M., and Vo-Dinh, T. 2005. Near-field scanning optical microscopy for bioanalysis at the nanometer resolution. In *Protein Nanotechnology*, T. Vo-Dinh, *Ed.*, Humana Press, Totowa, NJ, pp. 437.
3. Vo-Dinh, T., Alarie, J. P., Cullum, B., and Griffin, G. D. 2000. Antibody-based nanoprobe for measurements in a single cell. *Nat. Biotechnol.* 18, 764.
4. Vo-Dinh, T. 2003. Nanosensors: Probing the sanctuary of individual living cells. *J. Cell. Biochem. Suppl.* 39, 154.
5. Kasili, P. M., Song, J. M., and Vo-Dinh, T. 2004. Optical sensor for the detection of caspase-9 activity in a single cell. *J. Am. Chem. Soc.* 126, 2799.
6. Vo-Dinh, T. 1998. Surface-enhanced Raman spectroscopy using metallic nanostructures. *Trends Anal. Chem.* 17, 557.
7. Vo-Dinh, T., Allain, L. R., and Stokes, D. L. 2002. Cancer gene detection using surface-enhanced Raman scattering (SERS). *J. Raman Spectrosc.* 33, 511.
8. Frey, B., Weiss, E. M., Rubner, Y., Wunderlich, R., Ott, O. J., and Suaer, R. 2012. Old and new facts about hyperthermia-induced modulations of the immune system. *Int. J. Hyperth.* 28, 528–542. doi:10.3109/02656736.2012.677933.
9. Schildkopf, P., Ott, O. J., Frey, B., Wadepohl, M., Sauer, R., Fietkau, R., and Gaipl, U. S. 2010. Biological rationales and clinical applications of temperature controlled hyperthermia—implications for multimodal cancer treatments. *Curr. Med. Chem.* 17, 3045–3057.
10. Wust, P., Hildebrandt, B., Sreenivasa, G., Rau, B., Gellermann, J., Riess, H., Felix, R., and Schlag, P. M. 2002. Hyperthermia in combined treatment of cancer. *Lancet Oncol.* 3, 487–497. doi:10.1016/s1470-2045(02)00818-5.
11. Hildebrandt, B., Wust, P., Ahlers, O., Dieing, A., Sreenivasa, G., Kerner, T., Felix, R., and Hanno Riess, H. 2002. The cellular and molecular basis of hyperthermia. *Crit. Rev. Oncol./Hematol.* 43, 33–56. doi:10.1016/s1040-8428(01)00179-2.
12. O'Neal, D. P., Hirsch, L. R., Halas, N. J., Payne, J. D., and West, J. L. 2004. Photo-thermal tumor ablation in mice using near infrared-absorbing nanoparticles. *Cancer Lett.* 209(2), 171.
13. Hirsch, L. R., Stafford, R. J., Bankson, J. A., Sershen, S. R., Rivera, B., Price, R. E. et al. 2003. Nanoshell-mediated near-infrared thermal therapy of tumors under magnetic resonance guidance. *PNAS.* 100:13549–13554.
14. Huang, X., Jain, P. K., El-Sayed, I. H., and El-Sayed, M. A. 2008. Plasmonic photothermal therapy (PPTT) using gold nanoparticles. *Lasers Med. Sci.* 23(3), 217–228.
15. Yuan, H., Khoury, C. G., Wilson, C. M., Grant, G. A., Bennett, A. J., and Vo-Dinh, T. 2012. In vivo particle tracking and photothermal ablation using plasmon resonant gold nanostars. *Nanomed.: Nanotechnol. Biol. Med.* 8, 1255–1363.
16. Yuan, H., Fales, A. M., and Vo-Dinh, T. 2012. TAT Peptide-functionalized gold nanostars: Enhanced intracellular delivery and efficient NIR photothermal therapy using ultralow irradiance. *J. Am. Chem. Soc.* 134, 11358–11361. doi:10.1021/ja304180y.
17. Yuan, H., Khoury, C. G., Hwang, H., Wilson, C. M., Grant, G. A., and Vo-Dinh, T. 2012. Gold nanostars: Surfactant-free synthesis, 3D modelling, and two-photon photoluminescence imaging. *Nanotechnology* 23(7), Article number: 075102. doi:10.1088/0957-4484/23/7/075102.

18. Vo-Dinh, T., Fales, A. M., Griffin, G. D., Khoury, C. G., Liu, Y., Ngo, H., Norton, S. J., Register, J. K., Wang, H. N., and Yuan, H. 2013. Plasmonic nanoprobes: From chemical sensing to medical diagnostics and therapy. *Nanoscale* 5, 10127–10140.

19. Liu, Y., Ashton, J. R., Moding, E. J., Yuan, H., Register, J. K., Fales, A. M. et al. 2015. A Plasmonic gold nanostar theranostic probe for *in vivo* tumor imaging and photothermal therapy. *Theranostics* 5, 946–960. doi:10.7150/thno.11974.

20. Liu, Y., Chang, Z., Yuan, H., Fales, A. M., and Vo-Dinh, T. 2013. Quintuple-modality (SERS-MRI-CT-TPL-PTT) plasmonic nanoprobe for theranostics. *Nanoscale* 5, 12126–12131. doi:10.1039/c3nr03762b.

21. Liu, Y., Maccarini, P., Palmer, G., Etienne, W., Zhao, Y., Lee, K., Ma, X., Inman, B. A., and Vo-Dinh, T. 2017. Synergistic immuno photothermal nanotherapy (SYMPHONY) to treat unresectable and metastatic cancers, submitted for publication.

22. Wabuyele, M. and Vo-Dinh, T. 2005. Detection of HIV type 1 DNA sequence using plasmonics nanoprobes. *Anal. Chem.* 77, 7810.

23. Wang, H. N., Fales, A. M., and Vo-Dinh, T. 2015. Plasmonics-based SERS nanobiosensor for homogeneous nucleic acid detection. *Nanomed.: Nanotechnol. Biol. Med.* 11(4), 511–520.

24. Ngo, H. T., Wang, H.-N., Fales, A., Nicholson, B. P., Woods, C. W., and Vo Dinh, T. 2014. DNA bioassay-on-chip using SERS detection for dengue diagnosis. *Analyst* 139(22), 5656–5660.

25. Wang, H. N., Crawford, B. M., Fales, A. M., Bowie, M. L., Seewaldt, V. L., and Vo-Dinh, T. 2016. Multiplexed detection of microRNA biomarkers using SERS-based inverse molecular sentinel (iMS) nanoprobes. *J. Phys. Chem. C* 120(37), 21047–21055.

26. Askin, A., Dziedzic, J. M., and Yamane, T. 1987. Optical trapping and manipulation of single cells using infrared laser beam. *Nature* 330, 769.

27. Kojima, H., Muto, E., Higuchi, H., and Yanagido, T. 1997. Mechanics of single kinesin molecules measured by optical trapping nanometry. *Biophys. J.* 73(4), 2012.

28. Maragò, O. M., Jones, P. H., Gucciardi, P. G., Volpe, G., and Ferrari, A. C. 2013. Optical trapping and manipulation of nanostructures. *Nat. Nanotechnol.* 8, 807–819.

29. Dudley, J. M., Genty, G., and Coen, S. 2006. Super continuum generation in photonic crystal fiber. *Rev. Mod. Phys.* 78, 1135–1184.

30. Rotenberg, N. and Kuipers, L. 2014. Mapping nanoscale light fields. *Nat. Photonics* 8, 919–926.

31. Wabuyele, M. B., Culha, M., Griffin, G. D., Viallet, P. M., and Vo-Dinh, T. 2005. *Near-Field Scanning Optical Microscopy for Bioanalysis at the Nanometer Resolution in Protein Nanotechnology*, T. Vo-Dinh, Ed., Humana Press, Totowa, NJ, pp. 437.

32. Backer, A. S., Lee, M. Y., and Moerner, W. E. 2016. Enhanced DNA Imaging using super-resolution microscopy and simultaneous single-molecule orientation measurements. *Optica* 3, 659–666.

33. Willing, K., Harke, B., Medda, R., and Hell, S. W. 2007. STED microscopy with continuous wave beams. *Nat. Methods* 4, 915–918.

34. Betzig, E. 2014. 3D live fluorescence imaging of cellular dynamics using Bessel beam plane illumination microscopy. *Nat. Protoc.* 9, 1083–1101.

35. Gustafsson, M. G. L. 2005. Nonlinear structured-illumination microscopy: Wide-field fluorescence imaging with theoretically unlimited resolution. *Proc. Natl. Acad. Sci.* 102, 13081–13086.

36. McNally, B., Singer, A., Yu, Z., Sun, Y., Weng, Z., and Meller, A. 2010. Optical recognition of converted DNA nucleotides for single- molecule DNA sequencing using nanopore arrays. *Nano Lett.* 10(6), 2237–2244.

Section I

Nanomaterials, Nanostructures, and Nanotools

2

Self-Assembled Organic Nanotubes: Novel Bionanomaterials for Orthopedics and Tissue Engineering

Rachel L. Beingessner, Baljit Singh, Thomas J. Webster, and Hicham Fenniri

CONTENTS

2.1 Introduction

The intricacies and elegant self-organization of tiny elements into well-defined functional architectures found in nature have been an invaluable source of inspiration for both scientists and engineers. We can all agree that biological materials have evolved complex structures yet simple processes to fit their purpose of durability, multifaceted functionality, programmability, self-assembly, information processing ability, and biodegradability, all of which surpass the current state of the art of materials industries.

The optimized properties and adaptability enjoyed by natural systems arise from their ability to sense their environment, integrate and process information in a controlled fashion, and adapt to new and evolving conditions. Such complexity has attracted multidisciplinary teams of materials engineers, chemists, biologists, and physicists alike, all qualified in one facet of nature, to design materials with similar capabilities. Research in supramolecular engineering of functional biomaterials sprouted from the need for (a) medicine for replacement materials and/or prosthetic devices with mechanical properties of soft and hard tissues such as skin, tendons, and bone, (b) agriculture and forestry for better crops and high value-added wood products, (c) food industries for improving production, quality, texture, processing, and manufacturing,[1] and (d) the biomedical and human health area where there is an insatiable need for ultrasensitive detection methods, diagnostic tools, more effective therapies, and separation technologies.

In the last decade or so, many technologies based on nanomaterials and nanodevices have emerged.[2–5] The purpose of this chapter, however, is to focus specifically on the hierarchical architecture of bone tissue and how we can tailor the next generation of orthopedic implant materials using current knowledge in supramolecular engineering. We will begin by reviewing biological systems in the context of nanoscale materials. For an appreciation of the usefulness of nanoscale materials in the effective repair of the skeletal system, we will review the architectural schemes of bone as a supramolecular bionanomaterial. We will then examine the versatility of a class of self-assembling organic nanotubes called rosette nanotubes (RNTs) and their potential in orthopedic implantology and nanomedicine.

2.2 Bionanosciences: The Art of Replicating the Structure and Function of Biological Systems

We understand that a living cell contains a number of reacting chemicals orchestrated by a complex network of feedback loops and sensing mechanisms, within a finite space that allows various forms of energy to transit across its boundaries. We also understand that the cell is a dynamic structure; self-replicating, energy dissipating, and adaptive. Yet, we have very little idea on how to connect these two sets of characteristics: how does life emerge from a system of chemical reactions? It is accepted today that the transition from the inanimate world of chemical reactions to that of living systems requires a new level of molecular and supramolecular organization. At the commencement of this process, biological systems build their structural components, such as microtubules, microfilaments, and chromatin in the range of 1–100 nm, a range that falls in-between what can be manufactured through conventional microfabrication and what can be synthesized chemically. The associations maintaining these components and the associations of other cellular components seem relatively simple when examined at the atomic scale: shape complementarity, electroneutrality, hydrogen bonding, and hydrophobic interactions are at the heart of these processes. A key property of biological nanostructures, however, is molecular recognition, leading to self-assembly and to the templating of molecular and higher-order architectures. For instance, complementary strands of DNA will pair to form a double helix (diameter = 2 nm). Then, an octamer of histone proteins coils the DNA helix to generate the nucleosome (diameter = 11 nm). The latter forms a "bead-on-a-string" ensemble that folds into higher-order fibers (diameter = 30 nm), which in a few more self-assembly/templating steps lead to the familiar X-shaped chromosomes.[6] This example illustrates three features of self-assembly: (a) the DNA strands recognize each other, (b) they form a predictable structure upon association, and (c) they undergo a hierarchical and templated self-organization process leading to a functional chromosome. The process does not end here; the chromosomes are the repository of the genetic information and are thus in a constant and dynamic relationship with the cellular maintenance and replication machinery.

A comparison of synthetic self-assembling nanoscale materials and biological materials reveals some key differences. First, many biological materials possess well-defined hierarchical architectures organized into increasing size levels adapted to meet the functional requirements of the material. If we zoom in and out of these structural entities, we observe recognizable architectures ordered as substructures with scales spanning many orders of magnitude from whole organisms to subnanometer components. Such pervasive tendency for biological materials to undergo a hierarchical organization confers unique physical and chemical properties rarely paralleled in man-made materials. For example, bones are organized into finer structures made of cells, collagen, and minerals. This arrangement confers strength to bone and a mechanism for active bone regeneration. Collagen itself self-assembles from procollagen molecules into triple-helical collagen fibrils and fibers that play important roles in the overall structure of various body tissues.[7,8]

Second, self-assembly and order in biological systems are driven by function.[9] For example, integral proteins aggregate to form focal points only when the cell begins to anchor on a surface. Filament structures responsible for cell repair appear only when a defect in the cell membrane exists, after which they disappear and/or cease to function. In contrast, synthetic materials require stepwise preparation, often irreversible, to generate the desired structure and to incorporate functionality.[9]

Third, biological systems are dynamic. Channel proteins, for example, enter "on" and "off" states to allow select ions to pass through depending on the chemical environment and cellular needs. Finally, biological systems are responsive, adaptive, and restorative. Classic examples are the directed response in muscle tissues to loads and the many repair mechanisms in DNA.[6,10] As Jeronimidis elegantly pointed out, design is the expression of function, which very often includes achieving compromises between conflicting requirements while extracting maximum benefit from the materials used.[11]

2.3 Supramolecular Engineering

2.3.1 Intermolecular Forces

Traditional organic chemists have for the past two centuries examined the *reactions* of molecules rather than their *interactions*.[12–22] Supramolecular chemists are interested in both because the synthesis of molecular assemblies requires designing and synthesizing building blocks capable of undergoing self-organization through intermolecular bonds akin to how nature holds itself. Driven by thermodynamics, self-assembling systems form spontaneously from their components. This also implies that they are in a dynamic equilibrium between associated and dissociated entities.[12–43] This feature confers a built-in capacity for error correction, a feature not available in fully covalent systems.

Noncovalent interactions include hydrogen bonding (H-bond) and $\pi-\pi$ interactions. Dispersion, polarization, and charge transfer interactions, combinations of which make up van der Waals forces, also play a role. The term H-bonds was used to describe the special structure of water. Consider molecules, A-H and B, where A in A-H and B are electronegative atoms (e.g., O, N, S, F, Cl). H-bonds occur when the hydrogen atom bonded to A (H-bond donor) is electronically attracted to B (H-bond acceptor). H-bonds can occur intramolecularly or intermolecularly. Individual H-bonds tend to be weak. However, collectively, they can contribute significant strength to a system. For neutral species, H-bond strengths are typically in the order of 5–60 kJ/mol. A distinct feature of H-bonds is their inherent directionality, which is well suited for achieving structural complementarity in supramolecular systems as will be seen in RNTs (Section 2.7). van der Waals interactions are long-range inductive or dispersive intermolecular forces. These interactions occur between nonpolar molecules at distances larger than the sum of their van der Waals radii. Although the magnitude of these forces varies as an inverse power of distance between the interacting species, and are thus weak, their effects are also additive. The inductive forces include attractive permanent dipole–dipole and induced dipole–dipole interactions. The dispersion forces (also known as London dispersion forces), on the other hand, result from fluctuations of electronic density within molecules.

$\pi-\pi$ interactions involve London dispersion forces and the hydrophobic effect. This form of stabilizing interaction is commonly found in DNA where the vertical base stacking contributes a significant stabilizing force to the double helix. In an aqueous environment, an unfavorable entropy effect occurs as a result of polar solvent molecules trying to order themselves around apolar (or hydrophobic) molecules. This unfavorable entropy provides a driving force for hydrophobic solute aggregation to reduce the total hydrophobic surface area accessible to polar solvent molecules. This form of binding can thus be described as the association of nonpolar regions of molecules in polar media, resulting from the tendency of polar solvent molecules to assume their thermodynamically favorable states. The hydrophobic effect is a salient force in, for instance, micelle formation, protein–protein interactions, and protein folding.

2.3.2 From Molecular to Supramolecular Chemistry

The heart of supramolecular chemistry lies in the increasing complexity beyond the molecule through intermolecular interactions. It is the creation of large, discrete, and ordered structures from molecular synthons. Since Wöhler's synthesis of the first organic molecule, urea in 1828,[44] organic chemists have masterfully developed a cache of synthetic methods for constructing molecules by making and/or breaking *covalent* bonds between atoms in a controlled and precise fashion.[19] However, nature's way of organizing and transforming matter from elementary particles to sophisticated functional structures has prompted chemists to think beyond the covalent bond and the molecule. Supramolecular chemists are thus concerned with forming increasingly complex molecules that are held together by *noncovalent* interactions. Lehn defined supramolecular chemistry as a sort of "molecular sociology," where the noncovalent interactions define the intercomponent bond, action, reaction, and behavior of individual and populations of molecules.[19] Supermolecules are thus ensembles of molecules having their own organization, stability, dynamics, and reactivity.

Because the collective properties and function of materials depend on both the nature of its constituents and the interactions between them, it is anticipated that the art of building supermolecules will pave the way for designing artificial abiotic systems capable of displaying evolutive processes with high efficiency and selectivity, similar to natural systems. As we go further down in the scales, due to the difficulty in manipulating individual molecules and atoms, scientists and engineers developed self-assembly and supramolecular synthesis as new tools to overcome this challenge.[45]

Self-assembly and self-organization processes are the thread that connects the reductionism of chemical reactions to the complexity and emergence of a dynamic living system. Understanding life will therefore require understanding these processes. Broadly defined, self-assembly[12,13,23–26,46–51] is the autonomous organization of matter into patterns or structures without human intervention. The principles of artificial self-assembly are derived from nature and its processes, and an understanding of these principles allows us to design nonbiological mimics with new types of function. Large molecules (e.g., histones), molecular aggregates (e.g., chromosomes), and complex forms of organized matter (e.g., cells) cannot be synthesized bond by bond. Rather, a new type of synthesis based on noncovalent forces is necessary to generate functional entities from the bottom up. This field of chemistry, termed supramolecular synthesis,[14–21] is at the heart of nanoscale science and technology.

The organization of matter brought about through supramolecular synthesis makes feats of molecular engineering possible that are virtually unthinkable from a covalent perspective. The challenge lies both in the chemical design and synthesis: the conceptualization of an organized state of matter is intimately linked with the chemical information embedded in the molecules in the form of charges, dipoles, and other functional elements necessary to translate chemical information into substances. Much of the research endeavor has been devoted to the use of noncovalent bonds as the alphabet for chemical information encoding, and the structures expressed have spanned the range of dimensions and shapes, from discrete[13–17,24,27–35] to infinite[12,20,21,23,25,36–42] networks. A step forward toward harnessing the noncovalent interaction is not only instructing the molecules to generate well-defined static assemblies but also designing them so that the ultimate entity displays a dynamic relationship with its environment, the ability to adapt, evolve, and self-replicate.

Given that supramolecular engineering can generate materials with tunable chemical, physical, and mechanical properties, this field has enormous potential in musculoskeletal tissue engineering and nanomedicine. The main goal of this review is to describe a novel approach for bone implant design based on supramolecular engineering. We will also briefly describe the potential of supramolecular materials in nanomedicine.

2.4 Why Are New Orthopedic Implant Materials Needed?

In the United States alone, an estimated 11 million people have received at least one medical implant device. In 1992, of these implants, orthopedic fractures, fixation, and artificial joint devices accounted for 51.3%.[52] If we examine the growth rate of joint replacements, surgery rates increased by 101% between 1988 and 1997 (Figure 2.1). The use of shoulder replacement increased by 126% and knee replacement rates increased by 120%.[53] Since 1990, the total number of hip replacements, which is the replacement of both the femoral head and acetabular cup with synthetic materials, has been steadily increasing. In fact, the 152,000 total hip replacements in 2000 is a 33% increase from the number performed in 1990 and a little over half of the projected number of total hip replacements (272,000) by 2030. These numbers attest to the increasing demand in orthopedic replacement and fixation devices.

Due to surgery, hospital care, physical therapy costs, and recuperation time, implanting devices is not only very costly but also involves considerable patient discomfort.[54] If postimplantation surgical revision becomes necessary, due to material failure under physiological loading condition, insufficient integration of implant to juxtaposed bone, and/or host tissue rejection, both cost and patient discomfort increase steeply. For instance, in 1997, 12.8% of the total hip arthroplasties were simply due to revision surgeries of previously implanted failed hip replacements.[54] Furthermore, since revision surgeries require the removal of large amounts of healthy bone, most people can undergo only one such revision. This finite number of surgical revision calls for implants that can last for 20–60 years or more, especially for younger and more active patients with joint and bone complications. Bone nonunions,

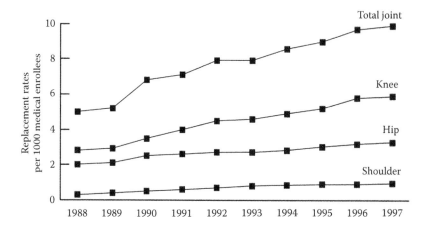

FIGURE 2.1
Growth in rates of joint replacement (1988–1997). Joint replacement surgery rates increased by 101% between 1988 and 1997. The use of shoulder replacement increased from 0.35 to 0.79 per 1000 enrollees (126%) and knee replacement rates increased from 2.7 to 5.9 per 1000 enrollees (120%). (Adapted from Praemer, A. et al. 1992. *Musculoskeletal Conditions in the United States*, The American Academy of Orthopaedic Surgeons, Park Ridge, IL; the original figure is copyrighted by the Trustees of Dartmouth College.)

implant loosening owing to poor osseointegration of implant and osseodegradation of bone surrounding the implant, are all difficult clinical problems. All of these conditions lead to acute pain and poor mobility. These problems are reasons why careful design is necessary to improve the functional lifetime of implants and to promote new bone growth on the surface of an orthopedic implant material (osseointegration) in order to reduce costs associated with prostheses retrieval and reimplantation. These problems have driven engineers and scientists to re-examine and investigate improvements in the design and formulations of current orthopedic implant technology. In order to develop new strategies for fabricating materials useful in the repair of our skeletal system, we need to identify the hierarchical and supramolecular organization of bone responsible for its unique load-bearing properties.

2.5 Bone Architectural Hierarchy, Function, and Adaptability

Our body is supported by the skeletal system, which has evolved an optimum design adapted for mechanical requirements, nutritional reserves, muscular power, and a compromise between size and weight. As Ontanon suggested, the direction of evolution was based on two premises: (a) macroscopic and microscopic features that are present to minimize working stresses and (b) appropriate distribution of material to achieve optimal density.[55]

Bone is one of the many specialized connective tissues in our body that is rich in extracellular matrix (ECM) made up of various proteins and minerals. It is a highly organized and strong structure with efficient modulation of scaffolding and mineral crystal arrangement at the molecular level. The synergy between molecular, cellular, and tissue arrangement provides excellent tensile strength for such a support structure.[56–58] Like most other complex systems in nature, bone evolves from irregularly organized structures (woven bone) at the time of birth to more patterned architectures in adulthood (lamellar bone) (Figure 2.2). In the embryonic skeleton, collagen fibers and vascular spaces are irregularly

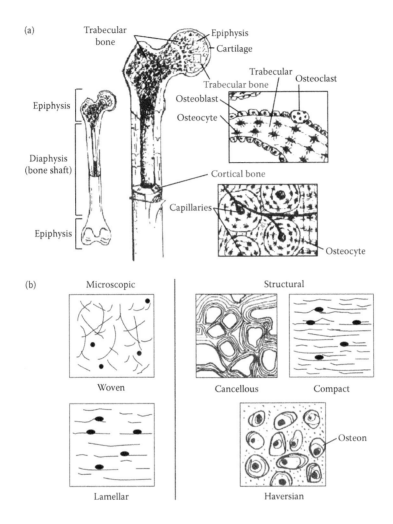

FIGURE 2.2
(a) Bone anatomy. Normal adult bone is organized into cortical (compact) and trabecular (cancellous) bone. Cortical bone is found in the diaphysis (bone shaft) of long bones and serves as protective surfaces for all bones. Trabecular bone is found in the epiphysis of bones and exists as interconnected networks of rods and plates with the lattices oriented toward the direction of principal stress. Osteoblasts (OBs) are bone-depositing cells; osteoclasts are bone-resorbing cells; osteocytes are bone-maintaining cells. Osteocytes form cellular networks that are believed to respond to mechanical deformation and loading in bone, forces that exert strong influences on bone shape and remodeling. (Reprinted from *Advances in Chemical Engineering. Nanostructured Materials*, Webster, T. J., Nanophase ceramics: The future orthopaedic and dental implant material, pp. 125–166, Copyright 2001, with permission from Elsevier.) (b) Types of bone. Embryonic bone is characteristically woven bone. By age 4, most immature woven bone is replaced by mature lamellar bone. Lamellar bone is found throughout the mature skeleton in both cancellous and compact bone. Lamellar bone has anisotropic properties due to the highly organized and stress-oriented collagen fibers. Compact bone has four times the mass of cancellous bone though the latter has a metabolic turnover eight times greater than the former due to the high surface area for cellular activity. Cancellous bone is found in the metaphysis and epiphysis of long bones. Cancellous bone is subjected predominantly to compression forces while compact bone is subjected to bending, torsional, and compressive forces. The most complex type of compact bone is the Haversian bone, which is composed of vascular channels surrounded by lamellar bone, forming what is known as osteons (major structural units of compact bone oriented in the long axis of bone). (Adapted from Bostrom, M. P. et al. 2000. Form and function of bone. In *Orthopaedic Basic Science: Biology and Biomechanics of the Musculoskeletal System*, 2nd ed., J. A. Buckwalter, T. A. Einhorn, and S. R. Sheldon, Eds., The American Academy of Orthopaedic Surgeons, Park Ridge, IL.)

arranged in the form of interlacing networks.[59] With time, successive layers of bone are deposited in areas that were previously vascular channels to form an organized structure possessing anisotropic mechanical properties.[54] Remodeling continues as initiated by osteoclastic resorption to create longitudinally oriented tubular channels known as Haversian systems in adult bone.[59]

Normal adult bone is structurally organized into 80% cortical (or compact) and 20% trabecular (or cancellous) bone (Figure 2.2).[60] Cortical bone is found in the diaphysis of long bones and comprises the outside protective surfaces of all bones. Due to its dense nature (80%–90% calcified) and low porosity (30% with small pore size of up to 1 mm in diameter), cortical bone is suited for mechanical, structural, and protective functions.[54,60] Trabecular bone, on the other hand, is found in the metaphyses and epiphyses of bones and exists as three-dimensional interconnected networks of rods and plates (trabeculae)[59] that are organized into branching lattices oriented toward the direction of principal stress.[54] Trabecular bone is less dense (5%–20% calcified), has a higher porosity (90% with large pores up to several millimeters in diameter), and as a result, a higher metabolic activity and water content compared with cortical bone.

The precise composition of bone varies with age, species, gender, bone type, and bone health. Bone is composed of approximately 70% inorganic material, 20% organic, and 10% water (Chart 2.1).[54] The organic phase, which is primarily collagen type I (90%) and amorphous ground substance (10%), makes up the protein scaffold in which the inorganic phase resides.[59]

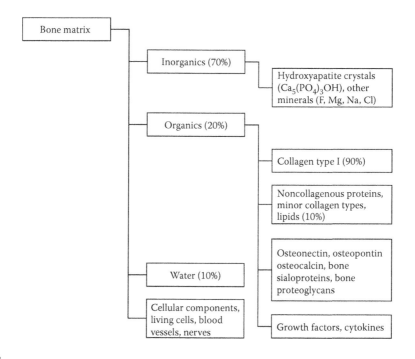

CHART 2.1
Bone components serving structural and regulatory functions. Bone matrix is approximately 70% inorganic material, 20% organic material, and 10% water. The main component of the organic phase is type I collagen. Mineral salts of calcium phosphate exist in the form of hydroxyapatite (HA) crystals and are arranged within the collagenous organic matrix. Bone function is accomplished by OBs, osteoclasts, and osteocytes, which are supplied as progenitor cells by blood vessels.

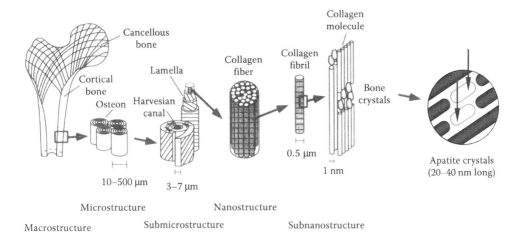

FIGURE 2.3

Hierarchical levels of bone. At the scale of 100s of microns, we see osteons, which are composed of cylindrical lamella surrounding longitudinal vascular canals called Haversian canals. Lamella exists at the scale of 10s of microns. They are repeating patterns of collagen fibroarchitecture arranged in an antiparallel manner to confer strength. Osteocytes (bone-maintaining cells) exist in the spaces of the lamella. As we proceed into the nanostructure and subnanostructure regime, we see the hierarchical framework of collagen going from fibers and fibrils to molecules. Within the collagen molecules reside the bone apatite crystals. Minerals make bone rigid and proteins (mainly collagen) provide strength and elasticity. Of main interest in current orthopedic biomaterials research are the nanostructure and subnanostructure levels of bone, which have yet to be accurately replicated in current orthopedic implants. (Adapted from Cowin, S. C. et al. 1987. *Handbook of Bioengineering*, McGraw-Hill, New York.)

The inorganic phase is composed of mineral salts of calcium phosphate and calcium carbonate in the form of HA crystals (20–80 nm in length and 4–6 nm thick in the human femur).[54] These crystals are arranged in an orderly pattern within the collagenous organic matrix, which is composed of ~300 nm long collagen fibrils.[54,55,61] In addition to the mineralization of bone, rigidity and strength also arise from the hierarchical arrangement of bone constituents (Figure 2.3). Bone signaling and function (e.g., bone matrix synthesis, bone remodeling, and mineral deposition) are modulated by bone cells (OB, osteoclast, and osteocyte) and various macromolecules embedded within the mineralized organic matrix. Some macromolecules include growth factors, cytokines, bone-inductive proteins such as osteonectin, osteocalcin, and osteopontin, lipids, and adhesive proteins such as vitronectin, laminin, and fibronectin.[54]

We can now appreciate how bone itself is a nanobiomaterial and in order to generate efficacious implants that possess cementitious, regenerative, or pseudoplastic[63] properties, material formulations should identify with natural systems to promote integration at a biologically relevant length scale.

2.6 Nanostructured Materials: The Next Generation of Orthopedic Implants

For effective functioning of implant materials, some characteristics must be met. Implants must display: (a) no toxicity, (b) resistance to corrosion in body fluids, (c) sufficient strength for normal and involuntary motion and loading, (d) resistance to fatigue, (e) an ability to promote cell adhesion, which is a key step toward subsequent cell function of anchorage-dependent cells, and (f) biocompatibility with the host tissue or organs.[64] Biomaterials used in orthopedic surgery have been reviewed previously.[65–68]

Healing characteristic of unsuccessful implants include fibrous encapsulation and chronic inflammation (Table 2.1).[69] The healing response, which is affected by the implant's chemical and physical characteristics following implantation, determines its lifetime.[64] Therefore, our task is to control these responses by (a) improving compatibility to decrease unwanted fibrous tissue formation; (b) investigating novel material formulations that can elicit specific, chronological, and desirable responses from surrounding cells and tissues to support osseointegration and enhance deposition of a mineralized matrix; and (c) developing a better match in mechanical (flexural strength, bending, modulus of elasticity, toughness, ductility) and electrical characteristics (resistivity, piezoelectricity) to bone tissue.

With this goal in mind and the knowledge that naturally occurring bone ECM is a nanostructured entity, nanomaterials have been envisaged as a potential solution to the obstacles described above.[64] Progressing into such dimensions is very significant because we are approaching the approximate size of the largest biological molecules: proteins[70] and DNA.[6,10,43] Compelling *in vitro* data have shown enhanced OB (bone-forming cells) function on nanostructured surface of carbon nanofibers,[71–73] ceramics,[54,74–77] poly(L-lactide-block-acrylic acid) (PLLA),[78,79] and organoapatite nanocrystals.[80–82] Yet, orthopedic materials such as titanium (Ti) do not possess desirable nanometer surface

TABLE 2.1

Biomaterial–Tissue Interaction

Effect of Implant on the Host	
Local	• *Blood–material interactions*: Protein adsorption, coagulation, platelet adhesion, activation and release, fibrinolysis, complement activation, hemolysis
	• Infection
	• Toxicity
	• *Modified healing*: Encapsulation, foreign body reaction, pannus formation
	• Tumorigenesis
Systemic	• *Embolization*: Thrombus formation
	• Hypersensitivity
	• Elevation of implant elements in the blood
	• Lymphatic particle transport
Effect of Host on Implant	
Physical	• Abrasive wear
Mechanical	• Fatigue
	• Stress–corrosion cracking
	• Corrosion (especially metals)
	• Degeneration and dissolution
Biological	• Adsorption of substances from tissues
	• Enzymatic degradation
	• Calcification

Source: Adapted from Anderson, J. M. et al. 1996. *Biomaterials Science: An Introduction to Materials in Medicine,* Academic Press, New York, pp. 165–214.

Note: In considering biomaterial–tissue interactions, it is important to know both the effect of implant on the host and the effect of host on the implant. The former can have local and systemic effects while the latter can have physical/mechanical and biological effects.

feature, which is believed to be a reason why these materials sometimes fail clinically. The knowledge that cells *in vivo* interact with nanometer-sized structures led us to study the impact of RNTs, a nanotubular assemblage with biologically inspired chemistries, on bone cell adhesion (*vide infra*).[83]

2.7 Rosette Nanotubes: Self-Assembling Organic Nanotubes with Tunable Properties

Unidimensional nanotubular objects have captivated the minds of the scientific community over the past two decades because of their boundless potential in nanoscale science and technology. The strategies developed to achieve the synthesis of these materials spanned the areas of inorganic[84–97] and organic[98–118] chemistry and resulted in, for instance, carbon nanotubes,[84] peptide nanotubes,[38,102] as well as surfactant-derived tubular architectures.[119–133] While inorganic systems benefit from the vast majority of the elements of the periodic table and the rich physical and chemical properties associated with them, organic systems inherited the power of synthetic molecular[134,135] and supramolecular[14–21] chemistry. As such, the latter approach offers limitless possibilities in terms of structural, physical, and chemical engineering. Following is a brief description of the design, synthesis, and investigation of a class of *adaptive* nanotubular architectures resulting from the self-assembly and self-organization of biologically inspired materials. The key advantages of this strategy are its compatibility with physiological conditions, and its generality, as it relies on a constant scaffold that can be tailored to achieve different functions.

2.7.1 Design

For a system based on hydrogen bonds to self-assemble in water, one has to balance the enthalpic loss (H-bonds) with a consequent entropic gain (stacking interactions and hydrophobic effect). If preorganized, ionic H-bonds could also add to the enthalpic term. Nature has ingeniously taken advantage of these design principles to compartmentalize the cell (membranes), and to create thermodynamically favorable pathways for protein and nucleic acid folding.

With this in mind, the heterobicyclic base G∧C (Figure 2.4a) was designed and synthesized to be a hydrophobic base unit possessing the Watson–Crick donor–donor–acceptor (DDA) H-bond array of guanine and acceptor–acceptor–donor (AAD) of cytosine. Because of the asymmetry of its hydrogen bonding arrays, their spatial arrangement, and the hydrophobic character of the bicyclic system, G∧C undergoes a hierarchical self-assembly process fueled by hydrophobic effects in water to form a six-membered supermacrocycle maintained by 18 H-bonds (rosette, Figure 2.4b). The resulting and substantially more hydrophobic aggregate self-organizes into a linear stack defining an open central channel 1.1 nm across, running the length of the assembly, and up to several millimeters long (Figure 2.4c). A covalently attached functional group, such as an amino acid moiety, dictates the supramolecular chirality of the resulting assembly. This tri-block design endows the modules with elements essential for the sequential self-assembly into stable nanotubular architectures (Figure 2.4c). The inner diameter is directly related to the distance separating the hydrogen bonding arrays within the G∧C motif while the peripheral diameter and its chemistry are dictated by the choice of the functional groups appended to this motif.

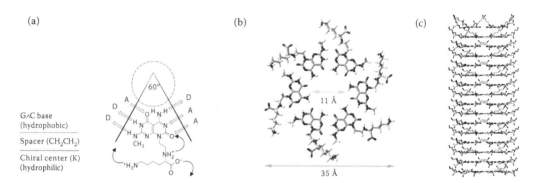

FIGURE 2.4

(a) Design features of a self-assembling module leading to the helical RNTs. (b) The G∧C motif self-assembles spontaneously in water to form a six-membered supermacrocycle (rosette) maintained by 18 H-bonds. (c) Second level of organization involves several rosettes stacking up to form a nanotube 3.5 nm in diameter, with a hollow core 1.1 nm across and up to several millimeters long.

COSY, ROESY, NOESY, and TOCSY 2D ^{1}H-NMR spectroscopy (600 MHz, 90% H_2O/D_2O) and electrospray ionization mass spectrometry (ESI-MS) established the self-assembly of the G∧C motif into six-membered supermacrocycles in water.[115] Circular dichroism (CD) spectroscopy, variable temperature UV-visible melting studies, dynamic light scattering (DLS), and small angle x-ray scattering (SAXS) provided additional evidence in support of this and of the formation of chiral tubular assemblies. Finally, transmission electron microscopy (TEM, Figure 2.5a,b), atomic force microscopy (AFM, Figure 2.5c–h), and scanning tunneling microscopy provided us with visual evidence of the formation of the proposed nanotubular assemblies, of their hollow nature, and corroborated the spectroscopic investigations.[115–117]

2.7.2 What Is Novel and Versatile about RNT?

As outlined earlier, orthopedic implants with surface properties that promote cell and tissue interactions for improved implant osseointegration are needed. Properties, which include surface area, charge, and topography, depend on the dimensions of the material. In this respect, nanostructured materials by their very nature possess higher surface area and exhibit enhanced magnetic, catalytic, electrical, and optical properties over conventional formulations of the same material, and thus have an advantage in biomedical applications.[54]

In principle, upon self-assembly, any functional group covalently attached to the G∧C motif could be expressed on the surface of the nanotubes, thereby offering a general "built-in" strategy for tailoring the physical and chemical properties of RNTs. As a proof of concept, we have covalently modified the G∧C motif with L-Lys and D-Lys and demonstrated that the nanotubes' helicity was dictated by the chirality of the amino acid moiety.[115]

We have synthesized over 150 G∧C derivatives to self-assemble into RNTs that express structurally and electronically diverse functional groups including aromatic[137] and alkyl moieties,[138] nanoparticles,[139] inorganic complexes,[140] carbohydrates,[141] peptides, and amino acids.[115–117,142,143] We have also developed a strategy whereby the nanotubes' properties can be altered after self-assembly. In this "dial-in" approach, the G∧C motif was designed so that the resulting nanotubes would express evenly distributed anchor points on their

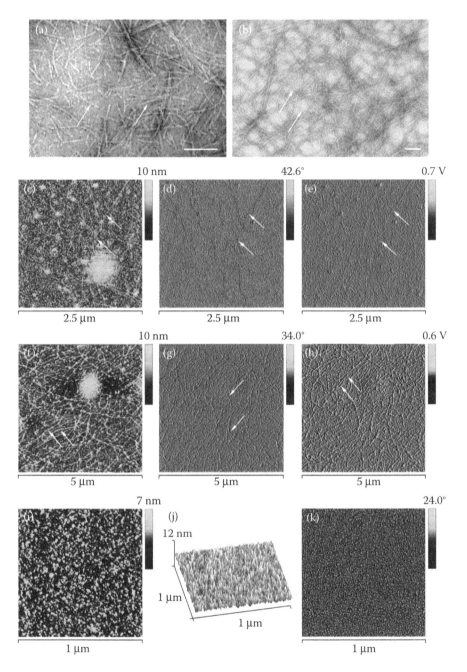

FIGURE 2.5
TEM and tapping-mode AFM (TM-AFM) images showing the effect of temperature on the degree of aggregation of RNT-K1. TEM image of negatively stained[115–117] (a) unheated (–T) and (b) heated (+T) RNT-K1 showing the formation of extensive networks, sheets, and bundles of long RNT-K1 (scale bar = 50 nm). TM-AFM of –T RNT-K1 (c: height; d: phase; e: amplitude) and +T RNT-K1 (f: height; g: phase; h: amplitude) corroborating the TEM data. Uncoated Ti (i: height, j: 3D height, k: phase) shows a relatively smooth surface with a maximum peak height of 10.5 nm. (Reproduced from Chun, A. L. et al. 2005. *Biomaterials* 26, 7304–7309. With permission.)

outer surface for further modification with external molecules (promoters).[116] This system was developed to establish RNTs as stable, yet noncovalent scaffolds for the self-assembly of helical nanotubes with tunable chiroptical properties.[117] This strategy offers a powerful approach to literally "dial-in" the desired properties by simply selecting promoters featuring the desired property. In a related study, we also reported the synthesis and characterization of a chiral G∧C motif that self-assembles into RNTs in MeOH and undergoes mirror image supramolecular chirality inversion upon the addition of very small amounts of water.[144] Physical and computation studies established that the mirror image RNTs obtained (called *chiromers*) result from thermodynamic (in water) and kinetic (in MeOH) self-assembly processes involving two conformational isomers of the parent G∧C motif.

Along with functionalization, another key aspect of the RNTs' flexible design is the tunability of their channel diameters and properties. An extended tricyclic version of the G∧C scaffold has been engineered, for example, to feature the same G and C face H-bonding arrays, but with a pyridine ring positioned in between (Figure 2.6).[145,146] Upon self-assembly, RNTs have a channel diameter of 1.4 nm, which is larger than that of the self-assembled bicyclic G∧C motif (1.1 nm).[115] Extending the G∧C core also leads to unique RNT J-type properties,[147] indicative of stronger dipole transitions between adjacent molecules. In addition to this tricyclic module, a tetracyclic version has also been synthesized in order to generate RNTs with an even larger inner diameter of 1.7 nm[148] (Figure 2.6, right).

Due to their mechanism of formation and the flexible synthetic scheme employed,[115] these nanotubular constructs are novel in many ways. First, such a scaffold is well suited for anchorage-dependent cells such as OB. The noncovalent nature of these materials also permits association/dissociation under the right thermodynamic conditions, which is an important criterion in replacement therapy. Second, the synthetic accessibility of RNT allows tailoring of the surface functionalities for different applications. Third, RNTs possess nanometer features that resemble naturally occurring nanostructured constituent components

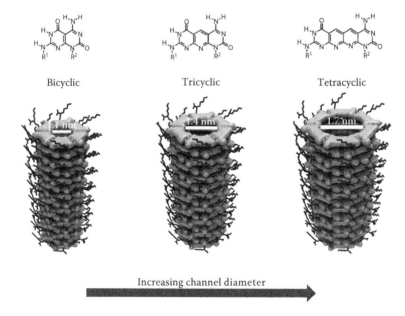

FIGURE 2.6
Self-complementary bicyclic, tricyclic, and tetracyclic (top, left to right) G∧C motifs and their respective self-assembly into RNTs (bottom).

in bone (such as collagen fibers and HA crystals) that bone cells are accustomed to interacting with. Much of the literature has revealed the relevance of surface nanotopography as a major selective parameter in promoting OB adhesion and its subsequent function.[54,64,77–79]

For example, one can attach to the G∧C motif, growth factors,[149] and/or specific bone recognition peptide sequences that will attract bone cell adhesion[150] (in place of lysine in Figure 2.4). Lysine and arginine functionalities on RNT (RNT-K1 and RNT-R1, respectively), two positively charged amino acids, have been shown to increase OB adhesion on RNT-coated Ti surfaces by about 50% compared with uncoated Ti (Figure 2.7).[83] RNT-K1 similarly enhances endothelial cell density and thus holds promise for vascular stent applications as well.[151] Activation of OB membrane integrin receptors using a coassembled mixture of RNT-RGD and RNT-K1 in a 95:5 molar ratio (RNT-K1[95]/RGD[5]) further enhances OB adhesion by 124% relative to the uncoated Ti substrate.[152] Likewise, RNTs expressing the cell-adhesive domain KRSR improve OB adhesion by about the same amount (122% relative to uncoated Ti) and, importantly, are selective relative to fibroblast (soft-tissue forming cells) and endothelial (cells that line the vasculature) cell adhesion.[152] Moreover, OB adhesion on Ti coated with RNT-KRSR is 84% higher than Ti coated with the KRSR peptide alone, which highlights the unique biomimetic features of RNTs for promoting OB functions.

Because these materials undergo extensive self-assembly at higher temperatures resulting in networks of long nanotubes and higher aggregation states of bundles and sheets (Figure 2.5), the effect of heated versus unheated RNT on OB adhesion in the presence and absence of proteins has also been examined (Figure 2.8).[136]

Figure 2.8 shows that (a) under both −S and +S conditions, +T and −T RNT-K1-treated Ti performed better than both uncoated Ti and glass. (b) In the case of −T RNT-K1-coated Ti, enhanced OB adhesion was observed regardless of the presence/absence of proteins. (c) In the case of +T RNT-K1-coated Ti, +S conditions led to a significant enhancement compared with −S conditions. (d) When comparing +T and −T RNT-K1-coated Ti under −S

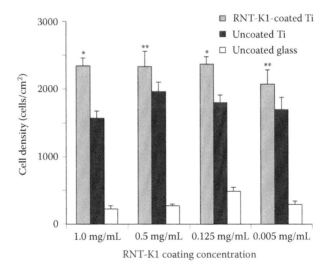

FIGURE 2.7
Human osteoblast adhesion on RNT-K1-coated Ti. RNT-K1-coated Ti showed a higher cell density than uncoated Ti (control) and glass (reference). RNT-K1 concentration did not show any trend. No significant differences between test groups indicate consistency of experimental method. (Data are mean ± SEM; $n = 3$; *$p < 0.01$; **$p < 0.10$ when compared to uncoated Ti; adhesion time = 1 h). (Reproduced from Chun, A. L. et al. 2004. *Nanotechnology*, 15, S234–S239. With permission.)

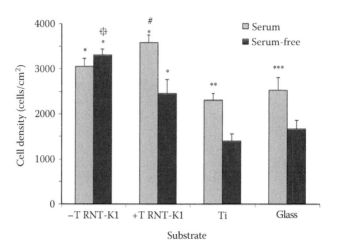

FIGURE 2.8

Human osteoblast (OB) adhesion on +T (heated) and −T (unheated) RNT-K1-coated Ti substrates under serum (+S) and serum-free (−S) conditions. Data are mean ± SEM; $n = 3$. *$p < 0.01$ when compared with uncoated Ti under +S and −S conditions, respectively. #$p < 0.01$ when compared with +T RNT-K1 under −S condition. ⊕$p < 0.05$ when compared with +T RNT-K1 under −S condition. **$p < 0.01$ when compared with uncoated Ti under −S condition. ***$p < 0.01$ when compared with glass under −S condition. (Reproduced from Chun, A. L. et al. 2005. *Biomaterials*, 26, 7304–7309. With permission.)

conditions, the latter displayed better adhesion than the former ($p < 0.05$), whereas under +S conditions, this trend was inverted. (e) Both uncoated Ti (negative control) and glass (reference) showed better OB adhesion under +S conditions versus −S conditions.

TEM images of −T and +T RNT-K1-coated Ti (Figure 2.5a,b) show a greater density of nanotubes and extensive networks in the latter case. The diameter measured from these images (3.4 ± 0.3 nm) was in agreement with the computed value (3.5 nm). TM-AFM of −T RNT-K1 (Figure 2.5, c: height, d: phase, e: amplitude) and +T RNT-K1 (Figure 2.5, f: height, g: phase, h: amplitude) corroborate the TEM results. These images resulted in a tube cross-section of 3.1 nm, in agreement with the TEM data. Uncoated Ti (Figure 2.5, i: height, j: 3D height, k: phase) showed a relatively smooth surface with a maximum peak height of 10.5 nm.

Figure 2.8 showed that proteins are necessary in the heated samples for enhanced OB adhesion, while they had no effect on the unheated RNT samples. This suggests an active role played by unheated RNT in promoting OB adhesion. Furthermore, in the absence of proteins, heated RNT displayed the same level of OB adhesion as uncoated Ti in the presence of proteins, which imply that RNT may either (a) possess certain signaling properties resembling those found in proteins that are known to enhance OB adhesion, or (b) embody a unique disposition that induces a different signaling mechanism. Although the underlying molecular basis for enhanced OB adhesion and its relationship to the presence/absence of proteins remains unclear at this stage, it is reasonable to postulate that lysine clusters on the RNT surface may act like certain lysine-rich bovine proteins known to promote OB adhesion, proliferation, and differentiation.[153] This hypothesis is further substantiated by the importance of lysine in cross-linking of bone collagen, which is significant in bone matrix formation, fracture healing, and bone remodeling.[154–156]

The act of heating RNTs-K1 to higher temperatures (+T) or treating with the serum-supplemented media (+S) described in these experiments causes the solution to undergo a visible-phase transition from a liquid to a viscous gel.[136] In taking advantage of this property, RNTs have also been embedded (or coat) within widely used pHEMA (poly(2-hydroxyethyl methacrylate)[157,158] hydrogels,[159,160] in order to create a more optimal material for OB functions. Indeed, pHEMA hydrogels embedded with RNTs-K1[90]/RGD[10] or RNT-K1 increase OB adhesion by 197% (after 4 h) and OB density by 115% (after 3 d), respectively, relative to pHEMA hydrogel controls.[160] Further elaborating the pHEMA-RNTs-K1 composites with 2 wt% nanocrystalline HA not only supports greater tensile stresses and greater comprehensive and tensile moduli,[161,162] but also increases the OB densities by 235% and 99% relative to hydrogel controls and 2 wt% nanocrystalline HA containing hydrogels, respectively.[162] Interestingly, RNTs have also been found to behave as a biomimetic template for HA deposition.[161] $CaCl_2$ and Na_2HPO_4 solutions rapidly form HA crystals on RNTs, in a pattern similar to collagen fibers in the natural bone. Since the deposition, nucleation, and growth of HA in pores of the extracellular collagen matrix are key steps for *in vivo* bone remodeling, K-RNT/HA nanocomposites are anticipated to be valuable in promoting this natural mineralization process.

Within these discussions, we have highlighted the RNTs' ability to promote OB cell growth as a result of their nanobiomimetic features and expression of bioactive peptide/amino acid sequences. RNTs may also be able to deliver hydrophobic drug molecules to further promote this healthy cell growth process. This is because they have a hydrophobic core and a hydrophilic outer surface and thus could incorporate water-insoluble drugs into their tubular architectures by hydrophobic interactions with the core. The outer hydrophilic surface, alternatively, could shield the drugs in a physiological environment for prolonged release.[163,164] Preliminary experiments investigating the loading of RNTs-K1 with dexamethasone (DEX), which is an anti-inflammatory agent widely used in orthopedic applications,[165,166] and the anticancer drug tamoxifen,[167] have been very promising.[163,164] In the case of DEX, ^{1}H-NMR studies showed approximately 32% loading of the drug at a molar ratio of 5:3 for DEX:RNTs-K1. TM-AFM and DOSY further corroborated the formation of a complex between DEX and RNTs, while UV–visible spectroscopy suggested this is via intercalation (stacking) interactions. Cell studies also showed that DEX was released from RNTs for up to 9 days and had a positive impact on OB functions compared with DEX added to cell culture media alone.[163]

Because the ultimate goal for a drug delivery system is to target certain cells to maximize the therapeutic activity and minimize adverse effects, we have also explored and demonstrated that the RNT surface functional groups can be used to target particular cell receptors. The $\alpha v \beta 3$ integrin on human adenocarcinoma cells (Calu-3), for example, recognizes the arginine-glycine-aspartic acid (RGD) peptide sequence.[142] By treating these Calu-3 cells with coassembled RNTs-K1/RGDSK, cell inflammation and apoptosis can be triggered through the P38 MAPK pathway (Figure 2.9).[142] This is significant because the $\alpha v \beta 3$ integrin is generally not expressed in normal and resting endothelial cells but is present in proliferating endothelial cells in tumors. The importance of the RNT functional groups in initiating this cellular response is apparent by comparing with RNTs-K1, which, in contrast, are biocompatible and do not promote a robust inflammatory response or cytotoxicity in the same Calu-3 cells[168] as well as in U937 human macrophage cells[169] and upon pulmonary exposure of C57/BL mice *in vivo*.[170]

In other related studies, RNTs-K1/RGDSK have also targeted neutrophil integrins to promote acute lung inflammation in a mouse and when used along with lipopolysaccharides (LPS), induce an exaggerated immune response in the lung.[171] As well, it has been shown that these coassembled RNTs can target the $\alpha v \beta 3$ integrin receptor of bovine neutrophils to inhibit chemotaxis.[172]

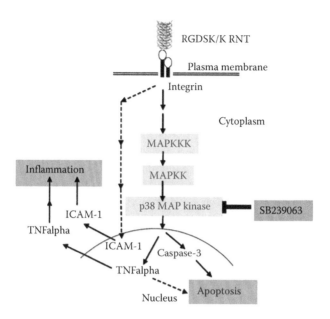

FIGURE 2.9
Proposed action of Calu-3 cells treated with RNTs-K1/RGDSK. (Reproduced from Suri, S. S. et al. 2009. *Biomaterials*, 30, 3084–3090. With permission.)

In general, these studies provide an excellent foundation for RNT-targeted delivery applications. Notably, the RNTs' ability to incorporate drug molecules via noncovalent interactions in combination with targeting capabilities could mean boundless therapeutic potential of these materials as stand-alone injectables or when used as coatings on medical devices.

2.8 Conclusion and Prospects

RNTs have proved to be a promising alternative orthopedic implant material that requires further exploration. There are many questions to be addressed: (a) How does RNT mediate OB adhesion? What are the underlying *mechanisms* that promote OB adhesion and function on RNT-coated surfaces? (b) In the case of heating, how does the aggregation state of RNT affect OB adhesion? (c) What are the long-term effects of RNT? The list of questions continues to grow as we learn more about RNTs and their interaction with OB.

There are a few areas under current investigation. First, determination of protein profiles on RNT-coated substrates versus uncoated substrates. It is necessary to identify whether there is selective adsorption on RNT-coated substrates of at least four proteins known to enhance OB adhesion and function on nanostructured materials—fibronectin, vitronectin, laminin, and collagen. Second, in addition to protein identification, since much of biomaterial interfacial interactions depend on individual protein properties (e.g., size, shape, stability, and surface activity) and surface characteristics (e.g., material chemistry and surface properties),[70,173] it is necessary to determine the concentration, conformation/organization, and bioactivity of these proteins on RNT-coated surfaces. Once adsorbed, these proteins

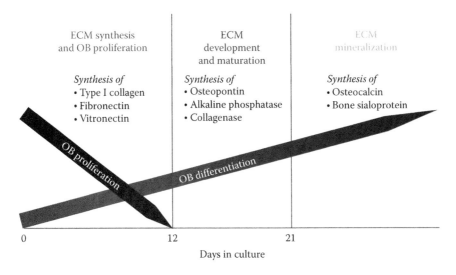

FIGURE 2.10
Time course of osteoblast (OB) differentiation and synthesis of extracellular matrix (ECM) proteins on newly implanted biomaterials. Three distinct stages are observed: (a) cell proliferation and ECM synthesis, (b) ECM development and maturation, and (c) ECM mineralization. The proteins synthesized at each stage are shown in the respective columns. (Reprinted from *Advances in Chemical Engineering. Nanostructured Materials*, Webster, T. J., Nanophase ceramics: The future orthopaedic and dental implant material., pp. 125–166, Copyright 2001, with permission from Elsevier.)

can exist either in native or in denatured form, intact or fragmented, which will ultimately affect how cells interact with them. Third, to determine the long-term functions of OB on RNT-coated substrates. *In vitro* functions of OBs leading to synthesis and deposition of bone on newly implanted prostheses are important first indications of good prostheses. Three distinct periods of OB differentiation at the genetic level were observed when progressive development of bone cell phenotype was examined *in vitro* (Figure 2.10).[54,174,175] Measures of the products of OBs (e.g., OB-specific matrix protein expression) is thus a good marker for bone metabolism as it can provide valuable information on OB physiology in response to the material in question.[176,177]

Fourth, elucidating the mechanism of OB adhesion on RNT-coated surfaces. Since OB adhesion can be mediated by a variety of mechanisms, it is worthwhile identifying which pathway is favored in the case of RNT. A common mechanism involves cell-membrane integrin receptors that bind preferentially to select peptide sequences (e.g., RGD[150]) found in ECM proteins. Other mechanisms that regulate OB adhesion include cell-membrane heparin sulfate proteoglycan interaction with heparin-binding sites on ECM proteins such as fibronectin and collagen.[178–180] Fifth, optimizing and understanding the mechanisms of encapsulation, release kinetics and bioavailability of drug-loaded RNTs, and targeted delivery. In this regard, RNTs have the potential to serve as a multifunctional scaffold that delivers osteogenic peptides and creates a biocompatible surface to promote healthy cell and tissue growth and also slowly release encapsulated drugs into the targeted areas.

In conclusion, supramolecular chemistry with self-assembly as a strategy is a powerful tool in generating well-defined and functional assemblies of molecules similar to those found in nature. Knowing the collection of structures and constituents in biology tells us about the architecture of the tissue or organ. This helps us fabricate materials with similar dimensions and properties. The challenge lies, however, in designing programmable building blocks that can self-organize into functional structures. In the case of orthopedic applications, there is an insatiable need for innovative development of similar self-assembling matter that will promote swift bone deposition in addition to closely matching the mechanical properties of bone. As described in this chapter, RNTs have emerged as a promising two-dimensional coating for Ti substrates and for fabricating three-dimensional constructs (hydrogels) that could be used for the repair of bone fractures.

Acknowledgments

This research was supported by Canada's National Research Council, Canada's Natural Science and Engineering Research Council, and the University of Alberta.

References

1. Jeronimidis, G. 2000. Structure-property relationships in biological materials. In *Structural Biological Materials: Design and Structure-Property Relationships*, M. Elices, Ed., Pergamon Press, Oxford, pp. 3–16.
2. Klefenz, H. 2004. Nanobiotechnology: From molecules to systems. *Eng. Life Sci.* 4, 211–218.
3. Alivisatos, P. A. 2001. Less is more in medicine. *Sci. Am.* 285, 67–73.
4. Whitesides, G. M. and Love, C. J. 2001. The art of building small. *Sci. Am.* 285, 39–47.
5. Lieber, C. M. 2001. The incredible shrinking circuit. *Sci. Am.* 285, 59–64.
6. Alberts, B., Bray, D., Lewis, J., Raff, M., Roberts, K., and Watson, J. D. 1989. *Molecular Biology of the Cell*, 2nd ed., Garland Publishing, Inc., New York.
7. Eyre, D. R. 1980. Collagen: Molecular diversity in the body's protein scaffold. *Science* 207, 1315–1322.
8. Kadler, K. E., Hojima, Y., and Prockop, D. J. 1988. Assembly of Type I collagen fibrils de novo. *J. Biol. Chem.* 263, 10517–10523.
9. Zhang, J. Z., Wang, Z. L., Liu, J., Chen, S., and Liu, G. Y. 2003. *Self-Assembled Nanostructures*, Kluwer Academic/Plenum Publishers, New York, pp. 53–75.
10. Lehninger, A. L., Nelson, D. L., and Cox, M. M. 1993. *DNA Repair, Principles of Biochemistry*, 2nd ed., Worth Publishers, New York, pp. 831–839.
11. Jeronimidis, G. 2000. Design and function of structural biological materials. In *Structural Biological Materials: Design and Structure-Property Relationships*, M. Elices, Ed., Pergamon, Oxford, pp. 19–29.
12. Etter, M. C. 1990. Encoding and decoding hydrogen-bond patterns of organic compounds. *Acc. Chem. Res.* 23, 120–126.
13. Müller, A., Reuter, H., and Dillinger, S. 1995. Supramolecular inorganic chemistry: Small guests in small and large hosts. *Angew. Chem. Int. Ed.* 34, 2328–2361.
14. Whitesides, G. M., Simanek, E. E., Mathias, J. P., Seto, C. T., Chin, D. N., Mammen, M., and Gordon, D. M. 1995. Noncovalent synthesis: Using physical-organic chemistry to make aggregates. *Acc. Chem. Res.* 28, 37–44.
15. Prins, L. J., Reinhoudt, D. N., and Timmerman, P. 2001. Non-covalent synthesis using hydrogen bonding. *Angew. Chem. Int. Ed.* 40, 2382–2426.

16. Reinhoudt, D. N. and Crego-Calama, M. 2002. Synthesis beyond the molecule. *Science* 295, 2403–2407.

17. Stoddard, J. F. and Tseng, H.-R. 2002. Chemical synthesis gets a fillip from molecular recognition and self-assembly processes. *Proc. Natl. Acad. Sci. U.S.A.* 99, 4797–4800.

18. Lehn, J.-M. 1996. Supramolecular chemistry and chemical synthesis. From molecular interactions to self-assembly. *NATO ASI Ser. Ser. E Appl. Sci.* 320, 511–524.

19. Lehn, J.-M. 1995. *Supramolecular Chemistry: Concepts and Perspectives*, VCH, Weinheim.

20. Desiraju, G. R. 1995. Supramolecular synthons in crystal engineering—A new organic synthesis. *Angew. Chem. Int. Ed.* 34, 2311–2327.

21. Mascal, M. 1994. Noncovalent design principles and the new synthesis. *Contemp. Org. Synth.* 1, 31–46.

22. Lindoy, L. F. and Atkinson, I. M. 2000. *Self-Assembly in Supramolecular Systems*, Royal Society of Chemistry Press, Cambridge.

23. Philp, D. and Stoddart, J. F. 1996. Self-assembly in natural and unnatural systems. *Angew. Chem. Int. Ed.* 35, 1155–1196.

24. MacGillivray, L. R. and Atwood, J. L. 1999. Structural classification and general principles for the design of spherical molecular hosts. *Angew. Chem. Int. Ed.* 38, 1018–1033.

25. Melendéz, R. E. and Hamilton, A. D. 1998. Hydrogen-bonded ribbons, tapes and sheets as motifs for crystal engineering. *Top. Curr. Chem.* 198, 97–129.

26. Lindsey, J. S. 1991. Self-assembly in synthetic routes to molecular devices. Biological principles and chemical perspectives: A review. *New J. Chem.* 15, 153–180.

27. Hof, F., Craig, S. L., Nuckolls, C., and Rebek, J. Jr. 2002. Molecular encapsulation. *Angew. Chem. Int. Ed.* 41, 1488–1508.

28. Hill, D. J., Mio, M. J., Prince, R. B., Hughes, T. S., and Moore, J. S. 2001. A field guide to foldamers. *Chem. Rev.* 101, 3893–4011.

29. Lawrence, D. S., Jiang, T., and Levett, M. 1995. Self-assembling supramolecular complexes. *Chem. Rev.* 95, 2229–2260.

30. Zimmerman, S. C. and Lawless, L. J. 2001. Supramolecular chemistry of dendrimers. *Top. Curr. Chem.* 217, 95–120.

31. Zeng, F. and Zimmerman, S. C. 1997. Dendrimers in supramolecular chemistry: From molecular recognition to self-assembly. *Chem. Rev.* 97, 1681–1712.

32. Albrecht, M. 2001. Let's twist again—Double-stranded, triple-stranded, and circular helicates. *Chem. Rev.* 101, 3457–3497.

33. Leininger, S., Olenyuk, B., and Stang, P. J. 2000. Self-assembly of discrete cyclic nanostructures mediated by transition metals. *Chem. Rev.* 100, 853–908.

34. Matile, S. 2001. Bioorganic chemistry a la baguette: Studies on molecular recognition in biological systems using rigid-rod molecules. *Switz. Chem. Rec.* 1, 162–172.

35. Saalfrank, R. W. and Demleitner, B. 1999. Ligand and metal control of self-assembly in supramolecular chemistry. *Perspect. Supramol. Chem.* 5, 1–51.

36. Menger, F. M. 2002. Supramolecular chemistry and self-assembly. *Proc. Natl. Acad. Sci. U.S.A.* 99, 4818–4822.

37. Kato, T. 2000. Hydrogen-bonded liquid crystals: Molecular self-assembly for dynamically functional materials. *Struct. Bond.* 96, 95–146.

38. Bong, D. T., Clark, T. D., Granja, J. R., and Ghadiri, M. R. 2001. Self-assembling organic nanotubes. *Angew. Chem. Int. Ed.* 40, 988–1011.

39. Hartgerink, J. D., Zubarev, E. R., and Stupp, S. I. 2001. Supramolecular one-dimensional objects. *Curr. Opin. Solid State Mater. Sci.* 5, 355–361.

40. MacDonald, J. C. and Whitesides, G. M. 1994. Solid-state structures of hydrogen-bonded tapes based on cyclic secondary diamides. *Chem. Rev.* 94, 2383–2420.

41. Brunsveld, L., Folmer, B. J. B., Meijer, E. W., and Sijbesma, R. P. 2001. Supramolecular polymers. *Chem. Rev.* 101, 4071–4097.

42. Cornelissen, J. J. L. M., Rowan, A. E., Nolte, R. J. M., and Sommerdijk, N. A. J. M. 2001. Chiral architectures from macromolecular building blocks. *Chem. Rev.* 101, 4039–4070.

43. Zhang, J. Z., Wang, Z. L., Liu, J., Chen, S., and Liu, G. Y. 2003. Synthetic self-assembled materials: Principles and practice. In *Self-Assembled Nanostructures*, Kluwer Academic/Plenum Publishers, New York, pp. 7–39.

44. Wöhler, F. 1828. *Ann. Phys.* 12, 253.

45. Lehn, J.-M. 2002. Toward complex matter: Supramolecular chemistry and self-organization. *Proc. Natl. Acad. Sci. U.S.A.* 99, 4763–4768.

46. Cramer, F. 1993. *Chaos and Order*, VCH, Weinheim.

47. 2002. Special issue on supramolecular chemistry and self-assembly. *Proc. Natl. Acad. Sci. U.S.A.* 99(8), 4762–5188.

48. Whitesides, G. M. and Grzybowski, B. 2002. Self-assembly at all scales. *Science* 295, 2418–2421.

49. Seeman, N. C. and Belcher, A. M. 2002. Emulating biology: Building nanostructures from the bottom up. *Proc. Natl. Acad. Sci. U.S.A.* 99, 6451–6455.

50. Lehn, J.-M. 2002. Toward self-organization and complex matter. *Science* 295, 2400–2403.

51. Sauvage, J.-P. and Hosseini, M. W., Eds. 1996. *Comprehensive Supramolecular Chemistry, Volume 9: Templating, Self-Assembly, and Self-Organization*, Pergamon, Oxford.

52. Praemer, A., Furner, S., and Rice, S. D. 1992. *Musculoskeletal Conditions in the United States*, The American Academy of Orthopaedic Surgeons, Park Ridge, IL.

53. Weinstein, J. 2000. *The Dartmouth Atlas of Musculoskeletal Health Care*, AHA Press, Chicago, IL.

54. Webster, T. J. 2001. Nanophase ceramics: The future orthopaedic and dental implant material. In *Advances in Chemical Engineering. Nanostructured Materials*, J. Y. Ying, Ed., Academic Press, San Diego, pp. 125–166.

55. Ontanon, M., Aparicio, C., Ginbera, M. P., and Planell, J. A. 2000. Structure and mechanical properties of cortical bone. In *Structural Biological Materials: Design and Structure-Property Relationships*, 1st ed., M. Elices, Ed., Pergamon, Amsterdam, pp. 33–71.

56. Bostrom, M. P., Boskey, A., Kaufman, J. K., and Einhorn, T. A. 2000. Form and function of bone. In *Orthopaedic Basic Science: Biology and Biomechanics of the Musculoskeletal System*, 2nd ed., J. A. Buckwalter, T. A. Einhorn, and S. R. Sheldon, Eds., The American Academy of Orthopaedic Surgeons, Park Ridge, IL.

57. Moore, K. L. and Dalley, A. F. 1992. *Clinically Oriented Anatomy*, 4th ed., Lippincott Williams & Wilkins, Philadelphia.

58. Cormack, D. H. 2001. Dense connective tissue, cartilage, bone and joints. In *Essential Histology*, 2nd ed., Lippincott Williams and Wilkins, Philadelphia.

59. Hayes, W. C. 1991. Biomechanics of cortical and trabecular bone: Implications for assessment of fracture risk. In *Basic Orthopaedic Biomechanics*, V. C. Mow and W. C. Hayes, Eds., Raven Press, Ltd., New York, pp. 93–99.

60. Christenson, R. H. 1997. Biochemical markers of bone metabolism: An overview. *Clin. Biochem.* 30, 573–593.

61. Tirrell, M., Kokkoli, E., and Biesalski, M. 2002. The role of surface science in bioengineered materials. *Surf. Sci.* 500, 61–83.

62. Cowin, S. C., Van Buskirk, W. C., and Ashman, R. B. 1987. The properties of bone. In *Handbook of Bioengineering*, R. Skalak and S. Chien, Eds., McGraw Hill, New York.

63. Stupp, S. I., Hanson, J. A., Ciegler, G. W., Mejicano, G. C., Eurell, J. A., and Johnson, A. L. 1992. Organoapatites: New materials for regenerative artificial bone. *Mat. Res. Soc. Symp. Proc.* 252 (Tissue-inducing biomaterials), 93–98.

64. Ejiofor, J. and Webster, T. J. 2004. Biomedical implants from nanostructured materials. In *Encyclopedia of Nanoscience and Nanotechnology*, J. A. Schwarz, C. Contescu, and K. Putyera, Eds., Marcel Dekker, Inc., New York.

65. Dunn, M. G. and Maxian, S. H. 1994. Biomaterials used in orthopaedic surgery. In *Implantation Biology. The Host Response and Biomedical Devices*, R. S. Greco, Ed., CRC Press, Boca Raton, pp. 229–252.

66. Kuhn, K.-D. 2000. *Bone Cements. Up-to-Date Comparison of Physical and Chemical Properties of Commercial Materials*, Springer-Verlag, Berlin.

67. Hench, L. L. and Ethridge, E. C. 1982. Orthopaedic implants. In *Biomaterials an Interfacial Approach*, A. Noordergraaf, Ed., Academic Press, Inc., New York, pp. 225–252.

68. Unwin, P. S. 2000. The recent advancements in bone and joint implant technology. In *Cost-Effective Titanium Component Technology for Leading-Edge Performance*, M. Ward-Close, Ed., Professional Engineers Publishing Ltd. for the Institution of Mechanical Engineers, Bury St. Edmunds & London.

69. Anderson, J. M., Gristina, A. G., Hanson, S. R., Harker, L. A., Johnson, R. J., Merritt, K., Naylor, P. T., and Schoen, F. J. 1996. Host reactions to biomaterials and their evaluation. In *Biomaterials Science: An Introduction to Materials in Medicine*, B. D. Ratner, A. S. Hoffman, F. J. Schoen, and J. E. Lemons, Eds., Academic Press, New York, pp. 165–214.

70. Horbett, T. A. 1996. Proteins: Structure, properties and adsorption to surfaces. In *Biomaterials Science: An Introduction to Materials in Medicine*, B. D. Ratner, A. S. Hoffman, F. J. Schoen, and J. E. Lemons, Eds., Academic Press, New York, pp. 133–140.

71. Price, R. L., Waid, M. C., Haberstroh, K. M., and Webster, T. J. 2003. Selective bone cell adhesion on formulations containing carbon nanofibers. *Biomaterials* 24, 1877–1887.

72. Price, R. L., Haberstroh, K. M., and Webster, T. J. 2003. Enhanced functions of osteoblast on nanostructured surfaces of carbon and alumina. *Med. Biol. Eng. Comput.* 41, 372–375.

73. Webster, T. J., Waid, M. C., McKenzie, J. L., Price, R. L., and Ejiofor, J. U. 2004. Nano-biotechnology: Carbon nanofibres as improved neural and orthopaedic implants. *Nanotechnology* 15, 48–54.

74. Webster, T. J., Ergun, C., Doremus, R. H., Segel, R. W., and Bizios, R. 2000. Enhanced functions of osteoblasts on nanophase ceramics. *Biomaterials* 21, 1803–1810.

75. Webster, T. J., Siegel, R. W., and Bizios, R. 1999. Osteoblast adhesion on nanophase ceramics. *Biomaterials* 20, 1221–1227.

76. Webster, T. J., Siegel, R. W., and Bizios, R. 2001. Nanoceramic surface roughness enhances osteoblast and osteoclast functions for improved orthopaedic/dental implant efficacy. *Scr. Mater.* 44, 1639–1642.

77. Webster, T. J. 2003. Nanophase ceramics as improved bone tissue engineering materials. *Am. Cer. Soc. Bull.* 82, 23–28.

78. Kyung, M. W., Chen, V. J., and Ma, P. X. 2003. Nano-fibrous scaffolding architecture selectively enhances protein adsorption contributing to cell attachment. *J. Biomed. Mater. Res.* 67A, 531–537.

79. Ma, P. X. and Zhang, R. 1999. Synthetic nanoscale fibrous extracellular matrix. *J. Biomed. Mater. Res.* 46, 60–72.

80. Stupp, S. I. and Ciegler, G. W. 1992. Organoapatites: Materials for artificial bone. I. Synthesis and microstructure. *J. Biomed. Mater. Res.* 26, 169–183.

81. Stupp, S. I., Hanson, J. A., Eurell, J. A., Ciegler, G. W., and Johnson, A. L. 1993. Organoapatites: Materials for artificial bone. III. Biological testing. *J. Biomed. Mater. Res.* 27, 301–311.

82. Stupp, S. I., Mejicano, G. C., and Hanson, J. A. 1993. Organoapatites: Materials for artificial bone. II. Hardening reactions and properties. *J. Biomed. Mater. Res.* 27, 289–299.

83. Chun, A. L., Moralez, J. G., Fenniri, H., and Webster, T. J. 2004. Helical rosette nanotubes: A more effective orthopaedic implant material. *Nanotechnology* 15, S234–S239.

84. Iijima, S. 1991. Helical microtubules of graphitic carbon. *Nature* 354, 56–58.

85. Hamilton, E. J. M., Dolan, S. E., Mann, C. M., Colijin, H. O., McDonald, C. A., and Shore, S. G. 1993. Preparation of amorphous boron nitride and its conversion to a turbostratic, tubular form. *Science* 260, 659–661.

86. Brumlik, C. J. and Martin, C. R. 1991. Template synthesis of metal microtubules. *J. Am. Chem. Soc.* 113, 3174–3175.

87. Brorson, M., Hansen, T. W., and Jacobsen, C. J. H. 2002. Rhenium(IV) sulfide nanotubes. *J. Am. Chem. Soc.* 124, 11582–11583.

88. Miyaji, F., Davis, S. A., Charmant, J. P. H., and Mann, S. 1999. Organic crystal templating of hollow silica fibers. *Chem. Mater.* 11, 3021–3024.

89. Hsu, W. K., Chang, B. H., Zhu, Y. Q., Han, W. Q., Terrones, H., Terrones, M., Grobert, N., Cheetham, A. K., Kroto, H. W., and Walton, D. R. M. 2000. An alternate route to molybdenum disulfide nanotubes. *J. Am. Chem. Soc.* 122, 10155–10158.

90. Seddon, A. M., Patel, H. M., Burkett, S. L., and Mann, S. 2002. Chiral templating of silica-lipid lamellar mesophase with helical tubular architecture. *Angew. Chem. Int. Ed.* 41, 2988–2991.
91. Raez, J., Barjovanu, R., Massey, J. A., Winnik, M. A., and Manners, I. 2000. Self-assembled organometallic block copolymer nanotubes. *Angew. Chem. Int. Ed.* 39, 3862–3865.
92. Shenton, W., Douglas, T., Young, M., Stubbs, G., and Mann, S. 1999. Inorganic-organic nanotubes composites from template mineralization of tobacco mosaic virus. *Adv. Mater.* 11, 253–256.
93. Chopra, N. G., Luyken, R. J., Cherrey, K., Crespi, V. H., Cohen, M. L., Louie, S. G., and Zettl, A. 1995. Boron nitride nanotubes. *Science* 269, 966–967.
94. Kasuga, T., Hiramatsu, M., Hoson, A., Sekino, T., and Niihara, K. 1998. Formation of titanium oxide nanotubes. *Langmuir* 14, 3160–3163.
95. Raez, J., Manners, I., and Winnik, M. A. 2002. Nanotubes from self-assembly of asymmetric crystalline-coil poly(ferrocenylsilane-siloxane) block copolymers. *J. Am. Chem. Soc.* 124, 10381–10395.
96. Mitchell, D. T., Lee, S. B., Trofin, L., Li, N., Nevanen, T. K., Soderlund, H., and Martin, C. R. 2002. Smart nanotubes for bioseparations and biocatalysis. *J. Am. Chem. Soc.* 124, 11864–11865.
97. Nath, M. and Rao, C. N. R. 2002. Nanotubes of group IV metal disulfides. *Angew. Chem. Int. Ed.* 41, 3451–3454.
98. Harada, A., Li, J., and Kamachi, M. 1993. Synthesis of a tubular polymer from threaded cyclodextrins. *Nature* 364, 516–518.
99. Drager, A. S., Zangmeister, A. P., Armstrong, N. R., and O'Brien, D. F. 2001. One-dimensional polymers of octasusbtituted phtalocyanines. *J. Am. Chem. Soc.* 123, 3595–3596.
100. Nelson, J. C., Saven, J. G., Moore, J. S., and Wolynes, P. G. 1997. Solvophobically-driven folding of non-biological oligomers. *Science* 277, 1793–1796.
101. Ashton, P. R., Brown, C. L., Menzer, S., Nepogodiev, S. A., Stoddart, J. F., and Williams, D. J. 1996. Synthetic cyclic oligosaccharides: Synthesis and structural properties of a cyclo[(1→4)-α-L-rhamnopyranosyl-(1→4)-α-D-mannopyranosyl]trioside and -tetraoside. *Chem. Eur. J.* 2, 580–591.
102. Fernandez-Lopez, S., Kim, H.-S., Choi, E. C., Delgado, M., Granja, J. R., Khasanov, A., Kraehenbuehl, K. et al. 2001. Antibacterial agents based on the cyclic D,L-α-peptide architecture. *Nature* 412, 452–455.
103. Leevy, W. M., Donato, G. M., Ferdani, R., Goldman, W. E., Schlessinger, P. H., and Gokel, G. W. 2002. Synthetic hydraphile channels of appropriate length kill *Escherichia coli. J. Am. Chem. Soc.* 124, 9022–9023.
104. Gauthier, D., Baillargeon, P., Drouin, M., and Dory, Y. L. 2001. Self-assembly of cyclic peptides into nanotubes and then into highly anisotropic crystalline materials. *Angew. Chem. Int. Ed.* 40, 4635–4638.
105. Ranganathan, D., Lakshmi, C., and Karle, I. L. 1999. Hydrogen-bonded self-assembled peptide nanotubes from cystine-based macrocyclic bisureas. *J. Am. Chem. Soc.* 121, 6103–6107.
106. Engelkamp, H., Middlebeek, S., and Nolte, R. J. M. 1999. Self-assembly of disk-shaped molecules to coiled-coil aggregates with tunable helicity. *Science* 284, 785–788.
107. Baumeister, B. and Matile, S. 2000. Rigid-rod β-barrels as lipocalin models: Probing confined space by carotenoid encapsulation. *Chem. Eur. J.* 6, 1739–1749.
108. Biron, E., Voyer, N., Meillon, J.-C., Cormier, M.-E. and Auger, M. 2000. Conformational and orientation studies of artificial ion channels incorporated into lipid bilayers. *Biopolymers (Pept. Sci.)* 55, 364–372.
109. Gokel, G. W. and Murillo, O. 1996. Synthetic organic chemical models for transmembrane channels. *Acc. Chem. Res.* 29, 425–432.
110. Das, G. and Matile, S. 2002. Transmembrane pores formed by synthetic *p*-octiphenyl β-barrels with internal carboxylate clusters: Regulation of ion transport by pH and Mg^{2+}- complexed 8-aminonaphthalene-1,3,6-trisulfonate. *Proc. Natl. Acad. Sci. U.S.A.* 99, 5183–5188.
111. Ranganathan, D., Lakshmi, C., Haridas, V., and Gopikumar, M. 2000. Designer cyclopeptides for self-assembled tubular structures. *Pure Appl. Chem.* 72, 365–372.

112. Cuccia, L. A., Lehn, J.-M., Homo, J.-C., and Schmutz, M. 2000. Encoded helical self-organization and self-assembly into helical fibers of an oligoheterocyclic pyridine-pyridazine molecular strand. *Angew. Chem. Int. Ed.* 39, 233–237.

113. Seebach, D., Matthews, J. L., Meden, A., Wessels, T., Baerlocher, C., and McCusker, L. B. 1997. Cyclo-β-peptides: Structure and tubular stacking of cyclic tetramers of 3-aminobutanoic acid as determined from powder diffraction data. *Helv. Chim. Acta* 80, 173–181.

114. Shimizu, L. S., Smith, M. D., Hughes, A. D., and Shimizu, K. D. 2001. Self-assembly of bis-urea macrocycle into a columnar nanotubes. *Chem. Commun.* 1592–1593.

115. Fenniri, H., Mathivanan, P., Vidale, K. L., Sherman, D. M., Hallenga, K., Wood, K. V., and Stowell, J. G. 2001. Helical rosette nanotubes: Design, self-assembly and characterization. *J. Am. Chem. Soc.* 123, 3854–3855.

116. Fenniri, H., Deng, B.-L., Ribbe, A. E., Hallenga, K., Jacob, J., and Thiyagarajan, P. 2002. Entropically-driven self-assembly of multi-channel rosette nanotubes. *Proc. Natl. Acad. Sci. U.S.A.* 99, 6487–6492.

117. Fenniri, H., Deng, B.-L., and Ribbe, A. E. 2002. Helical rosette nanotubes with tunable chiroptical properties. *J. Am. Chem. Soc.* 124, 11064–11072.

118. Fenniri, H., Packiarajan, M., Ribbe, A. E., and Vidale, K. E. 2001. Toward self-assembled electro- and photo-active organic nanotubes. *Polym. Prepr.* 42, 569–570.

119. Schnur, J. M. 1993. Lipid tubules: A paradigm for molecularly engineered structures. *Science* 262, 1669–1676.

120. Georger, J. H., Singh, A., Price, R. R., Schnur, J. M., Yager, P., and Schoen, P. E. 1987. Helical and tubular, microstructures formed by polymerizable phosphatidylcholines. *J. Am. Chem. Soc.* 109, 6169–6175.

121. Vauthey, S., Santoso, S., Gong, H., Watson, N., and Zhang, S. 2002. Molecular self-assembly of surfactant-like peptides to form nanotubes and nanovesicles. *Proc. Natl. Acad. Sci. U.S.A.* 99, 5355–5360.

122. Nakashima, N., Asakuma, S., and Kunitake, T. 1985. Optical microscope study of helical super-structures of chiral bilayer membranes. *J. Am. Chem. Soc.* 107, 509–510.

123. Fuhrhop, J.-H., Spiroski, D., and Boettcher, C. 1993. Molecular monolayer rods and tubules made of α-(L-lysine),ω-(amino)bolaamphiphiles. *J. Am. Chem. Soc.* 115, 1600–1601.

124. Frankel, D. A. and O'Brien, D. F. 1994. Supramolecular assemblies of diacetylenic aldonamides. *J. Am. Chem. Soc.* 116, 10057–10069.

125. Mueller, A. and O'Brien, D. F. 2002. Supramolecular materials via polymerization of meso-phases of hydrated amphiphiles. *Chem. Rev.* 102, 727–757.

126. Imae, T., Takahashi, Y., and Maramatsu, H. 1992. Formation of fibrous molecular assemblies by amino acid surfactants in water. *J. Am. Chem. Soc.* 114, 3414–3419.

127. Kimizuka, N., Fujikawa, S., Kuwahara, S., Kunitake, T., Marsh, A., and Lehn, J.-M. 1995. Mesoscopic supramolecular assembly of a "janus" molecule and a melamine derivative via complementary hydrogen bonds. *Chem. Commun.* 2103–2104.

128. Kimizuka, M., Kawasaki, T., Hirata, K., and Kunitake, T. 1995. Tube-like nanostructures composed of networks of complementary hydrogen bonds. *J. Am. Chem. Soc.* 117, 6360–6361.

129. Klok, H.-A., Joliffe, K. A., Schauer, C. L., Prins, L. J., Spatz, J. P., Möller, M., Timmerman, P., and Reinhoudt, D. N. 1999. Self-assembly of rodlike hydrogen-bonded nanostructures. *J. Am. Chem. Soc.* 121, 7154–7155.

130. Choi, I. S., Li, X., Simanek, E. E., Akaba, R., and Whitesides, G. M. 1999. Self-assembly of hydro-gen-bonded polymeric rods based on the cyanuric acid-melamine lattice. *Chem. Mater.* 11, 684–690.

131. Marchi-Artzner, V., Artzner, F., Karthaus, O., Shimomura, M., Ariga, K., Kunitake, T., and Lehn, J.-M. 1998. Molecular recognition between 2,4,6-triaminopyrimidine lipid monolayers and complementary barbituric acid molecules at the air/water interface: Effects of hydrophilic spacer, ionic strength, and pH. *Langmuir* 14, 5164–5171.

132. Marchi-Artzner, V., Lehn, J.-M., and Kunitake, T. 1998. Specific adhesion and lipid exchange between complementary vesicle and supported or langmuir films. *Langmuir* 14, 6470–6478.

133. Kawasaki, T., Tokuhiro, M., Kimizuka, N., and Kunitake, T. 2001. Hierarchical self-assembly of chiral complementary hydrogen-bond networks in water: Reconstitution of supramolecular membranes. *J. Am. Chem. Soc.* 123, 6792–6800.

134. Corey, E. J. and Cheng, X.-M. 1989. *The Logic of Chemical Synthesis*, John Wiley & Sons, New York.

135. Nicolaou, K. C. and Sorensen, E. J. 1996. *Classics in Total Synthesis*, VCH, New York.

136. Chun, A. L., Moralez, J. G., Webster, T. J., and Fenniri, H. 2005. Helical rosette nanotubes: A biomimetic coating for orthopaedics? *Biomaterials* 26, 7304–7309.

137. Beingessner, R. L., Deng, B.-L., Fanwick, P. E., and Fenniri, H. 2008. A regioselective approach to trisubstituted 2(or 6)-arylaminopyrimidine-5-carbaldehydes and their application in the synthesis of structurally and electronically unique G∧C base precursors. *J. Org. Chem.* 73, 931–939.

138. Tikhomirov, G., Oderinde, M., Makeiff, D., Mansouri, A., Lu, W., Heirtzler, F., Kwok, D. Y., and Fenniri, H. 2008. Synthesis of hydrophobic derivatives of the G∧C base for rosette nanotube self-assembly in apolar media. *J. Org. Chem.* 73, 4248–425.

139. Chhabra, R., Moralez, J. G., Raez, J., Yamazaki, T., Cho, J.-Y., Myles, A. J., Kovalenko, A., and Fenniri, H. 2010. One-pot nucleation, growth, morphogenesis and passivation of 1.4 nm Au nanoparticles on self-assembled rosette nanotubes. *J. Am. Chem. Soc.* 132, 32–33.

140. Alsbaiee, A., St. Jules, M., Beingessner, R. L., Cho, J.-Y., Yamazaki, T., and Fenniri, H. 2012. Synthesis of rhenium chelated MAG$_3$ functionalized rosette nanotubes. *Tet. Lett.* http://dx.doi.org/10.1016/j.tetlet.2012.01.090.

141. Beingessner, R. L., Diaz, J. A., Hemraz, U. D., and Fenniri, H. 2011. Synthesis of a β-glycoside functionalized G∧C motif for self-assembly into rosette nanotubes with predefined length. *Tet. Lett.* 52, 661–664.

142. Suri, S. S., Rakotondradany, F., Myles, A. J., Fenniri, H., and Singh, B. 2009. The role of RGD-tagged helical rosette nanotubes in the induction of inflammation and apoptosis in human lung adenocarcinoma cells through the P38 MAPK pathway. *Biomaterials* 30, 3084–3090.

143. Moralez, J. G., Raez, J., Yamazaki, T., Motkuri, R. K., Kovalenko, A., and Fenniri, H. 2005. Helical rosette nanotubes with tunable stability and hierarchy. *J. Am. Chem. Soc.* 127, 8307–8309.

144. Johnson, R. S., Yamazaki, T., Kovalenko, A., and Fenniri, H. 2007. Molecular basis for water-promoted supramolecular chirality inversion in helical rosette nanotubes. *J. Am. Chem. Soc.* 129, 5735–5743.

145. Borzsonyi, G., Johnson, R. S., Myles, A. J., Cho, J.-Y., Yamazaki, T., Beingessner, R. L., Kovalenko, A., and Fenniri, H. 2010. Rosette nanotubes with 1.4 nm inner diameter from a tricyclic variant of the Lehn-Mascal G∧C base. *Chem. Comm.* 46, 6527–6529.

146. Borzsonyi, G., Beingessner, R. L., Yamazaki, T., Cho, J.-Y., Myles, A. J., Malac, M., Egerton, R. et al. 2010. Water-soluble J-type rosette nanotubes with giant molar ellipticity. *J. Am. Chem. Soc.* 132, 15136–15139.

147. Kuhn, H. and Kuhn, C. 1996. In *J-Aggregates*, T. Kobayashi, Ed., World Scientific, Singapore.

148. Borzsonyi, G., Alsbaiee, A., Beingessner, R. L., and Fenniri, H. 2010. Synthesis of a tetracyclic G∧C scaffold for the assembly of rosette nanotubes with 1.7 nm inner diameter. *J. Org. Chem.* 75, 7233–7239.

149. Gittens, S. A. and Uludag, H. 2001. Growth factor delivery for bone tissue engineering. *J. Drug Target* 9, 407–429.

150. Dee, K., Thomas, T., and Andersen, R. B. 1998. Design and function of novel osteoblast-adhesive peptides for chemical modification of biomaterials. *J. Biomed. Mater. Res.* 40, 371–377.

151. Fine, E., Zhang, L., Fenniri, H., and Webster, T. J. 2009. Enhanced endothelial cell functions on rosette nanotube-coated titanium vascular stents. *Int. J. Nanomed.* 4, 91–97.

152. Zhang, L., Hemraz, U. D., Fenniri, H., and Webster, T. J. 2010. Tuning cell adhesion on titanium with osteogenic rosette nanotubes. *J. Biomed. Mater. Res.* 95A, 550–563.

153. Zhou, H.-Y., Ohnuma, Y., Takita, H., Fujisawa, R., Mizuno, M., and Kuboki, Y. 1992. Effects of a bone lysine-rich 18 kDa protein on osteoblast-like MC3T3-E1 cells. *Biochem. Biophys. Res. Commun.* 186, 1288–1293.

154. Oxlund, H., Barckman, M., Ortoft, G., and Andreassen, T. T. 1995 Reduced concentrations of collagen cross-links are associated with reduced strength of bone. *Bone* 17, 365S–371S.

155. Fini, M., Torricelli, P., Giavaresi, G., Carpi, A., Nicolini, A., and Giardino, R. 2001. Effect of L-lysine and L-arginine on primary osteoblast cultures from normal and osteopenic rats. *Biomed. Pharmacother.* 55, 213–220.

156. Torricelli, P., Fini, M., Giavaresi, G., and Giardino, R. 2003. Human osteopenic bone-derived osteoblasts: Essential amino acids treatment effects. *Artif. Cells Blood Substit. Immobil. Biotechnol.* 31, 35–46.

157. Lee, K. Y. and Mooney, D. J. 2001. Hydrogels for tissue engineering. *Chem. Rev.* 101, 1869–1879.

158. Hoffmann, A. S. 2002. Hydrogels for biomedical applications. *Adv. Drug. Deliv. Rev.* 54, 3–12.

159. Zhang, L., Ramsaywack, S., Fenniri, H., and Webster, T. J. 2008. Enhanced osteoblast adhesion on self-assembled nanostructured hydrogel scaffolds. *Tissue. Eng. A.* 14, 1353–1364.

160. Zhang, L., Rakotondradany, F., Myles, A. J., Fenniri, H., and Webster, T. J. 2009. Arginine-glycine-aspartic acid modified rosette nanotube-hydrogel composites for bone tissue engineering. *Biomaterials* 30, 1309–1320.

161. Zhang, L., Chen, L. Y., Rodriguez, J., Fenniri, H., and Webster, T. J. 2008. Biomimetic helical rosette nanotubes and nanocrystalline hydroxyapatite coatings on titanium for improving orthopedic implants. *Int. J. Nanomed.* 3, 323–333.

162. Zhang, L., Rodriguez, J., Raez, J., Myles, A. J., Fenniri, H., and Webster, T. J. 2009. Biologically inspired rosette nanotubes and nanocrystalline hydroxyapatite hydrogel nanocomposites as improved bone substitutes. *Nanotechnology* 20, 1–12.

163. Chen, Y., Song, S., Yan, Z., Fenniri, H., and Webster, T. J. 2011. Self-assembled rosette nanotubes encapsulate and slowly release dexamethasone. *Int. J. Nanomed.* 6, 1035–1044.

164. Song, S., Chen, Y., Yan, Z., Fenniri, H., and Webster, T. J. 2011. Self-assembled rosette nanotubes for incorporating hydrophobic drugs in physiological environments. *Int. J. Nanomed.* 6, 101–107.

165. Guzmán-Morales, J., El-Gabalawy, H., Pham, M. H., Tran-Khanh, N., McKee, M. D., Wu, W., Centola, M., and Hoemann, C. D. 2009. Effect of chitosan particles and dexamethasone on human bone marrow stromal cell osteogenesis and angiogenic factor secretion. *Bone* 45, 617–626.

166. Beule, A. G., Steinmeier, E., Kaftan, H., Biebler, K. E., Göpferich, A., Wolf, E., and Hosemann, W. 2009. Effects of a dexamethasone-releasing stent on osteoneogenesis in a rabbit model. *Am. J. Rhinol. Allergy* 23, 433–436.

167. Macgregor, J. and Jordan, V. C. 1998. Basic guide to the mechanisms of antiestrogen action. *Pharmacol. Rev.* 50, 151–196.

168. Journeay, W. S., Suri, S. S., Moralez, J. G., Fenniri, H., and Singh, B. 2008. Low inflammatory activation by self-assembling rosette nanotubes in human calu-3 pulmonary epithelial cells. *Small* 4, 817–823.

169. Journeay, W. S., Suri, S. S., Moralez, J. G., Fenniri, H., and Singh, B. 2009. Macrophage inflammatory response to self-assembling rosette nanotubes. *Small* 5, 1446–1452.

170. Journey, W. S., Suri, S. S., Moralez, J. G., Fenniri, H., and Singh, B. 2008. Rosette nanotubes show low acute pulmonary toxicity *in vivo*. *Int. J. Nanomed.* 3, 373–383.

171. Suri, S. S., Mills, S., Aulakh, G. K., Rakotondradany, F., Fenniri, H., and Singh, B. 2011. RGD-tagged helical rosette nanotubes aggravate acute lipopolysaccharide-induced lung inflammation. *Int. J. Nanomed.* 6, 3113–3123.

172. Le, M. H. A., Suri, S. S., Rakotondradany, F., Fenniri, H., and Singh, B. 2010. Rosette nanotubes inhibit bovine neutrophil chemotaxis. *Vet. Res.* 41, 75.

173. Webster, T. J. 2004. Proteins: Structure and interaction patterns to solid surfaces. In *Encyclopedia of Nanoscience and Nanotechnology*, J. A. Schwarz, C. Contescu, and K. Putyera, Eds., Marcel Dekker, Inc., New York, pp. 1–16.

174. Stein, G. and Lian, J. B. 1993. Molecular mechanisms mediating proliferation/differentiation interrelationships during progressive development of the osteoblast phenotype. *Endocr. Rev.* 14, 424–441.

175. Cooper, L. F., Masuda, T., Yliheikklika, P. K., and Felton, D. A. 1998. Generalizations regarding the process and phenomenon of osseointegration. Part II. In vitro studies. *Int. J. Oral Maxillofac. Implants* 13, 163–174.

176. Puleo, D. A., Preston, K. E., Shaffer, J. B., and Bizios, R. 1993. Examination of osteoblast-orthopaedic biomaterial interactions using molecular techniques. *Biomaterials* 14, 111–114.

177. Puleo, D. A., Holleran, L. A., Doremus, R. H., and Bizios, R. 1991. Osteoblast responses to orthopedic implant materials in vitro. *J. Biomed. Mater. Res.* 25, 711–723.

178. Dalton, B. A., McFarland, C. D., Underwood, P. A., and Steele, J. G. 1995. Role of the heparin-binding domain of fibronectin in attachment and spreading of human bone-derived cells. *J. Cell Sci.* 108, 2083–2092.

179. Nakamura, H. and Ozawa, H. 1994. Immunohistochemical localization of heparan sulfate proteoglycan in rat tibiae. *J. Bone Min. Res.* 9, 1289–1299.

180. Puleo, D. A. and Bizios, R. 1992. Mechanisms of fibronectin-mediated attachment of osteoblasts to substrates in vivo. *Bone Miner.* 18, 215–226.

3

Gold Nanoparticles with Organic Linkers for Applications in Biomedicine

Olga Shimoni and Stella M. Valenzuela

CONTENTS

3.1 Introduction

Gold nanoparticles (Au NPs) have a rich history of drawing the attention, at first, of alchemists followed by scientists from the fields of chemistry, physics, photonics, and, in more recent times, biology and medicine. Such wide scientific attention has arisen due to the unique physical and chemical properties of these nanoparticles, including their optical and electronic virtues dependent on size, shape, and high surface-to-volume area with facile surface chemistry. Added to these favorable physicochemical properties is their highly desirable and now increasingly accepted biocompatible profile. These factors converge to provide the rationale for the current growing interest in their pursuit as new breakthrough therapeutic and prophylactic agents. The range of applications are being explored, which include their use as drug (Cesbron et al. 2015, Danesh et al. 2015, Khandelia et al. 2015) and gene delivery agents (Li et al. 2015); photothermal therapeutic agents (Pissuwan et al. 2006, 2008); contrast dyes for *in vivo* imaging (Chen et al. 2015); adjuvants in vaccine development (Safari et al. 2012); and antibacterial (Huo et al. 2014), antiparasitic (Pissuwan et al. 2007, 2009), and antiviral agents (Paul et al. 2014).

Our focus in this chapter is the use of organic linkers to modify Au NPs and the subsequent interactions and effects these modified particles have, when used within biological systems.

3.2 Synthesis

The earliest scientific mentioning of the preparation of spherical Au NPs appeared in the mid-nineteenth century. Nonetheless, their use has been known for over two millenniums. Most of the modern synthesis of spherical Au NPs is based on the modified Turkevich (one-phase method) (Turkevich et al. 1951, Frens 1973) or Brust–Schiffrin (two-phase method) (Brust et al. 1994) methods (Figure 3.1). In both methods, chloroauric acid is reduced by means of a reducing agent in the presence of capping agents. In the case of the Turkevich method, citrate molecules play a double role as a reducing and as a stabilizing agent. The size of the resultant spherical Au NPs can be controlled by varying the concentration or ratio of gold ions, reducing agent, or stabilizing agent. In the Brust–Schiffrin method, surfactant acts as a capping or protecting layer and sodium borohydride is usually used as a reducing agent. Production of Au NPs via the Turkevich method results in their formation in an aqueous environment, which is convenient for subsequent bioapplications. When synthesis occurs via the Brust–Schiffrin method, Au NPs are found in the organic phase (e.g., toluene, hexane, etc.), which produces smaller-sized NPs with narrow size distribution.

Preparation of molecular-sized gold nanoclusters [$Au_{55}(PPh_3)_{12}Cl_6$] was reported by Schmid et al., where phosphine-stabilized Au NPs were prepared by reduction of PPh_3AuCl with diborane gas in organic solvents (Schmid et al. 1981). Later, the same group described formation of water-soluble nanoclusters by exchange of PPh_3 with monosulfonated triphenylphosphane ($Ph_2PC_6H_4SO_3Na$) (Schmid et al. 1988). This discovery has led to the investigation of these nanoclusters in bioapplications, discussed later in this chapter.

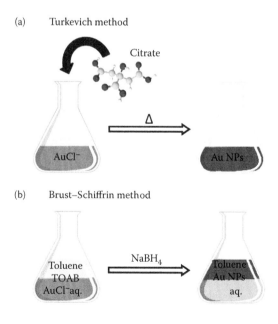

(a)   Turkevich method

Citrate

Δ

AuCl⁻   Au NPs

(b)   Brust–Schiffrin method

NaBH₄

Toluene   Toluene
TOAB   Au NPs
AuCl⁻aq.   aq.

FIGURE 3.1
Two most common methods of spherical gold nanoparticle synthesis. (a) Turkevich method, where chloroauric ions are reduced and stabilized by citrate molecules. (b) Brust–Schiffrin method involves reduction of chloroauric ions with sodium borohydride (NaBH₄) and produced gold nanoparticles are stabilized by the surfactant, tetraoctylammonium bromide (TOAB). Formed and stabilized gold nanoparticles are then eventually dispersed in the organic phase.

Over the last decade, there have been multiple reports on "green" synthesis of Au NPs, a process that avoids the use of harsh or undesirable molecules, such as surfactants, organic solvents, and others (Mukherjee et al. 2001, Raveendran et al. 2006, Narayanan and Sakthivel 2008, Xie et al. 2009, Park et al. 2011). The "green" method involves a use of naturally occurring molecules, including starch, proteins, biopolymers, plant extracts, and more, to play a role as capping and reducing agents simultaneously. Although "green" synthesis is somewhat desirable for application of Au NPs in biomedicine, there are still some concerns about reproducibility of high-quality nanoparticles.

From the 1990s, Brumlik and Martin began a new era of shape diversity of Au NPs (Brumlik and Martin 1991). They were the first to show the possibility of obtaining porous Au NPs with high aspect ratio by templating from porous alumina. To date, numerous articles have been published on the synthesis of Au NPs of various shapes, including rods, shells, cubes, triangular bipyramids, octahedra, hollow cubes, or spheres, and many more (Oldenburg et al. 1998, Jana et al. 2001, Sun and Xia 2002, Liang et al. 2003, Kim et al. 2004, Shankar et al. 2004, Personick et al. 2011). Synthetic strategies to produce different shapes of Au NPs include use of surfactants (cetyltrimethylammonium bromide [CTAB] and cetyltrimethylammonium chloride [CTAC]), template-assisted electrochemical deposition, galvanic displacement, or vapor phase deposition (Dreaden et al. 2012).

3.3 Gold-Linker Chemistry

As discussed previously, one of the most popular methods to synthesize Au NPs is based on citrate-assisted chloroauric ionic reduction method. Despite the fact that this method has been in use for almost 60 years, it was not until recently that the exact conformational structure of citrate molecules on the surface of Au NPs was elucidated (Park and Shumaker-Parry 2014). Specifically, it was established that partially protonated citrate molecules (dihydrogen citrate anions) are adsorbed onto the gold surface through a coordination complex with a central carboxylate group. Citrate molecules create a layer with a thickness of ~10 Å through hydrogen bonding and van der Waals forces with adjacent and adsorbed molecules.

One of the most abundant techniques to obtain organic linkers on the surface of Au NPs is through thiol–gold reaction. This reaction originates from bonding between gold atoms sitting on the surface of Au NPs (adatoms) and sulfur atoms in thiol (Jadzinsky et al. 2007). Previously, it was assumed that the bond between adatoms and sulfur is purely ionic; however, more recently it has been established by Riemers et al. via density function theory (DFT) calculations that the bonding is essentially covalent and relatively strong (Reimers et al. 2010).

A plethora of publications now report the use of a thiol–gold chemistry to functionalize surface of Au NPs with different species (Figure 3.2). Regardless of the initial capping agent on the surface of the Au NPs (citrate, surfactant, or other thiols), tethered molecules can be attached via a ligand exchange reaction (Yeh et al. 2012). Specifically, introduced molecules with thiol end groups displace the existing species of Au NPs in an equilibrium process. Using this method, researchers have successfully functionalized the surface of Au NPs with organic molecules (PEG, fluorophores, drugs, or analytes) (Gu et al. 2003, Lytton-Jean and Mirkin 2005, Hwu et al. 2008, Perni et al. 2011), biomolecules (proteins, peptides, DNA, iRNA, antibodies, or antigens) (Thanh and Rosenzweig 2002, Giljohann et al. 2010,

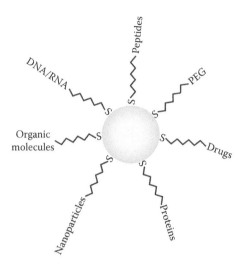

FIGURE 3.2
Schematic representation of attachment of different molecules using gold–thiol bond.

Rana et al. 2012, Zhang et al. 2012), and various nanoparticles (Kinoshita et al. 2007, Gandra et al. 2012) (Figure 3.2).

Although gold–thiol bonding is considered to be relatively stable, thiol is able to dissociate from the gold surface under physiological conditions with changing redox conditions. For some time, ligands with dithiolates have been considered superior in their chemical stability compared with mono-thiol ligands, due to their multivalent binding ability. However, it was found that dithiolates are in fact more susceptible to oxidative desorption as a result of inefficient packing and ease of disulfide formation (Hou et al. 2009).

An additional approach to the attachment of organic molecules onto the surface of Au NPs involves using the favorable interaction between gold and amine groups. Amine groups create a complex with gold atoms, but the stability of this interaction is rather weak. For instance, bond strength of gold–amine is ~6 kcal/mol, while thiol–gold is 47 kcal/mol (Hoft et al. 2007). Although the bonding affinity is significantly lower, it has the advantage of easy "release," which can be exploited as a mechanism for the delivery of drug in therapeutic applications (Vigderman and Zubarev 2013).

3.4 Gold Nanoparticle Cellular Interactions

Despite numerous potential uses of Au NPs and their great promise as a wonder drug, there still remains the obligation by researchers to thoroughly examine their biokinetics and biocompatibility. This includes characterization of the interactions and effects of Au NPs on different cell and tissue types, ranging from changes to protein and gene expression profiles, for understanding how they impact and are distributed within and across the various body systems.

There is growing evidence in the literature that when used within defined concentrations (typically ~10–100 μM) (Shukla et al. 2005) Au NPs appear to be inert and demonstrate low levels of cell and tissue toxicity; however, this is also highly dependent upon particle size, particle shape, and surface-coating chemistries (Shukla et al. 2005, Pan et al. 2007). Moreover, attention is now turning to unraveling the specific effects of various surface coatings on the Au NPs, which greatly influence uptake, tissue retention, and circulation half-life (Wang et al. 2015). Great promise is being given to the organic polymer linkers, such as polyethylene glycol (PEG) and PEG variants, with growing support for their apparent ability to confer increased biocompatibility properties to Au NPs (Bogdanov et al. 2014). The following sections will focus on the role of different coatings and other critical parameters, such as particle size and charge, in nanoparticle biodistribution and retention within biological systems.

3.5 *In Vitro* Cytotoxicity Studies

The common approaches used to assess cell toxicity in *in vitro* systems are via cell viability assays, such as MTT assay, which assesses cell metabolic activity. The MTT assay uses the drug 3-(4,5-dimethylthiazol-2-yl)-2,5-diphenyltetrazolium bromide (MTT), which is readily taken up by cells grown in culture (Mosmann 1983). Within viable cells, the MTT is enzymatically reduced to formazan, resulting in its conversion from a yellow to purple color that is readily detected at 570 nm using a UV–Vis microplate reader. It should, however, be noted that assay systems such as MTT and others employ similar wavelengths to detect color changes by the converted drug of interest (e.g., 570 nm for formazan), which can overlap with the peak emission spectra for the Au NPs, dependent on their size, shape, and concentration (Pan et al. 2007, Kroll et al. 2012). Therefore, the imperative for inclusion of relevant controls in these assays is to exclude interference and background attributable to the Au NPs themselves.

Cellular uptake of particulate material can occur via a variety of processes ranging from phagocytosis to receptor-mediated endocytic uptake. A number of excellent review articles discuss Au NPs cellular interactions in detail (Vonarbourg et al. 2006, Alkilany and Murphy 2010, Zhao et al. 2011). As expected, Au NP cellular cytotoxicity is greatly influenced by nanoparticle size and surface charge (Zhao et al. 2011). A study of varying sized Au NPs ranging between 0.8 and 15 nm in diameter, coated with monosulfonated triphenylphosphane, demonstrated that particles of 1.4 nm diameter were the most toxic with an IC_{50} of 46 μM. Similarly sized particles were found to be less toxic with 0.8, 1.2, 1.8, and larger 15 nm Au NPs having IC_{50} values of 250, 140, 230, and 6300 μM, respectively (Pan et al. 2007). This same study also showed that the cellular effects of the toxicity differed between the particles, with the 1.4 nm particles conferring rapid cell death via necrosis within 12 hours, while the similar-sized 1.2 nm Au NPs induced death via an apoptotic pathway (Pan et al. 2007).

A more recent report showed that 4.5 nm PEG-coated Au NPs added to cultured mouse myoblastoma cells were also found to be noncytotoxic up to a concentration of 5×10^{13} particles/mL (Leite et al. 2015). However, when the same Au NPs were added to cells in addition to the drug staurospaurin, they induced a greater cell susceptibility to the effects of the drug. The authors concluded that the PEG-coated Au NPs potentiate the effects of the drug via apoptotic means (Leite et al. 2015).

3.5.1 *In Vitro* Studies Reveal PEG-Coating Density Regulates Cellular Uptake of Au NPs

As mentioned previously, coating Au NPs with PEG has been shown to reduce *in vitro* cellular uptake and improve biocompatibility of the particles (Vonarbourg et al. 2006).

Attention has now turned to studying the underlying mechanisms responsible for this biocompatibility, in order to better understand how these effects arise, which will ultimately allow for the fine-tuning and rational design of tailored nanoparticles destined for biological applications. Some recent developments are outlined below.

The cellular uptake of PEG-coated Au NPs and the effect of PEG-coating density were studied in detail using murine macrophage cell line J774A.1 (Walkey et al. 2012). The study found that the degree of serum protein adsorption onto the particles directly correlated with particle uptake by the cells (Walkey et al. 2012). The researchers showed that as they increased PEG-coating density (0–1.25 PEG/nm^2), the amount of adsorbed serum proteins onto the particles decreased. They then demonstrated that not only was the amount of bound protein different, but also there was a concomitant change in the types of serum proteins bound to the particles. Detailed analysis revealed that they could group the proteins in clusters that correlated with the various PEG-coating densities and these clusters provided insight into the cellular uptake mechanisms and efficiencies of uptake. Overall, they demonstrated that not only Au NPs size but also PEG-grafting density ultimately controls the cellular uptake mechanism and uptake efficiency (Walkey et al. 2012). Even at high PEG-coating density that greatly reduced nonspecific serum protein binding, it did not totally eliminate Au NPs uptake by the cells.

Comparison of the cellular uptake of PEG-coated nanospheres (50 nm, negatively charged) compared to nanorods (10 × 45 nm, near neutral charge) by macrophage cell line RAW264.7 demonstrated that shape and surface charge remain critical factors in cellular interactions (Janát-Amsbury et al. 2011). Specifically, the researchers found that there were around four times as many PEG-coated nanospheres compared to nanorods in comparable cell samples (Janát-Amsbury et al. 2011). Their *in vivo* studies also demonstrated a significant reduction of nanorod accumulation in the liver and a longer blood circulation half-life compared to the spherical particles. They explained these difference largely in terms of the varied surface charge of the particles and indicated that particle geometry is also likely to influence these processes; however, they did not expound a mechanism for this, instead concluding that more work needs to be done (Vonarbourg et al. 2006).

Stability of the PEG surface coatings on the Au NPs is another critical aspect of the biocompatibility characterization process. Research has now emerged to indicate that physiological concentrations of cysteine can displace methoxy-PEG-thiol molecules from the Au NPs surface (Larson et al. 2012). Researchers showed that once the PEG particles were placed in cell culture media containing serum, the PEG coating was displaced by cysteine molecules, allowing for the adsorption of proteins onto the particle surface. To overcome this problem, the group included a small alkyl chain, which they refer to as a "hydrophobic shield," between the Au NPs surface and the outer hydrophilic PEG layer. In doing so, they found that mPEG–alkyl-thiol-coated Au NPs had greatly reduced protein adsorption and this also led to reduced uptake *in vitro*, by macrophage cells presumably mediated via the protein corona that normally forms on the mPEG–thiol particles (Larson et al. 2012).

Similarly, a combined *in vivo* and *in vitro* study of 5 nm polymer-coated Au NPs revealed that the polymer coating was partially removed following introduction of the particles to biological system (Kreyling et al. 2015). *In vivo*, the injected Au NPs were largely retained within the liver of the rats, while fragments of the coatings were excreted through the kidney (Kreyling et al. 2015). Complementary *in vitro* studies suggested that the polymer coating was degraded via enzymatic proteolytic cleavage within cells of the liver (Kreyling et al. 2015). This was supported by demonstration that the internalized particles were localized to lysosomal and endosomal compartments within HUVEC and Kuppfer macrophage cells (Kreyling et al. 2015).

3.6 *In Vivo* Biokinetic and Biodistribution Studies

Controlling *in vivo* Au NP localization has obvious benefits, with active and passive targeting approaches becoming better understood and exploited. An increasing number of studies demonstrate that coating of Au NPs with molecules, such as PEG, results in reduced surface fouling of the particles and steric stabilization (Shao et al. 2011). This has been suggested to be due to the inhibition or blocking of nonspecific attachment by serum proteins onto the Au NPs surface preventing the formation of a protein corona and reduced opsonization. In turn, it is believed that PEG coating results in a reduced incidence of uptake and clearance of these PEG-coated Au NPs by cells of the reticulo-endothelial system (Shao et al. 2011, Walkey et al. 2012). The coating appears to act as a "stealth" or "invisibility cloak" around the Au NPs, increasing their circulation half-life within the body and bloodstream. Given the advantages provided by these coatings in making Au NPs more biocompatible and better targeted, a clearer understanding of their mechanism of action is obviously needed.

In vivo studies using animal models to determine Au NPs biocompatibility, clearance from the body, and accumulation within discrete cell and tissue types have usually been assessed by isolation of the tissues of interest, followed by approaches to quantify the presence of Au NPs. Detection methods include scanning and transmission electron microscopy (Pissuwan et al. 2009, Chen et al. 2013), along with quantification methods, such as inductively coupled plasma mass spectrometry (ICP-MS) (Chen et al. 2013) and radioactive labeling of the particles themselves (Hirn et al. 2011, Kreyling et al. 2014). The following section summarizes a collection of studies carried out principally by a team based at the Helmholtz Zentrum Munich, Germany, that demonstrate the power of well-designed systematic research that can begin to form the basis for developing models to accurately predict the behavior of Au NPs when introduced into the body.

The three, 24-hour biokinetic studies were carried out in female rats, using monodispersed, spherical Au NPs across a range of core diameter sizes (monosulfonated triphenylphosphane [TPPMS]-coated 1.4, 5, 18, 80, 100 nm Au NPs as negatively charged surface), as well as 2.8 nm size (either carboxyl-coated negative surface or amino-coated positive surface), administered by following methods:

1. Intravenous (IV) injection into the tail vein, and following the movement of the particles from the bloodstream into body organs, tissues, and excrement (Hirn et al. 2011)
2. Oral ingestion by delivery of the particles intra-oesophageally, and following the absorption and uptake of the particles through the gastrointestinal tract (GIT) into the body systems and excrement (Schleh et al. 2012)
3. Inhaled Au NPs uptake, through the lung air–blood barrier, delivered by intratracheal instillation to determine accumulation of the particles in the body organs, tissues, and excrement of the animals (Kreyling et al. 2014)

The three studies confirmed the importance of size and surface charge on the *in vivo* distribution and accumulation of the particles, as well as demonstrating that the administration route was also a critically important determining factor. Specifically, the particles administered orally demonstrated the lowest uptake of all three procedures with <0.4% present in the internal organs and tissues, while between 99.63% and 99.99% of the different-sized and charged particles were found to be located in the GIT and feces at the 24-hour time point post-ingestion (Schleh et al. 2012). Overall, the study showed that the smaller-sized and more negatively charged particles tended to have higher levels of absorption and accumulation in tissues and organs. However, the researchers take care to point out that the 18-nm particles were more readily absorbed compared to the 5-nm particles and were the most highly accumulated particle in the brain of all the Au NPs tested. Moreover, they discussed the role that the types of adsorbed proteins forming the "corona" also likely influence particle accumulation. They speculated that the profile of adsorbed proteins onto the particle surface would be different based on the route of administration and, hence, also led to different tissue localization and accumulation sites. They concluded that individual tailoring and bespoke design of Au NPs for specific organ targeting is necessary (Schleh et al. 2012).

The study of body distribution 24 hours post-IV injection of the various Au NPs showed a rapid movement of the particles from the blood circulation, predominantly into the liver (Hirn et al. 2011). The researchers found that the vast majority of 5–200 nm particles ended up in the liver (91.9%–96.9%), while 81.6% of the negatively charged 2.8 nm particles, 72% for the positively charged 2.8 nm particles, and only 51.3% of the 1.4 nm particles localized to the liver. They demonstrated the existence of a linear relationship between the particles sized 1.4–5 nm in diameter, when plotted against liver retention at 24 hours post-IV injection. They, however, found little size dependency of accumulation in other tissues for particles sized between 18 and 200 nm.

Moreover, some differences have been identified in the accumulation of positively versus negatively charged particles. The results obtained for the spleen, for example, had a 2% accumulation of all particles ranging in size from 1.4 to 200 nm that had a negative TPPMS-coated surface, while the 2.8-nm particles surface coated either carboxyl-negative or amino-positive showed accumulation rates of 8.6% and 11.4%, respectively. This was considered as an unexpected result, given that the spleen, which is part of the reticulo-endothelial system, has previously been shown to accumulate other particles in a size-dependent fashion (Moghimi et al. 1991, Hirn et al. 2011). Again, surface charge was raised as a potential difference resulting in distinct proteins binding onto the particle to form the "corona." Others have also shown variation in organ distribution of 15-nm PEG-coated Au NPs with differing surface charges (Lee et al. 2015). This "dynamic protein binding and exchange" in turn is suggested to influence the ultimate site of particle accumulation and retention (Hirn et al. 2011).

The final study of this trilogy was aimed at examining the Au NPs distribution and accumulation within the body post-inhalation (Kreyling et al. 2014). The researchers measured biodistribution at 1, 3, and 24 hours post-intratracheal administration of the various-sized particles. They found a strong size dependency, with smaller-sized particles much more likely to cross the air–blood barrier compared to the larger-sized particles up to 80 nm in diameter (Kreyling et al. 2014). The percentage of Au NPs translocated across the air–blood barrier at 24 hours was ∼6%–7% of 1.4 nm Au NPs, <0.1% of 80 nm Au NPs, and ∼0.2%–0.3% of the 200 nm Au NPs. Therefore, of the total 100% administered, only a relatively small percentage of particles had in fact entered the body past the air–blood barrier within a 24-hour time frame. Interestingly, the 200 nm Au NPs did not follow this size-dependency relationship, as they translocated much more readily compared with the 80-nm particles. However, relative to the amount of Au NPs that crossed the air–blood barrier, retention within secondary organs and tissues was found to be independent of the Au NPs size. The researchers also importantly point out that the processes of Au NP translocation from the air into the bloodstream and subsequent accumulation and retention within organs and tissues should be viewed as two distinct processes, with each influenced independently by Au NP size and surface charge (Kreyling et al. 2014).

An additional important finding of the researchers was that Au NP retention in the rat carcass (comprising skeleton, soft tissue, and fat) was greater than that of all the secondary organ retention across all different-sized Au NPs (Kreyling et al. 2014). Of these total carcass retained Au NPs, 10%–20% were within the skeleton and presumed to be located within the bone marrow, as they most likely arrived there via the bloodstream. It is highlighted that these Au NPs particles are, thus, in direct contact with the pluripotent stem cells located within the bone marrow, which are known to be highly sensitive to exogenous stimuli (Kreyling et al. 2014). These findings serve to remind us of the complexities such studies face in attempting to unravel the myriad of processes involved in particle accumulation within tissues and organs.

Independent studies by researchers at the University of Technology, Sydney, Australia, also support the dependency between *in vivo* accumulation and size distribution of Au NPs, upon their initial route of administration. A study of 20–30 nm citrate-coated Au NPs administered via intraperitoneal injection in mice resulted in significant accumulation of the particles in the abdominal fat pad as well as in the liver, with no evidence of accumulation in brain, kidney, and heart tissues, at 72 hours post-injection (Chen et al. 2013). Another group of researchers using 25-nm-sized Au NPs (polyvinyl alcohol stabilized) compared particle distribution following oral or intravenous administration in rats over a relatively long period (Bednarski et al. 2015). After 10 days post-administration, researchers showed the injected Au NPs accumulated principally in the liver (>50%), with relatively smaller amounts in the lungs (5.7%) and spleen (2%) and only minor amounts recovered from the urine and feces during the 10-day period. The orally administered Au NPs showed a completely different distribution profile, with the majority of the particles excreted in the feces (55.8%) within the first 4 days of administration, and ~50-fold fewer particles detected within internal organs (Bednarski et al. 2015). The particles were principally located within the heart (0.6%), serum (0.3%), and brain (0.23%) (Bednarski et al. 2015). Recovery from the organs, of the initial dose of Au NPs, post 10 days, was 60% from the injected rats and only 1.4% from the orally administered rats (Bednarski et al. 2015).

Results of an *in vivo* study examining Au NPs toxicity undertaken in mice further extend this correlation in support of the importance of administration route on Au NPs activity and fate (Zhang et al. 2010). The study used unmodified 13.5 nm Au NPs that were administered via three different routes: oral, peritoneal injection, and tail vein injection, over a range of concentrations. The study found that of the three administration routes, the least toxic was the intravenous injection route (Zhang et al. 2010).

It is clear that a much deeper understanding of how Au NPs interact within biological systems is needed before they can be accepted as medical and mainstream drug and therapeutic agents. However, this journey of discovery is also proving to be highly valuable for equally important reasons of basic biological and medical research. *In vivo* and cell *in vitro* studies of these particles have already led to a number of novel and serendipitous findings (Vonarbourg et al. 2006, Chen et al. 2013), which in the long run can only serve to enrich our fundamental understanding of the natural and physical world, as well as providing new possibilities in the treatment of disease.

References

Alkilany, A. M. and Murphy, C. J. 2010. Toxicity and cellular uptake of gold nanoparticles: What we have learned so far? *Journal of Nanoparticle Research* 12 (7):2313–2333.

Bednarski, M., Dudek, M., Knutelska, J., Nowiński, L., Sapa, J., Zygmunt, M., Nowak, G., Luty-Błocho, M., Wojnicki, M., and Fitzner, K. 2015. The influence of the route of administration of gold nanoparticles on their tissue distribution and basic biochemical parameters: In vivo studies. *Pharmacological Reports* 67 (3):405–409.

Bogdanov, A. A. Jr., Gupta, S., Koshkina, N., Corr, S. J., Zhang, S., Curley, S. A., and Han, G. 2014. Gold nanoparticles stabilized with MPEG-grafted poly (l-lysine): In vitro and in vivo evaluation of a potential theranostic agent. *Bioconjugate Chemistry* 26 (1):39–50.

Brumlik, C. J. and Martin, C. R. 1991. Template synthesis of metal microtubules. *Journal of the American Chemical Society* 113 (8):3174–3175.

Brust, M., Walker, M., Bethell, D., Schiffrin, D. J., and Whyman, R. 1994. Synthesis of thiol-derivatised gold nanoparticles in a two-phase liquid–liquid system. *Journal of the Chemical Society, Chemical Communications* (7):801–802.

Cesbron, Y., Shaheen, U., Free, P., and Levy, R. 2015. TAT and HA2 facilitate cellular uptake of gold nanoparticles but do not lead to cytosolic localisation. *PLoS One* 10 (4):e0121683.

Chen, H., Dorrigan, A., Saad, S., Hare, D. J., Cortie, M. B., and Valenzuela, S. M. 2013. In vivo study of spherical gold nanoparticles: Inflammatory effects and distribution in mice. *PLoS One* 8 (2):e58208.

Chen, C. H., Lin, F. S., Liao, W. N., Liang, S. L., Chen, M. H., Chen, Y. W., Lin, W. Y. et al. 2015. Establishment of a trimodality analytical platform for tracing, imaging and quantification of gold nanoparticles in animals by radiotracer techniques. *Analytical Chemistry* 87 (1):601–608.

Danesh, N. M., Lavaee, P., Ramezani, M., Abnous, K., and Taghdisi, S. M. 2015. Targeted and controlled release delivery of daunorubicin to T-cell acute lymphoblastic leukemia by aptamer-modified gold nanoparticles. *International Journal of Pharmaceutics* 489 (1–2):311–317.

Dreaden, E. C., Alkilany, A. M., Huang, X., Murphy, C. J., and El-Sayed, M. A. 2012. The golden age: Gold nanoparticles for biomedicine. *Chemical Society Reviews* 41 (7):2740–2779.

Frens, G. 1973. Controlled nucleation for the regulation of the particle size in monodisperse gold suspensions. *Nature* 241 (105):20–22.

Gandra, N., Abbas, A., Tian, L., and Singamaneni, S. 2012. Plasmonic planet–satellite analogues: Hierarchical self-assembly of gold nanostructures. *Nano Letters* 12 (5):2645–2651.

Giljohann, D. A., Seferos, D. S., Weston, L. D., Matthew, D. M., Patel, P. C., and Mirkin, C. A. 2010. Gold nanoparticles for biology and medicine. *Angewandte Chemie International Edition* 49 (19):3280–3294.

Gu, H., Ho, P. L., Tong, E., Wang, L., and Xu, B. 2003. Presenting vancomycin on nanoparticles to enhance antimicrobial activities. *Nano Letters* 3 (9):1261–1263.

Hirn, S., Semmler-Behnke, M., Schleh, C., Wenk, A., Lipka, J., Schaffler, M., Takenaka, S. et al. 2011. Particle size-dependent and surface charge-dependent biodistribution of gold nanoparticles after intravenous administration. *European Journal of Pharmaceutics and Biopharmaceutics* 77 (3):407–416.

Hoft, R. C., Ford, M. J., McDonagh, A. M., and Cortie, M. B. 2007. Adsorption of amine compounds on the Au (111) surface: A density functional study. *The Journal of Physical Chemistry C* 111 (37):13886–13891.

Hou, W., Dasog, M., and Scott, R. W.-J. 2009. Probing the relative stability of thiolate-and dithiolate-protected Au monolayer-protected clusters. *Langmuir* 25 (22):12954–12961.

Huo, D., Ding, J., Cui, Y. X., Xia, L. Y., Li, H., He, J., Zhou, Z. Y., Wang, H. W., and Hu, Y. 2014. X-ray CT and pneumonia inhibition properties of gold-silver nanoparticles for targeting MRSA induced pneumonia. *Biomaterials* 35 (25):7032–7041.

Hwu, J. R., Lin, Y. S., Josephrajan, T., Hsu, M.-H., Cheng, F-Y., Yeh, C.-S., Su, W.-C., and Shieh, D.-B. 2008. Targeted paclitaxel by conjugation to iron oxide and gold nanoparticles. *Journal of the American Chemical Society* 131 (1):66–68.

Jadzinsky, P. D., Calero, G., Ackerson, C. J., Bushnell, D. A., and Kornberg, R. D. 2007. Structure of a thiol monolayer-protected gold nanoparticle at 1.1 Å resolution. *Science* 318, 5849:430–433.

Jana, N. R., Gearheart, L., and Murphy, C. J. 2001. Seed-mediated growth approach for shape-controlled synthesis of spheroidal and rod-like gold nanoparticles using a surfactant template. *Advanced Materials* 13 (18):1389.

Janát-Amsbury, M. M., Ray, A., Peterson, C. M., and Ghandehari, H. 2011. Geometry and surface characteristics of gold nanoparticles influence their biodistribution and uptake by macrophages. *European Journal of Pharmaceutics and Biopharmaceutics* 77 (3):417–423.

Kim, F., Connor, S., Song, H., Kuykendall, T., and Yang, P. 2004. Platonic gold nanocrystals. *Angewandte Chemie* 116 (28):3759–3763.

Kinoshita, T., Seino, S., Mizukoshi, Y., Nakagawa, T., and Yamamoto, T. A. 2007. Functionalization of magnetic gold/iron-oxide composite nanoparticles with oligonucleotides and magnetic separation of specific target. *Journal of Magnetism and Magnetic Materials* 311 (1):255–258.

Khandelia, R., Bhandari, S., Pan, U. N., Ghosh, S. S., and Chattopadhyay, A. 2015. Gold nanocluster embedded albumin nanoparticles for two-photon imaging of cancer cells accompanying drug delivery. *Small*. doi: 10.1002/smll.201500216.

Kreyling, W. G., Abdelmonem, A. M., Ali, Z., Alves, F., Geiser, M., Haberl, N., Hartmann, R. et al. 2015. In vivo integrity of polymer-coated gold nanoparticles. *Nature Nanotechnology* 10 (7):619–623.

Kreyling, W. G., Hirn, S., Moller, W., Schleh, C., Wenk, A., Celik, G., Lipka, J. et al. 2014. Air-blood barrier translocation of tracheally instilled gold nanoparticles inversely depends on particle size. *ACS Nano* 8 (1):222–233.

Kroll, A., Pillukat, M. H., Hahn, D., and Schnekenburger, J. 2012. Interference of engineered nanoparticles with *in vitro* toxicity assays. *Archives of Toxicology* 86 (7):1123–1136.

Larson, T. A., Joshi, P. P., and Sokolov, K. 2012. Preventing protein adsorption and macrophage uptake of gold nanoparticles via a hydrophobic shield. *ACS Nano* 6 (10):9182–9190.

Lee, J. K., Kim, T. S., Bae, J. Y., Jung, A. Y., Lee, S. M., Seok, J. H., Roh, H. S., Song, C. W., Choi, M. J., and Jeong, J. 2015. Organ-specific distribution of gold nanoparticles by their surface functionalization. *Journal of Applied Toxicology* 35 (6):573–580.

Leite, P. E., Pereira, M. R., Santos, C. A., Campos, A. P., Esteves, T. M., and Granjeiro, J. M. 2015. Gold nanoparticles do not induce myotube cytotoxicity but increase the susceptibility to cell death. *Toxicology in Vitro* 29 (5):819–827.

Li, M., Li, Y., Huang, X., and Lu, X. 2015. Captopril-polyethyleneimine conjugate modified gold nanoparticles for co-delivery of drug and gene in anti-angiogenesis breast cancer therapy. *Journal of Biomaterials Science, Polymer Edition* 26 (13):813–827.

Liang, Z., Susha, A., and Caruso, F. 2003. Gold nanoparticle-based core-shell and hollow spheres and ordered assemblies thereof. *Chemistry of Materials* 15 (16):3176–3183.

Lytton-Jean, A. K. R. and Mirkin, C. A. 2005. A thermodynamic investigation into the binding properties of DNA functionalized gold nanoparticle probes and molecular fluorophore probes. *Journal of the American Chemical Society* 127 (37):12754–12755.

Moghimi, S. M., Porter, C. J., Muir, I. S., Illum, L., and Davis, S. S. 1991. Non-phagocytic uptake of intravenously injected microspheres in rat spleen: Influence of particle size and hydrophilic coating. *Biochemical and Biophysical Research Communications* 177 (2):861–866.

Mosmann, T. 1983. Rapid colorimetric assay for cellular growth and survival: Application to proliferation and cytotoxicity assays. *Journal of Immunological Methods* 65 (1–2):55–63.

Mukherjee, P., Ahmad, A., Mandal, D., Senapati, S., Sainkar, S. R., Khan, M. I., Ramani, R., Parischa, R., Ajayakumar, P. V., and Alam, M. 2001. Bioreduction of AuCl4$^-$ ions by the fungus, *Verticillium* sp. and surface trapping of the gold nanoparticles formed. *Angewandte Chemie International Edition* 40 (19):3585–3588.

Narayanan, K. B. and Sakthivel, N. 2008. Coriander leaf mediated biosynthesis of gold nanoparticles. *Materials Letters* 62 (30):4588–4590.

Oldenburg, S. J., Averitt, R. D., Westcott, S. L., and Halas, N. J. 1998. Nanoengineering of optical resonances. *Chemical Physics Letters* 288 (2):243–247.

Pan, Y., Neuss, S., Leifert, A., Fischler, M., Wen, F., Simon, U., Schmid, G., Brandau, W., and Jahnen-Dechent, W. 2007. Size-dependent cytotoxicity of gold nanoparticles. *Small* 3 (11):1941–1949.

Park, J-W. and Shumaker-Parry, J. S. 2014. Structural study of citrate layers on gold nanoparticles: Role of intermolecular interactions in stabilizing nanoparticles. *Journal of the American Chemical Society* 136 (5):1907–1921.

Park, Y.-S., Hong, Y. N., Weyers, A., Kim, Y. S., and Linhardt, R. J. 2011. Polysaccharides and phytochemicals: A natural reservoir for the green synthesis of gold and silver nanoparticles. *Nanobiotechnology, IET* 5 (3):69–78.

Paul, A. M., Shi, Y., Acharya, D., Douglas, J. R., Cooley, A., Anderson, J. F., Huang, F., and Bai, F. 2014. Delivery of antiviral small interfering RNA with gold nanoparticles inhibits dengue virus infection *in vitro*. *Journal of General Virology* 95 (Pt. 8):1712–1722.

Perni, S., Prokopovich, P., Pratten, J., Parkin, I. P., and Wilson, M. 2011. Nanoparticles: Their potential use in antibacterial photodynamic therapy. *Photochemical & Photobiological Sciences* 10 (5):712–720.

Personick, M. L., Langille, M. R., Zhang, J., Harris, N., Schatz, G. C., and Mirkin, C. A. 2011. Synthesis and isolation of {110}-faceted gold bipyramids and rhombic dodecahedra. *Journal of the American Chemical Society* 133 (16):6170–6173.

Pissuwan, D., Valenzuela, S. M., and Cortie, M. B. 2006. Therapeutic possibilities of plasmonically heated gold nanoparticles. *Trends in Biotechnology* 24 (2):62–67.

Pissuwan, D., Valenzuela, S., and Cortie, M. B. 2008. Prospects for gold nanorod particles in diagnostic and therapeutic applications. *Biotechnology & Genetic Engineering Reviews* 25:93–112.

Pissuwan, D., Valenzuela, S. M., Miller, C. M., and Cortie, M. B. 2007. A golden bullet? Selective targeting of *Toxoplasma gondii* tachyzoites using antibody-functionalized gold nanorods. *Nano Letters* 7 (12):3808–3812.

Pissuwan, D., Valenzuela, S. M., Miller, C. M., Killingsworth, M. C., and Cortie, M. B. 2009. Destruction and control of *Toxoplasma gondii* tachyzoites using gold nanosphere/antibody conjugates. *Small* 5 (9):1030–1034.

Rana, S., Bajaj, A., Mout, R., and Rotello, V. M. 2012. Monolayer coated gold nanoparticles for delivery applications. *Advanced Drug Delivery Reviews* 64 (2):200–216.

Raveendran, P., Fu, J., and Wallen, S. L. 2006. A simple and "green" method for the synthesis of Au, Ag, and Au–Ag alloy nanoparticles. *Green Chemistry* 8 (1):34–38.

Reimers, J. R., Wang, Y., Cankurtaran, B. O., and Ford, M. J. 2010. Chemical analysis of the superatom model for sulfur-stabilized gold nanoparticles. *Journal of the American Chemical Society* 132 (24):8378–8384.

Safari, D., Marradi, M., Chiodo, F., Th Dekker, H. A., Shan, Y., Adamo, R., Oscarson, S. et al. 2012. Gold nanoparticles as carriers for a synthetic Streptococcus pneumoniae type 14 conjugate vaccine. *Nanomedicine* 7 (5):651–662.

Schleh, C., Semmler-Behnke, M., Lipka, J., Wenk, A., Hirn, S., Schaffler, M., Schmid, G., Simon, U., and Kreyling, W. G. 2012. Size and surface charge of gold nanoparticles determine absorption across intestinal barriers and accumulation in secondary target organs after oral administration. *Nanotoxicology* 6 (1):36–46.

Schmid, G., Klein, N., Korste, L., Kreibig, U., and Schönauer, D. 1988. Large transition metal clusters—VI. Ligand exchange reactions on Au 55 (PPh 3) 12 Cl 6—The formation of a water soluble Au 55 cluster. *Polyhedron* 7 (8):605–608.

Schmid, G., Pfeil, R., Boese, R., Bandermann, F., Meyer, S., Calis, G. H. M., and van der Velden, J. W. A. 1981. Au55[P(C6H5)3]12CI6—ein Goldcluster ungewöhnlicher Größe. *Chemische Berichte* 114 (11):3634–3642.

Shankar, S. S., Rai, A., Ankamwar, B., Singh, A., Ahmad, A., and Sastry, M. 2004. Biological synthesis of triangular gold nanoprisms. *Nature Materials* 3 (7):482–488.

Shao, X., Agarwal, A., Rajian, J. R., Kotov, N. A., and Wang, X. 2011. Synthesis and bioevaluation of (1)(2)(5)I-labeled gold nanorods. *Nanotechnology* 22 (13):135102.

Shukla, R., Bansal, V., Chaudhary, M., Basu, A., Bhonde, R. R., and Sastry, M. 2005. Biocompatibility of gold nanoparticles and their endocytotic fate inside the cellular compartment: A microscopic overview. *Langmuir* 21 (23):10644–10654.

Sun, Y. and Xia, Y. 2002. Shape-controlled synthesis of gold and silver nanoparticles. *Science* 298, 5601:2176–2179.

Thanh, N. T. K. and Rosenzweig, Z. 2002. Development of an aggregation-based immunoassay for anti-protein A using gold nanoparticles. *Analytical Chemistry* 74 (7):1624–1628.

Turkevich, J., Stevenson, P. C., and Hillier, J. 1951. A study of the nucleation and growth processes in the synthesis of colloidal gold. *Discussions of the Faraday Society* 11:55–75.

Vigderman, L. and Zubarev, E. R. 2013. Therapeutic platforms based on gold nanoparticles and their covalent conjugates with drug molecules. *Advanced Drug Delivery Reviews* 65 (5):663–676.

Vonarbourg, A., Passirani, C., Saulnier, P., and Benoit, J.-P. 2006. Parameters influencing the stealthiness of colloidal drug delivery systems. *Biomaterials* 27 (24):4356–4373.

Walkey, C. D., Olsen, J. B., Guo, H., Emili, A., and Chan, W. C. 2012. Nanoparticle size and surface chemistry determine serum protein adsorption and macrophage uptake. *Journal of the American Chemical Society* 134 (4):2139–2147.

Wang, J., Bai, R., Yang, R., Liu, J., Tang, J., Liu, Y., Li, J., Chai, Z., and Chen, C. 2015. Size-and surface chemistry-dependent pharmacokinetics and tumor accumulation of engineered gold nanoparticles after intravenous administration. *Metallomics* 7 (3):516–524.

Xie, J., Zheng, Y. and Ying, J. Y. 2009. Protein-directed synthesis of highly fluorescent gold nanoclusters. *Journal of the American Chemical Society* 131 (3):888–889.

Yeh, Y.-C., Creran, B., and Rotello, V. M. 2012. Gold nanoparticles: Preparation, properties, and applications in bionanotechnology. *Nanoscale* 4 (6):1871–1880.

Zhang, X., Servos, M. R., and Liu, J. 2012. Instantaneous and quantitative functionalization of gold nanoparticles with thiolated dna using a pH-assisted and surfactant-free route. *Journal of the American Chemical Society* 134 (17):7266–7269.

Zhang, X. D., Wu, H. Y., Wu, D., Wang, Y. Y., Chang, J. H., Zhai, Z. B., Meng, A. M., Liu, P. X., Zhang, L. A., and Fan, F. Y. 2010. Toxicologic effects of gold nanoparticles *in vivo* by different administration routes. *International Journal of Nanomedicine* 5:771–781.

Zhao, F., Zhao, Y., Liu, Y., Chang, X., Chen, C., and Zhao, Y. 2011. Cellular uptake, intracellular trafficking, and cytotoxicity of nanomaterials. *Small* 7 (10):1322–1337.

4

Nucleoprotein-Based Nanodevices in Drug Design and Delivery

Elizabeth Singer, Katarzyna Lamparska-Kupsik, Jarrod Clark, Kristofer Munson, Leo Kretzner, and Steven S. Smith

CONTENTS

4.1 Introduction

Bionanotechnology is a new field based on chemistry, physics, and molecular biology. It is largely concerned with the development of nanoscale bioassemblies that do not occur in nature. The assemblies produced in this field form devices that are under 100 nm in their largest dimension, monodisperse, and soluble in aqueous media. Although chemical conjugation can be used in the assembly of several components into a device, naturally occurring biospecificities are often helpful in that they permit self-assembly. Several of these new technologies afford novel approaches to drug design and delivery. An essential component of this third generation of drugs [1] is their capacity for selective targeting.

4.1.1 Bionanotechnology for Molecular Targeting

4.1.1.1 Dendrimer-Based Targeting Assemblies

Dendrimers are synthetic scaffolds that can carry bioconjugates. The core of a dendrimer anchors subsequent polymerization. It carries multiple linking functionalities that permit the sequential addition of branching monomers. After the first round of synthesis (generation G0) the system carries twice as many linking functionalities; after the second round (generation G1) it carries four times as many, and so on. Dendrimers synthesized in this fashion are now commercially available with generation levels as high as G10. The larger versions adopt a roughly spherical shape [2] that permits the attachment of nucleic acids [3] and small molecules suitable for molecular targeting [4,5]. Dendrimers appear to have the capacity to flatten out so as to conform to an irregular surface, which may contribute to their capacity for strong binding at cell surfaces [6]. The details of dendrimer targeting are reviewed in Ref. [7].

4.1.1.2 Protein-Based Targeting Assemblies

Proteins can also be used as molecular scaffolds. Protein scaffolds are generally based on protein–protein interactions. This form of self-assembly permits the formation of extended interlocked structures or interlocked closed shells like the viral capsid. The rules for assembly of icosahedral viral capsids were developed by Caspar and Klug [8]. These naturally occurring systems generally form from one or more capsid proteins that can assemble into triangular facets that aggregate into larger deltahedra.

In cowpea mosaic virus capsids, each triangular facet carries a lysine residue with enhanced chemical reactivity that has permitted the conjugation of a variety of species including fluorescein, rhodamine, biotin, and 900 nm diameter gold particles [9–11]. Although cell surface targeting has not been demonstrated with these capsids, this remains a possibility [7].

Extended tubes, filaments, and vesicles that self-assemble from proteins are not monodisperse, and have not been adapted to bionanotechnological applications. However, self-assembling fusion proteins [12] and multisubunit bioconjugates [13] are well suited for these applications.

4.1.1.3 Protein–Nucleic Acid Based Targeting Assemblies

Bionanotechnological designs for self-assembling nucleoprotein biostructures suggest that nanoscale devices capable of site-specific molecular targeting can be constructed. These tools are uniquely equipped to augment immunohistochemistry or drug targeting techniques where increased sensitivity is required, or where antibodies are unavailable. The system we are developing is a nanotechnological implementation of both molecular biology and chemistry. It uses DNA-methyltransferase-directed covalent addressing of fusion proteins to chemically synthesize DNA scaffolds. It offers a practical approach to the construction of bionanostructures in a wide variety of designs that may make it possible to match cell surface structures in a lock and key fashion. The fundamental principle of this technology lies in the capacity of DNA methyltransferases to form covalent linkages with DNA (Figure 4.1). This property, coupled with their selectivity for defined

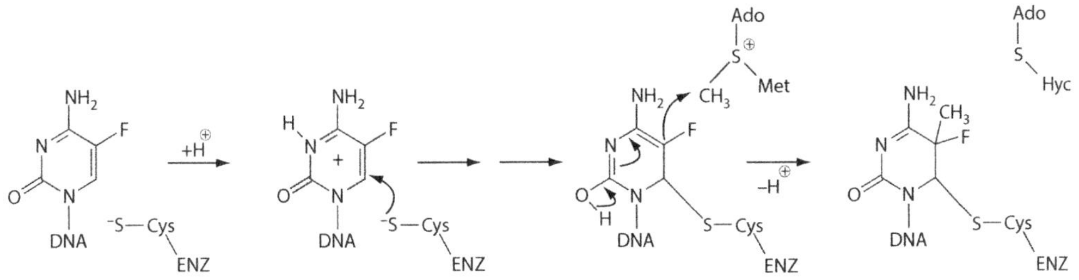

FIGURE 4.1

DNA methyltransferase mechanism of action. During catalysis cytosine methyltransferases make a nucleophilic attack on C6 of cytosine or 5-fluorocytosine. This breaks the 5–6 double bond in the ring and activates C5 for methyltransfer. After the methyl group is transferred from *S*-adenosylmethionine (AdoMet) to C5, the normal progress of the reaction is to remove the hydrogen at C5 and the enzyme nucleophile from C6 by β-elimination. However, when fluorine is present at C5 this cannot occur because of the strength of the fluourine–carbon bond. Thus the progress of the reaction is stalled and a covalent complex forms between the enzyme and the cytosine ring targeted by the enzyme. In the case of M·*Eco*RII this is the second cytosine in the CCWGG recognition sequence.

nucleic sequences, allows them to serve as targeting agents for fusion proteins directed to specific sites on a DNA scaffold [14,15].

The technology used here differs from the adaptor concept described by Gibson and Lamond [16]. That method for ordering components was first developed by Niemeyer et al. [17–19] and later studied by Alivisados et al. [20]. In the adaptor concept, elements are ordered on a template of single-stranded nucleic acid by coupling them to the adaptor single-strands having Watson–Crick complementarity to juxtaposed regions of the template. In the adaptor system, one uses the selectivity of the DNA–DNA hybridization to obtain a set order for desired elements (e.g., biotinylated proteins [17] or gold nanocrystals [20]). These methods have been used successfully in a number of applications. However, the elements to be ordered must be capable of surviving mild chemical reactions (e.g., biotinylization for proteins) and moderate temperatures needed for hybridization of the adaptor-coupled elements.

As can be seen from the schematic illustration given in Figure 4.2, the use of DNA methyltransferases as targeting devices for fusion proteins obviates these problems. Labile

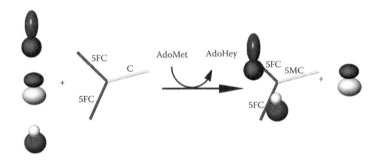

FIGURE 4.2

Schematic of the nanoscale assembly system. Fusion proteins are represented schematically. Linked ovoid and spherical shapes indicate the methyltransferase fusion proteins. The ovoid and spherical models at the base of each fusion represent the methyltransferase portion of the fusion. Each of these is colored so as to match the DNA arm of the Y-junction containing its recognition sequence. Only 5-fluorocytosine-substituted recognition sequences can form covalent links with the methyltransferases. Thus, sites activated by fluorocytosine substitution retain fusion proteins in a preselected order set down during DNA synthesis. Those with cytosine in their recognition sequence are methylated at the recognition site but the fusion protein is not linked to the DNA.

proteins are formed as fusions, and gentle conditions (37°C, neutral pH) can be used in the ordered assembly reaction to covalently couple the fusion protein to a preselected site on a duplex DNA scaffold. In what follows, we focus on the implementation of the methyltransferase-based technology in three-address systems like that depicted in Figure 4.2.

4.1.2 Assembly of Three-Address Nucleoprotein Arrays

The DNA Y-junction has proven to be a very useful DNA scaffold for the construction of nanoscale targeting devices employing ordered methyltransferase-fusions. The Y-junction can have symmetry; however, unlike the four-way Holliday junction [21], it is not capable of branch migration via Watson–Crick base pairing. The presence of mispairs near the center of the system tends to force the system into an asymmetric Y with two of the arms stacking on one another while forcing the third arm into an obtuse angle relative to the linear helix formed by the other two arms so as to accommodate the mispaired bases in the resulting space [22]. The system can also adopt a T conformation with the mispaired bases accommodated in the two arms that stack on each other and oppose the third arm [23]. Although such systems may prove useful in certain applications, three-address nucleoprotein arrays that have thus far been exploited do not contain mispairs and are symmetric. Thus, a system with complete Watson–Crick homology to the center of the junction should be roughly Y-shaped [24] when viewed from above (Figure 4.3). Studies of the arrangement of arms around the central junction using resonance energy transfer show that the mean interarm distance is similar for each arm [25]. The structure is dynamic and the interarm distances are rather broadly distributed around the mean, with a range of about 30% of the measured mean encompassing 50% of the measured values. The values are consistent with a system that is in rapid equilibrium with a planar T conformation, in which any two arms can stack on each other at random, while possibly undergoing a planar and pyramidal interconversion with an equilibrium favoring the trigonal pyramid [26]. Torsional constraints have not been studied extensively. Decoration of the arms with recombinant proteins is expected to slow or hinder planar and pyramidal interconversions and restrict the lateral and torsional range of motion of the arms.

4.2 Molecular Models

It is generally valuable to prepare molecular models of the final device designs. This is often quite difficult because structural information on the proteins involved may be unavailable. For designs involving well-characterized structures, models have been constructed in Insight II (Accelrys, San Diego, CA). In those cases [15,27] a model of the DNA scaffold (linear DNA or Y-junction) was constructed, and the DNA present in the three dimensional structure of the bacterial methyltransferase–DNA complex [28,29] was spliced into the DNA model at the appropriate site. After constraining the protein and the DNA at the protein binding sites, the structures were then minimized in molecular mechanics using the Dreiding force field in Biograf (MSI, San Diego, CA) or the AMBER force field in Insight II until the RMS force was less that 0.1 (kcal/mol)/Å, followed by 1000 steps of simple dynamics. Models of the M·*Eco*RII containing devices have not yet been prepared, however inspection of the homology models of this protein [30] suggests that the parameters used in the construction of the devices containing M·*Hha*I [1,27] would carry over to the devices described here.

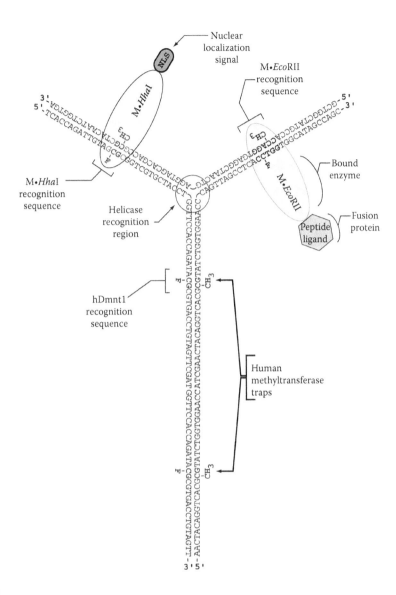

FIGURE 4.3

A targeted molecular device. Selective targeting to the nucleus requires that a nanoscale device locate and target a specific subset of cells, gain access to the cytoplasm and be targeted to the nucleus. The Y-junction depicted in Figure 4.2 provides a scaffold for protein signals provided as fusions to the DNA methyltransferases, and a linking arm that can be utilized to inhibit nuclear systems. DNA methyltransferase traps are depicted in this example as nuclear process inhibitors.

4.3 Oligodeoxynucleotide Preparation

Oligodeoxynucleotides were synthesized using standard phosphoramidite chemistry. The TMP-F-dU-CE convertible phosphoramidite (Glen Research, Sterling, VA) was used to introduce 5-fluorodeoxycytidine. In certain cases fluorescent labels have been introduced. For this purpose the 5′ fluorescein phosphoramidite (Glen Research, Sterling, VA) was used. For experiments in which the DNA Y-junction is to be exposed to cultured cells or crude extracts, the DNA can be synthesized using protective phosphothiolate backbones. These backbones do not interfere with methyltransferase binding but do protect against cleavage by nucleases. However, we have found that this is unnecessary when the devices are used only to target the cell surface. Oligodeoxynucleotide concentrations were measured by absorbance spectroscopy at 260 nm. In order to form duplexes and Y-junctions, oligodeoxynucleotides were mixed in equimolar amounts in a buffer containing 10 mM Tris-HCl at pH 7.2, 1 mM EDTA, and 100 mM NaCl, to a final concentration of 6 μM. They were then annealed at 95°C for 5 min, 50°C for 60 min, room temperature for 10 min, and then put on ice for 10 min. Each of these procedures has been used routinely in the laboratory [14,15,27]. Representative oligodeoxynucleotide sequences employed in previous work are given below.

Y-junction oligodeoxynucleotide sequences:

5'-GCTGGCTATGCCACMAGGTGAGGCTAACTGAGGTAGCACGACCGFGCTACAATCTGGTGA-3'

5'-TCACCAGATTGTAGMGCGGTCGTGCTACCTGGTTCCACCAGATGFGCGTGACCTGTAGTT-3'

5'-AACTACAGGTCACGMGCATCTGGTGGAACCCAGTTAGCCTCACFTGGTGGCATAGCCAGC-3'

where M indicates 5-methyl and F indicates a 5-fluoro moiety on cytosine.

4.3.1 Cloning, Expression, and Purification of Fusion Proteins

Cloning and expression of ligand sequences can be performed by more or less routine methods. The current method used for the M·*Eco*RII–peptide–ligand fusions can be summarized as follows. M·*Eco*RII was cloned into the vector pET28b(+) (EMD Biosciences, Inc. Novagen Brand, Madison, WI) by PCR amplification. The PCR product contained the added restriction sites *Nco*I and *Bam*HI. The vector was cut with *Nco*I and *Bam*HI to remove the vector's His-tag, thrombin, and T7 tags and to create compatible ends for ligation of the M·*Eco*RII product. This new vector, pET28-M·*Eco*RII, has its start ATG in the same position as the native pET28b start ATG. The *Nco*I site containing the start ATG codon was used for cloning of the ligand, in this case Thioredoxin (Trx). Ligands have been PCR amplified from an appropriate clone or cell line with an *Nco*I site added to both ends of the PCR product. After in-frame ligation of the peptide–ligand sequence into the pET28-M·*Eco*RII construct, the correct sequence and orientation were verified by sequencing. The pET28-ligand-M·*Eco*RII vector was transformed into BL21-DE3 cells and protein expression was induced with 1 mM IPTG. Following the induction and expression period, the cells were lysed by treatment with lysozyme followed by sonication. Debris was pelleted by centrifugation and the supernatant fluid was applied to a phosphocellulose (P-11) column. The column was eluted with a linear gradient, and then applied to a DEAE (DE-52) column, and likewise gradient eluted, resulting in a high-specific activity enzyme [31].

4.3.2 Y-junction Device Assembly

4.3.2.1 Fusion Protein Coupling

Once the fusion proteins were purified, they were coupled to the Y-junction as follows. Annealed Y-junctions, at 0.6 μM, were exposed to 5.3 μg/mL of M·*Eco*RII-fusions (carrying the peptide–ligand) in a binding buffer containing 50 mM of Tris-HCl at pH 7.8, 10 mM of EDTA, 5 mM of β-mercaptoethanol and 80 μM of *S*-adenosyl-L-Methionine (AdoMet). The final volume of the reaction depends on the desired scale of the preparation. The reaction was incubated for 2.5 h at 37°C.

4.3.2.2 Monitoring Final Assembly with Microfluidics Chip–Based Protein Mobility Shift

Electrophoretic mobility shift analysis (EMSA) is a well-characterized and widely employed technique for the analysis of protein–DNA interaction and the analysis of transcription factor combinatorics. As currently implemented, EMSA generally involves the use of radio-labeled DNA and polyacrylamide gel electrophoresis. We noted [27] that this technique could be effectively implemented with microfluidics chips designed for the separation of DNA fragments. To accomplish this, samples were run on a 2100 Bioanalyzer (Agilent Technologies, Palo Alto CA) using a DNA 500 LabChip or a DNA 7500 LabChip (Caliper Technologies, Mountain View CA) according to the manufacturer's instructions (see below).

4.3.3 Applications of Ordered Arrays in Smart Drug Design

The design progression for DNA methyltransferase inhibitors is depicted in Figure 4.4. Here we see examples of small molecules that must be metabolized in order to be effective as first generation inhibitors. Preformed mimics of a complex substrate or structure are examples of the second generation inhibitors. Ordered protein arrays provide an example of the third generation inhibitors selectively targeted to cells for intracellular delivery of a drug (Figure 4.5).

4.3.3.1 Thioredoxin-Targeted Fluorescent Nanodevices

To test the functioning of a system of this type, an *Eco*RII-methyltransferase–thioredoxin fusion protein (M·*Eco*RII-Trx) was cloned, purified, and covalently coupled to the target

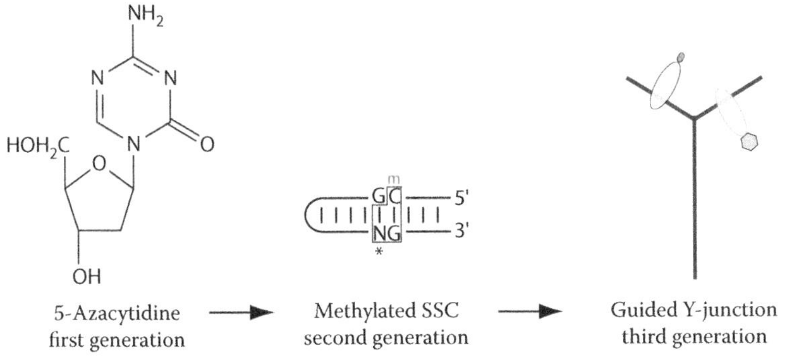

5-Azacytidine   Methylated SSC   Guided Y-junction
first generation   second generation   third generation

FIGURE 4.4

Design progression. First generation inhibitors are small molecules. For methyltransferases these are represented by 5-azacytidine. This drug can be metabolized and incorporated into DNA where it serves to trap methyltransferases. Second generation inhibitors are pre-formed mimics of a complex substrate or structure. For methyltransferases they are represented by single-strand conformers (SSCs) that carry a methyltransferase trap. In this case the recognition motif characteristic of hDnmt1 is incorporated into the trap and is marked with an L-shaped box. Third generation inhibitors are devices with a delivery system (hexagon), allowing specific cell targeting and intracellular delivery. Nuclear localization signals (oval) guide them to the cell nucleus, where they can deliver a lethal payload (long stem of the Y).

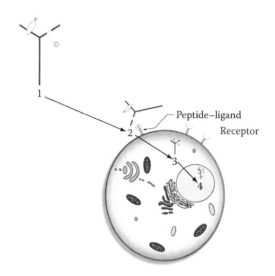

FIGURE 4.5
Surface binding, internalization and nuclear targeting. In (1), the device binds to a cell surface receptor protein specific for the protein ligand carried by the nanodevice. In (2), the device is internalized by the receptor. In (3), transport via the nuclear localization signal carries the device to the cell nucleus (4).

sequence in the Y-junction DNA. In this test case the fusion was ligated into the multicloning site of pET32a. Expression and purification were as described above.

The Y-junction DNA used in this application has three target sequences for M·*Eco*RII-Trx fusion proteins bound to one Y-junction. We evaluated the assembly of the Y-junction coupled products by using a microfluidics-based EMSA described above and developed for this purpose in our laboratory [27]. This allowed us to distinguish between free Y-junctions and Y-junction coupled products with one, two, or three M·*Eco*RII fusion proteins bound. We observed a greater mobility shift with the Y-junction M·*Eco*RII-Trx-coupled product than with the Y-junction M·*Eco*RII-coupled product as predicted from the difference in the molecular weight (Figure 4.6). We observed only one product with a Y-junction that has only one M·*Eco*RII binding site, whereas we observed three products with a Y-junction having three M·*Eco*RII binding sites. The Y-junction coupled to M·*Eco*RII-Trx generates three products, with di- and tri-substituted forms being the most prevalent.

The nanodevices tested in the preliminary work display bacterial thioredoxin [32] as the targeting ligand. This peptide is structurally homologous to human thioredoxin, although it shares little sequence homology with its human counterpart. In human cells, thioredoxin is thought to be an inhibitor of apoptosis [33]. It is also exported [34], where it can serve as a cytokine [35]. Thus, its continued expression, and that of its associated receptors (e.g., thioredoxin reductase) and transporters [36], suggests that it may be a hallmark of certain aggressive forms of prostate cancer. The ability to detect this marker may be of significant value in tumor classification. To test this possibility, MCF-7 cells were exposed to solutions containing the nanodevice and were evaluated by fluorescence microscopy. The cells exposed to the Y-junction M· *Eco*RII-Trx had a high overall fluorescence and localized fluorescent signals around the cell surface (Figure 4.7a). This suggests that the nanodevice may be binding to receptors in the cell membrane. Cells exposed to the Y-junction linked to M·*Eco*RII but lacking the fused Trx (Figure 4.7b) had a similar level of fluorescence as cells exposed to the Y-junction DNA alone (Figure 4.7c) and cells exposed to phosphate-buffered saline (Figure 4.7d).

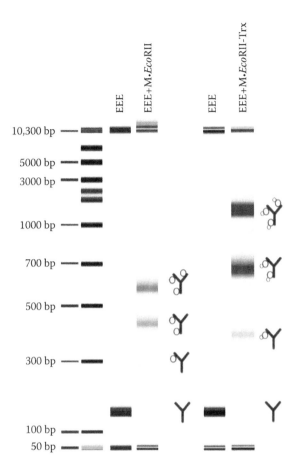

FIGURE 4.6

Microfluidics monitoring of Y-junction coupling. Lane EEE: Y-junction DNA containing three M·*Eco*RII recognition sites. Lane EEE + M·*Eco*RII: Y-junction DNA coupled to the fusion protein containing M·*Eco*RII lacking fused Thioredoxin. Lane EEE + M·*Eco*RII-Trx: Y-junction DNA coupled to the fusion protein containing M·*Eco*RII and thioredoxin domains. Di- and trisubstituted forms dominate the products. Note that the additional molecular weight of the thioredoxin peptide relative to the control M·*Eco*RII gives a significantly greater retardation of the Y-junction. An illustration representing each of the forms present in the virtual gel is given to the right of each set of lanes.

4.3.3.2 Designs Expected to Be Internalized and Localized to the Nucleus

The thioredoxin targeted Y-junction provides a proof-of-concept demonstration that nanodevices of this type can be targeted to the cell surface. However, the design depicted in Figure 4.5 is expected not only to target the cell surface but also to be transported across the cell membrane by ligand receptor internalization and then transported to the nucleus. In general, nuclear localization signals are recognized in the cytoplasm by the importin system [37] that mediates transport to the nucleus. However, this system operates only on proteins that have gained entry to the cytoplasm. Selective transport of peptide–ligands is mediated by cell surface receptors. Several well-characterized cell surface receptors are members of the epidermal growth factor receptor (EGFR) family. Our expectation is that receptors of this type will internalize [38,39] the nanodevice, and hand it off to the importin system for transport to the nucleus.

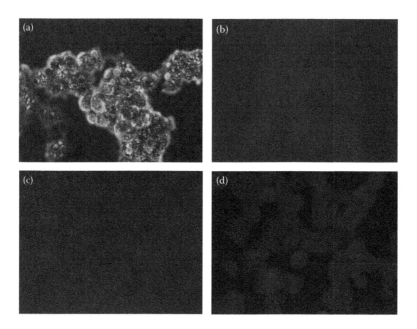

FIGURE 4.7
Fluorescent images of MCF7 cells exposed to the thioredoxin-targeted device. The cells were observed at 400×
under fluorescent light with a FITC filter at the City of Hope Cytogenetics Core Laboratory using high-quality
fluorescent photomicroscopes and computerized imaging system. MCF7 cells exposed to the Y-junction linked
to M·*Eco*RII-Trx (a) had a high overall fluorescence and localized fluorescence around the cell surface while
cells exposed to Y-junction linked to M·*Eco*RII (b) and Y-junction DNA (c) had a similar level of fluorescence to
cells exposed to PBS (d).

4.3.4 Molecular Payloads

Given a nanoscale targeting device, it is important to consider the types of payload that
can be delivered and how they might work. Payloads generally fall into two classifications:
those causing general damage to an important cellular system and those that are designed
to selectively attack a given metabolic pathway. Delivery of nanoparticles for subsequent
electromagnetic energy capture, e.g., light energy capture by carbon nanotubes [40] or neutron energy capture by boron cages [41], can be effective with selective targeting to the cell
surface or cytoplasm and do not require delivery to the nucleus. Devices that carry small-
molecule based lethal poisons that act in the cell nucleus (e.g., α-amanitin) have the disadvantage that premature or nonspecific release can result nonspecific cell killing, because
these poisons are generally transported to the nucleus without the aid of a nanodevice. We
focus here on DNA methyltransferase traps and DNAzymes because these systems can be
easily built into the nanodevice we describe, cannot act until selectively internalized by
the cell, and are most effective in the cell nucleus.

4.3.4.1 Biology of DNA Methylation

Selective gene activation and repression in normal cells is not fully understood, however these processes appear to center on promoter activation or repression mediated by
protein–DNA interaction combinatorics. These patterns are tissue specific and stably
maintained in a given cell lineage. Once a transcription state is established, it appears

to be stably maintained by a self-reinforcing network of protein and DNA modifications involving histone methylation, histone acetylation, and cytosine methylation in DNA [42].

In general, gene expression patterns are randomized during tumorigenesis by genetic damage and natural selection during tumor progression. Hallmarks of this process are the establishment of patterns of ectopic gene expression and ectopic gene silencing that adapt them for their role as invasive tumors. This has led to the development of drugs directed at the disruption of stable patterns of gene expression, in the hope that selective delivery to tumor cells will inhibit growth or induce cell death [43]. Among the drugs that have been discovered are a variety of histone deacetylase inhibitors and DNA (cytosine-5) methyltransferase inhibitors [44] that tend to act synergistically [45,46] to disrupt these gene silencing systems. These drugs can also be effective alone. In principle, DNA methyltransferases can be inhibited by either the direct interaction of the inhibitor with the enzyme active site or its protein targeting signals, or by selectively interfering with methyltransferase synthesis. The DNA Y-junction can be used to target DNA methyltransferase traps (noncompetitive inhibitors of the enzyme) or DNAzymes targeting the messenger RNA for the methyltransferase itself to the nucleus.

4.3.4.2 First Generation DNA Methyltransferase Traps

Since the target of methylation is deoxycytidine, the first inhibitors of MT activity to be developed were analogs of this nucleoside [47,48]. The compounds in this group are structurally based upon 5-azacytidine. These drugs are phosphorylated in cells and incorporated into RNA and DNA, with the deoxyribose analog (Figure 4.4) preferentially incorporated into DNA. Both of these compounds have a nitrogen atom at the 5-position [43], and since N5 does not allow for the β-elimination step in the methyltransferase enzymatic mechanism, the enzyme becomes trapped in a covalent intermediate with its substrate [49]. This mechanistic prediction was confirmed by in vitro studies with 5-fluorocytosine for the human enzyme [50]. Since that time, a wide variety of bases have been shown to operate similarly when incorporated into DNA [51]. Of these, only 5-azacytidine and 2-pyrimidinone (Table 4.1) have been touted as first generation methyltransferase inhibitors in cancer chemotherapy [52,53].

4.3.4.3 Second Generation Methyltransferase Traps

Second generation methyltransferase inhibitors are DNA substrate analogs (Figure 4.4). They can take a variety of forms (e.g., simple hairpins or annealed duplexes), and are generally synthesized as short oligodeoxynucleotides [51]. Since one of the key properties of the human enzyme is its response to 5-methylcytosine (5mC), this property can be exploited in second-generation inhibitor design. For the major form of the human DNA methyltransferase (hDnmt1) the presence of a 5mC focuses the enzyme active site so that it probes the symmetrically placed base (normally an unmethylated cytosine residue) in the three-nucleotide motif (L-shaped box in Figure 4.4) recognized by the enzyme [54,55]. Electronic structure–activity relations for target bases at this site have been developed [51]. Table 4.1 lists the targets. They fall into three categories: (1) productive targets, those that actually permit nucleophic attack, subsequent methyltransfer, and β-elimination; (2) nonproductive targets, those expected to undergo nucleophilic attack at a negligible rate, and therefore merely slow down the enzyme by forcing it to unstack an unproductive base [51];

TABLE 4.1

Electronic Structure Classification of DNA Methyltransferase Targets

Productive Target[a]		Nonproductive Target		Trapping Target	
Attacked Target	Intermediate	Target	Intermediate	Attacked Target	Intermediate
Cyt+	Cyt Enol	4-ThioU	–	2-Pyrimidinone+	2-Pyrimidinone Enol
		5-BrU	–	5-FCyt+	5-FCyt Enol
		5-FU	–	5-AzaCyt+	5-AzaC Enol
		U	–		
		Pseudo U (ψ)	–		
		T	–		
		8-Oxo-G	–		
		A	–		
		G	–		
		7-Deaza-G	–		

[a] Electronic structure–activity relations allow the classification of bases at a site targeted for attack by the methyltransferase into the three categories listed in the table: productive, nonproductive, and trapping bases. LUMO, HOMO, and frontier orbital differences for model compounds used in the calculations are given in Ref. [51].

and (3) trapping targets, those that undergo nucleophilic attack and methyltransfer but not β-elimination. This latter group is represented by 5-fluorocytosine (Figure 4.1), which forms a dead-end complex between the enzyme nucleophile (in general a cysteine residue) and the DNA because β-elimination cannot occur.

In the model device depicted in Figure 4.5, 5-fluorocytosine has been used with the spacing between trapping sites oriented so that the device is capable of trapping two methyltransferase molecules [15]. In this case an odd multiple of 5 bp places the two enzyme molecules on opposite sides of the DNA and at the required intersite spacing of about 22 bp [56], so that they need not compete with one another during binding [15].

A second inhibitor is possible with 5mC on one strand targeting nonproductive nucleotides. Here the density of the 5mC residues is important. For example an oligodeoxynucleotide carrying a single hemi-methylated productive target surrounded by hemi-methylated nonproductive targets is a strong inhibitor of the enzyme when the local density of nonproductive targets is 0.40 5mC/nt [51], but is not an effective inhibitor when the density is only 0.15 5mC/nt [31].

4.3.4.4 DNAzymes

Deoxyribozymes, or DNAzymes, have been in existence for 10 years [57–59]. They combine the catalytic efficiency of their predecessors, the ribozymes [60], with the stability and ease of synthesis of DNA. Certain catalytic cores, flanked by single-stranded arms available for base pairing with a desired target RNA, can be generated by multiple rounds of selection interspersed with intervening PCR amplification steps [57,59,61]. However, de novo selection of an active DNAzyme is not strictly necessary in each case, as certain core sequence motifs have been identified with given properties, for example, RNA cleavage [61], synthetic capability [58,59], or fluorescence reporting activity [62].

The most studied among DNAzyme activities, and one of obvious therapeutic interest, is that resulting in targeted cleavage of a desired RNA substrate. The prototype of this is the so-called 10–23 DNAzyme of Santoro and Joyce [61]. This term is still used for the same or very similar DNA catalytic cores [63], although it designated a particular clone isolated from the 10th round of amplification in the original study [61]. This oligo-nucleotide has an almost invariant 15-nucleotide single-stranded loop, flanked by highly variable arms. These latter nucleotides are capable of base-pairing with a variety of RNA substrates, and this hybridization results in cleavage of the substrates at an invariant, characteristic site of the RNA: immediately 3′ of a single unpaired purine residue of the target [61]. The 10–23 design combines seemingly limitless target specificities of the free arms with an acceptable catalytic phosphoesterase activity of the core: $k_{cat} = 3.4$ min^{-1}; $K_m = 0.76$ nM; k_{cat}/K_m (catalytic efficiency) $\cong 10^9$ M^{-1}, min^{-1}, which is comparable to nat-ural and designed ribozymes [61]. These values were obtained in vitro under simulated physiological conditions of 2 mM MgCl$_2$ and 150 mM KCl at pH 7.5, and 37°C, although the enzyme was maximally active with an apparent K_m for Mg^{2+} of 180 mM at pH 8.0 and 37°C.

In the intervening years, 10–23 DNAzymes have been obtained that possess improved stability and activity [63], and can be used to cleave a wide variety of RNA targets. They have been shown to inhibit cell proliferation and migration in bioassays, both of cell cultures but also, more importantly, of tumor xenotransplants in athymic (nude) mice [64]. An easily conceivable hDnmt1-specific DNAzyme that can be linked to the nanode-vice is depicted in Figure 4.8. This sequence surrounds the translation start site of the human methyltransferase and has been shown to be available for DNAzyme hybrid-ization [61]. It is also noteworthy that hDnmt1 is cleavable in cells with characterized ribozymes [65].

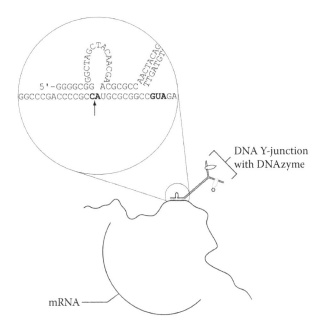

FIGURE 4.8
A nanodevice-targeted DNAzyme. A nanodevice targeting the messenger RNA of the human DNA methyl-transferase (hDnmt1) is depicted. The linked single-stranded DNAzyme is pictured bound to the messenger RNA target. The predicted RNA cleavage site is marked with an arrow.

4.4 Conclusion

Progress in bionanotechnology is rapidly generating new devices that can be selectively targeted to cell structures. Improvements in our understanding of dendrimer, viral capsid, and nucleoprotein assembly are providing new approaches to rational drug delivery. Of these, the nucleoprotein assemblies may be particularly well suited for peptide–ligand targeting, internalization, and nuclear localization. Nanodevices of the type described here are suited to the delivery of a variety of nucleic acid based payloads including second-generation DNA methyltransferase inhibitors and DNAzymes.

References

1. Clark, J., T. Shevchuk, P.M. Swiderski, R. Dabur, L.E. Crocitto, Y.I. Buryanov, and S.S. Smith. 2005. Construction of ordered protein arrays. *Methods Mol Biol* 300: 325–348.
2. Ballauff, M. and C.N. Likos. 2004. Dendrimers in solution: Insight from theory and simulation. *Angew Chem Int Ed Engl* 43: 2998–3020.
3. Striebel, H.M., E. Birch-Hirschfeld, R. Egerer, Z. Foldes-Papp, G.P. Tilz, and A. Stelzner. 2004. Enhancing sensitivity of human herpes virus diagnosis with DNA microarrays using dendrimers. *Exp Mol Pathol* 77: 89–97.
4. Shukla, S., G. Wu, M. Chatterjee, W. Yang, M. Sekido, L.A. Diop, R. Muller, J.J. Sudimack, R.J. Lee, R.F. Barth, et al. 2003. Synthesis and biological evaluation of folate receptor-targeted boronated PAMAM dendrimers as potential agents for neutron capture therapy. *Bioconjug Chem* 14: 158–167.
5. Choi, Y., T. Thomas, A. Kotlyar, M.T. Islam, and J.R. Baker, Jr. 2005. Synthesis and functional evaluation of DNA-assembled polyamidoamine dendrimer clusters for cancer cell-specific targeting. *Chem Biol* 12: 35–43.
6. Mecke, A., I. Lee, J.R. Baker Jr., M.M. Holl, and B.G. Orr. 2004. Deformability of poly(amidoamine) dendrimers. *Eur Phys J E Soft Matter* 14: 7–16.
7. Clark, J. and S.S. Smith. 2005. Application of nanoscale bioassemblies to clinical laboratory diagnostics. *Adv Clin Chem* 41:23–48.
8. Caspar, D.L.D. and A. Klug. 1962. Physical principles in the construction of regular viruses. *Cold Spring Harb Symp on Quant Biol* 27: 1–24.
9. Wang, Q., E. Kaltgrad, T. Lin, J.E. Johnson, and M.G. Finn. 2002. Natural supramolecular building blocks. Wild-type cowpea mosaic virus. *Chem Biol* 9: 805–811.
10. Wang, Q., T. Lin, J.E. Johnson, and M.G. Finn. 2002. Natural supramolecular building blocks. Cysteine-added mutants of cowpea mosaic virus. *Chem Biol* 9: 813–819.
11. Cheung, C.L., J.A. Camarero, B.W. Woods, T. Lin, J.E. Johnson, and J.J. De Yoreo. 2003. Fabrication of assembled virus nanostructures on templates of chemoselective linkers formed by scanning probe nanolithography. *J Am Chem Soc* 125: 6848–6849.
12. Deyev, S.M., R. Waibel, E.N. Lebedenko, A.P. Schubiger, and A. Pluckthun. 2003. Design of multivalent complexes using the barnase*barstar module. *Nat Biotechnol* 21: 1486–1492.
13. Kipriyanov, S.M., M. Little, H. Kropshofer, F. Breitling, S. Gotter, and S. Dubel. 1995. Affinity enhancement of a recombinant antibody: Formation of complexes with multiple valency by a single-chain Fv fragment-core streptavidin fusion. *Protein Eng* 9: 203–211.
14. Smith, S.S., L. Niu, D.J. Baker, J.A. Wendel, S.E. Kane, and D.S. Joy. 1997. Nucleoprotein-based nanoscale assembly. *Proc Natl Acad Sci U S A* 94: 2162–2167.

15. Smith, S.S. 2001. A self-assembling nanoscale camshaft: Implications for nanoscale materials and devices constructed from proteins and nucleic acids. *Nano Lett* 1: 51–55.
16. Gibson, T.J. and A.I. Lamond. 1990. Metabolic complexity in the RNA world and implications for the origin of protein synthesis. *J Mol Evol* 30: 7–15.
17. Niemeyer, C.M., T. Sano, C.L. Smith, and C.R. Cantor. 1994. Oligonucleotide-directed self-assembly of proteins: Semisynthetic DNA–streptavidin hybrid molecules as connectors for the generation of macroscopic arrays and the construction of supramolecular bioconjugates. *Nucleic Acids Res* 22: 5530–5539.
18. Niemeyer, C.M., M. Adler, B. Pignataro, S. Lenhert, S. Gao, L. Chi, H. Fuchs, and D. Blohm. 1999. Self–assembly of DNA–streptavidin nanostructures and their use as reagents in immuno-PCR. *Nucleic Acids Res* 27: 4553–4561.
19. Niemeyer, C.M., J. Koehler, and C. Wuerdemann. 2002. DNA-directed assembly of bienzymic complexes from *in vivo* biotinylated NAD(P)H:FMN oxidoreductase and luciferase. *Chembiochem* 3: 242–245.
20. Alivisatos, A.P., K.P. Johnsson, X. Peng, T.E. Wilson, C.J. Loweth, M.P. Bruchez Jr., and P.G. Schultz. 1995. Organization of 'nanocrystal molecules' using DNA. *Nature* 382: 609–611.
21. Zhang, S., T.J. Fu, and N.C. Seeman. 1993. Symmetric immobile DNA branched junctions. *Biochemistry* 32: 8062–8067.
22. Wu, B., F. Girard, B. van Buuren, J. Schleucher, M. Tessari, and S. Wijmenga. 2004. Global structure of a DNA three-way junction by solution NMR: Towards prediction of 3H fold. *Nucleic Acids Res* 32: 3228–3239.
23. Assenberg, R., A. Weston, D.L. Cardy, and K.R. Fox. 2002. Sequence-dependent folding of DNA three-way junctions. *Nucleic Acids Res* 30: 5142–5150.
24. Stuhmeier, F., J.B. Welch, A.I. Murchie, D.M. Lilley, and R.M. Clegg. 1997. Global structure of three-way DNA junctions with and without additional unpaired bases: A fluorescence resonance energy transfer analysis. *Biochemistry* 36: 13530–13538.
25. Yang, M. and D.P. Millar. 1996. Conformational flexibility of three-way DNA junctions containing unpaired nucleotides. *Biochemistry* 35: 7959–7967.
26. Shlyakhtenko, L.S., V.N. Potaman, R.R. Sinden, A.A. Gall, and Y.L. Lyubchenko. 2000. Structure and dynamics of three-way DNA junctions: Atomic force microscopy studies. *Nucleic Acids Res* 28: 3472–3477.
27. Clark, J., T. Shevchuk, P.M. Swiderski, R. Dabur, L.E. Crocitto, Y.I. Buryanov, and S.S. Smith. 2003. Mobility-shift analysis with microfluidics chips. *Biotechniques* 35: 548–554.
28. Klimasauskas, S., S. Kumar, R.J. Roberts, and X. Cheng. 1994. HhaI methyltransferase flips its target base out of the DNA helix. *Cell* 76: 357–369.
29. Berman, H.M., J. Westbrook, Z. Feng, G. Gilliland, T.N. Bhat, H. Weissig, I.N. Shindyalov, and P.E. Bourne. 2000. The protein data bank. *Nucleic Acids Res* 28: 235–242.
30. Schroeder, S.G., and C.T. Samudzi. 1997. Structural studies of EcoRII methylase: Exploring similarities among methylases. *Protein Eng* 10: 1385–1393.
31. Shevchuk, T., L. Kretzner, K. Munson, J. Axume, J. Clark, O.V. Dyachenko, M. Caudill, Y. Buryanov, and S.S. Smith. 2005. Transgene-induced CCWGG methylation does not alter CG methylation patterning in human kidney cells. *Nucleic Acids Res* 33: 6124–6135.
32. Huber, D., D. Boyd, Y. Xia, M.H. Olma, M. Gerstein, and J. Beckwith. 2005. Use of thioredoxin as a reporter to identify a subset of *Escherichia coli* signal sequences that promote signal recognition particle-dependent translocation. *J Bacteriol* 187: 2983–2991.
33. Saitoh, M., H. Nishitoh, M. Fujii, K. Takeda, K. Tobiume, Y. Sawada, M. Kawabata, K. Miyazono, and H. Ichijo. 1998. Mammalian thioredoxin is a direct inhibitor of apoptosis signal-regulating kinase (ASK) 1. *EMBO J* 17: 2596–2605.
34. Nickel, W. 2003. The mystery of nonclassical protein secretion. A current view on cargo proteins and potential export routes. *Eur J Biochem* 270: 2109–2119.
35. Pekkari, K., R. Gurunath, E.S. Arner, and A. Holmgren. 2000. Truncated thioredoxin is a mitogenic cytokine for resting human peripheral blood mononuclear cells and is present in human plasma. *J Biol Chem* 275: 37474–37480.76

36. Rubartelli, A., A. Bajetto, G. Allavena, E. Wollman, and R. Sitia. 1992. Secretion of thiore-doxin by normal and neoplastic cells through a leaderless secretory pathway. *J Biol Chem* 267: 24161–24164.

37. Yoneda, Y. 2000. Nucleocytoplasmic protein traffic and its significance to cell function. *Genes Cells* 5: 777–787.

38. Jiang, X. and A. Sorkin. 2003. Epidermal growth factor receptor internalization through clath-rin-coated pits requires Cbl RING finger and proline-rich domains but not receptor polyubiq-uitylation. *Traffic* 4: 529–543.

39. Jiang, X., F. Huang, A. Marusyk, and A. Sorkin. 2003. Grb2 regulates internalization of EGF receptors through clathrin-coated pits. *Mol Biol Cell* 14: 858–870.

40. Teker, K., R. Sirdeshmukh, K. Sivakumar, S. Lu, E. Wickstrom, H.-N. Wang, T. Vo-Dinh, and B. Panchapakesan. 2005. Applications of carbon nanotubes for cancer research. *NanoBiotechnology* 1: 171–182.

41. Hawthorne, M.F. and M.W. Lee. 2003. A critical assessment of boron target compounds for boron neutron capture therapy. *J Neurooncol* 62: 33–45.

42. Richards, E.J. and S.C. Elgin. 2002. Epigenetic codes for heterochromatin formation and silenc-ing: Rounding up the usual suspects. *Cell* 108: 489–500.

43. Goffin, J. and E. Eisenhauer. 2002. DNA methyltransferase inhibitors-state of the art. *Ann Oncol* 13: 1699–1715.

44. Szyf, M. and N. Detich. 2001. Regulation of the DNA methylation machinery and its role in cellular transformation. *Prog Nucleic Acid Res Mol Biol* 69: 47–79.

45. Chiurazzi, P., M.G. Pomponi, R. Pietrobono, C.E. Bakker, G. Neri, and B.A. Oostra. 1999. Synergistic effect of histone hyperacetylation and DNA demethylation in the reactivation of the FMR1 gene. *Hum Mol Genet* 8: 2317–2323.

46. Cameron, E.E., K.E. Bachman, S. Myohanen, J.G. Herman, and S.B. Baylin. 1999. Synergy of demethylation and histone deacetylase inhibition in the re-expression of genes silenced in cancer. *Nat Genet* 21: 103–107.

47. Sorm, F., A. Piskala, A. Cihak, and J. Vesely. 1964. 5-Azacytidine, a new, highly effective can-cerostatic. *Experientia* 20: 202–203.

48. Sorm, F. and J. Vesely. 1964. The activity of a new antimetabolite, 5-azacytidine, against lym-phoid leukaemia in Ak mice. *Neoplasma* 11: 123–130.

49. Santi, D.V., A. Norment, and C.E. Garrett. 1984. Covalent bond formation between a DNA-cytosine methyltransferase and DNA containing 5-azacytosine. *Proc Natl Acad Sci U S A* 81: 6993–6997.

50. Smith, S.S., B.E. Kaplan, L.C. Sowers, and E.M. Newman. 1992. Mechanism of human methyl-directed DNA methyltransferase and the fidelity of cytosine methylation. *Proc Natl Acad Sci U S A* 89: 4744–4748.

51. Clark, J., T. Shevchuk, M.R. Kho, and S.S. Smith. 2003. Methods for the design and analysis of oligodeoxynucleotide-based DNA (cytosine-5) methyltransferase inhibitors. *Anal Biochem* 321: 50–64.

52. Egger, G., G. Liang, A. Aparicio, and P.A. Jones. 2004. Epigenetics in human disease and pros-pects for epigenetic therapy. *Nature* 429: 457–463.

53. Marquez, V.E., J.J. Barchi, Jr., J.A. Kelley, K.V. Rao, R. Agbaria, T. Ben-Kasus, J.C. Cheng, C.B. Yoo, and P.A. Jones. 2005. Zebularine: A unique molecule for an epigenetically based strategy in cancer chemotherapy. The magic of its chemistry and biology. *Nucleosides Nucleotides Nucleic Acids* 24: 305–318.

54. Smith, S.S., J.L. Kan, D.J. Baker, B.E. Kaplan, and P. Dembek. 1991. Recognition of unusual DNA structures by human DNA (cytosine-5)methyltransferase. *J Mol Biol* 217: 39–51.

55. Smith, S.S., T.A. Hardy, and D.J. Baker. 1987. Human DNA (cytosine-5)methyltransferase selec-tively methylates duplex DNA containing mispairs. *Nucleic Acids Res* 15: 6899–6915.

56. Laayoun, A. and S.S. Smith. 1995. Methylation of slipped duplexes, snapbacks and cruciforms by human DNA (cytosine-5) methyltransferase. *Nucleic Acids Res* 23: 1584–1589.

57. Breaker, R.R. and G.F. Joyce. 1994. A DNA enzyme that cleaves RNA. *Chem Biol* 1: 223–229.

58. Cuenoud, B. and J.W. Szostak. 1995. A DNA metalloenzyme with DNA ligase activity. *Nature* 375: 611–614.
59. Li, Y. and D. Sen. 1995. A catalytic DNA for porphyrin metallation. *Nat Struct Biol* 3: 743–747.
60. Fedor, M.J. and J.R. Williamson. 2005. The catalytic diversity of RNAs. *Nat Rev Mol Cell Biol* 6: 399–412.
61. Santoro, S.W. and G.F. Joyce. 1997. A general purpose RNA-cleaving DNA enzyme. *Proc Natl Acad Sci U S A* 94: 4262–4265.
62. Stojanovic, M.N. and D. Stefanovic. 2003. A deoxyribozyme-based molecular automaton. *Nat Biotechnol* 21: 1069–1074.
63. Schubert, S., D.C. Gul, H.P. Grunert, H. Zeichhardt, V.A. Erdmann, and J. Kurreck. 2003. RNA cleaving "10–23" DNAzymes with enhanced stability and activity. *Nucleic Acids Res* 31: 5982–5992.
64. Mitchell, A., C.R. Dass, L.Q. Sun, and L.M. Khachigian. 2004. Inhibition of human breast carcinoma proliferation, migration, chemoinvasion and solid tumour growth by DNAzymes targeting the zinc finger transcription factor EGR-1. *Nucleic Acids Res* 32: 3065–3069.
65. Scherr, M., M. Reed, C.F. Huang, A.D. Riggs, and J.J. Rossi. 2000. Oligonucleotide scanning of native mRNAs in extracts predicts intracellular ribozyme efficiency: Ribozyme-mediated reduction of the murine DNA methyltransferase. *Mol Ther* 2: 26–38.

5

Bimetallic Nanoparticles: Synthesis and Characterization

Tarasankar Pal, Anjali Pal, and Sudipa Panigrahi

CONTENTS

5.1 Introduction

The intense research in the field of nanoparticles by chemists, physicists, and materials scientists has gained tremendous momentum by the search for new materials of dimension <100 nm to further miniaturize electronic devices,[1–6] and improve the idea in catalysis,[7] spectroscopy,[8–10] etc. While nanomaterials become fascinating, the fundamental question of how molecular electronic properties evolve with increasing size in this intermediate region between molecular and solid-state physics becomes intriguing. The collective electronic, optical, and magnetic properties of organized assemblies of size-selective monodispersed nanocrystals are increasingly being the subjects of investigation.[11] Control over the spatial arrangement of these building blocks often leads to new materials with chemical, mechanical, optical, or electronic properties distinctly different from their bulk component.[12] A variety of metal,[13] metal oxide,[14] semiconductor[15] nanoparticles/nanorod, carbon dots,[16] and carbon nanotubes[17] have been synthesized and proposed as potential entities for optical and electronic devices.[17,18] Recently, metallic clusters with fractal structure have sparked much interest because of the localization of dynamical excitations in these fractal objects, which plays important roles in many physical processes.[19] In particular, the localization of resonant dipolar eigenmodes can lead to a dramatic enhancement of many optical effects in fractals.[19] Their origin is attributed to the collective oscillation of the free conduction electrons induced by an interacting electromagnetic field. These resonances are also denoted as surface plasmons (SPs). Mie[20] was the first to explain this phenomenon by applying classical electrodynamics to spherical particles and solved Maxwell's equations for the appropriate boundary conditions. The explanation looks apparently much simpler for spherical monometallics but becomes more complicated while one moves to a system containing two metals with definite interaction in the closest proximity and geometry deviates from spheroid.

It has been established that the particle size of noble metals is reduced and approaches the Fermi wavelength of electrons (i.e., the electron de Broglie wavelength at the Fermi level: 0.5 nm for Au and Ag); the continuous density of states breaks up into discrete energy levels leading to the observation of dramatically different optical, electrical, and chemical properties compared with nanoparticles.[21] Their molecule-like properties result in discrete electronic transitions leading to strong fluorescence. Free electron model and Jellium model are usually employed to explain the stability of the metal clusters, that is, subnanometer particle containing magic number of atoms.[22] Specific scaffolds-like polymer, thiols, DNA, etc. are used for synthesis purpose.[8,21,22] Although silver clusters synthesized from same scaffold become brighter than those of gold, spontaneous oxidation and subsequent toxicity of silver ions out of silver clusters limited its application. Thus, research on suitable scaffold to tune emission and minimize cytotoxicity has become very promising.[23] Recent reports regarding multiple practical applications like toxic metal, anions, detection of trace amount of solvent molecules in a miscible solvent, and bioimaging make the field very promising.[24–27]

5.2 Bimetallic Nanoparticles

The concept of the coordination complex enabled Alfred Werner to make a great sense of discovery but there was no scheme in the concept for direct bonding between metal atoms. Later on, in complex compounds the existence of metal-to-metal bonding has been proved beyond doubt and so is the case in metallic clusters in the nanoregime. Single-crystal x-ray diffraction (XRD) analysis provides concrete evidence for metal–metal bonding. Bimetallic nanoparticles, composed of two different elements, have been reported to show outstanding characters different from the corresponding monometallic ones.[28–30] In the nanoregime when two metals are combined within as a single nanoparticle (bimetallic nanoparticles), the optical, electronic, and magnetic properties of the bimetallic particles are directed by a combination of the properties (dielectric constants) of both metals. Such a combination strongly depends on the microscopic arrangement of the metals within the particle, that is, whether an alloy, a perfect core–shell structure, or something in-between is obtained, but in any of these cases there is direct interaction between the metals. These features may be enhanced, modified, or suppressed in the case of bimetallic and multimetallic nanoparticles because of intermetallic interactions arising from their constitutional and morphological combinations. Totally new functions may be created by overcoming disadvantages of single-component nanoparticles. Unique features expected for multimetallic nanoparticles may include (1) physical and chemical interactions among different atoms and phases that lead to novel functions, (2) altered miscibility and interactions unique to nanometer dimension (macroscopic phase property may not apply), and (3) morphological variations that are related to new properties.

Bimetallic clusters and colloids are of special interest as their chemical and physical properties may be tuned by varying the atomic ordering, composition, and size,[31] and eventually fetches superior catalytic properties[32–35] than those of single-metallic components. They may serve as models to study the formation of different alloys in the nanoregime at a much lower temperature. It is now possible to save precious metals, by optimizing the synthetic conditions so that only very thin surface layers occur. Bimetallic nanoparticles have played an important role in improving the catalyst quality,[33–36] changing the SP band,[37–40] and regulating the magnetic properties.[41,42] Because of all the above special properties that are brought about by the changes on surface and structure caused by alloying, control of composition distribution of bimetallic nanoparticles is crucial to the improvement of particle properties. It has been found that bimetallic structures with fivefold symmetry have a number of structural variants and some unexpected pentagonal shapes that cannot be obtained by simply rotating the main symmetry axis at all, but by displacements of the fivefold axis from the center of the particle, which generates asymmetric structures.[43] Perfectly monodispersed nanoparticles are of course ideal but special properties are to be expected even if the ideality is not perfectly realized. Mass production of these nanoparticles with uniformity is most important to realize, as most chemical or physical properties of nanosize materials have not been elucidated yet. Much attention is now being paid to this area.

One interesting platform has been reported for alloying of metals. Keeping the lowering of melting points of metals with size and also the lattice spacing of the participating metal alloy materials have been produced in high boiling silicon oil.[44] The study can also be extended to metalloids even.[45] Recent development is the synthesis in confined environments, that is, using soft or hard templates.[46,47]

In general, size-selective synthesis of particles is carried out mainly to show enhanced catalytic activity of nanoparticles, which depends on the surface-to-volume ratio. Now the time has come to elaborate morphology-dependent nanoparticle properties. It may be monometallic or multimetallic particles. There, surface structure plays a role and even the porosity has been considered greatly, especially to catalysis. In recent times, facet-selective particle synthesis is coming in a big way. It has been experimentally verified that because of capping agents' preferential affinity for facet selection, high index faceted particles can be synthesized by etching, dissolution, redox transformation, etc. Then more number of surface atoms in the edges, kink, corners, etc. endorse shapes which eventually enhance catalytic property of metallic particles leaving aside low index facets.

The recent development in material science has now made it possible to tailor-made metal to obtain controllable size and shape in nanoregime.[48,49] This gives a new opening to recognize the nature of the active sites, the origin of the structure-dependent reactivity, and the mechanism of interaction. To do this, careful design and fabrication of size- and shape-controlled catalyst nanoparticle synthesis becomes useful. In general, size-selective synthesis of particles is carried out mainly to show enhanced catalytic activity of nanoparticles, which depends on the surface-to-volume ratio.[50,51] Now the time has come to elaborate morphology-dependent nanoparticle properties.[52,53] There surface structure plays a role mainly when porosity is taken into consideration. Furthermore, the shapes of the nanocatalyst are determined by their crystal structures, including terminating facets, crystallinity, and anisotropy. In recent times, facet selective particle synthesis is coming in big way. Generally, the shape of a nanoparticle controls the facet(s) and thus the surface structure as well as the fraction of atoms at corners and edges happens to control the reaction.[54] During the syntheses of faceted nanomaterials, if the growth process comes under thermodynamic control, the as-obtained product will be bounded by low-index facets, that is, lower surface energy.[55] In contrast, when the growth process is under kinetic control, then the product will be bound by high-index facets, that is, high surface energy, and the product yield is drastically different from the thermodynamically favored structure.[56] It has been experimentally verified that because of capping agents' preferential affinity for facet selection, high index faceted particles can be synthesized by etching, dissolution, redox transformation etc. Then more number of surface atoms in the edges, steps, and kinks endorses shapes which eventually enhance catalytic property of metallic particles.[57]

Again, surface atomic arrangement and bond length have been found to play a dominant role for adsorption leading to effective catalysis.

Interestingly, formation of Pd–Rh bimetallic core–frame concave nanocubes (Cb) as the growth of Rh on the corners and edges of the Pd Cb is continued; formation of Rh cubic nanoframes by selectively etching takes away the Pd cores.[58]

5.3 Preparation of Bimetallic Nanoparticles

Metallic nanoparticles of definite size are easily synthesized via a "bottom-up" approach that too in gram quantities and can be surface modified with special functional groups. Nanocomposites, that is, alloy and core–shell particles, are an attractive subject mainly because of their composition-dependent optical, magnetic, and catalytic properties.

In general, bimetallic nanoparticles can be prepared by simultaneous reduction or by successive reduction of two metal ions through suitable stabilization strategy (capping/template),[59] combating steric hindrance and static-electronic repulsive force. The former reduction methods may obtain a particle structure between core–shell and homogeneous alloy depending on the reduction condition,[60–62] while the latter methods are used for the purpose of production of core–shell particles.[63] Alloy particles may be conveniently synthesized by simultaneous reduction of two or more metal ions with proper lattice matching.[39,64] Growth of core–shell structures may be accomplished by the successive reduction of one metal ion over the preformed metallic core of another metal,[65] generally by weak reducing agent. Otherwise, latter process often leads to the formation of fresh nuclei of the second metal in solution, in addition to a shell around the first metal core,[66] and is clearly undesirable from the application point of view. Another possible strategy to overcome this drawback could be based on immobilization of a reducing agent on the surface of the core metal which, when exposed to the second metal ions, would reduce them, thereby leading to the formation of a thin metallic shell. However, control of the reduction, nucleation, and aggregation rates of the two components may be effective to control the size, structure, and composition distribution of bimetallic nanoparticles. A facile synthetic strategy comes from the judicious selection of commercial ion-exchange resins. Considering cationic or anionic metal complexes or simple compounds of metal precursor compounds, one can reduce one metal over the other. Thus, one can tactfully separate the mono-, bi-, and even multimetallic nanoparticles leaving aside resin matrix outside. Instead, metal resin composite can be produced with ease for some useful applications.[67]

Simultaneous reduction of two kinds of precious metal ions usually give core–shell structured bimetallic nanoparticles[33] in which atoms of the first element form a core and the atoms of the second cover the core to form a shell. Successive reduction of two metal salts can be considered as one of the most suitable methods to prepare core–shell structured bimetallic particles. The deposition of one metal onto preformed monometallic nanoparticles of another metal seems to be very effective. For this purpose, however, the second element must be deposited on the surface of preformed particles, and the preformed monometallic nanoparticles must be chemically surrounded by the deposited element. The core–shell structure has been considered to be controlled by the order of redox potentials of both the ions and the coordination ability of both atoms in relation to the reducing agent. However, some difficulties are to be overcome for the preparation of bimetallic nanoparticles having a controllable core–shell structure. For instance, oxidation of the preformed core to metal ions often takes place when the other types of metal ions, added for making the shell, have a higher redox potential. This redox potential may result in the production of large islands of the shell metal on the preformed metal core and the rereduction of core metal ions produced. Alternate adsorptions of one of the excess ions and further reductions of the complex by radiolytic radicals progressively build the alloyed cluster. In most of these bimetallic cluster systems, segregation occurs during the reduction so that the nobler metal constitutes the core and the less noble metal the shell of a bilayered cluster. This structure stems from an intermetallic electron transfer occurring with the mixture according to the respective redox potentials of the two metals. Initially, the reduction may be equiprobable but then the less noble atoms behave as electron relays toward the other metal ions up to the complete reduction of the latter, this favoring the formation of the clusters of the more noble metal first. A few exceptions observed in intimately alloyed metal clusters, prepared either by chemical reduction or by low dose rate irradiation, result from an extremely slow electron transfer, which allows the simultaneous reduction of both ions to occur under nonequilibrium kinetics.

Various methods have been reported so far for the preparation of bimetallics, for example, alcohol reduction,[33–40] citrate reduction,[39,68] polyol process,[69] solvent extraction reduction,[40,70] sonochemical method,[71] photolytic reduction,[72] decomposition of organometallic precursors,[73] and electrolysis of a bulk metal.[74] Belloni and Henglein also proposed γ radiolysis to produce bimetallic nanoparticles from two different noble metals.[1,75]

Gold–silver system is the most interesting bimetallics since both metals are miscible in the bulk phase, owing to very similar lattice constants (0.408 for Au and 0.409 for Ag).[76] Especially, the size effect on the plasmon absorption in connection with the Mie theory and its modifications has been of major interest.[77] Those nanoparticles also showed two SP absorption bands originating from the individual gold and silver domains.[76] One can obtain gold over silver (inverted core–shell) nanoparticle or the usual normal core–shell nanoparticles. This has been reported for the first time considering kinetic control of the reaction, which supersedes thermodynamic control.[31] A new reagent, beta-cyclodextrin has been employed for the synthesis of both types of gold/silver particles. This can also be done just by employing commercial ion-exchange resins as discussed earlier. The optical properties (plasmon absorption in the visible range) are examined and compared with the calculated absorption spectra using the Mie theory. Recently, a strategy for extracting optical constants of the core and/or the shell material of bimetallic Ag–Au nanoparticles from their measured SP extinction spectra has been reported.[78] Freeman et al.[79] and Morriss and Collins[80] prepared nanoparticles consisting of gold core and silver shell. Mulvaney et al.[81] deposited gold onto radiolytically prepared silver seeds by irradiation of $KAu(CN)_2$ solution. Treguer et al.[60] prepared layered nanoparticles by radiolysis of mixed Au^{III}/Ag^I solution. Silver colloid with gold reduced in the surface layer was prepared by Chen and Nickel[82] by mixing a solution of $HAuCl_4$ with a silver colloid and addition of a reductant (*p*-phenylenediamine) in the second step. A two-step wet radiolytic synthesis resulting in a size-dependent spontaneous alloying within Au_{core}–Ag_{shell} nanoparticles and a photochemical approach to Au_{core}–Ag_{shell} nanoparticle preparation were reported.[72,83] Since Ag and Au are miscible in all proportions, but differ in both redox potentials and surface energies, the results of a particular preparative strategy with respect to formation of either the alloyed or layered nanoparticle composition are not always readily predictable, and characterization of the composition of the resulting nanoparticles is thus of key importance. Liz-Marzan et al.[84] used inorganic fibers in aqueous solution for the stabilization of gold–silver particles with diameters of 2–3 nm after the simultaneous reduction of gold and silver salts by sodium borohydride. Mulvaney et al.[81] and Sinzig et al.[85] prepared silver nanoparticles coated with an overlayer of gold (core–shell nanoparticles). These particles have two distinct plasmon absorption bands and their relative intensities depend on the thickness of the shell. However, alloy formation within the shell was suggested on the basis of the optical absorption spectra. Similarly, gold–silver composite colloids (30–150 nm in diameter) consisting of mixtures of gold and silver domains were obtained by irradiating aqueous solutions of gold and silver ions with 253.7 nm UV light.[86] Alloy nanoparticles, on the other hand, have mainly been studied because of their catalytic effects.[87]

However, for numerous other couples of metal ions, the radiation-induced reduction of mixed ions does not eventually produce solid solutions but a segregation of the metals in core–shell structure, such as for the case of Ag_{core}–Cu_{shell}[88] or of Au_{core}–Pt_{shell}.[61] The absorption spectrum changes in all these systems from the SP spectrum of the first metal to that of the second one, suggesting that the latter is coated on the clusters formed first. Since both ions have almost the same initial probabilities of encountering the radiolytic reducing radicals and of being reduced, the results were interpreted as a consequence of an electron transfer from the less noble metal, as soon as it is reduced to one of its lower valency states, to the ions of the more noble metal, which is obviously favored by this displacement and reduced first.[1,2,61,88] The role of nuclei played by more noble metals such as Pt, Pd, or Cu is efficient to reduce the metal ions of Ni, Co, Fe, Pb, and Hg, which otherwise do not easily yield stable monometallic clusters through irradiation.[61] The bimetallic character of Fe–Cu clusters is attested by their ferromagnetic properties and the change of the optical spectrum relative to pure copper clusters. Similarly, a chemical reduction of a mixture of two ions among Au [III], Pt [IV], and Pd [II] (with decreasing order of redox potentials) yields bilayered clusters of Au–Pd, Au–Pt, or Pt–Pd.[89] Composite clusters of Ag–Pd have been also characterized.[92,93] Bilayered Au–Pt, Ag–Pt, and Ag–Au supported on megalith fibers have been studied by optical absorptions.[84] Photochemical reduction of precursor salts in the presence of suitable surfactant has now become worthy addition to this field of research.

As gold has low catalytic activity compared with platinum or palladium, the structural and catalytic changes have been examined for the admixture of platinum or palladium to gold.[33,92–94] Turkevich et al.[95] have synthesized the Au–Pd bimetallic particles and described their morphologies. Toshima et al.[33] have described the catalytic activity and analyzed the structure of the poly(N-vinyl-2-pyrrolidone)-protected Au–Pd bimetallic clusters prepared by the simultaneous reduction of $HAuCl_4$ and $PdCl_2$ in the presence of poly(N-vinyl-2-pyrrolidone). Other groups have reported the formation of the Au–Pd bimetallic particles with a palladium-rich shell by the simultaneous alcoholic reduction method.[94] In contrast, successive alcoholic reduction did not give the core–shell products but, instead, "cluster-in-cluster" based on the coordination number of the constituents.[36] Mizukoshi et al.[96] reported the preparation and structure of gold–palladium bimetallic nanoparticles by sonochemical reduction of the gold(III) and palladium(II) ions. Au–Pt bimetallic nanoparticles were prepared by citrate reduction by Miner et al. from the corresponding two metal salts.[68] Following the similar method, citrate-stabilized Pd–Pt bimetallic nanoparticles can also be prepared.[68] Colloidal dispersions of Pd–Pt bimetallic nanoparticles can be prepared by refluxing the alcohol water mixture in the presence of PVP.[97] Toshima et al. have succeeded in the preparation of polymer-protected various nanoscopic bimetallic colloids of noble metals by coreduction of the corresponding metal ions in a refluxing mixture of water and alcohol, as well as the preparation of Cu–Pd, Cu–Pt bimetallic colloids with well-defined alloy structures by a cold alloying process.[29,33,92,97,98] The resulting bimetallic crystal particles depend on the ratio of ionic precursors; the structures Cu–Pd and Cu_3–Pd,[89,99] Ni–Pt,[99,100] Cu–Au, and Cu_3–Au[99,100] have been found. Alloyed Ag–Pt clusters in ethylene glycol have also been prepared by chemical reduction of silver bis(oxalato)platinate and characterized by a homogeneous (111) lattice spacing.[101] The properties of both the metals may be somewhat different in relation to their redox reactions that control their formation through reduction. The alloying takes place upon irradiation[102] through fast association between atoms or clusters and excess ions, which in the case of mixed solutions yield bimetallic complexes.

Silver particles having a gold layer were prepared and the UV–Vis absorption spectra of these bimetallic nanoparticles were intensively investigated. Several metal ions were deposited onto silver sols to produce bimetallic nanoparticles.[1,103] Hg ions can be reduced in the presence of silver sol, which results in the formation of Ag_{core}–Hg_{shell} bimetallic nanoparticles.[104] Ligand-stabilized Au–Pd[63] and Au–Pt[105] bimetallic nanoparticles were prepared by Schmid et al. by successive reduction. In an earlier study, Turkevich and Kim proposed gold-layered palladium nanoparticles.[95] Three types of Au–Pd bimetallic nanoparticles such as Au_{core}–Pd_{shell}, Pd_{core}–Au_{shell}, and random alloyed particles were prepared by the application of successive reduction technique.[106]

Reduction of the corresponding double salts is one of the important techniques for the synthesis of bimetallic nanoparticle. Torigoe and Esumi proposed silver(I) bis(oxalato) palladate(II) as a precursor of Ag–Pd bimetallic nanoparticles stabilized by PVP.[95] Preparation of PVP-stabilized Ag–Pt bimetallic nanoparticles by borohydride reduction from silver(I) bis(oxalato)platinate(I) was also reported.[101]

Instead of chemical reduction, an electrochemical process can be used to create metal atoms from bulk metal. Reetz and Helbig proposed an electrochemical method including both oxidation of bulk metal and reduction of the metal ions for the size-selective metal nanoparticles.[107] The particle size can be controlled by the current density. The Pd–Pt, Ni–Pd, Fe–Co, and Fe–Ni bimetallic nanoparticles have also been obtained by this method.[108]

5.4 Characterization

For more than a decade, efforts have been made for the preparation and characterization of nanosized materials. To understand the special properties of nanoparticle systems, it has become increasingly important to develop techniques for characterizing such materials at the nanometer level. Their prominence stems from the recognition that nanophase systems often possess dramatically different properties because they may show different characteristics compared with conventional bulk materials or to atoms, the smallest units of matter. Characterization is the most important for the bimetallic nanoparticles. Characterization of colloids (hydrosols) of bimetallic nanoparticles constituting various combinations of noble metals was the subject of numerous articles, for example, Au–Pd,[95,109] Au–Pt,[105,109] Ag–Pd,[92,110] Ag–Pt,[84,101] and Ag–Au.[60,80,81,111]

The first question asked about metal nanoparticles is concerned with aggregation state, size, and morphology. Among the technique commonly used, transmission electron microscopy (TEM) is indispensable for metal nanoparticle studies. Structural information is obtained from TEM and high-resolution TEM (HRTEM). TEM is a key tool in the quest to understand nanophase systems even though there remain many challenges in applying modern microscopy techniques to small particles.[112] The TEM emphasizes the intensity contrast. Further, HRTEM can now provide information not only on the particle size and shape but also on the crystallography of the monometallic and bimetallic nanoparticles. High-resolution phase contrast microscopy is well suited to determine the lattice spacing of thin crystals. This method enables one to distinguish core–shell structure on the basis of observing different lattice spacing. For large crystalline metal nanoparticles, HRTEM suggests the area composition by the fringe measurement, using crystal information of nanoparticles observed in the particle images.[105,113,114] It is easier to measure the lattice spacing on particles of well-defined shape in exact zone-axis orientations. Furthermore, in the case of supported metal nanoparticles, particle growth can be directly seen by *in situ* TEM observation. It is also necessary to measure the lattice changes and surface relaxations in nanoparticles, which may cause significant changes in the properties of nanophase materials. In bimetallic systems, small changes in lattice spacing may be related to the formation of alloy phases. Accurate determination of the lattice spacing in tiny crystallites is one area of importance for nanoparticle systems. When energy dispersive x-ray microanalysis (EDX) is used in conjunction with TEM, localized elemental information can be obtained.[105,113] Urban et al. have conducted a number of studies on small well-defined clusters in zone-axis orientations.[116]

Gold, silver, and copper (group IB metal) nanoparticles all have characteristic colors (rich plasmon absorption) related with their particle size. Thus, for these metals, observation of UV–Vis spectra can be a useful complement to other methods in characterizing metal particles. Optical properties of bimetallic nanoparticles comprising Ag and Au are thus the subject of considerable interest. Comparison of the calculated and measured SP extinction spectra was frequently employed as one of the criteria of distinguishing between an alloyed and/or layered (core–shell) structure of the bimetallic Ag–Au nanoparticles. Comparison of spectra of bimetallic nanoparticles with the spectra of physical mixture of the respective monometallic particle dispersions can confirm a bimetallic structure for the nanoparticles.[68,117] The UV–Vis spectral changes during the reduction can provide quite important information.[90] Moreover, the UV–Vis absorption spectra of the Au–Ag bimetallic particles show substantial differences between the alloy and a core–shell structure, which has been studied explicitly by experimental and theoretical measurements.[118]

Extended x-ray absorption fine structure (EXAFS) has been especially useful in providing structural informations about the nanocrystalline, as well as the crystalline materials. The analysis of EXAFS allows determination of local structural parameters, such as interatomic distance and coordination number, which are difficult to measure by any other methods. In contrast, colloidal dispersions of nanoparticles can be made at a low concentration of metal and the particles in the dispersions can be small and uniform. EXAFS samples can be obtained by concentrating the dispersions without aggregation, to provide high-quality data. Sinfelt et al. carried out major studies on the EXAFS of supported bimetallic nanoparticles.[119,120] Bradley et al. prepared Cu–Pd bimetallic nanoparticles by deposition of zerovalent Cu atoms onto preformed Pd nanoparticles.[121]

Infrared spectroscopy (IR) has been widely applied to the investigation of the surface chemistry of adsorbed small molecules. By comparison of IR spectra of CO on a series of bimetallic nanoparticles at various metal compositions, one can elucidate the surface microstructure of bimetallic nanoparticles.[35,121]

X-ray methods are informative for nondestructive elemental and structural analyses. XRD gives structural information of nanoparticles, including qualitative elemental information. For monometallic nanoparticles, the phase changes with increasing diameter of nanoparticles can be investigated with XRD. The presence of bimetallic particles opposed to a mixture of monometallic particles can also be demonstrated by XRD since the diffraction pattern of the physical mixtures consists of overlapping lines of the two individual monometallic nanoparticles and is clearly different from that of the bimetallic nanoparticles. The structural model of bimetallic nanoparticles can be proposed by comparing the observed XRD spectra and the computer-simulated ones.[122]

For a rationalization of their catalytic properties, the surface composition and structure are indispensable information and quantitative x-ray photoelectron spectroscopy (XPS) analysis is a powerful tool in the elucidation of the surface composition. From the quantitative analysis of XPS data of bimetallic nanoparticles, one can note which elements are present in the surface region. For the core–shell structure, the XPS data give the peak of the binding energy corresponding to the metal in the shell. On the other hand, for alloy nanoparticles a separate peak is assigned that is different from both the constituting metal.

One of the most revealing analytical methods for the composition of bimetallic nanoparticles is energy-dispersive x-ray spectroscopy (EDX), which is usually coupled with a TEM with high resolution.[92,105,123] EDX is a kind of electron probe microanalysis (EPMA) or x-ray microanalysis (XMA) method, which has higher sensitivity than the usual EPMA or XMA techniques. This method provides analytical data that cannot be obtained by the other three methods mentioned above.

NMR spectroscopy of metal isotopes[124] is a powerful technique for understanding the electronic environment of metal atoms in metallic particles by virtue of the NMR shifts caused by free electrons (Knight shifts).[125] The NMR spectra of metal nanoparticles, having Pauli paramagnetic properties, are governed both by the density of energy levels at the Fermi energy and by the corresponding wave function intensities of each site: the local density of states (LDOS). Furthermore, the electronic properties of metal nanoparticles may be informative for investigating catalytic properties of the metal nanoparticles.

All the tools at hand comfortably provide enough information about the nanostructural material ordering but quantitative evaluation of the thickness of the add-layer over the core structure still remains a debate.

5.5 Application of Bimetallics

The alteration in collective electronic, optical, and magnetic properties of organized nanoensembles for bimetallics over their monometallic counterpart is increasingly the subject of investigation,[126] leaving catalytic applications aside.[3,126–128] It has the potential to be used as a nonlinear optical material. Possible future applications include the areas of ultrafast data communication and optical data storage.[3,4,129] Such materials would be useful in developing nanodevices (e.g., integrated circuits, quantum dots, etc.)[130–132] and sensors,[133,134] even for DNA sequencing.[135] Studies on metal alloy nanocatalysts have long been of interest for investigating the relationship between catalytic activity and the electronic structure of metals.[94,98,105,136] Sometimes it has been found that the catalytic activity of bimetallic nanoparticles has been found to be superior to the activities of monometallic nanoparticles. Moreover, it has been found that catalytic effect of bimetallic nanoparticles is sensitive to the composition of the particle components. This could easily be attributed to the electronic effect and the segregation behavior of the materials.

Gold, which was not thought as an active catalyst,[137] can be used to improve the catalytic activity of other precious metal nanoparticles. Now, the relationship between catalytic activity and electronic structure of metals can be understood to some extent. For example, Au–Pt,[30] Au–Pd,[28,36] and Au–Rh[138] core–shell bimetallic nanoparticles have shown higher catalytic activity for hydrogenation of water than the corresponding monometallic nanoparticles. Nanoparticles composed of free-electron-like metals such as Ag and Au are known to provide strong resonance optical responses to irradiation by light, which results in amplification of light-induced processes undergone by molecules localized on their surfaces, such as Raman scattering, giving rise to surface-enhanced Raman scattering (SERS).[139] Thus, bimetallic nanostructures have a future to become ubiquitous and might be critical for the future development of electro-optical communications.

Attempts have been directed to transfer metallic nanoparticles from aqueous solution to nonpolar organic solvent as organosol to study localized surface plasma resonance (LSPR).[140] The shift of the LSPR is a quantitative measure of the electronic parameters for the incoming compounds in nonpolar organic solvent. The work is important to sense biomolecules using even bimetallics dispersed in organic solvent and this might be an alternative to nanosphere lithography in future.

5.6 Bimetallics

Catalytic process involving monometallic particles was developed by Bergilious. At present, bimetallic or even multimetallic particles have shown much promise. In this respect, noble metal bimetallic particles are very important. Recently, Cu–Ag, Ag–Au, Ni–Pd, Pt–Pd, Co–Pd, Co–Pt, etc. have been used for catalytic purposes.

For the fuel cell application, that is, from the electro-oxidation point of view, Co/Pt_3 has recently been exploited for methanol oxidation.[141] It has been described that morphologically different particles use different proportion of stabilizing agents for electro-catalytic methanol oxidation. The catalyst bearing (111) facets for nanoflower morphology has the highest surface energy than any other facets, even though there is size variation of the nanocrystals of Co/Pt_3.

Coreduction of Pt and Au precursor salts has been done to obtain high-energy {100}-faceted Pt/Au alloy Cb for 4-NP reduction using a capping agent that tunes the crystal facet.[142]

Bimetallic alloy nanoparticles of Pt–Pd-bearing nanotetrahedron (Td) and Cb have been meticulously synthesized and exploited in catalysis.[143] Using different capping agents, (Td) and Cb different morphologies are obtained. It has been shown that {100}-facet-bound cube has higher electrocatalytic activity than {111}-faceted Td toward methanol oxidation. On the other hand, {111}-facet enclosed Pt–Pd Td exhibits better durability.

Xia and coworkers reported the preparation of distinctive shaped Pd–Ag bimetallic nanocrystals through neat nucleation and growth-controlled processes.[144]

Considering the importance of bimetallic catalyst particles, Xia et al. have also considered Rh deposition on selected facets of Pd nanocrystals by seed-mediated growth.[145] Again capping agent, kinetic control, and selected exploitation of etching agent Pd–Rh cubic structures have been obtained.

Hollow structure or concave faces have received noteworthy attention due to the presence of high-index facets as well as their unique optical and catalytic properties.[146–154] Xia et al. have reported the synthesis of Pd–Rh core–frame concave nanocubes through site-specific overgrowth.[155] Rh is highly resistant to oxidative corrosion than from Pd and has also been realized to obtain Rh nanoframes with a unique open structure through selective etching of the Pd cores.

Researchers focused on synthesis of shape- and size-dependent synthesis of nanocrystals from less costly metals without losing catalytic performance.[156]

References

1. Henglein, A., Physicochemical properties of small metal particles in solution: "Microelectrode" reactions, chemisorption, composite metal particles, and the atom-to-metal transition, *J. Phys. Chem.*, 97, 5457, 1993.
2. Henglein, A., Small-particle research: Physicochemical properties of extremely small colloidal metal and semiconductor particles, *Chem. Rev.*, 89, 1861, 1989.
3. Schmid, G., *Clusters and Colloids: From Theory to Application*, VCH Verlagsgesellschaft, Weinheim, 1994.
4. Kamat, P. V. and Miesel, D., *Studies in Surface Science and Catalysis Semiconductor Nanoclusters—Physical, Chemical, and Catalytic Aspects*, Vol. 103, Elsevier, Amsterdam, 1997.
5. Alivisatos, A. P., Perspectives on the physical chemistry of semiconductor nanocrystals, *J. Phys. Chem.*, 100, 13226, 1996.
6. Graetzel, M., In: *Electrochemistry in Colloids and Dispersions*, Mackay, R. A., Texter, J., *J. Chem. Educ.*, 1993, 70(5), A146.
7. Aditya, T., Pal, A., and Pal, T., Nitroarene reduction: A trusted model reaction to test nanoparticle catalysts, *Chem. Commun.*, 2015, DOI: 10.1039/C5CC01131K
8. Ganguly, M., Pal, A., Negishi, Y., and Pal, T. Photoproduced fluorescent Au(I)@(Ag2/Ag3)-thiolate giant cluster: An intriguing sensing platform for DMSO and Pb(II). *Langmuir*, 29, 2033, 2013.
9. Sarkar, S., Sinha, A. K., Pradhan, M., Basu, M., Negishi, Y., and Pal, T., Redox transmetalation of prickly nickel nanowires for morphology controlled hierarchical synthesis of nickel/gold nanostructures for enhanced catalytic activity and SERS responsive functional material, *J. Phys. Chem. C.*, 115, 1659, 2011.
10. Pradhan, M., Maji, S., Sinha, A. K., Dutta, S., and Pal, T., Sensing trace arsenate by surface enhanced Raman scattering using a FeOOH doped dendritic Ag nanostructure, *J. Mater. Chem. A.*, 3, 10254, 2015.
11. Alivisatos, A. P., Semiconductor clusters, nanocrystals, and quantum dots, *Science*, 271, 933, 1996.
12. Marinakos, S. M., Schultz, D. A., and Feldheim, D. L., Gold nanoparticles as templates for the synthesis of hollow nanometer-sized conductive polymer capsules, *Adv. Mater.*, 11, 34, 1999.
13. Toshima, N. et al. In: Sugimoto T. (Ed.) *Fine Particles Synthesis, Characterization and Mechanisms of Growth*, Marcel Dekker, New York, Chapter 9, 2000.
14. Gonsalves, K. E. et al., *Nanostructured Materials and Nanotechnology*, Academic Press, California, Chapter 1, 2002.
15. Jana, J., Ganguly, M., and Pl, T., Intriguing cysteine induced improvement of the emissive property of carbon dots with sensing applications, *Phys. Chem. Chem. Phys.*, 17, 2394, 2015.
16. Talapin, D. V. et al., Synthesis and surface modification of amino-stabilized CdSe, CdTe and InP nanocrystals, *Colloids Surf. A.*, 202, 145, 2002.
17. Ajayan, P. M. In: Nalwa H. S. (Ed.), *Nanostructured Materials and Nanotechnology*, Academic Press, California, 2002, Chapter 8.
18. Natan, M. J. and Lyon, L. A. In: Feldheim, D. L. and Foss, C. A. Jr. (Eds.), *Metal Nanoparticle Synthesis, Characterization, and Applications*, Marcel Dekker, New York, 2002, Chapter 8.
19. Schmidt-Winkel, P. et al., Mesocellular siliceous foams with uniformly sized cells and windows, *J. Am. Chem. Soc.*, 121, 254, 1999.
20. Mie, G., Contributions to the optics of turbid media, especially colloidal metal solutions, *Ann. Phys.*, 25, 377, 1908.
21. Xu, H. and Suslick, K. S., Water-soluble fluorescent silver nanoclusters, *Adv. Mater.*, 22, 1078, 2010.
22. Zheng, J., Nicovich, P. R., and Dickson, R. M., Highly fluorescent noble-metal quantum dots. *Annu. Rev. Phys. Chem.*, 58, 409, 2007.

23. Ganguly, M., Pal, J., Das, S., Mondal, C., Pal, A., Negishi, Y., and Pal, T., Green synthesis and reversible dispersion of a giant fluorescent cluster in solid and liquid phase, *Langmuir*, 25, 10945, 2013.

24. Ganguly, M., Mondal, C., Pal, J., Pal, A., Negishi, Y., and Pal, T., Fluorescent Au(I)@Ag2/Ag3 giant cluster for selective sensing of mercury(II) ion, *Dalton. Trans.*, 43, 11557, 2014.

25. Ganguly, M., Mondal, C., Jana, J., Pal, A., and Pal, T., Photoproduced fluorescent Au(I)@(Ag2/Ag3)-thiolate giant cluster: An intriguing sensing platform for DMSO and Pb(II). *Langmuir*, 30, 348, 2014.

26. Ganguly, M., Jana, J., Das, B., Dhara, S., Pal, A., and Pal, T., Orange-red silver emitters for sensing application and bio-imaging, *Dalton Trans.*, 44, 11457–11469, 2015.

27. Ganguly, M., Jana, J., Mondal, C., Pal, A., and Pal, T., Fluorescent Au(I)@Ag2/Ag3 giant cluster for selective sensing of mercury(II) ion, *Phys. Chem. Chem. Phys.*, 16, 18185, 2014.

28. Lee, A. F. et al., Structural and catalytic properties of novel Au/Pd bimetallic colloid particles: EXAFS, XRD, and acetylene coupling, *J. Phys. Chem.*, 99, 6096, 1995.

29. Toshima, N. and Wang, Y., Polymer-protected cu/pd bimetallic clusters, *Adv. Mater.*, 6, 245, 1994.

30. Harriman, A., Making photoactive molecular-scale wires, *Chem. Commun.*, 2, 1990.

31. Pande, S., Ghosh, S. K., Praharaj, S., Panigrahi, S., Basu, S., Jana, S., Pal, A., Tsukuda, T., and Pal, T., Synthesis and size-selective catalysis by supported gold nanoparticles: Study on heterogeneous and homogeneous catalytic process, *J. Phys. Chem. C.*, 111, 10806, 2007.

32. Nath, S., Ghosh, S. K., and Pal, T., Solution phase evolution of AuSe nanoalloys in Triton X-100 under UV-photoactivation, *Chem. Commun.*, 966, 2004.

33. Toshima, N. et al., Catalytic activity and structural analysis of polymer-protected Au-Pd bimetallic clusters prepared by the simultaneous reduction of $HAuCl_4$, and $PdCl_2$, *J. Phys. Chem.*, 96, 9927, 1992.

34. Zhang, H., Watanabe, T., Okumura, M., Haruta, M., and Toshima, N., Catalytically highly active top gold atom on palladium nanocluster, *Nat. Mater.*, 11, 49, 2012.

35. Ghosh, S. K. et al., Bimetallic Pt–Ni nanoparticles can catalyze reduction of aromatic nitro compounds by sodium borohydride in aqueous solution, *J. Appl. Catal. A.*, 286, 61, 2004.

36. Harada, M. et al., Catalytic activity and structural analysis of polymer-protected Au/Pd bimetallic clusters prepared by the successive reduction of $HAuCl_4$ and PdCl2, *J. Phys. Chem.*, 97, 5103, 1993.

37. Ghosh, S. K. and Pal, T., Interparticle coupling effect on the surface plasmon resonance of gold nanoparticles: From theory to applications. *Chem. Rev.*, 107, 4797, 2007.

38. Mallik, K., Mandal, M., Pradhan, N., and Pal, T., Seed mediated formation of bimetallic nanoparticles by UV irradiation: A photochemical approach for the preparation of "core–shell" type structures, *Nano. Lett.*, 1, 319, 2001.

39. Link, S., Wang, Z. L., and El-Sayed, M. A., Spectral properties and relaxation dynamics of surface plasmon electronic oscillations in gold and silver nanodots and nanorods, *J. Phys. Chem. B.*, 103, 3529, 1999.

40. Han, S. W., Kim, Y., and Kim, K., Dodecanethiol-derivatized Au/Ag bimetallic nanoparticles: TEM, UV/VIS, XPS, and FTIR analysis, *J. Colloid. Interface. Sci.*, 208, 272, 1998.

41. Sun, S. et al., Spin-dependent tunneling in self-assembled cobalt-nanocrystal superlattices, *Science*, 287, 1989, 2000.

42. Bian, B. et al., Core(Fe)shell(Au) nanoparticles obtained from thin Fe/Au bilayers employing surface segregation, *J. Electron Microsc.*, 48, 753, 1999.

43. Srnova-Sloufova, I. et al., Core–shell (Ag)Au bimetallic nanoparticles: Analysis of transmission electron microscopy images, *Langmuir*, 16, 9928, 2000.

44. Pande, S., Sarkar, A. K., Basu, M., Jana, S., Sinha, A. K., Sarkar, S., Pradhan, M., Saha, S., Pal, A., and Pal, T., Gram level synthesis of lead-free solder in the nanometer length scale obtained from tin and silver compounds using silicone oil, *Langmuir*, 24, 8991, 2008.

45. Sinha, A. K., Sasmal, A. K., Mehetor, S. K., Pradhan, M., and Pal, T., Evolution of amorphous selenium nanoballs in silicone oil and their solvent induced morphological transformation, *Chem. Commun.*, 50, 15733, 2014.
46. Liang, Y. N., Yu, K., Yan, Q., and Hu, X., Improved elevated temperature performance of Al-intercalated V_2O_5 electrospun nanofibers for lithium-ion batteries, *ACS. Appl. Mater. Interfaces*, 5, 4100, 2013.
47. Takai, A., Saida, T., Sugimoto, W., Wang, L., Yamauchi, Y., and Kuroda, K., Preparation of mesoporous Pt-Ru alloy fibers with tunable compositions via evaporation-mediated direct templating (EDIT) method utilizing porous anodic alumina membranes, *Chem. Mater.*, 21, 3414, 2009.
48. Semagina, N. and Kiwi-Minsker, L., Recent advances in the liquid-phase synthesis of metal nanostructures with controlled shape and size for catalysis, *Catal. Rev.*, 51, 147–217, 2009.
49. Goesmann, H. and Feldmann, C., Nanoparticulate functional materials, *Angew. Chem. Int. Ed.*, 49, 1362–1395, 2010.
50. Baldyga, L. M., Blavo, S. O., Kuo, C.-H., Tsung, C.-K., and Kuhn, J. N., Universal sulfide-assisted synthesis of M–Ag heterodimers (M=Pd, Au, Pt) as efficient platforms for fabricating metal–semiconductor heteronanostructures, *ACS Catal.*, 2, 2626–2629, 2012.
51. Wu, K.-L., Yu, R., and Wei, X.-W., Monodispersed $FeNi_2$ alloy nanostructures: Solvothermal synthesis, magnetic properties and size-dependent catalytic activity, *Cryst. Eng. Commun.*, 14, 7626–7632, 2012.
52. Pradhan, M., Sarkar, S., Sinha, A. K., Basu, M., and Pal, T., Morphology controlled uranium oxide hydroxide hydrate for catalysis, luminescence and SERS studies, *Cryst. Eng. Comm.*, 13, 2878–2889, 2011.
53. Zhang, X., Yin, H., Wang, J., Chang, L., Gao, Y., Liu, W., and Tang, Z., Shape-dependent electrocatalytic activity of monodispersed palladium nanocrystals toward formic acid oxidation, *Nanoscale*, 5, 8392–8397, 2013.
54. Lim, B., Jiang, M. J., Tao, J., Camargo, P. H. C., Zhu, Y. M., and Xia, Y., Shape-controlled synthesis of Pd nanocrystals in aqueous solutions, *Adv. Funct. Mater.*, 19, 189, 2009.
55. Sun, Y. and Xia, Y., Shape-controlled synthesis of gold and silver nanoparticles, *Science*, 298, 2176–2179, 2002.
56. Zhang, H., Jin, M., and Xia, Y., The science and art of carving metal nanocrystals, *Angew. Chem., Int. Ed.*, 51, 7656–7673, 2012.
57. Van Santen, R. A., Complementary structure sensitive and insensitive catalytic relationships, *Acc. Chem. Res.*, 42, 57–66, 2008.
58. Xia, Y. et al., On the role of surface diffusion in determining the shape or morphology of noble-metal nanocrystals, *Angew. Chem. Int. Ed.*, 51, 10266–10270, 2012.
59. Mandal, M. et al., Synthesis and characterization of superparamagnetic Ni–Pt nanoalloy, *Chem. Mater.*, 15, 3710, 2003.
60. Treguer, M. et al., Dose rate effects on radiolytic synthesis of gold–silver bimetallic clusters in solution, *J. Phys. Chem.*, 102, 4310, 1998.
61. Remita, S., Mostafavi, M., and Delcourt, M. O., Bimetallic Ag-Pt and Au-Pt aggregates synthesized by radiolysis, *Radiat. Phys. Chem.*, 47, 275, 1996.
62. Belloni, J. et al., Radiation-induced synthesis of mono- and multi-metallic clusters and nanocolloids, *New J. Chem.*, 22, 1239, 1998.
63. Schmid, G. et al., Catalytic properties of layered gold–palladium colloids, *Chem. Eur. J.*, 2, 1099, 1996.
64. Mallin, M. P. and Murphy, C. J., Solution-phase synthesis of Sub-10 nm Au–Ag alloy nanoparticles, *Nano. Lett.*, 2, 1235, 2002.
65. Ah, C. S., Hong, S. D., and Jang, D. J., Preparation of $Au_{core}Ag_{shell}$ nanorods and characterization of their surface plasmon resonances, *J. Phys. Chem. B.*, 105, 7871, 2001.
66. Ryan, D. et al., Heterosupramolecular chemistry: Recognition initiated and inhibited silver nanocrystal aggregation by pseudorotaxane assembly, *J. Am. Chem. Soc.*, 122, 6252, 2000.

67. Praharaj, S., Nath, S., Ghosh, S. K., Kundu, S., and Pal, T., Immobilization and recovery of Au nanoparticles from anion exchange resin: Resin-bound nanoparticle matrix as a catalyst for the reduction of 4-nitrophenol, *Langmuir*, 20, 9889, 2004.
68. Miner, R. S., Namba, S., and Turkevich, J., In *Proceedings of the Seventh International Congress on Catalysis*, Tokyo, 1981, 160.
69. Silvert, P. Y., Synthesis and characterization of nanoscale Ag-Pd alloy particles, *Nanostruct. Mater.*, 7, 611, 1996.
70. Esumi, K. et al., Preparation of bimetallic Pd-Pt colloids in organic solvent by solvent extraction-reduction, *Langmuir*, 7, 457, 1991.
71. Mizukoshi, Y. et al., Sonochemical preparation of bimetallic nanoparticles of gold/palladium in aqueous solution, *J. Phys. Chem. B.*, 101, 7033, 1997.
72. Mandal, M. et al., *Nano. Lett.*, 1, 319, 2001.
73. Pan, C. et al., A new synthetic method toward bimetallic ruthenium platinum nanoparticles; composition induced structural changes, *J. Phys. Chem. B.*, 103, 10098, 1999.
74. Reetz, M. T. and Quaiser, S. A., *Angew. Chem., Int. Ed. Engl.*, 34, 2240, 1995.
75. Belloni, J., Metal nanocolloids, *Curr. Opin. Colloid Interface Sci.*, 1, 184, 1996.
76. Kittel, C., *Introduction to Solid State Physics*, Wiley, New York, 1996.
77. Kreibig, U. and Vollmer, M., *Optical Properties of Metal Clusters*, Springer, Berlin, 1995.
78. Moskovits, M., Srnova-Sloufova, I., and Vlckova, B., Bimetallic Ag–Au nanoparticles: Extracting meaningful optical constants from the surface-plasmon extinction spectrum. *J. Chem. Phys.*, 116, 10435, 2002.
79. Freeman, R. G. et al., Ag-Clad Au nanoparticles: Novel aggregation, optical, and surface-enhanced Raman scattering properties. *J. Phys. Chem.*, 100, 718, 1996.
80. Morriss, R. H. and Collins, L. F., Optical properties of multilayer colloids, *J. Chem. Phys.*, 41, 3357, 1964.
81. Mulvaney, P., Giersig, M., and Henglein, A., Electrochemistry of multilayer colloids: Preparation and absorption spectrum of gold-coated silver particles, *J. Phys. Chem.*, 97, 7061, 1993.
82. Chen, Y. H. and Nickel, U., Superadditive catalysis of homogeneous redox reactions with mixed silver–gold colloids, *J. Chem. Soc. Faraday Trans.*, 89, 2479, 1993.
83. Shibata, T. et al., Size-dependent spontaneous alloying of Au-Ag nanoparticles. *J. Am. Chem. Soc.*, 124, 11989, 2002.
84. Liz-Marzan, L. M. and Philipse, A. P., Stable hydrosols of metallic and bimetallic nanoparticles immobilized on imogolite fibers, *J. Phys. Chem.*, 99, 15120, 1995.
85. Sinzig, J. et al., Z. Binary clusters: Homogeneous alloys and nucleus-shell structures, *Phys. D.*, 26, 242, 1993.
86. Sato, T. et al., Photochemical formation of silver-gold (Ag–Au) composite colloids in solutions containing sodium alginate, *Appl. Organomet. Chem.*, 5, 261, 1991.
87. Schwank, J., Catalytic gold, *Gold. Bull.*, 16, 98, 1983.
88. Sosebee, T. et al., The nucleation of colloidal copper in the presence of poly(ethyleneimine), *Ber. Bunsen-Ges. Phys. Chem.*, 99, 40, 1995.
89. J. L. et al., Microaggregates of non-noble metals and bimetallic alloys prepared by radiation-induced reduction Marignier, *Nature*, 317, 344, 1985.
90. Yonezawa, T. and Toshima, N., Mechanistic consideration of formation of polymer-protected nanoscopic bimetallic clusters, *J. Chem. Soc. Faraday Trans.*, 91, 4111, 1995.
91. Yala, F. et al., Silver-palladium submicronic powders Part II structural characterization, *J. Mater. Sci.*, 30, 1203, 1995.
92. Torigoe, K. and Esumi, K., Preparation of bimetallic silver-palladium colloids from silver(I) bis(oxalato)palladate(II), *Langmuir*, 9, 1664, 1993.
93. Turkevich, J., Colloidal gold. Part I, *Gold. Bull.*, 18, 86, 1985.
94. Liu, H., Mao, G., and Meng, S., Preparation and characterization of the polymer-protected palladium—gold colloidal bimetallic catalysts, *J. Mol. Catal.*, 74, 275, 1992.
95. Turkevich, J., and Kim, G., Palladium: Preparation and catalytic properties of particles of uniform size, *Science*, 169, 873, 1970.

96. Mizukoshi, Y. et al., Characterization and catalytic activity of core–shell structured gold/palladium bimetallic nanoparticles synthesized by the sonochemical method, *J. Phys. Chem. B.*, 104, 6028, 2000.

97. Toshima, N. et al., Colloidal dispersions of palladium–platinum bimetallic clusters protected by polymers. Preparation and application to catalysis, *Chem. Lett.*, 1769, 1989.

98. Toshima, N. and Wang, Y., Preparation and catalysis of novel colloidal dispersions of copper/noble metal bimetallic clusters, *Langmuir*, 10, 4574, 1994.

99. Belloni, J. et al., U.S. Patent 4, 629, 709, December 16, 1986.

100. Marignier, J. L., *These de Doctoral d'Etat*, Universite Paris-Sud, Orsay, 1987.

101. Torigoe, K., Nakajima, Y., and Esumi, K., Preparation and characterization of colloidal silver-platinum alloys, *J. Phys. Chem.*, 97, 8304, 1993.

102. Belloni, J. et al., In: Hedwig, P., Nyikos, L., and Schiller, R., (Eds.). *Radiation Chemistry*, Akademia Kiado, Budapest, 1987, 89.

103. Henglein, A. et al., Electrochemistry of colloidal silver particles in aqueous solution: Deposition of lead and indium and accompanying optical effects, *Ber. Bunsenges. Phys. Chem.*, 96, 2411, 1992.

104. Henglein, A. and Brancewicz, C., Absorption spectra and reactions of colloidal bimetallic nanoparticles containing mercury, *Chem. Mater.*, 9, 2164, 1997.

105. Schmid, G. et al., Ligand-stabilized bimetallic colloids identified by HRTEM and EDX, *Angew. Chem. Int. Ed. Engl.*, 30, 874, 1991.

106. Degani, Y. and Willner, I., Photoinduced hydrogenation of ethylene and acetylene in aqueous media: The functions of palladium and platinum colloids as catalytic charge relays, *J. Chem. Soc. Perkin Tans.*, 2, 37, 1986.

107. Reetz, M. T. and Helbig, W., Size-selective synthesis of nanostructured transition metal clusters, *J. Am. Chem. Soc.*, 116, 7401, 1994.

108. Reetz, M. T., Helbig, W., and Quasier, S. A., Electrochemical preparation of nanostructural bimetallic clusters, *Chem. Mater.*, 7, 2227, 1995.

109. Schmid, G. et al., Hydrosilation reactions catalyzed by supported bimetallic colloids. *Inorg. Chem.*, 36, 891, 1997.

110. Esumi, K., Wakabayashi, M., and Torigoe, K., Preparation of colloidal silver—Palladium alloys by UV-irradiation in mixtures of acetone and 2-propanol, *Colloids Surf. A.*, 109, 55, 1996.

111. Rivas, L. et al., Mixed silver/gold colloids: A study of their formation, morphology, and surface-enhanced Raman activity, *Langmuir*, 16, 9722, 2000.

112. Gallezot, P. and Leclereq, C., Characterization of catalysts by conventional and analytical electron microscopy, In: Imelik, B. and Vedrine, J. C. (Eds.), *Catalyst Characterization*, Plenum, New York, 1994, 509.

113. Duff, D. G. et al., The microstructure of colloidal silver: Evidence for a polytetrahedral growth sequence, *J. Chem. Soc. Chem. Commun.*, 1264, 1987.

114. Curtis, A. C. et al., Preparation and structural characterization of an unprotected copper sol, *J. Phys. Chem.*, 92, 2270, 1988.

115. Harada, M., Asakura, K., and Toshima, N., Structural analysis of polymer-protected platinum/rhodium bimetallic clusters using extended x-ray absorption fine structure spectroscopy. Importance of microclusters for the formation of bimetallic clusters. *J. Phys. Chem.*, 98, 2653, 1994.

116. Urban, J., Sack-Kongehl, H., and Weiss, K., Z. HREM studies of the structure and the oxidation process of copper clusters created by inert gas aggregation, *Phys. D.*, 36, 73, 1996.

117. Toshima, N. et al., Synthesis of highly coheterotactic poly(methyl methacrylate-alt-styrene), *Chem. Lett.*, 2157, 1990.

118. Mulvaney, P., Surface plasmon spectroscopy of nanosized metal particles. *Langmuir*, 12, 788, 1996.

119. Sinfelt, J. H., Structure of bimetallic clusters, *Acc. Chem. Res.*, 20, 134, 1987.

120. Sinfelt, J. H., Via, G. H., and Lytle, F. W., Structure of bimetallic clusters. Extended x-ray absorption fine structure (EXAFS) studies of Re–Cu, Ir–Cu, and Pt–Cu clusters, *J. Chem. Phys.*, 76, 2779, 1982.

121. Bradley, J. S. et al., Infrared and EXAFS study of compositional effects in nanoscale colloidal palladium–copper alloys, *Chem. Mater.*, 8, 1895, 1996.
122. Zhu, B. et al., Bimetallic Pd–Cu catalysts: X-ray diffraction and theoretical modeling studies, *J. Catal.*, 167, 412, 1998.
123. Touroude, R., Preparation of colloidal platinum/palladium alloy particles from non-ionic microemulsions: Characterization and catalytic behaviour, *Colloids Surf.*, 67, 9, 1992.
124. Bucher, J. P. and van der Link, J. J., Electronic properties of small supported Pt particles: NMR study of 195Pt hyperfine parameters, *Phys. Rev. B.*, 38, 11038, 1988.
125. Tong, Y. Y., Martin, G. A., and van der Link, J. J., ^{195}Pt NMR observation of local density of states enhancement on alkali-promoted Pt catalyst surfaces, *J. Phys. D.*, 6, L533, 1994.
126. Ascencio, J. A. et al., A truncated icosahedral structure observed in gold nanoparticles. *Surf. Sci.*, 447, 73, 2000.
127. Lewis, L. N., Chemical catalysis by colloids and clusters, *Chem. Rev.*, 93, 1693, 1993.
128. Kavanagh, K. E. and Nord, F. F., Systematic studies on palladium-synthetic high polymer catalysts, *J. Am. Chem. Soc.*, 65, 2121, 1943.
129. Edelstein, A. S. and Cammarata, R. C., *Nanoparticles: Synthesis, Properties and Applications*, Institute of Physics Publishing, Bristol, 1996.
130. Gomez-Romero, P., Hybrid organic–inorganic materials—in search of synergic activity, *Adv. Mater.*, 13, 163, 2001.
131. Hickman, J. J. et al., Molecular self-assembly of two-terminal, voltammetric microsensors with internal references. *Science*, 252, 688, 1991.
132. Elghanian, R., Selective colorimetric detection of polynucleotides based on the distance-dependent optical properties of gold nanoparticles, *Science*, 277, 1078, 1997.
133. Willner, I. and Willner, B., Molecular and biomolecular optoelectronics, *Pure Appl. Chem.*, 73, 535, 2001.
134. Markovich, G. et al., Architectonic quantum dot solids, *Acc. Chem. Res.*, 32, 415, 1999.
135. Cao, Y. W., Jin, R., and Mirkin, C. A., DNA-modified core–shell Ag/Au nanoparticles, *J. Am. Chem. Soc.*, 123, 7961, 2001.
136. Kolb, O. et al., Investigation of tetraalkylammonium bromide stabilized palladium/platinum bimetallic clusters using extended X-ray absorption fine structure spectroscopy, *Chem. Mater.*, 8, 1889, 1996.
137. Haruta, M. et al., Low-temperature oxidation of CO over gold supported on TiO_2, α-Fe_2O_3 and Co_3O_4, *J. Catal.*, 114, 175, 1993.
138. Toshima, N. and Hirakawa, K., Polymer-protected bimetallic nanocluster catalysts having core/shell structure for accelerated electron transfer in visible-light-induced hydrogen generation, *Polym. J.*, 31, 1127, 1999.
139. Mandal, M. et al., Synthesis of Au_{core}–Ag_{shell} type bimetallic nanoparticles for single molecule detection in solution by SERS method, *J. Nanoparticle Res.*, 6, 53, 2004.
140. Ghosh, S. K. et al., Solvent and ligand effects on the localized surface plasmon resonance (LSPR) of gold colloids, *J. Phys. Chem. B.*, 108, 13963, 2004.
141. Luo, X., Liu, Y., Zhang, H. and Yang, B., Shape-selective synthesis and facet-dependent electro-catalytic activity of $CoPt_3$ nanocrystals, *Cryst. Eng. Comm.*, 14, 3359–3362, 2012.
142. Fu, G., Ding, L., Chen, Y., Lin, J., Tang, Y., and Lu, T., Facile water-based synthesis and catalytic properties of platinum–gold alloy nanocubes, *Cryst. Eng. Comm.*, 16, 1606–1610, 2014.
143. Yin, A.-X., Min, X.-Q., Zhang, Y.-W., and Yan, C.-H., Shape-selective synthesis and facet-dependent enhanced electrocatalytic activity and durability of monodisperse Sub-10 nm Pt–Pd tetrahedrons and cubes, *J. Am. Chem. Soc.*, 133, 3816–3819, 2011.
144. Zeng, J., Zhu, C., Tao, J., Jin, M., Zhang, H., Li, Z-Y., Zhu, Y., and Xia, Y., Controlling the nucleation and growth of silver on palladium nanocubes by manipulating the reaction kinetics, *Angew. Chem. Int. Ed.*, 51, 2354–2358, 2012.
145. Xie, S., Peng, H.-C., Lu, N., Wang, J., Kim, M. J., Xie, Z., and Xia, Y., Confining the nucleation and overgrowth of Rh to the {111} facets of Pd nanocrystal seeds: The roles of capping agent and surface diffusion, *J. Am. Chem. Soc.*, 135, 16658–16667, 2013.

146. Zhang, H., Jin, M., and Xia, Y., Noble-metal nanocrystals with concave surfaces: Synthesis and applications, *Angew. Chem. Int. Ed.*, 51, 7656, 2012.
147. Hu, M., Chen, J., Li, Z. Y., Au, L., Hartland, G. V., Li, X., Marquez, M., and Xia, Y., Gold nanostructures: Engineering their plasmonic properties for biomedical applications, *Chem. Soc. Rev.*, 35, 1084, 2006.
148. Yavuz, M. S. et al., Gold nanocages covered by smart polymers for controlled release with near-infrared light, *Nat. Mater.*, 8, 935, 2009.
149. Peng, Z., You, H., Wu, J., and Yang, H., Electrochemical synthesis and catalytic property of sub-10 nm platinum cubic nanoboxes, *Nano Lett.*, 10, 1492, 2010.
150. Metraux, G. S., Cao, Y. C., Jin, R., and Mirkin, C. A., Triangular nanoframes made of gold and silver, *Nano Lett.*, 3, 519, 2003.
151. Lu, X., Au, L., McLellan, J., Li, Z. Y., Marquez, M., and Xia, Y., Fabrication of cubic nanocages and nanoframes by dealloying Au/Ag alloy nanoboxes with an aqueous etchant based on $Fe(NO_3)_3$ or NH_4OH, *Nano Lett.*, 7, 1764, 2007.
152. Au, L., Chen, Y., Zhou, F., Camargo, P. H. C., Lim, B., Li, Z. Y., Ginger, D. S., and Xia, Y., Synthesis and optical properties of cubic gold nanoframes, *Nano Res.*, 1, 441, 2008.
153. McEachran, M., Keogh, D., Pietrobon, B., Cathcart, N., Gourevich, I., Coombs, N., and Kitaev, V., Ultrathin gold nanoframes through surfactant-free templating of faceted pentagonal silver nanoparticles, *J. Am. Chem. Soc.*, 133, 8066, 2011.
154. Fan, N., Yang, Y., Wang, W., Zhang, L., Chen, W., Zou, C., and Huang, S.,: Selective etching induces selective growth and controlled formation of various platinum nanostructures by modifying seed surface free energy, *ACS. Nano.*, 6, 4072, 2012.
155. Xie, S., Lu, N., Xie, Z., Wang, J., Kim, M. J., and Xia, Y., Synthesis of Pd-Rh core–frame concave nanocubes and their conversion to Rh cubic nanoframes by selective etching of the Pd cores, *Angew. Chem., Int. Ed.*, 51, 10266–10270, 2012.
156. Jin, R., The impacts of nanotechnology on catalysis by precious metal nanoparticles, *Nanotechnol. Rev.*, 1, 31, 2012.

6

Nanotube-Based Membrane Systems

Lane A. Baker and Charles R. Martin

CONTENTS

6.1 Introduction

We have been investigating membranes with pores of controllable size, geometry, and surface chemistry [1–3]. The pores of these membranes can be modified to create nanometer scale tubes that retain the geometry of the original pores, but impart functionality to the membrane. These nanotube membranes show promise for use in the fields of bioanalysis and biotechnology. Specifically, these membranes and materials prepared therefrom can be used for template synthesis of biofunctionalized materials, chemical and biochemical separations, and as platforms for biochemical sensing. In this chapter, we will review the materials and techniques used to create, manipulate, and interrogate nanotube-based

Template synthesis

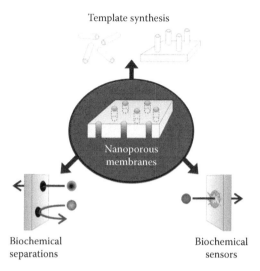

Nanoporous
membranes

Biochemical
separations

Biochemical
sensors

FIGURE 6.1
Illustration of the uses of nanotube-based membrane systems in biotechnology. Clockwise from top, template synthesis, biochemical sensors, and biochemical separations.

membrane systems. We will review applications of nanotube-based membrane systems to problems that are both basic and applied in nature.

Nanotube membranes are an attractive platform for nanotechnology in large part due to the simple, yet effective, manner in which they can be used. Membrane approaches to nanotechnology offer a facile manner to handle and manipulate nanomaterials without the use of highly specialized equipment. Further, homogenous pores ensure homogenous nanomaterials, a characteristic that is often not easily achieved at these small scales. Appropriate membranes can be purchased commercially, or can be fabricated by the user, and relatively simple techniques can be used to chemically or physically modify the membrane properties.

There are three general membrane-based strategies that have been used to prepare nanomaterials. These strategies are illustrated in Figure 6.1. In the first strategy, template synthesis, nanometer scale pores are used to synthesize and modify materials in which at least one dimension is nanometer in scale. In the second strategy, nanometer scale pores are used to separate species that translocate a nanotube membrane. In the third approach, nanotube membranes are used as sensors. In this chapter, we will discuss the uses of nanotube-based membranes in the context of biotechnology. We will briefly review the materials and methods of nanotube membrane technology and then discuss biochemically oriented research and applications of these membranes with respect to template synthesis, separations, and sensing.

6.2 Materials and Methods of Nanotube-Based Membrane Systems

6.2.1 Porous Alumina Membranes

Alumina membranes are obtained through the electrochemical growth of a thin, porous layer of aluminum oxide from aluminum metal in acidic media. Membranes of this type may be obtained commercially with a variety of pore sizes, or can be grown by using well-established procedures. Pores with dimensions from 200 to 5 nm can be obtained in millimeter-thick membranes. The pores created are nominally arranged in a hexagonally packed array. Pore densities can be as high as 10^{11} pores/cm^2. An example of such a membrane is shown in Figure 6.2 [4].

6.2.2 Track Etched Membranes

Membranes prepared by the track-etch procedure are created by bombarding (or tracking) a thin film of the material of interest with high-energy particles, creating damage tracks. The damage tracks are then chemically developed (or etched) to produce pores. A variety of membrane materials are compatible with this technique, however, polymer films have shown the greatest utility. Porous poly(carbonate), poly(ethylene terephthalate), and poly(imide) membranes are all commonly produced with this method. Track-etch membranes are available from commercial sources or can be fabricated using tracked material. Pore dimensions can be controlled by development conditions, including pH, temperature, and time. The density of pores can range from a single pore to millions of pores, and is

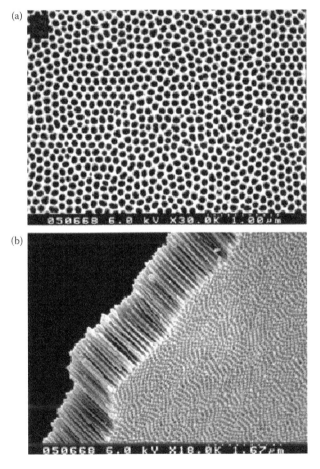

FIGURE 6.2
Scanning electron micrographs, (a) top, (b) cross section, of an anodically etched alumina membrane.

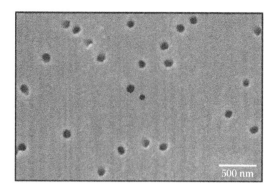

FIGURE 6.3
Scanning electron micrograph of polycarbonate membrane.

controlled by the fluence of impinging particles in the tracking process. An example of a poly(carbonate) membrane produced using this method is shown in Figure 6.3 [5].

6.2.3 Electroless Plating

We have found electroless plating of materials in and on these nanoporous membranes to be a powerful method for both synthesizing materials and providing a membrane that is amenable to surface modification. We have described our method for electroless plating in detail previously [6]. Briefly, template membranes are sensitized by soaking in a solution of Sn(II)Cl, which results in adsorption of Sn(II) to the surface of the membrane. The Sn(II)-coated membrane is then soaked in a solution of Ag(NO$_3$). The Sn(II) adsorbed to the surface of the membrane is oxidized by the Ag(I) present in solution (Equation 6.1), resulting in the deposition of Ag(0) nanoparticles on the membrane surfaces.

$$Sn(II)_{surf} + 2Ag(I)_{aq} \rightarrow Sn(IV)_{aq} + 2Ag(0)_{surf} \tag{6.1}$$

The membrane with Ag particles adsorbed to the surface is then soaked in a commercial Au plating solution. The silver particles at the surface reduce Au(I) present in plating solution, resulting in the deposition of Au(0) nanoparticles on the membrane surfaces (Equation 6.2).

$$Ag(0)_{surf} + Au(I)_{aq} \rightarrow Ag(I)_{aq} + Au(0)_{surf} \tag{6.2}$$

The deposited Au particles serve as autocatalysts for further Au deposition using formaldehyde as a reducing agent. The Au films deposited on the membranes cover both the pore walls and the surface of the membrane, but do not close the pore mouths. Further, the gold surface layer can be selectively removed, leaving Au nanotubes present in the pores. Electroless plating affords additional control over two critical parameters in nanotube-based membrane systems, pore size and surface chemistry. By controlling the plating time and conditions, the amount of Au deposited can be controlled. This translates into more precise control over the pore diameter. The use of Au allows facile Au–thiol chemistry to be utilized to control the surface chemistry of the pores. Adsorption of charged thiols, thiolated DNAs, or functional thiols (which can undergo further chemical modification) allows the incorporation of appropriate chemistries for separations and sensing into the pores.

6.2.4 Sol–Gel Deposition

Another method we have found useful for materials preparation using membranes is sol–gel chemistry [7]. This method has been extremely versatile and can be used to produce nanotubes or nanowires, as desired. In the sol–gel method, a sol of tetraethyl orthosilicate is formed in an acidic ethanol solution. A template membrane is then sonicated in the sol solution, is removed, and is then dried and cured overnight at 150°C. Adsorption of the sol on the membrane produces a thin film of silica on the pore walls and membrane surface. Control of time and sol concentration during deposition allows control of thickness at the nanometer level. Deposited silica can be easily further modified using silane chemistry, allowing the incorporation of almost any functionality. Further, by mechanical polishing, the silica present at either or both faces of the membrane can be removed. When the membrane is dissolved, a silica negative of the original template membrane is obtained. Additionally, other inorganic materials, such as TiO_2, ZnO, and WO_3 can be templated using this sol–gel method.

6.2.5 Membrane Measurements

Measuring the transport or separation of an analyte with a nanotube membrane typically requires the use of a U-tube permeation cell or a conductivity cell. Membranes are initially mounted in a holder to ensure a tight seal, with the membrane separating the two halves of the cell. A schematic diagram of a typical configuration for mounting a membrane using parafilm spacers and glass slides is shown in Figure 6.4 [8]. The membrane is then placed between two half-cells of a U-tube cell (schematically illustrated in Figure 6.5) and the entire assembly is held together with a clamp. By placing a solution with species to be separated in the half-cell on one side of the membrane (feed side) and monitoring the concentration of species present in the opposite half-cell (permeate side) as a function of time, the flux of a species across the membrane may be determined. Typically, the flux is monitored using UV–vis spectroscopy or chromatographic methods. Flux of species across the membrane may be modulated by applying a voltage between the two half-cells, resulting in electrophoretic movement, or by applying an anisotropic pressure between the two half-cells, resulting in pressure-driven flow.

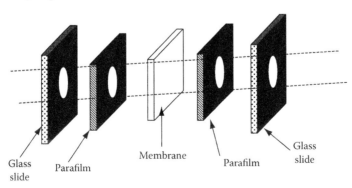

FIGURE 6.4
Membrane assembly.

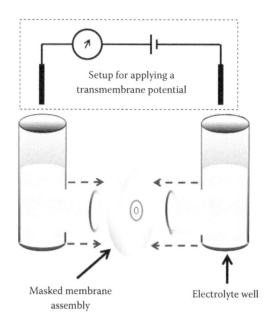

Setup for applying a
transmembrane potential

Masked membrane
assembly

Electrolyte well

FIGURE 6.5
Illustration of a U-tube cell used for preparing and measuring the properties of nanotube membranes.

6.3 Template Synthesis

Template synthesis is a powerful and elegant method capable of producing nanometer scale materials in a controlled fashion. Template synthesis involves the use of a template, or master, with nanometer scale features. Membranes, such as those described in the experimental section, have been our templates of choice because they are convenient, versatile, and robust. A general scheme for template synthesis is shown in Figure 6.6. Synthesis in the template involves the growth or deposition of materials inside the pores. The surrounding membrane material is then selectively removed, leaving nanomaterials that are negatives of the original membrane template. Depending on the conditions of membrane removal, the templated material can form a surface-bound array, or can be freed from the template to form individual nanoparticles. By controlling the parameters involved in the synthesis of a given material, a variety of geometries can be obtained. For instance, wires, tubes, and cones can be prepared with ease. Additionally, multicomponent structures, such as segmented wires or coaxial tubes can be prepared by modulating the materials templated. A diverse range of materials are amenable to template synthesis, including metals, semiconductors, and polymers. Furthermore, we have demonstrated that template-synthesized nanomaterials can be modified with or constructed using biochemical species.

6.3.1 Enzymatic Nanoreactors

Materials prepared through template synthesis can be used as nanometer scale test tubes. One example of our use of these nano test tubes is the immobilization of enzymes [9]. In

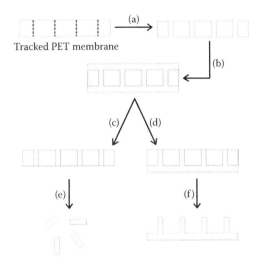

FIGURE 6.6
Schematic of a method for template synthesis with nanoporous membranes. (a) Chemical etch of damage tracks, (b) deposition of material to be templated (i.e., gold, silica), (c) removal of material templated at the membrane faces through a tape stripping or mechanical abrasion, (d) removal of material templated at one membrane face through tape stripping or mechanical abrasion, (e) dissolution of the membrane and filtration of nanotubes, (f) membrane removal through dissolution or oxygen plasma etch.

Figure 6.7, a schematic of the process used to create nanometer scale enzymatic bioreactors is shown. This method makes use of a combination of electrochemical, chemical, and physical deposition methods. A polycarbonate membrane is first sputtered with a thin layer of gold (~50 nm) (Figure 6.7a). This gold film serves as an electrode to electropolymerize a thin polypyrrole film across the membrane. A short polypyrrole plug is deposited in the pores, as well (Figure 6.7b). Additional polypyrrole is then chemically polymerized at the nanopore walls, forming a closed nanotubule, in effect a nano test tube (Figure 6.7c). The thickness of the polypyrrole deposited can be controlled through the reaction conditions. Thickness is an important parameter, as the entrapment ability and permeability of the film depends greatly on the films' thickness.

These features are used to encapsulate an enzyme using the electropolymerized film as a filter. An enzyme is then loaded into the nano test tubes by filtering a solution of the enzyme through the polypyrrole-modified membrane (Figure 6.7d). The solvent can pass through the polymer coating, but the enzyme is too large to pass and is retained. After the nano test tube is loaded with the enzyme, a layer of Torrseal epoxy is applied to the membrane (Figure 6.7e). The membrane is then dissolved in dichloromethane (Figure 6.7f), leaving the enyzme-loaded nano test tubes affixed to the epoxy backing in a random array.

Using this method, glucose oxidase (GOD), catalase, subtilisin, trypsin, and alcohol dehydrogenase have been successfully encapsulated. An example of the activity of GOD-filled nano test tubes is shown in Figure 6.8. The enzymatic activity was evaluated using a standard *o*-dianisidine–peroxidase assay. In curves a and b, the catalytic activities of two different capsule arrays with different enzyme loadings are shown. In curves c and d, a competing encapsulation method, incorporation into a thin film of polypyrrole is shown. In curve e, nano test tubes with no enzyme are shown. This work demonstrated that biochemical activity of templated materials could be retained, and that certain advantages, such as high surface area and volume ratio, can be obtained using template synthesis.

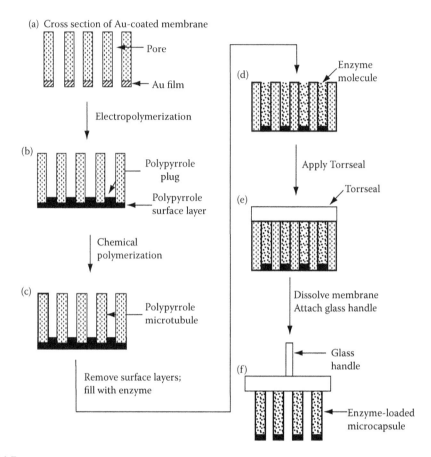

FIGURE 6.7
Schematic diagram of methods used to synthesize and enzyme-load the capsule arrays. (a) Au-coated template membrane. (b) Electropolymerization of polypyrrole film. (c) Chemical polymerization of polypyrrole tubules. (d) Loading with enzyme. (e) Capping with epoxy. (f) Dissolution of the template membrane.

6.3.2 Nano Test Tubes

Another use of template synthesis with implications for biotechnology is the synthesis of nano test tubes free of a solid support [10]. Such materials have potential applications in drug delivery. For instance, if the void region of a nano test tube could be loaded with a specific payload, the open end could then be capped, forming a nanometer scale delivery vehicle. Molecular recognition chemistry could then be incorporated on the exterior of the capped nano test tube that would direct the nano test tube and payload to a specific portion of a cell. The cap could then be selectively released, or the nano test tube could degrade, releasing the payload at the targeted site.

Nano test tubes comprised of silica have been synthesized using an anodically oxidized alumina as the template. A schematic of the synthetic process is shown in Figure 6.9. In the first step, Figure 6.9a, an aluminum–alumina template is produced by partial anodization of the aluminum substrate. This creates a template with one end open and the other end closed, an appropriate configuration for the formation of nano test tubes. Silica is then deposited on the pore walls and surface of the template using a sol–gel method (Figure 6.9b). The surface silica film is removed through a mechanical and chemical step with

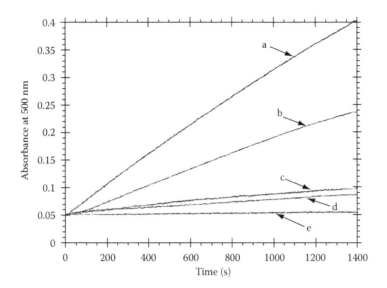

FIGURE 6.8
Evaluation of the enzymatic activity of GOD-loaded capsules (curves a and b) and empty capsules (curve e). The standard *o*-dianisidine–peroxidase assay was used. A larger amount of GOD was loaded into the capsules used for curve a than in the capsules used for curve b. Curves c and d are for a competing GOD-immobilization methods, entrapment within a polypyrrole film.

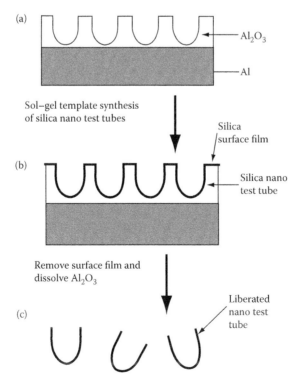

FIGURE 6.9
Schematic of the template synthesis method used to prepare the nano test tubes.

ethanol and polishing. The template membrane is then dissolved in a 25% (w/w) solution of H_3PO_4, liberating the templated nano test tubes (Figure 6.9c). Transmission electron micrographs of silica nano test tubes prepared by using this method are shown in Figure 6.10. By varying the pore size and depth, the diameter and length of the prepared test tubes can be controlled. In Figure 6.10a through c, nano test tubes prepared from membranes of differing geometries are shown. In the inset of Figure 6.10a, it is clear that one end of the nano test tube is closed, as expected.

We have also shown that open-ended silica nanotubes can be prepared with a strategy similar to that used to prepare nano test tubes [11]. These open-ended silica nanotubes can be selectively functionalized with different chemistries on the interior or exterior of the tubes. Template-synthesized, open-ended silica nanotubes are prepared by the sol–gel method previously described using either 60 or 200 nm diameter alumina membranes. While still in the membrane, the tubes were exposed to a solution of a silane to selectively modify the interior of the tubes. Silica at the membrane faces is then removed by mechanically polishing the membrane on each side. The alumina membrane is then dissolved, liberating the silica nanotubes with the interiors selectively silylated. The liberated nanotubes with interior chemical modification are then exposed to a second solution of silane with a different chemical functionality, which only attaches to the previously unexposed nanotube exterior. In this manner, silanes with hydrophobic–hydrophilic character or silanes with molecular recognition capabilities can be selectively placed on the interior or exterior of a nanotube. This creates a functionalized nanotube with a specific chemistry on the tube interior and a potentially different chemistry on the tube exterior. These differentially functionalized nanotubes could be used for bioseparations and biocatalysis.

We have demonstrated how the use of these materials as a smart nanophase extractor to remove molecules from solution was demonstrated. In these experiments, 5 mg of nanotubes having hydrophobic octadecyl silane coatings on the interior of the tubes and hydrophilic bare silica exteriors are suspended in a 1.0×10^{-5} M solution of aqueous 7,8-benzoquinoline (BQ). BQ is hydrophobic and has an octanol–water partition coefficient of 10. The suspension is stirred for 5 min and then filtered to recover the nanotubes. UV–vis spectroscopy of the filtrate solution showed as much as 82% of the BQ could be removed from solution. Control nanotubes with no coating showed less than 10% extraction of BQ.

The use of hydrophobic–hydrophilic interactions for extraction–separation is a general but nonspecific example of the use of functionalized nanotubes. The ability to use functionalized nanotubes for bioseparations in a highly specific manner has also been demonstrated. In these experiments, enantiomers of the drug 4-[3-(4-fluorophenyl)-2-hydroxy-1-[1,2,4]triazol-1-yl-propyl]-benzonitrile (FTB, Figure 6.11) could be separated from a racemic mixture using RS enantiomer-specific Fab antibody fragments. The Fab fragments are attached to the interior and exterior of the nanotubes using an aldehyde-terminated silane. The nanotubes are then suspended in a racemic mixture of FTB and stirred. Nanotubes are collected by filtration and the filtrate is assayed for the presence of the two enantiomers using chiral high-performance liquid chromatography (HPLC). Chromatograms of the filtrate are shown in Figure 6.11. The top chromatogram (A) is a solution 20 μM in SR and RS enantiomers of FTB. The middle chromatogram (B) is the same solution as that present in (A), but after exposure to the Fab-functionalized nanotubes. From integration of the peak ratios, 75% of the RS enantiomer, but none of the SR enantiomer is removed. When the concentration of the initial racemic solution is lowered from 20 to 10 μM, all of the RS enantiomer could be removed (chromatogram C). Unfunctionalized nanotubes do not remove any appreciable quantity of either enantiomer. Differential modification of the nanotube interiors with the RS-specific Fab fragments also removed only RS FTB from solution, but at lower concentrations.

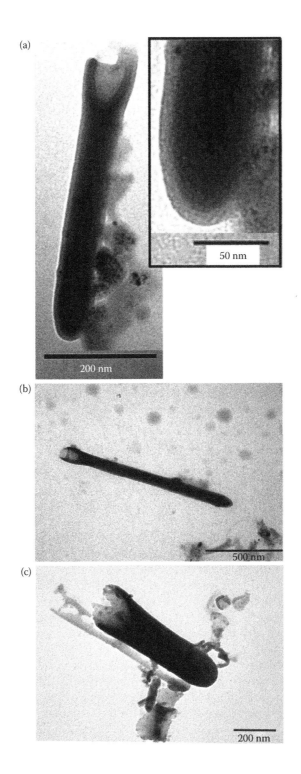

FIGURE 6.10
(a) Transmission electron micrograph of a prepared nano test tube. The inset shows a close-up of the closed end of this nano test tube. (b, c) Transmission electron micrographs of nano test tubes prepared in membranes with different pore dimensions, demonstrating the variability in nano test tube size that can be templated.

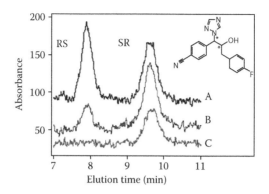

FIGURE 6.11
Chiral HPLC chromatograms for racemic mixtures of FTB before (A) and after (B, C) extraction with 18 mg/
mL of 200 nm Fab-containing nanotubes. Solutions were 5% dimethyl sulfoxide in sodium phosphate buffer,
pH 8.5.

We have further demonstrated the ability to effect biocatalytic transformations with
these modified nanotubes. The enzyme GOD is immobilized on the interior and exterior of the silica nanotubes using the same aldehyde silane coupling procedure used for
the Fab fragments. The GOD nanotubes are suspended in a solution of glucose (90 mM)
and the activity is assayed using a standard dianisidine-based assay. A GOD activity of
0.5 ± 0.2 units/mg is determined. When the nanotubes are filtered from solution, oxidation stopped, indicating that the enzyme does not leach from the nanotubes and that the
enzyme retains biochemical activity when immobilized on the nanotubes.

6.3.3 Self-Assembly with Nano- and Microtubes

We have also reported a method for preparing materials using template synthesis that can
be self-assembled through modification with biomolecular recognition elements, namely
streptavidin and biotin [12]. Using a modified deposition procedure, nanowires comprised
of poly[N-(2-aminoethyl)-2,5-di(2-thienyl)pyrrole] (poly(AEPy) or of poly(AEPy)) coated with
a thin film of gold could be produced. The membrane was then soaked in a solution of bioti-
nyl-N-hydroxysuccinimide, resulting in coupling of biotin to the amine-containing polymer
through amide bonds only at the tip of the nanowires. The array was then soaked in a solution
of polystyrene beads coated with streptavidin (Spherotech). Scanning electron micrographs
(SEMs) of such experiments are shown in Figure 6.12. In Figure 6.12a and b, membranes
with nanowires of poly(AEPy) before membrane dissolution are shown. In Figure 6.12a, the
membrane has been biotinylated, whereas in Figure 6.12b the membrane has not. In both
instances, the membranes are exposed to streptavidin-coated spheres, but only in the case
of the biotinylated membrane, specific irreversible adsorption is observed. In Figure 6.12c,
free standing (membrane dissolved) gold-coated biotinylated poly(AEPy) nanowires with
streptavidin particles assembled specifically at the tip of the tubes are shown.
    The examples given above demonstrate several important aspects of template synthe-
sis in the context of biotechnology. First, the activity of templated biochemical species,
such as enzymes, can be retained. Second, biochemical recognition can be used for diverse
functions, such as the assembly of templated materials or the separation of stereoisomers.
Finally, templated materials of sizes appropriate for drug delivery applications can be
synthesized. We believe that the template synthesis method affords the ability to prepare
materials with unique properties, such as biodegradability, biocompatibility, ruggedness,

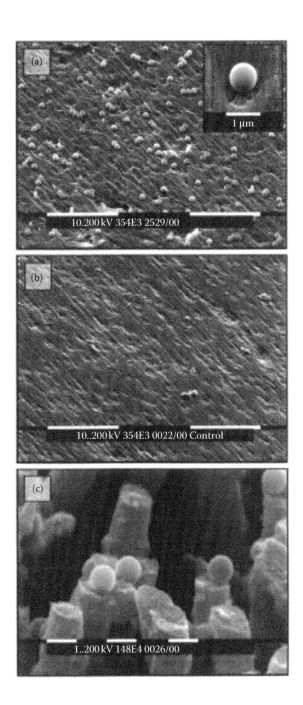

FIGURE 6.12
Scanning electron micrographs: (a) the surface of a poly(AEPy) microwire-containing membrane after self-assembly of the latex particles to the ends of the microwires; the inset shows a higher magnification image of a single microwire/latex assembly; (b) the surface of an analogous membrane treated in the same way as in panel a but omitting the biotinylation step; and (c) Au/poly(AEPy) concentric tubular microwires after dissolution of the template membrane and self-assembly.

size, and functionality. The ability to control these and other biomaterial design parameters will allow the investigation of new nanometer scale materials for biological applications.

6.4 Biochemical Separations with Nanotube Membranes

Nanotube membranes may be used in biochemical separations by placing a mixture of molecules on one side of a membrane (feed side); selected molecules then translocate the membrane to the opposite side (permeate side) either by passive diffusion or a diffusion-assisted process, such as electrophoresis. Selectivity of nanotube membranes for species in the feed solution can be achieved through several means, including the size of the nanotubes in the membrane, the surface charge of the nanotubes in the membrane, and the addition of selective complexing agents to the nanotubes of the membrane. We have used these methods, and others, to create membranes for separating species of biochemical interest, including ions, small molecules, proteins, and nucleic acids. We have used both alumina and polymer membranes in a variety of configurations. Experiments related to biochemical separations will be discussed in this section.

6.4.1 Separation of Proteins by Size

The simplest and most straightforward use of nanotube membranes for separation uses the size of the inner diameter (i.d.) of the nanotube to physically select for a molecule based on its hydrodynamic radius, in this case the molecules being separated are proteins [13]. Polycarbonate membranes of 6 μm thick with pores either 30 or 50 nm in diameter are plated with gold using electroless deposition. After deposition, the i.d. of the gold-plated 50 nm pore membrane is 45 nm. Membranes with smaller diameter pores are obtained after plating the 30 nm pore membranes. The gold-plated membranes are soaked 6 days in a 1 mM solution of a thiol-terminated poly(ethylene glycol) (PEG-thiol, MW = 5000 kDa). Previous measurements have shown that the thin film formed from this PEG-thiol is approximately 2.4 nm. The PEG-thiol film is used to prevent nonspecific adsorption to (and the resulting pore blockage of) the membrane surfaces.

Gold nanotube membranes with PEG-thiol monolayers are placed in a U-tube permeation cell and solution from the feed side of the membrane is forced through the cell by applying 20 psi pressure. The concentration of the protein is monitored by periodically sampling the permeate side of the U-tube using UV–vis spectroscopy. The results of permeation experiments with single protein solutions of lysozyme (Lys, MW = 14 kDa) and bovine serum albumin (BSA, MW = 67 kDa) through a 40 nm i.d. nanotube membrane are shown in Figure 6.13.

The Stokes radii for BSA and Lys are 3.6 and 2 nm, respectively. Using the Stokes–Einstein equation, the calculated diffusion coefficient for Lys should be 1.8 times higher than BSA. From the data in Figure 6.13, a much higher flux for Lys relative to BSA is observed. The higher flux is a consequence of the smaller size of Lys relative to BSA, as BSA is physically hindered from translocating the membrane.

Experiments with two proteins in the feed solution were also performed. In this case, the experiments are carried out in a similar fashion, except the feed side of the membrane contained an equimolar ratio of Lys and BSA, and the results were monitored using HPLC. The results for the two protein Lys and BSA permeation experiments as a function of nanotube

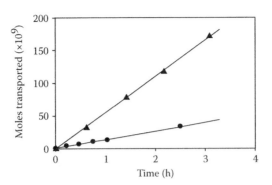

FIGURE 6.13
Plots of moles transported versus time for Lys (upper) and BSA (lower) across a 40 nm Au i.d. nanotube membrane.

i.d. are shown in Figure 6.14. In Figure 6.14a, the HPLC data of the initial feed solution is shown. Figure 6.14b shows HPLC data of the permeate solution after transport through a 45 nm i.d. nanotube membrane. In analogy to the single-protein permeation experiments shown in Figure 6.13, the permeation of BSA is hindered relative to Lys, as observed by the diminished BSA peak. Figure 6.14c shows the HPLC of the permeate solution of a 30 nm i.d. nanotube membrane; in this case, the BSA peak is highly attenuated, indicating a small relative amount of BSA has permeated. In the case of a 20 nm i.d. nanotube membrane, Figure 6.14d, there is no detectable permeation of BSA. (While it is unlikely there is no BSA present in the permeate solution, the amount of BSA present is below the detection limit.) The selectivity coefficient, defined as the ratio of the concentration of Lys to the concentration of BSA present in the permeate solution, is a quantifiable measurement of nanotube membrane

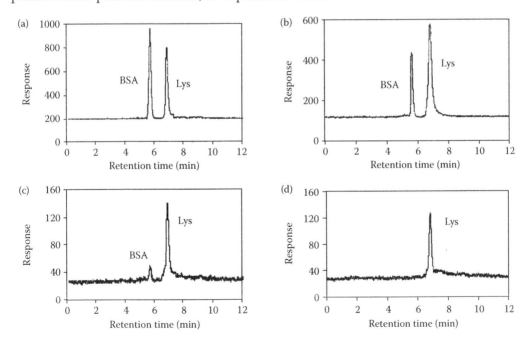

FIGURE 6.14
HPLC data for two-protein (Lys and BSA) permeation experiments. (a) Feed solution. Permeate solutions after transport through (i.d.) 45 nm (b), 30 nm (c), and 20 nm (d) nanotube membranes.

performance. The selectivity coefficient increases with decreasing nanotube diameter, from ≥ 20 to 13 to 2.2 for nanotube i.d.s of 20, 30, and 45 nm, respectively. There is a trade-off for increased selectivity, however, as flux decreases with decreasing nanotube i.d., resulting higher selectivity but lower productivity for the smaller i.d. nanotube membranes.

6.4.2 Charge-Based Separation of Ions

In the case of the experiments just discussed, we have used a chemisorbed thiol to prevent nonspecific adsorption to the membrane and nanotube walls. We have also demonstrated this same strategy, adsorption of a functional thiol can induce chemical selectivity to gold-plated nanotube membranes. Early experiments showed that the adsorption of a hydrophobic or hydrophilic thiol could promote transport of a chemical species based on the hydrophobicity of the permeate molecule [6,14]. In a similar manner, we have demonstrated that the chemisorption of L-cysteine (through the sulfur-bearing side chain) to the inside of the nanotube walls affords a method of separating ionic species based on charge, creating a pH-switchable ion-transport selective membrane [15]. A schematic representation of this is shown in Figure 6.15. Polycarbonate membranes of 6 μm thick with 30 nm diameter pores are plated with gold to varying inner diameters. L-Cysteine is chemisorbed to the gold nanotube walls by soaking the plated membranes in a 2 mM solution overnight. The chemical structures and molecular volumes of species separated, methylviologen (MV^{2+}), 1,5-naphthalene disulfonate (NDS^{2-}), 1,4-dimethylpyridinium iodide (DMP^+), and picric acid (Pic^-), are shown in Figure 6.16. Cysteine-modified gold nanotube membranes are mounted in a U-tube for permeation experiments. The transport of species across the membrane as a function of time is monitored using UV–vis spectroscopy at molecule-appropriate wavelengths.

The effect of pH on the transport of cations MV^{2+} and DMP^+ through the L-cysteine-modified gold nanotubes is shown in Figure 6.17. At pH $= 2$, the nanotube walls are positively charged, resulting in low cation flux due to electrostatic repulsion of the like-charged cations. In the case of pH $= 12$, the nanotube walls are negatively charged and a high cation flux is observed. At pH $= 6$, close to the isoelectric point of cysteine, an intermediate flux is observed. In the case of transport of anions, NDS^{2-} and Pic^-, the opposite effects are observed (not shown). At pH $= 12$, the nanotube walls are negatively charged, resulting in low anion flux due to electrostatic repulsion of the like-charged anions. In the case of pH $= 2$, the nanotube walls are positively charged and a high anion flux is observed. Again at pH $= 6$, close to the isoelectric point of cysteine, an intermediate flux is observed. The results of these experiments in terms of nanotube i.d. and flux (Table 6.1) and selectivity coefficient, as defined by the ratio permeate transported as a function of pH (Table 6.2), are shown. A detailed investigation of the mechanism of the observed pH transport properties determined two electrostatic effects responsible for the selectivity observed. One electrostatic effect, an electrostatic accumulation effect, occurs when the permeate ion has a charge opposite to the charge on the nanotube wall. A second electrostatic effect, an electrostatic rejection effect, occurs when the permeate ion has the same charge as the charge on the nanotube walls. These experiments clearly demonstrate the ability to design membranes selective for ionic species, in this case, using an chemisorbed amino acid.

6.4.3 Separations Using Molecular Recognition

6.4.3.1 Enzymatic Molecular Recognition

One of the earliest examples of biochemical separations with a nanotube membrane uses enzymes immobilized in a polymeric membrane as a selective molecular recognition

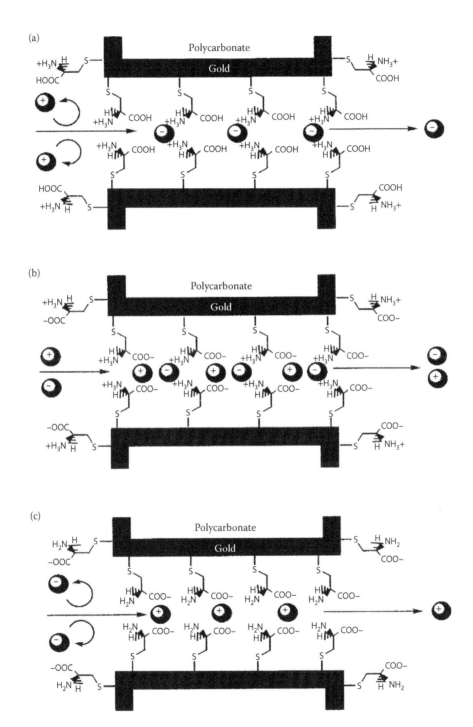

FIGURE 6.15
Schematic representation showing the three states of protonation and the resulting ion-permselectivity of the chemisorbed cysteine. (a) Low pH, cation-rejecting/anion-transporting state. (b) pH 6.0, non-ion-permselective state. (c) High pH, anion-rejecting/cation-transporting state. *Note*: Ion transport in one direction (e.g., anions from left to right in A) is balanced by an equal flux of the same charge in the opposite direction, so that electroneutrality is not violated in the two solution phases.

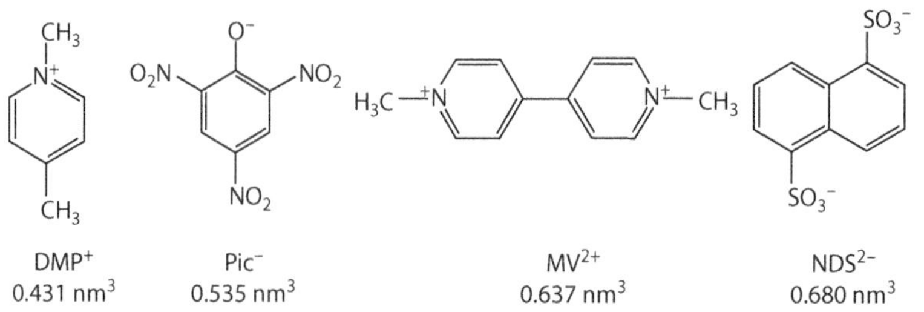

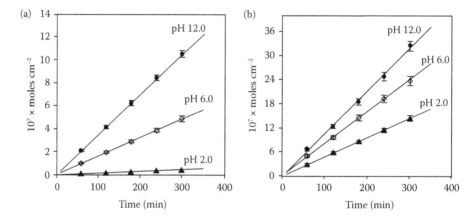

FIGURE 6.16
Chemical structures and molecular volumes for the permeate ions that were investigated.

FIGURE 6.17
Permeation data for (a) MV^{2+} and (b) DMP^+ at various pH values for the 1.4 nm i.d. Au nanotubule membrane that was modified with L-cysteine. The error bars represent the maximum and minimum values obtained from three replicate measurements.

TABLE 6.1

Flux Data

Permeate Ion	Nanotubule i.d. (nm)	Flux ($\times 10^7$ mol/cm²/h)		
		pH 2.0	pH 6.0	pH 12.0
DMP^+	0.9	0.25	0.66	1.6
Pic^-	0.9	0.89	0.44	0.22
MV^{2+}	0.9	0.018	0.18	0.40
NDS^{2-}	0.9	0.12	0.042	0.030
DMP^+	1.4	2.9	4.8	6.5
Pic^-	1.4	4.7	3.6	2.6
MV^{2+}	1.4	0.086	0.98	2.1
NDS^{2-}	1.4	1.7	0.75	0.14
MV^{2+}	1.9	1.5	6.8	15
NDS^{2-}	1.9	14	6.3	2.0
MV^{2+}	3.0	3.1	11	21
NDS^{2-}	3.0	20	11	3.4

TABLE 6.2

$\alpha_{pH\ 12/pH\ 2}$ and $\alpha_{pH\ 2/pH\ 12}$ Values

		Selectivity Coefficient	
Permeate Ion	Nanotubule i.d. (nm)	$\alpha_{pH\ 12/pH\ 2}$	$\alpha_{pH\ 2/pH\ 12}$
DMP⁺	0.9	6.4	
Pic⁻	0.9		4.0
MV²⁺	0.9	22	
NDS²⁻	0.9		4.0
DMP⁺	1.4	2.2	
Pic⁻	1.4		1.8
MV²⁺	1.4	24	
NDS²⁻	1.4		12
MV²⁺	1.9	10	
NDS²⁻	1.9		6.8
MV²⁺	3.0	6.9	
NDS²⁻	3.0		5.9

agent [16]. The membrane used for this separation is a 10 μm thick polycarbonate membrane with 400 nm diameter pores. A cartoon of the final-modified membrane is shown in Figure 6.18. To modify the membrane for biochemical separations, a thin gold film is then sputtered across one face of the membrane. This sputtered film is too thin to close the membrane pores, but is thick enough to provide a conductive electrode layer. This electrode is then used to electropolymerize a thin (ca. 100 nm) polypyrrole layer, forming plugs of polypyrrole that were porous enough for solvent molecules to permeate, but were not porous enough for larger enzymes to permeate. A thin gold film was then sputtered on the other side of the membrane and a solution of an apoenzyme is vacuum filtered through the membrane from the open to closed end. An apoenzyme is chosen as a molecular recognition agent, because without the addition of a cofactor, substrate molecules would not be catalyzed by the apoenzyme, allowing the substrate to be selected for without chemical conversion. After the membrane is loaded with an apoenzyme, a layer of polypyrrole is

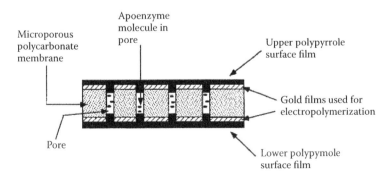

FIGURE 6.18
Schematic cross section of the polypyrrole–polycarbonate–polypyrrole sandwich membrane with the apoenzyme entrapped in the pores. The membrane is drawn as coming out of the plane of the paper. The various components are not drawn to scale.

electropolymerized across the top layer of the membrane, encapsulating the apoenzyme in a porous matrix permeable to solvent and substrate molecules.

Membranes modified with alcohol dehydrogenase apoenzyme (apo-ADH) are mounted in a U-tube permeation cell. The membranes are then subjected to pure and mixed solutions of ethanol and phenol. (Ethanol is a substrate for apo-ADH, but phenol is not.) The results of transport experiments with this membrane and a control membrane with no apoenzyme loaded are shown in Figure 6.19. In Figure 6.19a, the control versus the apo-modified membrane, it is clear that the amount of ethanol transported by the apo-ADH-modified membrane is higher than the unmodified membrane. In Figure 6.19b, the transport of ethanol and phenol with the apo-ADH-modified membrane is shown. The ratio of the slopes of the flux of ethanol and phenol yields a selectivity coefficient of 9.2 for ethanol. The selectivity from a mixed solution (shown in Figure 6.19b) is analogous to the selectivity obtained when transport experiments of the individual molecules were performed. Membranes are also modified with a variety of other apoenzymes, including

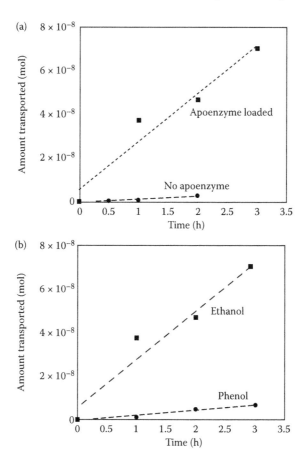

FIGURE 6.19
(a) Plots of amount of ethanol transported from the feed solution through the membrane and into the permeant solution versus time for a membrane loaded with apo-ADH and for an apo-ADH-free membrane. The feed solution was 0.5 mM in both ethanol and phenol. The slopes of these lines provide the ethanol flux across the membrane. (b) Plots of amount of ethanol (substrate) and phenol (nonsubstrate) transported versus time for the apo-ADH-loaded membrane. Feed solution as (a). The ratio of the slopes provides the selectivity coefficient for ethanol versus phenol transport.

apo-aldehyde dehydrogenase and apo-ᴅ-amino acid oxidase (apo-ᴅ-AAO). Apo-ᴅ-AAO binds only to ᴅ-amino acids, allowing us to interrogate the ability of this type of membrane to separate enantiomers. In these enantioselective membranes, a selectivity coefficient of 3.3 is obtained for ᴅ-phenylalanine versus ʟ-phenylalanine. By using smaller pores, this selectivity coefficient could be increased to 4.9, due to an increase in the amount of permeate transported through facilitated mechanisms as compared to permeate transported through passive diffusion.

6.4.3.2 Chiral Separation

We have also used antibody-modified alumina membranes to perform enantiomeric separations of a drug molecule [17]. In these experiments, alumina membranes with initial pore diameters of 20 and 35 nm are used. A sol–gel method (similar to that described previously) is used to deposit silica nanotubes in the pores of the membrane. The silica nanotubes are then modified with an aldehyde silane. Biochemical recognition is then incorporated into this membrane by coupling the aldehyde groups of the silanes with primary amines present in antibody Fab fragments. Antibodies are selected that bind the drug 4-[3-(4-fluorophenyl)-2-hydroxy-1-[1,2,4]triazol-1-yl-propyl]-benzonitrile, an inhibitor of aromatase enzyme activity. This molecule has two chiral centers, yielding four possible isomers: RR, SS, SR, and RS. Fab fragments (anti-RS) of this antibody that selectively bind the RS relative to the SR form of the drug are used to modify membranes.

A racemic mixture of the drug molecule is placed on one side of a U-tube permeation cell and the flux of each species is monitored as a function of time by periodically monitoring the concentration of each enantiomer present in the permeate solution with a chiral chromatographic method. A selectivity of 2 was obtained for the RS relative to the SR enantiomer, indicating that the membranes transport the RS form twice as fast as the SR form. A facilitated transport mechanism was determined to be responsible for transport in these membranes. As in the case of the apoenzyme-modified membranes, by decreasing the pore diameter the selectivity coefficient is increased to 4.5 (at the expense of lower total flux). It was also found that by adding dimethyl sulfoxide (DMSO) to the feed and permeate solutions in concentrations from 10% to 30%, the rate of transport for the RS form of the drug could be regulated. This occurs because DMSO weakens the affinity of the anti-RS Fab fragment for the RS enantiomer. Thus, at 30% DMSO content the relative transport rates for the RS and SR enantiomers were essentially equal. Because antibodies can be developed for a wide variety of species of biochemical interest, this method should be highly adaptable to a wide variety of targets.

6.4.3.3 Separation of Nucleic Acids

We have also used nanotube membranes to perform separation of DNA with single-base mismatch selectivity [18]. In these experiments, 6 μm thick polycarbonate membranes with 30 nm diameter pores are coated with gold using electroless deposition. The diameter of the pores after gold deposition is determined to be 12 ± 2 nm. Linear DNA or hairpin DNA are used as the molecular recognition agent in these experiments. DNA hairpins contain complementary sequences at each end of the molecule, and under appropriate conditions form a stem–loop structure. As a result of this structure, hybridization of complementary DNA is very selective, in optimal cases a single-base mismatch will not hybridize. A 30 base DNA hairpin with a thiol modification at the 5′ end allowed facile chemisorption of the molecular recognition agent to the gold-coated nanotubes. The six bases at each end of the DNA strand were complementary, forming the stem, with the loop comprised of the remaining 18 bases in the middle of the DNA strand. The thiol-modified linear DNA molecular recognition modifiers used the same 18 bases in the middle of the molecule, but the six bases at each end were not complementary, thus these linear sequences do not form the stem–loop structure. DNA molecules to transport are 18 bases long and are either perfect complements to the bases in the loop or contained one or more mismatches.

DNA-modified membranes are mounted in a U-tube permeation cell and molecules to transport are added to the feed side of the membrane. Transport is monitored by measuring the UV–vis absorbance of the permeate solution as a function of time. These systems also demonstrated a facilitated transport mechanism for complementary sequences of DNA. In the case of linear DNA, the selectivity coefficient for perfect complement DNA (PC-DNA) versus single-base mismatch DNA is 1, that is to say there was no selectivity. PC-DNA versus a seven-base mismatch showed a selectivity coefficient of 5. In the case of hairpin DNA-modified membranes, transport plots of PC-DNA through a modified and unmodified membrane are shown in Figure 6.20. In Figure 6.20a, the flux of DNA through an unmodified membrane is significantly lower than transport through the membrane modified with a perfectly complementary hairpin DNA. In Figure 6.20b, the Langmuirian

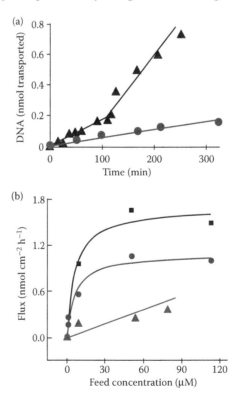

FIGURE 6.20
(a) Transport plots for PC-DNA through gold nanotube membranes with (blue triangles) and without (red circles) the immobilized hairpin-DNA transporter. The feed solution concentration was 9 μM. (b) Flux versus feed concentration for PC-DNA. The data in red and blue were obtained for a gold nanotube membrane containing the hairpin-DNA transporter. At feed concentrations of 9 μM and above, the transport plot shows two linear regions. The data in blue (squares) were obtained from the high slope region at longer times. The data in red (circles) were obtained from the low slope region at shorter times. The data in pink (triangles) were obtained for an analogous nanotube membrane with no DNA transporter.

shape characteristic of facilitated transport is observed for the PC-DNA, whereas diffusive flux is observed for the membrane with no DNA modification. In the case of hairpin DNA molecular recognition elements, a selectivity coefficient of 3 is obtained for a PC-DNA sequence versus a single-base mismatch sequence. A selectivity coefficient of 7 is obtained for a PC-DNA sequence versus a seven-base mismatch.

Nanotube membranes have shown the ability to separate an amazingly diverse field of biochemical species, from DNA to proteins to drug molecules. The selectivity in each of these separations is governed by the inherent selectivity in the immobilized biochemical species used to effect recognition or through physical properties of the nanotubes themselves.

6.5 Toward Nanotube Membranes for Biochemical Sensors

Many of the principles of biochemical sensing with nanotube membranes are inspired by the results obtained with separations using such membranes. The small, often molecular, sizes of the nanotubes prepared offer new approaches to bioanalytical chemistry at the nanometer scale. We have previously described composite membranes with thin polymer skins that function as chemical sensors. In this chapter, we will discuss our results with nanotube membranes that function as ion channel mimics. These experiments are our first step toward constructing nanotube-based biochemical sensors that function in a manner analogous to biological channels.

6.5.1 Ligand-Gated Membranes

Ligand-gated ion channels in biochemical systems respond to an external chemical stimulus by switching between an off (no current or low current) and on (high current) state [8]. We have created synthetic nanotube membranes that can mimic the function of natural ligand-gated ion channels. Our ion channel mimics start in the low or no current state and convert to the high current state in the presence of the appropriate analyte, in analogy to the functioning of acetylcholine-gated channels found in nature. Alumina membranes of 60 μm thick with 200 nm diameter pores were modified with an octadecyl silane or gold-coated alumina membranes were modified with an octadecyl thiol. This creates a highly hydrophobic membrane that does not wet when placed in water. When the membrane is mounted in a U-tube permeation cell and a transmembrane potential is applied, the hydrophobicity of the pores results in the passage of zero or very low currents, effectively an off state. Initial experiments using an ionic surfactant, dodecylbenzene sulfonate, showed that when 10^{-6}–10^{-5} M were added to one side of the membrane it partitions into the pore. This creates a more hydrophilic environment inside the pore, allowing the pores to wet. This wetting results in a dramatic drop in resistance and the passage of a measurable current, effectively an on state.

These ligand-gated ion channel mimics can also be used to detect drug molecules. In these experiments the effects of the hydrophobicity of three drug molecules, bupivacaine, amiodarone, and amitriptyline on the observed transmembrane resistances were investigated. The hydrophobicity of these molecules, a function of molecular weight and polarity, increases in the following order: bupivacaine < amitriptyline < amiodarone. If the hydrophobic nature of these molecules is responsible for the partitioning of these molecules into the membrane, and thus turning on the current, then transition from the off to on state of the membrane would occur at the lowest concentration of amiodarone. This is experimentally observed (Figure 6.21). Bupivacaine is the least hydrophobic of these compounds, and it is also observed experimentally that bupivacaine requires the highest concentration to effect gating from off to on.

6.5.2 Voltage-Gated Conical Nanotube Membranes

In addition to ligand-gated ion channels, we have also mimicked the properties of voltage-gated ion channels [19]. In these studies, we have used polymer membranes with a single pore. The single pore membranes are prepared either by isolating individual pores in low-density tracked films, or by using films with a single damage track. This approach allows us to investigate the properties of a single nanopore rather than the ensemble of pores present in conventional membranes. By applying a transmembrane current, we are able to monitor the flow of ionic currents through the pore analogous to ion channels in lipid bilayers using traditional patch clamp techniques. Current can be monitored as a function of time or as a function of applied voltage. Pores used for these studies are anisotropically etched to create conical pores, rather than cylindrical pores. The use of conical pores lowers the total resistance of the pore, allowing higher currents to flow, while retaining the nanometer dimension at the tip of the conical pore. Single conical pore membranes used in these studies have been plated with gold through electroless deposition to permit the chemisorption of functionalized thiols that enable us to control the surface chemistry of the nanotube walls.

In the first study, 12 μm thick poly(ethylene terephthalate) membranes with a single damage track were obtained from GSI (Darmstadt, Germany). The track was anisotropically etched using a basic solution on one side and an acidic stopping medium on the other side. This results in the formation of a conical pore. By controlling the etching time and concentrations of base and acid, pores with nominal cone tips 20 nm in diameter and cone bases 600 nm in diameter can be obtained. Conical pores are then plated with gold, forming

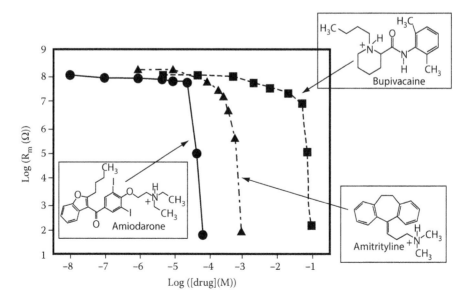

FIGURE 6.21
Plots of log membrane resistance versus log[drug] for the indicated drugs and a C18-modified alumina membrane.

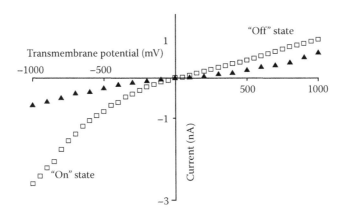

FIGURE 6.22
I–V curves in 0.1 M KCl (□) and 0.1 M KF (▲).

conical gold nanotubes. After plating, the small diameter (cone tip) of the pore was nominally 10 nm. The membrane was then mounted in a conductivity cell with a solution of 0.1 M KCl on both sides of the half-cell. In the case of a bare gold membrane (Figure 6.22), ionic currents are rectified, creating a two state system. At negative potentials, the pore is *on* whereas at positive potentials the pore is *off*. This phenomenon is observed due to the adsorption of Cl⁻ to the walls of the gold nanotube, creating a high negative charge at the nanotube surface. When the solution is changed from 0.1 M KCl to 0.1 M KF, rectification is not observed (Figure 6.22). This is due to the fact that F⁻ does not adsorb to gold, as Cl⁻ does.

The effect of charge on the nanotube walls was further investigated by measuring current–voltage curves of nanotubes with chemisorbed 2-mercaptopropionic acid (Figure 6.23) or mercaptoethyl ammonium (not shown) to respective nanotube membranes. In the case of 2-mercaptopropionic acid, the carboxylate group can be protonated or deprotonated by varying the solution pH. At pH = 6.6, the carboxylic acids are deprotonated, resulting in a negatively charged nanotube surface. Current–voltage curves at this pH showed rectification, similar to that observed in the case of Cl⁻ adsorbed to bare gold nanotubes. When the pH was lowered to 3.5, the carboxylic acid groups are protonated, removing the negative charge at the surface. Current–voltage curves at this pH showed no rectification, as observed when KF was used as the electrolyte. By using mercaptoethyl ammonium, a positively charged cation, current rectification can be reversed, meaning that at positive potentials higher current is passed, the nanotube is *on* and at negative potentials, low current is passed, the nanotube is *off*. A detailed model of the mechanism of rectification based on the formation of an electrostatic trap that arises due to the inherent asymmetry in charged conical pores was developed to explain the observed current–voltage curves and rectification.

6.5.3 Electromechanically Gated Conical Nanotube Membranes

In an effort to design more sophisticated biomimetic conical nanotubes, we have constructed single conical nanotubes with a built-in electromechanical mechanism that controls rectification of ionic currents based on the movement of charged DNA strands [20]. In these experiments, low-density tracked polycarbonate membranes were anisotropically etched to form conical nanopores. Membranes were masked in a manner that allowed the isolation and characterization of a single conical nanotube. Figure 6.24 shows SEM images

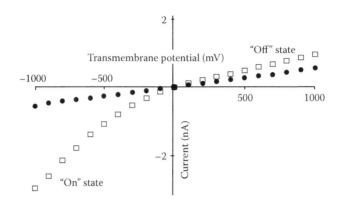

FIGURE 6.23
I–V curves in 0.1 M KF for gold nanotubes modified with 2-mercaptopropionic acid; pH = 6.6 (□) and pH = 3.5 (•).

of the large opening of a pore (a) and the small opening of a pore (b). In Figure 6.24c, a SEM image of a gold replica of a prepared pore is shown that demonstrates the conical geometry of the conical nanopore. Conical nanopores were plated with gold through electroless deposition, forming membranes possessing a single conical gold nanotube. After plating, conical gold nanotubes with small-diameter radii between 13 and 100 nm were obtained (Table 6.3). Thiolated DNA strands of varying base pair length and sequence were then chemisorbed to the surface of the gold nanotube. The DNA nanotubes prepared show an *off* state (low currents at positive potentials) and an *on* state (high currents at negative potentials) (Figure 6.25). We propose the rectification observed is due to electrophoretic movement of the DNA chains into (off state, Figure 6.25c) and out of (on state, Figure 6.25d) the nanotube mouth. The movement of the DNA chains into the nanotube mouth results in occlusion of the nanotube orifice, resulting in a higher ionic resistance. In Figure 6.25, the effect of chain length on rectification can be clearly observed. That is to say, as DNA chain length increases, the extent of rectification increases. It was found that an optimal length of DNA induces rectification based on the diameter of the small end of the nanotube. This work demonstrated the first example of a simple chemical (DNA chain length) or physical (nanotube pore size) method to control the extent of rectification of an artificial ion channel.

Studies of nanotubes and conical nanotubes that function as artificial ion channels are a relatively new endeavor in bioanalytical chemistry. We expect future applications of

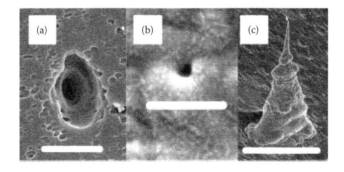

FIGURE 6.24
Electron micrographs showing (a) large-diameter (scale bar = 5.0 μm) and (b) small-diameter (scale bar = 333 nm) opening of a conical nanopore, and (c) a liberated conical Au nanotube (scale bar = 5.0 μm).

TABLE 6.3

Nanotube Mouth Diameter (d), DNA Attached, r_{max}, Radius of Gyration of DNA (r_g), and Extended Chain Length (l)

d (nm)	DNA Attached	r_{max}	r_g (nm)	l (nm)[a]
41	12-mer	1.5	1.4	5.7
46	15-mer	2.2	1.6	6.9
42	30-mer	3.9	2.9	12.9
38	45-mer	7.1	4.0	18.9
98	30-mer	1.1	2.9	12.9
59	30-mer	2.1	2.9	12.9
39	30-mer	3.9	2.9	12.9
27	30-mer	11.5	2.9	12.9
13	30-mer	4.7	2.9	12.9
39	30-mer hairpin	1.4	n.a.	6.9

[a] Includes the $(CH_2)_6$ spacer.

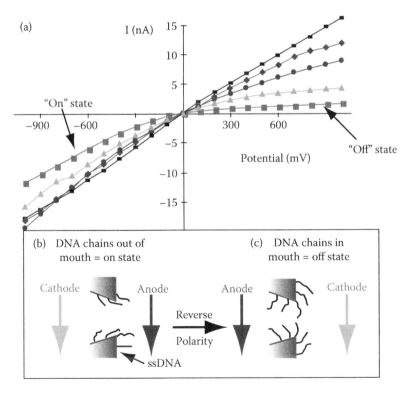

FIGURE 6.25
(a) *I–V* curves for nanotubes with a mouth diameter of 40 nm containing no DNA (black) and attached 12-mer (blue), 15-mer (red), 30-mer (green), and 45-mer (orange) DNAs. (b and c) Schematics showing electrode polarity and DNA chain positions for on (b) and off (c) states.

nanotube membranes to include highly sensitive and selective chemical sensors based on the design principles of mother nature.

6.6 Future Outlook

In this chapter, we have described our work related to nanotube-based membrane systems in biologically oriented or inspired settings. The ability to tune the material, size, and surface chemistries of the nanotubes affords a flexible venue to address a host of questions and problems at the forefront of bionanotechnology. Smart nanodelivery systems, artificial ion channels, and separations platforms with unique selectivity are some of the immediate questions we and others seek to address. An ultimate goal of nanotube membranes is to match or exceed the performance of transmembrane proteins found in living systems.

While these technologies are largely tools available to the experimental nanotechnologist, mass production of templated materials is certainly possible. There is much to be done to optimize and expand the techniques of membrane-based nanotubes. Eventually, we hope to transition these important nanoscale systems to the biotechnology community in general.

Acknowledgments

C.R.M. and L.A.B. wish to acknowledge past and present group members whose work has contributed to this chapter. Aspects of this work have been funded by the National Science Foundation, Office of Naval Research, and DARPA.

References

1. Bayley, H., and C.R. Martin. 2000. *Chem Rev* 100:2575–2594.
2. Martin, C.R., and P. Kohli. 2003. *Nat Rev Drug Discov* 2:29–37.
3. Martin, C.R. 1994. *Science* 266:1961–1966.
4. Kang, M., S. Yu, N. Li, and C.R. Martin. 2005. *Small* 1:69–72.
5. Sides, C.R., and C.R. Martin. Unpublished results.
6. Martin, C.R., M. Nishizawa, K. Jirage, and M. Kang. 2001. *J Phys Chem B* 105:1925–1934.
7. Hulteen, J.C., and C.R. Martin. 1997. *J Mater Chem* 7:1075–1087.
8. Steinle, E.D., D.T. Mitchell, M. Wirtz, S.B. Lee, V.Y. Young, and C.R. Martin. 2002. *Anal Chem* 74:2416–2422.
9. Parthasarathy, R., and C.R. Martin. 1994. *Nature* 369:298–301.
10. Gasparac, R., P. Kohli, M.O.M. Paulino, L. Trofin, and C.R. Martin. 2004. *Nano Lett* 4:513–516.
11. Mitchell, D.T., S.B. Lee, L. Trofin, N. Li, T.K. Nevanen, H. Soederlund, and C.R. Martin. 2002. *J Am Chem Soc* 124:11864–11865.
12. Sapp, S.A., D.T. Mitchell, and C.R. Martin. 1999. *Chem Mater* 11:1183–1185.

13. Yu, S., S.B. Lee, M. Kang, and C.R. Martin. 2001. *Nano Lett* 1:495–497.
14. Hulteen, J.C., K. Jirage, and C.R. Martin. 1997. *J Am Chem Soc* 120:6603–6604.
15. Lee, S.B., and C.R. Martin. 2001. *Anal Chem* 73:768–775.
16. Lakashmi, B.B., and C.R. Martin. 1997. *Nature* 338:758–760.
17. Lee, S.B., D.T. Mitchell, L. Trofin, T.K. Nevanen, H. Soederlund, and C.R. Martin. 2002. *Science* 296:2198–2200.
18. Kohli, P., C.C. Harrell, Z. Cao, R. Gasparac, W. Tan, and C.R. Martin. 2004. *Science* 305:984–986.
19. Siwy, Z., E. Heins, C.C. Harrell, P. Kohli, and C.R. Martin. 2004. *J Am Chem Soc* 126:10850–10851.
20. Harrell, C.C., P. Kohli, Z. Siwy, and C.R. Martin. 2004. *J Am Chem Soc* 126:15646–15647.

7

Nanoimaging of Biomolecules Using Near-Field Scanning Optical Microscopy

Musundi B. Wabuyele and Tuan Vo-Dinh

CONTENTS

7.1 Introduction

In recent years, there have been significant advances in the development of microscopic techniques with high spatial resolution that are essential for a wide range of biological applications. The need for tools that are capable of locating, characterizing, and distinguishing features at a nanoscale level has been the driving force in the growing field of nanoimaging. The advent of scanning probe microscopy, such as scanning tunneling microscopy (STM) (1,2), atomic force microscopy (AFM) (3–10), scanning confocal microscopy, and near-field scanning optical microscopy (NSOM) (11–16), has enabled imaging of biomolecules at nanometer resolutions.

Although SPM techniques provide atomic-scale resolution, which can be used to manipulate atoms and nanostructures, they lack the chemical specificity for molecular characterization. For this reason, they are not suitable for observation of spectral and dynamic properties required in imaging. Therefore, only topographical information can be obtained from AFM since the acquired images are based on the force interaction between the tip and the surface of the sample. This limitation, however, has recently been addressed with the development of AFM probes modified with bioreceptors that are specific for certain molecules. Alternatively, STM techniques can be used but samples must be conductive.

A promising instrumental approach that combines a high-resolution scanning probe and optical microscopy has been developed.

7.1.1 Nanoimaging

Over the past decade, AFM and NSOM have evolved into new frontiers of science with significant impact in various areas of research. Several investigations applying these techniques have been extensively reported in literature. AFM produces high-resolution topography with unique features applicable to biological systems (17–20), and can be used to obtain two- and three-dimensional images of a wide variety of biological samples, such as living cells, DNA molecules, and proteins, as shown in Figure 7.1. In addition, the relative heights of the structural features on the surfaces of objects obtained by AFM enable quantitative study of surface modifications (21–24). NSOM on the other hand is an imaging technique that combines high-resolution scanning probe microscopy and fluorescence microscopy (25–27). Recently, NSOM has increasingly been used for biological applications to visualize biological structures (28–32) and monitor interaction or intermolecular dynamics at the single molecule level on cells (26,33). Localization of proteins within the substructure and cellular organelle allows one to have a better understanding of protein structure and their functions. Enderle

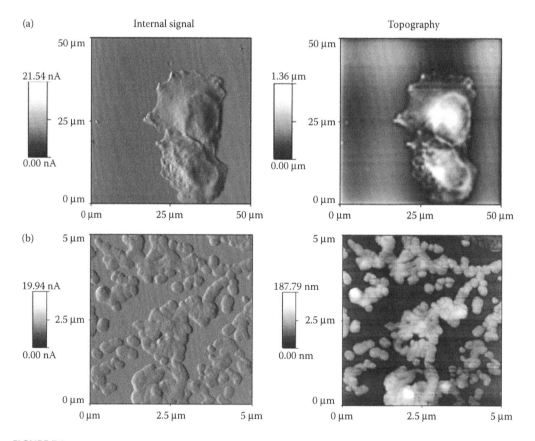

FIGURE 7.1
AFM images (noncontact mode) of (a) CHO cells and (b) dsDNA molecule adsorbed onto mica.

and coworkers demonstrated the application of NSOM to simultaneously map and detect colocalized malarial and host skeletal proteins that were indirectly labeled with immunofluorescence antibodies (34).

7.1.2 Near-Field Scanning Optical Microscopy

NSOM offers the advantages of subdiffraction limited optical resolution, specificity, and the sensitivity of fluorescence-based techniques (7,8). In addition, NSOM probes do not come in contact with the sample; hence, it is noninvasive in nature and does not perturb the sample. Besides topography, optical information can be acquired from the NSOM images. The illumination light from a tapered optical fiber is scanned close to the sample surface. This light passes through an aperture (typically, 2–120 nm in diameter, providing a resolution beyond the diffraction limit) with an exponential attenuation away from the tip. The intensity of the evanescent light waves decays exponentially to insignificant levels ~100 nm from the tip. As shown in Figure 7.2, near-field illumination of the sample at a distance h_1 smaller than the wavelength (λ) of incident light yields high-resolution images compared to those obtained from far-field illumination where the distance $h_2 \gg \lambda$ and greater than the aperture. In this latter case, subwavelength details of the image are lost.

7.1.3 Multidrug Resistance

We have exploited the optical detection sensitivity and the high resolution of NSOM to detect the cellular localization and effect of ABC (ATP-binding cassette) proteins associated with multidrug resistance (MDR) (35–39). Drug resistance can be associated with several cellular mechanisms ranging from reduced drug uptake to reduction of drug sensitivity caused by genetic alterations. MDR is therefore a phenomenon that indicates a variety of strategies that cancer cells are able to develop in order to resist the cytotoxic effects of anticancer drugs. Decades of studies have demonstrated that

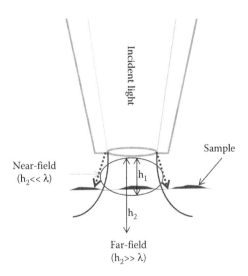

FIGURE 7.2
Schematic diagram showing NSOM principle. A subwavelength-sized aperture confines the laser light and illuminates the sample in close proximity (typically <10 nm) with a depth of h_1 (near field) and h_2 (far field).

there are different ways in which tumor cells can develop resistance. MDR can result from (1) decreased influx of cytotoxic drugs (40); (2) overexpression of drug transporters that belong to the ABC family of proteins including the P-glycoprotein (Pgp), MDR-associated protein (MRP1), and the breast cancer resistance protein 1 (BCRP1); and (3) changes in cellular physiology affecting the structure of the plasma membrane, the cytosolic pH, and the rates and extent of intracellular transport through membranes (41–43).

In particular, Pgp (a transmembrane glycoprotein of 170 kDa) was the first protein found to be associated with MDR. This protein is strongly homologous to a family of ABC protein membrane transporters, which are capable of translocating drugs and other xenobiotic compounds out of the cell. Pgp is a broad-spectrum multidrug efflux pump that has 12 transmembrane regions and 2 ATP-binding sites (44). The transmembrane regions bind hydrophobic drug substrates that are either neutral or positively charged, and are probably presented to the transporter directly from the lipid bilayer. Two ATP hydrolysis events, which do not occur simultaneously, are needed to transport a drug molecule (45). Binding of substrate to the transmembrane regions stimulates the ATPase activity of Pgp, causing a conformational change that releases substrate to either the outer leaflet of the membrane (from which it can diffuse into the medium) or the extracellular space (46). Hydrolysis at the second ATP site is required to "reset" the transporter so that it can bind substrate again, completing one catalytic cycle.

One of the salient features of Pgp is its broad substrate recognition pattern. Over the past decade, the substrate list expanded from the original description of Pgp conferring resistance to the vinca alkaloids and anthracyclines, to the current extensive list of compounds, which includes structurally unrelated anticancer agents, antihuman immunodeficiency virus (HIV) agents, and fluorophores. A classification of the drug interactions with Pgp in four categories has been established: agonists, partial agonist, antagonists, and nonsubstrates (47). An *agonist* would be both an ATPase activator and a transport substrate. Some typical examples are the "classical" substrates of Pgp: the anthracyclines and the vinca alkaloids. A *partial agonist* would be a molecule, which stimulates the Pgp ATPase activity but which does not exhibit any significant transport substrate features. To this group belongs verapamil and progesterone, both of which activate Pgp at the catalytic level, but inhibit at the transport level.

An *antagonist* would inhibit the action of Pgp and inhibit both at the ATPase and at the transport level. An example of a well-known antagonist is vanadate, which inhibits the ATPase activity of Pgp by binding to the catalytic site, thus acting as a noncompetitive inhibitor of transport. Another group of drugs, which inhibit the Pgp-associated ATPase activity, are the cyclosporins. Nonsubstrates would simply be drugs that do not appear to interact with Pgp, neither at the ATPase level nor at the transport site. Methotrexate belongs to this group.

For a better understanding of the localization of the MDR proteins and their association with the MDR substrates, a technique that is capable of mapping the distribution of the MDR proteins and monitoring their effect on the localization of therapeutic regimes in cells is required. NSOM, which is a highly sensitive and specific technique with nanoscale resolution, is the most appropriate technique to determine the localization of the MDR proteins and their substrates in cancer cells. Unlike confocal microscopy, which is suitable for imaging inside a cell, NSOM imaging is capable of simultaneously acquiring fluorescence and topography images, thus allowing real-space mapping with subwavelength resolution of cellular components localized mainly on the surface of the cell.

We have used NSOM to investigate the distribution and the localization of MDR proteins (Pgp) and MDR substrates (doxorubicin and verapamil) in living and fixed malignant rat prostate tumor cells, AT3B-1 and MLLB-2, developed from the AT-3 and MAT-LyLu cell lines, respectively (48). Chinese ovarian hamster (CHO) cells that are non-MDR expressing were used as a control.

7.2 Material and Methods

Drugs: Doxorubicin, tetramethyl rhodamine ester (TMRE), and verapamil were obtained from Sigma Chemical Company (St Louis, MO). Stock solutions (1 mM) were prepared by dissolving the drugs in sterile double-distilled water and then storing them under appropriate temperature conditions. The concentrations of the drugs used for incubating the cells were verapamil (5 μM), doxorubicin (10 μM), and TMRE (0.2 μM).

Cell Lines: CHO cells and MDR-expressing malignant rat prostate tumor cells, AT3B-1 and MLLB-2, were purchased from the American Type Culture Collection (ATCC), Manassas, VA, and grown according to the specified protocols, to confluency. The AT3B-1 and MLLB-2 cells were developed from the AT-3 and MAT-LyLu cell lines, respectively, which were derived from an adult malignant rat prostate tumor. The CHO cells were used as a control.

Cell Culture: AT3B-1 and MLLB-2 cells were propagated in T-25 culture flasks in RPMI 1640 medium supplemented with 1 μM doxorubicin (Aldrich Sigma), 10% fetal bovine serum (Invitrogen, Carlsbad, CA), L-glutamine (2 mM), 1.5 g/L sodium bicarbonate, 4.5 g/L glucose, 10 mM HEPES, and 1.0 mM sodium pyruvate. CHO cells (non-MDR-expressing control cells) were grown in T-25 or T-75 flasks (Corning, Corning, NY) using Ham's F-12 medium (Invitrogen) containing 1.5 g/L sodium bicarbonate and 2 mM L-glutamine, and supplemented with 10% fetal bovine serum (Gibco, Grand Island, NY). The stock cultures were kept in a 5% CO_2 cell culture incubator at 37°C with 95% relative humidity. When cells reached 70%–80% confluence, they were subcultured at 1:20 split ratio. For experiments, cells were seeded onto glass chamber slides from Nalge Nunc International (Naperville, IL).

The MDR-expressing (AT3B-1 and MLLB-2) and non-expressing (CHO) cells were incubated with drugs for 2 h in a 5% CO_2 incubator at 37°C. Prior to NSOM studies, cells were washed three times, in order to remove the drugs, with 1× phosphate-buffered saline (PBS) and fixed with 4% methanol-free formaldehyde in PBS buffer for 20 min at 37°C and dehydrated in an ethanol series. Fluorescence microscopy experiments were performed on cells that were unfixed or mildly fixed in 1% methanol-free formaldehyde in PBS buffer for 30 min on ice.

Immunofluorescence Labeling: Cells were cultured in two-chamber glass slides and fixed using the above procedure. The fixed cells were permeabilized with 0.02% Triton X-100 for 5 min, and then blocked with 10% normal goat serum (NGS) for 30 min at room temperature. After thorough washing with 1× PBS, the cells were incubated for 5 h at 37°C with primary monoclonal antibodies, mouse anti-Pgp, and rat anti-MRP1 diluted 1:50 in PBS/1% NGS. Cells were washed three times in PBS and reacted for 2 h with Alexa Fluor 488 conjugated goat anti-mouse or goat anti-rat immunoglobulin G (H+L) (Molecular Probes) diluted 1:1000 in PBS/1% NGS buffer at 37°C.

7.3 Instrumentation

7.3.1 Fluorescence Microscopy

Fluorescence microscopy experiments were performed using a Nikon Diaphot 300 Inverted microscope (Nikon, Melville, NY) using 60×, 0.85NA objective and a thermoelectrically cooled intensified charged-coupled device (ICCD) containing a front-illuminated chip with a 512×512 two-dimensional array of $19 \times 19 \ \mu m^2$ (PI-Max:512 GEN II, Roper Scientific, Trenton, NJ). The ICCD was computer-controlled with Win View software. Fluorescent dyes Alexa Fluor 488 and verapamil were excited using a mercury arc lamp with the following filter sets: $\lambda_{ex,}$ 470–490 nm, λ_{em} 520–560 nm and $\lambda_{ex,}$ 510–560 nm, λ_{em} 580 band-pass filter sets for doxorubicin and TMRE, respectively.

7.3.2 Near-Field Scanning Optical Microscopy

A TopoMetrix Aurora-2 near-field scanning optical microscope was used for our experiments. A schematic of the NSOM system is shown in Figure 7.3. The fiber-optic probe (aperture size ~60–100 nm) attached to a piezoelectric tuning fork was mounted on the removable Aurora-2 microscope head and positioned above the sample. A resonating frequency ranging between 90 and 100 kHz was selected with <1 nm lateral amplitude at the probe end. Samples were mounted directly beneath the tip on an XY piezo scanner that is

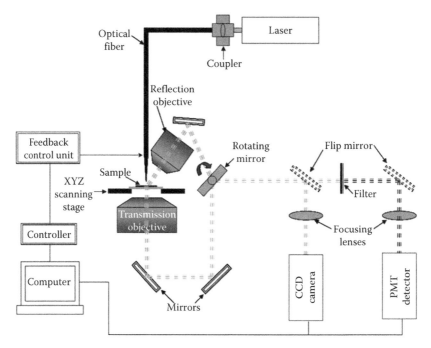

FIGURE 7.3

Schematic diagram of a near-field scanning optical microscope. The sample is mounted on a scanning stage, which is controlled by a XYZ-piezo scanner. The NSOM optical fiber probe is mounted on the removable Aurora-2 microscope head and positioned above the sample. A constant probe-sample distance is maintained at <10 nm using an electronic feedback system. The fluorescence is collected by a microscope objective through a filter set and imaged onto a photomultiplier tube (PMT) detector.

used to scan the sample under near field. The 488-nm line of the argon-ion laser was used as an excitation source for Alexa Fluor 488 immunolabeled Pgp and MRP1 proteins and the MDR inhibitor drug, verapamil, while a 532-nm laser was used to excite doxorubicin and TMRE drugs. The fluorescence emitted light was collected via the transmission mode through a 40×, 0.65 NA objective, and a band-pass filter (520LP10: for Alexa 488 and verapamil; 580LP10: for doxorubicin and TMRE), obtained from Omega Optical. The fluorescence signal was then detected by a photomultiplier tube and analyzed with commercial software (SPMLab).

7.4 Cellular and Intracellular Localization of MDR Proteins

Understanding the effect and localization of the MDR proteins will facilitate a more targeted strategy for chemotherapeutic agents or drugs being developed for clinical application. The MDR activity has been thought to be mediated at the plasma membrane by several proteins. However, studies have shown that the intracellular localization of these proteins may also play a major role in the MDR activity. Therefore, for our localization studies, drug-sensitive CHO cells and drug-resistant AT3B-1 and MLLB-2 cells that overly expressed Pgp were used. Immunofluorescence labeling against Pgp was carried out using monoclonal antibodies, mouse anti-Pgp together with a secondary antibody, goat anti-mouse IgG that were conjugated with Alexa 488 fluorophore. Confocal microscopy images of AT3BB-1 cells shown in Figure 7.4a–c were recorded to investigate the immunofluorescence distribution of Pgp. Control experiments using drug-sensitive CHO cells resulted in minimal or no fluorescence. The composite images of the AT3BB-1 stained cells revealed that Pgp was localized evenly in the plasma membrane of the cell (Figure 7.4c).

To further determine the distributional pattern of the MDR proteins in AT3B-1 cells, NSOM experiments were carried out using sets of cells similar to the confocal microscopy experiments. The NSOM tip was positioned over Pgp-expressing cell and scanned over an area of 30×30 μm². Figure 7.5 shows the internal feedback signal (a), shear force topography (b), NSOM fluorescence (c), and composite (d) (topography and NSOM fluorescence) images of AT3B-1 cells labeled with anti-Pgp antibodies, allowing the visualization of the distribution of MDR cell surface proteins. In agreement with the confocal imaging results, we observed a homogeneous distribution of MRP1 in the plasma membrane and the localization of Pgp around the perinuclear region of the cells. The composite images

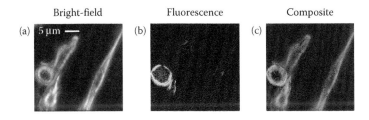

FIGURE 7.4
Fluorescence and bright-field images of MDR-expressing malignant rat prostate tumor cells, AT3B-1 labeled with primary monoclonal antibodies to Pgp, and mouse anti-Pgp (a–c). The primary antibodies were detected with a goat anti-mouse or goat anti-rat immunoglobin G conjugated with Alexa 488. The fluorescence and bright-field images were merged to form a composite image.

in Figure 7.5c and d clearly show the correlation of the fluorescence and topographical features obtained from a cell. A closer examination of a smaller region of an immunolabeled cell shown in Figure 7.6 revealed that the proteins Pgp are unevenly distributed in the cell, forming membrane patches or clusters with an average size of ∼50 nm as seen from the cross section through the NSOM image in Figure 7.6c. Conventional microscopy cannot reveal these observed significant differences. Figure 7.6d and e shows the internal feedback signal and shear force topography images of the AT3B-1 cell. The random distribution of the protein Pgp in the plasma membrane would most likely increase activity of the proteins to mediate drug resistance by lowering the accumulation of chemotherapeutic regimens. Alternatively, clustering of the Pgp within the perinuclear regions and other intracellular organelles would increase the protein activity to efflux MDR substrates from the nucleus of the cell.

Since NSOM can also detect a fluorescence signal in a near-field mode (<100 nm) with diminishing efficiency, the fluorescence signal detected from surface proteins is significantly strong, allowing space mapping of the NSOM image with the topography image. Moreover, in addition to the high sensitivity (minimal autofluorescence) and spatial

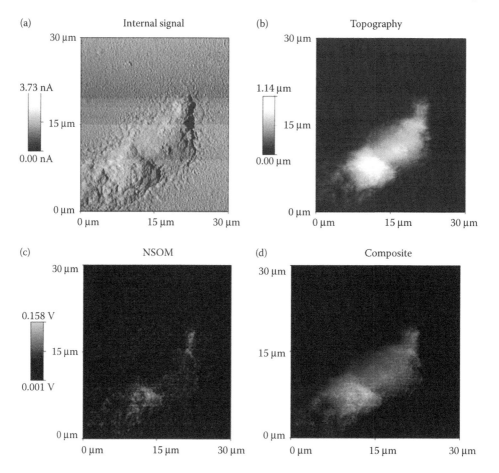

FIGURE 7.5
Images of AT3B-1 cells labeled with primary monoclonal antibodies to Pgp (a–d). Primary antibodies were detected using Alexa 488 conjugated goat anti-mouse or goat anti-rat immunoglobin G. Internal signal, topography, NSOM, and their composite images are shown.

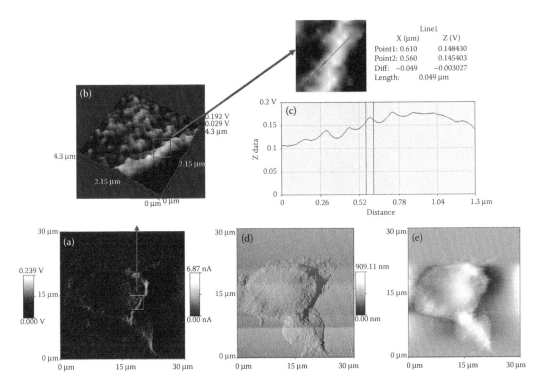

FIGURE 7.6
(a) NSOM image of an AT3B-1 cell labeled with anti-Pgp antibodies. (b) A $4.3 \times 4.3 \, \mu m^2$ 3-D NSOM image obtained by zooming into the indicated region of the cell. (c) A cross section through the NSOM image revealed optical features having a step resolution of ~49 nm; (d) and (e) show the internal feedback signal and shear force topography images of the AT3B-1 cell.

resolution exhibited by NSOM imaging, topography and internal signal information allows real-space mapping of the proteins at nanoscale resolution. Also the small excitation volume of the NSOM tip can allow the interrogation of individual protein molecules or a clustered type of organization at a single-cell level, which is not possible with diffraction-limited imaging techniques.

7.5 Effect of MDR Expression on Drug Accumulation

Overexpression of MDR proteins is associated with the decreased intracellular accumulation and increased efflux of various chemotherapeutic drugs. Consequently, it is necessary for researchers to study and understand how the activity and distribution of the MDR proteins affect the accumulation and localization of therapeutic drugs. In our study, confocal microscopy and NSOM were used to monitor the effect of MDR protein Pgp on the intracellular accumulation of MDR substrates and anthracycline doxorubicin in both drug-sensitive and drug-resistant cells.

To determine the activity of the MDR proteins Pgp, we studied their effect on the intracellular accumulation of TMRE, a fluorescent MDR substrate that does not accumulate

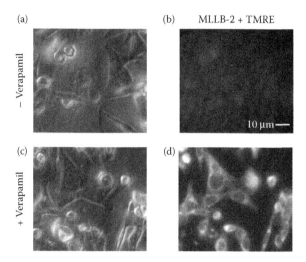

(a)   (b)   MLLB-2 + TMRE

− Verapamil

+ Verapamil

(c)   (d)

10 μm ———

FIGURE 7.7
Confocal microscopy images of MLLB-2 cells incubated with 0.2 μM of a fluorescent MDR substrate, tetramethyl rhodamine ester (TMRE). Cells were treated with TMRE in the absence of an MDR inhibitor, verapamil (a and b), and in the presence of 5 μm verapamil (c and d) for 2 h, mildly fixed with 1% methanol-free formaldehyde in PBS buffer for 30 min on ice.

in drug-resistant cells (49) and anthracycline doxorubicin. Drug-sensitive cells, however, accumulate TMRE in the mitochondria. Confocal microscopy was used to assay the effect of the proteins on the accumulation of TMRE (0.2 μM) by treating drug-resistant MLLB-2 cells with 5 μM of the MDR inhibitor verapamil. Figure 7.7 shows the bright-field (a and c) and fluorescence (b and d) images of the cells. Upon treating the drug-resistant cells with verapamil, the protein activity at the plasma membrane was reversed, and therefore, both drug-sensitive and drug-resistant cells accumulated the TRME drug into the cytoplasm, resulting in staining of the mitochondria (Figure 7.7d). On the other hand, cells that were not treated with verapamil accumulated little to none of the drug (Figure 7.7b). MLLB-2 cells expressed drug-resistance protein, Pgp, thus lowering the accumulation of TMRE.

Unlike the MDR activity on TMRE drug, doxorubicin accumulates in the nucleus and perinuclear regions of drug-sensitive and drug-resistant cells, respectively. This indicates that the presence of MDR proteins in the plasma membrane has a minimal to negligible effect on doxorubicin drug influx. However, the distribution of doxorubicin in drug-resistant cells is due to the intracellular localization of MDR proteins on the membrane of cellular organelles located around the nucleus.

We therefore tested the effect of Pgp expression on anthracycline doxorubicin in drug-sensitive CHO and drug-resistant AT3B-1 cells using confocal microscopy (Figure 7.8a–h). The confocal images show non-MDR-expressing CHO cells and MDR-expressing AT3B-1 cell treated with 5 μM verapamil (green) and 10 μm doxorubicin (red). Verapamil, a Ca^{2+}-channel blocker, is one of several molecules known to inhibit MRP1- or Pgp-mediated drug efflux, and is located mainly in the plasma membrane (50). Both CHO cells incubated with verapamil and those that were not accumulated a substantial amount of doxorubicin in the nucleus. Therefore, non-Pgp-expressing CHO cells have no effect on the accumulation of doxorubicin drug in the cell (Figure 7.8e and f). However, Pgp-expressing cells AT3B-1 showed no detectable drug accumulation (Figure 7.8c and g) due to the rapid efflux of the drug from the cellular compartments. Conversely, verapamil reversed the activity

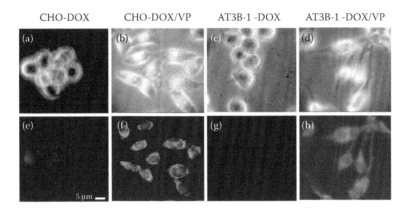

FIGURE 7.8
Images of a fixed CHO and AT3B-1 cells incubated with 10 μm doxorubicin and 5 μm verapamil for 2 h. Confocal microscopy images, bright-field (a–d) and fluorescence (e–h), were obtained using a mercury arc lamp with the following filter sets: λ_{ex}, 470–490 nm, λ_{em} 520–560 nm for verapamil and λ_{ex}, 510–560 nm, λ_{em} 580 band-pass, for doxorubicin.

of Pgp in AT3B-1 cells allowing doxorubicin drug to accumulate in the nucleus of the cells (Figure 7.8d and h).

Unlike the drug-sensitive CHO cells, the expression of MDR proteins in the drug-resistant cell lines changed the intracellular distribution of doxorubicin in the MDR cells. We used NSOM to demonstrate the intracellular distribution of doxorubicin in CHO drug-sensitive (data not shown) and MLLB-2 drug-resistant cells (Figure 7.9). The composite (topography and NSOM combined) images showed localization of doxorubicin in the

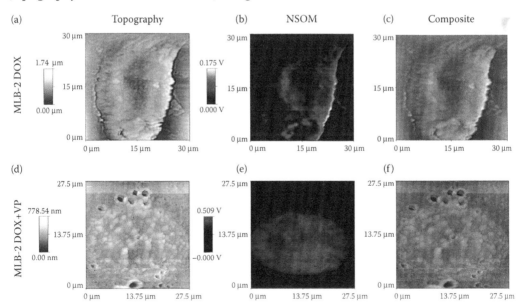

FIGURE 7.9
Images of MLLB-2 cells incubated with chemotherapeutic drugs. Topography (a and d), NSOM (b and e), and composite (c and f) images of fixed cells. The MLLB-2 cells treated with doxorubicin in the absence of MDR inhibitor verapamil (a–c) and in the presence of 5 μm verapamil (d–f). The composite images were obtained by merging topography and NSOM images.

nucleus and other intracellular organelles in the cytoplasmic compartment of verapamil-treated and -untreated MLLB-2 and CHO cells. We investigate the internal signal, topography, and the corresponding NSOM images of fixed anthracycline-resistant MLLB-2 cells that were treated with only doxorubicin (a–c), doxorubicin and verapamil (a–f), and CHO cells treated with only doxorubicin (data not shown). After the cells were incubated with drugs for 2 h, verapamil rendered the drug-resistant cell MLLB-2 incapable of extruding doxorubicin (Figure 7.9d–f) from the nucleus. These effects were similar to the confocal images observed in drug-sensitive CHO cells (Figure 7.8f). However, MDR-expressing cells resulted in the redistribution of doxorubicin from the nucleus into the perinuclear regions. These results are comparable to a previously reported observation where confocal microscopy was applied (49,51). We also observed that the NSOM intensity signal from the MDR-resistant cells treated with MDR reverser verapamil had higher (~4 times) concentrations of doxorubicin than those of the untreated MDR cells (Figure 7.9b and e).

As shown in Figure 7.10a, doxorubicin is extruded from the nucleus and is concentrated in the organelles throughout the cytoplasm and the perinuclear regions. Studies have shown that, in MDR cells, weak-base chemotherapeutic drugs accumulated only within certain organelles (such as the trans-Golgi network [TGN], lysosomes, secretion vesicles, or endosomes) with almost none in the nucleus (52,53). Using NSOM high-resolution imaging

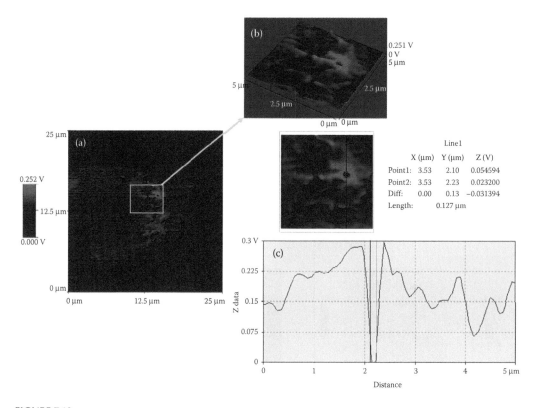

FIGURE 7.10
(a) NSOM image of an MLLB-2 cell treated with 10 μm doxorubicin drug. (b) A 3-D NSOM image of cells showing intracellular doxorubicin localization of doxorubicin the perinuclear organelles. Line trace through the subcellular structure revealed nanosized features (~127 nm) in the subcellular compartments loaded with doxorubicin. (c) Cross-section profile along the line drawn revealing nanosized saclike features.

capability, we were able to demonstrate the accumulation of doxorubicin in the TGN and other organelles. An optical image of TGN localized with doxorubicin is shown in Figure 7.10b. A cross-sectional profile along the line drawn in Figure 7.10c revealed nanosized saclike features (with a FWHM of ~127 nm), which are characteristic of the Golgi apparatus. Such small features revealed by NSOM are almost impossible to visualize using confocal microscopy.

Our studies therefore indicate that activity of Pgp located on the intracellular organelles sequestered doxorubicin from the nucleus of drug-resistant cells that were treated with verapamil. The localization of the drug in this organelle was also visualized using NSOM imaging, which confirmed the pattern distribution of doxorubicin localized in certain organelles in the endomembrane system. A line trace through the subcellular structure revealed nanosized features of TGN loaded with doxorubicin. Moreover, doxorubicin was also observed to accumulate in other organelles located in the perinuclear regions of the MLLB-2 drug-resistant cells.

7.6 Conclusion

We have used NSOM to demonstrate that the MDR protein transporters, Pgp, located in the plasma membrane and other intracellular compartments confer increased resistance to chemotherapeutic drugs and other MDR substrates. The high spatial resolution and surface sensitivity of NSOM revealed the distribution of the MDR proteins in small patches and membrane domains on a scale from 10 to 100 nm in size. In this work, we demonstrate that the activity and distribution of the MDR proteins have a significant effect on the uptake and localization of therapeutic regimes in cancer cells. Surface plasma membrane proteins confer resistance by mediating drug efflux from the cells at the plasma membrane, while intracellular membrane proteins located on internal organelles sequester drugs from the cytoplasm and cell nucleus. For this reason, understanding the cellular localization and activity of these proteins will enable the development of highly specific MDR modulators and anti-tumor drugs. In addition, nanoimaging techniques such as NSOM provide a great potential to visualize and study individual membrane domains and protein clusters at the nanoscale level. The application of NSOM in nanobiology is certainly becoming a method of choice due to the high sensitivity and high resolution information of this emerging nanoimaging technology.

Acknowledgments

This work was sponsored by the National Institutes of Health (R01 EB 006201), the Office of Biological and Environmental Research, Department of Energy (DOE), under Contract DE-AC05-00OR22725 with UT-Battelle, LLC, and by the Oak Ridge National Laboratory LDRD Program (Advanced Plasmonics Sensors). This research was also supported, in part, by the appointments of M. B. Wabuyele to the U.S. DOE Laboratory Cooperative Postdoctoral Research Training Program administered by Oak Ridge Institute for Science and Education.

References

1. Pettinger, B., Picardi, G., Schuster, R., and Ertl, G. 2002. Surface-enhanced and STM-tip-enhanced Raman spectroscopy at metal surfaces. *Single Mol.* 3, 285–294.
2. Otero, R., Federico, R., and Flemming, B. 2006. Scanning tunneling microscopy manipulation of complex organic molecules on solid surfaces. *Annu. Rev. Phys. Chem.* 57, 497–525.
3. Daniel, J. M., and Yves, F. D. 2011. Atomic force microscopy: A nanoscopic window on the cell surface. *Trends Cell Biol.* 21(8), 461–469.
4. Hinterdorfer, P., and Dufrene, Y. F. 2006. Detection and localization of single molecular recognition events using atomic force microscopy. *Nat. Methods* 3, 347–355.
5. Moffitt, J. R., Chemla, Y. R., Smith, S. B., and Bustamante, C. 2008. Recent advances in optical tweezers. *Annu. Rev. Biochem.* 77, 205–228.
6. Neuman, K. C., and Nagy, A. K. 2008. Single-molecule force spectroscopy: Optical tweezers, magnetic tweezers and atomic force microscopy. *Nat. Methods* 5, 491–505.
7. Müller, D. J., Helenius, J., Alsteens, D., and Dufrêne, Y. F. 2009. Force probing surfaces of living cells to molecular resolution. *Nat. Chem. Biol.* 5, 383–390.
8. Hansma, H. G. 2001. Surface biology of DNA by atomic force microscopy. *Annu. Rev. Phys. Chem.* 52, 71–92.
9. Niemeyer, C. M., Adler, M., Pignataro, B., Lenhert, S., Gao, S., Chi, L., Fuchs, H., and Blohm, D. 1999. Self-assembly of DNA-streptavidin nanostructures and their use as reagents in immuno-PCR. *Nucleic Acids Res.* 27, 4553–4561.
10. Smith, G. C., Cary, R. B., Lakin, N. D., Hann, B. C., Teo, S. H., Chen, D. J., and Jackson, S. P. 1999. Purification and DNA binding properties of the ataxia-telangiectasia gene product ATM. *Proc. Natl. Acad. Sci. U.S.A.* 96, 11134–11139.
11. Ianoul, A., Street, M., Grant, D., Pezacki, J., Taylor, R. S., and Johnston, L. J. 2004. Near-field scanning fluorescence microscopy study of ion channel clusters in cardiac myocyte membranes. *Biophys. J.* 87, 3525–3535.
12. Burgos, P., Lu, Z., Ianoul, A., Hnatovsky, C., Viriot, M. L., Johnston, L. J., and Taylor, R. S. 2003. Near-field scanning optical microscopy probes: A comparison of pulled and double-etched bent NSOM probes for fluorescence imaging of biological samples. *J. Microsc.* 211, 37–47.
13. Dickenson, N. E., Armendariz, K. P., and Dunn, R. C. 2010. Near-field scanning optical microscopy: A tool for nanometric exploration of biological membranes. *Anal. Bioanal. Chem.* 396, 31–43.
14. Lereu, A. L., Passian, A., and Dumas, P. H. 2012. Near field optical microscopy: A brief review. *Int. J. Nanotechnol.* 9, 488–501.
15. Hinterdorfer, P., Garcia-Parajo, M. F., and Dufrêne, Y. F. 2012. Single-molecule imaging of cell surfaces using near-field nanoscopy. *Acc. Chem. Res.* 45, 327–336.
16. Edidin, M. 2001. Near-field scanning optical microscopy, a siren call to biology. *Traffic* 2, 797–803.
17. Sannohe, Y., Endo, M., Katsuda, Y., Hidaka, K., and Sugiyama, H. 2010. Visualization of dynamic conformational switching of the G-quadruplex in a DNA nanostructure. *J. Am. Chem. Soc.* 132, 16311–16313.
18. Noy, A., Vezenov, D. V., and Lieber, C. M. 1997. Chemical force microscopy. *Annu. Rev. Mater. Sci.* 27, 381–421.
19. Lehenkari, P. P., Charras, G. T., Nykanen, A., and Horton, M. A. 2000. Adapting atomic force microscopy for cell biology. *Ultramicroscopy* 82, 289–295.
20. Fotiadis, D., Scheuring, S., Muller, S. A., Engel, A., and Muller, D. J. 2002. Imaging and manipulation of biological structures with the AFM. *Micron* 33, 385–397.
21. Hamon, L., Pastré, D., Dupaigne, P., Le Breton, C., Le Cam, E., and Piétrement, O. 2007. High-resolution AFM imaging of single-stranded DNA-binding (SSB) protein–DNA complexes. *Nucleic Acids Res.* 35, e58.

22. Dammer, U., Popescu, O., Wagner, P., Anselmetti, D., Güntherodt, H.-J., and Misevic, G. N. 1995. Binding strength between cell adhesion proteoglycans measured by atomic force microscopy. *Science* 267, 1173–1175.

23. Shi, D., Somlyo, A. V., Somlyo, A. P., and Shao, Z. 2001. Visualizing filamentous actin on lipid bilayers by atomic force microscopy in solution. *J. Microsc.* 201, 377–382.

24. Rotsch, C., and Radmacher, M. 2000. Drug-induced changes of cytoskeletal structure and mechanics in fibroblasts: An atomic force microscopy study. *Biophys. J.* 78, 520–535.

25. Pohl, D. W., Denk, W., and Lanz, M. 1984. Optical stethoscopy: Image recording with resolution 1/20. *Appl. Phys. Lett.* 44, 651–653.

26. Koopman, M., Cambi, A., de Bakker, B. I., Joosten, B., Figdor, C. G., van Hulst, N. F., and Garcia-Parajo, M. F. 2004. Near-field scanning optical microscopy in liquid for high resolution single molecule detection on dendritic cells. *FEBS Lett.* 573, 6–10.

27. Betzig, E., and Chichester, R. J. 1993. Single molecules observed by near-field scanning optical microscopy. *Science* 262, 1422–1425.

28. Meixner, A. J., and Kneppe, H. 1998. Scanning near-field optical microscopy in cell biology and microbiology. *Cell. Mol. Biol.* 44, 673–688.

29. Huckabay, H. A., Armendariz, K. P., Newhart, W. H., Wildgen, S. M., and Dunn, R. C. 2013. Near-field scanning optical microscopy for high-resolution membrane studies. *Methods Mol. Biol.* 950, 373–394.

30. Dickenson, N. E., Armendariz, K. P., Huckabay, H. A., Livanec, P. W., and Dunn, R. C. 2010. Near-field scanning optical microscopy: A tool for nanometric exploration of biological membranes. *Anal. Bioanal. Chem.* 396, 31–43.

31. Van Hulst, N. F., Veerman, J. A., Garcia-Parajo, M. F., and Kuipers, L. 2000. Analysis of individual (macro)molecules and proteins using near-field optics. *J. Chem. Phys.* 112, 7799–7810.

32. Nagy, P., Jenei, A., Kirsch, A. K., Szollosi, J., Damjanovich, S., and Jovin, T. M. 1999. Activation-dependent clustering of the erbB2 receptor tyrosine kinase detected by scanning near-field optical microscopy. *J. Cell. Sci.* 112, 1733–1741.

33. Ruiter, A. G., Veerman, J. A., Garcia-Parajo, M. F., and Van Hulst, N. F. 1997. Single molecule rotational and translational diffusion observed by near-field scanning optical microscopy. *J. Phys. Chem.* 101, 7318–7323.

34. Enderle, T., Ha, T., Ogletree, D. F., Chemla, D. S., Magowan, C., and Weiss, S. 1997. Membrane specific mapping and colocalization of malarial and host skeletal proteins in the *Plasmodium falciparum* infected erythrocyte by dual-color near-field scanning optical microscopy. *Proc. Natl. Acad. Sci. U.S.A.* 94, 520–525.

35. Piddock, L. J. 2006. Multidrug-resistance efflux pumps—Not just for resistance. *Nat. Rev. Microbiol.* 4, 629–636.

36. Lage, H. 2008. An overview of cancer multidrug resistance: A still unsolved problem. *Cell. Mol. Life Sci.* 65, 3145–3167.

37. Juliano, R. L., and Ling, V. 1976. A surface glycoprotein modulating drug permeability in Chinese hamster ovary cell mutants. *Biochim. Biophys. Acta* 455, 152–162.

38. Borst, P., Evers, R., Kool, M., and Wijnholds, J. 2000. A family of drug transporters: The multidrug resistance-associated proteins. *J. Natl. Cancer Inst.* 92, 1295–1302.

39. Gottesman, M. M., Fojo, T., and Bates, S. E. 2002. Multidrug resistance in cancer: Role of ATP-dependent transporters. *Nat. Rev. Cancer* 2, 48–58.

40. Shen, D., Pastan, I., and Gottesman, M. M. 1998. Cross-resistance to methotrexate and metals in human cisplatin-resistant cell lines results from a pleiotropic defect in accumulation of these compounds associated with reduced plasma membrane binding proteins. *Cancer Res.* 58, 268–275.

41. Cole, S. P. 2014. Multidrug resistance protein 1 (MRP1, ABCC1), a "multitasking" ATP-binding cassette (ABC) transporter. *J. Biol. Chem.* 289, 30880–30888.

42. Amaral, L., Engi, H., Viveiros, M., and Molnar, J. 2007. Review. Comparison of multidrug resistant efflux pumps of cancer and bacterial cells with respect to the same inhibitory agents. *In Vivo* 21, 237–244.

43. Simon, S. M., and Schindler, M. 1994. Cell biological mechanisms of multidrug resistance in tumors. *Proc. Natl. Acad. Sci. U.S.A.* 91, 3497–3504.

44. Chen, C. J., Chin, J. E., Ueda, K., Clark, D. P., Pastan, I., Gottesman, M. M., and Roninson, I. B. 1986. Internal duplication and homology with bacterial transport proteins in Mdr-1 (P-glycoprotein) gene from multidrug-resistant human cells. *Cell* 47, 371–380.

45. Senior, A. E., and Bhagat, S. 1998. P-glycoprotein shows strong catalytic cooperativity between the two nucleotide sites. *Biochemistry* 37, 831–836.

46. Liu, R., and Sharom, F. J. 1996. Site-directed fluorescence labeling of P-glycoprotein on cysteine residues in the nucleotide binding domains. *Biochemistry* 35, 11865–11873.

47. Litman, T., Druley, T. E., Stein, W. D., and Bates, S. E. 2001. From MDR to MXR: New understanding of multidrug resistance system, their properties and clinical significance. *Cell. Mol. Life Sci.* 58, 931–959.

48. Isaacs, J. T., Isaacs, W. B., Feitz, W. F., and Scheres, J. 1986. Establishment and characterization of seven Dunning rat prostatic cancer cell lines and their use in developing methods for predicting metastatic abilities of prostatic cancers. *Prostate* 9, 261–281.

49. Rajagopal, A., and Simon, S. M. 2003. Subcellular localization and activity of multidrug resistance proteins. *Mol. Biol. Cell.* 14, 3389–3399.

50. Hindenburg, A. A., Baker, M. A., Gleyzer, E., Stewart, V. J., Case, N., and Taub, R. N. 1987. Effect of verapamil and other agents on the distribution of anthracyclines and on reversal of drug resistance. *Cancer Res.* 47, 1421–1425.

51. Zhou, Y., Sridhar, R., Shan, L., Sha, W., Gu, X., and Sukumar, S. 2012. Loperamide, an FDA-approved antidiarrhea drug, effectively reverses the resistance of multidrug resistant MCF-7/MDR1 human breast cancer cells to doxorubicin-induced cytotoxicity. *Cancer Invest.* 30, 119–125.

52. Altan, N., Chen, Y., Schindler, M., and Simon, S. M. 1998. Defective acidification in human breast tumor cells and implications for chemotherapy. *J. Exp. Med.* 187, 1583–1598.

53. Altan, N., Chen, Y., Schindler, M., Simon, S. M. 1999. Tamoxifen inhibits acidification in cells independent of the estrogen receptor. *Proc. Natl. Acad. Sci. U.S.A.* 96, 4432–4437.

8

Development and Modeling of a Novel Self-Assembly Process for Polymer and Polymeric Composite Nanoparticles

B. G. Sumpter, J.-M. Y. Carrillo, S.-K. Ahn, M. D. Barnes,
W. A. Shelton, R. J. Harrison, and D. W. Noid

CONTENTS

8.1 Introduction

The confinement of materials at the nanoscale and the resulting alterations of their properties and chemistry have been and continue to be a subject of considerable excitement and interest in numerous science and technology sectors.[1] Recent results have shown dramatic effects in this regard including profound changes in kinetics and reaction products of organic molecules in porous media,[2] altered structures and enhanced melting points of proteins and other macromolecules,[3] new types of fluid dynamics,[4] modified fluorescence lifetimes,[5] and the modification of the structure of liquid water.[6] For many macromolecules, in particular polymers, three-dimensional (3-D) confinement at a nanometer-size scale is often comparable to a polymer's radius of gyration and is of special interest in the context of the so-called collapse transitions associated with semiconducting polymers and how the confinement affects intra- and interchain organization. For a mixed polymer system, there is the interesting question of spinodal decomposition (sudden phase segregation) of such systems under 3-D confinement, and the possibility of deeply quenched single-phase (homogeneous) polymer-blend particles or polymer alloy[7] with specially tailored electronic,[8,9] optical, or mechanical[10] properties.[11–13] However, commonly used techniques capable of producing these types of structures such as thin-film or self-assembly processes can suffer from substrate interactions, which may dominate or obscure the underlying polymer physics. In order to minimize these complexities, we have recently explored ink-jet printing methods for producing polymer particles with arbitrary size and composition. This method is based on using droplet-on-demand generation to create a small drop consisting of a very dilute polymer mixture in a solvent.[14,15] As the solvent evaporates, a polymer particle is produced whose size is defined by the initial size of the droplet (typically between 5 and 30 μm), and the weight fraction of polymer (or other nonvolatile species) in solution. Because the droplets are produced with small excess charge during ejection from the nozzle, this approach lends itself naturally to spatial manipulation of micro- and nanoparticles using electrodynamic focusing techniques.[16] Polymeric particles in the micro- and nanometer-size range provide many unique properties due to size reduction to the point where critical length scales of physical phenomena become comparable to or larger than the size of the structure. Applications of these types of particles take advantage of high surface area and confinement effects, leading to interesting nanostructures with different properties that cannot be produced using conventional methods. Clearly, there is extraordinary potential for developing new materials in the form of bulk, composites, and blends that can be used for coatings, optoelectronic components, magnetic media, ceramics and special metals, micro- or nanomanufacturing, and bioengineering. The key to beneficially exploiting these interesting materials is a detailed understanding of the connection of nanoparticle technology to atomic and molecular origins of the process.

The question of morphological control of individual macromolecules is an important one from the standpoint of fundamental physical understanding of interchain interactions, as well as of polymer-based optoelectronic device applications.[17] Stiff-chain polymers possessing structural defects may adopt different morphologies (e.g., toroids or rods) depending on the structural nature of the polymer (number of defects, persistence length, etc.) and the strength of the interchain interactions. The case of conducting polymers is particularly intriguing as these species possess both interesting structural and luminescence properties where spectral or polarization signatures carry information on morphological properties of single-polymer chains.

Since the discovery of conducting polymers in the late 1970s, an enormous literature has evolved on the structural, photophysical, spectroscopic, and charge transport properties of these materials. There has been considerable interest in these organic-conjugated systems because they provide the basis of novel materials that combine optoelectronic properties of semiconductors with the mechanical properties and processing advantages of plastics. It is relatively easy to functionalize the backbone with a variety of flexible side groups, which can make the materials soluble in organic solvents and thus easily processed into thin films and, as we show in the present chapter, precise nanoparticles. As such, conjugated polymers offer new possibilities for use in optoelectronic devices such as organic LEDs, flexible displays, photovoltaics, transistors, as well as a number of biomedical imaging possibilities. However, despite the enormous versatility for practical applications, the fundamental physics underlying the optimization of the optoelectronic properties has remained elusive. Much of the problem in this regard is centered about a poor understanding of the interactions between conjugated polymer chains in solutions and films. It is clear that the electronic structure of conjugated polymers depends sensitively on the physical conformation of the polymer chains and the way the chains pack together. In recent single-molecule studies, evidence for coil–rod "collapse transitions" of conducting polymers in dilute thin films was presented where polarization-modulated fluorescence indicated compact chain morphologies characterized as "defect rods" with a broad distribution of morphologies and varying degree of stiff-chain alignment. However, the nature of the substrate and host–polymer interactions and whether these conspire to arrest collapse transitions at various points in-between random coil and ideal rod morphologies in thin-film preparations are still not understood.

Of the large class of conjugated polymers, poly(para-phenyelene vinylene) (PPV) derivatives have received a great deal of attention in the context of polymer-based optoelectronic devices because of its efficient luminescence and charge transport properties. PPV macromolecules are described structurally as a large number of stiff-chain (conjugated) segments—typically between 50 and 200—that are linked by the so-called tetrahedral defects. Each conjugated segment has a length of between 6 and 12 monomer units and can act as a local optical chromophore within the molecule where the final chain morphology depends sensitively on solvent and film-processing parameters.[18–22] Currently, much insight has been gained through single-molecule spectroscopy into exciton dynamics, photochemical stability, and chain organization of PPV-based polymer molecules isolated in dilute thin films. Polarization spectroscopy of single-poly(2-methoxy-5-[2′-ethyl-hexyloxy]-1,4-phenylene) vinylene (MEH-PPV) molecules in polycarbonate host films have shown evidence of (partial) coil–rod collapse transitions during spin casting of the film, where the distribution of single-molecule polarization anisotropy parameters could be correlated with simulations based on different chain morphologies.[17] However, these results also suggested a wide variation in chain morphologies, and the role of substrate and host polymer in affecting the degree of collapse is unclear.

In this chapter, we describe our previous work on investigating the effects of 3-D confinement of single molecules of conjugated organic polymers. We updated the second edition of this chapter to reflect some of our recent work on polythiophenes, which is another class of conjugated organic polymers that have potential applications in organic photovoltaics, field-effect transistors, electroluminescent devices, and sensors. By using a combination of the state-of-the-art experimental and computational techniques, we provide new insight into the organization of stiff-chain polymers confined to nanoscale domains, and unambiguously show that the photophysical properties of these systems are profoundly altered as a result. We also looked into the effect of polymer architecture on the nanoscale assembly process. These developments have important ramifications for biomedical applications. While medicinal uses of polymeric materials have been fairly broad, such as using biodegradable polymers for sutures, artificial skin, and materials for covering wounds, other medical and biochemical applications include possible use of polymer particles in absorbents, latex diagnostics, affinity bio-separators, and drug and enzyme carriers. In particular, the use of nanoparticles as drug delivery vehicles has enjoyed significant recent activity and research. Drugs or other biologically active molecules have been dissolved, entrapped, encapsulated, absorbed onto surfaces, and chemically attached to polymeric particles as a means for delivery. Some other important biomedical applications that may depend/benefit on new advances in polymer particle technology are medical imaging, bioassays, and biosensors. The incorporation of functional nanoparticles can be highly advantageous for the performance of numerous bioassays. The tremendous increase in surface area offers the ultimate ability for binding to target molecules such as proteins and enzymes and these same particles offer complementary advantages for the production of highly sensitive biosensors. With future advances in the functionalization of luminescent particles, noninvasive biomedical imaging could be substantially improved.

8.2 Summary of Experimental Results

8.2.1 PPV-Based Polymers and Oligomers

In the first part of this section, we describe the experimental approach that has been used in our laboratory to investigate structural and photophysical properties of PPV-based polymeric particles. It is important to lay this background before proceeding to the computational and theoretical methods and results, as much of those calculations are directed by what we have found through extensive experimental studies.

In our more recent experiments, we used inkjet printing methods to isolate single-conducting polymer chains in microdroplets in various organic solvents (tetrahydrofuran, toluene, etc.) typically <5 μm initial diameter.[23,24] The droplets evaporate en route to the cover glass substrate thus allowing self-organization of the polymer chain in the absence of host polymer or substrate interactions. The probe polymer used in our experiments was MEH-PPV—with an average molecular weight of 250,000 (polydispersity $M_n/M_w = 4$), where each chain comprises ≈100 conjugated segments (8–12 monomer units long). Droplets of dilute MEH-PPV solution (10^{-11}–10^{-12} M) in doubly distilled tetrahydrofuran (THF) were generated from both piezoelectric on-demand droplet generators or nebulized from a 2-μm glass nozzle, and the dry particles were deposited on clean glass coverslips. We found concentration-dependent nanoparticle coverage at MEH-PPV concentrations as low as 10^{-14} M indicating clearly that the polymer nanoparticles probed in our experiments are single-MEH-PPV chains. To probe structural organization in individual particles, we used a combination of dipole emission pattern imaging, polarization-modulated fluorescence, and atomic force microscopy (AFM). Together, these measurements provide a clear picture of stiff-chain organization within individual macromolecules adsorbed on the glass surface.

Details of the fluorescence microscopy are given in Reference 23. Fluorescence images were acquired on a Nikon TE300 inverted microscope with a 1.4 NA × 100× oil objective combined with a 4× expander. The imager was a high-speed back-illuminated frame-transfer camera (Roper Scientific EEV57). The "donut-like" spatial intensity patterns seen in the fluorescence image (Figure 8.1) are characteristic of single-dipole emitters oriented parallel to the optic (Z) axis. The slight circular asymmetry is ascribed to small tilt angles (≈3° nom.) with respect to the surface normal. We find a high degree of orientational uniformity for MEH-PPV from most droplet-generated samples. Depending on the mode of sample production, there may be a fraction of the population that shows in-plane orientation; however, these species tend to be very short-lived photochemically, and only the z-oriented species remain fluorescent for longer than a few seconds. This unusual transition dipole orientation is precisely the opposite of that seen for spin-coated films where all molecules appear to show spatial intensity patterns in fluorescence characteristic of in-plane (parallel to the glass substrate) emitters. It is highly surprising to find uniform transition moment orientation of these species in the nonintuitive z-direction (perpendicular

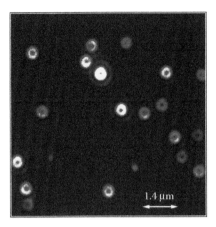

FIGURE 8.1
High-resolution image of z-oriented MEH-PPV single molecules acquired under 514.5 nm excitation. The image was acquired for 20 s for better signal-to-noise ratio. Note that the size of the emission pattern does not represent the actual size or morphology of the nanoparticle, only the antenna image.

to the support substrate). The high spatial resolution fluorescence images (real-space distance per pixel is 35 nm) from single molecules of MEH-PPV deposited on a clean glass coverslip, all look similar to that shown in Figure 8.1. It is known that a single molecule, or single dipole, with a transition moment orientation fixed in space that emits light with a sine-squared angular distribution about the dipole axis, μ, generates an optical emission pattern that is uniquely defined by its orientation.[25] Since emission is forbidden at angles along μ, dipoles oriented perpendicular to the substrate show "donut"-like emission patterns that are seen in focus as well as for small defocusing. Quantitative fitting of many patterns from different particles indicates an extraordinary uniformity in the polar angle, θ, to within a few percent. However, the observation of dipolar emission patterns does not, of itself, provide information on intramolecular structure, if the radiative recombination site is localized within the molecule. If the radiative recombination site within the molecule is localized, some local 3-D orientation of the molecule can be perceived experimentally from the spatial intensity pattern (or linear dichroism measurement). However, in a random coil structure approximation, each local orientation should logically be expected to be different with no net orientation in the sample. The fact that our ensemble sample shows uniform Z orientation is significant in that this particular orientation can only be explained if the molecules possess a high degree of alignment between conjugated segments. This is due to the fact that the transition moment for the (^1B_u) optically pumped exciton is polarized collinearly with the conjugation axis.[26] Thus, the only way for the ensemble of polymer chains to show the same transition moment is for them all to possess the same up–down/left–right structural identity with respect to the surface.

Further evidence of a nanocylindrical geometry was seen by scanning the contact with a modified Digital Instruments Bioscope/Dimension scanner with a Nanoscope III controller (Figure 8.2). All measurements were made in tapping mode. Particle heights ranging from 7 to 12 nm were seen, in good agreement with the persistence length of MEH-PPV ($\approx$10 monomer units), with a small minority of larger ($>$20 nm) particles. AFM surface scans in tapping mode of the same sample region revealed a lateral broadening (along the direction of the scan angle) that depends *inversely* on the contact force. At low contact force (high set point), the cantilever interacts with the particle near the turning point of its

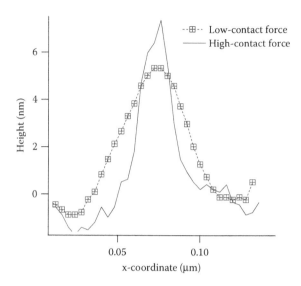

FIGURE 8.2

Representative AFM surface-height image of a selected MEH-PPV nanoparticle. The lateral surface-height profile of the particle is shown for both low- and high-contact force modes of operation.

oscillatory motion revealing an attractive particle–cantilever tip interaction that results in
lever-like motion of the nanorod manifested as a lateral broadening in the surface height
image. This is the opposite of what would be expected for a globular particle.

To further probe the intramolecular organization on individual z-oriented nanoparticles,
we have used polarization spectroscopy. The z component of the evanescent excitation
field at the air–cover glass interface was modulated by rotating between S (transverse electric) and P (transverse magnetic) input polarizations with a half-wave plate. The known
orientation of the emission moment for a given nanoparticle (from fitting the spatial fluorescence intensity patterns) allows the comparison of the measured fluorescence polarization anisotropy to geometrical approximations in some limiting possible cases. In these
measurements, polarization modulation in fluorescence intensity is a function of (1) the
structural organization within a given nanoparticle and (2) the projection of the absorption moment in the x,y plane. If the absorption moment is approximately equal to the emission, then excitation polarization anisotropy can be estimated. Histograms of anisotropy
parameters compared with approximation of simulated distortions for different possible
single-molecule morphologies including thin films compared to the experimental histogram of the z-oriented single-molecule nanoparticles differ significantly in peak value
(M = 0.92) from random coil (M = 0.1), thin film (blue curve), defect cylinder (M = 0.5), and
rod shaped (M = 0.7). This difference indicates a structure with more organization than
the other models, one that has a structure similar to that which is shown as computed from
molecular dynamics (MD) and mechanics simulations. This structure has a high degree of
chain organization with parallel chain segments located within a relatively short distance
(3.5 Å) stacked into a cylindrical morphology (full details of the simulations are given in
Section 8.3.

The single-molecule polymer nanoparticle species that we observed clearly show strongly
enhanced photochemical stability relative to similar molecules in spun-cast thin films, as
well as significantly higher modulation contrast in polarization anisotropy, and a narrow

bandwidth emission. These novel luminescence features appear to be derived from a high degree of structural order within each molecule, as well as decoupling from electronic perturbations or surface trap states by virtue of its orientation. In addition, our recent measurements using photon antibunching provide definitive experimental evidence on the high structural order and on the nature of the emissive site.[27] This study clearly indicates that luminescence occurs from an individual rod-shaped, oriented nanostructure and not from multiple sites or a coherent combination (super-radiant) of multiple sites.

The size range (5–15 nm) associated with dimensions of a single chain of a conducting polymer provides an interesting proving ground for theoretical models, as we are now in a position to directly compare experimental measurements with structural calculations. One further important piece of experimental evidence that sheds some light on the driving force behind the highly order and cylindrical morphology of these systems comes from fluorescence correlation spectroscopy (FCS).[28] The FCS measurements were carried out using the same types of freshly prepared solutions, and data were acquired periodically up to 100 h to study the polymer chain dynamics. FCS is based on fluctuations in the fluorescence intensity of the molecules diffusing in and out of a laser focal volume. FCS measurements were carried out using Nikon TE2000 inverted microscope operating in epi-illumination mode. A 514.5 nm line of an argon ion laser with a power of 100 μW at the objective was used as an excitation source. A high numerical aperture oil immersion objective (Nikon 100×, 1.3 NA) was used for high collection efficiency. Fluorescence from the sample was collected using the same objective and was directed into a high-efficiency avalanche photodiode (APD). A long pass filter (550 nm, Melles-Griot) and interference bandpass filter (Omega optical Inc.) were used to reduce background fluorescence. The signal from the APD was then sent to a correlator card (ALV-6010, ALV-Laser, Germany), which calculated the correlation function. The dilute MEH-PPV and CN-PPV solutions were loaded into a sample cuvette, and a rubber stopper was inserted into the top to prevent solvent evaporation. The fluorescence from the solutions was collected for 30 s, and the correlation curve was fit to a single-diffusion coefficient model. From extensive FCS studies, we have been able to show a clear correlation of solution-phase morphologies with the luminescence properties (dipole orientation, photostability, and photoluminescence spectra) of single molecules of various PPV systems. The initial nucleation of the compact-structured PPV systems is strongly favored by poorer solvents. In fact, a good solvent does not lead to the formation of a structured system as will also be demonstrated in Section 8.4 via MD methods. It appears that solution-phase molecular organization is the dominate factor underlying the generation of the highly structured PPV systems in single-molecule nanoparticles with some secondary contributions due to the droplet generation. The dynamical dependence of the structure and morphology on the type of solvent[29] and the secondary structural changes induced from confinement of a single PPV molecule to a nanoparticle are examined in more detail using computational and theoretical methods. In addition, the effects of structural changes, both intramolecular and overall morphological shape, on the excited-state electronic structure can be examined.

8.2.2 Poly(norbornene)-*g*-poly(3-hexylthiophene) (PNB-*g*-P3HT) Molecular Bottlebrush

In this section, we summarize our work on the synthesis of poly(3-hexylthiophene) (P3HT) molecular bottlebrushes composed of a poly(norbornene) (PNB) backbone and

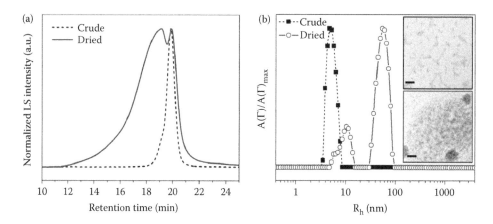

FIGURE 8.3
Experimental evidence for physical aggregation of PNB_{25}-g-$P3HT_{12}$ upon freeze-drying. (a) SEC for crude and dried samples in THF (LS detector). (b) Scattering amplitude distribution of crude and dried samples in THF at 25°C obtained by DLS (inset: TEM images for crude [top] and dried [bottom], where scale bar is 200 nm). (Reprinted with permission from S.-K. Ahn et al., *ACS Macro Lett.* 2, 761. Copyright 2013. American Chemical Society.)

regioregular P3HT side chains (denoted PNB_x-g-$P3HT_y$, where x and y represent the degree of polymerization [DP] of the backbone and side chain, respectively).[30] This work was designed to explore the implication of complex polymer architecture, specifically that of a molecular bottlebrush, on the assembly of conjugated polymers. Interestingly, the P3HT molecular bottlebrush displays a markedly strong physical aggregation that was first observed by size exclusion chromatography (SEC) after the P3HT molecular bottlebrushes were purified by precipitation, filtration, and freeze-drying from benzene. SEC traces of the crude (solution after the polymerization, but prior to any drying process) and dried solutions of PNB_{25}-g-$P3HT_{12}$ in THF are compared in Figure 8.3a. The signal from the light-scattering detector shows that dispersity of the crude solution broadens considerably upon freeze-drying: the appearance of a broad distribution of products emerging at lower retention time indicates the presence of high molecular weight material. The appearance of the high molecular weight materials is attributed to aggregation caused by the freeze-drying process. Moreover, this aggregation proved to be strong: redissolution in THF did not completely disperse the chains and aggregate sizes were only reduced when a strong ultrasonic wave was applied directly through tip sonication to the solution.

To further explore this behavior, dynamic light scattering (DLS) was performed on crude and dried solutions at ~1 mg/mL in THF. A single distribution ($R_h = 6$ nm) was observed from the crude solution while two discrete distributions ($R_h = 10$ and 48 nm) were observed from the dried solution (Figure 8.3b). The DLS results clearly suggest that the freeze-drying step causes aggregation of P3HT molecular bottlebrushes and agrees with the SEC results. The aggregation behavior of P3HT molecular bottlebrushes was further examined using transmission electron microscopy (TEM) to probe the thin-film structures, which are related to the aggregated structure adopted by the macromolecules in solution. TEM images (Figure 8.3b, inset) show that the P3HT molecular bottlebrushes deposited from the crude solution are more dispersed in contrast to observed aggregated clusters found in the dried product.

8.3 Computational and Theoretical Methods

The interplay between chemical composition, atomic arrangements, microstructures, and macroscopic behavior makes computational modeling and materials design extremely difficult. Even with the fundamental laws of quantum mechanics and statistical physics, the availability of high-performance computers, and with the growing range of sophisticated software systems accessible to an increasing number of scientists and engineers, the goal of designing novel materials from first principles continues to elude most attempts. On the other hand, computational experiments have led to increased understanding of atomistic origins of molecular structure and dynamics. In particular, Monte Carlo, MD, and molecular mechanics methods have yielded a wealth of knowledge on the structural behavior of various polymeric materials and their dynamic behavior as a function of temperature and pressure. Quantum chemistry methods, although generally difficult to directly apply to large molecular systems, allow *ab initio* determination of many-body molecular interactions, which can be used in classically based methods. In addition, these methods are now becoming applicable for systems containing several hundreds of atoms, making quantum mechanics-based prediction of structure and properties feasible. The combination of all of these computational chemistry methods clearly provides a good framework to examine some of the fundamental questions concerned with the role of solvation in controlling molecular self-organization of macromolecules and the resulting photophysical properties.

8.3.1 MD, Monte Carlo, and Molecular Mechanics

The MD, Monte Carlo, and molecular mechanics methods are well-reviewed and proven modeling methods and we only discuss our particular implementations. These methods require the specification of molecular potential energy functions to describe the various many-body interactions common in molecular systems. Since our interest here is to understand the structure and morphology in solution and in dry single-molecule nanoparticles, we have elected to use potential functions with a proven record of accurate prediction of structure and electronic spectra for conjugated organic molecules. These potentials are harmonic or Morse oscillators for the bond-stretching and angle-bending terms (both in and out of plane), truncated Fourier series for the torsion interactions (regular dihedral and improper), and Lennard-Jones (LJ) 6-12 plus Coulomb potentials for the nonbonded interactions. A number of standard force fields fall into this category, such as the MM2, MM3, MM4, Dreiding, UFF, MMFF, CHARMM, AMBER, GROMOS, TRIPOS, and OPLS models.[31] In the present study, we have used parameters defined within the MM3 model as this particular parameterization has proven to give very accurate results for structural optimizations of many conjugated organic molecules.[32] Of particular importance is the capability of the MM3 model to account for intermolecular interactions of the π-electron densities through the dependence of the stretching and torsion terms on iterative self-consistent field evaluations for the relevant π-conjugated bonds. The overall reliability of this model for structural calculations has continually been demonstrated for numerous aromatic compounds (benzene, biphenyl, annulene)[33] and conjugated systems (t-stilbene and even multiple oligomers of PPV).[34] The structures were also verified by comparing the results to those obtained from identical simulations using calculations with no assumed potentials via semiempirical and *ab initio* quantum mechanics.

Monte Carlo methods used in macromolecular science generally begin by constructing a Markov chain generated by the Metropolis algorithm (i.e., sampling of states according to their thermal importance: Boltzmann distribution for the ensemble under consideration, usually the canonical ensemble).[35] As the chain length of a simulated system becomes longer, it quickly becomes necessary to introduce a series of biased moves in which additional information about the system is incorporated into the Monte Carlo selection process in such a way as to maintain detailed balance. The most commonly used biased sampling techniques are the continuum and concerted rotation moves. These modern algorithms or slightly modified versions can efficiently generate dense fluid polymer systems for chain lengths of 30–100 monomers. Longer polymer chain lengths pose additional convergence problems and often require the use of other types of biased moves, in particular the double-bridging moves.[36] We are primarily interested in understanding the morphology and molecular structure of polymeric molecules composed of PPV-based molecules in dilute solution. The applicability of the Monte Carlo methods to this situation is somewhat different than in dense fluids, but the biased moves developed for that regime are still valid. Addition of solvent via continuum models (discussed in Section 8.3.2) or explicit atoms can also be easily implemented. One of our primary interests in using the Monte Carlo method is to insure adequate equilibration of longer chain polymers, that is, those with hundreds of monomers. Molecular mechanics and MD methods can also be used but generally require considerably longer times to equilibrate. For shorter chain lengths, however, we use these methods to obtain temporal data on the dynamical processes of chain self-organization.

Molecular mechanics methods use the laws of classical physics to predict structures and properties of molecules by optimizing the positions of atoms based on the energy derived from an empirical force field describing the interactions between all nuclei (electrons are not treated explicitly).[37] As such, molecular mechanics can determine the equilibrium geometry in a much more computationally efficient manner than *ab initio* quantum chemistry methods, yet the results for many systems are often comparable. However, since molecular mechanics treats molecular systems as an array of atoms governed by a set of potential energy functions, the model nature of this approach should always be noted.

MD simulations essentially consist of integrating Hamilton's equations of motion over small time steps. Although these equations are valid for any set of conjugate positions and momenta, Cartesian coordinates greatly simplifies the kinetic energy term. In our MD simulations, the integrations of the equations of motion are carried out in Cartesian coordinates, thus giving an exact definition of the kinetic energy and coupling and the classical equations of motion are formulated using our geometric statement function approach, which reduces the number of mathematical operations required by a factor of ~60 over many traditional approaches.[17] These coupled first-order ordinary differential equations are solved using novel symplectic integrators developed in our laboratory that conserve the volume of phase space and robustly allow integration for virtually any timescale.[17]

In investigating the self-assembly process of polymers, there is a timescale and length-scale gap between MD simulations with atomistic details and experimental methods. Atomistic molecular dynamics (AMD) simulations integrate the equations of motion at a given time step, typically femtoseconds. A characteristic AMD simulation using today's high-performance computers consists of about 10 million integration steps, which is equivalent to 10 ns consisting of $\sim10^5$ atoms and a simulation box size on the order of 10 nm. This is inadequate to sample self-assembly processes that occur in the order of 10–1000 ns, consisting of a mole of particles ($\sim10^{23}$) and domain sizes in the order of 10 nm. In order to bridge these gaps, coarse-graining is often used where an agglomerate of atoms (super atoms)[38] or the polymer is simply represented as beads and springs.[39] When combined with MD, this method is called coarse-grained molecular dynamics (CGMD).

Together with CGMD, we use efficient sampling methods such as umbrella sampling (US) to investigate the aggregation of P3HT molecular bottlebrushes to make comparison with experiments discussed in the prior section. The US method consists of a series of independent but related MD simulations. It can be used to calculate the free energy and potential of mean force (PMF). It employs a harmonic biasing function where the minimum of the bias is shifted to a desired location of a given reaction coordinate across the free-energy landscape. In so doing, the reaction coordinate is sampled. And for each of these simulations, the probability distribution can be estimated and then combined to provide the free energy across the reaction coordinate. One method of combining the probability distribution is the weighted histogram analysis method (WHAM),[40] which determines an optimal set of weights for each probability distribution derived from an independent simulation run.

8.3.2 Continuum Solvation Models

It is well known that solvation effects are critical to the structural and dynamical properties of macromolecules. In the present case, the polymeric nanoparticles are produced from a very dilute solution and show strong solvent dependences of the observed photophysical properties. Therefore, modeling the structure of these PPV-based nanoparticles must include the influence of the solvent. Although a "full" microscopic description of solvation is possible using MD or mechanics with explicit solvent molecules (there are still approximations in the many-body electrostatic interactions), this approach can be computationally time-consuming. As such, considerable research has previously been devoted toward developing reliable implicit solvent models in which the solvent molecules are generally replaced by a structureless dielectric continuum.[41,42] These models greatly increase the speed of the calculation and often avoid some of the convergence problems in explicit models (where longer simulations or different solvent starting geometries yield different final energies). The continuum models generally divide the solvation effect into nonpolar contributions treated in terms of the amount of solvent accessible surface area, and electrostatic contributions computed based on the Poisson–Boltzmann equation or one of its many simplifications (in particular the generalized Born [GB] models). Most schemes for evaluating the nonpolar components of the solvation-free energy are ad hoc. As such, a simple model is generally used, where the free energy associated with the nonpolar solvation of any atom is assumed to be characteristic for that atom and proportional to its solvent-exposed surface area, where SA is the exposed surface area and sigma is the characteristic "surface tension" (this is not the surface tension of the solvent but simply a parameter with units of energy per area) associated with the same atom. The solvent-exposed surface area can be computed by a variety of different procedures but one common method uses a spherical probe molecule rolling over the van der Waals surface of the solute atoms. The atomic "surface tension" parameters are generally taken from fits to collections of experimental data for the free energy of solvation in a specific solvent minus the electrostatic part computed via the GB method. These types of data fits are generally available for water, carbon tetrachloride, chloroform, and octanol solvents. Crammer and Truhlar[22,43] have developed the SMx models, which generalized the computation of the "surface tension" parameters to any solvent by making these values a function of more quantifiable solvent properties such as macroscopic surface tension, index of refraction, relative percent composition of aromatic carbon atoms and halogen atoms, and hydrogen-bonding acidity and basicity. These models also attempt to better account for other contributions to solvation such as the cavitation energy (making a "hole" in the solvent for the solute), attractive dispersion forces between the solute and solvent molecules, and local structural changes in the solvent such as changes in the extent of hydrogen bonding.

In principle, the atom-centered monopoles used by this GB model generate all of the multipoles required to represent the true electronic distribution. Currently, there are several different GB models, differing mainly in how the Born radii are computed. In the present study, we have implemented the analytical techniques that use a pairwise atomic summation to give the volume integration for the Born radii as described by Still et al.[44] and also by Hawkins et al. (the pairwise descreening method).[45] We have also used the Eisenberg–McLachlan Atomic Solvation Parameter model,[46] the original numerical integration of the solvent-accessible area implemented by Still et al. (ONION),[47] and the analytical continuum electrostatics solvation method of Karplus et al.[48] In the present case, we are interested in other solvents such as toluene, tetrahydrofurane (THF), and dichloromethane (DCM). These solvents can be considered to span the range of good (good solubility or highly solvated system) to bad (low solubility or low solvation) solvents for the PPV-based polymers of this study. Appropriate dielectric constants and surface probe radii were used for these cases.

The principle reason for examining the various models was to determine whether there were any qualitative changes in the dynamics and resulting structure due to the assumed continuum solvation model. Since we are not directly concerned with quantitatively accurate structures at this point in the polymer particle formation, the particular computational details of the GB model should not cause many large changes. From our studies using the various implementations of continuum solvation, the qualitative solvated morphologies are indeed quite similar—there is a collapse of a PPV-based molecule into an "organized" folded structure, as shown in Figure 8.5. The results we report in the present chapter are therefore only given based on those obtained from the GB/SA model as described by Still et al.

8.3.3 Single-Molecule Nanoparticle Formation Procedure

The procedure we used to model the overall experimental process for producing single-molecule PPV polymers was to start a GB/SA-MD simulation at a randomly chosen configuration of the PPV polymer, where we used substitution on PPV backbone to give 2-methoxy-5-(2′-ethyl-hexyloxy)-p-phenylene vinylene (MEH-PPV) and 2,5,2′,5′-tetrahexyloxy-7,8′-dicyano-p-phenylene vinylene (CN-PPV). MD simulations are performed until the geometry of the PPV systems reaches an "equilibrium structure" as measured by the fluctuation in the total nonbonded energy, end-to-end distance, and an orientational autocorrelation function of a unit vector oriented along the main chain backbone. This typically requires a trajectory on the order of nanoseconds, somewhat dependent on the solvation model but mainly on the nature of the substituted PPV polymer. Figure 8.4 shows the final geometry of a MEH-PPV polymer consisting of 28 monomers and three sp^3 (tetrahedral) defects located every 7 monomers. In this particular figure, the time evolution of the positions of the chain ends is marked as a solid red and blue lines (the side chains are not shown for better clarity) and the green and yellow lines show the progression of the positions for two of the sp^3 defects. Following these computations, which are taken as giving the initial qualitative but crucial stages of folding of the polymer in solvent, the polymer structure obtained is used to start a second series of computations, which include explicit solvent molecules (Figure 8.5). This approach is used in an attempt to reduce any "minor" structural dependence on the continuum model as well as to better account for minor differences between the solvents (THF, DCM, toluene). MD and molecular mechanics simulations are used in the full system to obtain a new structure, called MEH-PPV-sol and CN-PPV-sol. These new structures are used to make correlations to experimental observations in solvent. Similar

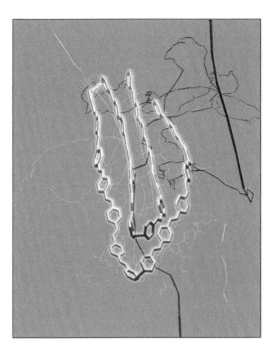

FIGURE 8.4
A final structure for a 28-monomer (with three tetrahedral defects) MEH-PPV molecule in a bad solvent. The side groups are made transparent and the solid colored lines indicate the progression of the folding dynamics: red is one of the chain ends and blue is the other. The green and yellow lines mark the progression of two of the tetrahedral defect sites.

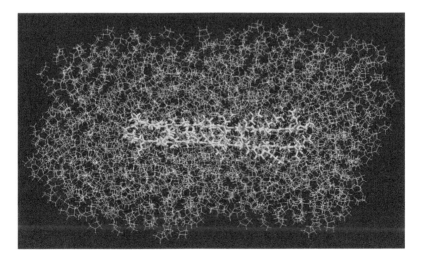

FIGURE 8.5
An explicit atom representation and the optimized geometry for a MEH-PPV molecule (highlighted in green) in tetrahydrofurane (THF). The initial geometry of the MEH-PPV molecule was obtained from GB/SA-MD simulations.

computations are also performed using the Monte Carlo approach with a combination of reptation, coupled-decoupled configuration-bias, concerted rotation, pivot, translation, and aggregation volume Monte Carlo moves.

To emulate the production of solvent-free single-molecule nanoparticles from solution, the MEH-PPV-sol and CN-PPV-sol structures are minimized in a vacuum using a combination of molecular mechanics with simulated annealing. Simulated annealing is used to take the room temperature solvated structures quickly through a temperature decrease as occurs during evaporative cooling in the experimental nanoparticle generation procedure. Combining this with soft boundary conditions to impose an overall cylindrical shape (this was only used for relatively long polymers >70 monomers long and is based on the experimental evidence discussed above) gives the final PPV polymer structure for a dry single-molecule polymer nanoparticle. Similarly, one might obtain some useful information on multimolecule (thin films and solutions are composed of many molecules) structural organization in thin films or to probe concentration dependences in solution by using periodic boundary conditions. For thin films, only two dimensions are periodic but with a substrate placed parallel to the periodic imaged axes. In the present chapter, we do not present any results for thin-film molecular structures but are currently performing computations in this area. To investigate concentration dependencies of the structure and morphologies in solution, 3-D periodic images are used where the size of the imaged box is varied to emulate different concentrated solutions. These results can be compared to the continuum solvation and nonperiodic explicit solvent model results, which emulate very dilute solutions thereby giving two concentration extremes, very concentrated and very dilute.

Figure 8.6 shows a typical MEH-PPV single-molecule nanoparticle structure obtained from the modeling and simulations procedures just described. A rod-shaped, compact structure with conjugated chain stacking (π stacking) is readily apparent.

8.3.4 *Ab Initio* and Semiempirical Quantum Chemistry

In order to eliminate any fundamental dependence on the assumed potential model, one must resort to the more complicated and time-consuming methods of quantum chemistry. While clearly these techniques in theory provide the correct treatment of molecular

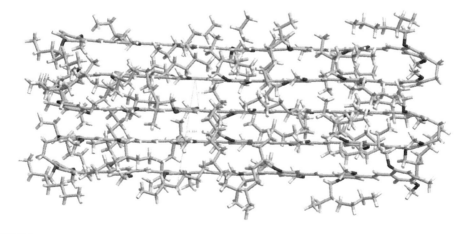

FIGURE 8.6
Final structure of a 35-monomer (four tetrahedral defects) MEH-PPV molecule obtained through sequences of continuum and explicit solvent molecular dynamics and mechanics simulations combined with simulated annealing.

and atomic systems, incomplete basis sets and limitations due to electron correlation or exchange for the methods capable of treating reasonably sized molecules can present considerable complications. In the present case, we are primarily interested in using quantum chemistry methods to evaluate structural geometries as determined by the classical deterministic (molecular mechanics and dynamics) and stochastic methods (Monte Carlo) described above. The quantum methods also provide an avenue for computing electronic absorption and emission spectra for comparison to observed experimental luminescence spectroscopy. Since the excited-state structure depends on the ground state, it is extremely important to have the correct optimization of the structure for the ground state. A structure taken from other computations will most likely not be the optimal one for the representation of the wavefunction in quantum chemistry calculations and likewise computations of excited-state properties will not conform to any meaningful result. As such we feel that it is necessary to go through some of the fundamental details of the various quantum chemistry methods so as to clearly define what we have done and why.

Quantum mechanical methods are most generally based on the wavefunction approach,[49] although density functional theory (DFT), which determines the ground-state electronic structure as a functional of the electron density,[50] offers an equivalently powerful method (Equation 8.7). The wavefunction approach to electronic structure determination begins with the Schroedinger equation in which the Hamiltonian operator is composed of the kinetic energy of the nuclei and the electrons, the electrostatic (Coulombic) interactions between the nuclei and the electrons, and the internuclear and interelectronic repulsions. Simplifications are generally made by imposing the Born–Oppenheimer approximation, which assumes that the nuclei do not move, and hence have no kinetic energy. This removes the terms for the nuclear kinetic energy and the nuclear–nuclear repulsions become a constant that is simply added to the electronic energy to get the total energy of the system. In order to determine a practical representation of the unknown wavefunction, the many-electron wavefunction is replaced by a product of one-electron wavefunctions. The simplest such replacement, called the Hartree–Fock (or single determinant) wavefunction, involves a single determinant of products of one-electron functions, called *spin orbitals*. Each spin orbital is written as a product of a space part, Φ, describing the location of a single electron, and one of two possible spin parts, α or β. The space part is essentially a molecular orbital, which can be occupied by only two electrons of opposite spin. By using a linear combination of atomic orbitals (LCAO) approximation (the molecular wavefunctions are composed of atomic wavefunctions, combined linearly), one obtains what are called the Roothaan–Hall equations. These equations can be solved numerically using a self-consistent iterative procedure, in which each molecular orbital is evaluated under the influence of an average potential field from the other electrons. The derived MO then contributes to the field used for the next MO, and the process repeats until it is internally consistent within specified limitations (SCF-HF).

The total Hartree–Fock energy then is given by

$$E^{HF} = E^{nuclear} + E^{core} + E^{Coulomb} + E^{exchange} \tag{8.1}$$

The four terms are defined as follows:

- Nuclear term

$$E^{nuclear} = \sum_{A<B}^{nuclei} \frac{Z_A Z_B}{R_{AB}} \tag{8.2}$$

- Core term

$$E^{core} = \frac{1}{2} \sum_{\mu}^{\text{basis functions}} \sum_{\nu}^{\text{functions}} P_{\mu\nu} H_{\mu\nu}^{core} \qquad (8.3)$$

- Coulomb term

$$E^{Coulomb} = \frac{1}{2} \sum_{\mu}^{\text{basis functions}} \sum_{\nu}^{\text{functions}} P_{\mu\nu} F_{\mu\nu}^{Coulomb} \qquad (8.4)$$

- Exchange term

$$E^{exchange} = \frac{1}{2} \sum_{\mu}^{\text{basis functions}} \sum_{\nu}^{\text{functions}} P_{\mu\nu} F_{\mu\nu}^{exchange} \qquad (8.5)$$

where the density matrix, $P_{\mu\nu}$, the elements of which are the squares of the molecular orbital coefficients summed over all occupied orbitals, is

$$P_{\mu\nu} = \sum c_{\mu i}^* c_{\nu i}^* \qquad (8.6)$$

Computations based on these approximations give quite good structures for molecules containing main group elements. However, even in the limit of a complete basis set, the Hartree–Fock energy is not equal to the experimental energy of a molecule, largely because of the error introduced by using the SCF model (electron correlation being one problem). These methods also scale somewhat poorly as a function of the number of basis functions $\sim N^4$, which tend to limit calculations to systems with less than 100 atoms. On the other hand, DFT, which is based on the Hohenberg–Kohn theorem, scales more like N^3 and in theory accounts for electron correlation (although this assumes correct exchange-correlation functionals).[51] In DFT, the minimum energy of a collection of electrons under the influence of an external field (the nuclei) is a functional of the electron density. The energy includes the same nuclear, core, and Coulomb terms as the Hartree–Fock energy. However, the HF exchange energy is replaced by an exchange-correlation functional, $E^{XC}(\rho)$, leading to the Kohn–Sham equation:

$$E^{DFT} = E^{nuclear} + E^{core} + E^{Coulomb} + E^{XC}(\rho) \qquad (8.7)$$

This unique functional valid for all systems can be formally proved, but an explicit form of the potential has been difficult to define, hence the large number of density functional (DFT) methods. Functional forms are often chosen based on which one give the best fit to a certain body of experimental data, which makes DFT more like a semiempirical method.

One advantage of using electron density is that the integrals for Coulomb repulsion need to be done only over the electron density, which leads to the N^3 scaling. The other advantage is that use of electron density automatically includes at least some of the electron correlation. Among the simplest models are those called local density models, such as the SVWN (Slater, Vosko, Wilk, Nusair)[52] functional. The basis of these models is the assumption of a many-electron gas of uniform density. This model is not generally expected to be satisfactory for molecular systems, in which the electron density is nonuniform. However, for band structure it is quite satisfactory. Substantial improvement can often be obtained by introducing explicit dependence on the gradient of the electron density, as well as the density itself. Such procedures are called gradient-corrected, or nonlocal DFT models. The most popular of these models are the so-called BP (Becke, Perdew),[53] BLYP (Becke, Lee, Yang, Parr),[54] and PBE (Perdew, Burke, Ernzerhof).[55] Another class of DFT models combines the exact Hartree–Fock exchange with a DFT exchange term, and adds a correlation functional.[56] In general, DFT-based methods can treat larger systems and even provide a route to achieve O(N) scaling. However, the current state of the art is still somewhat premature to be used as a black-box.

Semiempirical molecular orbital methods lie somewhere between classical molecular mechanics and *ab initio* quantum mechanics methods.[57] This approach makes use of a number of experimentally determined parameters but like fundamental *ab initio* quantum methods is based on a SCF-Hartree–Fock solution with atomic orbital basis functions. Semiempirical models use s orbitals and the p_x and p_y orbitals for the valence shell. The remaining part of each atom is treated as a "core," with a net or effective charge equal to the atomic number Z minus the number of inner shell electrons. The mathematical representation of each of the four basis orbitals used in constructing the LCAO is a representation that reflects the spatial distribution of the electrons occupying the orbitals. The integrals are divided into two sets, H and S, according to whether they contained the Hamiltonian operator. Semiempirical methods represent the H-type integrals as a sum of five terms:

1. One-center, one-electron integrals, which represent the sum of the kinetic energy of an electron in an AO on atom X and its potential energy due to attraction to its own core

2. One-center, two-electron repulsion integrals

3. Two-center, one-electron core resonance integrals

4. Two-center, one-electron attractions between an electron on atom X and the core of atom Y

5. Two-center, two-electron repulsion integrals

The integrals of types (1) and (2) are replaced by numerical values obtained by fitting spectroscopic values of electron energies in various valence states. Because these parameters are based on experimental data from real molecules, in which electron correlation is a fact of nature, some correlation is built into semiempirical methods. This makes up, in part, for the failure of semiempirical methods to consider correlation explicitly. Integrals of type (3) represent the main contribution to the bonding energy of the molecule. Semiempirical methods treat these as proportional to the overlap integral, S:

$$\beta = f_x S_{ij} \tag{8.8}$$

where f_x is an adjustable parameter. Note that none of these terms are set to zero as in Hückel theory; however, they do not contain any explicit dependence on interatomic distance. A modified electrostatic treatment is used to evaluate the integrals of type (4). They are taken as proportional to eZ_{eff}/r, with an adjustable parameter for the proportionality constant. They also include a term of the form exp-αR, where α is an adjustable parameter. This function is included to ensure that the net repulsion between neutral atoms vanishes as their separation goes to infinity.

The remaining integrals, type (5), represent the energy of interaction between the charge distribution at atom X and that at atom Y. They are calculated as the sum over all multipole interactions, again with adjustable parameters. STOs are used to represent the spatial distribution of the charges. The overlap integrals, S, are evaluated analytically; however, the orbital exponents are treated as adjustable parameters.

Core repulsions (the core of one atom interacting with the core of the next) are the remaining item to be evaluated. This is done with two-center repulsion integrals, the cores being represented as Gaussians. In order to reduce excessive repulsions at large atomic separations, which arise in a more classical electrostatic treatment, attractive Gaussians are added to the repulsive ones for most elements. Finally, the total energy of the molecule is represented as the sum of the electronic energy (net negative) and the core repulsions (net positive).

With this general structure of the model completed, a so-called training set of molecules is selected, chosen to cover as many types of bonding situations as possible. A nonlinear least squares optimization procedure is applied with the values of the various adjustable parameters as variables and a set of measured properties of the training set as constants to be reproduced. The measured properties include heats of formation, geometrical variables, dipole moments, and first ionization potentials.

Depending upon the choice of training sets, the exact numbers of types of adjustable parameters, and the mode of fitting to experimental properties, different semiempirical methods have been developed, ranging from the AM1 (Austin Model 1) method of M. J. S. Dewar and the PM3 (Parameter Model 3) method of J. J. P. Stewart, to variants of the older MNDO models.

8.3.5 Excited States: CIS-ZINDO/s

In order to obtain information pertinent to the electronic excitation such as optical absorption (vertical transitions) and emission (adiabatic transitions), computations that probe the electronic excited states are required. Given a good approximation of the electronic ground state, which clearly requires optimization of the structure within a reasonable level of theory and for a reasonably complete basis set, there are a number of possible methods that can be used to probe the electronic excited states. However, most of the more rigorous methods such as coupled clusters (CC), configuration interactions (CI), and other multireference methods (CASSCF, etc., generalized valence bond) scale very poorly with the number of basis functions and therefore cannot be effectively applied to the relatively large systems of the present study. DFT combined with the response due to a linear electric field that is fluctuating (linear response or RPA for HF), referred to as time-dependent density function theory (TDDFT), can in principle overcome the scaling problems. However, we have found through extensive computations that this approach is particularly inaccurate for the current systems, which tend to exhibit considerable charge-transfer character of the wavefunction (there is considerable literature on this topic[58]). While TDHF (time-dependent Hartree–Fock, the RPA) can also be used, the effect of electron correlation is neglected. Another approach is to resort back to the semiempircal methods, in particular those of Zerner et al., which have parameterized the INDO method based on fairly extensive spectroscopic data (sometimes referred to as INDO/s or ZINDO/s).[59] This approach, in principle, does take into account some of the effects of electron correlation in that it is fit to experimental data. When combined with configuration interaction singles (CIS), it can be very successful for a variety of transitions, including $\pi \rightarrow \pi^*$ transitions.[60] Since this approach is based on a semiempirical model, it is reasonably computational-efficient when compared to the fully *ab initio* methods. Bredas and coworkers[52] have shown how this method can be very successful for conjugated organic systems. As such, the results discussed for electronic transitions are only given for those calculations based on CIS-ZINDO/s.

8.4 Results and Discussion

Molecular modeling can often provide an efficient method for visualizing processes at a submacromolecular level, which also can be used to connect theory and experiment. Particularly attractive from a computational point of view is that polymer nanoparticles are very close to the size scale where a complete atomistic model can be studied without using artificial constraints such as periodic boundary conditions, yet these particles are too small for traditional experimental structure/property determination. Polymeric particles in the nano- and micrometer size range have shown many new and interesting properties due to size reduction to the point where critical length scales of physical phenomena become comparable to or larger than the size of the structure itself. This size-scale mediation of the properties (mechanical, physical, electrical, etc.) opens a facile avenue for the production of materials with pre-designed properties.[61] It is therefore extremely important to develop an understanding of these phenomena.

It is important to realize that details of the molecular structure for these types of conjugated organic systems are fundamental for determining photophysical properties involving electron excitations since subtle variations of the π-conjugation can strongly influence the outer valence electron levels. In order to gain more insight into the structural organization of MEH-PPV and CN-PPV, a combination of MD and molecular mechanics (using the MM3 potential), simulated annealing (using a linear annealing schedule over a 1-ns trajectory), semiempirical quantum mechanics (AM1 and INDO/s levels),[62] and *ab initio* quantum mechanics (Hartree–Fock, second-order Moller–Plesset perturbation theory, DFT, and CIS)[63] were performed as described above in the methods section. The minimum energy configuration of single-molecule systems consisting of 14–112 monomers with tetrahedral defects located every 7 monomers, as well as similar computations with more random locations, was determined for both MEH-PPV and CN-PPV. These simulations were performed for isolated single-chain systems (no solvent, in order to make contact to what has typically been done for these types of systems) and with inclusion of solvent via the continuum model of the generalized Born approximation (GB/SA) as well as with explicit solvent molecules (THF, toluene, dichloromethane). It is important to realize that details of the molecular structure for these types of conjugated organic systems are fundamental for determining photophysical properties involving electron excitations since subtle variations of the π-conjugation can strongly influence the outer valence electron levels. As such, the majority of the present chapter is devoted toward describing results obtained from the determination of the structure and morphologies of the various PPV systems produced by our specific experimental-droplet generation technique discussed above.

8.4.1 Structure of PPV-Based Systems

Based on molecular mechanics energy minimization in a vacuum, a single MEH-PPV or CN-PPV molecule containing tetrahedral defects has a lower total potential energy when folded into cofacial stacked chain segments (Figure 8.6) than in an extended chain configuration: on the order of 10s of kcal/mol determined from the MM3 model for the all *trans-anti* configuration. For MEH-PPV, an all *trans-syn* configuration lies at ~3 kcal/mol higher in potential energy than the *anti* configuration due to steric interactions of the side groups. For CN-PPV the *syn* configuration is ~4 kcal/mol lower in potential energy mainly due to the strong intermolecular interactions of the CN groups. The *syn* configuration is also lower in energy for PPV, ~2 kcal/mol relative to *anti*. Similar results can be found for various isomeric forms, ranging from having the vinyl linkages all *cis* to random amounts of *cis/trans*. For molecules with shorter oligomer segments between tetrahedral defects, steric repulsion can become significant and folding is not energetically favorable for less than four monomers.

The intermolecular interactions in all of the PPV-based systems are strong enough to lead to efficient nucleation of multiple oligomer chains into aggregates with varying degrees of crystalline order. For a PPV system consisting of seven chains with eight monomers, based on molecular mechanics optimization using the MM3 model (converged to a root mean square gradient of 10^{-5}), a herringbone-type packing arrangement is found with identifiable crystallographic parameters of a = 5.2 Å, b = 8.0 Å, and c = 6.4 Å, and a setting angle of 58°. This is in good agreement with experimental determination and with recent molecular mechanics simulations.[64] For an MEH-PPV system of the same size, aggregate formation occurs but the packing does not conform to a herringbone-type arrangement but instead more to a cofacial staking of the oligomers with an average interchain separation of ~3.7 Å. CN-PPV behaves quite similar to MEH-PPV except with a closer interchain separation of 3.4 Å and a greater degree of helical twisting (the phenyl rings remain cofacial). This type of π stacking of the conjugated oligomers leads to increased charge carrier mobility by generating large valence or conduction bandwidths proportional to the orbital overlap of adjacent oligomers. Quantum chemistry computations show increased transport properties and indicate decreased luminescence quenching for π-staked structures[65] as well as substantial self-solvation affects that lead to greater enhancement of orbital overlap. A herringbone packing arrangement destroys the orbital overlap between oligomers (see discussion below on electronic spectra), and due to the two symmetry-inequivalent oligomers in a unit cell, leads to a splitting of the bands (Davydov splitting), which subsequently can cause luminescent quenching. Thus, it is desirable to generate PPV-based systems with a π-stacked arrangement.[66] The single-molecule nanoparticles generated by our experimental procedure provide this unique capability by using a combination of 3-D confinement and solvent-induced morphologies in the absence of an interacting substrate.

We have also carried out extensive semiempirical quantum mechanics calculations to determine single-molecule structures. For these calculations, the initial geometry was taken from an MM3 optimized structure. The results obtained from the AM1 calculations (PM3 results were very similar) show larger interchain separation for both MEH-PPV and CN-PPV molecules with a fold (as much as 0.5 Å). The optimized AM1 structures

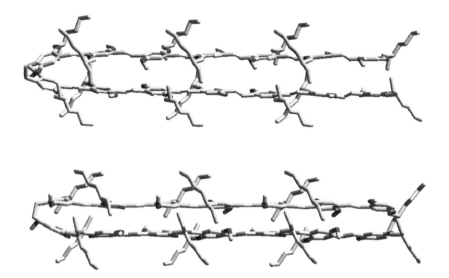

FIGURE 8.7
Structure of a 14-monomer MEH-PPV molecule with one tetrahedral defect: top determined with AM1, bottom determined with MM3. Note the increased interchain separation (about 1 Å larger) and backbone twisting for the AM1 structure.

for folded 14-monomer MEH-PPV and CN-PPV molecules have larger torsional rotations about the bonds adjacent to the vinyl C=C, which tends to force the oligomer chains farther apart (Figure 8.7).

On the other hand, for just PPV (no side groups), the structures for small oligomers (up to four monomers) and even multiple cofacial oligomers are reasonably similar to the MM3 results. However, as the number of monomers increases, deviations from planarity occur even for PPV. Again, the result is a significantly different structure than that obtained from classical MM3 molecular mechanics. Either the enhanced intermolecular interactions caused by the cofacial arrangements of the substituted PPV oligomer chains (dispersive van der Waals forces) cause considerable changes that are not accounted for in the MM3 model or the semiempirical AM1 parameterization or model (perhaps inadequate electron correlation) is not appropriate for this type of conjugated system. Since the MM3 generated multi-oligomer aggregate structure is particularly accurate compared to experimental determinations for PPV, the source of the difference in structures would appear to be in the semiempirical models. In addition, the AM1 optimized geometric structures for MEH-PPV do not give electronic structure results for the vertical transitions that agree with experiment, while the vertical transitions computed based on the MM3-optimized structures conform quite closely to experimental results. Full quantum calculations using wavefunction or DFT also give approximately planar structures with interchain distance on the order of 3.5 Å for both PPV and MEH-PPV. As such, geometry optimizations of multi-oligomer PPV-based systems using the semiempirical AM1 and PM3 models should be questioned. We recommend using the much faster, and apparently for this particular case, accurate, classical molecular mechanics minimization of the MM3 force-field for the PPV-based systems. In addition, structural differences between semiempirical AM1, DFT, and HF calculations have also been previously observed for small oligomers of PPV.[67]

One consistent structural observation obtained from the various optimizations using the MM3 and AM1 models is that while small oligomers of PPV with no substitution

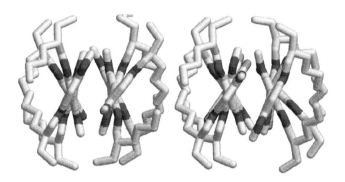

FIGURE 8.8
Structures of MEH-PPV (left) and CN-PPV (right) obtained from AM1 semiempirical calculations. The systems consist of 2–4 monomer oligomers without a fold. The twist is more pronounced in these types nonfolded oligomer systems.

tend to form cofacial-planar geometries with a shift along the chain axis, MEH-PPV and CN-PPV tend to have some "helical" twisting of the PPV backbone (Figure 8.8), with a larger angle for CN-PPV. The phenyl rings in these structures are rotated about the backbone by about ±6° for MEH-PPV and ±10° for CN-PPV but maintain an approximately cofacial orientation with respect to the phenyl rings on the neighboring chain. Addition of a fold and larger oligomers causes the twist angle to decrease (depending on the oligomer length between the folds). For example, a twist is present in structures determined from MM3 calculations with folds and longer oligomer segments but not for those with <4 monomers. Quantum calculations based on Hartree–Fock SCF theory using a modest basis set (6-31G*) gives a structure for MEH-PPV consisting of eight monomers and one tetrahedral defect that has an interchain separation of $d^{IC}=3.53$ Å and has very little helical twisting. The intermolecular interactions of a multiply folded MEH-PPV molecule will tend to decrease any backbone twisting as well as decrease the interchain separation. This influence, often referred to as self-solvation, is discussed in more detail in Section 8.4.3.

The values for interchain separations are probably not as accurate as the general trend of decreasing interchain separation for CN-PPV as compared to MEH-PPV. The interchain distance reported from X-ray diffraction studies of MEH-PPV thin films[68] is $d^{IC} = 3.56$ Å and is good accord with our results from MM3 (discussed above) and full *ab initio* quantum calculations carried out for short oligomer chains of PPV (no side chains) as well as folded MEH-PPV molecules at the MP2/3-21++G level of theory. For short bi-oligomer PPV systems, the syn conformation is of lower energy than the anti conformation by ∼0.96 kcal/mol (side chains change this result as noted earlier). An optimized structure for four-monomer PPV systems has an interchain distance of $d^{IC} = 3.5$ Å and there is a shift of 0.98 Å along the chain backbone axis and one of 0.7 Å along the remaining axis. Addition of side groups to produce MEH- or CN-PPV leads to geometries that do not show the shifts and $d^{IC} = 3.53$ Å for MEH-PPV and $d^{IC} = 3.4$ Å for CN-PPV. Whether the interchain distance and the shifts (for PPV) about the remaining axes depend on the number of monomers in the oligomer segments or to chain folds is clearly of interest. Unfortunately, the only way the shifts can be quantified is through full quantum calculations using many-body perturbation theory, which scales like N^5. The largest system that we have been able to treat at this level of theory and get converged results is for two oligomers consisting of four monomers, which does maintain the shifts for PPV. While we believe the shift gives a true structural minimum, the values we obtain are significantly smaller than those typically used for stacked PPV molecules (generally half a unit cell length, ∼3.3 Å). As will be shown below, the degree of orbital overlap is strongly dependent on both the inter-chain distance and the shifts along the other axes and it is therefore important to obtain an accurate initial structure. The degree the shifts along the two axes change as the system is taken from the very small isolated cluster to the bulk can be examined reasonably effectively by using periodic DFT-LDA calculations. Here we have carefully calibrated the particular implementation (plane wave norm-conserving pseudopotentials) of the DFT-LDA method to the geometry of the MP2 study to ensure consistency. We emphasize the importance of doing this calibration as many geometry optimizations for short oligomers (from three to four monomers per oligomer) based on either DFT or SCF-HF theory can generate optimal (lowest energy) geometries that are "T shaped." Inclusions of electron correlation and dispersion appear to be crucial in obtaining the cofacial geometries for a given basis set ranging from STO-3G * to cc-aug-pVTZ. We mention this in passing only to point out some of the many problems with electronic structure geometry optimization when used in the black-box context. Errors from the incompleteness of the basis sets and from electron correlation tend to cause significant variation in the optimized geometries and likewise for the vertical transitions. On the other hand, a plane-wave basis function representation is a "complete" one (there is no BSSE) and with careful selection of the representation of the core via pseudopotentials, these calculations are considerably quicker yet often provide very accurate results. From the periodic DFT-LDA calculations of four-monomer PPV oligomers, we find that interchain separation decreases somewhat but the shifts about the two other axes are not significantly altered. The decreased interchain distance going toward a more bulk-like phase is assignable to a self-solvation effect that we discuss in more detail below and the persistence of the shift about the other axes means these structural details should be noted in any excited-state calculation for these types of stacked PPV oligomers.

8.4.2 The Effects of Solvent on Structure and Morphology

On the basis of the above results, it would not appear entirely unreasonable to begin geometry optimization with an organized (stacked oligomer segments) folded or a cofacial oligomer structure in vacuum (the most common starting point assumed in previously reported calculations for PPV[40,43,69]). On the other hand, it should also be mentioned that previous Monte Carlo simulations for simplified models consisting of beads on a chain (no atomic structure or interactions were present in the models) suggest random coil-like geometries (defect-coil structures since little to no rotation happens about the unsaturated C=C bonds) would be preferential.[11] Indeed, these types of results appear more in accord with the standard interpretation within polymer science for the structure of macromolecules in dense fluids.[70] However, in past molecular mechanics models the determination of the minimum energy configuration for PPV-based oligomers has been performed assuming a cofacial arrangement of very short oligomers without folds. This implies highly organized initial structures. While either modeling approach is certainly not unreasonable, neither they provide information directly related to the solution-phase morphologies, which are clearly important as indicated by the QM/MM results discussed above, nor do they provide any details on the dynamical processes leading to such structures. Since there is fairly substantial evidence from experiment as discussed/shown in the previous section that the solution-phase morphologies are crucial to those of thin films and in particular to those of single-molecule nanoparticles, it is clear that one needs to directly take into account solvent effects. In the computations discussed in the next paragraph, it is shown that solvent indeed provides the key to producing self-organization into compact and structured morphologies.

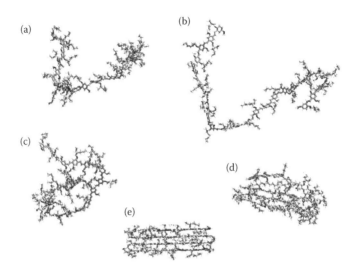

FIGURE 8.9
Snapshots taken during a 1-ns GB/SA-MD simulation for a model of a bad solvent.

Figure 8.9 (also Figure 8.4) shows the progression of a typical GB/SA MD simulation over a 1-ns trajectory (the total trajectory time to reach and "equilibrated" conformation depends on the number of monomers in the molecule and on the nature of the side groups). The initial configuration (Figure 8.9a) was obtained by propagating an MD simulation of a linear chain of MEH-PPV consisting of 28 monomers and 3 tetrahedral defects at an elevated temperature (800 K) for 10 ps. This allows the system to sample some of the possible phase space available from which an individual geometry is randomly selected. The next set of snapshots (Figure 8.9b–d) show how the MEH-PPV chain folds in a bad solvent (treated as a continuum dielectric) at the tetrahedral defects during the course of a 1-ns MD simulation. The structure and morphology obtained from this simulation (Figure 8.9d) is one that shows considerable folding into a reasonably compact structure with a rod-shaped morphology. While the oligomer chains between the tetrahedral defects do not stack perfectly cofacial (interchain distances range from 3.5 to 4 Å and there are shifts of about one phenyl ring along the chain axis of one oligomer with respect to another as well as some backbone twisting along the chains), there is a definite preference for this organization. Determination of the final or lowest energy configuration (Figure 8.9e) was obtained by using a combination of simulated annealing and molecular mechanics without any solvent interactions (an attempt to emulate a dry particle as is obtained from the experimental generation) starting with the structure obtained from the MD simulation with solvent (Figure 8.9d). Some secondary organization is notable, in particular the preponderance toward cofacial oligomer chain stacking (the interchain distance becomes more uniform) and reduced backbone twisting (Figures 8.9e and 8.6). Similar simulations with inclusion of shorter oligomer segments (three to six monomer defects) only show folding at the defects sites separated by at least four monomers. The single-molecule systems appear to nucleate at a particular oligomer length and the remaining oligomer segments conform very closely to this nucleated size even if there is oligomer segments composed of fewer monomers. This structural arrangement tends to maximize the intermolecular interactions and still leads to a rod-shaped structure of near-uniform length. While this type of a single-molecule nanoparticle is composed of multiple conjugated oligomer segments, the shape and length conform to one particular size and due to extensive orbital overlap induced by this structure, it has luminescence properties like a single chromophore of fixed length. Longer MEH-PPV molecules, those with more than 100 monomers, with regularly spaced tetrahedral defects tend to have a much slower equilibration time and go through a number of complicated geometric and structural changes on a timescale of 10 ns. Multiple nucleation sites can occur, which leads to self-organization into a number of different regions of structural regularity. This results in a single-molecule system that is composed of several regularly stacked oligomer segments separated by several angstroms. This type of solution-phase system should not exhibit the luminescent properties we have measured experimentally for the PPV-based nanoparticles, and the secondary structural changes induced by nanoparticle formation process become an essential component of driving these conjugated systems to the required rod-like structures. Finally, there is a third component to generating the high amount of structural regularity unique to the single-molecule nanoparticles. This comes from the excess electric charge on the nanoparticles and is also key for achieving z-orientation on the deposition substrate. Based on semiempirical quantum calculations for a MEH-PPV structure with one excess electron, we note a clear preference for the z-orientation as well as some enhancement in the structural organization toward π-stacks with reduced interchain separation.

The same type of MD simulations but in a continuum model of a good solvent such as DCM does not lead to folded or compact structures but to extended chain conformations with random amounts of folding at the tetrahedral defects. The final structure obtained from this particular type of simulation does not have a compact morphology and is more accurately described as a defect-extended chain. There is clearly a substantial difference in the dynamics and resulting structure and morphology of a MEH-PPV molecule in the different solvents. In a later part of this section, we discuss results obtained by using an explicit solvent model, which allows us to account more accurately for the differences between three solvents: toluene, THF, and DCM. The solution-phase structural differences are small enough not to induce any large alterations in the final dry-single-molecule nanoparticle structures or morphologies, although there are some details that may be important to the solution-phase spectral measurements.

The effects of explicit solvent (full inclusion of all of the atoms) on the structure of MEH-PPV and CN-PPV were also examined. In these calculations, explicit solvent molecules were added to an MD box to achieve the appropriate density followed by aggressive energy minimization using molecular mechanics. In order to reduce the number of required solvent molecules, we started all calculations with a polymer molecule that had already been equilibrated in a continuum solvent simulation (such as that shown in Figure 8.9d). This approach also allows us to examine solvents that have very similar dielectric constants. It should be stressed that these simulations began with an initial structure determined from a continuum model for toluene. The structure is already folded. A continuum model for DCM does not lead to a folded structure but one that has very little folding and is more accurately described as a defect-extended chain. The final structures obtained were qualitatively similar to those obtained for the continuum solvent systems but with some notable quantitative differences. For MEH-PPV, the interchain distance increased from $d^{IC} = 3.7$ to 4 Å in THF and to $d^{IC} = 4.1$ Å for DCM but decreased to $d^{IC} = 3.4$ Å in toluene. The interchain distance for CN-PPV did not seem to be as strongly dependent on the solvent but did show an increased helical backbone structure over the continuum model. The PPV structures in DCM were clearly much more disorganized compared to the other solvents, with very little alignment of the chain segments, even though the simulation started from a prefolded and compact structure. Interestingly, there were no strong transitions to the first excited state observed using the CIS-ZINDO/s for the PPV systems in DCM. The disorganization of the chain segments coupled with the increased separation clearly has dramatic effects on the vertical transitions.

The average interchain distance, d^{IC}, obtained for the optimized single-molecule MEH-PPV nanoparticle created from the solution-phase bad solvent morphology (Figure 8.9d) is $d^{IC} = 3.7$ Å (determined between the two center chains). For CN-PPV of the same backbone length, very similar chain dynamics (although the rate of folding was ~2 times slower mainly due to the increased steric hindrance about the sp^3 C–C bond due to the relatively large CN groups) was found and the minimum energy configuration had an interchain distance of $d^{IC} = 3.5$ Å, a relatively large decrease in the separation. These values are somewhat different than that determined by Conwell et al. using the MM2 model and assuming cofacial initial geometries. Here we are using slightly different potential energy functions, MM3, have explicitly included chain folds and all of the atoms of the system for longer oligomer segments and larger number of "stacked oligomers" (some self-solvation effects are thus possible), and we have accounted for the influence of solvent on the folding process and resulting geometry as well as emulated the process of single-molecule nanoparticle production from the solution phase. The "secondary" role of nanoparticle formation from the dilute solution-phase structures induces some notable changes as mentioned above. In particular, the 3-D confinement coupled with the extremely rapid evaporation of the solvent tends to cause much more compact and regular stacks of oligomer segments to form.

The effects of solution concentration on the structure of MEH-PPV were also investigated by using 3-D periodic images of the explicit atom solvent model (Figure 8.5). This approach was used to emulate a concentrated solution (since the size of the imaged box is constrained to be small) and can be compared to the dilute solution structure obtained from the results discussed above (which are for extremely dilute solutions as no boundary conditions were used). The results obtained clearly revealed how the interchain structure becomes disrupted but maintains some cofacial ordering between the oligomer segments. The degree of the interchain disorganization is dependent on the size of the MD box, with smaller or higher concentrated MEH-PPV solutions having less structural organization. These types of more disorganized folded structures do not have a transition dipole oriented along the chain axis, and as noted in the experiments, do not exhibit the donut-like emission patterns.

8.4.3 Structural Effects Induced by Interchain Interactions: Self-Solvation

The general trend observed from the MM3 results is a decrease in the interchain separation as larger numbers of oligomer segments are added. This decrease is on the order of 0.3 Å for oligomer segments in the center of a particle and indicates some type of chain–chain self-solvation effect. In order to obtain a better idea of the change in the distance due to self-solvation, we performed limited multiscale modeling (QM/MM) simulations. The simulations were set up by modeling the inner-folded MEH-PPV molecule with semiempirical quantum mechanics (AM1 model) and the outer chains with molecular mechanics (MM3 model). The outer chains were fixed at a distance of $d^{IC} = 3.7$ Å from the center of MEH-PPV molecule in accordance with the MM3 results and the MEH-PPV molecule was optimized using AM1. Since we are only comparing differences here instead of absolute structure, the AM1 model for the molecule should be reasonable. The results of these QM/MM calculations show a significant decrease in the interchain separation of about 0.9 Å. The resulting structure of the MEH-PPV molecule was actually very similar to that obtained from the MM3 calculations (interchain separation of $d^{IC} \sim 3.5$ Å and much smaller torsional rotations about the bonds adjacent to the vinyl group) but differed substantially from the AM1 vacuum results. This provides some interesting evidence that interchain separation of the rod-shaped morphologies tends to decrease toward the center of the single-molecule nanoparticle. As we will show below, the electronic structure depends quite strongly on this interchain distance, with enhanced singlet-to-singlet transition moments for shorter distances. This might provide some rationalization of the definitive experimental results that show photon anti-bunching for single-molecule z-oriented nanoparticles.[26] These results show that the z-oriented single-molecule nanoparticles act as single-photon emitters but multi-chromophore absorbers. The self-solvated inner core structure, where the interchain distance becomes closer and where there is much higher degree of structural organization, is probably acting as the primary emission site. From our semiempirical results as well as others, we know that the HOMO–LUMO bandgaps in molecular systems are strongly dependent on the degree of confinement, generally decreasing with increasing confinement. For the PPV-based systems, we have observed the following HOMO–LUMO gap dependences:

1. A decreasing HOMO–LUMO gap with increasing numbers of folded oligomer segments for PPV, CN-PPV, and MEH-PPV.
2. For a fixed oligomer segment length between folds, the magnitude of the HOMO–LUMO gap decrease becomes smaller with increasing number of segments.
3. For a fixed number of oligomer segments but a varying number of monomers in each oligomer, there is also a general decrease in the HOMO–LUMO bandgap for the PPV systems.
4. The computed HOMO–LUMO gap dependence on self-solvation as determined for a 14 (1 tetrahedral defect)-monomer MEH-PPV oligomer is a decrease of ∼0.1 eV and a increase in the wavelength for the first excited-state transition ∼23 nm (*this is now a red shift from quantum results*). The change in the electronic structure comes primarily from a LUMO lowering of ∼0.09 eV. The HOMO, which often shows greater sensitivity to confinement, does not change as much as the LUMO, increasing by ∼0.02 eV.

The self-solvation effects noted in the present simulations lead to reasonably large changes in the electronic structure and related optical transitions. It seems plausible to consider these single-molecule nanoparticles as dielectric core–shell systems, with the emissive core being composed of the self-solvated more ordered and tightly packed chains and the shell as less tightly packed chains (larger interchain separation but still close enough to allow orbital delocalization and nonradiative Förster energy transfer processes). This interpretation, which is backed by the self-solvation computational results, would also allow direct implementation of classical electrostatics arguments due to vacuum field interactions attenuating the fluorescence lifetimes, a property which has been observed experimentally.

8.4.4 The Effects of Polymer Architecture on Self-Assembly and Miscibility

Using CGMD simulation for modeling P3HT grafted to PNB resulting into molecular bottlebrush architecture, we were able to mimic the SEC, DLS, and TEM experimental results showing the physical aggregation of bottlebrushes after drying—despite dissolving the dried system using a good solvent (THF) (Figure 8.3). The CGMD simulations were performed assuming a short-range attractive interaction between P3HT side chain beads that were described by an LJ potential with well depth $\varepsilon_{LJ} = 0.43\ k_B T$. The CGMD results were

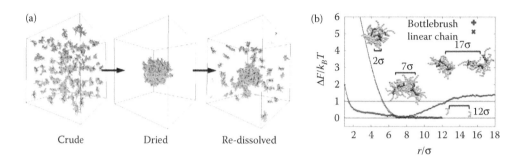

FIGURE 8.10

(a) Snapshots for bottlebrush polymers at three different states. (b) Potential of mean force for separating two free chains (blue) and two bottlebrush polymers (red). The bottlebrushes and macromoners are modeled using LJ beads with diameter, σ. (Reprinted with permission from S.-K. Ahn et al., *ACS Macro Lett.* 2, 761. Copyright 2013. American Chemical Society.)

qualitatively similar to the observed aggregation behavior, as illustrated by the simulation snapshots in Figure 8.10a. (The details of the CGMD can be found in Reference 30.) As a reference, we used the same interaction parameters, molecular weights, and simulation protocol used in the simulations in Figure 8.10a but without the PNB backbone. And, physical aggregation was absent in the reference simulation. We were able to quantify the aggregation behavior on the basis of umbrella sampling (US) simulation procedure and potential of mean force (PMF) calculations through the weighted histogram analysis method (WHAM) described in Section 8.3. These calculations show that an energy barrier of approximately 1 $k_B T$ (thermal energy) is required to separate two molecular bottlebrushes. This energy barrier is not present in separating two linear chains (see Figure 8.10b). An increase in inter-brush contact will compound this energy barrier; for example, it is more difficult to separate a bottlebrush surrounded with other bottlebrushes in comparison to a bottlebrush that is in contact with only one. This energy barrier can be attributed to localized side chain attractions around a backbone and we speculate that the strong aggregation of P3HT bottlebrushes is primarily driven by the increase in the number of attractive π–π interactions. Although, we caution that other types of noncovalent interactions or physical factors may also contribute to the aggregation and warrants further investigation. However, all of these interactions are grouped into an LJ type of interaction in the CGMD.

8.5 Summary

By combining experimental observations and developments with extensive computational chemistry studies, we have presented substantial evidence suggesting highly ordered rod-shaped structures for single-molecule MEH-PPV and CN-PPV systems. The chain organization is crucial to the photophysical properties and can be controlled to a large extent by the solvent. A relatively bad solvent leads to a structure that is tightly folded into a rod-shaped morphology while a good solvent produces structures with little or random folding and more of a defect extended-chain morphology. Secondary structural regularity is also induced by producing single-molecule nanoparticles from micro-droplets of the solution without the presence of a substrate. For toluene and THF solvent preparations of MEH-PPV and CN-PPV, the resulting solvent-free single-molecule nanoparticle structures show a very high level of organization consisting of π-stacked folded chains. Due to this structural organization, which imposes a rod-shaped morphology, single-molecule nanoparticle orientation on an appropriate surface occurs such that the particle stands on its end with near-perfect z-orientation. This orientational alignment is caused by excess charges that are induced on the particle during its production, which tend to localize at the surface,[71] causing the rod-shaped particle to minimize the Coulombic interactions with the surface (Si-O)$^-$ groups. The excess charge also appears to increase the structural organization into π-stacks. These z-oriented single-molecule nanoparticles appear to act something like a core–shell system, where the inner core is a self-solvated PPV system with interchain distances ~0.3 Å closer than the surrounding chains. Since there is still orbital overlap throughout the system, Forster energy transfer can occur. In addition, this type of interchain distance anisotropy might exhibit behavior like a single quantum emitter in a dielectric medium at the nanometer scale (Raleigh scattering regime) and thus show strong effects due to classical electromagnetic interactions (vacuum fluctuations are altered from boundary reflections), which lead to the observed altered fluorescence lifetimes.

In our recent work of systems consisting of P3HT, we studied the effect of polymer architecture on the nanoscale assembly and aggregation of polythiophenes grafted to a polymer chain forming a molecular bottlebrush. The combination of macromolecular architecture and the inherent intermolecular interactions between the P3HT side chains results in the strong physical aggregation of the bottlebrushes upon drying. This effect is absent in linear chains having the same molecular weight as the grafted chains.

The overall ramifications of developing this fundamentally new processing technique for generating optoelectronic materials are far-reaching. By achieving uniform orientation perpendicular to the substrate with enhanced luminescence lifetimes and photostability under ambient conditions, the door is now open for major developments in molecular photonics, display technology, and bioimaging, as well as new possibilities for optical coupling to molecular nanostructures and for novel nanoscale optoelectronics devices.

Disclaimer

This manuscript has been authored by UT-Battelle, LLC under contract no. DE-AC05-00OR22725 with the U.S. Department of Energy. The United States Government retains and the publisher, by accepting the article for publication, acknowledges that the United States Government retains a non-exclusive, paid-up, irrevocable, world-wide license to publish or reproduce the published form of this manuscript, or allow others to do so, for United States Government purposes. The Department of Energy will provide public access to these results of federally sponsored research in accordance with the DOE Public Access Plan (http://energy.gov/downloads/doe-public-access-plan).

Acknowledgments

This work was supported in part by the ORNL Laboratory Directed Research and Development and the Materials Science and Engineering Division, Office of Basic Energy

Sciences, U.S. Department of Energy. The extensive computational work was performed using resources of the Oak Ridge Leadership Computing Facility (OLCF) at ORNL. Work on P3HT was performed at the Center for Nanophase Materials Sciences, a DOE Office of Science user facility.

References

1. M.A. El-Sayed, Some interesting properties of metals confined in time and nanometer space of different shapes, *Acc. Chem. Res.* 34, 257, 2001; C.M. Niemer, Nanoparticles, proteins, and nucleic acids: Biotechnology meets materials science, *Angew. Chem. Int. Ed.* 40, 4128, 2001; D.M. Adams et al., Charge transfer on the nanoscale: Current status, *J. Phys. Chem. B* 107, 6668, 2003; M. Antonietti, K. Landfester, Single molecule chemistry with polymers and colloids: A way to handle complex reactions and physical processes? *ChemPhysChem* 2, 207, 2001.

2. C.H. Turner, J.K. Brennan, J.K. Johnson, K.E. Gubbins, Effect of confinement by porous materials on chemical reaction kinetics, *J. Chem. Phys.* 116, 2138, 2002; C.H. Turner, J.K. Johnson, K.E. Gubbins, Effect of confinement on chemical reaction equilibria: The reactions $2NO \Leftrightarrow (NO)_2$ and $N_2 + 3H_2 \Leftrightarrow 2NH_3$ in carbon micropores, *J. Chem. Phys.* 114, 1841, 2001; S.J. Stuart, B.M. Dickson, B.G. Sumpter, D.W. Noid, Hydrocarbon reactions in carbon nanotubes: Pyrolysis, *Proc. MRS* 651, T7.15, 2001; M.K. Kidder, P.F. Britt, Z. Zhang, S. Dai, A.C. Buchannan, Confinement effects on product selectivity in the pyrolysis of phenethyl phenyl ether in mesoporous silica, *ChemComm* 2804, 2003.

3. D.K. Eggers, J.S. Valentine, Molecular confinement influences protein structure and enhances thermal protein stability, *Protein Sci.* 10, 250, 2001; H.-X. Zhou, K.A. Dill., Stabilization of proteins in confined spaces, *Biochemistry* 40, 11289, 2001; R. Rittg, A. Huwe, G. Fleischer, J. Karger, F. Kremer, Molecular dynamics of glass-forming liquids in confining geometries, *Phys. Chem. Phys.* 1, 519, 1999.

4. G. Hummer, J.C. Rasaiah, J.P. Noworyta, Water conduction through the hydrophobic channel of a carbon nanotube, *Nature* 414, 188, 2001; R.E. Tuzun, D.W. Noid, B.G. Sumpter, R.C. Merkle, Dynamics of flow inside carbon nanotubes, *Nanotechnology* 8, 112, 1997; R.E. Tuzun, D.W. Noid, B.G. Sumpter, R.C. Merkle, Dynamics of fluid flow inside carbon nanotubes, 7, 241, 1996.

5. W.L. Barnes, Fluorescence near interfaces: The role of photonic mode density, *J. Mod. Opt.* 45, 661, 1998; H. Schniepp, V. Sandoghdar, Spontaneous emission of europium ions embedded in dielectric nanospheres, *Phys. Rev. Lett.* 89, 257403, 2002; H. Chew, Radiation and lifetimes of atoms inside dielectric particles, *Phys. Rev. A* 38, 3410, 1988.

6. A. Striolo, A.A. Chialvo, P.T. Cummings, K.E. Gubbins, Water adsorption in carbon-slit nanopores, *Langmuir* 19, 8583, 2003; R. Allen, J.-P. Hansen, S. Melchionna, Molecular dynamics investigation of water permeation through nanopores, *J. Chem. Phys.* 119, 3905, 2003; K. Koga, G.T. Gao, H. Tanaka, X.C. Zeng, Formation of ordered ice nanotubes inside carbon nanotubes, *Nature* 412, 802, 2001; Y. Maniwa, H. Kataura, M. Abe, S. Suzuki, Y. Achiba, H. Kira, K. Matsuda.

7. P. Eisenberg, J.C. Lucas, R.J.J. Williams, Hybrid organic-inorganic polymer networks based on the copolymerization of methacryloxypropyl-silsesquioxanes and styrene, *Macromol. Symp.* 189, 1–13, 2002.

8. A.J. Heeger, Semiconducting and metallic polymers: The fourth generation of polymeric materials, *J. Phys. Chem. B* 105, 8475, 2001.

9. J.F. Feller, I. Linossier, Y. Grohens, Conductive polymer composites: comparative study of poly(ester)-short carbon fibres and poly(epoxy)-short carbon fibres mechanical and electrical properties, *Mater. Lett.* 57, 64–71, 2002.

10. R. Gensler, P. Groppel, V. Muhrer, N. Muller, Improving the dielectric properties of high density polyethylene by incorporating clay-nanofiller, *Part. Part. Syst. Charact.* 19(5), 293, 2002.

11. R.H. Friend et al., Electroluminescence in conjugated polymers, *Nature* 397, 121–128, 1999.

12. F. Hide, M.A. Diazgarcia, B.J. Schwartz, and A.J. Heeger, New developments in the photonic applications of conjugated polymers, *Acc. Chem. Res.* 30, 430, 1997.

13. M. Srinivasarao, D. Collings, A. Philips, and S. Patel, Three-dimensionally ordered array of air bubbles in a polymer film, *Science* 292, 79, 2001.

14. C.-Y. Kung, M.D. Barnes, N. Lermer, W.B. Whitten, and J.M. Ramsey, Single-molecule analysis of ultradilute solutions with guided streams of 1-μm water droplets, *Appl. Opt.* 38, 1481, 1999.

15. M.D. Barnes, K.C. Ng, K. Fukui, B.G. Sumpter, and D.W. Noid, Probing phase-separation behavior in polymer-blend microparticles: Effects of particle size and polymer mobility, *Macromolecules* 32, 7183, 1999.

16. K.C. Ng, J.V. Ford, S.C. Jacobson, J.M. Ramsey, and M.D. Barnes, Polymer microparticle arrays from electrodynamically focused microdroplet streams, *Rev. of Sci. Instrum.* 71, 2497, 2000.

17. B.J. Schwartz, Conjugated polymers as molecular materials: How chain conformation and film morphology influence energy transfer and interchain interactions, *Annu. Rev. Phys. Chem.* 54, 141, 2003.

18. D. Hu, L. Yu, P.F. Barbara, Single-molecule spectroscopy of the conjugated polymer MEH-PPV, *J. Am. Chem. Soc.* 121, 6936, 1999; D. Hu, J. Yu, B. Bagchi, P.J. Rossky, P.F. Barbara, Collapse of stiff conjugated polymers with chemical defects into ordered, cylindrical conformations, *Nature* 405, 1030, 2000.

19. D.A. Vandenbout et al., PF discrete intensity jumps and intramolecular electronic energy transfer in the spectroscopy of single conjugated polymer molecules, *Science* 277, 1074, 1997.

20. J. Yu, D. Hu, and P.F. Barbara, Unmasking electronic energy transfer of conjugated polymers by suppression of O2 quenching, *Science* 289, 1327, 2000.

21. G. Padmanaban, S. Ramakrishnan, Conjugation length control in soluble poly[2-methoxy-5-((2'-ethylhexyl)oxy)-1,4-phenylenevinylene] (MEHPPV): Synthesis, optical properties, and energy transfer, *J. Am. Chem. Soc.* 122, 2244, 2000.

22. C.L. Gettinger, A.J. Heeger, J.M. Drake, D.J. Pine, A photoluminescence study of poly(phenylene vinylene) derivatives: The effect of intrinsic persistence length, *J. Chem. Phys.* 101, 1673, 1994.

23. A. Mehta, P. Kumar, M. Dadmun, J. Zheng, R.M. Dickson, T. Thundat, B.G. Sumpter, and M.D. Barnes, Oriented nanostructures from single molecules of a semiconducting polymer: Polarization evidence for highly aligned intramolecular geometries, *Nano Let.* 3(5), 603, 2003.

24. P. Kumar, A. Mehta, M. Dadmun, J. Zheng, L. Peyser, R.M. Dickson, T. Thundat, B.G. Sumpter, and M.D. Barnes, Narrow-bandwidth spontaneous luminescence from oriented semiconducting polymer nanostructures, *J. Phys. Chem. B* 107, 6252, 2003.

25. E.H. Hellen, D. Axelrod, Fluorescence emission at dielectric and metal-film interfaces, *J. Opt. Soc. Am. B* 4, 337, 1987.

26. J.L. Bredas, D. Beljonne, J. Cornil, J.P. Calbert, Z. Shuai, R. Silbey, Electronic structure of π-conjugated oligomers and polymers: A quantum–chemical approach to transport properties, *Synth. Met.* 125, 107, 2001.

27. P. Kumar , T.H. Lee, A. Mehta , B.G. Sumpter, R.M. Dickson, M.D. Barnes, Photon antibunching from oriented semiconducting polymer nanostructures, *J. Am. Chem. Soc.* 126, 3376, 2004.

28. P. Kumar, A. Mehta, S.M. Mahurin, S. Dai, M.D. Dadmun, T. Thundat, B.G. Sumpter, The role of solvent on formation of oriented nanostructures from single molecules of conjugated polymers, Formation of oriented nanostructures from single molecules of conjugated polymers in microdroplets of solution: The role of solvent, *Macromolecules* 37, 6132, 2004.

29. T.-Q. Nguyen, V. Doan, B.J. Schwatz, Understanding the electronic properties of conjugated polymers, *J. Chem. Phys.* 110, 4068, 1999; C.W. Hollars, S.M. Lane, T. Huser, Controlled non-classical photon emission from single conjugated polymer molecules, *Chem. Phys. Lett.* 370, 393, 2003.

30. S.-K. Ahn et al., Poly(3-hexylthiophene) molecular bottlebrushes via ring-opening metathesis polymerization: Macromolecular architecture enhanced aggregation, *ACS Macro Lett.* 2, 761, 2013.

31. See for example, C.J. Cramer, *Essentials of Computational Chemistry: Theory and Models* (Wiley, 2002).

32. N.L. Allinger, Y.H. Yuh, J.-H. Lii, Molecular mechanics. The MM3 force field for hydrocarbons. 1, *J. Am. Chem. Soc.* 111, 8551, 1989.

33. J.L. Tai, N.L. Allinger, Effect of inclusion of electron correlation in MM3 studies of cyclic conjugated compounds, *J. Comp. Chem.* 19, 475, 1998; J.-L. Lii, N.L. Allinger, Molecular mechanics. The MM3 force field for hydrocarbons. 3. The van der Waals' potentials and crystal data for aliphatic and aromatic hydrocarbons, *J. Am. Chem. Soc.* 111, 8576, 1989.

34. A.V. Fratini, K.N. Baker, T. Resch, H.C. Knachel, W.W. Adams, E.P. Socci, B.L. Farmer, *Polymer* 43, 1571, 1993; N. Nevins, J.-H. Lii, N.L. Allinger, Molecular mechanics (MM4) calculations on conjugated hydrocarbons, *J. Comp. Chem.* 17, 695, 1996; L. Claes, M.S. Deleuze, J.-P. Francois, Comparative study of the molecular structure of stilbene using molecular mechanics, Hartree–Fock and density functional theories, *J. Mol. Struct. (THEOCHEM)* 549, 63, 2001.

35. See for example, D. Frenkel, B. Smit, *Understanding Molecular Simulations: From Algorithms to Applications* (Academic Press, New York, 2002).

36. N.C. Karayiannis, A.E. Giannousaki, V.G. Mavrantzas, D.N. Theodorou, Atomistic Monte Carlo simulation of strictly monodisperse long polyethylene melts through a generalized chain bridging algorithm, *J. Chem. Phys.* 17, 5465, 2002; N.C. Karayiannis, A.E. Giannousaki, V.G. Mavrantzas, An advanced Monte Carlo method for the equilibration of model long-chain branched polymers with a well-defined molecular architecture: Detailed atomistic simulation of an H-shaped polyethylene melt, *J. Chem. Phys.* 118, 2451, 2003.

37. U. Burkert, N.L. Allinger, *Molecular Mechanics*, ACS Monograph 177 (ACS, Washington D.C., 1992).

38. F. Müller-Plathe, Coarse-graining in polymer simulation: From the atomistic to the mesoscopic scale and back, *Chem Phys Chem.* 3, 754, 2002.

39. K. Kremer, G. Grest, Dynamics of entangled linear polymer melts: A molecular-dynamics simulation, *J. Chem. Phys.* 92, 5057, 1990.

40. S. Kumar, J.M. Rosenberg, D. Bouzida, R.H. Swendsen, P.A. Kollman, The weighted histogram analysis method for free-energy calculations on biomolecules. I. The method, *J. Chem. Phys.* 13, 1011, 1992.

41. D. Bashford, D.A. Case, Generalized born models of macromolecular solvation effects, *Annu. Rev. Phys. Chem.* 51, 129, 2000.

42. C.J. Cramer, D.G. Truhlar, Implicit solvation models: Equilibria, structure, spectra, and dynamics, *Chem. Rev.* 99, 2161, 1999.

43. J. Li, C.J. Cramer, D.G. Truhlar, Application of a universal solvation model to nucleic acid bases: comparison of semiempirical molecular orbital theory, ab initio Hartree-Fock theory, and density functional theory, *Biophys. Chem.* 78, 147, 1999.

44. D. Qiu, P.S. Shenkin, F.P. Hollinger, W.C. Still, The GB/SA continuum model for solvation. A fast analytical method for the calculation of approximate Born radii, *J. Phys. Chem. A* 101, 3005, 1997.

45. G.D. Hawkins, C.J. Cramer, D.G. Truhlar, Pairwise solute descreening of solute charges from a dielectric medium, *Chem. Phys. Lett.* 246, 122, 1995; G.D. Hawkins, C.J. Cramer, D.G. Truhlar, Parametrized models of aqueous free energies of solvation based on pairwise descreening of solute atomic charges from a dielectric medium, *J.Phys. Chem.* 100, 19824, 1996.

46. D. Eisenberg, A.D. McLachlan, Solvation energy in protein folding and binding, *Nature* 319, 199, 1986.

47. W.C. Still, A. Tempczyk, R.C. Hawley, T. Hendrickson, Semianalytical treatment of solvation for molecular mechanics and dynamics, *J. Am. Chem. Soc.* 112, 6127, 1990.

48. M. Schaefer, M. Karplus, A Comprehensive analytical treatment of continuum electrostatics, *J. Phys. Chem.* 100, 1578, 1996; M. Schaefer, C. Bartels, M. Karplus, Solution conformations and thermodynamics of structured peptides: Molecular dynamics simulation with an implicit solvation model, *J. Mol. Biol.* 284, 835, 1998.

49. T. Helgaker, P. Jorgensen, J. Olsen, *Molecular Electronic-Structure Theory* (Wiley, New York, 2000).

50. R.G. Parr, W. Yang, *Density-Functional Theory of Atoms and Molecules* (Oxford University Press, New York, 1989).

51. W. Koch, M.C. Holthausen, *A Chemist's Guide to Density Functional Theory* (Wiley-VCH, New York, 2001).

52. S.H. Vosko, L. Wilk, N. Nusair, Accurate spin-dependent electron liquid correlation energies for local spin density calculations: A critical analysis, *Can. J. Phys.* 58, 1200, 1980.

53. A.D. Becke, Density functional calculations of molecular bond energies, *J. Chem. Phys.* 84, 4524, 1986; J.P. Perdew, Density-functional approximation for the correlation energy of the inhomogeneous electron gas, *Phys. Rev. B* 33, 8822, 1986.

54. C. Lee, W. Yang, R.G. Parr, Development of the Colle-Salvetti correlation-energy formula into a functional of the electron density, *Phys. Rev. B* 37, 785, 1988.

55. J.P. Perdew, M. Ernzerhof, K. Burke, Rationale for mixing exact exchange with density functional approximations, *J. Chem. Phys.* 105, 9982, 1996.

56. A.D. Becke, Density-functional thermochemistry. III. The role of exact exchange, *J. Chem. Phys.* 98, 5648, 1993.

57. F. Jensen, *Introduction to Computational Chemistry* (Wiley, New York, 1999).

58. A. Dreuw, J.L. Weisman, M. Head-Gordon, Long-range charge-transfer excited states in time-dependent density functional theory require non-local exchange, *J. Chem. Phys.* 119, 2943, 2003; D.J. Tozer, Relationship between long-range charge-transfer excitation energy error and integer discontinuity in Kohn–Sham theory, *J. Chem. Phys.* 119, 12697, 2003; H. Iikura, T. Tsuneda, T. Yanai, K. Hirao, A long-range correction scheme for generalized-gradient-approximation exchange functionals, *J. Chem. Phys.* 115, 3540, 2001; Z.-L. Cai, K. Sendt, J.R. Reimers, Failure of density-functional theory and time-dependent density-functional theory for large extended π systems, *J. Chem. Phys.* 117, 5543, 2002; S. Grimme, M. Parac, Substantial errors from time-dependent density functional theory for the calculation of excited states of large pi systems, *ChemPhysChem* 4, 292, 2003; S.J.A. van Gisbergen, P.R.T. Schipper, O.V. Gritsenko, E.J. Baerends, J.G. Snijders, B. Champagne, B. Kirtman, Electric field dependence of the exchange-correlation potential in molecular chains, *Phys. Rev. Lett.* 83, 694, 1999.

59. M.C. Zerner, in *Reviews in Computational Chemistry*, ed K.B. Lipkowitz, D.B. Boyd, p. 313 Vol. II (VCH, New York, 1991); G.M. Pearl, M.C. Zerner, A. Broo, J. McKelvey, Method of calculating band shape for molecular electronic spectra, *J. Comp. Chem.* 19, 781, 1998.

60. I.G. Hill, A. Kahn, J. Cornil, D.A. dos Santos, J.L. Bredas, Occupied and unoccupied electronic levels in organic π-conjugated molecules: Comparison between experiment and theory, *Chem. Phys. Lett.* 317, 444, 2000; Y. Han, S.U. Lee, Molecular orbital study on the ground and excited states of methyl substituted tris(8-hydroxyquinoline) aluminum(III), *Chem. Phys. Lett.* 366, 9, 2002; J. Cornil, D. Belojonne, C.M. Heller, I.H. Campbell, B.K. Laurich, D.L. Smith, D.D.C. Bradley, K. Mullen, J.L. Bredas, Photoluminescence spectra of oligo-paraphenylenevinylenes: A joint theoretical and experimental characterization, *Chem. Phys. Lett.* 278, 139, 1997; J. Cornil, A.J. Heeger, J.L. Bredas, Effects of intermolecular interactions on the lowest excited state in luminescent conjugated polymers and oligomers, *Chem. Phys. Lett.* 272, 463, 1997; J. Cornil, D. Beljonee, J.L. Bredas, Nature of optical transitions in conjugated oligomers. I. Theoretical characterization of neutral and doped oligo(phenylenevinylene)s, *J. Chem. Phys.* 103, 834, 1995; J. Cornil, J.Ph. Calbert, D. Beljonne, R. Silbey, J.L. Bredas, Interchain interactions in π-conjugated oligomers and polymers: A primer, *Syn. Met.* 119, 1, 2001; A. Pogantsch, A.K. Mahler, G. Hayn, R. Saf, F. Stelzer, E.J.W. List, J.L Bredas, E. Zojer, Excited-state localization effects in alternating meta- and para-linked poly(phenylene-vinylene)s, *Chem. Phys.* 297, 143, 2004.

61. C. Hayashi, R. Uyeda, A. Tasaki, *Ultra Fine Particles Technology* (Noyes, New Jersey, 1997).

62. Semiempirical quantum calculations were performed using GAMESS, M.W. Schmidt, K.K. Baldridge, J.A. Boatz, S.T. Elbert, M.S. Gordon, J.H. Jensen, S. Koseki, N. Matsunaga, K.A. Nguyen, S.J. Su, T.L. Windus, M. Dupuis, J.A. Montogemery, General atomic and molecular electronic structure system, *J. Comput. Chem.* 14, 1347–1363, 1993; and Hyperchem7.0, Hypercube, Inc., Gainesville, FL, 2003.

63. *Ab initio* quantum calculations were performed using NWChem Version 4.5, as developed and distributed by Pacific Northwest National Laboratory, P.O. Box 999, Richland, Washington 99352, USA, and funded by the U.S. Department of Energy.

64. D. Chen, M.J. Winokur, M.A. Masse, F.E. Karaz, Structural phases of sodium-doped polyparaphenylene vinylene, *Phys. Rev. B* 41, 6759, 1990; P.F. van Hutten, J. Wildeman, A. Meetsma, G. Hadziioannou, Molecular packing in unsubstituted semiconducting phenylenevinylene oligomer and polymer, *J. Am. Chem. Soc.* 121, 5910, 1999; T. Grainer, E.L. Thomas, D.R. Gagnon, F.E. Karasz, R.W. Lenz, Structure investigation of poly(p-phenylene vinylene), *J. Polym. Sci. B* 24, 2793, 1986; L. Claes, J.-P. Francois, M.S. Deleuze, Molecular packing of oligomer chains of poly(p-phenylene vinylene), *Chem. Phys. Lett.* 339, 216, 2001.

65. A. Ferretti, A. Ruini, E. Molinari, Electronic properties of polymer crystals: The effect of interchain interactions, *Phys. Rev. Lett.* 90, 086401-1, 2003; A. Ruini, M.J. Caldas, G. Bussi, E. Molinari, Solid state effects on exciton states and optical properties of PPV, *Phys. Rev. Lett.* 88, 206403, 2002.

66. M.D. Curtis, J. Cao, J.W. Kampf, Solid-state packing of conjugated oligomers: From pi-stacks to the herringbone structure, *J. Am. Chem. Soc.* 126, 4318, 2004.

67. F.C. Grozema, L.P. Candeias, M. Swart, P.Th. Van Duijnen, J. Wildeman, G. Hadziioanou, L.D.A. Siebbeles, J.M. Warman, Theoretical and experimental studies of the opto-electronic properties of positively charged oligo(phenylene vinylene)s: Effects of chain length and alkoxy substitution, *J. Chem. Phys.* 117, 11366, 2002.

68. C.Y. Yang, F. Hide, M.A. Diaz-Garcia, A.J. Heeger, Y. Cao, Microstructure of thin films of photoluminescent semiconducting polymers, *Polymer 39*, 2299–2304, 1998.

69. E.M. Conwell, J. Perlstein, S. Shaik, Interchain photoluminescence in poly(phenylene vinylene) derivatives, *Phys. Rev. B* 54, R2308, 1996; J. Cornil, A.J. Heeger, J.L. Bredas, Effects of intermolecular interactions on the lowest excited state in luminescent conjugated polymers and oligomers, *Chem. Phys. Lett.* 272, 463, 1997.

70. V.A. Ivanov, W. Paul, K. Binder, Finite chain length effects on the coil–globule transition of stiffchain macromolecules: A Monte Carlo simulation, *J. Chem. Phys.* 109, 5659, 1998; K.S. Schweizer, Order–disorder transitions of π-conjugated polymers in condensed phases. I. General theory, *J. Chem. Phys.* 85, 4181, 1986; B.E. Kohler, I.D.W. Samuel. Experimental determination of conjugation lengths in long polyene chains, *J. Chem. Phys.* 103, 6248, 1995.

71. B.G. Sumpter, P. Kumar, A. Mehta, W.A. Shelton, R.J. Harrison, Computational study of the structure, dynamics, and photophysical properties of conjugated polymers and oligomers under nanoscale confinement. *J. Phys. Chem. B* 109, 7671, 2005.

9

Cellular Interfacing with Arrays of Vertically Aligned Carbon Nanofibers and Nanofiber-Templated Materials

Timothy E. McKnight, Anatoli V. Melechko, Guy D. Griffin, Michael A. Guillorn, Vladimir I. Merkulov, Mitchel J. Doktycz, M. Nance Ericson, and Michael L. Simpson

CONTENTS

9.1 Preface

The self-assembly and controlled synthesis properties of vertically aligned carbon nanofibers (VACNFs) have been exploited to provide parallel, subcellular probes to manipulate and monitor biological matrices. This chapter provides a summary of efforts to fabricate, characterize, and biologically integrate several embodiments of VACNF-based systems, including electrophysiological probing arrays, material delivery vectors, and nanoscale fluidic elements. Incorporating nanoscale functionality in multiscale devices, these systems feature elements that exist at a size scale, which enable cellular integration in tissue matrices, within individual cells, and even within subcellular compartments, including the mammalian nucleus. Ultimately, these approaches and their refinement may provide effective strategies for interfacing to cells and cellular processes at the molecular scale.

9.2 Background: The Evolution of Ever Finer Tools for Biological Manipulation

The evaluation of cause and effect is perhaps the most fundamental of investigative techniques, and it is one whose application is ubiquitous across the disciplines of science. It is no wonder, then, that some of the earliest tools in a discipline are designed to manipulate local regions of a system, or more specifically, to introduce causal influences to a system and thereby elucidate the function or effect of a system's parts. Early efforts in physiology provide some striking examples.[1] Jan Swammerdam (1637–1680) was one of the first students of microscopy. His fascination with the form and function of insects and plants led to the development of miniaturized tools, such as fine glass tubes used for injection of air and inks to probe local regions of his specimens. Albrecht von Haller (1708–1777) provided an early framework for moving these studies into living systems and thereby enabling causal analyses of physiological response. He defined physiology as *animated anatomy* and sought to understand the nervous system through systematic stimulus or destruction of specific parts of the whole organism. Shortly thereafter, Luigi Galvani (1737–1798) demonstrated conclusively that electrical stimulation, applied through wire probes, caused the isolated system of the frog leg to twitch.

As technology advanced, so did the investigator's ability to influence more spatially refined biological targets. Swammerdam's early efforts with hollow glass tubes perhaps helped inspire even finer microneedles and microcapillaries for introducing and manipulating more refined regions of biological matrices and even single cells. Microscale glass lances were being used by Chabry in 1887 to perforate and destroy individual blastomeres of the ascidian egg. Shortly thereafter, Schouten (1899) and Barber (1904) began to implement pulled glass capillaries for manipulating individual bacterial cells.[2] Almost 100 years ago, Barber was generating micropipettes with tip openings of $\sim$1 μm and applying these microscale implements to direct the delivery of microorganisms into plant tissue and protozoans to observe the impact of these deterministic infections on the host organism.

Since these early, but remarkable demonstrations, solid needles, metal wires, hollow glass microcapillaries, and their evolving and often intertwined embodiments have become established as workhorses of biological investigation. Material delivery via hollow microneedles has now enabled the genetic manipulation of virtually all cell types and is routinely applied clinically for *in vitro* fertilization. The foundations of modern electrophysiology are also built on the application of fine wires and capillaries for stimulus and monitoring of excitable cells. By 1949, Ling and Gerard had combined a cell-penetrant pulled capillary with a metallic electrode to observe the resting membrane potential of individual frog sartorius fibers.[3] Shortly thereafter, Hodgkin and Huxley used similar embodiments to hold the membrane potential constant in individual giant squid axons in order to isolate the measurement of current flow and provide a comprehensive analysis of the action potential. In 1976, Neher and Sakmann refined tools and techniques even further with their development of the patch clamp. This refinement of the pulled capillary and its application enabled measurement of current through even more isolated regions of cells, specifically isolated ion channels of muscle cell membranes. In the 1980s and 1990s, techniques to entrain electrochemical electrodes in the fine tips of microcapillaries enabled the observation of highly localized subcellular phenomena in and around living cells, including exocytosis during evoked depolarization and localized molecular probing of easily oxidized or reduced species.[4]

With the advent of micromachining and the more recent focus on nanostructured materials, technology continues to augment the researcher's ability to probe ever smaller regions of biological matrices. The *merging* of recent advances in the synthesis of nanostructured materials with the mature technology of microfabrication is beginning to provide complex devices capable of interfacing with biological systems at the molecular scale. This advance is driven largely by the ability to incorporate nanoscale (and ultimately molecular-scale) functionality into practical, multiscale physical devices. In this overview, we focus on one embodiment of nanoscale science, the vertically aligned carbon nanofiber, or VACNF, and investigate how this recently discovered embodiment of an old material is providing new approaches to cellular probing, electrophysiology, and gene-delivery applications.

9.3 Vertically Aligned Carbon Nanofiber

The VACNF described in this work refers to a catalytically derived, nanostructured material comprising predominantly of carbon and growth oriented on a substrate of typically silicon or quartz (Figure 9.1). VACNFs are cylindrical or conical in shape and span multiple length scales, featuring nanoscale tip radii (10–200+ nm) and lengths up to tens of microns.

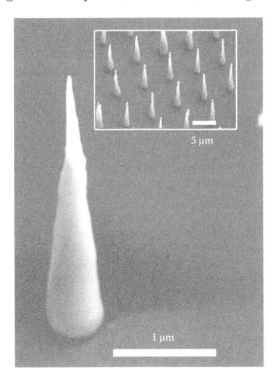

FIGURE 9.1
The deterministic synthesis of vertically aligned carbon nanofibers provides for the generation of structures at a size scale well suited to cellular interfacing. Here, a VACNF array is shown featuring nanofibers at a 5-μm pitch, with lengths of 6 μm, and tip diameters of ~100 nm. This array was synthesized using photolithographically defined dots of Ni catalyst.

Nanofibers are frequently compared and contrasted with the more familiar *nanotube*. Both comprise hexagonal networks of covalently bonded carbon (graphene) and, thus, both feature exceptional mechanical strength. However, whereas nanotubes comprise *concentrically rolled* graphene sheets, the internal structure of carbon nanofibers consists of stacked curved graphite layers that form cones or "cups."[5,6] This configuration lends properties to nanofibers that are highly advantageous to cellular interfacing applications. Most apparent is the potential for a conical morphology. Similar to the pulled capillaries of electrophysiology and microinjection, conical morphology of VACNFs provides base strength and rigidity while also featuring nanoscale tip radii that facilitate tissue penetration and probing. The stacked plane configuration also results in *edge plane* formation during nanofiber synthesis. These edge planes provide sites for electron transfer, which allow the nanofiber to serve as a faradaic sensor or actuator for electrochemical processes. These sites also provide rich surface chemistry, facilitating the postsynthesis modification of the nanofiber with coatings and chemical derivatizations that augment the nanofiber's biological utility.

Awareness of the carbon nanofiber, or closely related materials, has existed for over a century. In a comprehensive overview of nanofiber synthesis, Melechko provides a historical perspective that reveals that a patent as early as 1889[7] taught that carbon "filaments" can be grown from carbon-containing gases using a hot, iron crucible.[8] However, for almost the next century, detailed studies of these materials were conducted largely from the viewpoint of them being an undesirable byproduct of industrial processes.[9] This viewpoint changed dramatically with the discovery and appeal of C_{60}[10] and the carbon nanotube.[11] Realizing the merits of nanoscale materials, the 1990s directed considerable effort toward controlling the synthesis of these and related structures. One of the most significant advancements during this time was the introduction of plasma-enhanced chemical vapor deposition (PECVD) for the formation of nanofibers.[12,13] In contrast to earlier synthesis methods, including arc discharge, laser ablation, and other types of chemical vapor deposition that produce nonaligned, entangled ropes of nanotubes, PECVD allowed for synthesis of *vertically aligned* nanofibers, where the nanofiber growth is oriented with respect to the underlying substrate.[14,15] Subsequent refinements in PECVD parameters, in turn, have provided additional levels of determinism in the synthesis of nanofibers, including the ability to position exactly where an individual nanofiber grows upon a substrate, how long it grows, its diameter, its morphology, and to some extent its chemical composition.[16–25]

In the devices described in this chapter, nanofibers are catalytically synthesized in a dc-PECVD process and can be produced on a large scale, for example, using standard 75 or 100 mm silicon or fused silica wafers as the growth substrate. The sites of fiber growth are defined by lithography and by deposition of a thin-film of nickel as catalyst, typically 50–1000 Å thick. When the substrate is heated to 700°C and dc plasma is initiated, the thin nickel film nucleates into isolated nanoparticles, each of which will initiate growth of a single fiber. Nickel nucleation varies based on the interaction of the nickel solution with the underlying substrate and often a buffering layer, such as titanium, is employed to aid in catalyst nucleation. Following particle coalescence, carbon nanofibers grow from the defined catalyst dots. Carbonaceous species decompose at the surface of and diffuse through the catalyst particle, and are incorporated into a growing nanostructure between the particle and the substrate. Such nanofibers, with the catalyst particle at the tip, grow oriented along the electric field lines, which are usually oriented perpendicular to the substrate. The radial dimensions of the nanofibers are influenced by the size of the catalytic particle. Nanofiber length can be precisely controlled by the growth time, which in case of PECVD is determined by the time the plasma is on (i.e., the growth stops immediately after the plasma is extinguished). Additionally, there is some control over the shape of the carbon nanofibers as they can be made more conical or more cylindrical depending on growth conditions, particularly gas composition and substrate composition.

9.4 Device Integration

As we will demonstrate later in this chapter, deterministic synthesis provides the ability to generate nanofibers and nanofiber arrays with morphologies and in patterns that facilitate their integration with cellular and tissue matrices. To provide additional levels of functionality, however, further processing is often desired. For example, to use a nanofiber as an electrophysiological probe, one must use postsynthesis processing techniques to enable the interconnection of the nanofiber electrode to external electronics. Toward this end, carbon nanofibers are compatible with a large number of microfabrication techniques including lithographic processing, material lift-off techniques, ultrasonic agitation, high-temperature processing, wet and dry etching, wafer dicing, and chemical/mechanical polish. Standard microfabrication techniques may, therefore, be readily employed to incorporate nanofibers, and therefore nanoscale functionality, into functional multiscale devices.

Guillorn et al. demonstrated that by using standard microfabrication techniques, single VACNFs can be synthesized on electrical interconnects and implemented as individually addressable electrochemical electrodes where only the extreme nanoscale tip of the fiber is electrochemically active.[26] In this work, nanofiber growth sites were defined by patterning catalyst with electron-beam lithography upon a layer of tungsten deposited on thermally oxidized silicon wafers. Electron beam lithography provided catalyst dots of ~100 nm diameters, from which individual nanofibers could be grown. Following nanofiber synthesis, the tungsten layer was lithographically patterned and etched with an SF_6/ CF_4-based refractory metal reactive ion etch to form interconnects between peripheral bonding pads and the individual nanofibers. The entire substrate was then covered with an insulating silicon dioxide layer that was deposited using PECVD. Subsequently, an etch process was used to selectively remove the oxide from the bonding pads and the tips of the nanofiber elements, providing a probing system where only the extreme tip was electrochemically active. McKnight et al. continued this work by incorporating a process developed by Melechko[24] to enable single nanofiber growth from larger, photolithographically defined, catalyst films of ~400–500 nm in diameter, thereby eliminating the infrastructural and processing demands imposed by electron beam lithography.[27] This embodiment also implemented additional passivation layers upon the interconnect structure to reduce capacitive coupling of the interconnect with overlying solution and to eliminate faradaic response from pinholes in the interconnect PECVD oxide. A photolithographically patternable epoxy, SU-8, was used for these passivation layers allowing simultaneous definition of structural elements in the devices. These included microfluidic channels and mechanical barriers that facilitated incorporation of the nanofiber arrays into standard semiconductor packages. The resultant devices featured up to 40 individually addressable nanofiber elements on a 5-mm square chip where up to 100 of these chips were fabricated on a single 100-mm silicon wafer substrate (Figure 9.2).

In both of these embodiments, the nanofiber served to elevate the electroanalytical measurement volume above the planar substrate, thereby providing a structure that could

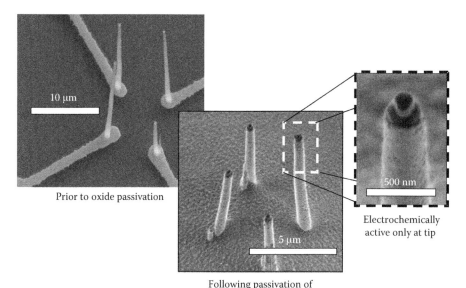

FIGURE 9.2
The compatibility of VACNFs with standard microfabrication processes enables the integration of nanoscale features into functional multiscale devices. Here, a four-element electrophysiological probing array is shown featuring cellular-scale, individually addressed VACNF elements with passivated sheaths and electroactive tips of <100 nm diameter.

penetrate into tissue matrices and potentially into individual cells. Furthermore, the nanofiber served as an electrical bridge between the nanoscale dimensions of the electroactive nanofiber tip and the microscale dimensions of the electrical interconnects of the substrate. By limiting the size of this exposed tip, electroanalytical probing could be achieved in extremely small volumes (<100 nm diameter, <500 zeptoliter volume). This enables the quantification of electroactive species in this volume, as well as the direct manipulation of this local environment via oxidation, reduction, pH variation, and application of electric field. Ultimately, this provides a means of manipulating exceptionally small volumes within and around living, functioning cells. Moreover, by exploiting the inherent parallel nature of microfabrication techniques, these devices demonstrated how highly parallel arrays can be fabricated to provide simultaneous probing and actuation of many positions throughout a cellular matrix.

9.5 Electrochemical Response

The application of nanofiber-based devices for electrophysiological interrogation and manipulation of cellular matrices requires an understanding of their electrochemical behavior. Carbon nanofibers and nanofiber devices, including those described in the previous sections, exhibit reproducible electron transfer *as-synthesized*, that is, without requiring specific pretreatments to electrochemically activate the surface. The electrochemical properties of photolithographically defined, individually addressed nanofiber elements (with electrochemically active lengths of 4–12 μm) and nanofiber tips (with active tips of <250 nm diameter) have been characterized against a variety of quasireversible, outer sphere redox species including $Fe(CN)_6^{3-/4-}$, $Ru(NH_3)_6^{2+/3+}$, and $IrCl_6^{2-/3-}$ (for example, see Figure 9.3).[27] The faradaic response of nanofibers was found to be dependent on the surface area of exposed nanofiber material, thus indicating that electron transfer occurs along the entire length of these vertically aligned structures. Nanofiber/solution interfacial specific capacitance was determined to be ~140 μF/cm² in 100 mM KCl by performing voltammetry of various length fibers (2–12 μm) at multiple scan rates. The potential window of these electrodes was also evaluated in several aqueous solvents, and was found to be 1.2–1.3 V vs. Ag/AgCl (3 M KCl) in the physiological buffer solution, phosphate-buffered saline (150 mM NaCl, 10 mM phosphate). Tip exposed (semihemispherical) and electrodes with an overall surface area of less than 10 μm² adhered well to classic theory of ultramicroelectrode steady-state response.[28] In millimolar concentrations of ruthenium hexamine trichloride and 100 mM KCl, 200 mV/s scan rates resulted in deviation from ultramicroelectrode behavior for longer electrodes. For example, fibers with lengths longer than 10 μm and surface areas greater than 10 μm² began to indicate the quasi-steady-state responses of cylindrical microelectrodes, which experience slight diffusion limited peaking of the oxidation and reduction curves at higher scan rates.

A unique aspect of microfabricated nanofiber electrode systems is that nonplanar, individually addressable elements can be synthesized in close proximity to one another. This provides the ability to simultaneously probe multiple positions within a biological sample, or to electrically or electrochemically excite one location while observing adjacent regions. This can be particularly advantageous with respect to electrophysiological probing, where pulled capillary probes have been manipulated as individual elements to provide multiple

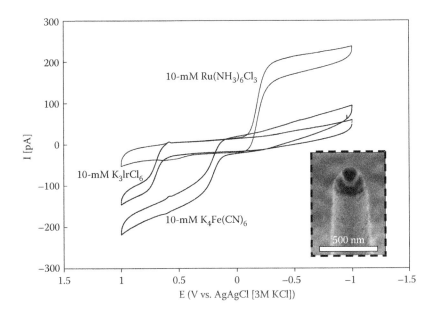

FIGURE 9.3
VACNFs feature reproducible electrochemical properties; demonstrated here by the oxidation and reduction of several outer-sphere, quasi-reversible redox species at the extreme electroactive tip of an individually addressed nanofiber probe, and by the oxidation and reduction of several electroactive neurotransmitters at nanofiber probes that have been fully released from their oxide passivation.

probing of tissue samples. The performance of closely spaced nanofibers has been investigated by generating $Ru(NH_3)_6^{2+}$ at one electrode and reoxidizing this species back to $Ru(NH_3)_6^{3+}$ at an adjacent electrode 10 μm away.[27] This demonstration of interelectrode communication, that is, generator–collector feedback, validated that even with closely spaced electrode arrays, individual responses of electrodes can be resolved. The demonstration of these mechanisms with vertical nanofiber elements provides exciting potential for using these techniques *within* tissue matrices. Communication between *penetrant* nanofiber structures may provide information regarding the fate of electrogenerated species as they travel and interact with the biological matrix between the paired electrodes.

9.6 Chemical Functionalization

Nanofibers provide an electroactive structure for probing and manipulating subcellular scale volumes, but they also provide the opportunity for scaffolding biologically active materials and introducing these materials into inter- and intra-cellular domains. The same functional groups that provide sites of reproducible electron transfer can also serve as chemical handles for postsynthesis chemical modification of the fiber surface. These surfaces can be modified using the same well-established techniques that have been developed for other carbon-based materials, including physical adsorption, covalent coupling, and molecular capture in polymer coatings. One strategy that is often employed for modifying carbon electrodes with active enzymes is carbodiimide-mediated coupling. For example, 1-ethyl-3-(3-dimethylaminopropyl)carbodiimide (EDC) may be used to conjugate enzymes to nanofibers via a condensation reaction between enzyme amines and carboxylic acid sites on fibers. Figure 9.4 provides the result of such a coupling between soybean peroxidase (SBP) and as-synthesized nanofibers. After the coupling reaction, tetramethylbenzidine and hydrogen peroxide were used to form an insoluble precipitate at local areas of active SBP. The result as observed with scanning electron microscopy reveals a dense region of SBP activity located near, but not upon, the tips of VACNFs, apparently due to a localized abundance of carboxylic acid sites on the nanofibers in these locations.

Using similar chemistries, the direct immobilization of glucose oxidase on the tips of vertical nanostructured carbon elements for the purpose of mediatorless electrochemical sensing of glucose has also been demonstrated.[29] Rather than relying on the presence of carboxylic acid sites on *as-synthesized* nanofibers, carboxylic acid sites were generated by clamping the nanofiber at 1.5 V for 90 s (vs. AgAgCl) and electrochemically oxidizing the nanofiber surface in 1 M NaOH. Other techniques to increase COOH coverage have included potassium permanganate oxidation and exposure of nanofibers to water, air, acetylene, and/or oxygen plasmas. Pretreatment of a nanofiber array with the latter results in a much more uniform coverage of soybean peroxidase over the entire fiber during EDC-mediated condensation, as visualized by the tetramethylbenzidine reaction (Figure 9.4), putatively due to a more homogenous distribution of COOH over the nanofiber surface.

Nanofiber surfaces can also be modified with functional oligonucleotides. Again using EDC, amine- and dye-terminated single-stranded oligonucleotides have been immobilized onto the tips of vertically aligned carbon nanostructures, and used for hybridization and detection of subattomole quantities of DNA targets using $Ru(bpy)_3^{2+}$-mediated guanine oxidation.[30–32] Much larger DNA sequences have also been immobilized to VACNF arrays, including 5000+ bp DNA plasmids[33] and even larger viral sequences.[34] As opposed

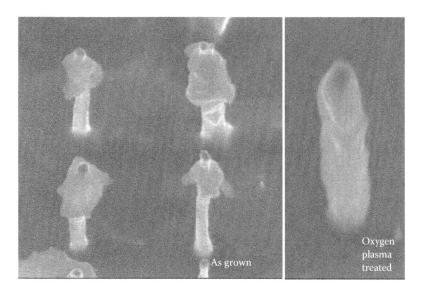

FIGURE 9.4
The surface chemistry of carbon nanofibers provides for functionalization to increase their biological utility. Carbodiimide chemistry was used to modify the surface of this nanofiber array with soybean peroxidase. Following immobilization, visualization of enzymatic activity was achieved by forming an insoluble precipitate from a peroxidase substrate. At left, active enzymes appear confined to a small portion of the nanofiber surface. Following surface activation via exposure to oxygen plasma, immobilization appears more uniform across the entire length of the nanofibers.

to using amine-terminated DNA constructs, double-stranded DNA (dsDNA) was bound putatively at guanine amines[35] available either via stochastic denaturation in local regions of the otherwise dsDNA or at the overhanging ends of linearized fragments.

Photoresist protection combined with the aforementioned chemistries provides a convenient means of derivitizing nanofiber arrays in spatially discrete patterns. For example, thick layers of diazonium-based photoresists (Shipley SPR220 and 1800 series photoresists) may be patterned over nanofiber arrays such that only regions that have been cleared of photoresist are accessible to subsequent chemical derivitization (Figure 9.5). Often following photoresist development, a brief oxygen plasma should be used to ensure that residual traces of the photoresist have been removed from the cleared regions of the sample. Following derivitization, the photoresist block can be removed with a variety of solvents, including acetone, sodium hydroxide, or methanol. Choice of solvent must be compatible with the nature of the bound material so as to not impact its biological functionality.

Photoresist templating strategies may also be implemented to derivitize nanofibers heterogeneously over their length. Using either resist layers with thicknesses less than the height of nanofibers, or by subsequently etching the resist layer back with oxygen plasma, the resist layer can be used to block the bases of nanofibers while allowing derivitization of their tips. This is a convenient means, for example, of ensuring the nanofiber substrate does not participate during derivitization. Figure 9.6 provides a scanning electron micrograph of a photoresist-blocked VACNF array derivitized with gold-labeled streptavidin using EDC condensation. Here, the base of the nanofibers and the substrate were protected in Shipley 1813 photoresist and etched with an oxygen plasma to clear the VACNF tips of spun up photoresist. Following an overnight reaction, the resist was removed with acetone, leaving behind gold-streptavidin on the nanofiber tips, as well as in threads between the nanofibers. The latter are likely due to the formation of carboxylic acid groups on the

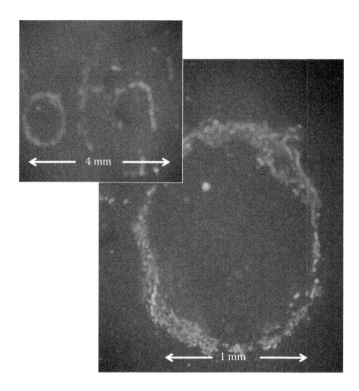

FIGURE 9.5
Photoresist-based protective schemes can be used to limit immobilization of material to specific regions of a nanofiber array. Here, a crude lithography mask was used to define the pattern "ORNL" in a layer of photoresist on a nanofiber array. DNA was then covalently tethered to the exposed elements of the nanofiber array using carbodiimide chemistry. Following the immobilization procedure, the photoresist layer was removed with acetone, leaving behind a VACNF array spatially patterned with tethered DNA.

surface of the photoresist layer during the oxygen plasma etch. As such, immobilization of gold-streptavidin occurred at COOH sites on both the fibers and the photoresist surfaces between fibers.

In addition to direct functionalization of the nanofiber surface with biologically active molecules, nanofiber surfaces may also be physically coated with nanometer thicknesses of other materials including metals, polymers, silicon nitride, and silicon dioxide. As with direct functionalization, these coating techniques can be combined with lithographic processing such that only local regions of a nanofiber array are modified. Similarly, the same techniques used for emancipating the nanofiber tips from blocking resist or passivating oxide can also be used for coating nanofibers heterogeneously over their length. For example, the nanofiber may be encapsulated in silicon dioxide and then processed such that only the extreme tip is exposed carbon. These discrete surfaces (oxide and carbon) may subsequently be modified independently, such as by using organosilane reactions to derivitize the silicon dioxide sheath and using EDC chemistry to derivitize the carbon tip. Similarly, by burying the base of a nanofiber array in photoresist, the tips of the array can be coated with physical vapor deposited gold and subsequently derivitized with thiolized species.

Electrochemical strategies provide additional VACNF derivitization schemes and the ability to direct functionalization at specific electrodes. For example, metals can be electrodeposited both as discrete nucleated islands and as continuous films on the surfaces of nanofibers by application of reducing potentials to the addressed nanofiber electrode.[26]

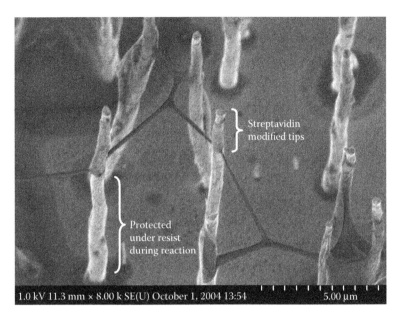

FIGURE 9.6
Protective schemes can be used to also derivitize nanofiber arrays heterogeneously over their length. This scanning electron micrograph shows an array of VACNFs modified with tethered streptavidin. Limitation of this material to the tops of nanofibers was achieved by burying the base in a layer of photoresist, using carbodiimide chemistry to bind the streptavidin, and subsequently removing the photoresist with acetone.

Gold or platinized nanofiber electrodes generated thus can be used for advanced electrochemical detection schemes such as pulsed amperometric detection. This technique was developed to enable long-term measurements via regeneration of the active electrode surface in fouling biological environments. Here, the applied waveform not only quantitates electroactive species in the probing volume but also provides for oxidative cleaning and reductive regeneration of the metal islands on the electrode surface.[36] Metallized sites can also be used to provide for affinity binding. The capture of thiolated species onto electrodeposited gold serves as an excellent example of this process. Lee et al. have demonstrated another approach for the capture of thiolated species at individually addressed VACNFs entailing the electrochemical activation of a surface-bound species.[37] First, VACNF surfaces were modified with 4-nitrobenzenediazonium tetrafluoroborate (36 mM in 1% sodium dodecyl sulfate) to provide coverage with nitrophenyl groups. These nitrophenyls were then reduced to primary amines at specific electrodes by application of -1.0 to -1.5 vs. AgAgCl. These primary amines could then be modified using a variety of conjugation chemistries. In the Lee study, the amines were made reactive to 5′-thiol terminated oligonucleotides by immersion in the heterobifunctional cross linker, sulfo-succinimidyl 4-(*N*-maleimidomethyl)cyclohexan-1-carboxylate (SSMCC).

Electrodeposited polymer films, for example, polypyrrole, can also be used to modify discretely addressed nanofiber electrodes. Chen et al. electropolymerized conformal polypyrrole films onto each individual element of a bulk-addressed vertical forest of carbon fibers from a solution of 17.3 mM pyrrole monomer in 0.1 M LiClO$_4$ at anodic potentials of >0.4 V vs. SCE.[38] The electroanalytical techniques used for these depositions allow for controlling the thickness of polypyrrole and incorporation of anionic species into the polypyrrole matrix as it forms. Often, these techniques have been used for immobilization of enzymes for enzyme-mediated redox reactions on the carbon electrode.

9.7 Compatability with Mammalian Cell Culture

Nanofiber arrays, as well as the materials used to passivate nanofiber-based devices, that is, SU-8 and PECVD oxide, provide for attachment and proliferation of many adherent mammalian cell types. These include Chinese hamster ovary (CHO K_1BH_4), human breast epithelial (MVLN), mouse macrophage (J774a.1), rat thoracic aorta (A-10), quail neuroretina (QNR), rat pheochromocytoma (PC-12), and day 18 embryonic rat hippocampal cells. Figure 9.7 provides a scanning electron micrograph after 48 h of culture of Chinese hamster ovary cells on an as-synthesized nanofiber array. Here, the nanofibers are spaced at 5 μm intervals and are grown to be ~10-μm tall, with diameters of <100 nm at the tip and ~1 μm at the base. CHO was seeded onto the array as a suspension of individual cells with diameters of 7–10 μm. As shown, CHO attaches and stretches out on the nanofiber array, using the nanofibers as anchorage points. Closer inspection reveals that CHO attachment is typically localized to a portion of the nanofiber slightly below the tip, corresponding perhaps to the same location where soybean peroxidase was found to preferentially immobilize during EDC-mediated condensation. It is anticipated that this region of the nanofiber is rich in functional groups, including those used for cellular attachment and carboxylic acids for peptide bonding.

Cell proliferation is often used as a metric to evaluate the suitability of a substrate, or a surface modification to a substrate, for cell culture. Cell proliferation on nanofiber arrays can be evaluated using several techniques. Sacrificial techniques may be used where individual VACNF chips are seeded with cells at a known density and are then treated with lysing agents after specific periods of growth. The number of cells at the time of sacrifice can be determined by assaying the lysate for biomarkers including, for example, DNA or

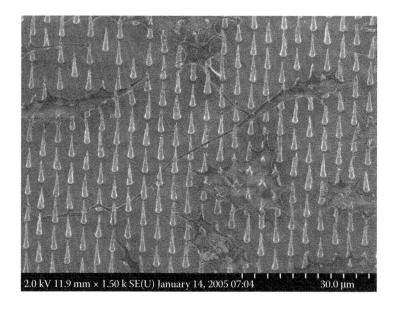

FIGURE 9.7
Adherent cell lines can use the elements of a nanofiber array for adhesion. In this scanning electron micrograph, Chinese hamster ovary cells are shown to attach to various locations along the nanofiber length. As-synthesized nanofibers, without treatment with specific cellular attachment matrices, provide adhesion points for a variety of cell types.

lactone dehydrogenase content. Cell proliferation may also be directly observed by monitoring colony formation and expansion, that is, clonal growth rate. Here, cells are seeded onto VACNF chips at low densities in order to provide for attachment of individual cells and formation of colonies well separated from neighbors. By tracking the formation of clonal colonies from individual cells, and their subsequent expansion, one can directly determine the doubling time of that strain on the VACNF array. Figure 9.8 provides a sequence of images depicting the clonal growth of a strain of CHO K_1BH_4 CIA (clonal isolate A) on the same nanofibers, as depicted in Figure 9.7. This strain was genetically modified to constitutively express a cyan fluorescent protein via stable insertion of the *pd2eCFP-N1* CFP gene (Clontech) and was chosen to facilitate imaging. Since the VACNF arrays are typically grown on an opaque substrate, that is, silicon, transmission microscopy, including phase contrast techniques, cannot be used. The expression of cyan fluorescent protein, however, allows the colonies to be tracked with fluorescent microscopy. In this case, the clonal colonies were imaged on an inverted epifluorescent microscope by placing the seeded nanofiber array face down in a culture dish during imaging. In this set of experiments, the clonal growth rate, or doubling time, of the CFP expressing cells was found to vary between 18 h and 34 h, with an average doubling time of 23.9 ± 6.9 h for n = 9 colonies. These rates are similar to the doubling time of this particular strain on conventional culture surfaces, including nunc-delta culture dishes (Nunc) and T-25 culture flasks (Falcon). As such, the morphology and chemistry of this particular *as-grown* VACNF array is very well suited to culture of this strain of CHO, even without specific surface treatments to promote attachment of these cells.

Cell attachment and proliferation also occurs on a variety of materials used in the fabrication of nanofiber-based devices, including SU-8 and PECVD silicon oxide, both of which are the surface materials of the devices depicted in Figure 9.2. To promote adhesion on these surfaces, a 30-s oxygen plasma etch can be used to both roughen the passivation

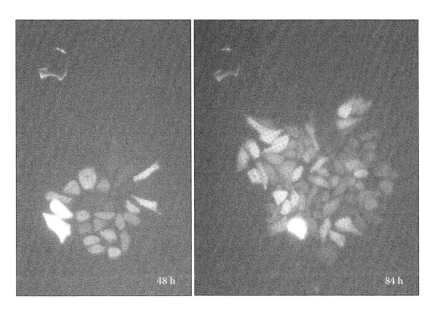

FIGURE 9.8
Cell proliferation on nanofiber arrays proceeds much like on planar surfaces. Here, a cyan fluorescent protein-expressing strain of Chinese hamster ovary cells is shown proliferating on a sample of the nanofiber array depicted in Figure 9.7.

surface as well as provide surface charges. Chemical treatments, such as the application of poly-L-lysine (PLL) or poly-D-lysine (PDL) and/or fibronectin, can also be used and facilitate improved attachment for most cell types. However, if the nanofiber devices are to be used for electrochemical probing or manipulation, caution must be applied so as to not coat and passivate the electrochemically active surface of the nanofiber elements. Even without such treatments, culture of a variety of cell types including quail neuroretina (QNR), rat thoracic aorta (A10), and rat embryonic hippocampal cells (E18) has been maintained upon the devices depicted in Figure 9.2 for periods as long as 8 weeks, at which point experiments were terminated.

The attachment and residence of cells upon nanofiber-based devices facilitate long-term monitoring of these cellular matrices with the nanofiber device, provided that the nanofiber-based probing arrays can remain electrochemically responsive during these prolonged periods of cell culture (weeks). To validate that nanofiber electrodes can remain electrochemically active after long periods of cell culture, quail neuroretina cells (QNR[39]) were seeded and cultured onto individually addressable nanofiber arrays (Figure 9.2) for a period of 4 weeks. Culture media (DMEM + 10% FBS + 4 mM glutamine) was changed daily. At days 23, 25, and 27, media was aspirated, cells were bathed in Tyrodes solution, and electrophysiological stimuli and measurement were performed at individually addressed nanofiber electrodes to evoke excitable activity in the confluent QNR monolayer. Evoked activity of these neuronal cells could be achieved by applying excitatory waveforms to nanofibers located in close proximity to cells. This induced current spiking behavior at the electrodes due putatively to both capacitive coupling of the electrode with the depolarization of the cell and oxidation of exocytosed electroactive materials from these cells. These demonstrations provide evidence that, even after long-term cell culture in high protein containing media, nanofiber electrodes remain sufficiently unfouled and electrochemically active to both induce and monitor the activity of excitable cells.

9.8 Electrical–Neural Interfacing

One of the most widely studied areas of cellular integration of VACNFs is their application as both growth substrates and electrodes for interfacing with neuronal cells and excitable tissue. Nguyen-Vu et al. presented one of the seminal articles anticipating VACNFs as a neural interface.[40] Anticipating that the capacitive coupling of high-surface area VACNF electrodes would be effective for functional electrical stimulation (FES) of tissue, forests of closely spaced VACNFs were grown using PECVD from nickel catalyst on a chrome-coated silicon wafer. To promote the capacitive potential of the VACNF electrode, polypyrrole was then electrodeposited onto the surfaces of the nanofiber forest and this was subsequently coated with collagen to promote cellular attachment. The result provided a robust, high-capacitance electroactive polymer electrode upon which rat pheochromocytoma cells (PC-12) were cultured. McKnight et al. subsequently demonstrated a series of silicon-based microelectrode arrays, which exploited individually addressable VACNFs as the active electrodes for electrophysiological and electroanalytical coupling with both PC-12 cells and dissociated cells from rat hippocampal tissue.[41] These studies demonstrated that microfabricated VACNF devices with SU-8 and oxide passivation could be implemented for culture of excitable cells for at least 16 days, with the electrodes remaining electrophysiologically and electroanalytically active over this entire time period. Both extracellular field potentials and oxidative current bursts due to neurotransmitter exo-cytosis from cells could be observed on individual VACNF electrodes. Zhu et al. shortly thereafter followed up this study using the same devices to stimulate and record activity from slices of hippocampal tissue from day 8 to 11 rat pups.[42] Spontaneous spiking activity from the hippocampal slice could be recorded at individual electrodes, with a bicuculline cocktail inducing epileptiform oscillatory behavior and tetrodotoxin suppressing sponta-neous spiking activity. Evoked potentials could be induced by injecting current via indi-vidual elements of a linear array of VACNF electrodes, and resulted in highly reproducible evoked responses of the tissue, putatively due to the improved coupling of the high aspect, penetrant electrode into the hippocampal slice. De Asis et al. conducted a systematic study of various coatings on VACNF "brush" electrodes for efficacy of electrical stimulation of hippocampal slices and found that polypyrrole-coated VACNF "microbrush arrays" could be used to safely stimulate neural tissue while reducing the electrolytic decomposition of water that normally results from such stimulation with more conventional electrodic materials, providing a path forward for more long-term implantation of VACNF devices for chronic neurostimulation.[43]

9.8.1 Intracellular Interfacing

In addition to cell culture and growth *on and around* carbon nanofibers, nanofiber arrays may also be integrated with the *intracellular* domain of cell and tissue matrices. The shape and resilience of nanofiber arrays enable them to efficiently penetrate the plasma mem-brane of cells using a variety of integration techniques including centrifugation of cells onto arrays and directly pressing arrays into cellular matrices.[33,44] The covalent-bonding structure of the VACNF provides considerable resilience to mechanical strain, enabling the fiber to deform under stress, and recover from this deformation when the stress is removed. Figure 9.9 provides a composite of two scanning electron micrographs of a nano-fiber array after the array was pressed into a suspension culture of typsinized Chinese hamster ovary cells. Prior to imaging, the cells were fixed with 2% gluteraldehyde and methanol dehydration. As evidenced in this image, either the integration process or the subsequent subsidence occasioned by dehydration caused many of the nanofibers to tortu-ously bend as a consequence of interactions with the CHO cell. One particular fiber at the left of the cell was imaged under high magnification, which caused the attachment with the cell to break. Upon doing so, the nanofiber immediately recovered from its deforma-tion and reassumed its vertical orientation on the substrate. Subsequent release by focus-ing the electron beam was achieved at several additional nanofibers on the substrate. This mechanical resilience makes the nanofiber an ideal structure for direct integration with cellular matrices.

In addition to survival of nanofibers during these integration procedures, cells can also recover and proliferate following nanofiber interfacing. Probably due to the nanoscale diameter of VACNFs, the plasma membranes of mammalian cells can apparently reseal around the penetrant nanostructure following the penetration event. This enables the recovery and continued proliferation of the interfaced cell. These results were observed by centrifuging cells onto nanofiber arrays either in the presence of a membrane impermeant stain, propidium iodide (PI), or with PI added to the solution ~5 min *after* centrifugation. PI allows fluorescent visualization of membrane compromise, as its fluorescence yield increases significantly when internalized due to intercalation with intracellular nucleic acids. Therefore, if dye is present during cell penetration, staining is observed and can be used as an indicator of membrane penetration. When dye is added several minutes after

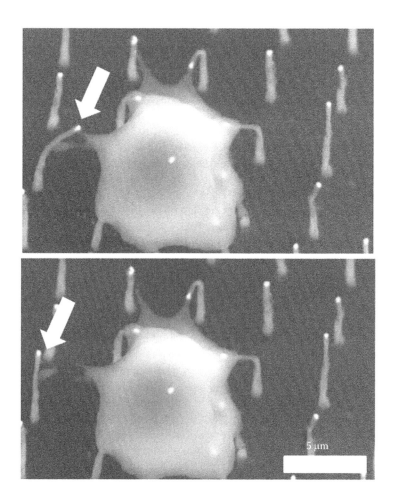

FIGURE 9.9
The covalent bonding structure of VACNFs provides a mechanically robust material that is well suited to cellular integration. This scanning electron micrograph shows a bed of nanofiber elements following integration with Chinese hamster ovary cells. In the top pane, a single fiber can be seen that is bent over due to interaction with the cell. Upon imaging with higher magnification, the cell/nanofiber interaction broke, causing the nanofiber to spring up and recover to its vertical orientation, as shown in the bottom pane.

the penetration event, however, the dye can be excluded, indicative that the membrane can reseal around the penetrant nanofiber.

Deterministically synthesized VACNFs modified with biological material can be inserted into cells to impact intracellular biochemistry. For example, arrays of nanofibers modified with either adsorbed or covalently linked plasmid DNA can be inserted into viable cells in a parallel manner to impart new genetic information.[33,44] In these experiments, nanofibers were modified with fluorescent reporter genes such that successful delivery could be evidenced by expression of the fluorescent gene product in penetrated cells. Furthermore, by continued observation of the manipulated cells, the long-term survival of these cells to nanofiber penetration could be determined.

Following synthesis, nanofiber arrays were surface-modified with plasmid DNA. The predominant plasmid used in these experiments was pGreenLantern-1, which contains an enhanced green fluorescent protein (eGFP) gene under the CMV immediate early enhancer/promoter, the SV40 t-intron and polyadenylation signal, and no mammalian origin of replication. Other reporter constructs, *pd2eYFP-N1* and *pd2eCFP-N1* (Clontech), which encode a yellow and a cyan fluorescent protein, respectively, have also been successfully used in similar experiments. Plasmid DNA at various concentrations (5–500 ng/μL) was either spotted onto the chips as 0.5–1 μL aliquots and allowed to dry or covalently tethered to the nanofibers using the carbodiimide coupling reaction, as discussed in earlier sections.

The cell line used predominantly for these experiments was a subclone of the Chinese hamster ovary (CHO) designated K_1-BH_4 and provided to us by Dr. A.W. Hsie,[45] although the technique has also been validated in quail neuroretina (QNR), human breast cancer (MVLN), and mouse macrophage (J774a.1). CHO cells were routinely grown in Ham's F-12 nutrient mixture supplemented with 5% fetal bovine serum and 1 mM glutamine. Cell cultures were grown in T-75 flasks and passaged at 80% confluency by trypsinization using 0.025% trypsin-EDTA. In preparation for fiber-mediated plasmid delivery, adherent cells were trypsinized from T-75 flasks, quenched with 10 mL of Ham's F12 media, pelleted at 100g for 10 min, resuspended in phosphate-buffered saline (PBS), counted, and diluted in PBS to a desired density ranging from 50,000 to 600,000 cells/mL.

The nanofiber arrays used for these cell-interfacing experiments were similar to those previously described for cell attachment and proliferation experiments. Arrays were synthesized with nanofibers at a 5-μm pitch, with tip diameters of ideally <100 nm and nanofiber lengths of several microns. The pitch of the nanofibers was chosen based on the morphology of mammalian cells in suspension, which are typically spheroids of 5–10 μm diameter. With these dimensions, a suspended cell will likely interact with only 1 nanofiber as it is centrifuged down onto a 5-μm pitched nanofiber array. Nanofiber length is an important parameter with respect to cellular penetration. If nanofibers are too short (<∼1 μm), penetration is reduced, possibly due to the compliance of the cell membrane and its ability to deform around the vertical nanofiber. Fiber lengths of at least 50% of the diameter of the suspended cell are effective for penetration and gene delivery. Tip diameter is an important parameter with respect to cell survival following the penetration event. Traditional microinjection literature indicates that pulled capillaries should ideally be <100 nm to improve survival rates of microinjected cells.[46] While nanofibers up to ∼300 nm have been successful with cell penetration and gene delivery, minimizing the nanofiber diameter likely minimizes trauma, and increases survival of penetrated cells. One of the most effective means of penetrating cells with VACNFs is to centrifuge the cells out of suspension onto the nanofiber array and then subsequently press them against a smooth, compliant surface to increase the penetration. For the centrifugation step, one must place the nanofiber array in a tube that orients the array appropriately in the centrifuge, such that cells impinge directly down onto the nanofibers. One approach is to fill a microcentrifuge tube with a castable material and spin the tube while the material gels or cures. Agarose or polydimethylsiloxane (PDMS) can be used for this purpose. Sylgard 184 is a two-part PDMS matrix that will cure to form a compliant solid in ∼12 h at room temperature when mixed. If the tubes are spun in a centrifuge during this time, the resultant slant in the PDMS will enable proper positioning of a nanofiber array for subsequent spins, provided the tube is always inserted into the centrifuge with the same orientation as used during PDMS cure. These tubes can also be autoclaved for subsequent reuse.

Immediately following centrifugation, cells retain their rounded shape and remain loosely coupled with VACNFs. If an optional press step is used to increase the probability and depth of nanofiber penetration into the cells, adherent cell types tend to deform from their spherical shape and attach to the nanofibers and interfiber surfaces of the substrate.

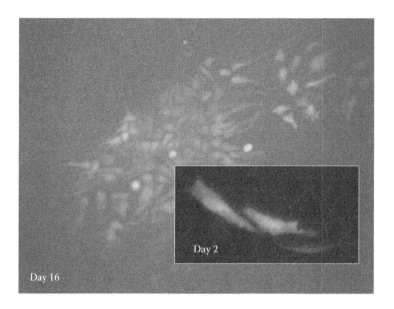

FIGURE 9.10
VACNF arrays are an effective vector for material delivery into mammalian cells and tissue. In these fluorescent micrographs, Chinese hamster ovary cells were impaled upon nanofiber arrays that had been modified by dehydrating a green-fluorescent protein encoding plasmid DNA onto the array. Following the impalement, cells can recover and proliferate, and often express the DNA cargo of the penetrant nanofiber.

Surviving cells eventually stretch out on the substrate and continue to proliferate, with those that received DNA during the interfacing going on to express nanofiber-delivered DNA over long time periods (up to weeks in some experiments). Figure 9.10 is an example of fluorescent protein expression following penetration into CHO cells of nanofibers that had been modified with physically adsorbed DNA. The fluorescent image in the background shows a large colony of GFP expressing cells and was photographed 16 days following nanofiber/cell interfacing.

9.8.2 Nuclear Penetration

DNA that is covalently tethered to penetrant nanofiber arrays can also be expressed within some nanofiber impaled cells for extended periods (several weeks).[33] Since the nucleus is the site of eukaryotic transcription, the first step of gene expression, this implies that nanofibers can achieve nuclear residence during or following the penetration event. Moreover, long-term expression of covalently tethered DNA implies that nanofibers and their tethered DNA cargo can *maintain* this nuclear presence, even apparently through the mitotic events of prometaphase nuclear membrane breakdown and telophase nuclear membrane reformation.

To directly visualize nuclear residence of nanofibers, CHO cells were centrifuged and pressed onto a nanofiber array and subsequently exposed to a nuclear extraction protocol developed by Butler.[47] In brief, VACNF-interfaced cells were exposed to a hypotonic solution in the presence of ethylhexadecyldimethylammonium bromide. This effectively caused the plasma membrane to rupture, spilling the contents of the cytoplasm, and leaving behind an intact nucleus. Following nuclear stabilization and cell lysis, the extracted nuclei on the nanofiber platform were fixed with 1% paraformaldehyde in PBS

FIGURE 9.11
Penetration of nanofibers can extend into the nucleus of interfaced mammalian cells. Chinese hamster ovary cells were impaled onto a nanofiber array and subsequently exposed to a nuclear extraction protocol. The "extracted nuclei" of the interfaced cells remained resident on the nanofiber platform. Subsequent fracture analysis of these nuclei demonstrated that nanofibers can penetrate into the mammalian nucleus, thereby, providing a nanostructured interface to the transcriptional control center of the cell.

and dehydrated with methanol. Subsequently, fixed nuclei resident on the VACNF array were subjected to fracture forces by application and removal of carbon tape to the array and its captured nuclei. The array was then inspected with scanning electron microscopy. Figure 9.11 provides a compelling picture of a nucleus, putatively fractured by the tape adhesion and peeling process. At the point of fracture, an individual VACNF can be observed, putatively resident within the nucleus during the nuclear extraction and fixation protocol. In retrospect, this phenomenon should not be that surprising. The nucleus of most *suspended* mammalian cells is a significant portion of the overall cell cross section. As such, penetration of the cellular membrane of a suspended cell should often result in co-penetration of the nucleus.

The penetration and residence of DNA-modified nanofibers within the nuclear domain offers exciting possibilities for both gene delivery applications and the fundamental study of gene expression and transcriptional phenomenon. For example, nuclear delivery, such as that provided by nuclear microinjection, is attributed to being more efficient per unit of delivered DNA vs. non-nuclear targeting methods. This is because the delivered material is protected from extranuclear degradative pathways, including cytosolic- and extracellular-nuclease activity. For similar reasons, nuclear targeting also facilitates the simultaneous multiplexed delivery of many genes. As delivered material need not survive degradative gauntlets, the likelihood of successful nuclear delivery and expression of multiplexes of template is improved.

Delivery of nanofiber-tethered DNA into the nucleus may also offer a higher level of control over the fate of introduced genes, including the potential to remove these genes from a system after a period of transient expression simply by removing the cells from the nanofiber array. In experiments conducted with covalently tethered DNA, it is often observed that a nanofiber-impaled cell maintains expression of the tethered gene, but its progeny do not. Since the DNA is tethered to the penetrant nanofiber scaffold, only those cells that maintain nuclear residence of the penetrant structure can access its DNA for transcription and expression. When a cell divides, at most only one of the mitotic couple can maintain this residence and continue expression of the tethered gene. With the reporter gene system of tethered plasmid pd2eYFP-N1, this is manifested by the impaled progenitor continuing to express yellow fluorescent protein while progeny cells only receive an aliquot of cytosolic YFP during mitosis. In time, this aliquot of protein is degraded and progeny no longer display yellow fluorescence. This scenario provides interesting opportunities for a variety of applications where protein-modified cells are desired, but genetic manipulation is prohibited. Tethered DNA arrays enable the production of transgene protein-filled progeny that maintain wild-type genotype.

Nuclear delivery via VACNF microinjection provides the opportunity for very rapid transgene expression of the delivered DNA template. In our laboratory, transgene expression can be visualized using a fluorescent microscope as early as 27 min following the delivery of both pd2eYFP-n1 and mVenus (both yellow fluorescent proteins of high brightness) into retinal pigment epithelial cells and African green monkey (Cos-7) cells. Even more slowly maturing fluorescent proteins, such as GFP, can typically begin to be visualized within an hour of cellular interfacing. The ease and rapidity of such transgene manipulation, therefore, opens the doorway toward application of VACNF-mediated gene delivery for diagnostic, critical care, and point-of-care opportunities. Nuclear delivery of tethered DNA on a parallel basis may also provide more efficient methods for studying the impact of template length and topology on transcriptional activity, which has traditionally been investigated through application of the serial method of microinjection.[48–50]

9.9 VACNF-Templated Devices

In previous sections, we have focused predominantly on the fabrication, modification, and cellular interaction of *solid* nanofiber elements. However, the same attributes that make the nanofiber attractive for these applications may also be exploited to provide even further utility of the nanofiber-based device. Similar to the historical perspective of solid glass agonizers evolving into pulled glass capillaries, techniques have been developed to transform the scaffolding of solid nanofibers into more elaborate structures that we refer to as "nanopipes" and "partial nanopipes."[51,52] These structures are fabricated by first synthesizing arrays of nanofiber templates that are then coated with thin (<100 nm) layers of other materials (e.g., silicon dioxide). The precursor carbon-based nanofiber scaffold is then selectively etched from the structure. The etching is performed by opening the tips of each coated nanofiber (e.g., using an oxide reactive ion etch), and then reactive ion etching, thermally decomposing, or electrochemically etching the internal carbon material. The result is a hollow pipe with nanoscale diameters (Figure 9.12). As these pipes are derived from nanoscale nanofiber precursors, they also exist on a size scale appropriate for interfacing to cellular matrices. Like VACNFs, these structures can be fabricated as vast arrays providing a highly parallel interface to cells. Nanopipes may be constructed on thin membranes and used for fluidic transport of macromolecules (propidium iodide and plasmid DNA) across the membrane, using both diffusive and electrokinetic transport

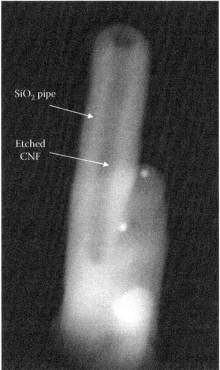

FIGURE 9.12
Nanofibers may serve as templates for other nanostructured devices. Here, a nanofiber was coated with oxide and subsequently etched, forming a nanoscale pipe. The piped region of this device features an inner diameter of ~40 nm.

techniques.[48] Figure 9.13 demonstrates the electrophoretic accumulation of DNA in partial pipes, which are structures similar to that in Figure 9.12, where the removal of the carbon nanofiber is not complete, but it rather forms just a partial pipe on one end of the silica tube. By applying potential to the nanopipe substrate, and thereby to the partial carbon core within these nanopipes, propidium iodide–labeled DNA could be accumulated within and upon the silica nanopipes from an overlying solution. These demonstrations show that nanopipes can provide an effective means of essentially providing highly parallel arrays of fine-tipped microcapillaries. We anticipate that they may be employed for similar applications of their conventional forebear, including electrophysiological probing, patch-clamping, material delivery via micro/nanoinjection, and potentially even material sampling from cellular and intracellular matrices.

9.10 Summary

The self-assembling and controlled synthesis properties of vertically aligned carbon nanofibers (VACNF) have been exploited to provide parallel subcellular and molecular-scale probes for biological manipulation and monitoring. VACNFs possess many attributes that make them very attractive for implementation as functional, nanoscale interfaces to

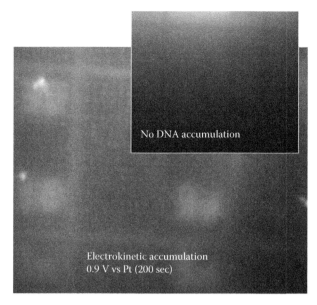

No DNA accumulation

Electrokinetic accumulation
0.9 V vs Pt (200 sec)

FIGURE 9.13
Nanopipe arrays may be electrokinetically loaded with material. Here, propidium iodide labeled DNA was electrophoretically accumulated in and upon an array of electrically addressed nanofibers, similar to that shown in Figure 9.12.

cellular processes. They can be synthesized at precise locations upon a substrate, can be grown many microns long, and feature sharp, nanodimensioned tips that enable their penetration into cellular environments. Nanofibers are highly compatible with most microfabrication processes and can, therefore, be deterministically integrated into functional microfabricated devices. As carbon-based electrodes, nanofibers exhibit characteristic electrochemical responses similar to conventionally studied materials such as the edge plane of pyrolytic graphite and surface-activated glassy carbon. Unlike these conventional materials, however, nanofibers appear to require very little conditioning to activate their electrochemical response. Nanofibers can be synthesized to feature rich surface chemistries, providing the opportunity for postsynthesis chemical modification to increase their biological utility. Finally, the combination of these features provides the ability to efficiently couple nanofiber-based systems with intact cells on a highly parallel basis for measurement, manipulation, and control of subcellular and molecular scale processes within and around live cells.

Acknowledgments

The authors acknowledge the many talented students, technicians, and colleagues who have assisted with the development of VACNF-based biological interfacing systems, including Teri Subich, Dale Hensley, Darrell Thomas, Rich Kasica, Pam Fleming, Jenny Morrell, Stephen Jones, Chorthip Peeraphatdit, David GJ Mann, Francisco Serna, Tyler Sims, Derek Austin, Kate Klein, Stephen Randolph, Seung Ik Jun, Jason Fowlkes, Phillip

Rack, Chris Culbertson, Stephen Jacobson, and Doug Lowndes. These works were supported, in part, by the National Institute for Biomedical Imaging and Bioengineering under assignments 1-R01EB000433-01, 1-R21EB004066, and 1-R01EB000440-01 and through the Laboratory Directed Research and Development funding program of the Oak Ridge National Laboratory, which is managed for the U.S. Department of Energy by UT-Battelle, LLC. MLS acknowledges support by the BES Material Sciences and Engineering Division Program of the DOE Office of Science under contract DE-AC05-00OR22725 with UT-Battelle, LLC.

References

1. Magner, L.N., *A History of the Life Sciences*, 3rd ed., Marcel Dekker, Inc., New York, 2002, Chap. 4.
2. Korzh, V. and Straehle, U., Marshall Barber and the century of microinjection: From cloning of bacteria to cloning of everything, *Differentiation*, 70, 221, 2002.
3. Aidley, D.J., *The Physiology of Excitable Cells*, 4th ed., Cambridge University Press, Cambridge, 1998.
4. Wightman, R.W., Jankowski, J.A., Kennedy, R.T., Kawagoe, K.T., Schroeder, T.J., Leszczyszyn, D.J., Near, J.A., Diliberto, E.J., and Viveros, O.H., Temporally resolved catecholamine spikes correspond to single vesicle release from individual chromaffin cells. *Proc. Natl. Acad. Sci. USA*, 88, 10754, 1991.
5. Krishnan, A., Dujardin, E., Treacy, M.M., Hugdahl, J., Lynum, S., and Ebbesen, T.W., Graphitic cones and the nucleation of curved carbon surfaces, *Nature*, 388, 6641, 451, 1997.
6. Endo, M., Kim, Y.A., Hayashi, T., Fukai, Y., Oshida, K., Terrones, M., Yanagisawa, T., Higaki, S., Dresselhaus, M.S., Structural characterization of cup-stacked-type nanofibers with an entirely hollow core, *Appl. Phys. Lett.*, 80, 7, 1267, 2002.
7. Hughes, T.V. and Chambers, C.R., *Manufacture of Carbon Filaments*, USA, US Patent #405480A, 1889.
8. Melechko, A.V., Merkulov, V.I., McKnight, T.E., Guillorn, M.A., Klein, K.L., Lowndes, D.H., and Simpson, M.L., Vertically aligned carbon nanofibers and related structures: Controlled synthesis and directed assembly, *J. Appl. Phys.*, 97, 141301–1, 2005.
9. Baker, R.T.K., Catalytic growth of carbon filaments, *Carbon*, 27, 3, 315, 1989.
10. Kroto, H.W., Heath, J.R., Obrien, S.C., Curl, R.F., and Smalley, R.E., C-60—Buckminsterfullerene, *Nature*, 318, 6042, 162, 1985.
11. Iijima, S., Helical microtubules of graphitic carbon, *Nature*, 354, 6348, 56, 1991.
12. Chen, Y., Guo, L.P., Johnson, D.J., and Prince, R.H., Plasma-induced low-temperature growth of graphitic nanofibers on nickel substrates, *J. Cryst. Growth*, 193, 3, 342, 1998.
13. Ren, Z.F., Huang, Z.P., Xu, J.W., Wang, J.H., Bush, P., Siegal, M.P., and Provencio, P.N., Synthesis of large arrays of well-aligned carbon nanotubes on glass, *Science*, 282, 5391, 1105, 1998.
14. Ren, Z.F., Huang, Z.P., Wang, D.Z., Wen, J.G., Xu, J.W., Wang, J.H., Calvet, L.E., Chen, J., Klemic, J.F., and Reed, M.A., Growth of a single freestanding multiwall carbon nanotube on each nanonickel dot, *Appl. Phys. Lett.*, 75, 8, 1086, 1999.
15. Merkulov, V.I., Lowndes, D.H., Wei, Y.Y., Eres, G., and Voelkl, E., Patterned growth of individual and multiple vertically aligned carbon nanofibers, *Appl. Phys. Lett.*, 76, 24, 3555, 2000.
16. Merkulov, V.I., Melechko, A.V., Guillorn, M.A., Lowndes, D.H., and Simpson, M.L., Sharpening of carbon nanocone tips during plasma-enhanced chemical vapor growth, *Chem. Phys. Lett.*, 350, 5, 381, 2001.
17. Merkulov, V.I., Melechko, A.V., Guillorn, M.A., Lowndes, D.H., and Simpson, M.L., Alignment mechanism of carbon nanofibers produced by plasma-enhanced chemical-vapor deposition, *Appl. Phys. Lett.*, 79, 18, 2970, 2001.

18. Merkulov, V.I., Guillorn, M.A., Lowndes, D.H., Simpson, M.L., and Voelkl, E., Shaping carbon nanostructures by controlling the synthesis process, *Appl. Phys. Lett.*, 79, 8, 1178, 2001.
19. Merkulov, V.I., Lowndes, D.H., and Baylor, L.H., Scanned-probe field-emission studies of vertically aligned carbon nanofibers, *J. Appl. Phys.* 89, 3, 1933, 2001.
20. Merkulov, V.I., Hensley, D.K., Melechko, A.V., Guillorn, M.A., Lowndes, D.H., and Simpson, M.L., Control mechanisms for the growth of isolated vertically aligned carbon nanofibers, *J. Phys. Chem. B*, 106, 10570, 2002.
21. Merkulov, V.I., Melechko, A.V., Guillorn, M.A., Lowndes, D.H., and Simpson, M.L., Growth rate of plasma-synthesized vertically aligned carbon nanofibers, *Chem. Phys. Lett.*, 361, 492, 2002.
22. Merkulov, V.I., Melechko, A.V., Guillorn, M.A., Lowndes, D.H., Simpson, M.L., Whealton, J.H., and Raridon, R.J., Controlled alignment of carbon nanofibers in a large-scale synthesis process, *Appl. Phys. Lett.*, 80, 4816, 2002.
23. Merkulov, V.I., Melechko, A.V., Guillorn, M.A., Lowndes, D.H., and Simpson, M.L., Effects of spatial separation on the growth of vertically aligned carbon nanofibers produced by plasma-enhanced chemical vapor deposition, *Appl. Phys. Lett.*, 80, 476, 2002.
24. Melechko, A.V., McKnight, T.E., Hensley, D.K., Guillorn, M.A., Borisevich, A.Y., Merkulov, V.I., Lowndes, D.H., and Simpson, M.L., Large-scale synthesis of arrays of high-aspect-ratio rigid vertically aligned carbon nanofibers, *Nanotechnology*, 14, 9, 1029, 2003.
25. Melechko, A.V., Merkulov, V.I., Lowndes, D.H., Guillorn, M.A., and Simpson, M.L., Transition between "base" and "tip" carbon nanofiber growth modes, *Chem. Phys. Lett.*, 356, 527, 2002.
26. Guillorn, M.A., McKnight, T.E., Melechko, A.V., Merkulov, V.I., Austin, D.W., Lowndes, D.H., and Simpson, M.L., Individually addressable vertically aligned carbon nanofiber-based electrochemical probes, *J. Appl. Phys.*, 91, 6, 3824, 2002.
27. McKnight, T.E., Melechko, A.V., Austin, D.W., Sims, T., Guillorn, M.A., and Simpson, M.L., Microarrays of vertically aligned carbon nanofiber electrodes in an open fluidic channel, *J. Phys. Chem. B*, 108, 7115, 2004.
28. Wightman, R.M. and Wipf, D.O., Voltammetry at ultramicroelectrodes, In *Electroanalytical Chemistry*, Bard A.J., Ed. vol. 15. Marcel Dekker, New York, 1988, pp. 267–351.
29. Lin, Y.H., Lu, F., Tu, Y., and Ren, Z.F., Glucose biosensors based on carbon nanotube nanoelectrode ensembles, *Nano Lett.*, 4, 2, 191, 2004.
30. Nguyen, C.V., Delzeit, L., Cassell, A.M., Li, J., Han, J., and Meyyappan, M., Preparation of nucleic acid functionalized carbon nanotube arrays, *Nano Lett.*, 2, 10, 1079, 2002.
31. Li, J., Ng, H.T., Cassell, A., Fan, W., Chen, H., Ye, Q., Koehne, J., Han, J., and Meyyappan, M., Carbon nanotube nanoelectrode array for ultrasensitive DNA detection, *Nano Lett.*, 3, 5, 597, 2003.
32. Koehne, J., Li, J., Cassell, A.M., Chen, H., Ye, Q., Ng, H.T., Han, J., and Meyyappan, M., The fabrication and electrochemical characterization of carbon nanotube nanoelectrode arrays, *J. Mater. Chem.*, 14, 4, 676, 2004.
33. McKnight, T.E., Melechko, A.V., Griffin, G.D., Guillorn, M.A., Merkulov, V.I., Serna, F., Hensley, D.K., Doktycz, M.J., Lowndes, D.H., and Simpson, M.L., Intracellular integration of synthetic nanostructures with viable cells for controlled biochemical manipulation, *Nanotechnology*, 14, 5, 551, 2003.
34. Dwyer, C., Guthold, M., Falvo, M., Washburn, S., Superfine, R., and Erie, D., DNA-functionalized single-walled carbon nanotubes, *Nanotechnology*, 13, 601, 2002.
35. Millan, K.M., Spurmanis, A.J., and Mikkelsen, S.R., Covalent immobilization of DNA onto glassy-carbon electrodes, *Electroanalysis*, 4, 10, 929, 1992.
36. Lau, Y.Y., Wong, D.K.Y., and Ewing, A.G., Intracellular voltammetry at ultrasmall platinum electrodes, D.K.Y. Wong, A.G. Ewing Ed. *Microchem. J.*, 47, 308–316, 1993.
37. Lee, C.S., Baker, S.E., Marcus, M.S., Yang, W.S., Eriksson, M.A., and Hamers, R.J., Electrically addressable biomolecular functionalization of carbon nanotube and carbon nanofiber electrodes, *Nano Lett.*, 4, 9, 1713, 2004.
38. Chen, J.H., Huang, Z.P., Wang, D.Z. et al., Electrochemical synthesis of polypyrrole films over each of well-aligned carbon nanotubes, *Synth. Met.*, 125, 3, 289, 2001.

39. Pessac, B., Girard, A., Crisanti, P., Lorinet, A.M., and Calothy, G., A neuronal clone derived from a Rous sarcoma virus-transformed quail embryo neuroretina established culture, *Nature*, 302, 616, 1983.

40. Nguyen-Vu, T.D.B., Chen, H., Cassell, A.M., Andrews, R., Meyyappan, M., and Li, J., Vertically aligned carbon nanofiber arrays: An advance toward electrical–neural interfaces, *Small* 2, 1, 89, 2006.

41. McKnight, T.E., Melechko, A.V., Fletcher, B.J., Jones, S.W., Hensley, D.K., Peckys, D.B., Griffin, G.D., Mann, D., Simpson, M.L., and Ericson, M.N., Resident neuroelectrochemical interfacing using carbon nanofiber arrays, *J. Phys. Chem. B*, 110, 15317, 2006.

42. Zhe, Y., McKnight, T.E., Ericson, M.N., Melechko, A.V., Simpson, M.L., and Morrison, B., Vertically aligned carbon nanofiber arrays record electrophysiological signals from hippocampal slices, *Nano Lett.*, 7, 8, 2188, 2007.

43. De Asis, E.D. Jr., Nguyen-Vu, T.D.B., Arumugam, P.U., Chen, H., Cassell, A.M., Andrews, R.J., Yang, C.Y., and Li, J., High efficient electrical stimulation of hippocampal slices with vertically aligned carbon nanofiber microbrush array, *Biomed. Microdevices*, 11, 801, 2009.

44. McKnight, T.E., Melechko, A.V., Hensley, D.K., Griffin, G.D., Mann, D., and Simpson, M.L., Tracking gene expression after dna delivery using spatially indexed nanofiber arrays, *Nano Lett.*, 4, 7, 1213, 2004.

45. Hsie, A.W., Casciano, D.A., Couch, D.B., Krahn, D.F., Oneill, J. P., and Whitfield, B.L., The use of chinese-hamster ovary cells to quantify specific locus mutation and to determine mutagenicity of chemicals—A report of the gene tox program, *Mutat. Res.*, 86, 193, 1981.

46. Proctor, G.N., Microinjection of DNA into mammalian cells in culture: Theory and practice. *Methods Mol. Cell Biol.*, 3, 209, 1992.

47. Butler, W.B., Preparation of nuclei from cells in monolayer cultures suitable for counting and following synchronized cells through the cell cycle, *Anal. Biochem.*, 141, 70, 1984.

48. Harland, R.M., Weintraub, H., and McKnight, S.L., Transcription of DNA injected into *Xenopus* oocytes is influenced by template topology, *Nature*, 301, 38, 1983.

49. Krebs, J.E. and Dunaway, M., DNA length is a critical parameter for eukaryotic transcription in vivo, *Mol. Cell Biol.*, 16, 10, 5821, 1996.

50. Weintraub, H., Cheng, P.F., and Conrad, K., Expression of transfected DNA depends on DNA topology, *Cell*, 46, 115, 1986.

51. Melechko, A.V., McKnight, T.E., Guillorn, M.A., Austin, D.W., Ilic, B., Merkulov, V.I., Doktycz, M.J., Lowndes, D.H., and Simpson, M.L., Nanopipe fabrication using vertically aligned carbon nanofiber templates, *J. Vac. Sci. Tech. B*, 20, 6, 2730, 2002.

52. Melechko, A.V., McKnight, T.E., Guillorn, M.A., Merkulov, V.I., Ilic, B., Doktycz, M.J., Lowndes, D.H., and Simpson, M.L., Vertically aligned carbon nanofibers as sacrificial templates for nanofluidic structures, *Appl. Phys. Lett.*, 82, 976, 2003.

10

Single-Molecule Detection Techniques for Monitoring Cellular Activity at the Nanoscale Level

Kui Chen and Tuan Vo-Dinh

CONTENTS

10.1 Introduction

Single-molecule detection (SMD) represents the ultimate goal in analytical chemistry and is of great scientific interest in many fields [1–4]. In particular, SMD and activity monitoring in fixed and living cells have become a fascinating topic of a wide variety of research activities [5–8]. Many of the initial applications of SMD have been in the area of extremely sensitive imaging and analyte detection [9,10]. Whereas these applications will undoubtedly continue to be important areas, the more intriguing aspect of SMD lies in the investigation of the dynamics and spectroscopy of single molecules and the interactions with their molecular environments, by monitoring the chemical and structural changes of individual molecules [11–13]. Real-time observation of single-molecule activities in living cells is another important aspect of single-molecule studies [14,15]. These investigations on the single-molecule level have the potential of offering important perspectives and providing fundamentally new information about intracellular processes.

Several advantages are offered by studying cellular activities and their dynamics at the single-molecule level. First and foremost, single-molecule measurement provides information free from ensemble averaging. It allows the examination of individual molecules in a complicated heterogeneous system so that differences in the structure or function of each molecule can be identified and related to its specific molecular environment. The distribution of a given molecular property among the members of a system, rather than the statistical ensemble-averaged property, can be revealed. As a result, rare events that are otherwise hidden can be captured and cellular processes can be directly visualized at the molecular level. Another benefit of SMD is that the need for synchronization of many molecules undergoing a time-dependent process is eliminated. This feature is because, at a given time, any single molecule of a system exists in only one particular conformational state. Therefore, the intermediates and paths of time-dependent chemical reactions can be directly measured and followed. Finally, SMD has the potential of providing spatial and temporal distribution information. This possibility would enable in vivo monitoring of dynamic movements of single molecules in intracellular space and the observation of their behavior over an extended period of time.

In this chapter, the basic requirements for SMD will be outlined followed by a discussion of how these requirements can be fulfilled in different ways using the state-of-the-art optical techniques. Finally, selected examples of SMD applications, particularly for monitoring of cellular events and activities, are presented.

10.2 Basic Requirements for Single-Molecule Detection

SMD techniques have mostly involved fluorescence and, more recently, surface-enhanced Raman scattering (SERS). In order to achieve single-molecule sensitivity, close attention needs to be paid to two critical issues: (a) excellent signal-to-noise ratio (SNR) and signal-to-background ratio (SBR) and (b) ensuring that the observed signal comes from single molecule.

10.2.1 Signal-to-Noise Ratio and Signal-to-Background Ratio

SMD is essentially an SNR and SBR issue. SNR determines the ability to detect the signal from a single molecule compared to the noise fluctuations that may appear as originating from a single molecule. SBR, on the other hand, is a measure of the signal of the molecule compared to the overall quality of the sample and the ability of the detection system to reduce background. In other words, the signal characteristic of a single molecule must be detected on top of the background associated with the surrounding media. SNR and SBR both depend on a number of parameters of the system such as incident intensity, collection and detection efficiencies, quality of sample, etc., whereas the SNR also depends on the detection bandwidth determined by the integration time. Attaining adequate SNR and SBR for SMD requires experimental efforts to maximize signal while minimizing noise from unwanted sources as much as possible.

10.2.1.1 Maximizing Signal Level

Since the "native" fluorescence of biomolecules is relatively weak in most cases, it is a common practice to label the biomolecules with fluorescent probes, which can be covalently and site-specifically attached to biomolecules [16,17]. To maximize signal using fluorescence-based techniques, fluorophores with high quantum yields and favorable photophysical properties such as large absorption cross sections and high photostability need to be used. Some commonly used classes of fluorescent probes used in biological labeling include rhodamines, cyanines, oxazines, etc. [7,18]. More recently, a new class of single-molecule fluorophores has been found among molecules originally optimized for nonlinear optical properties [19]. Alternatively, variants of green fluorescent protein (GFP) can be fused onto proteins as fluorescent tags, which have been successfully used to monitor motor proteins [20,21]. In principle, any protein can be fluorescently labeled by constructing cDNAs of desired proteins fused to the genes of GFPs and expressing them in living cells. However, dramatic blinking of GFP is a problem that needs to be further addressed [22]. Strong fluorescing semiconductor quantum dots have also been proposed as biological fluorescent labels. Advantages offered by quantum dots include narrow emission lines and resistance to photobleaching. However, their application has been limited by poor attachment of the quantum dots to biomolecules, which has been the subject of intense ongoing research [23]. Biocompatible quantum dots have been developed by encapsulating quantum dots in an organic disguise that prevents them from coming into direct contact with the aqueous biological environment.

For Raman-based techniques, efficient and reproducible substrates for surface enhancement need to be prepared to increase the cross section for Raman scattering. Different schemes have been developed to produce highly uniform and reproducible localized surface plasmon resonance (LSPR) nanostructures from several different materials [24–27]. Silver nanostructures are the most common media and provide the largest enhancements. Gold is another material frequently used in various SERS applications because of their large enhancement, biocompatibility, chemical robustness, and established functionalization chemistry. The production schemes include vapor deposition through self-assembled monolayer masks [25], electrochemical etching procedures [26], templated self-assembly of colloidal crystals [24,27,28], and annealing of vapor-plated metal islands [29].

10.2.1.2 Signal-to-Noise Considerations

To probe a single molecule, an SNR for the single-molecule signal greater than unity for a reasonable averaging time is a prerequisite. Assuming that noise factors limiting detection are the intrinsic photon shot noise fluctuations of the single-molecule signal, the background signal and the dark counts, Basche et al. [30] have developed the following equation to estimate the SNR for fluorescence-based SMD:

$$\text{SNR} = \frac{D\Phi_F \left(\frac{\sigma_p}{A} \right) \left(\frac{P_0}{h\nu} \right) T}{\sqrt{\left(\frac{D\Phi_F \sigma_p P_0 T}{A h \nu} \right) + C_b P_0 T + N_d T}}$$

where Φ_F is the fluorescence quantum yield of the fluorophore, σ_p is the absorption cross section, T is the detector counting interval, A is the beam area, $P_0/h\nu$ is the number of incident photons per second, C_b is the background count rate per watt of excitation power, and N_d is the dark count rate, and D is an instrument-dependent collection factor. According to this equation, several parameters must be chosen carefully in order to maximize the SNR. First, a fluorophore with a large quantum yield (Φ_F) and large absorption cross section (σ_p) must be used. The laser spot should be as small as possible. Higher power produces higher SNR values, but the power (P_0) cannot be increased arbitrarily because saturation causes the absorption cross section to decrease. In cellular applications, the excitation power also has to be kept low enough not to interfere with the functions of a cell.

10.2.1.3 Background Suppression

Despite much effort to reduce background, almost all single-molecule experiments are background limited. Therefore, suppression of the background signal must be one of the primary goals in single-molecule experiment design. Most background photons arise from Rayleigh scattering, interfering Raman scattering, impurity fluorescence, and, in the case of cells, autofluorescence of cellular components. One of the most efficient ways to minimize the effect of the background is to reduce the probe volume [31,32]. This is due to the fact that the signal from a single molecule is independent of the probe volume whereas the background signal is proportional to the probe volume: the Rayleigh and Raman scattering intensities decrease significantly due to the presence of less scattering molecules in a smaller interrogated volume, and background fluorescence from the solvent or other media is also decreased due to the ability to illuminate specific sites of interest with a small probe volume. Other approaches to suppress the background signal include elimination of background fluorescence by using ultrapure solvent for sample preparation, photobleaching the impurities in the solvent, and employing low-fluorescing optics.

Reduction of the probe volume has been accomplished optically by laser excitation in the confocal, two-photon excitation (TPE), evanescent schemes as well as near-field optics [4,16,32–35]. Confocal excitation and detection is a simple yet effective optical approach to attain subfemtoliter probe volumes. A schematic representation of a typical confocal microscope setup is shown in Figure 10.1. In a confocal setup, a laser beam is focused down to the diffraction limit using an objective with a high numerical aperture. A pinhole positioned in the primary focal plane of the objective rejects all light except that originating from the focal point, thus restricting the probe volume to the close vicinity of the focal point. A typical probe volume in a confocal setup is estimated to be ~1 fL. TPE is another scheme that has been employed to successfully confine the sample volume and reduce the background signal for SMD. In a two-photon scheme, the analyte molecules are excited by simultaneous absorption of two photons with a total energy corresponding to the excitation energy of the molecule. The reason for the superior ability of TPE to reduce background is twofold. First, the efficiency of TPE has a quadratic dependence on the laser intensity. As a result, only the immediate vicinity of the focal spot receives sufficient intensity for significant excitation to occur, thus forming a tremendously reduced excitation volume (Figure 10.2). In addition, because of the large separation between the excitation and detection wavelengths, Rayleigh and Raman scattering of the excitation laser beam from the sample can be easily removed with high-efficiency optical filters. Evanescent excitation with high numerical aperture collection has also been used to suppress background in SMD. Evanescent excitation is normally achieved when the excitation beam is passed from a high refractive index media to a lower refractive index media at or beyond the critical angle. When total internal reflection (TIR) of the excitation beam occurs at the interface under these conditions, an electromagnetic field known as an "evanescent wave"

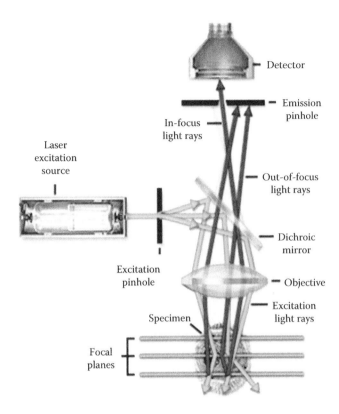

FIGURE 10.1
Schematic representation of a typical confocal microscope setup. (From http://www.microscopyu.com)

is generated and propagates into the media with lower refractive index. Since the strength of evanescent wave decays exponentially as a function of the distance from the interface, the effective probing depth is generally limited to ~300 nm beyond the interface, resulting in drastically reduced probe volumes.

Although optical reduction of the detection volume results in excellent SNRs, the detection efficiencies are generally low. Alternatively, probe volume for SMD can be reduced by using nanofabricated devices such as microcapillary tubes and microchannels in combination with hydrodynamic focusing. Since this review focuses mainly on optical techniques, this scheme will only be briefly discussed and further information can be obtained from several excellent reviews on this topic [36,37]. The effective observation volumes created by these channels are ~100 times smaller than the observation volumes using conventional confocal optics and thus enable SMD at higher concentrations. Furthermore, the use of microcapillaries and microstructures to force the molecules to pass through the detection volume has resulted in larger molecule detection efficiencies. Detection efficiencies of single-molecule events of up to 60% have been shown in these microstructures [38]. Additional advantages include increased rate of detection and reduced data acquisition time. However, the requirements on the geometries of these nanofabricated devices are demanding and critical for the success of this scheme. Characterization of the size, shape, and geometry of the capillary as well as a description of important optical characteristics in order to achieve SMD has been carried out by Lundqvist et al. [39].

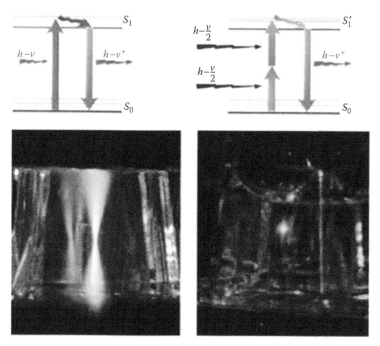

FIGURE 10.2
Comparison of the excitation profiles of the focused laser beam with one-photon (left) and two-photon (right) excitation. The energy diagrams of one-photon and two-photon excitation are also illustrated. (Adapted from Dittrich, P.S. et al., *Appl. Phys. B*, 2001, 73(8), 829–837.)

10.2.2 Ensure That the Signal Actually Originates from a Single Molecule

Another experimental challenge in SMD is to ensure that the observed signal arises from a single molecule. Intensity fluctuations observed in single-molecule experiments could be due to changes in single-molecule dynamics or the molecular environment, but they could also originate from emissions of nearby molecules. It is crucial to eliminate this source of noise uncertainty. One way to accomplish this involves a combination of focusing the laser to a small excitation volume and working at an ultra low concentration of the molecule of interest so that the average number of fluorescent molecules residing in the probe volume is one or less [3,40]. Alternatively, a molecule-specific excitation and detection procedure can be used so that only the target molecule in the probed volume is in resonance with the laser. Even if the laser beam excites a large area, no molecules other than the target molecule in the excitation volume would be excited because they are out of resonance.

10.3 Optical Techniques for Single-Molecule Detection

SMD has been made possible in the past decade by the advances in optical spectroscopic techniques, such as laser-induced fluorescence (LIF), near-field scanning optical microscopy (NSOM), SERS, and "optical tweezers" techniques. Due to their noninvasiveness and high sensitivity to changes in molecular conformation and environment, optical spectroscopic techniques are especially suitable for applications in SMD. The availability of these techniques not only allows us to detect and image single molecules, but also to conduct spectroscopic measurements and monitor dynamic processes as well. In this chapter, we will focus on the discussion of these optical spectroscopic techniques. The principle of each technique will be briefly described followed by a discussion of the benefits offered for SMD.

10.3.1 Laser-Induced Fluorescence

LIF is an extremely valuable tool for the study of molecular phenomena because of its superb sensitivity, high information content, noninvasiveness, and the availability of a large pool of excellent fluorescent probes. Various properties of fluorescent probes, such as polarization and fluorescence lifetime, can be utilized to provide information on conformational dynamics, reaction kinetics, and changes in chemical microenvironment.

10.3.1.1 Fluorescence Correlation Spectroscopy

Fluorescence correlation spectroscopy (FCS) is a solution-phase optical technique used to detect the random Brownian motion of fluorescent molecules by monitoring the time-dependent fluctuation in the fluorescence intensity of molecules that pass through the focus of a laser beam [22,41,42]. Figure 10.3 illustrates the principle of fluctuation correlation analysis and representative data of fluctuating signals induced by random motion of fluorophore molecules through the detection volume. FCS utilizes small spontaneous signal fluctuations in an ensemble of fluorescent molecules to extract dynamic information about a system in thermodynamic equilibrium. Any molecular process causing a change in the emission characteristics of the fluorophore can be monitored with high precision. Among a number of physical parameters that are, in principle, accessible by FCS, are the determination of local concentrations, mobility coefficients, and characteristic rate constant of reactions of fluorescently labeled biomolecules at very low concentrations. Furthermore, from the characteristic correlation time, time-dependent dynamic process can be followed in detail.

In recent years, the analytical and diagnostic potential of FCS in life sciences has been discussed and demonstrated [43–48]. FCS has been successfully applied for studying reaction kinetics of nucleic acids and proteins [45,46]. Based on fluctuations in the fluorescence yield of single-dye molecules, electron transfer, ion concentrations, and conformational changes of nucleic acid oligomers could be monitored [47,48]. The potential of FCS was further illustrated when substantial improvements in SNR were made by defining extremely small probe volumes using confocal and two-photon excitation. The potential of confocal FCS with high temporal resolution for rapid enzyme screening has been reported [49,50]. The tiny volume of confocal FCS in which the measurements are performed also makes it possible to evaluate molecular processes at the cell membrane. An increasing number of intracellular applications involving the study of molecular mobility of proteins and DNA in different locations inside cells have been developed using FCS [45,51,52]. Combined with TPE, FCS yields substantially improved signal quality in turbid samples such as deep cell layers in tissue [33]. At comparable signal levels, TPE minimizes photobleaching in spatially restrictive cellular compartments thereby preserving long-term signal stability. Since the definition of small volume is purely of an optical nature and no mechanical constraints are involved, FCS is relatively noninvasive and ideally suited for both in vitro and in vivo measurements. Potential artifacts that interfere with FCS measurements for intracellular applications include cellular autofluorescence, reduced signal quality due to

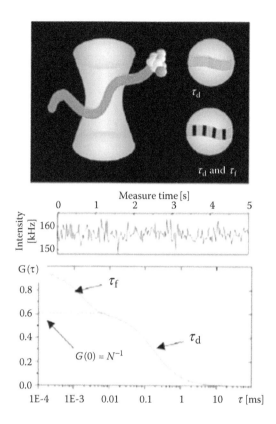

FIGURE 10.3

Principle of fluctuation correlation analysis (top). Representative data of fluctuating fluorescence signals induced by random motion of dye molecules through the detection volume (middle). Representative autocorrelation curve $G(\tau)$ (bottom), describing the temporal decay function of fluctuations. The characteristic decay times for molecular residence times in the volume (τ_d) and internal intensity fluctuations (τ_f) are indicated. (Adapted from Medina, M.A. et al., *BioEssays*, 2002, 24(8), 758–764.)

light absorption and scattering, and dye depletion resulting from photobleaching in the restricted compartments inside cells.

10.3.1.2 Fluorescence Resonance Energy Transfer

Fluorescence resonance energy transfer (FRET) is a nonradiative transfer of energy between two fluorophores that are placed within close vicinity ($\sim$20 to $\sim$100 Å) of each other in a proper angular orientation [13,17]. In a typical FRET experiment, a biological macromolecule is labeled at two different positions with a donor and an acceptor fluorophore having spectral overlap. Single-pair FRET between the two fluorophores has been measured to determine the distance between two fluorophores and has been suggested to be a useful tool for investigating the dynamic processes of proteins. Intramolecular FRET (Figure 10.4a) is based on a change in the distance between different parts of proteins as a result of folding or conformational changes. FRET can be employed to report and monitor such changes by correlating the variations in the donor and acceptor fluorescence intensities due to the distance-dependent energy transfer between the fluorophores. Intermolecular FRET (Figure 10.4b), in which two protein molecules are labeled with a donor and an acceptor fluorophore, respectively, could be used to detect protein–protein interactions.

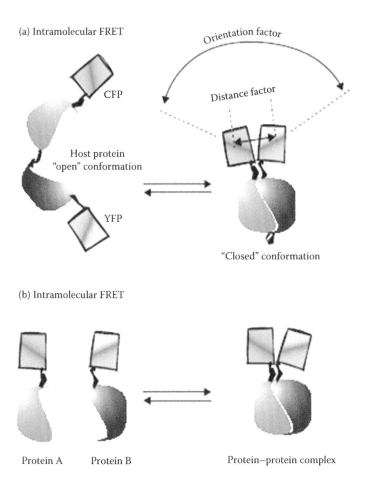

FIGURE 10.4

Intra- and intermolecular FRET: (a) Intramolecular FRET occurs when both the donor and acceptor fluorophores (in this case cyan fluorescent protein [CFP] and yellow fluorescent protein [YFP], respectively) are on the same host molecule, which undergoes a transition between "open" and "closed" conformations. The amount of FRET transferred strongly depends on the relative orientation and distance between the donor and acceptor fluorophores. (b) Intermolecular FRET occurs between one molecule fused to the donor (CFP) and another molecule (protein B) fused to the acceptor (YFP). When the two proteins bind to each other, FRET occurs. When they dissociate, FRET diminishes. (From Truong, K. and Ikura, M., *Curr. Opin. in Struct. Biol.*, 2001, 11(5), 573–578.)

The introduction of the fluorescent probes into desired locations on the biomolecule is crucial for the success of FRET. One can utilize the existing native fluorophore of protein together with an extrinsic label covalently attached to the protein. However, since the native fluorescence of most proteins is relatively weak, it is more common that both the donor and acceptor fluorophores are extrinsic probes covalently attached to the protein. Sometimes, it is possible to label two sites on the biomolecule with the same probe molecule that is capable of transferring energy between them, significantly simplifying the labeling step. Covalent attachment of the probe is most often achieved by using natural or engineered cysteine residues and thiol-reactive fluorescence probes. Fusion of the target protein with GFP variants can also be used to constitute a donor–acceptor pair. GFP is a spontaneously fluorescent polypeptide of 238 amino acid residues from the jellyfish *Aequorea victoria* that absorbs UV-blue light and emits in the green region of the spectrum [53]. Its structure and potential uses for studying living cells have been the subject of several excellent reviews [54–56]. Variants of the GFP with different excitation and emission spectra have been engineered to better match available light sources. When fused to proteins, these variants generally retain their fluorescence without affecting the functionality of the tagged protein, therefore making them candidates for intracellular reporters. The labeled protein can be expressed in cells and their conformation can be followed in the native cellular environment [57]. Fusions with GFP variants have also been used to design several fluorescent indicators of cellular events or signaling molecules [58–60]. However, some limitations exist for the use of GFP variants for probing protein conformational changes. First, these variants are fused to the N- and C-terminals of the studied protein, which are generally not the best location to detect conformational changes resulting from the binding to other proteins or enzyme substrates. Furthermore, their relatively large size makes them unusable for tagging small proteins.

10.3.1.3 *Total Internal Reflection Fluorescence Microscopy*

Total internal reflection fluorescence microscopy (TIR-FM), a general term for any spectroscopic or microscopic technique based on the evanescent field created by TIR of light, has been established as an important tool for studying near-surface phenomena. In TIR-FM, fluorophores are excited by an evanescent wave generated in the optically less dense medium when TIR occurs at the interface between two media having different refractive indices. The emission intensity and spectral profiles of the evanescent-wave-induced fluorescence can be related to the concentration and conformation of species in the evanescent regime [61–64]. This technique is generally nondestructive, making it suitable for single-molecule applications in cells and tissues under ambient conditions. The strength of evanescent field rapidly decays beyond the interface where it is generated. Another primary benefit of TIR-FM is the extremely low background as a result of a significantly reduced illumination volume. A reduction of the nonspecific fluorescence from inside the cell by a factor of ~20 has been reported when using TIR excitation [65]. TIR-FM is the method of choice used to excite sufficiently thin layers in order to reduce background luminescence. Only molecules within a few hundred nanometers of the interface will be interrogated using TIR-FM. Two widely used configurations for TIR-FM are "prism-based" TIR-FM and "through the objective" TIR-FM (Figure 10.5). In the prism-based configuration (Figure 10.5a), a prism on the side of the sample opposing the objective is used to generate the evanescent field. As the excitation light is totally reflected away from the detector, very high SNR values can be obtained and even weakly fluorescent molecules can be detected. In the through the objective configuration (Figure 10.5b), the excitation laser beam illuminates the sample through a high-numerical-aperture objective lens and generates the evanescent wave at the near-side boundary between a glass coverslip and the sample. In general, prism-based TIR-FM is found to offer better SNR whereas the through the objective configuration is better for obtaining the largest number of photons before photobleaching.

Single-molecule imaging of living cells by TIR-FM was first reported by Sako and Uyemura [7]. In particular, the epidermal growth factor (EGF) and its receptor (EGFR) located on the cell surface were of some interest in SMD [66]. Llobet et al. [67] successfully combined TIR-FM with interference reflection microscopy to image cell membranes and to detect changes during endocytosis or exocytosis at the synaptic terminal of retinal cells. In another TIR-FM application, microtubule ends of fibroblasts were examined close to the cell surface after transfection with GFP-tagged tubulin [68].

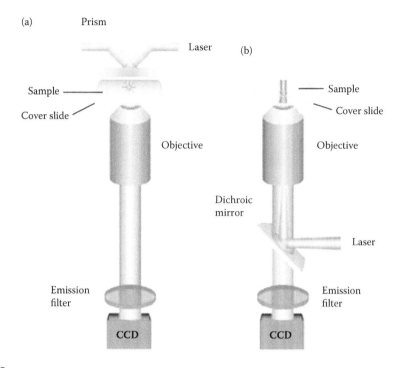

FIGURE 10.5
(a) Prism-based total internal reflection fluorescence microscopy (TIR-FM). The laser light is passed through a prism on the side of the sample opposing a microscope objective to generate the evanescent field. The emitted fluorescence is collected using the same objective and detected by a CCD camera. (b) Through the objective TIR-FM. Collimated laser light is directed via a dichroic mirror into a microscope objective. The emitted fluorescence is collected using the same objective and detected by a CCD camera. (Adapted from Haustein, E. et al., *Curr. Opin. Struct. Biol.*, 2004, 14(5), 531–540.)

10.3.2 Near-Field Scanning Optical Microscopy

Although LIF is a powerful tool for single-molecule studies, a drawback is the fundamental limit in spatial resolution that can be achieved by a far-field optical spectroscopic technique such as LIF, where the laws of diffraction dictate that the resolution limit is approximately half of the wavelength used. NSOM has been developed to break this diffraction limit and allows optical measurements with subwavelength resolution. A typical NSOM is illustrated in Figure 10.6. In NSOM, a sharp probe scans across a sample surface at a close and constant distance from the sample. The most generally applied NSOM probe consists of a small aperture of subwavelength dimensions, typically 50–100 nm in diameter, at the end of a metal-coated pulled single-mode fiber. The light emitted from the probe is predominantly composed of evanescent waves that only effectively excite molecules within a layer of ∼200 nm from the tip of the probe. Subdiffraction-limited spatial resolutions can be achieved within this optical near-field of the aperture because photons do not have enough distance to experience diffraction. Other advantages offered by NSOM include: (a) lower background signal due to an extremely small excitation/detection volume as a result of the aperture dimension as well as the penetration depth of the evanescent wave and (b) the ability to simultaneously obtain both optical and topographic information of the sample. The combination of topographical information, high optical resolution, and single-molecule sensitivity makes NSOM a unique tool for

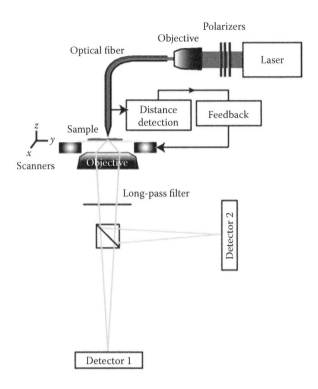

FIGURE 10.6

Schematic diagram of a typical near-field scanning optical microscope (NSOM). The NSOM probe is a tapered optical fiber. Laser light is coupled into the fiber and is used to excite fluorophores as the probe scans across the sample surface. The probe-sample distance is maintained constant at <10 nm during scanning by shear-force-based distance detection in combination with an electronic feedback system controlling the piezoelectric scan stage. Fluorescence is collected by a conventional inverted microscope. Dual-channel optical detection allows wavelength and polarization discrimination. (From de Lange, F. et al., *J. Cell Sci.*, 2001, 114(23), 4153–4160.)

biological applications [69–71]. For example, NSOM has been used to investigate multi-drug resistance (MDR) processes in single cells treated with cancer drugs [72]. However, several technical challenges remain for NSOM. First of all, the increase in spatial resolution comes at the price of increased experimental complexity compared to far-field microscopy. In addition, the typical transmission efficiency through a pulled fiber is very low ($\sim 10^{-5}$–10^{-6}). In addition, it is not possible to image molecules far from the sample surface as the near-field regime rapidly disappears with distance, preventing the use of NSOM to probe the interior of a cell. As a potential solution to some of these issues, apertureless NSOM probes have been proposed [73–75] in which an ultrasharp tip is used as an antenna to localize the excitation region.

10.3.3 Surface-Enhanced Raman Spectroscopy

SERS is another optical detection technique suited for single-molecule studies because of its trace analytical capabilities together with its high structural selectivity compared to other optical spectroscopies [76–79]. Strong enhancement in Raman signals can be observed from molecules attached to nanometer-sized architectures such as silver and gold nanoparticles [76,79–83]. Several potential schemes to prepare these nanostructured SERS-active

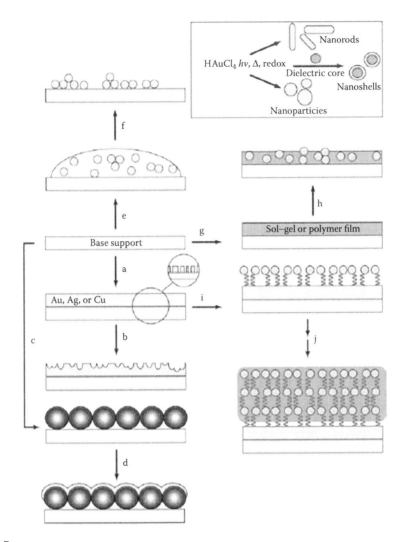

FIGURE 10.7

Schematic overview showing some potential routes to solid state or nanostructured SERS-active architectures: vapor or vacuum deposition of continuous noble metal films, "punctuated" films, or metal islands (a); corrosive etching or electrochemical roughening (b); microsphere or nanoparticle deposition and ordering (c); sputter coating over a regular or irregular surface (e.g., polymer latex, randomly arranged quartz posts, and colloidal crystal) (d); aqueous sol deposition (e); and subsequent aggregation, gravitational deposition, or assembly (f); polymer or sol–gel deposition (g); noble-metal nanoparticle entrapment in a matrix or in situ formation (h); metal colloid monolayer self-assembly (i); and layer-by-layer colloid multilayer formation and subsequent organosilane overlayer deposition via the surface sol–gel method (j). The boxed inset shows common building blocks used to prepare SERS-active materials beginning with noble metal precursor salts such as $HAuCl_4$, $AgNO_3$, and Ag_2SO_4. (From Baker, G.A. et al., *Anal. Bioanal. Chem.*, 2005, 382(8), 1751–1770.)

architectures are schematically illustrated in Figure 10.7. The significant increase in cross section in SERS has been associated primarily with the enhancement of the electromagnetic field surrounding small metal objects through the interaction with SPR, and the chemical enhancement due to specific interactions of the adsorbed molecule with the metal surface. SERS enhancement factors on the order of 10^{14} corresponding to effective SERS cross sections of about 10^{-16} cm^2/molecule allow Raman detection of single molecules. A further increase of the sensitivity can be obtained by coupling surface-enhanced resonance Raman scattering spectroscopy (SERRS) [84]. In addition to increased cross sections, SERRS has the advantage of higher specificity, because only the molecular vibrations associated with resonant electronic transition contribute to the SERRS spectrum [78,84]. Another interesting aspect of SERS is its spatial resolution. By taking advantage of the local optical fields of special metallic nanostructures, SERS can provide lateral resolutions better than 20 nm [85–88]. This is well below the diffraction limit and smaller than the resolution of common near-field microscopy.

The analytical capabilities of SERS and SERRS for ultratrace detection have been recently exploited for biophysical and biomedical applications at the single-molecule level [77,78,89]. This is of great interest because biologically relevant molecules available for characterization exist often in extremely small amounts. Examples of such applications include the detection and identification of neurotransmitters [90], SERS-based DNA and gene probes [91,92], and immunoassays [93]. More recently, SERS has been used in the analysis of single proteins such as myoglobin, horseradish peroxidase (HRP), and cytochrome *c* [84,94,95]. SERS studies have also been performed in living cells [96,97]. In these experiments, colloidal silver particles were incorporated inside the cells and SERS was employed to monitor the intracellular distribution of drugs in the whole cell and the interactions between drugs and nucleic acids.

10.3.4 Optical Tweezers

"Optical tweezers" is another technique of great interest in the context of SMD, especially for applications in cells. Optical tweezers, also known as optical trap, exploit the fact that refracted laser light exerts radiation force on matter [98–100]. For the radiation forces to be significant, a laser beam must be used and tightly focused with a high-numerical-aperture

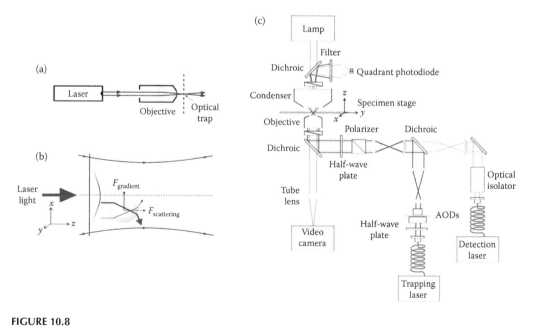

FIGURE 10.8
(a) Generation of an optical trap; (b) the basic principle of an optical trap; and (c) a typical optical trap setup. (From http://www.stanford.edu)

objective (Figure 10.8a). Although the forces might be only on the order of piconewtons, they can be dominant at the microlevel. For biological applications, usually an infrared laser is used to avoid damage to the biological sample, and absorption of light in the solution. Figure 10.8b provides a schematic diagram on how an optical trap works. When the incoming light is focused and interacts with a bead, the sum of the forces can be split into two components: $F_{scattering}$, the scattering force which tends to push the bead along the direction of the incident light, and $F_{gradient}$, the gradient force arising from the gradient in light intensity and pointing transversely toward the high-intensity region of the focused beam. A stable trap will exist when the gradient force $F_{gradient}$ overcomes the scattering force $F_{scattering}$. Optical tweezers have been used to trap dielectric spheres, viruses, bacteria, living cells, organelles, small metal particles, and even strands of DNA. Dielectric objects and biological samples ranging from tens of nanometers to tens of microns can be trapped and manipulated by tailoring the wavelength and other properties of the laser beam [98,100]. A typical optical tweezers setup is shown in Figure 10.8c.

Optical trapping and manipulation has attracted interest with its potential in various biological applications such as trapping and positioning of cells and bacteria, performing internal cell surgery, and investigation of motor molecules, selective chemical reactions and molecular assembly [101–103]. For SMD, two of the main uses for optical traps have been the studies of molecular motors [104–107] and the physical properties of DNA [108–110]. In both areas, a biological specimen is biochemically attached to a micron-sized glass or polystyrene bead that is then trapped. By attaching a single molecular motor and a piece of DNA to such a bead, experiments have been performed to probe various properties of molecular motors and to measure the elasticity of the DNA, as well as the forces under which the DNA breaks or undergoes a phase transition.

10.4 Applications in Fixed and Living Cells

SMD techniques are widely used in probing fixed cells as well as living cells, thus providing the potential for a wide variety of medical applications at the cellular level. Due to many adaptations by many researchers, we will not attempt to exhaustively review all applications but instead present some illustrative examples of different aspects of SMD applications in cells. Applications that are discussed include the studies of molecular motors, cell signaling, protein conformational dynamics, and ion channels.

10.4.1 Molecular Motors

Examples of molecular motors include biological machines that move in cell environments, such as protein tracks (actin filament or microtubules) and drive processes such as intracellular transport, cell division, and muscle contraction [111]. Several SMD techniques were first developed to study molecular motors such as myosin and kinesin [112–114]. Mechanisms of the movement of molecular motors have been successfully elucidated by monitoring the conformation changes, location of single motor proteins as they move along their tracks, and determining the speed and distance traveled before dissociating or pausing. The studies of conformational changes of motor proteins associated with their motions have greatly benefited from single-molecule fluorescence resonance energy transfer (smFRET) and single-molecule fluorescence polarization anisotropies (smFPA) [115,116]. Single-molecule nanomanipulation techniques such as optical trapping have also played an important role in the studies of molecular motors. Combined with single-molecule imaging techniques, important insights into the coupling between individual mechanical and chemical events of single myosin molecules can be observed directly [14]. Yanagida and coworkers reported the imaging of the motions of single fluorescently labeled myosin molecules and the detection of individual ATP turnover reactions using TIR-FM at the single-molecule level [117]. The ability to manipulate actin filaments with a microneedle or an optical trap combined with position-sensitive detectors has enabled direct measurements of nanometer displacements and piconewton forces exerted by individual myosin molecules [118,119]. Movement of single kinesin molecules along a microtubule has also been directly observed by TIR-FM and measured with nanometer accuracy by optical trapping nanometry [113,120]. Kinesin has been found to move along a microtubule with regular 8-nm steps, indicating a path along the α–β tubulin dimer repeat in a microtubule [121,122]. Vale and coworkers studied several kinesin mutants in an effort to understand which part of kinesin determines the direction of motion using TIR-FM [123,124]. Based on the findings of these studies, it is generally agreed that the "lever arm swinging model" and "hand-over-hand model" can be used to explain the motions of myosin and kinesin molecules, respectively [125,126], whereas the "biased Brownian motion" appears to play an essential role in the movement of some single-arm myosin and single-head kinesin molecules [127,128].

10.4.2 Cell Signaling

Cell signaling is one of the major target areas for single-molecule imaging investigations. The transduction of signals inside cells and among cells is a central and basic process in biological systems and often involves dynamic interactions among proteins. A comprehensive characterization of protein interaction during this process is critical for the understanding of the regulatory mechanisms that control cellular functions and can offer valuable information for new schemes for the development of drugs and new therapies for diseases. The high temporal and spatial resolution available in single-molecule spectroscopy makes it ideal to quantify conformational dynamics and localization of proteins during a cell-signaling event under physiological conditions [129,130]. Cell-signaling proteins, including membrane receptors, small G proteins, as well as small signaling compounds labeled with single fluorophore or GFPs have been visualized in living cells [131–134]. Using TIR-FM, binding of single molecules of fluorescently labeled EGF to its receptor EGFR has been observed in living cells [135]. The mechanism underlying the highly sensitive response to EGF was revealed by single-molecule studies and was attributed to the amplification of the EGF receptor signal by dynamic clustering, reorganization of the dimers, and lateral mobility of EGFR on the cell surface. Slaughter et al. [136] used FRET to probe the dynamics and conformational distributions of calmodulin, a calcium-signaling protein that mediates various cell-signaling pathways important to muscle contraction and energy metabolism. The fluctuations in FRET efficiency for Ca-calmodulin were monitored and used to resolve dynamics of calmodulin on the microsecond and millisecond timescales. Lu et al. reported a study on protein–protein noncovalent interactions in an intracellular-signaling protein complex, Cdc42–WASP, using single-molecule spectroscopy and molecular dynamic simulations [15]. Measurements of reaction kinetics and the activation of the cell-signaling molecules have also been reported [137].

10.4.3 Protein Conformational Dynamics

The conformational dynamics of proteins is crucial for their function. FRET has been extensively used to directly monitor conformational changes of proteins [138–142]. The ability to follow protein conformation at concentrations down to a single molecule in a variety of solution conditions including living cells and the ability to monitor the conformational status of proteins in real time are particularly useful features of FRET. The technique has seen numerous applications in the study of fluctuations and stability of proteins, protein folding and unfolding, and enzyme structural changes during catalysis [143–146]. Trakselis and coworkers utilized the real-time detection capability of FRET to study the rapid kinetics of functional conformational changes of T4 DNA polymerase holoenzyme assembly [140]. Jia et al. [146] measured conformational equilibrium distributions and fluctuations in the FRET signal as a dimeric peptide, GCN4, folds, and unfolds.

Studies of conformation changes in proteins at a single-molecule level in the native context of living cells are also possible. This was accomplished by fusion of the protein of interest with GFP, producing a donor–acceptor pair-labeled protein that can be expressed in cells. Fluorescence microscopy is used to follow different conformations in the native cellular environment. In one such experiment, functional conformational changes of MK2 protein were detected in vivo using FRET-based detection [139]. New ways of introducing fluorescence probes into proteins and the availability of new probes will further expand the applicability of FRET for protein conformation studies. smFPA is another useful technique for monitoring protein conformational dynamics, especially information on dynamic changes in orientation. Orientation information can be obtained by analyzing the emission polarization of the two chromophores excited by polarized light. Using smFPA, rotational motions of an actin filament and myosin have been observed at the single-molecule level [147–149]. Simultaneous measurements of the FRET or polarized fluorescence together with the mechanical properties of a myosin molecule during force generation can provide insight on how the conformational changes in myosin are involved in the force generation process.

10.4.4 Ion Channels

Ion channels are proteins that precisely regulate the ionic flow across cell membranes and generate ionic gradients that are responsible for nerve and muscle excitability [150,151]. Single-molecule spectroscopies, in combination with traditional patch-clamp electric current recording, have the potential to provide new information on the conformation changes between the open and closed states of an active ion channel during ion channel gating processes. Sonnleitner et al. [65] visualized single molecules of a voltage-gated K^+ channel conjugated to tetramethylrhodamine in living cells using TIR-FM in which changes of the fluorescence intensity and the membrane potential were detected at the same time. Protein motions that are not directly involved in opening or closing the ion channel were revealed. Harms and coworkers proposed an approach combining patch-clamp measurements with confocal fluorescence microscopy (PC-CFM) to probe single-molecule ion channel kinetics and conformational dynamics [152]. Using this technique, single-molecule ion channel kinetics and conformational dynamics can be probed using simultaneous ultrafast fluorescence spectroscopy and single-channel electric current recording. PC-CFM was used to determine single-channel conformational dynamics by probing smFRET, fluorescence self-quenching, or anisotropy of the dye-labeled gramicidin ion channel incorporated in an artificial lipid bilayer. Several "silent" intermediate states that have not been detected before by single-channel current recording were revealed by single-molecule fluorescence spectroscopy. Detection of conformational changes and chemical state of ion channels using SMD techniques has shed light on the complex mechanisms of ion channels as well as the kinetics and pharmacological properties of many ion channels [152–154]. Such analysis can also be carried out in vivo. In a recent report, human cardiac L-type Ca^{2+} ion channels labeled with yellow fluorescent protein (YFP) were expressed in a cell [155]. Their structure and dynamics in the plasma membrane were characterized using wide-field fluorescence microscopy at the level of individual channels in living cells.

10.4.5 Monitoring Reactions and Chemical Constituents in Living Cells

SMD has been effectively used both to quantify intracellular reactions and to allow the spatial analysis of such reactions with high resolution by using optical nanosensors [156–158]. One such optical nanobiosensor can be inserted into single living cells to monitor and measure, in vivo, molecules and chemicals of biomedical interest without disrupting normal cellular processes [157–159]. It consists of a biological recognition molecule coupled to the optical transducing nanometer-size optical fiber interfaced to a photometric

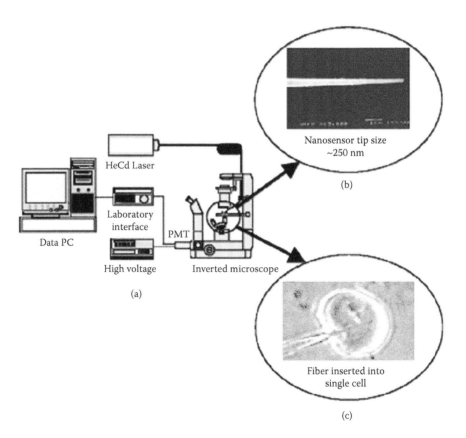

FIGURE 10.9

(a) Laser-induced fluorescent measurement system used for data acquisition and processing. (b) Scanning electron micrograph of a fiber-optic nanosensor after coating with 200 nm of silver. (c) Image of an antibody-based nanoprobe inserted into a single MCF-7 cell. (From Kasili, P.M. et al., *J. Am. Chem. Soc.*, 2004, 126(9), 2799–2806.)

detection system (Figure 10.9). Both quantitative and qualitative information are available using biological recognition elements (e.g., DNA, protein) that are in direct spatial contact with a solid-state optical transducer element. Such a nanobiotechnology-based device could potentially provide unprecedented insights into living cell function, allowing studies of molecular functions such as apoptosis-signaling process [160–162], DNA–protein interaction, and protein–protein interaction. This device also provides a novel and powerful tool for fundamental biological research, ultrahigh throughput drug screening and medical diagnostics applications. SERS studies with colloidal gold and silver nanoparticles for sensitive and structurally selective detection of native chemicals inside a cell and their intracellular distributions inside living cells have also been reported [163–165]. The colloidal nanoparticles were incorporated inside the cells by either incubating the cells with these nanoparticles or by fluid-phase uptake during the growth in a culture medium supplemented with the nanoparticles. Nabiev et al. [165] monitored the intracellular distribution of drugs in the whole cell and studied the antitumor drugs and nucleic acid complexes. An SERS image inside a living cell was obtained by measuring the SERS signal of the native constituents in the cell nucleus and cytoplasm. Imaging of labeled silver nanoprobes in single cells was performed using hyperspectral SERS-imaging techniques [166]

10.5 Conclusion

In this chapter, we present an overview of the latest developments in single-molecule spectroscopy with respect to important experimental aspects and applications. In the last two decades following the first demonstration of SMD measurements, optical studies of single molecules have matured to a level where SMD is conducted on a more regular basis. Techniques such as LIF, TIR-FM, NSOM, etc. have been adopted to provide a set of relatively standard techniques for SMD, both in vitro and in vivo. With the ultimate level of sensitivity offered by these techniques, we are able to catch glimpses of the inner life of cells, visualize, and gain insight into molecular mechanisms and dynamics of subcellular systems with unprecedented details. Single-molecule studies will most certainly continue to hold enormous potential for applications in intracellular processes and generate exciting possibilities in many disciplines for years to come.

References

1. Tamarat, P. et al., Ten years of single-molecule spectroscopy. *Journal of Physical Chemistry A*, 2000, 104(1), 1–16.
2. Peterman, E.J.G., H. Sosa, and W.E. Moerner, Single-molecule fluorescence spectroscopy and microscopy of biomolecular motors. *Annual Review of Physical Chemistry*, 2004, 55, 79–96.
3. Michalet, X. and S. Weiss, Single-molecule spectroscopy and microscopy. *Comptes Rendus Physique*, 2002, 3(5), 619–644.
4. Moerner, W.E. and D.P. Fromm, Methods of single-molecule fluorescence spectroscopy and microscopy. *Review of Scientific Instruments*, 2003, 74(8), 3597–3619.
5. Kobayashi, T. et al., Membrane dynamics and cell signaling as studied by single molecule imaging. *Seikagaku*, 2004, 76(2), 91–100.

6. Sako, Y. and T. Yanagida, Single-molecule visualization in cell biology. *Nature Cell Biology*, 2003, September (Suppl), SS1–SS5.
7. Sako, Y. and T. Uyemura, Total internal reflection fluorescence microscopy for single-molecule imaging in living cells. *Cell Structure and Function*, 2002, 27(5), 357–365.
8. Ishijima, A. and T. Yanagida, Single molecule nanobioscience. *Trends in Biochemical Sciences*, 2001, 26(7), 438–444.
9. Schmidt, T. et al., Imaging of single molecule diffusion. *Proceedings of the National Academy of Sciences of the United States of America*, 1996, 93(7), 2926–2929.
10. Beechem, J.M., Single-molecule spectroscopies and imaging techniques shed new light on the future of biophysics. *Biophysical Journal*, 1994, 67(6), 2133–2134.
11. Forkey, J.N., M.E. Quinlan, and Y.E. Goldman, Protein structural dynamics by single-molecule fluorescence polarization. *Progress in Biophysics and Molecular Biology*, 2000, 74(1–2), 1–35.
12. Haran, G., Single-molecule fluorescence spectroscopy of biomolecular folding. *Journal of Physics: Condensed Matter*, 2003, 15(32), R1291–R1317.
13. Weiss, S., Measuring conformational dynamics of biomolecules by single molecule fluorescence spectroscopy. *Nature Structural Biology*, 2000, 7(9), 724–729.
14. Oosawa, F., The loose coupling mechanism in molecular machines of living cells. *Genes to Cells*, 2000, 5(1), 9–16.
15. Tan, X. et al., Single-molecule study of protein–protein interaction dynamics in a cell signaling system. *Journal of Physical Chemistry B*, 2004, 108(2), 737–744.
16. Bohmer, M. and J. Enderlein, Fluorescence spectroscopy of single molecules under ambient conditions: Methodology and technology. *Chemphyschem*, 2003, 4(8), 793–808.
17. Heyduk, T., Measuring protein conformational changes by FRET/LRET. *Current Opinion in Biotechnology*, 2002, 13(4), 292–296.
18. Benes, M. et al., Coumarin 6, hypericin, resorufins, and flavins: Suitable chromophores for fluorescence correlation spectroscopy of biological molecules. *Collection of Czechoslovak Chemical Communications*, 2001, 66(6), 855–869.
19. Willets, K.A. et al., Novel fluorophores for single-molecule imaging. *Journal of the American Chemical Society*, 2003, 125(5), 1174–1175.
20. Pierce, D.W. et al., Single-molecule behavior of monomeric and heteromeric kinesins. *Biochemistry*, 1999, 38(17), 5412–5421.
21. Romberg, L., D.W. Pierce, and R.D. Vale, Role of the kinesin neck region in processive microtubule-based motility. *Journal of Cell Biology*, 1998, 140(6), 1407–1416.
22. Schwille, P., Fluorescence correlation spectroscopy and its potential for intracellular applications. *Cell Biochemistry and Biophysics*, 2001, 34(3), 383–408.
23. Pierce, D.W. and R.D. Vale, Single-molecule fluorescence detection of green fluorescence protein and application to single-protein dynamics. *Methods in Cell Biology*, 1999, 58, 49–73.
24. Tessier, P.M. et al., On-line spectroscopic characterization of sodium cyanide with nanostructured gold surface-enhanced Raman spectroscopy substrates. *Applied Spectroscopy*, 2002, 56(12), 1524–1530.
25. Litorja, M. et al., Surface-enhanced Raman scattering detected temperature programmed desorption: Optical properties, nanostructure, and stability of silver film over SiO_2 nanosphere surfaces. *Journal of Physical Chemistry B*, 2001, 105(29), 6907–6915.
26. Sylvia, J.M. et al., Surface-enhanced Raman detection of 1,4-dinitrotoluene impurity vapor as a marker to locate landmines. *Analytical Chemistry*, 2000, 72(23), 5834–5840.
27. Freeman, R.G. et al., Self-assembled metal colloid monolayers—an approach to SERS substrates. *Science*, 1995, 267(5204), 1629–1632.
28. Kneipp, K. et al., Near-infrared surface-enhanced Raman-scattering (NIR-SERS) of neurotransmitters in colloidal silver solutions. *Spectrochimica Acta. Part A, Molecular and Biomolecular Spectroscopy*, 1995, 51(3), 481–487.
29. Gupta, R. and W.A. Weimer, High enhancement factor gold films for surface enhanced Raman spectroscopy. *Chemical Physics Letters*, 2003, 374(3–4), 302–306.

30. Basche, T., W.P. Ambrose, and W.E. Moerner, Optical-spectra and kinetics of single impurity molecules in a polymer: Spectral diffusion and persistent spectral hole burning. *Journal of the Optical Society of America B. Optical Physics*, 1992, 9(5), 829–836.

31. Foquet, M. et al., Focal volume confinement by submicrometer-sized fluidic channels. *Analytical Chemistry*, 2004, 76(6), 1618–1626.

32. Hill, E.K. and A.J. de Mello, Single-molecule detection using confocal fluorescence detection: Assessment of optical probe volumes. *Analyst*, 2000, 125(6), 1033–1036.

33. Diaspro, A. and M. Robello, Two-photon excitation of fluorescence for three-dimensional optical imaging of biological structures. *Journal of Photochemistry and Photobiology. B, Biology*, 2000, 55(1), 1–8.

34. Weiss, S., Fluorescence spectroscopy of single biomolecules. *Science*, 1999, 283(5408), 1676–1683.

35. Dittrich, P.S. and P. Sohwille, Photobleaching and stabilization of fluorophores used for single-molecule analysis with one-and two-photon excitation. *Applied Physics B-Lasers and Optics*, 2001, 73(8), 829–837.

36. Shaqfeh, E.S.G., The dynamics of single-molecule DNA in flow. *Journal of Non-Newtonian Fluid Mechanics*, 2005, 130(1), 1–28.

37. Dittrich, P.S. and A. Manz, Single-molecule fluorescence detection in microfluidic channels—the Holy Grail in muTAS? *Analytical and Bioanalytical Chemistry*, 2005, 382(8), 1771–1782.

38. Dorre, K. et al., Highly efficient single molecule detection in microstructures. *Journal of Biotechnology*, 2001, 86(3), 225–236.

39. Lundqvist, A., D.T. Chiu, and O. Orwar, Electrophoretic separation and confocal laser-induced fluorescence detection at ultralow concentrations in constricted fused-silica capillaries. *Electrophoresis*, 2003, 24(11), 1737–1744.

40. Kohler, J., Optical spectroscopy of individual objects. *Naturwissenschaften*, 2001, 88(12), 514–521.

41. Schwille, P., J. Korlach, and W.W. Webb, Fluorescence correlation spectroscopy with single-molecule sensitivity on cell and model membranes. *Cytometry*, 1999, 36(3), 176–182.

42. Medina, M.A. and P. Schwille, Fluorescence correlation spectroscopy for the detection and study of single molecules in Biology. *Bioessays*, 2002, 24(8), 758–764.

43. Schwille, P., J. Bieschke, and F. Oehlenschlager, Kinetic investigations by fluorescence correlation spectroscopy: The analytical and diagnostic potential of diffusion studies. *Biophysical Chemistry*, 1997, 66(2–3), 211–228.

44. Maiti, S., U. Haupts, and W.W. Webb, Fluorescence correlation spectroscopy: Diagnostics for sparse molecules. *Proceedings of the National Academy of Sciences of the United States of America*, 1997, 94(22), 11753–11757.

45. Bacia, K. and P. Schwille, A dynamic view of cellular processes by in vivo fluorescence auto- and cross-correlation spectroscopy. *Methods*, 2003, 29(1), 74–85.

46. Dittrich, P. et al., Accessing molecular dynamics in cells by fluorescence correlation spectroscopy. *Biological Chemistry*, 2001, 382(3), 491–494.

47. Li, H.T. et al., Measuring single-molecule nucleic acid dynamics in solution by two-color filtered ratiometric fluorescence correlation spectroscopy. *Proceedings of the National Academy of Sciences of the United States of America*, 2004, 101(40), 14425–14430.

48. Cosa, G. et al., Secondary structure and secondary structure dynamics of DNA hairpins complexed with HIV-1NC protein. *Biophysical Journal*, 2004, 87(4), 2759–2767.

49. Koltermann, A. et al., Rapid assay processing by integration of dual-color fluorescence cross-correlation spectroscopy: High throughput screening for enzyme activity. *Proceedings of the National Academy of Sciences of the United States of America*, 1998, 95(4), 1421–1426.

50. Sterrer, S. and K. Henco, Fluorescence correlation spectroscopy (FCS)—A highly sensitive method to analyze drug/target interactions. *Journal of Receptor and Signal Transduction Research*, 1997, 17(1–3), 511–520.

51. Pramanik, A., Ligand–receptor interactions in live cells by fluorescence correlation spectroscopy. *Current Pharmaceutical Biotechnology*, 2004, 5(2), 205–212.

52. Elson, E.L., Fluorescence correlation spectroscopy measures molecular transport in cells. *Traffic*, 2001, 2(11), 789–796.

53. Prasher, D.C. et al., Primary structure of the *Aequorea victoria* green-fluorescent protein. *Gene*, 1992, 111(2), 229–233.
54. Chamberlain, C. and K.M. Hahn, Watching proteins in the wild: Fluorescence methods to study protein dynamics in living cells. *Traffic*, 2000, 1(10), 755–762.
55. Whitaker, M., Fluorescent tags of protein function in living cells. *BioEssays*, 2000, 22(2), 180–187.
56. Ludin, B. and A. Matus, GFP illuminates the cytoskeleton. *Trends in Cell Biology*, 1998, 8(2), 72–77.
57. Truong, K. and M. Ikura, The use of FRET imaging microscopy to detect protein–protein interactions and protein conformational changes in vivo. *Current Opinion in Structural Biology*, 2001, 11(5), 573–578.
58. Matsuoka, H., S. Nada, and M. Okada, Mechanism of Csk-mediated down-regulation of Src family tyrosine kinases in epidermal growth factor signaling. *Journal of Biological Chemistry*, 2004, 279(7), 5975–5983.
59. Itoh, R.E. et al., Activation of Rac and Cdc42 video imaged by fluorescent resonance energy transfer-based single-molecule probes in the membrane of living cells. *Molecular and Cellular Biology*, 2002, 22(18), 6582–6591.
60. Saito, K. et al., Direct detection of caspase-3 activation in single live cells by cross-correlation analysis. *Biochemical and Biophysical Research Communications*, 2004, 324(2), 849–854.
61. Wazawa, T. and M. Ueda, Total internal reflection fluorescence microscopy in single molecule nanobioscience. *Advances in Biochemical Engineering/Biotechnology*, 2005, 95, 77–106.
62. Schneckenburger, H., Total internal reflection fluorescence microscopy: Technical innovations and novel applications. *Current Opinion in Biotechnology*, 2005, 16(1), 13–18.
63. Mashanov, G.I. et al., Visualizing single molecules inside living cells using total internal reflection fluorescence microscopy. *Methods*, 2003, 29(2), 142–152.
64. Haustein, E. and P. Schwille, single-molecule spectroscopic methods. *Current Opinion in Structural Biology*, 2004, 14(5), 531–540.
65. Sonnleitner, A. et al., Structural rearrangements in single ion channels detected optically in living cells. *Proceedings of the National Academy of Sciences of the United States of America*, 2002, 99(20), 12759–12764.
66. Sako, Y., S. Minoghchi, and T. Yanagida, Single-molecule imaging of EGFR signalling on the surface of living cells. *Nature Cell Biology*, 2000, 2(3), 168–172.
67. Llobet, A., V. Beaumont, and L. Lagnado, Real-time measurement of exocytosis and endocytosis using interference of light. *Neuron*, 2003, 40(6), 1075–1086.
68. Krylyshkina, O. et al., Nanometer targeting of microtubules to focal adhesions. *Journal of Cell Biology*, 2003, 161(5), 853–859.
69. de Lange, F. et al., Cell biology beyond the diffraction limit: Near-field scanning optical microscopy. *Journal of Cell Science*, 2001, 114(23), 4153–4160.
70. Meixner, A.J. and H. Kneppe, Scanning near-field optical microscopy in cell biology and microbiology. *Cellular and Molecular Biology*, 1998, 44(5), 673–688.
71. Garcia-Parajo, M.F. et al., Near-field optical microscopy for DNA studies at the single molecular level. *Bioimaging*, 1998, 6(1), 43–53.
72. Wabuyele, M.B., M. Culha, G.D. Griffin, P.M. Viallet, and T. Vo-Dinh, Near-field scanning optical microscopy for bioanalysis at the nanometer resolution, in *Protein Nanotechnology*, T. Vo-Dinh (Ed.), Humana Press, Totowa, NJ, 2005, pp. 437–452.
73. Hamann, H.F. et al., Molecular fluorescence in the vicinity of a nanoscopic probe. *Journal of Chemical Physics*, 2001, 114(19), 8596–8609.
74. Sanchez, E.J., L. Novotny, and X.S. Xie, Near-field fluorescence microscopy based on two-photon excitation with metal tips. *Physical Review Letters*, 1999, 82(20), 4014–4017.
75. Zenhausern, F., Y. Martin, and H.K. Wickramasinghe, Scanning interferometric apertureless microscopy—optical imaging at 10 Angstrom resolution. *Science*, 1995, 269(5227), 1083–1085.
76. Kneipp, K. et al., Surface-enhanced Raman spectroscopy in single living cells using gold nanoparticles. *Applied Spectroscopy*, 2002, 56(2), 150–154.

77. Moskovits, M. et al., SERS and the single molecule, in *Optical Properties of Nanostructured Random Media*, Springer, Berlin, 2002, pp. 215–226.

78. Tolaieb, B., C.J.L. Constantino, and R.F. Aroca, Surface-enhanced resonance Raman scattering as an analytical tool for single molecule detection. *Analyst*, 2004, 129(4), 337–341.

79. Kneipp, K. et al., Surface-enhanced Raman scattering and biophysics. *Journal of Physics: Condensed Matter*, 2002, 14(18), R597–R624.

80. Vo-Dinh, T., Surface-enhanced Raman spectroscopy using metallic nanostructures. *TrAC: Trends in Analytical Chemistry*, 1998, 17(8–9), 557–582.

81. Vo-Dinh, T., F. Yan, and M.B. Wabuyele, Surface-enhanced Raman scattering for medical diagnostics and biological imaging. *Journal of Raman Spectroscopy*, 2005, 36(6–7), 640–647.

82. Wabuyele, M.B. and T. Vo-Dinh, Detection of human immunodeficiency virus type 1 DNA sequence using plasmonics nanoprobes. *Analytical Chemistry*, 2005, 77(23), 7810–7815.

83. Baker, G.A. and D.S. Moore, Progress in plasmomic engineering of surface-enhanced Raman-scattering substrates toward ultra-trace analysis. *Analytical and Bioanalytical Chemistry*, 2005, 382(8), 1751–1770.

84. Bizzarri, A.R. and S. Cannistraro, Surface-enhanced resonance Raman spectroscopy signals from single myoglobin molecules. *Applied Spectroscopy*, 2002, 56(12), 1531–1537.

85. Kneipp, K. et al., Surface-enhanced and normal Stokes and anti-Stokes Raman spectroscopy of single-walled carbon nanotubes. *Physical Review Letters*, 2000, 84(15), 3470–3473.

86. Safonov, V.P. et al., Spectral dependence of selective photomodification in fractal aggregates of colloidal particles. *Physical Review Letters*, 1998, 80(5), 1102–1105.

87. Zeisel, D. et al., Near-field surface-enhanced Raman spectroscopy of dye molecules adsorbed on silver island films. *Chemical Physics Letters*, 1998, 283(5–6), 381–385.

88. Deckert, V. et al., Near-field surface enhanced Raman imaging of dye-labeled DNA with 100-nm resolution. *Analytical Chemistry*, 1998, 70(13), 2646–2650.

89. Nie, S.M. and S.R. Emery, Probing single molecules and single nanoparticles by surface-enhanced Raman scattering. *Science*, 1997, 275(5303), 1102–1106.

90. Lee, N.S. et al., Surface-enhanced Raman-spectroscopy of the catecholamine neurotransmitters and related-compounds. *Analytical Chemistry*, 1988, 60(5), 442–446.

91. Graham, D., B.J. Mallinder, and W.E. Smith, Detection and identification of labeled DNA by surface enhanced resonance Raman scattering. *Biopolymers*, 2000, 57(2), 85–91.

92. Graham, D., B.J. Mallinder, and W.E. Smith, Surface-enhanced resonance Raman scattering as a novel method of DNA discrimination. *Angewandte Chemie-International Edition*, 2000, 39(6), 1061–1063.

93. Ni, J. et al., Immunoassay readout method using extrinsic Raman labels adsorbed on immuno-gold colloids. *Analytical Chemistry*, 1999, 71(21), 4903–4908.

94. Delfino, I., A.R. Bizzarri, and S. Cannistraro, Single-molecule detection of yeast cytochrome *c* by surface-enhanced Raman spectroscopy. *Biophysical Chemistry*, 2005, 113(1), 41–51.

95. Bjerneld, E.J. et al., Single-molecule surface-enhanced Raman and fluorescence correlation spectroscopy of horseradish peroxidase. *Journal of Physical Chemistry B*, 2002, 106(6), 1213–1218.

96. Sockalingum, G.D. et al., Characterization of island films as surface-enhanced Raman spectroscopy substrates for detecting low antitumor drug concentrations at single cell level. *Biospectroscopy*, 1998, 4(5), S71–S78.

97. Beljebbar, A. et al., Near-infrared FT-SERS microspectroscopy on silver and gold surfaces: Technical development, mass sensitivity, and biological applications. *Applied Spectroscopy*, 1996, 50(2), 148–153.

98. Ashkin, A., History of optical trapping and manipulation of small-neutral particle, atoms, and molecules. *IEEE Journal of Selected Topics in Quantum Electronics*, 2000, 6(6), 841–856.

99. Allaway, D., N.A. Schofield, and P.S. Poole, Optical traps: Shedding light on biological processes. *Biotechnology Letters*, 2000, 22(11), 887–892.

100. Ashkin, A., J.M. Dziedzic, and T. Yamane, Optical trapping and manipulation of single cells using infrared-laser beams. *Nature*, 1987, 330(6150), 769–771.

101. Calander, N. and M. Willander, Optical trapping of single fluorescent molecules at the detection spots of nanoprobes. *Physical Review Letters*, 2002, 89(14), 143603-1–143603-4.
102. Bennink, M.L. et al., Single-molecule manipulation of double-stranded DNA using optical tweezers: Interaction studies of DNA with RecA and YOYO-1. *Cytometry*, 1999, 36(3), 200–208.
103. Chiu, D.T. and R.N. Zare, Optical detection and manipulation of single molecules in room-temperature solutions. *Chemistry: A European Journal*, 1997, 3(3), 335–339.
104. Purcell, T.J., H.L. Sweeney, and J.A. Spudich, A force-dependent state controls the coordination of processive myosin V. *Proceedings of the National Academy of Sciences of the United States of America*, 2005, 102(39), 13873–13878.
105. Jeney, S. et al., Mechanical properties of single motor molecules studied by three-dimensional thermal force probing in optical tweezers. *Chemphyschem*, 2004, 5(8), 1150–1158.
106. Rief, M. et al., Myosin-V stepping kinetics: A molecular model for processivity. *Proceedings of the National Academy of Sciences of the United States of America*, 2000, 97(17), 9482–9486.
107. Tyska, M.J. et al., Two heads of myosin are better than one for generating force and motion. *Proceedings of the National Academy of Sciences of the United States of America*, 1999, 96(8), 4402–4407.
108. McDonald, M.E., Determining the physical properties of DNA in DNA microarrays using optical tweezers. *Biophysical Journal*, 2005, 88(1), 569A–570A.
109. Oana, H. et al., On-site manipulation of single whole-genome DNA molecules using optical tweezers. *Applied Physics Letters*, 2004, 85(21), 5090–5092.
110. Tessmer, I. et al., Mode of drug binding to DNA determined by optical tweezers force spectroscopy. *Journal of Modern Optics*, 2003, 50(10), 1627–1636.
111. Schliwa, M. and G. Woehlke, Molecular motors. *Nature*, 2003, 422(6933), 759–765.
112. Funatsu, T. et al., Imaging and nano-manipulation of single biomolecules. *Biophysical Chemistry*, 1997, 68(1–3), 63–72.
113. Svoboda, K. et al., Direct observation of kinesin stepping by optical trapping interferometry. *Nature*, 1993, 365(6448), 721–727.
114. Block, S.M., L.S.B. Goldstein, and B.J. Schnapp, Bead movement by single kinesin molecules studied with optical tweezers. *Nature*, 1990, 348(6299), 348–352.
115. Forkey, J.N. et al., Three-dimensional structural dynamics of myosin V by single-molecule fluorescence polarization. *Nature*, 2003, 422(6930), 399–404.
116. Sosa, H. et al., ADP-induced rocking of the kinesin motor domain revealed by single-molecule fluorescence polarization microscopy. *Nature Structural Biology*, 2001, 8(6), 540–544.
117. Funatsu, T. et al., Imaging of single fluorescent molecules and individual ATP turnovers by single myosin molecules in aqueous-solution. *Nature*, 1995, 374(6522), 555–559.
118. Ishijima, A. et al., Single-molecule analysis of the actomyosin motor using nano-manipulation. *Biochemical and Biophysical Research Communications*, 1994, 199(2), 1057–1063.
119. Finer, J.T., R.M. Simmons, and J.A. Spudich, Single myosin molecule mechanics—Piconewton forces and nanometer steps. *Nature*, 1994, 368(6467), 113–119.
120. Vale, R.D. et al., Direct observation of single kinesin molecules moving along microtubules. *Nature*, 1996, 380(6573), 451–453.
121. Hua, W. et al., Coupling of kinesin steps to ATP hydrolysis. *Nature*, 1997, 388(6640), 390–393.
122. Svoboda, K. and S.M. Block, Force and velocity measured for single kinesin molecules. *Cell*, 1994, 77(5), 773–784.
123. Thorn, K.S., J.A. Ubersax, and R.D. Vale, Engineering the processive run length of the kinesin motor. *Journal of Cell Biology*, 2000, 151(5), 1093–1100.
124. Case, R.B. et al., The directional preference of kinesin motors is specified by an element outside of the motor catalytic domain. *Cell*, 1997, 90(5), 959–966.
125. Spudich, J.A., How molecular motors work. *Nature*, 1994, 372(6506), 515–518.
126. Woehlke, G. and M. Schliwa, Walking on two heads: The many talents of kinesin. *Nature Reviews Molecular Cell Biology*, 2000, 1(1), 50–58.
127. Ait-Haddou, R. and W. Herzog, Brownian ratchet models of molecular motors. *Cell Biochemistry and Biophysics*, 2003, 38(2), 191–213.

128. Astumian, R.D. and I. Derenyi, A chemically reversible Brownian motor: Application to kinesin and Ncd. *Biophysical Journal*, 1999, 77(2), 993–1002.

129. Zhuang, X.W. et al., A single-molecule study of RNA catalysis and folding. *Science*, 2000, 288(5473), 2048–2051.

130. Lu, H.P., L.Y. Xun, and X.S. Xie, Single-molecule enzymatic dynamics. *Science*, 1998, 282(5395), 1877–1882.

131. Peleg, G. et al., Single-molecule spectroscopy of the beta(2) adrenergic receptor: Observation of conformational substates in a membrane protein. *Proceedings of the National Academy of Sciences of the United States of America*, 2001, 98(15), 8469–8474.

132. Scheel, A.A. et al., Receptor–ligand interactions studied with homogeneous fluorescence-based assays suitable for miniaturized screening. *Journal of Biomolecular Screening*, 2001, 6(1), 11–18.

133. Murakoshi, H. et al., Single-molecule imaging analysis of Ras activation in living cells. *Proceedings of the National Academy of Sciences of the United States of America*, 2004, 101(19), 7317–7322.

134. Haupts, U. et al., Single-molecule detection technologies in miniaturized high-throughput screening: Fluorescence intensity distribution analysis. *Journal of Biomolecular Screening*, 2003, 8(1), 19–33.

135. Sako, Y. et al., Optical bioimaging: From living tissue to a single molecule: Single-molecule visualization of cell signaling processes of epidermal growth factor receptor. *Journal of Pharmacological Sciences*, 2003, 93(3), 253–258.

136. Slaughter, B.D. et al., Single-molecule resonance energy transfer and fluorescence correlation spectroscopy of calmodulin in solution. *Journal of Physical Chemistry B*, 2004, 108(29), 10388–10397.

137. Ueda, M. et al., Single-molecule analysis of chemotactic signaling in Dictyostelium cells. *Science*, 2001, 294(5543), 864–867.

138. Mekler, V. et al., Structural organization of bacterial RNA polymerase holoenzyme and the RNA polymerase-promoter open complex. *Cell*, 2002, 108(5), 599–614.

139. Neininger, A., H. Thielemann, and M. Gaestel, FRET-based detection of different conformations of MK2. *Embo Reports*, 2001, 2(8), 703–708.

140. Trakselis, M.A., S.C. Alley, E. Abel-Santos, and S.J. Benkovic, Creating a dynamic picture of the sliding clamp during T4 DNA polymerase holoenzyme assembly by using fluorescence resonance energy transfer. *Proceedings of the National Academy of Sciences of the United States of America*, 2001, 98(15), 8368–8375.

141. Fa, M. et al., Conformational studies of plasminogen activator inhibitor type 1 by fluorescence spectroscopy—Analysis of the reactive centre of inhibitory and substrate forms, and of their respective reactive-centre cleaved forms. *European Journal of Biochemistry*, 2000, 267(12), 3729–3734.

142. Deniz, A.A. et al., Single-molecule protein folding: Diffusion fluorescence resonance energy transfer studies of the denaturation of chymotrypsin inhibitor 2. *Proceedings of the National Academy of Sciences of the United States of America*, 2000, 97(10), 5179–5184.

143. Schutz, G.J., W. Trabesinger, and T. Schmidt, Direct observation of ligand colocalization on individual receptor molecules. *Biophysical Journal*, 1998, 74(5), 2223–2226.

144. Brasselet, S. et al., Single-molecule fluorescence resonant energy transfer in calcium concentration dependent cameleon. *Journal of Physical Chemistry B*, 2000, 104(15), 3676–3682.

145. Ha, T. et al., Ligand-induced conformational changes observed in single RNA molecules. *Proceedings of the National Academy of Sciences of the United States of America*, 1999, 96(16), 9077–9082.

146. Jia, Y.W. et al., Folding dynamics of single GCN4 peptides by fluorescence resonant energy transfer confocal microscopy. *Chemical Physics*, 1999, 247(1), 69–83.

147. Warshaw, D.M. et al., Myosin conformational states determined by single fluorophore polarization. *Proceedings of the National Academy of Sciences of the United States of America*, 1998, 95(14), 8034–8039.

148. Sase, I. et al., Real-time imaging of single fluorophores on moving actin with an epifluorescence microscope. *Biophysical Journal*, 1995, 69(2), 323–328.
149. Sase, I. et al., Axial rotation of sliding actin filaments revealed by single-fluorophore imaging. *Proceedings of the National Academy of Sciences of the United States of America*, 1997, 94(11), 5646–5650.
150. Sakmann, B. and E. Neher, *Single Channel Recordings*, 2nd edn., Kluwer, New York, 1995.
151. Hille, B., *Ion channels of Excitable Membranes*, 3rd edn., Sinauer Associates, Inc., Sunderland, MA, 2001.
152. Harms, G., G. Orr, and H.P. Lu, Probing ion channel conformational dynamics using simultaneous single-molecule ultrafast spectroscopy and patch-clamp electric recording. *Applied Physics Letters*, 2004, 84(10), 1792–1794.
153. Milescu, L.S. et al., Hidden Markov model applications in QuB: Analysis of nanometer steps in single molecule fluorescence data and ensemble ion channel kinetics. *Biophysical Journal*, 2003, 84(2), 124A–124A.
154. Lougheed, T. et al., Fluorescent gramicidin derivatives for single-molecule fluorescence and ion channel measurements. *Bioconjugate Chemistry*, 2001, 12(4), 594–602.
155. Harms, G.S. et al., Single-molecule imaging of L-type Ca^{2+} channels in live cells. *Biophysical Journal*, 2001, 81(5), 2639–2646.
156. Nakane, J., M. Wiggin, and A. Marziali, A nanosensor for transmembrane capture and identification of single nucleic acid molecules. *Biophysical Journal*, 2004, 87(1), 615–621.
157. Cullum, B.M. and T. Vo-Dinh, The development of optical nanosensors for biological measurements. *Trends in Biotechnology*, 2000, 18(9), 388–393.
158. Vo-Dinh, T. et al., Antibody-based nanoprobe for measurement of a fluorescent analyte in a single cell. *Nature Biotechnology*, 2000, 18(7), 764–767.
159. Vo-Dinh, T., Nanobiosensors: Probing the sanctuary of individual living cells. *Journal of Cellular Biochemistry*, 2002, 39, 154–161.
160. Vo-Dinh, T., P.M. Kasili, and M.B. Wabuyele, Nanoprobes and nanobiosensors for monitoring and imaging individual living cells. *Nanomedicine*, 2006, 2(1), 22–30.
161. Song, J.M. et al., Detection of cytochrome *c* in a single cell using an optical nanobiosensor. *Analytical Chemistry*, 2004, 76(9), 2591–2594.
162. Kasili, P.M., J.M. Song, and T. Vo-Dinh, Optical sensor for the detection of caspase-9 activity in a single cell. *Journal of the American Chemical Society*, 2004, 126(9), 2799–2806.
163. Sijtsema, N.M. et al., Intracellular reactions in single human granulocytes upon phorbol myristate acetate activation using confocal Raman microspectroscopy. *Biophysical Journal*, 2000, 78(5), 2606–2613.
164. Morjani, H. et al., Molecular and cellular interactions between intoplicine, DNA, and topoisomerase-II studied by surface-enhanced Raman-scattering spectroscopy. *Cancer Research*, 1993, 53(20), 4784–4790.
165. Nabiev, I.R., H. Morjani, and M. Manfait, Selective analysis of antitumor drug-interaction with living cancer-cells as probed by surface-enhanced Raman-spectroscopy. *European Biophysics Journal*, 1991, 19(6), 311–316.
166. Wabuyele, M.B. et al., Hyperspectral surface-enhanced Raman imaging of labeled silver nanoparticles in single cells. *Review of Scientific Instruments*, 2005, 76(6), 063710-1–063710-7.

11

Optical Nanobiosensors and Nanoprobes

Tuan Vo-Dinh

CONTENTS

11.1 Introduction

Optical sensors provide significant advantages for *in situ* monitoring applications due to the optical nature of the excitation and detection modalities. Fiberoptic sensors are not affected by electromagnetic interferences from static electricity, strong magnetic fields, or surface potentials. Another advantage of fiberoptic sensors is the small size of optical fibers, which allow sensing intracellular/intercellular physiological and biological parameters in microenvironments. Biosensors, which use biological probes coupled to a transducer, have been developed during the last two decades for environmental, industrial, and biomedical diagnostics. Extensive research and development activities in our laboratory have been devoted to the development of a variety of fiberoptics chemical sensors and biosensors (1–9). Recent advances in nanotechnology have led the development of fiberoptics-based nanosensor systems having nanoscale dimensions suitable for intracellular measurements. The possibilities to monitor *in vivo* processes within living cells could dramatically improve our understanding of cellular function, thereby revolutionizing cell biology. The application of a submicron fiberoptics chemical sensor has been reported (10,11). Submicron tapered optical fibers with distal diameters between 20 and 500 nm have been employed to study the submicron spatial resolution achievable using near-field scanning optical microscopy (NSOM). The combination of NSOM and surface-enhanced Raman scattering (SERS) has been demonstrated to detect chemicals on solid substrates with subwavelength 100-nm spatial resolution (12,13). Submicron optical fiber probes have been developed for chemical analyses (14,15). Nanosensors with antibody probes have been developed and used to detect physiological and biochemical targets inside single cells (16–28).

This chapter describes the principle, development, and applications of fiberoptics nanobiosensor systems using antibody-based probes. The chapter provides background information on biosensors, a description of the fabrication methods for fiberoptics nanosensors and detection systems, and applications in single-cell analysis. The usefulness and potential of fiberoptics nanosensor technology in biological research and applications are discussed. Applications of nanobiosensors in medical applications are further discussed in Chapter 20.

11.2 Basic Components of Biosensors

A biosensor generally consists of a probe with biological recognition element, often called a bioreceptor, and a transducer. The interaction of the analyte with the bioreceptor is designed to produce an effect measured by the transducer, which converts the information into a measurable effect, for example, an electrical signal. Figure 11.1 illustrates the operating principle of a typical biosensing system. Based on antibody–antigen interactions, immunoassay techniques are very powerful monitoring tools because of their excellent

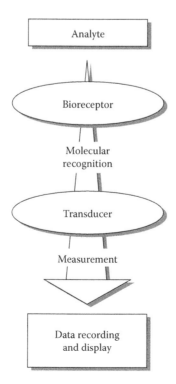

FIGURE 11.1
Operating principle of biosensor systems.

specificity and reasonable sensitivity. The immunological principle can be combined with laser and fiberoptics technology to develop a new generation of nanosensors for intracellular measurements.

11.2.1 Bioreceptors

The specificity of biosensors is based on the bioreceptors used. A bioreceptor is a biological molecular species (e.g., an antibody, an enzyme, a protein, or a nucleic acid) or a living biological system (e.g., cells, tissue, or whole organisms) that utilizes a biochemical mechanism for recognition. Bioreceptors allow binding the specific analyte of interest to the sensor for the measurement with minimum interference from other components in complex mixtures. The sampling component of a biosensor contains a biosensitive layer. The layer can either contain bioreceptors or be made of bioreceptors covalently attached to the transducer. The most common forms of bioreceptors used in biosensing are based on (1) antibody/antigen interactions, (2) nucleic acid interactions, (3) enzymatic interactions, (4) cellular interactions (i.e., microorganisms, proteins), and (5) interactions using biomimetic materials (i.e., synthetic bioreceptors). Detection techniques can use various schemes: optical, electrochemical, and mass-sensitive methods, etc. This chapter discusses various biosensors that use chemical-sensitive ligands or antibody probes, which are often called immunosensors.

11.2.2 Antibody Probes

Antigen–antibody (Ag–Ab) binding reaction, which is a key mechanism by which the immune system detects and eliminates foreign matter, provides the basis for specificity of immunoassays. Antibodies are complex biomolecules, made-up of hundreds of individual amino acids arranged in a highly ordered sequence. Antibodies are produced by immune system cells (B cells) when such cells are exposed to substances or molecules, which are called antigens. The antibodies called forth following antigen exposure have recognition/binding sites for specific molecular structures (or substructures) of the antigen. The way in which antigen and antigen-specific antibody interact is analogous to a lock-and-key fit, in which specific configurations of a unique key enable it to open a lock. In the same way, an antigen-specific antibody fits its unique antigen in a highly specific manner, so that the three-dimensional structures of antigen and antibody molecules are complementary. Due to this three-dimensional shape fitting, and the diversity inherent in individual antibody make-up, it is possible to find an antibody that can recognize and bind to any one of a large variety of molecular shapes. This unique property of antibodies is the key to their usefulness in immunosensors; this ability to recognize molecular structures allows one to develop antibodies that bind specifically to chemicals, biomolecules, microorganism components, etc. One can then use such antibodies as specific probes to recognize and bind to an analyte of interest that is present, even in extremely small amounts, within a large number of other chemical substances. Another property of great importance to antibodies' analytical role in immunosensors is the strength or avidity/affinity of the Ag–Ab interaction. Because of the variety of interactions that can take place as the Ag–Ab surfaces lie in close proximity one to another, the overall strength of the interaction can be considerable, with correspondingly favorable association and equilibrium constants. What this means in practical terms is that the Ag–Ab interactions can take place very rapidly (for small antigen molecules, almost as rapidly as diffusion processes can bring antigen and antibody together), and that, once formed, the Ag–Ab complex has a reasonable lifetime.

The production of antibodies requires the use of immunogenic species. For a substance to be immunogenic (i.e., capable of producing an immune response), a certain molecular size and complexity are necessary: proteins with molecular weights >5000 Da are generally immunogenic. Radioimmunoassay (RIA), which utilizes radioactive labels, has been one of the most widely used immunoassay methods. RIA has been applied to a number of fields including pharmacology, clinical chemistry, forensic science, environmental monitoring, molecular epidemiology, and agricultural science. The usefulness of RIA, however, is limited by several shortcomings, including the required use of radioactive labels, the limited shelf life of radioisotopes, the potential deleterious biological effects inherent to radioactive materials, and the cost of radioactive waste disposal. For these reasons, there are extensive research efforts aimed at developing simpler, more practical immunochemical techniques and instrumentation, which offer comparable sensitivity and selectivity to RIA. In the 1980s, advances in spectrochemical instrumentation, laser miniaturization, biotechnology, and fiberoptics research have provided opportunities for novel approaches to the development of sensors for the detection of chemicals and biological materials of environmental and biomedical interest. Since the first development of a remote fiberoptics immunosensor for *in situ* detection of the chemical carcinogen benzo[a]pyrene (BaP) (1), antibodies have become common bioreceptors used in biosensors today.

11.3 Fiberoptics Nanosensor System

11.3.1 Development of Fiberoptics Nanoprobes

This section discusses the protocols and instrumental systems involved in the fabrication of fiberoptics nanoprobes. The fabrication of near-field optical probes is a crucial prerequisite for the development of nanosensors. There are two methods for preparing the nanofiber tips. The most frequently used technique is the so-called heat and pull method. It is based on local heating of a glass fiber using a laser or a filament and subsequently pulling the fiber apart. The resulting tip shapes depend strongly on the temperature and the timing of the procedure. The second method is based on chemical etching of glass fibers (29,30).

Figure 11.2 illustrates the experimental steps involved in the fabrication of nanosensors using the heat and pull method (21). Fabrication of nanosensors involves techniques capable of making optical fibers with submicron-size diameter core. Since these nanoprobes are not commercially available, they have to be fabricated in the laboratory. One procedure consists of pulling from a larger silica optical fiber using a special fiber-pulling device (Sutter Instruments P-2000). This method yields fibers with submicron diameters. One end of a 600-μm silica/silica fiber is polished to a 0.3 μm finish with an Ultratec fiber polisher. The other end of the optical fiber is then pulled to a submicron length using a fiber puller. Figure 11.3 shows a scanning election microscopy photograph of one of the fiber probes fabricated for our preliminary studies. The scale on the photograph of this sample indicates that the distal end of the fiber is ~30 nm.

The sidewall of the tapered end is then coated with a thin layer of silver, aluminum, or gold (100–300 nm) to prevent light leakage of the excitation light on the tapered side of the fiber. The coating procedure is designed to leave the distal end of the fiber free for subsequent binding with bioreceptors. Such a coating system is illustrated in Figure 11.4. The fiber probe is attached on a rotating plate inside a thermal evaporation chamber (3,19,21).

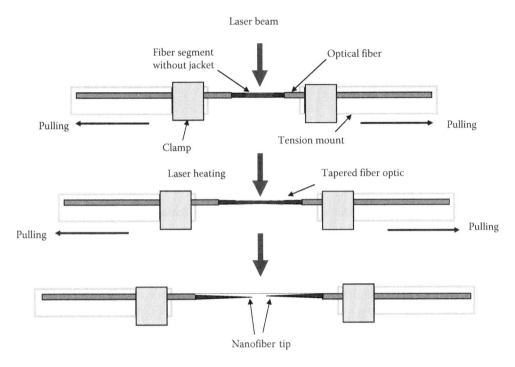

FIGURE 11.2
Method for the fabrication of nanofibers.

FIGURE 11.3
Scanning electron photograph of nanofiber (the size of the fiber tip diameter is ∼30 nm). Insert: Fiberoptic nanosensor used to monitor a single cell. (Adapted from Vo-Dinh T., Alarie J. P., Cullum B., and Griffin G. D. 2000. *Nat. Biotechnol.* 18, 76.)

The fiber axis and the evaporation direction formed an angle of ∼45°. While the probe is rotated, the metal is allowed to evaporate onto the tapered side of the fiber tip to form a thin coating. Since the fiber tip is pointed away from the metal source, it remains free from any metal coating. The tapered end is coated with 300–400 nm of silver in a Cooke Vacuum Evaporator system using a thermal source at 10^{-6} Torr. With the metal coating, the size of the probe tip is ∼250–300 nm.

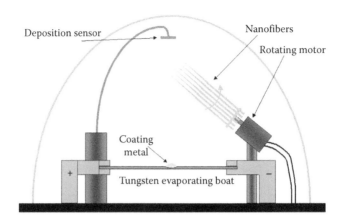

FIGURE 11.4
Instrumental setup for coating the nanofiber tips with silver.

The next step in the preparation of the biosensor probes involves covalent immobilization of receptors onto the fiber tip. Antibodies can be immobilized onto the fiberoptics probes using standard chemical procedures. Briefly, the fiber is derivatized in 10% GOPS in H_2O (v/v) at 90°C for 3 h. The pH of the mixture is maintained below 3 with concentrated HCl (1 M). After derivatization, the fiber is washed in ethanol and dried overnight in a vacuum oven at 105°C. The fiber is then coated with silver as described previously. The derivatized fiber is activated in a solution of 100 mg/mL 1,1′ carbonyldiimidazole (CDI) in acetonitrile for 20 min followed by rinsing with acetonitrile and then phosphate-buffered saline (PBS). The fiber tip is then incubated in a 1.2-mg/mL anti-BPT solution (PBS solvent) for 4 days at 4°C and then stored overnight in PBS to hydrolyze any unreacted sites. The fibers are then stored at 4°C with the antibody immobilized tips stored in PBS. This procedure has been shown to maintain over 95% antibody activity for BPT (21).

11.3.2 Experimental Protocol

This section discusses the procedures to grow cell cultures for analysis using the nanosensors. Cell cultures were grown in a water-jacketed cell culture incubator at 37°C in an atmosphere of 5% CO_2 in air. Clone 9 cells, a rat liver epithelial cell line, were grown in Ham's F-12 medium (cat# 11765-054, Gibco/BRL, Grand Island, NY) supplemented with 10% fetal bovine serum and an additional 1 mM glutamine (cat #15039-027 Gibco, Grand Island, NY). In preparation for an experiment, 1×10^5 cells in 5 mL of medium were seeded into 60-mm diameter dishes (cat #25010, Corning Costar Corp., Corning, NY). The growth of the cells was monitored daily by microscopic observation, and when the cells reached a state of confluence of 50%–60%, BPT was added and left in contact with the cells for 18 h (i.e., overnight). The growth conditions were chosen so that the cells would be in log phase growth during the chemical treatment, but would not be so close to confluence that a confluent monolayer would form by the termination of the chemical exposure. Benzopyrene tetrol (BPT) was prepared as a 1-mM stock solution in reagent grade methanol and further diluted in reagent grade ethanol (95%) prior to addition to the cells. The final concentration of BPT in the culture medium of the dish was 1×10^{-7} M and the final alcohol concentration (combination of methanol and ethanol) was 0.1%. Following chemical treatment, the medium containing BPT was aspirated and replaced with standard growth medium, prior to the nanoprobe procedure.

A unique application of nanosensor involves monitoring single cells *in vivo* (18–28). Monitoring BPT in single cells using the nanoprobe was carried out in the following way. A culture dish of cells was placed on the prewarmed microscope stage, and the nanoprobe, mounted on the micropipette holder, was moved into position (i.e., in the same plane of the cells), using bright-field microscopic illumination, so that the tip was outside the cell to be probed. The total magnification was usually 400×. All room light and microscope illumination light were extinguished, the laser shutter was opened, laser light was allowed to illuminate the optical fiber, and excitation light was transmitted into the fiber tip. Usually, if the silver coating on the nanoprobe was appropriate, no light leaked out of the sidewall of the tapered fiber. Only a faint glow of laser excitation at the tip could be observed on the nanoprobe. A reading was taken with the nanoprobe outside the cell and the laser shutter was closed. The nanoprobe was then moved into the cell, inside the cell membrane and extending a short way into the cytoplasm, but care was taken not to penetrate the nuclear envelope. The laser was again opened, and readings were then taken and recorded as a function of time during which the nanoprobe was inside the cell.

11.3.3 Optical Instrumentation

The optical measurement system used for monitoring single cells using the nanosensors is illustrated in Figure 11.5. The 325-nm line of a HeCd laser (Omnichrome, 8 mW laser power) or the 488-nm line of an argon ion laser (Coherent, 10 mW) was focused onto a 600-μm delivery fiber, which terminated with an SMA connector. The antibody immobilized tapered fiber was coupled to the delivery fiber through the SMA connector and was secured to the micromanipulators on the microscope. The fluorescence emitted from the cells was collected by the microscope objective and passed through a 400-nm longpass

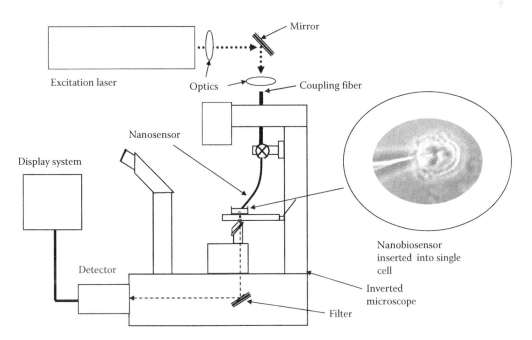

FIGURE 11.5
Instrumental system for fluorescence measurements of single cells using nanosensors.

dichroic mirror and then focused onto a photomultiplier tube (PMT) for detection. The output from the PMT was passed through a picoammeter and recorded on a strip chart recorder or a personal computer (PC) for further data treatment. The experimental setup used to probe single cells was adapted to this purpose from a standard micromanipulation/microinjection apparatus. A Nikon Diaphot 300 inverted microscope (Nikon, Inc, Melville, NY) with Diaphot 300/Diaphot 200 Incubator, to maintain the cell cultures at ~37°C on the microscope stage, was used for these experiments. The micromanipulation equipment used consisted of MN-2 (Narishige Co., LTD, Tokyo, Japan) Narishige three-dimensional manipulators for coarse adjustment, and Narishige MMW-23 three-dimensional hydraulic micromanipulators for final movements. The optical fiber nanoprobe was mounted on a micropipette holder (World Precision Instruments, Inc, Sarasota, FL). To record the fluorescence of BPT molecules binding to antibodies at the fiber tip, a Hamamatsu PMT detector assembly (HC125-2) was mounted in the front port of the Diaphot 300 microscope, and fluorescence was collected via this optical path (80% of available light at the focal plane can be collected through the front port).

11.4 Applications in Bioanalysis

11.4.1 Fluorescence Measurements Using Optical Nanoprobes

Nanofiber probes have been fabricated and used to detect fluorescence emission from species inside cells (17). In this study, mouse epithelial cells were incubated with a fluorescent dye by incubating the cells in the dye solution and allowing membrane permeabilization to take place. Another procedure for loading cells with fluorophores involved the method called "scrape loading." In this procedure, a portion of the cell monolayer was removed by mechanical means, and cells along the boundary of this "scrape" were transiently permeabilized, allowing the dye to enter these cells. The dye was subsequently washed away, and only permeabilized cells retain the dye molecules, as they were not internalized by cells with intact membranes. Following incubation, the fluorescence signal in single cells was detected using laser light excitation through the optical nanofibers fabricated using the procedures described previously. Micromanipulators were used to move the optical fiber into contact with the cell membrane. The fiber tip was then gently inserted just inside the cell membrane for fluorescence measurements. Light from an argon-ion laser (10 m; 488-nm line) was used and passed through the optical nanofiber. The dye inside the cell was excited, and fluorescence emission was collected and observed through the filter cube of a Nikon Microphot microscope, which effectively filter out the laser excitation light at 488 nm. Background measurements were performed with cells that were not loaded with the fluorophores. Fluorescence signals were successfully detected inside the fluorophore-loaded cells and not inside nonloaded cells. As another control, the optical fiber probe was then moved to an area of the specimen where there were no cells, and the laser light was again passed down the fiber for excitation. No visible fluorescence (at the emission wavelength) was detected for this control measurement, thus demonstrating the successful detection of the fluorescent dye molecules inside single cells. These results demonstrated the capability of optical nanofibers for measurements in intracellular environments of luminophores incorporated into cells.

11.4.2 Monitoring Single Living Cells Using Antibody-Based Nanoprobes

The success of intracellular investigations with single cells depends largely not only on the sensitivity of the measurement system, but also on the selectivity of the probe and on the small size of the probes. Biosensors can provide the required selectivity due to the Ab–Ag recognition process, which is one of the most selective molecular recognition processes. Several strategies for designing antibody-based nanosensors may be considered. For example, a membrane-based antibody sensor produced a concentration effect of 40 times greater than the nonantibody sensor (8,9). Membrane-based probes, however, increases the size of the measurement system, making it difficult to be inserted inside a cell. For biosensors used in intracellular measurements, covalent binding of antibody molecules directly on the tip of the bare nanofiber probe is the most straightforward procedure. This method does not significantly increase the probe sizes.

Our laboratory has developed and used nanosensors having antibody-based probes to measure fluorescent targets inside a single cell (18–21). The antibody probe was targeted against BPT, an important biological compound, which was used as a biomarker of human exposure to the carcinogen BaP, a polycyclic aromatic hydrocarbon (PAH) of great environmental and toxicological interest because of its mutagenic/carcinogenic properties and its ubiquitous presence in the environment. BaP has been identified as a chemical carcinogen in laboratory animal studies (31). The measurements were performed on rat liver epithelial cells (Clone 9) used as the model cell system. The cells had been previously incubated with BPT molecules prior to measurements. The results demonstrated the possibility of *in situ* measurements of BPT inside a single cell.

The small size of the probe allowed manipulation of the nanosensor at specific locations within the cells (Figure 11.6). To demonstrate proof of concept of single-cell measurements with antibody-based nanoprobes, experiments were performed on a rat liver epithelial Clone 9 cell line, which was used as the model cell system (17–21). The cells were first incubated with BPT prior to measurements. Interrogation of single cells for the presence of BPT was then carried out using antibody nanoprobes for excitation and a photometric system

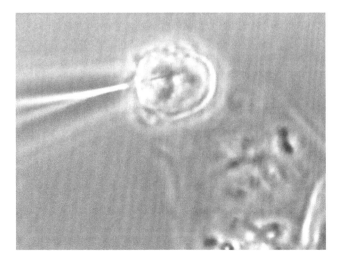

FIGURE 11.6
Photograph of single-cell sensing using the nanosensor system (the small size of the probe allowed manipulation of the nanoprobe at specific locations within a single cell).

for fluorescence signal detection. Multiple (e.g., five) recordings of the fluorescence signals could be taken with each measurement using a specific nanoprobe. We have made a series of calibration measurements of solutions containing different BPT concentrations in order to obtain a quantitative estimation of the amount of BPT molecules detected. For these calibration measurements, the fibers were placed in petri dishes containing solutions of BPT with concentrations ranging from 1.56×10^{-10} to 1.56×10^{-8} M. By plotting the increase in fluorescence from one concentration to the next versus the concentration of BPT, and fitting these data with an exponential function in order to simulate a saturated condition, a concentration of $(9.6 \pm 0.2) \times 10^{-11}$ M has been determined for BPT in the individual cell investigated (17–21). Optical nanobiosensors are capable of minimal-to-noninvasive analysis of single living cells as demonstrated in their applications in the measurement of carcinogenic compounds (31) and molecular pathways (23) within a single living cell.

11.4.3 Nanosensor for Monitoring pH in a Single Cell

A suitable plasmonics-active nanoprobe has been developed for intracellular measurement of pH in single living human cells using SERS detection (26–28). It is important to monitor intracellular pH values in cellular studies of protein structure, enzymatic catalysis, drug delivery and effects of cellular transport, and exposure to environmental toxicants. The fiberoptic nanoprobe was functionalized with thiolated ligands or labels, thereby imparting the molecular specificity required for intracellular biosensing detection. SERS-active nanoprobe insertion, interrogation, and subsequent removal from a cell (often <30 s) greatly reduced the possibility of silver nanoprobe degradation within the intracellular environment. The pH nanosensors were fabricated by tapering 400 mm core-diameter optical fibers using a commercially available pipette puller to produce nanoprobes smaller than 100 nm in diameter. These tapered optical fibers were then coated with a 6-nm mass thickness of silver using an electron beam evaporator. Following Ag island film (AgIF) deposition, the nanoprobes were functionalized in 10 mM paramercaptobenzoic acid (pMBA) dissolved in ethanol, which anchored pMBA to the AgIF via a silver-thiol covalent bond. The carboxyl group of pMBA is pH-sensitive, thereby rendering the nanoprobe pH-sensitive. The effectiveness and usefulness of the SERS nanoprobes were demonstrated in measurements of pH values in HMEC-15/hTERT immortalized "normal" human mammary epithelial cells and PC-3 human prostate cancer cells (26). The pH-sensitive nanoprobes were inserted into the cells using micromanipulators, the confocal Raman microscope was focused on the portion of the nanoprobe inserted into the cell, and single 10-s SERS spectra were acquired. SERS-active, pH-sensitive fiberoptic nanoprobes can be quickly and easily inserted into single living cells suspended in a supporting matrix or grown on microscope slides. The results indicate that nanoprobe insertion and interrogation provide a sensitive and selective means to monitor cellular microenvironments at the single-cell level.

11.5 Conclusion

Nanosensors could provide the tools to investigate important biological processes at the cellular level *in vivo*. Not only can antibodies be developed against specific epitopes, but also an array of antibodies can be established, so as to investigate the overall structural architecture of a given protein. For monitoring nonfluorescent analytes, the method of competitive binding may be used. Finally, the most significant advantage of the nano-sensors for cell monitoring is the minimal invasiveness of the technique. The integration of these advances in biotechnology and nanotechnology could lead to a new generation of nanosensor arrays with unprecedented sensitivity and selectivity to simultaneously probe subcompartments of living cells at the molecular level in a system approach. An important advantage of the optical sensing modality is the capability to measure biological parameters in a noninvasive or minimally invasive manner due to the very small size of the nanoprobe. Following measurements using the nanobiosensor, cells have been shown to survive and undergo mitosis. It was also shown in this study that the insertion of a nanobiosensor into a mammalian somatic cell not only appears to have no effect on the cell membrane, but also does not affect the cell's normal function. This was demonstrated by inserting a nanobiosensor into a cell that was just beginning to undergo mitosis and monitoring cell division following a 5-min incubation of the fiber in the cytoplasm and fluorescence measurement. The integration of advances in biotechnology and nanotech-nology could lead to a new generation of nanobiosensors with unprecedented sensitivity and selectivity to probe cells at the molecular level.

Acknowledgments

The author acknowledges the contribution of G. D. Griffin, J. P. Alarie, B. M. Cullum, P. Kasili, Y. Zhang, and J. Scaffidi.

References

1. Vo-Dinh, T., Tromberg, B. J., Griffin, G. D., Ambrose, K. R., Sepaniak, M. J., and Gardenshire, E. M. 1987. Antibody-based fiberoptics biosensor for the carcinogen benzo(a)pyrene. *Appl. Spectrosc.* 41, 735.
2. Vo-Dinh, T., Griffin, G. D., and Sepaniak, M. J. 1991. Fiberoptic immunosensors. In: *Chemical Sensors and Biosensors*, Wolfbeis, O. S., Ed., CRC Press, Boca Raton, Florida.
3. Vo-Dinh T., Sepaniak M. J., Griffin G. D., and Alarie J. P. 1993. Immunosensors: Principles and applications. *Immunomethods* 3, 85.
4. Alarie J. P. and Vo-Dinh T. 1996. An antibody-based submicron biosensor for BaP. *Polycycl. Aromat. Comp.* 8, 45.
5. Alarie, J. P. and Vo-Dinh, T. A. 1991. Fiberoptic cyclodextrin-based sensor. *Talanta* 38, 529.
6. Alarie, J. P., Sepaniak, M. J., and Vo-Dinh, T. 1990. Evaluation of antibody immobilization tech-niques for fiberoptics fluoroimmunosensor. *Anal. Chim. Acta* 229, 69.
7. Tromberg, B. J., Sepaniak, M. J., Alarie, J. P., Vo-Dinh, T., and Santella, R. M. 1998. Development of antibody-based fiberoptics sensor for the detection of benzo(a)pyrene metabolite. *Anal. Chem.* 60, 1901.
8. Alarie, J. P., Bowyer, J. R., Sepaniak, M. J., Hoyt, A. M., and Vo-Dinh, T. 1990. Fluorescence monitoring of benzo(a)pyrene metabolite using a regenerable immunochemical-based fiber-optic sensor. *Anal. Chim. Acta* 236, 237.
9. Bowyer, J. R., Alarie, J. P., Sepaniak, M. J., Vo-Dinh, T., and Thompson, R. Q. 1991. Construction and evaluation of regenerable, fluoroimmunochemical-based fiber optic biosensor. *Analyst* 116, 117.

10. Betzig, E., Trautman, J. K., Harris, T. D., Weiner, J. S., and Kostelak, R. L. 1991. Breaking the diffraction barrier—Optical microscopy on a nanometric scale. *Science* 251, 1468.
11. Betzig, E. and Chichester, R. J. 1993. Single molecules observed by near-field scanning optical microscopy. *Science* 262, 1422.
12. Zeisel, D., Deckert, V., Zenobi, R., and Vo-Dinh, T. 1998. Near-field surface-enhanced Raman spectroscopy of dye molecules adsorbed on silver island films. *Chem. Phys. Lett.* 283, 381.
13. Deckert, V., Zeisel, D., Zenobi, R., and Vo-Dinh, T. 1998. Near-field surface-enhanced Raman of DNA probes. *Anal. Chem.* 70, 2646.
14. Tan, W. H., Shi, Z. Y., and Kopelman, R. 1992. Development of submicron chemical fiber optic sensors. *Anal. Chem.* 64, 2985.
15. Tan, W. H., Shi, Z. Y., Smith, S., Birnbaum, D., and Kopelman, R. 1992. Submicrometer intracellular chemical optical fiber sensors. *Science* 258, 778.
16. Cullum, B., Griffin, G. D., Miller, G. H., and Vo-Dinh, T. 2000. Intracellular measurements in mammary carcinoma cells using fiberoptic nanosensors. *Anal. Biochem.* 277, 25.
17. Vo-Dinh, T., Griffin, G. D., Alarie, J. P., Cullum, B., Sumpter, B., and Noid, D. 2000. Development of nanosensors and bioprobes. *J. Nanopart. Res.* 2, 17–27.
18. Vo-Dinh, T. and Cullum, B. 2000. Biosensors and biochips, advances in biological and medical diagnostics. *Fresenius J. Anal. Chem.* 366, 540.
19. Cullum, B. and Vo-Dinh, T. 2000. Development of optical nanosensors for biological measurements. *Trends Biotechnol.* 18, 388.
20. Vo-Dinh T., Cullum, B. M., and Stokes, D. L. 2001. Nanosensors and biochips: Frontiers in biomolecular diagnostics. *Sens. Actuators* B74, 2.
21. Vo-Dinh, T., Alarie, J. P., Cullum, B., and Griffin, G. D. 2000. Antibody-based nanoprobe for measurements in a single cell. *Nat. Biotechnol.* 18, 76.
22. Kasili, P. M., Cullum, B. M., Griffin, J. D., and Vo-Dinh, T. 2002. Nanosensor for in-vivo measurement of the carcinogen benzo[a]pyrene in a single cell. *J. Nanosci. Nanotechnol.* 6, 653.
23. Kasili, P. M., Song, J. M., and Vo-Dinh, T. 2004. Optical sensor for the detection of caspase-9 activity in a single cell. *J. Am. Chem. Soc.* 126, 2799–2806.
24. Zhang, Y., Dhawan, A., and Vo-Dinh, T. 2010. Design and fabrication of fiberoptic nanoprobes for optical sensing. *Nanoscale Res. Lett.* 6. DOI: 10.1007/s11671-010-9744-5.
25. Vo-Dinh, T. and Zhang, Y. 2011. Single-cell monitoring using fiberoptic nanosensors. *Wiley Interdiscip. Rev. Nanomed. Nanobiotechnol.* 3, 79–85.
26. Scaffidi, J. S., Gregas, M., and Vo-Dinh, T. 2009. SERS fiberoptics nanoprobe for pH sensing in a single living cell. *Anal. Bioanal. Chem.* 393, 1135–1141.
27. Vo-Dinh, T., Scaffidi, J. S., Gregas, M., Zhang, Y., and Seewaldt, V. 2009. Applications of fiberoptics-based nanosensors to drug discovery. *Exp. Opin. Drug Discov.* 4, 889–900.
28. Vo-Dinh, T., Wang, H. N., and Scaffidi, J. S. 2010. Plasmonic nanoprobes for SERS biosensing and bioimaging. *J. Biophotonics* 3, 89–102.
29. Hoffmann, P., Dutoit, B., and Salathe, R. P. 1995. Comparison of mechanically drawn and protection layer chemically etched optical fiber tips. *Ultramicroscopy* 61(1–4), 165.
30. Lambelet, P., Sayah, A., Pfeffer, M., Philipona, C., and Marquis-Weible, F. 1998. Chemically etched fiber tips for near-field optical microscopy: A process for smoother tips. *Appl. Opt.* 37, 7289.
31. Vo-Dinh, T., Ed. 1989. *Chemical Analysis of Polycyclic Aromatic Compounds*, Wiley, New York.

12

Surface-Enhanced Fluorescence-Based Biosensors

Samuel Grésillon and Emmanuel Fort

CONTENTS

12.1 Introduction

Fluorescence imaging is today one of the most sensitive and widespread technique to observe living cells and organisms. Due to fast progresses in detection devices, light sources, optical lenses and filters, as well as in fluorescent probes, it is now possible to track minute concentrations of biomolecules, down to the single molecule, directly in living cells.

However, when tracking membranes processes, one really has to push fluorescence microscopy to its limit. The fluorescence coming from the inner part of the cell tends to blur the image and reduce the sensitivity. The diffraction-limited axial sectionings need to be significantly improved. Standard optical configurations such as wide field, two-photon, or even confocal microscopy are usually not sufficient to observe these membrane processes, which play crucial roles in most of the cell mechanisms.

Improving the sensitivity can be tackled by increasing the emission efficiency of the fluorophores at the membrane and reducing one of the fluorophores outside this region. This is called surface-enhanced fluorescence (SEF).

SEF is a fast-growing field aiming at controlling the local electromagnetic (EM) environment of the emitters to increase the detection efficiency of the fluorophores in 2D geometry. Controlling fluorescent processes and photon emission is not confined to bioapplications and is the subject of active research in numerous fields from solar cells (Li et al. 2015) to light sources (Lozano 2013) or integrated nanophotonics (Novotny and van Hulst 2011).

SEF is part of this general trend to engineer the emission of the fluorophore to improve its localization and detection. In the field of microscopy, it has permitted to break the diffraction limit and reach nanometric resolution. It has also enabled to dramatically enhance the detection efficiency. It has received much attention because it is one of its simple geometry and can still provide a significant enhancement in fluorescence detection.

One of the most promising aspect of SEF is its association with plasmonics (Fort and Grésillon 2008). Surface plasmons (SPs) are collective charge oscillation modes at the surface of conductors with unique light-interaction properties. Due to their high sensitivity to surface properties and surroundings, plasmonics allow to channel and concentrate light within subwavelength volumes (Barnes et al. 2003). These plasmonic surfaces have been used for a variety of optical applications where enhancement of the EM fields is of importance, in particular, nanosensors (Anker et al. 2008) and nonlinear spectroscopy phenomena such as surface-enhanced Raman scattering (Moskovits 1985, Schatz and Van Duyne 2002).

In this chapter, we present the general concepts of SEF before discussing in detail the effect of a simple continuous metallic thin film to improve detection and imaging at the interface. The understanding of the interplay between fluorophores, propagating photons, and the SP in its most basic geometry captures the essence of SEF. We show that this configuration provides significant improvements when applied to bioimaging.

This chapter is divided into further three sections. The first section introduces the principles of SEF and the second deals with SP coupling and its combined effect on fluorophores. The third section is devoted to analyzing the various cases of fluorescence enhancement near metallic surface for imaging applications.

12.2 SEF Principles

12.2.1 Principles of Fluorescence

Fluorescence is defined as the ability of any material to absorb light and subsequently undergo a radiative relaxation from an electronically singlet-excited state. The fluorescence cycle is usually illustrated by a Perrin–Jabłoński diagram. Figure 12.1a shows a simplified version of the diagram. The emitted fluorescent light is emitted at a longer wavelength. When a fluorophore absorbs light energy, it is usually excited to a higher vibrational energy level in the first S1 or second S2 electronically excited state before rapidly relaxing to the lowest exited energy level (*internal conversion*) in about a picosecond or less. Fluorescence lifetime Γ, defined as the average relaxation time to the electronic ground state, is typically four orders of magnitude slower than internal conversion, giving the fluorophore enough time to reach the thermally equilibrated lowest vibrational state of S1 (Lakowicz 1999).

From the EM point of view, the excited electronic states can be modeled as dipoles. The orientation of these dipoles is given by the electronic structure of a particular fluorophore. Hence, excitation and emission dipoles can have different orientation as shown for instance in Figure 12.1b, in the case of coumarin. Exciting the S2 level would result in two different excitation and emission dipole orientations while exciting the S1 level would result in a parallel configuration. As we will see in the next section, SEF processes depend dramatically on the orientation of both excitation and emission dipoles.

The fluorescence quantum yield Q is usually defined as the probability that a given excited fluorophore will produce a fluorescence photon (Lakowicz 1999). In this review, we will use a more general definition, which takes into account all the EM processes:

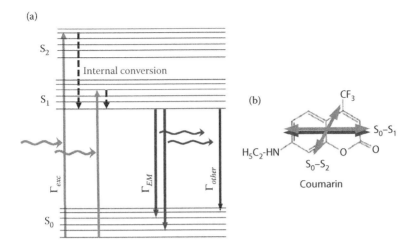

FIGURE 12.1
(a) Simplified Perrin–Jabłoński diagram illustrating the molecular processes involved during a fluorescence excitation and emission cycle. The molecular levels are represented as horizontal lines: S_0, S_1, and S_2 being, respectively, the fundamental electronic state, the first excited singlet state, and the second one. The vertical axis represents the energy of the levels. Γ_{exc}, Γ_{EM}, and Γ_{other} are the excitation rate, EM relaxation rate, and the non-EM relaxation rate, respectively. (b) Example of absorption and emission transition dipole orientations for coumarin molecule.

$$Q = \frac{\Gamma_{EM}}{\Gamma_{EM} + \Gamma_{other}} \tag{12.1}$$

where Γ_{EM} and Γ_{other} are the relaxation rate due to EM processes and to all other relaxation processes, respectively. EM processes include both radiative and nonradiative processes. Other processes are essentially molecular phenomena, induced for instance by collision with other molecules (Lichtman 2005). Note that in the cases of a free fluorophore in a homogeneous nonabsorbing medium, the upper definition of Q is recovered.

The presence of an interface in the vicinity of the fluorophore can dramatically alter the fluorescence processes through the modification of its local EM environment. Such changes induce modifications of the excitation as well as the emission processes. The internal conversion depending mainly on the fluorophore internal electronic structure is, in first approximation, insensitive to the modification of the local EM environment. Consequently, excitation and emission processes are independent and can be analyzed separately. In the following, we detail this influence.

12.2.2 Molecular Detection Efficiency

Molecular detection technique can be characterized by the relative molecular detection efficiency (MDE). MDE is defined by the number of detected photons versus the number of absorbed photons. It is given at position **r** and can be written as follows:

$$MDE(\mathbf{r}) = \underbrace{\Gamma_{exc}(\mathbf{r})}_{\text{excitation process}} \times \underbrace{Q(\mathbf{r}) \times MCE(\mathbf{r})}_{\text{emission process}} \tag{12.2}$$

The MDE function is the product of an excitation term given by the excitation rate $\Gamma_{exc}(\mathbf{r})$ and an emission term, which is the product of the fluorescence quantum yield $Q(\mathbf{r})$ by the molecular collection efficiency (MCE) function $MCE(\mathbf{r})$ at point $\mathbf{r}$. This latter quantity $MCE(\mathbf{r})$ is defined as the proportion of detected photons over all the emitted ones.

For the excitation, the SEF design must simply maximize the intensity at the fluorophore position $\mathbf{r}$ since $\Gamma_{exc}(\mathbf{r})$ increases with light intensity $I_{exc}(\mathbf{r})$. Besides, the electric field should be directed parallel to the excitation dipole of the fluorophore. However, in many cases, the fluorophore distribution is isotropic and this last requirement does not hold.

The ability for an interface to modify the relaxation processes, through the modification of Q, is more complex and less intuitive. Fluorescence emission is a spontaneous emission process hence its rate of relaxation is given by the coupling of the excited state of the molecule to the vacuum oscillations of the surrounding environment. This can be reformulated in classical terms: the probability to emit a photon is related to the local photonic mode density (PMD) (Barnes 1998). The fluorophore emission can consequently be controlled through the modification of the EM boundary conditions surrounding the fluorophore. The alteration of the optical mode structure available to a 3D-confined dipole in a small cavity has been studied as early as 1946 by Purcell (Purcell 1946). First experimental works were carried out by Drexhage in the early 1970s involving changes in the emission characteristics from fluorophores placed close to metallic interface (Drexhage 1974).

The fluorescence relaxation processes, which are associated with the radiative recombination of the excited electron–hole pair, compete with other nonfluorescent relaxation processes. SEF partly relies on the ability to tailor the local environment of a molecule to maximize the radiative relaxation rate as compared with one of the same fluorophore in free space. This is done while maintaining the competitive nonradiative processes to a low level. Basically, this ends up in reducing the fluorescence lifetime and increasing the quantum efficiency of the fluorophore. This is why SEF is sometimes abusively reduced to radiative decay engineering (RDE) (Lakowicz et al. 2003). Moreover, the efficient surface for SEF should also maximize the MDE through the redirection of the light toward the detector.

12.2.3 Radiative Decay Engineering Model

To evaluate the modification of the relaxation processes induced by the modification of the environment, a classical approach gives accurate results and simple understanding of the underlying principle. This model is sometimes designated as CPS model after Chance, Prock, and Silbey who introduced it 40 years ago (Chance et al. 1978). The excited fluorophore is modeled as an oscillating dipole with a dipolar moment parallel to one of the emission transition. The fluorophore spatial extension is thus neglected. The dynamics of the dipolar moment $\mathbf{p}$ is given by the fundamental law of dynamics applied to the excited electron:

$$\frac{d^2\mathbf{p}}{dt^2} + \Gamma_0 \frac{d\mathbf{p}}{dt} + \omega_0^2 \mathbf{p} = \frac{e^2}{m} \mathbf{E}_{loc}(\mathbf{r}_0) \tag{12.3}$$

where ω_0 and Γ_0 are the resonance frequency in free space and the oscillator damping rate in free space, respectively, and $\mathbf{E}_{loc}(\mathbf{r}_0)$ is the electric field at the dipole position $\mathbf{r}_0$ resulting from the feedback provided by the surrounding objects ($\mathbf{E}_{loc}(\mathbf{r}_0) = 0$ in free space). In the case of small damping, that is, $\Gamma_0 < 2\omega_0$, satisfying most of the experimental conditions, the solution is $\mathbf{p} \approx \mathbf{p}_0 \exp(-\Gamma t)\exp(i\omega t)$. The frequency shift is negligible $\omega \cong \omega_0$ and the emission rate is given by

$$\Gamma = \Gamma_0 + \frac{e^2}{m\omega_0 |\mathbf{p}|^2} \Im m(\mathbf{p}^* \cdot \mathbf{E}_{loc}(\mathbf{r}_0)) \tag{12.4}$$

where the subscript * stands for the complex conjugate.

This expression can be written using the electric field susceptibility, $\mathbf{S}(\mathbf{r}, \mathbf{r}_0, \omega_0)$ (Chance et al. 1978, Wylie and Sipe 1984, Carminati et al. 2006). The reflected electric field at position $\mathbf{r}$ is then defined by $\mathbf{E}_{loc}(\mathbf{r}) = \mathbf{S}(\mathbf{r}, \mathbf{r}_0, \omega_0)\,\mathbf{p}$. Moreover, using the energy conservation, the power radiated by an oscillating dipole is $\Gamma_0 = e^2\omega_0^2/6\pi m\varepsilon_0 c^3$. Hence, for a fluorophore with a quantum yield Q, the relative relaxation rate variation is

$$\frac{\Gamma}{\Gamma_0} = 1 + \frac{6\pi\varepsilon_0}{k^3} \Im m[\mathbf{u} \cdot \mathbf{S}(\mathbf{r}, \mathbf{r}_0, \omega_0) \cdot \mathbf{u}] \tag{12.5}$$

where $\mathbf{r}$ is the unit vector along the direction of the transition dipole $\mathbf{p}$. Hence, the modification of the emission rate only depends on the local surroundings through the tensor $\mathbf{S}$. It is noteworthy that the classical dipolar model gives similar relative variations of the emission rate to that obtained with a full quantum treatment (Barnes 1998). The main challenge of RDE for SEF is thus to control and design surface geometries to maximize the fluorescence enhancement by adjusting the local electric field $\mathbf{E}_{loc}$ (i.e., the PMD). The modified damping rate of the fluorophore Γ can be measured experimentally using fluorescence lifetime measurements since $\tau = 1/\Gamma$. Care should be taken in maximizing the EM relaxation processes since it includes evanescent components. Hence, the influence of the environment on the relaxation rate of a fluorophore must be analyzed in detail to determine what eventually contributes to the radiative emission of the fluorophore.

Using the CPS model, we can write the effect of a flat interface placed at a distance $\mathbf{d}$ of the fluorophore. As we will see, this simplest geometry has a dramatic effect on the fluorophore relaxation processes. It is convenient to evaluate the coupling as a function of the normalized in-plane component of the wavevector (u). The total modification of the spontaneous emission rate is evaluated by integrating the modification to the damping rate over all in-plane wavevectors (Chance et al. 1978). The modification to the relative damping rate for a dipole at distance d to the interface between two medium 1 and 2 strongly depends on the dipole orientation, parallel // or perpendicular $\perp$ to the surface. It is given by

$$\frac{\Gamma_\perp}{\Gamma_0} = 1 - \frac{3Q}{2} \Im m \int_0^\infty r_p e^{-\Delta\varphi l} \frac{u^3}{l_1} du \tag{12.6}$$

$$\frac{\Gamma_{//}}{\Gamma_0} = 1 - \frac{3Q}{4} \Im m \int_0^\infty \left[(1-u^2)r_p + r_s\right] e^{-\Delta\varphi l} \frac{u}{l_1} du \tag{12.7}$$

where r_p and r_s are the Fresnel reflection coefficients for the p- and s-polarization; $l_i = -i(\varepsilon_i/\varepsilon_1 - u^2)^{1/2}$ is the normalized component orthogonal to the surface; $\Delta\varphi_l = 2k_1l_1d$ is the phase shift associated with the light propagation from the fluorophore to the interface 1–2. In the following, we will study the simple case for which the surface is a metallic flat thin film that supports SP modes.

12.3 Surface Plasmon

12.3.1 Coupling with Propagative SP

A plasmon is a mode of collective excitation of free electrons in solids. SPs are the electron plasma oscillations at a metal surface that stem from the broken translational invariance in the direction perpendicular to the surface. The charge oscillations are perpendicular to the surface plane and induce an evanescent EM field at the surface with a transverse magnetic field. The association of the SP and its coupled EM field is called a surface plasmon polariton (SPP). Figure 12.2a shows a schematic of the SPP propagating at the metal–dielectric interface. $+$ and $-$ represent the regions with lower and higher electron density, respectively. SPPs are transverse magnetic modes, and the generation of surface charge requires an electric field normal to the surface. The EM field decreases exponentially with the distance $|z|$ from the surface on both sides of the interface.

Figure 12.2b shows the SPP dispersion curve, which gives the variation of the angular frequency ω as a function of their in-plane wavevector $(k_{//})$. For light propagating with an incident angle θ with the surface normal in a medium of relative permittivity ε_d, the dispersion relationship is simply given by $k_{//} = n_d \sin \theta\, \omega/c$ where c is the speed of light and $n_d = \sqrt{\varepsilon_d}$ is the refractive index of the medium. The dispersion relationship between the frequency and in-plane wavevector for SPPs propagating along the interface between a metal and a dielectric can be found by looking for surface mode solutions in Maxwell's equations under appropriate boundary conditions. The dispersion relation is given by the following equation (Raether 1988):

$$k_{SPP} = \frac{\omega}{c} \sqrt{\frac{\varepsilon_d\, \varepsilon_m}{\varepsilon_d + \varepsilon_m}} \tag{12.8}$$

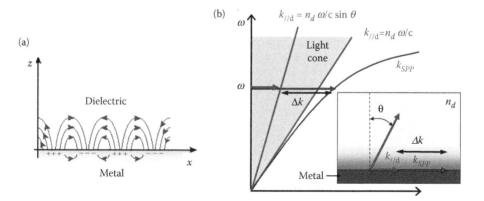

FIGURE 12.2
(a) Schematic representation of the SPP propagating along a metal–dielectric interface. $+$ and $-$ represent the regions with lower and higher electron density, respectively. SPPs are transverse magnetic modes; the generation of surface charge requires an electric field normal to the surface (represented by the red arrows). (b) The dispersion curve for an SP mode shows the momentum mismatch Δk that must be overcome in order to couple light and SP modes together. The SPP is always beyond the light line with a greater momentum (k_{SPP}) than a free-space photon $k_{//} \leq n_d \omega/c$ at the same frequency ω.

The dispersion curve for an SPP mode is always beyond the light line. SPP has a greater momentum (k_{SPP}) than a free-space photon (ω/c) of the same pulsation ω. The EM field associated with SP is evanescent and cannot propagate away from the surface.

This momentum mismatch must be overcome to couple together free-propagative light and SPP modes. This is of crucial importance in the use of SP for SEF for both excitation and relaxation processes. Indeed, SP offers an efficient way to excite fluorophores placed in their associated strong evanescent EM field. However, since fluorophores can also transfer a significant part of their energy to SPP through their near-field components, it will result in quenching of their fluorescence. Hence, energy flow from the fluorophore to the SPP modes is *a priori* detrimental to the SEF process. The need to ultimately recover this energy into light is thus essential.

In the following section, we will describe the various ways to couple SPP and propagative light.

12.3.2 SPP Excitation on Flat Metallic Surfaces

The coupling between propagative light and SPP on flat surfaces is essential for both exciting the fluorophore and enhancing the emission. The evanescent EM field associated with the SPP is maximum at the very surface of the metal and thus can be quite efficient for excitation processes. Conversely, the radiative recovery of the energy given to the SPP by near-field coupling with the fluorophore plays a crucial role in RDE.

SP on the front side of the sample can be excited by the rear side of the metallic thin film using the so-called Kretschmann–Raether configuration provided that the index of refraction of the material on the backside is higher than the one on the front side (Raether 1988). When impinging on the metallic thin film with an angle θ_0, the light in-plane wavevector is thus $k_{//} = n_s \sin\theta_0 \omega/c$. The dispersion relation for the SPPs propagating on the opposite side of the metallic thin film can thus be satisfied for a precise incident angle θ_{SP} as shown in Figure 12.3a. The excitation of the SPP can be recognized by a minimum in the reflected intensity (Raether 1988). At the resonant angle for SPP excitation, the reflected intensity equals zero for a specific thickness d_c of the metallic thin film (Raether 1988).

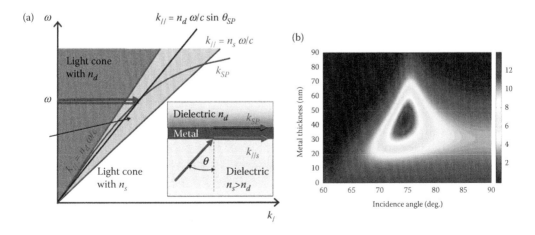

FIGURE 12.3
(a) Dispersion curve for the Kretschmann–Raether configuration showing the resonant coupling between light and SPP; (b) calculated enhancement.

The EM field reaches its maximum at the metallic surface on the opposite side. The field can be significantly enhanced compared with a similar configuration without metal (Enderlein 2005). The enhancement G is defined as the ratio of the transmitted EM intensity on the opposite surface of the metallic thin film T_{metal} over the one without metal $T_{no\ metal}$. It is obtained using Fresnel's equations for the three-layer system dielectric medium substrate (s)/metal (m)/dielectric medium (w) sample (Born and Wolf 1999):

$$G = \frac{T_{metal}}{T_{no\ metal}} = \frac{t_{sm}t_{mw}}{1 + r_{sm}r_{mw}e^{2ik_{zm}d}} \qquad (12.9)$$

where t_{ij} and r_{ij} are the p-polarized Fresnel's transmission and reflection coefficients, respectively, for one boundary between medium i and medium j for light incident from i to j (s: substrate; m: metal; w: second dielectric—usually water), d is the thickness of the metal thin film, and k_{zi} is the z component normal to the surface of the wave number of the light. The resonant excitation of the SPP on the opposite side of the metallic thin film induces a "cumulative" effect due to the charge excitation, which results in enhanced EM field at the metallic surface. The EM field increases throughout the films thickness to reach a maximum at the thin film surface (Raether 1988). The calculated enhancement values of the EM field are 14.3 for silver and 5.9 for gold, respectively, at 634 nm for an optimized metallic thin film thickness of about 50 nm. Far from the saturation regime of the fluorophore, this results in huge excitation rate enhancements (about 200-fold in the case of silver thin film).

It is of particular interest in nonlinear excitation processes, in particular for two-photon fluorescence (Goh et al. 2005).

12.4 Surface Enhancement Near Metal Surfaces

12.4.1 Relaxation Processes on Flat Metallic Thin Films

The vicinity of a fluorophore to a flat metallic surface offers additional loss channels due to the coupling with the SPP (Barnes 1998). As mentioned in the previous section, since the metal surface acts as a mirror, the emitter interferes with the reflected EM waves. The spontaneous emission rate oscillates with increasing distance depending on the relative phase of the reflected field.

For small emitter–surface distances, this simple picture needs to be refined since the excited molecule may decay nonradiatively via coupling to guided waves such as SPs and/or lossy waves through its near field. These three decay channels depend on the fluorophore–metal distance and on the dipole orientation. The classical model presented in the previous section allows to observe and evaluate quantitatively each relaxation channel through the amount of energy transfer Γ versus in-plane wavevector $k_{//}$ using Equations 12.6 and 12.7 in the case of a flat interface.

Figure 12.4 shows the relative energy flow versus the normalized in-plane wavevector $u = k_{//}/(n_d k_0)$ for various fluorophore/surface distances in the case of a silver/air interface, where n_d is the refraction index of the dielectric.

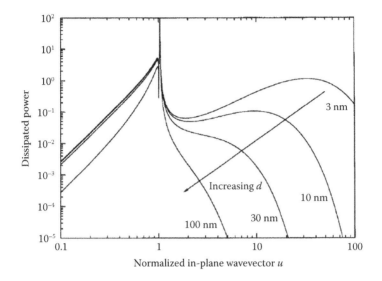

FIGURE 12.4
Power dissipated with an isotropic distribution of dipole orientation as function of the normalized in-plane wavevector u, calculated using Equations 12.6 and 12.7. The figure shows the power dissipated by the emitter for a range of emitter–surface separations, as indicated. The system consists of an emitter having a wavelength of emission 614 nm, immersed in vacuum and positioned above an Ag surface. (From Barnes W. L. 1998. *Journal of Modern Optics* 45:661–699. Reprinted with permission from Taylor & Francis Ltd.)

- *Emission of photons*: The process associated with the emission of photons in free space is given by the part of the curve $k_{//} < k_0$. Half of the total power is emitted in the upper half space and the other half is emitted toward the substrate. Hence, it can be divided into a reflected, transmitted, and an absorbed component, the relative amplitude of which depends on the nature of the substrate. In practice, when using an objective lens with numerical aperture *N.A.* placed above the substrate to collect the emitted light, the collected light power equals a sum of half the power represented by the curve, integrated over $(k_{//}/k_0) < N.A.$ plus the reflected part on the substrate (readily evaluated using Fresnel formulas). The angular emission of the fluorophore can also be obtained from Equations 6 and 7 since $k_{//} \cong \sin \theta$ (Drexhage 1974).

- *Coupling with SPP modes*: Excitation of SPP modes is clearly visible in Figure 12.4. It is associated to the sharp peak centered at the in-plane wavevector of the SPP k_{SPP} in the wavevector spectrum. The part of the energy flowing from the fluorophore to the SPP can be evaluated by integrating the curve around this value k_{SPP} (Chance et al. 1978). As the fluorophore–surface distance d decreases, its coupling with the SPP is getting increasingly important. SPP coupling is the prevailing process between 20 and 200 nm.

- *Nonradiative transition and exciton coupling*: For fluorophore–metal distances smaller than about 20 nm, fluorescence quenching originates from the transfer of energy from the excited dipole to the metallic surface through coupling with high in-plane wavevector values (Figure 12.4). The energy is transferred to electron–hole pairs of the metal (excitons), which act as energy acceptors. This energy is ultimately dissipated in the metal. This transfer is dipole–dipole in nature and yields a d^4 dependence for the transfer rate (when considering the planar nature of the surface). The relative part of this third channel can be evaluated by integrating the curve over $k_{//} > k_0$ subtracting the SPP channel.

Figure 12.5 shows the fraction of power dissipated into each channel for dipole orientation parallel and perpendicular to the surface. The dipoles that oscillate perpendicular to the surface are the one that can couple very efficiently to the SPP modes. In that case, up to 93% of the total dissipated power can be given to the SPP (Weber 1979).

Each relaxation channel competes with each other. The number of decaying processes increases as the fluorophore–surface distance decreases due to the presence of strong near-field coupling. This can be observed through lifetime measurements. The fluorescence lifetime drops dramatically for small metal–fluorophore distances. The lossy wave coupling is becoming the prevailing process for distances <20 nm even though the SPP coupling is getting stronger.

For fluorophores with $Q < 1$, the non-EM relaxation processes remain unaffected by the presence of the surface. Since the number of EM relaxation processes increases dramatically for small fluorophore–surface distance, the generalized quantum yield Q defined in Equation 12.1 increases toward unity (as the fluorescence lifetime τ tends toward zero). This is of practical importance since the photostability is related to the total time spent by the fluorophore in the excited state. Hence, for a given fluorophore, decreasing its lifetime permits to increase the number of cycles before photodestruction. This means collecting

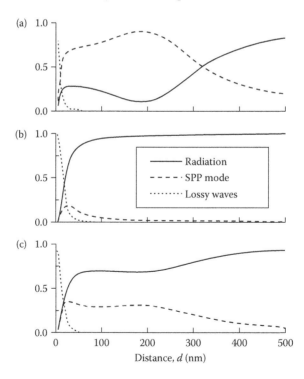

FIGURE 12.5
Normalized fraction of power dissipated in each relaxation channel, that is, radiation, SPP mode, and lossy waves, for various fluorophore/surface distances: (a) for a dipole-oriented normal to the surface, (b) parallel to the surface, and (c) for an isotropic distribution. (From Barnes W. L. 1998. *Journal of Modern Optics* 45:661–699. Reprinted with permission from Taylor & Francis Ltd.)

more emitted photon for proper experimental parameters before photodestruction. We must keep in mind, however, that these additional coupling processes through near-field components are *a priori* detrimental for fluorescence detection since they are related to nonradiative EM processes. Alteration in the photobleaching processes has been investigated both theoretically and experimentally (Barnes 1998, Vasilev et al. 2004). A good agreement between the classical EM model and the experimental results is found for metal/fluorophore distances larger than 10 nm. Essentially, for spacer layer below 20 nm, the strongly increased relaxation rate allows the fluorophores to undergo more excitation–emission cycles before photobleaching while the longer lifetime for thicker spacer layers is due to a decrease in excitation rate.

An important aspect of SEF is consequently to use proper surface structures to couple resonantly the SPP modes to radiative emission to recover the energy. Two main momentum-matching techniques are used to couple the SPP modes to photons. The first one consists in increasing the momentum of the photon by using high-index materials and specific optical geometry. The second technique is based on the braking of the translational invariance of the smooth metallic surface through the use of corrugated or roughened surfaces. Theses coupling techniques are the subject of the following sections.

For this excitation configuration, the fluorophore couples to the SPP resonantly through its near-field components with higher wavelengths. As mentioned before, it is necessary that the fluorophore be placed at a proper distance from the metallic surface to avoid nonradiative relaxation processes by use of a transparent dielectric spacer layer of typically a few tens of nanometer thick (Figure 12.5).

This optical configuration often designated as SP fluorescence spectroscopy (SPFS). It has been used in biological applications such as real-time monitoring of hybridization processes and for trace amount of PCR products in biosensors with detection limit of ~100 fmol or membrane processes (Liebermann and Knoll 2000, Yao et al. 2004, Morigaki and Tawa 2006).

The presence of the metallic thin films has a direct impact on the fluorescence excitation and fluorescence collection processes. A simple enhancement technique for fluorescent signals is to use simple interference effects induced by the mirror coating. The presence of standing waves can enhance the excitation process up to about fourfold (for a perfect mirror). The redirection of the emitted light on the mirror allows a more efficient collection which depends on the light detection geometry. The modification of the emission patterns due to planar boundaries was originally described by Drexhage (1974) and more recently exploited for semiconductor cavity design (Benisty et al. 1998). As mentioned previously, dielectric mirrors can also be used since the SP plays no role in this simple enhancement process (Choumane et al. 2005). The overall enhancement reaches typical values of 10–15-fold as compared with a standard glass slide.

The modification of the fluorescence lifetime and intensity on a single molecule as a function of the distance to a smooth metallic mirror has been investigated experimentally, recently (Buchler et al. 2005). The results agree with classical models showing a clear correlation between intensity and lifetime measurements.

Using SPP excitation on a flat metallic thin film, single-molecule real-time imaging of single fluorophores attached to protein molecules on metal surfaces in aqueous solution using SP resonance fluorescent microscopy has been reported (Yokota et al. 1998). Using such a configuration with silver thin film on Cyanin 5 dye marker, a signal enhancement of a factor 13 compared to glass substrate can be obtained. The enhancement is mainly due to the intensified evanescent field associated with the SP and the redirection of the emission signal toward the detector.

It is noteworthy to mention that fluorophores can also be used as local probes of the evanescent field associated with the SPP. SP fields can thus be imaged by detecting the fluorescence of molecular film close to the SP carrying metal surface, as shown by the field profile of SPP launched at lithographically designed nanoscopic defects (Ditlbacher et al. 2002).

12.4.2 Surface Plasmon Cross Emission

The core idea of surface plasmon cross emission (SPCE) is the recovery of the energy transferred by the fluorophore near field to the SP by coupling the SP to the radiating EM modes on the backside of the film.

As we mentioned previously, the fluorophore can couple resonantly to the SPP through its near-field components when placed in the vicinity of the metallic surface. The Kretschmann–Raether configuration permits to excite the SP efficiently by the rear side of the thin film. By use of reciprocity, the SPP couples back to the radiating EM on the rear side. For a particular emission wavelength, the light emitted is highly directional taking the shape of a cone with a half-angle of $\theta_{SPP}(\omega_{fluo})$ satisfying the SPP momentum conservation (Figure 12.3). The first observation of this coupling dates back to the years 1979 by the Weber's group (Weber and Eagen 1979) and has been more recently extensively studied by Lakowicz's group (Lakowicz 2004, Gryczynski et al. 2004).

The plasmon-coupled fluorescence emission exhibits a characteristic angular distribution and polarization (radial polarization), which are determined by the emission spectrum of the fluorophore and the dispersion relation of the SPs.

The excitation of the fluorophores can be performed either by the evanescent field associated with the SPP (excited in the Kretschmann–Raether configuration) or by shading light directly on the fluorophore (Gryczynski 2004). This latter geometry is less favorable since it does not take advantage of the huge SPP field enhancement and the EM near null at the metallic surface. The claimed experimental enhancements compare with standard optical geometry.

Stefani et al. (Stefani et al. 2005) have recently observed the fluorescence of single molecules in the vicinity of a thin gold film using an epi-illumination scanning confocal microscope with both excitation and emission mediated by the SPP. They show that the number of photons detectable from fluorophores perpendicular to the interface is enhanced by 140% by the presence of the metal while it is reduced to 26% for parallel molecules.

It was anticipated that SPCE could increase the fluorescence signals by a factor of up to 1000 (Lakowicz et al. 2003, Gryczynski 2004), which would be a revolutionary improvement for analytical assays. Recent experimental (Stefani 2005) and theoretical results (Calander 2004, Enderlein and Ruckstuhl 2005) contradict this claim. Enderlein and Ruckstuhl in a theoretical article (Enderlein and Ruckstuhl 2005) even claim that significantly less energy is coupled through a silver film into the glass than in the ideal case where no silver film is present. Note that, in the experiments, surface roughness is always present and may lead to additional enhancement through more enhanced electric fields and thus fluorescence excitation.

SPCE can be advantageous with respect to more conventional detection schemes because of the high directionality of fluorescence emission in SPCE, which may for instance help to better discriminate fluorescence, the strong wavelength-dependent angular position that makes it possible to use SPCE as a spectrally resolving technique and radial polarization of the emission (in contrast to emission generated at a glass/water interface).

Additionally, SPP can be coupled with more advanced fluorescence microscopy such as total internal reflection fluorescence (TIRF) (Axelrod 1981), a microscopy technique to illuminate selective emitters close to the surface using total internal reflection. TIRF is used in many biomedical applications (Groves et al. 2008). It is possible to replace the prism in TIRF by a high numerical aperture objective (He et al. 2010) coupled with plasmon to increase the fluorescence image. Through SPP, MDE is better than in standard TIRF. This technique is called SP-mediated fluorescence microscopy (SPMFM) and is particularly interesting to increase the signal-to-noise ratio. With the help of the metal film, photostability, signal, and confinement are improved (Balaa and Fort 2009).

12.4.3 SP Field-Enhanced Fluorescence on Textured and Corrugated Surfaces

In a typical interface between vacuum ($\varepsilon_d = 1$) and a metal with low losses ($\varepsilon_m = \varepsilon'_m + i\varepsilon''_m$ with $\varepsilon''_m << |\varepsilon'_m|$ and $|\varepsilon'_m| > 1$), the real part of the SPP effective refractive index $n_{SPP} = \sqrt{\varepsilon_d \varepsilon_m / \varepsilon_d + \varepsilon_m}$ is higher than that of the dielectric; SPPs are nonpropagative. The nonradiative nature of SPP prevents coupling with free photons but textured metal surface can bring the missing momentum to propagative wave. The momentum conservation is conserved with the breaking of the translational invariance, for instance, by adding a periodic wavelength scale corrugation to the metallic surface. For a corrugation of period a, the momentum conservation is given by

$$k'_{SPP} = k_{SPP} \pm n G_B$$

where $G_B = 2\pi / a$ is the grating wavevector and n is an integer.

It was demonstrated experimentally by Knoll et al. with a dye monolayers deposited on a silver grating (Knoll et al. 1981), followed by Adams et al. (1981). They confirmed the predictions of Aravind et al. 1981. Adams et al. (1982) and Sullivan et al. (1994) have later extended the study to thicker films to understand the scattering of waveguide modes.

Since the photonic mode density experienced by the emitters will be necessarily different from the planar case due to the different boundary conditions, the presence of the grating also affects the spontaneous emission. Both the lifetime dependence with distance and the spatial distribution of the emitted light are significantly changed upon the introduction of the corrugation, differently from the Bragg-scattered bound-mode features. It has been suggested that these perturbations arise from the interference of the dipole fields scattered by the grating with the propagating and reflected fields (Andrews and Barnes 2001).

The efficiency of the cross-coupling can also be enhanced by using corrugated metal surfaces. Wedge and Barnes showed that the fluorescence emission from a structure with a corrugated thin metal film is over 50 times greater than that from a similar planar structure and that samples containing a planar metal film with a corrugated dielectric overlayer have similar properties (Wedge 2004).

12.4.4 SP Field-Enhanced Fluorescence on Roughened Surfaces

The coupling between a corrugated surface and a fluorophore is also a simplified approach to study the coupling with a roughened surface. To a first approximation, for sufficiently small roughness, roughened surfaces can be considered as a multicorrugated surface (Raether 1988).

Okamoto et al. showed that rough metallic layers could enhance the light emission from InGaN quantum wells (Okamoto et al. 2004). Roughness and imperfections in evaporated metal coatings can efficiently scatter SPPs into radiative modes. Measuring the topography of an uncoated GaN surface and of a 50-nm-thick Ag film evaporated onto GaN using scanning electron microscopy (SEM), they obtained a modulation depth of the Ag surface of ∼30–40 nm with a length scale of a few hundred nanometers. Enhancement of light emission using a corrugated grating with similar parameters was found, while no enhancement at all for other grating parameters. It suggests that the size of the metal structures determines the SP–photon coupling and light extraction.

The metal nanostructures not only influence the excitation rate of the fluorophore but also have a dramatic effect on the redirection of the angular emission. The influence of nearby nanosized metal objects on the angular photon emission has been observed experimentally on a single molecule for various dipole orientations (Gersen et al. 2000). This property offers the possibility of enhancing the light collection by tailoring the local environment of a fluorophore.

For random media within nanometer scale structures, multiple light interferences induce strong localization of light (John 1990, Shalaev 2000) analogous to the Anderson localization of electron wave function (Gresillon et al. 1999). Furthermore, variations in the dielectric constant of the medium are large enough to realize sufficiently strong multiple scattering of light. Indeed, SPPs are scattered by surface roughness when the scattering is sufficiently strong (Bozhevolnyi et al. 2002).

Coupling between Si nanocrystals and a highly roughened surface made of nanoporous gold film was studied by Biteen et al. (2005). A fourfold enhancement in the fluorescence intensity at $\lambda = 780$ nm was demonstrated (Figure 12.6), related to the fourfold enhancement in radiative decay rate as a result of local field effects. Both the effective excitation cross section and quantum efficiency are enhanced. The role of strong localization of SPP on rough metallic films was confirmed by modeling the complex geometry of the nanoporous gold film by gold nanoparticles.

We showed how surface shaping, additional corrugation, and roughness improve fluorescence enhancement on flat surface. Structured films with desired properties amplify

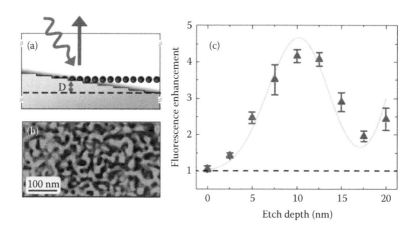

FIGURE 12.6

(a) Schematic cross section of the sample. Fluorescence excitation and measurements are made from the same side. Excitation wavelength is 488 nm. (b) SEM image of the nanoporous gold surface showing features. (c) Fluorescence enhancement, measured at ∼780 nm as a function of the etch depth, D (triangles). The solid line is a fit to the data using a model that accounts for the spatial distribution of Si nanocrystals and the enhanced local field. (Reprinted with permission from Biteen, J. S. et al. Enhanced radiative emission rate and quantum efficiency in coupled silicon nanocrystal-nanostructured gold emitters. *Nanoletters* 5:1768–1773. Copyright 2005 American Chemical Society.)

fluorescence through propagating SP modes, which act both on fluorescence emission rate and emission direction. It leads to advantageous solutions not only for biosensors but also for light emission devices or when improving collection efficiency is crucial. Additionally, using metal particles and their so-called localized plasmon resonance (Anker et al. 2008; Sepulveda et al. 2009) can further increase the ability to sense biomolecules.

References

Adams, A., J. Moreland, and P. K. Hansma. 1981. Angular resonances in the light emission from atoms near a grating. *Surface Science* 111:351.

Adams, A., J. Moreland, P. K. Hansma et al. 1982. Angular resonances in the emission from a dipole located near a grating. *Physical Review B* 25:3457–3461.

Andrew, P. and W. L. Barnes. 2001. Molecular fluorescence above metallic gratings. *Physical Review B* 64:125405.

Anker, J. N., W. P. Hall, O. Lyandres et al. 2008. Biosensing with plasmonic nanosensors. *Nature Materials* 7:442–453.

Aravind, P. K., E. Hood, and H. Metiu. 1981. Angular resonances in the emission from atoms near a grating. *Surface Science* 109:95–108.

Axelrod, D. 1981. Cell-substrate contacts illuminated by total internal reflection fluorescence. *Journal of Cell Biology* 89:141–145.

Balaa, K. and E. Fort. 2009. Surface plasmon enhanced TIRF imaging. *Imaging & Microscopy* 11:55–56.

Barnes, W. L. 1998. Fluorescence near interfaces: The role of photonic mode density. *Journal of Modern Optics* 45:661–699.

Barnes, W. L., A. Dereux, and T. W. Ebbesen. 2003. Surface plasmon subwavelength optics. *Nature* 424:824–830.

Benisty, H., R. Stanley, and M. Mayer. 1998. Methods of source terms for dipole emission modification in modes of arbitrary planar structures. *Journal of the Optical Society of America A* 15: 1192–1201.

Biteen, J. S., D. Pacifici, N. S. Lewis et al. 2005. Enhanced radiative emission rate and quantum efficiency in coupled silicon nanocrystal-nanostructured gold emitters. *Nanoletters* 5:1768–1773.

Born M. and E. Wolf. 1999. *Principles of Optics*. 7th edn. Cambridge: Cambridge University Press.

Bozhevolnyi, S. I., V. S. Volkov, and K. Leosson. 2002. Localization and waveguiding of surface plasmon polaritons in random nanostructures. *Physical Review Letters* 89:186801.

Buchler, B. C., T. Kalkbrenner, C. Hettich et al. 2005. Measuring the quantum efficiency of the optical emission of single radiating dipoles using a scanning mirror. *Physical Review Letters* 95:063003.

Calander, N. 2004. Theory and simulation of surface plasmon-coupled directional emission from fluorophores at planar structures. *Analytical Biochemistry* 76:2168–2173.

Carminati, R., J.-J. Greffet, and C. Henkel et al. 2006. Radiative and non-radiative decay of a single molecule close to a metallic nanoparticle. *Optic Communication* 261:368–375.

Chance, R. R., A. Prock, and R. Silbey. 1978. Molecular fluorescence and energy transfer near interfaces. In *Advances in Chemical Physics*, I. Prigogine and S. A. Rice, eds. New York: Wiley Interscience, 37:1–65.

Choumane, H., C. Nelep, N. Ha et al. 2005. Double interference fluorescence enhancement from reflective slides: Application to bi-color microarrays. *Applied Physics Letters* 87:031102.

Ditlbacher, H., J. R. Krenn, N. Felidj et al. 2002. Fluorescence imaging of surface plasmon fields. *Applied Physics Letters* 80:404–406.

Drexhage, K. H. 1974. Interaction of light with monomolecular dye layers. *Progress in Optics* 12:163–232.

Enderlein, J. and T. Ruckstuhl. 2005. The efficiency of surface-plasmon coupled emission for sensitive fluorescence detection. *Optics Express* 13:8856–8865.

Fort, E. and S. Grésillon. 2008. Surface enhanced fluorescence. *Journal of Physics D: Applied Physics* 41:013001.

Gersen, H., M. F. Garcia-Parajo, L. Novotny et al. 2000. Influencing the angular emission of a single molecule. *Physical Review Letters* 85:5312–5315.

Goh, J. Y. L., M. G. Somekh, C. W. See et al. 2005. Two-photon fluorescence surface wave microscopy. *Journal of Microscopy* 220:168–175.

Gresillon, S., L. Aigouy, A. C. Boccara et al. 1999. Experimental observation of localized optical excitation in random metal-dielectric films. *Physical Review Letters* 82:4520–4523.

Groves, J. T., R. Parthasarathy, and M. B. Forstner. 2008. Fluorescence imaging of membrane dynamics. *Annual Review of Biomedical Engineering* 10:311–338.

Gryczynski, I., J. Malicka, Z. Gryczynski et al. 2004. Radiative decay engineering: IV. Experimental studies of surface plasmon-coupled directional emission. *Analytical Biochemistry* 324:170–182.

He, R. Y., C. Y. Lin, Y. D. Su et al. 2010. Imaging live cell membranes via surface plasmon-enhanced fluorescence and phase microscopy. *Optics Express* 18:3649–3659.

John, S. 1990. The localization of waves in disordered media. In *Scattering and Localization of Classical Waves in Random Media,*. Ping Shen, ed. Singapore: World Scientific, 1–96.

Knoll, W., M. R. Philpott, J. D. Swalen et al. 1981. Emission of light from Ag metal gratings coated with dye monolayer assemblies. *Journal of Chemical Physics* 75:4795–4799.

Lakowicz, J. R. 1999. *Principle of Fluorescence Spectroscopy*. New York: Kluwer/Plenum.

Lakowicz, J. R. 2004. Radiative decay engineering: III. Surface plasmon-coupled directional emission. *Analytical Biochemistry* 324:153–169.

Lakowicz, J. R., J. Malicka, and I. Gryczynski et al. 2003. Radiative decay engineering: The role of photonic mode density in biotechnology. *Journal of Physics D: Applied Physics* 36:R240–R249.

Liebermann, T. and W. Knoll. 2000. Surface-plasmon field-enhanced fluorescence spectroscopy. *Colloids and Surfaces A* 171:115–130.

Li, J., S. K. Cushing, and F. Meng et al., 2015. Plasmon-induced resonance energy transfer for solar energy conversion. *Nature Photonics* 9:601–608.

Lichtman, J. W. and J.-A. Conchello. 2005. Fluorescence microscopy. *Nature Methods* 2:910–919.

Lozano, G., D. J. Louwers, S. R. K. Rodriguez et al. 2013. Plasmonics for solid-state lighting: Enhanced excitation and directional emission of highly efficient light sources. *Light: Science & Applications* e66.

Morigaki, K. and K. Tawa. 2006. Vesicle fusion studied by surface plasmon resonance and surface plasmon fluorescence spectroscopy. *Biophysics Journal* 91:1380–1387.

Moskovits, M. 1985. Surface-enhanced spectroscopy. *Review of Modern Physics* 57:783–826.

Novotny, L. and N. van Hulst. 2011. Antennas for light. *Nature Photonics* 5:83–90.

Okamoto, K., I. Niki, A. Shvartser et al. 2004. Surface-plasmon-enhanced light emitters based on InGaN quantum wells. *Nature Materials* 3:601–605.

Purcell, E. M. 1946. Spontaneous emission probabilities at radio frequencies. *Physical Review* 69:681.

Raether, H. 1988. Surface plasmon on smooth and rough surfaces and on gratings. In *Springer Tracts in Modern Physics*. Berlin: Springer, 111.

Schatz, C. C. and R. P. Van Duyne. 2002. *Handbook of Vibrational Spectroscopy*. New York: Wiley.

Sepulveda, B., P. C. Angelome, L. M. Lechuga et al. 2009. LSPR-based nanobiosensors. *Nano Today* 4:244–251.

Shalaev, V. M. 2000. *Nonlinear Optics of Random Media–Fractal Composites and Metal-Dielectric Films*. Berlin: Springer.

Stefani, F. D., K. Vasilev, N. Bocchio et al. 2005. Surface-plasmon-mediated single-molecule fluorescence through a thin metallic film. *Physical Review Letters* 94:023005.

Sullivan, K. G., O. King, C. Sigg et al. 1994. Directional enhanced fluorescence from molecules near a periodic surface. *Applied Optics* 33:2447–2454.

Vasilev, K., W. Knoll, and M. Kreiter. 2004. Fluorescence intensities of chromophores in front of a thin metal film. *Journal of Chemical Physics* 120:3439–3445.

Weber W. H. and C. F. Eagen. 1979. Energy transfer from an excited dye molecule to the surface plasmons of an adjacent metal. *Optics Letters* 4:236–238.

Wedge, S. and W. L. Barnes. 2004. Surface plasmon-polariton mediated light emission through thin metal films. *Optics Express* 12:3673.

Wylie, J. M. and J. E. Sipe. 1984. Quantum electrodynamics near an interface. *Physical Review A* 30:1185–1193.

Yao, D. F., F. Yu, J. Y. Kim et al. 2004. Surface-plasmon field-enhancement fluorescence spectroscopy in PCR product analysis by peptide acid probes. *Nucleic Acids Research* 32:177–192.

Yokota, H., K. Saito, and T. Yanagida. 1998. Single molecule imaging of fluorescently labeled proteins on metal by surface plasmons in aqueous solution. *Physic Review Letters* 80:4606.

[4]. Until today, this method has been the most popular technique for generating SPWs. Practical SPR systems for detecting chemical and biological agents were first demonstrated by Nylander and Liedberg in 1983 [5,6]. Since then, SPR-sensing techniques have attracted much attention in the scientific and instrumentation communities, especially for applications concerned with biological detection. The research for a biosensor that can measure molecular interactions of many types, like antibody–antigen, receptor–ligand, protein–DNA, and so on is always on the top of the medical healthcare list [7–9]. In recent years, R&D activities of SPR have been mainly directed toward biosensing applications, which include drug screening and clinical studies, food and environmental monitoring, and cell membrane mimicry. This is because of the potential of such sensors for applications in the health-related market. In the past 10 years, there have been an increasing number of companies offering commercial SPR biosensor systems targeting at customers conducting basic research in the field of life sciences. In fact, SPR biosensors have already become an important tool for characterizing and quantifying biomolecular interactions in many laboratories. Recent applications of the SPR-sensing technique have been expanding into the fields of environmental pollution, chemistry, theoretical physics, and experimental optics. Our literature search shows that the annual total number of research papers on SPR increased by almost 108-fold, from 6 to 651, during the period 1990–2002 [10–12]. PubMed, a search engine on biomedical literatures, with the keyword "SPR biosensor" resulted in >10,000 publications up to 2015. This clearly indicates the technological importance and application value of SPR sensors.

SPR sensors offer the capability of measuring very low levels of chemical and biological species near the sensing surface in real time, through monitoring the value refractive index within the vicinity of the sensor surface. Thus, any physical phenomenon at the surface that alters the refractive index will elicit a response. Moreover, since the sensor head is probed by an external optical beam, the front end of the system can operate in extreme environmental conditions such as high pressure and temperature. Until now, the SPR effect has already found application in a number of optoelectronic devices including light modulators [13,14], optical tunable filter [15,16], gas sensors [17,18], liquid sensors [19,20], biosensors [21,22], thin-film thickness detection, as well as SPR image (SPRi) [23–25].

In this chapter, we first describe the theory behind the SPR phenomenon and explain how effect may be used for sensing applications. Various optical-coupling schemes and their respective practical SPR-sensor designs for performing biomolecular sensing will be reviewed. Examples of application areas including drug discovery, clinical diagnostics, food testing and environmental monitoring, and cell membrane mimicry will also be presented. We must emphasize that a wide spectrum of applications is already in existence in the literature. Our list is by no means exhaustive.

13.2 SPR Phenomenon

The phenomenon of SPR is an excited charge density oscillation propagating along the boundary between a metal and a dielectric. The charge density oscillation can be induced by an optical wave, electromagnetic wave, or electron beam, etc. In this section, we shall discuss the coupling of SPR from an optical light beam and the conditions governing such coupling. The ATR scheme is a good example to illustrate the principle behind energy coupling from optical wave to SPR. The waveguide and grating coupling schemes will also be discussed.

13.2.1 Total Internal Reflection

Let us start from the phenomenon of total internal reflection (TIR) of light at the interface of two dielectric media. In this case, the phenomenon is described by Snell's law, as shown in Figure 13.1. Snell's law basically related the angles of incidence and refraction according to the following equations:

$$n_1 \sin\theta_i = n_2 \sin\theta_t, \text{ if } \theta_t = 90°, \text{ so } \theta_c = \sin^{-1}\left(\frac{n_2}{n_1}\right) \tag{13.1}$$

where θ_i and θ_t are the incident and transmitted angles, respectively. θ_c is the critical angle. There are three cases to be concerned with: (1) when the angle of incidence is less than the critical angle, the incoming light ray is split into two parts, the reflected ray and the refracted ray (Figure 13.1a); (2) when the angle of incidence is equal to the critical angle, the reflected light beam will propagate along the boundary between the two media; and (3) when the angle of incidence is larger than the critical angle, all of the incident light is reflected back into the high refractive index medium. And so, it is called TIR (Figure 13.1b).

During TIR, an interesting physical quantity called evanescent wave is also induced at the same time. The net energy of the light beam does not suffer any loss across the boundary of the two media when TIR occurs. But part of the electrical field intensity will continue to propagate into the lower refractive index medium. This evanescent wave has the same frequency as the incident light but its amplitude decreases exponentially with the distance from the boundary. Also interesting is the phenomenon that the evanescent wave can interact with a layer of conducting material deposited on the boundary interface if the layer is thin enough. In fact, the evanescent wave can penetrate into the metal layer and excite electromagnetic waves that propagate along the interface between the dielectric sample medium and the metallic layer. Such wave is due to oscillations of free electrons

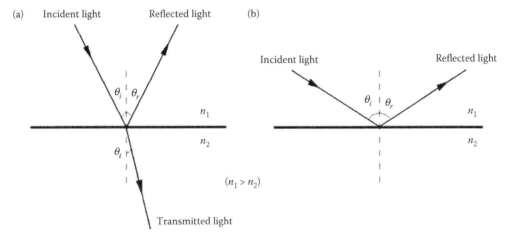

FIGURE 13.1
Light ray incident at an interface between two media, (a) reflection and refraction ($\theta_i < \theta_c$) and (b) total internal reflection ($\theta_i > \theta_c$).

in the surface of metal film. These electron oscillation waves are called SPWs, and their propagation characteristics are very similar to those of light waves.

13.2.2 Conditions for SPR

In order to efficiently couple the energy from photons to SPW, the orientation of the incident light of electric field vector must be equal to the orientation of free electrons oscillation on the metal film. This means that SPWs can only be excited by p-polarized light of the metal film. Figure 13.2 helps us understand the situation better. The optical wave with an electric field normal to the boundary and propagation along the x-axis is called transverse magnetic (TM) polarization. Others refer to it as p-polarization. The s-polarized, transverse electric (TE) polarization cannot couple into the plasmon mode since its electric field vector is oriented parallel to the metal film. For nonmagnetic metals such as gold and silver, the surface plasmon mode will also be p-polarized, and it will create an enhanced evanescent wave field. The energy becomes heat and then it disappears. In Figure 13.2, we show the excitation of evanescent wave, which has its intensity enhanced because of the presence of SP and penetrates into the dielectric medium.

To understand the conditions of optical excitation of SPR, the momentum of the incident light and the propagation constant of SPW must be considered. In mathematical concept, the momentum of the incident light can be illustrated in the form of a vector. This vector can be resolved conceptually into two components: one is parallel while the other is perpendicular to the metal-dielectric boundary (Figure 13.3). The magnitudes of these two components can be modified by varying the incident angle (θ_i). As for the SPW, its propagation is confined to the boundary. The momentum of this wave can be affected by factors such as thickness of the metal layer and dielectric constant of the metal film and its surrounding media. The mathematical treatment will be further described in Section 13.2.3.

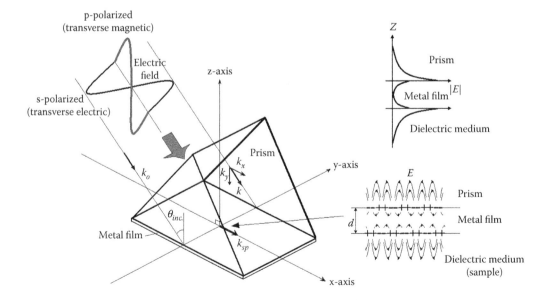

FIGURE 13.2
Definition of p- and s-polarization and the enhanced evanescent wave generated by SPR penetrates of the order of one wavelength into the medium on the opposite side, decaying exponentially with distance from the surface.

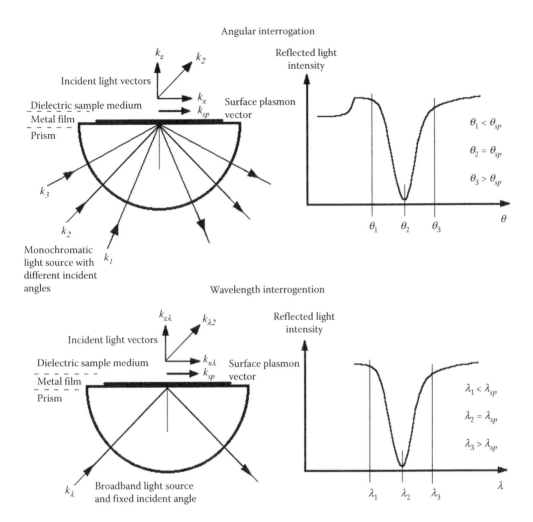

FIGURE 13.3
Angular and wavelength interrogation schemes are commonly used in practical SPR-sensing instruments.

SPW can be induced in the conducting materials (e.g., gold and silver) only when the momentum of the incident light vector parallel to the boundary is matched to the momentum of the SPWs. Under such conditions, the energy of the incident light is transferred to SPWs. Therefore, the intensity of the light reflected from the surface is reduced. For this reason, the amplitude of the wave vector in the plane of the metallic film depends on the angle (θ_{spr}) at which it strikes the interface. An evanescent (decaying) electrical field associated with the plasmon wave extends for a short distance into the medium from the metallic film. Because of this, the resonant frequency of the SPW (and thus θ_{spr}) depends on the refractive index of the medium. Therefore, it can be applied on sensing applications with extremely high sensitivity.

There are two simple ways to modify the momentum vector of the incident light parallel to the boundary matched to the momentum of the SP. One is to vary the incident angle, which modifies the relative magnitudes of the vector between the parallel and perpendicular components to the boundary; the other is to change the wavelength, which changes the incident photon energy and thus the momentum. As shown in Figure 13.3, these two approaches are commonly known as angular interrogation and wavelength interrogation. In SPR sensors based on angular interrogation, the light used has a fixed wavelength (monochromatic light). The reflected light beam experiences a dip when the incident angle of light is scanned. Here, the angle of the SPR absorption dip shift corresponds to the variation of dielectric constant of the sample medium near the metal surface. In wavelength interrogation, the spectral absorption response exhibits an absorption dip and the location of the dip are closely related to the refractive index of the sample medium (see Figure 13.3). Any refractive index variation in the sample medium will lead to a shift in the absorption dip.

13.2.3 Wave Vectors

To better illustrate the SPR phenomenon, resonance conditions are explained with the use of wave vectors. Wave vectors are mathematical expressions that describe the propagation of light and other electromagnetic phenomena. Under SPR, a high concentration of the electromagnetic field associated with the SPW exists in both media, as described in Figure 13.2. Therefore, the condition for resonant interaction between an optical wave and the SPW is very sensitive to the change of optical properties in the sample medium. The relationship between the wave vector of the incident light and that parallel to the interface (k_x) is given by the following equation:

$$k_x = k_o n_{glass} \sin(\theta_{inc}) \tag{13.2}$$

where k_o is the free space wavevector of the optical wave, n_{glass} the refractive index of the prism, and θ_{inc} the angle of incidence. An approximation of the SPW wave vector (k_{sp}) is given by

$$k_{sp} = k_o \sqrt{\frac{\varepsilon_{metal} \varepsilon_{sample}}{\varepsilon_{metal} + \varepsilon_{sample}}} \tag{13.3}$$

where ε_{metal} and ε_{sample} are, respectively, the dielectric constants of metal and the sample medium. When SPW excitation occurs, we have k_x equal to k_{sp}. Disregarding the imaginary portion of ε, k_{sp} can be rewritten as follows:

$$k_{sp} = \frac{2\pi}{\lambda} \sqrt{\frac{n_{metal}^2 \cdot n_{sample}^2}{n_{metal}^2 + n_{sample}^2}} \tag{13.4}$$

where $n_{metal} = (\varepsilon_{metal})^{1/2}$ is the refractive index of the metal and $n_{sample} = (\varepsilon_{sample})^{1/2}$ is the refractive index of the sample.

Since the dielectric constant of glass prism, metal film, and sample are functions of wavelength λ, of light, the conditions (dispersion relation) for surface plasmon excitation at the interface between metal and dielectric are given by

$$n_{glass} \cdot \sin \theta = \sqrt{\frac{\varepsilon_{metal} \cdot \varepsilon_{sample}}{\varepsilon_{metal} + \varepsilon_{sample}}} \tag{13.5}$$

The imaginary component of the complex refractive index term is represented by absorbance of light, the unit of measurement commonly encountered in Beer's law for transmission-based absorbance spectrophotometers. The sensing technique that is carried out in the majority of SPR applications detects the real refractive index change due to chemical or biochemical interactions. Therefore, the equations used here will neglect the imaginary component. The other properties of SPW are surface plasmon propagation and the decay length of the field into the metal film and a dielectric layer [26]. The propagation length of surface plasmon is related to the imaginary part of the surface plasmon wave vector, $k_{sp} = k_{sp}^{real} + ik_{sp}^{imag.}$, using the following equation:

$$\delta_{sp} = \frac{1}{2k_{sp}^{imag.}} = \frac{c}{\omega} \left(\frac{\varepsilon_{metal}^{real} + \varepsilon_{sample}}{\varepsilon_{metal}^{real} \varepsilon_{sample}} \right)^{3/2} \frac{\left(\varepsilon_{metal}^{real} \right)^2}{\varepsilon_{metal}^{imag.}} \tag{13.6}$$

where dielectric function of metal film can be expressed as $\varepsilon_{metal} = \varepsilon_{metal}^{real} + i\varepsilon_{metal}^{imag.}$. The decay length of the field penetrated into the dielectric medium is of the order of half the wavelength of light involved and the decay length into the metal film is determined by the skin depth.

13.2.4 SPR Described by Fresnel's Theory

In SPR sensors, a simple prism-coupling scheme, Kretschmann configuration, can be used to enhance wave vector momentum to permit coupling to the SPW. To study the reflected light in the Kretschmann configuration, one can start by analyzing the multiple reflections inside a simple three-layer system (prism–metal–sample), as illustrated in Figure 13.4.

For understanding the SPR reflection curve, a theoretical treatment based on Fresnel's theory of light reflection in a multilayered system is desirable [27]. For the present case, the reflection coefficient, r_{123}, of the optical light is given by

$$r_{123} = \frac{r_{12} + r_{23} \exp(2ik_{z1}d)}{1 + r_{12}r_{23} \exp(2ik_{z1}d)} \tag{13.7}$$

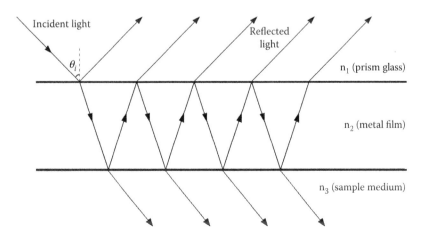

FIGURE 13.4
Ray tracing of multiple reflection of a three-layer system.

the efficiency of coupling between the SPs and the incoming light beam. Because of these factors, this configuration is not frequently used in real sensing system [4,28].

An alternative configuration was introduced by Erwin Kretschmann in 1971, and since then this has become a very common approach for optical excitation of SP (Figure 13.5b). The evanescent wave penetrates into the other side of a thin metal layer and directly excites SPs. The evanescent field created at the interface between the prism and the dielectric medium decays exponentially into the metal film and the SPs are excited at the boundary between the metal and the dielectric medium, as shown in Figure 13.2. The reasons of Kretschmann configuration being so widely used in real applications over the Otto configuration are due to its higher efficiency in SP coupling and the generated SP is in direct contact with the ambient medium. To keep the conditions of a thin metal film deposited on the prism base is much easier than to make a precise gap in nanometer scale as required by the Otto configuration. It is important to mention that unit until now the Kretschmann configuration plays an important role in the development of practical SPR measurement systems for chemical and biological sensing applications.

13.3.2 Coupling Scheme Using Grating Couplers

The operation of grating couplers may be explained by first showing the dispersion relations of an SPW propagating in a planar metal–dielectric interface, as shown in Figure 13.6. The wave vector of light is always less than the wave vector of the SPW for all frequencies. As previously explained, SPW can only be excited when the wave vector of the incident light is enhanced. Apart from using a high refractive index prism, which has been explained in previous sections, one can increase the value of wave vector using a scattering

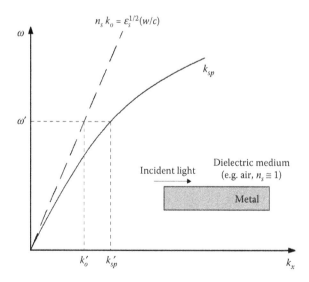

FIGURE 13.6
The dispersion curve showing that the momentum mismatch problem between incident light SPW propagating in a planar metal–dielectric interface, $k_{sp} = k_o(\varepsilon_m \varepsilon_S / \varepsilon_m + \varepsilon_S)^{1/2}$, where the permittivities of the dielectric and metallic media are $\varepsilon_S = n_S^2$ and ε_m, respectively, the wave vector of the surface plasmon wave is k_{sp}, the wave vector of the incident light is $k_o = (\varpi/c)$. The dashed line shows the maximum possible value of wave vector of an incoming photon propagating parallel to the interface. The momentum of SPW ($\hbar k_{sp}'$) is always larger than that of free space photon ($\hbar k o'$) of the same frequency (ω').

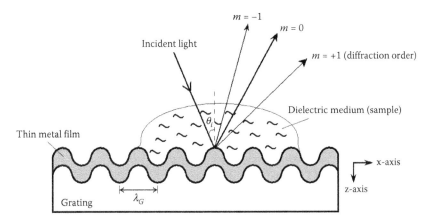

FIGURE 13.7
Typical grating coupler-based SPR configuration.

effect. When the optical beam is allowed to impinge on a periodically distorted surface, as shown in Figure 13.7, optical diffraction will occur. This will generate a series of diffracted beams, which exit the surface at angles very different from the angle of reflection, thus indicating that the wave vector has been modified by the grating structure.

Here, a portion of the wave vectors may propagate along the x-axis in which the grooves of a diffraction grating are oriented perpendicular to the plane of incidence [29]. The component of diffracted light along the interface (x-axis) is altered as follows:

$$k_{xm} = \frac{2\pi}{\lambda} n_S \sin\theta_i + mG = k_{sp} \tag{13.12}$$

where n_s is the refractive index of dielectric medium (sample), θ_i is the angle of incidence of the optical light, m is an integer, G is the grating wave vector ($G = 2\pi/\lambda_G$), λ_G is grating constant (the pitch of the grating), and k_{xm} is the wave vector of the diffracted optical wave. It is assumed that the dispersion properties of the SPW are not disturbed by the grating. Momentum conservation for an optical wave exciting an SPW via a diffraction grating may be rewritten as [30]

$$n_S \sin\theta_i + m\frac{\lambda}{\lambda_G} = \pm\sqrt{\frac{\varepsilon_m \varepsilon_S}{\varepsilon_m + \varepsilon_S}} \tag{13.13}$$

where for the sign "±" on right side of equation, '+' corresponds to the diffraction order $m > 0$ or "−" corresponds to diffraction order $m < 0$. Similar to prism coupler scheme, the energy of incoming photon coupled to the surface plasmon in grating-based optical structures can be observed by monitoring the variation of the minimum in reflection at a certain incident angle with a fixed wavelength [31], or the wavelength at a fixed angle of incidence [32–34], or the reflected light intensity variation at SPR [35].

13.3.3 Coupling Scheme Using Optical Waveguides

In optical waveguides, photons can propagate for a long distance with minimal loss. The use of optical waveguides in practical SPR-sensing system offers several advantages:

small size, ruggedness, and the ease to control the optical path within the sensor system. In principle, SPW excitation in an optical waveguide structure is similar to that in the ATR-coupler scheme. When a thin metal film is deposited on a waveguide, as shown in Figure 13.8, the evanescent wave generated by TIR may be able to penetrate through the metal layer and interact with the metal as well as the dielectric medium above. This means that it is possible to excite an SPW at the outer surface of the thin metal film if the SPW and the guided optical mode are phase-matched. Theoretically, the best achievable sensitivity factor of waveguide-based SPR devices can be the same as that of corresponding ATR configurations when they are put under similar operation conditions.

For practical systems, step index optical fibers may be used. A simple modification of the fiber is required for generating the required SPW. A short length of fiber cladding is polished away in order to expose the core. An evanescent wave may leak out from the core. When a thin metal layer is deposited on the polished surface, the evanescent wave will generate SPW on the outer surface, leading to a simple miniaturized fiber-based SPR-sensor device.

Despite its simplicity, the waveguide SPR sensor scheme does have its limitations. Since the incident angle at the metal–dielectric interface cannot be changed, measurement range (i.e., dynamic range) of such sensors may be limited. For minimizing this problem, performance parameters including metal film thickness, choice of material for the metallic layer (e.g., Au vs. Ag), and wavelength of the light source should be carefully selected. For this reason, the wavelength interrogation technique is commonly used in waveguide SPR-sensing devices. A broadband light source illuminates the fiber end and the SPR response can be obtained by analyzing the spectral attenuation characteristics of the exit beam [37]. It should be mentioned that when one uses a multimode instead of a single-mode optical fiber, the output signal tends to fluctuate because mechanical vibration can disturb the modal distribution of light in the fiber and this effect is particularly strong in the sensor surface [38,39]. Once the procedures are right, optical fibers can be easily adapted to SPR sensors. The first practical fiber-coupled SPR system was demonstrated by Jorgenson and Yee [40] in 1993.

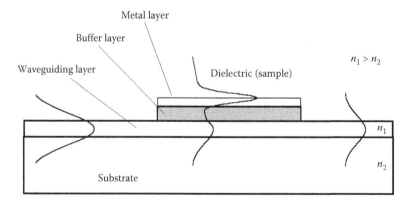

FIGURE 13.8
The concept of optical waveguides is based on the total internal reflection phenomenon. An optical wave propagates through the planar waveguide and couples some of its energy into a guided surface plasmon coupled mode and back. (Adapted from M. Weisser, B. Menges, and S.M. Neher, *Sensors and Actuators B* 56, 1999, 189–197.)

13.4 SPR Signal Detection Schemes

The reason why SPR is so useful for sensing applications is twofold. First, the resonance condition is extremely sensitive to any changes in refractive index. Second, the metal surface can be conveniently made to be in contact with a liquid medium so that any material bound to the surface may lead to a strong signal, thereby making the device very appropriate for detection affinity reactions between any target biomolecular species. The shift of resonance usually accompanies with a change in the coupling efficiency from photon energy to SPW. The theoretical aspects of this effect have been dealt with in Section 13.2. In this section, we shall present the various techniques for extracting the SPR information, namely, the angular [17,20], wavelength [19,24], or phase [41,42] interrogation schemes. Practical systems that are commercially available will also be reviewed.

13.4.1 Angular Interrogation

The angular interrogation scheme involves measuring the reflectivity variation with respect to incident angle under monochromatic light illumination. In Figure 13.9a, we show some numerical simulation results obtained from solving the Fresnel's equations for an SPR system when we vary the incident angle. The sample medium is water with refractive index 1.3333. The absorption dip where the reflected light intensity is at its minimum is often called the SPR-coupling angle (θ_{spr}) and θ_{spr} is shown to change in accordance with a change in the dielectric constant value of the sample medium from 1.3333 refractive index unit (RIU) to 1.3403 RIU (Figure 13.9b). In this simulation, we also need to input other parameters including (i) the light source being an He–Ne laser operating at 632.8 nm, the gold film being 50 nm thick, and the prism being made from BK7 glass.

13.4.1.1 Commercial Systems Based on Angular Interrogation

GE healthcare Biacore biosensor is the pioneer of commercialization of SPR-based biosensing devices. The first product based on angular interrogation scheme was launched in 1990

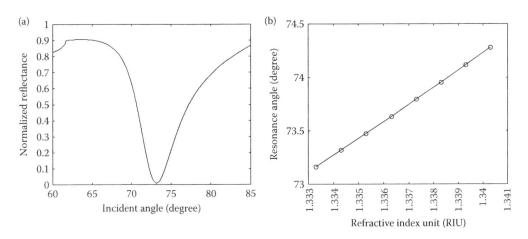

FIGURE 13.9
Typical angular SPR response curve and results: (a) angular SPR absorption dip observed in p-polarized reflected light; (b) shift of the resonant angle caused by variation of refractive index of sample medium.

[43–45]. Several generations of Biacore instruments including Biacore (series 3000, 4000) as well as the Biacore X, Biacore S200, and Biacore T200 are in existence in the market. They offer varying degrees of automation and parameter specifications. The optical design of Biacore [44] is shown in Figure 13.10. When a monochromatic light source is focused onto the metal film, a high-resolution photodetector array detects the variation of reflectivity with respect to angle. This can eliminate the use of high precise mechanical rotational stage. Most Biacore systems are designed to work in a refractive index range from 1.33 to 1.40 RIU, making them well suited for sensing of biomolecular interaction in aqueous media.

Many successful SPR-biosensing instruments are already in existence in the market. Table 13.1 provides a brief comparison between the SPR instruments supplied by different manufacturers. The performance of a SPR-biosensing system is concerned with its minimum resolvable refractive index change of sample medium. This value is mainly governed by the signal-to-noise (S/N) ratio and drift of components. Another figure of merit is concerned with minimum sample volume consumed by each measurement. Often biorelated materials do come with a very small quantity. In order to ensure minimal sample wastage, microfluidic system may be incorporated. Since the SPR effect itself is a temperature-sensitive phenomenon, it is also necessary to actively control the ambient temperature. Furthermore, by reducing the size of the entire system, unwanted drift may also be reduced. Some companies, like Biacore, also supply pretreated biosensor chips with well-defined surface chemistry to fit different applications

13.4.2 Wavelength Interrogation

The wavelength interrogation technique operates through the combined effect due to spectral dispersion (i.e., variation of dielectric permittivity in relation to wavelength) of the prism, metal film, and sample medium. In this case, one can use a fixed illumination angle and simply observe the variation of reflectivity at different wavelengths using a spectrometer. A spectral absorption dip signifies the presence of SPR. The spectral location of the dip is defined as the resonant wavelength, λ_{spr}. To perform the simulation of spectral

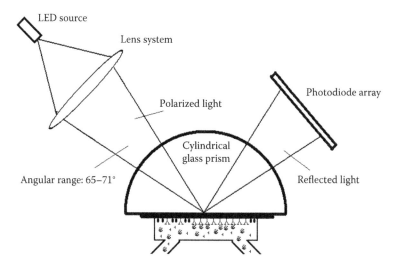

FIGURE 13.10
Diagrammatic illustration of biomolecule detection using a commercial SPR instrument (Biacore).

TABLE 13.1

Comparison of Several Commercially Available SPR Biosensors

Model	Biacore 3000 (Prism-Based SPR)	IBIS MX96 (Vibrating Mirror SPR)	Plasmoon (Broad-Range SPR)	Bio-Rad XSPR36 (Prism-Based SPR)	IASys (Resonant Mirror)
Flow-injection analysis (FIA) system	√	√	×	√	√
Temperature control	√	√	√	√	√
Multiplex	×	√	√	√	×
Microfluidics	√	√	×	√	×
Disposable sensing element	√	√	√	√	√
Refractive index range	1.33–1.40	1.33–1.43	1.33–1.48	1.33–1.37	–
Minimum sensitivity (RIU)	3×10^{-7}	2×10^{-6}	6×10^{-6}	3×10^{-7}	$>1 \times 10^{-6}$

Source: Adapted from J. Melendez et al., *Sensors and Actuators B* 35–36 (1996) 212–216; R.L. Rich, D.G. Myszka, *Journal of Molecular Recognition* 14, 2008, 355–400.

SPR response curve, one can solve the Fresnel's equations using dispersion characteristics of each of the materials involved. A set of simulated SPR response curves is shown in Figure 13.11a. The SPR system is the same as the one used in the previous case, as shown in Figure 13.9, and the angle of incidence is 70.0°. If we plot resonant wavelength, λ_{spr}, versus refractive index of the sample medium, as shown in Figure 13.11b, we can see that the spectral dip shifts toward longer wavelength as the refractive index of the sample medium gradually increases.

13.4.2.1 Practical Systems Based on Wavelength Interrogation

For the hardware configuration of wavelength interrogation, both ATR prism coupler and waveguide coupler schemes may be used. However, most of the reported systems are based on ATR prism coupler scheme. The difference is that the light source used a broadband

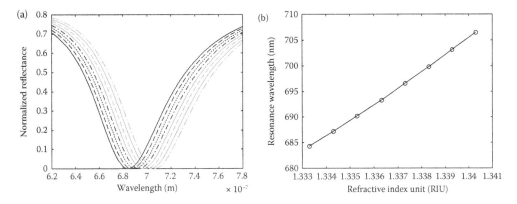

FIGURE 13.11
Simulation results of SPR wavelength interrogation: (a) typical spectral SPR response curves; (b) plot between resonant wavelength of absorption dips as obtained from simulation curves shown in (a) and refractive index of sample medium.

light source and the incident angle of light is fixed. This makes the optical instrumentation less complicated since there will be no need for any rotational stage for angular scanning. As for the broadband light source, a halogen lamp may be a good choice as long as proper calibration of the spectral characteristics has been conducted during the setup stage of the instrument. Typically, an optical fiber is used for carrying the light source to the entrance of the sensor head [21,24]. Apart from flexibility, optical fiber offers inherent beam profile shaping and isolation of heat from the lamp. One important point to be mentioned is that the beam entering the prism has to be collimated in order to ensure a sharp spectral dip, which provides minimum measurement error. The exit beam from sensor head is then analyzed using common spectrometer. The resolution of the spectrometer also defines the sensitivity limit of the instrument. As an effort to further simplify the optical design of wavelength interrogation scheme, we previously reported that a white light emitting diode may be taken as a direct replacement of the halogen lamp and fiber coupler system [48]. The approach has significantly reduced the size and cost of the instrument and hence improved the marketability of wavelength interrogation SPR systems. As for the spectrometer, high resolution ones is the obvious choice, but cost is an unfavorable factor. It has been suggested that low resolution spectrometers (16-channel) may be used instead of high resolution ones (1024-channel) without much compromise in performance. The rationale is that SPR response curves are fairly well-defined absorption curves that can be fitted with a set of invariant curve-fitting parameters. It has been shown that the switch to 16-channel spectrometers can still provide a measurement resolution of the resonant dip of 0.02 nm [49]. With 0.02 nm as the wavelength measurement resolution, simulation results indicate that the sensitivity limit of wavelength interrogation scheme is 2×10^{-5} RIU when the spectral window is around 630 nm [10]. Longer operation wavelength can provide a better sensitivity [50]. An improved sensitivity resolution of 1×10^{-6} RIU can be achieved by operating the system in a spectral region around 850 nm.

In fact, wavelength interrogation has an advantage over waveguide SPR-sensing devices since this scheme can be operated on fixed incident angle of light. Based on the linewidth of broadband light source wide enough, the wavelength interrogation method can provide a wide sensing range on waveguide coupler scheme. It is an important parameter for general applications' use of SPR-sensing devices. Fiber SPR-sensing devices can provide the highest degree of miniaturization as small as ~ 2 µm to be demonstrated [51]. The small physical size of SPR-sensing device can provide high potential on widely real application. The waveguide devices, optical fibers, and optical waveguide are commonly designed as SPR sensors by coating a thin metal film on around the exposed core. The sensitivity of these sensors based on wavelength interrogation typically can achieve the level as 10^{-5}–10^{-6} RIU with a wide sensing range [19,52,53]. One of the reasons for limited system sensitivity is that the inherent modal noise in multimode fibers causes the strength of the interaction between the fiber-guided light wave and SPW to fluctuate. In order to break through the limitation, a single-mode polarization-maintaining optical fiber was proposed as the substrate. This fiber optic SPR sensor is successfully demonstrated to resolve the change of refractive index as low as 4×10^{-6} RIU under moderate fiber deformations [53]. The waveguide coupler scheme is not only applied on fiber optics but also on multichannel planar light pipe sensing substrate [54], channel waveguide [52], and miniaturization of side active retroreflector [19].

13.4.3 Phase Interrogation

The fact that SPR is a resonant effect also means that as the system goes in and out of resonance, the phase of the incident optical wave experiences a massive jump. The steepness

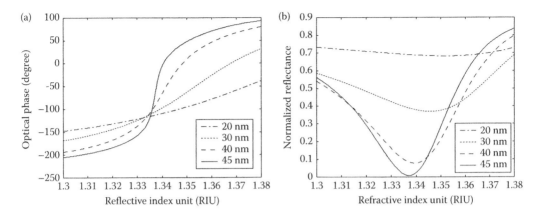

FIGURE 13.12
SPR phase and reflectivity responses versus refractive index with various thicknesses of gold layer from 20 to 45 nm: (a) phase; (b) amplitude.

and extent of the phase jump depends on the materials chosen and the thickness of the metal film. It is interesting to note that since SPR occurs only in p-polarization and not s-polarization, any phase change caused by SPR actually appears as retardation variation on the beam, that is, the optical beam goes from linearly polarized to elliptically polarized. Nonetheless, it has been reported that because of the very steep phase change across resonance [55,56], the measurement of SPR phase leads to the best sensitivity factor in comparison to angular and spectral techniques. Figure 13.12 shows a set of simulated response curves for different gold film thicknesses. The ATR prism substrate is BK7. The conditions of the simulation are $\lambda = 633$ nm and $\theta = 73.5°$, with the sensor structure being the same as the previous examples.

13.4.3.1 Practical Systems Based on Phase Interrogation

The first practical SPR phase sensing system reported by Nelson et al. in 1996 was based on a heterodyne phase detection scheme [42]. An acousto-optic modulator (AOM) frequency modulates a 45° polarized He–Ne laser at a frequency close to 100 MHz. In this setup, the AOM modulated input beam is split into two parts, one as the reference while the other as the signal beam, which goes through the ATR prism sensor head. The phase difference between the signals detected from the reference and signal beams provides the phase change induced by the SPR effect. The phase change can then be related to the refractive index of sample medium. The refractive index resolution of this system has been reported to be as high as 5×10^{-7} RIU. An alternative design based on the use of a Zeeman laser, in which the laser itself provides a built-in signal modulation, has simplified the system considerably [57]. The SPR phase-detection technique has also been applied to fiber SPR biosensor with encouraging success [58]. This system offers a sensitivity factor of 2×10^{-6} RIU. Other configurations based on the Mach–Zehnder interferometer have also been reported [59–63]. In particular, we recently demonstrated a differential phase system in which a sensitivity factor as high as 5.5×10^{-8} RIU is possible [61]. Figure 13.13 shows the optical setup of our design. The main contribution of this design is that both the reference and signal beams go through identical optical paths except for the short region between the output Wollaston prism and the two photodetectors. This ensures that much of the noise present in the system will be common to reference as well as signal channels. The

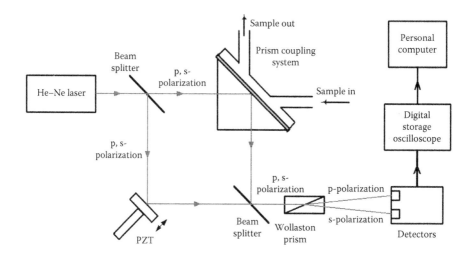

FIGURE 13.13
Differential phase SPR sensor based on the Mach–Zehnder interferometer configuration. (Adapted from H.P. Ho et al., *Biosensors and Bioelectronics* 20, 2005, 2177–2180.)

phase difference between the two channels will eliminate such common-mode fluctuations. Experimental stability of $\sim0.01°$ phase fluctuation over a period of 1 h has been demonstrated [62]. Apart from the much increased sensitivity factor, another very important benefit is the ease to perform imaging, which means that a large number of sensor sites may be monitored simultaneously. Finally, the use of Mach–Zehnder interferometer in a channel waveguide SPR sensor structure also highlights the promising possibility of building very compact biosensors on a planar waveguide device [63].

Other than the single-pass Mach–Zehnder and the double-pass Michelson configurations, a new multipass Fabry–Perot configuration with differential phase interrogation provides an intrinsic phase amplification based on the accumulation effect of the light beam hitting the sensor surface multiple times. In the Fabry–Perot configuration shown in Figure 13.14, the SPR sensor is placed in a signal arm of the interferometer, the interrogating optical beam will traverse the sensor surface an infinite number of times [64].

However, the high-detection sensitivity factor of 10^{-8} RIU is restricted to a narrow dynamic range 10^{-4} RIU of measurement because of the highly nonlinear nature of SPR phase jump, which limits its biomolecular sensing in clinical samples of wide dynamic concentration range [65,66]. Therefore, white light source for SPR excitation can expand the dynamic range of phase-sensitive region without any compromise in phase detection resolution [67–69]. The corresponding SPR phase change at the optimized coupling wavelength with fixed angle of incidence across the visible spectrum can be recorded to offer the optimal sensitivity factor of 10^{-7} RIU over a dynamic range of 10^{-2} RIU [67].

Other than Mach–Zehnder configuration, common-path configuration in Figure 13.15 is a simple optical setup for phase interrogation SPR based on the Kretschmann configuration [70,71]. Spectral oscillation is resulted by the optical path difference (OPD), determined by the thickness of birefringent crystal, between the p- and s-polarizations traversing the system in common-path SPR spectral interferometry. To further enhance its sensitivity, a temporal carrier in kilohertz such as liquid crystal modulator is employed to introduce spectral-temporal signal to solve the difficulty that phase jump at optimal wavelength is sharp to cause a discontinuity in spectral phase, where true phase signal cannot be computed. With this setup, the practical sensitivity factor reaches 2×10^{-8} RIU, which is an

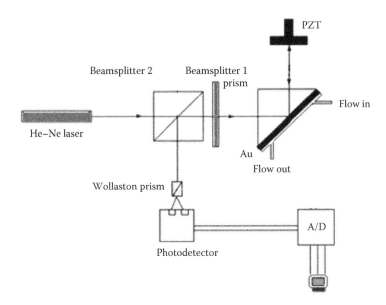

FIGURE 13.14
Schematic of the multipass Fabry–Perot interferometer. (Adapted from H.P. Ho et al., *Optics Communications* 276, 2007, 491–496.)

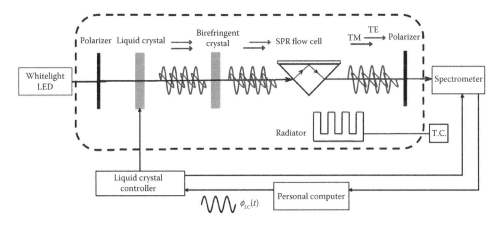

FIGURE 13.15
Schematic of the common-path SPR spectral interferometer with the liquid crystal modulator to generate the temporal carrier. (Adapted from S.P. Ng et al., *Optics Express* 17, 2013, 20268–20273.)

order of magnitude more sensitive over Michelson configuration setup where the sensitivity factor is 1×10^{-7} RIU [71].

13.5 Biomolecule Sensing Applications

Application of SPR-based sensors to biomolecular interaction monitoring was first demonstrated in 1983 [5]. The first biospecific interaction real-time analysis method appeared

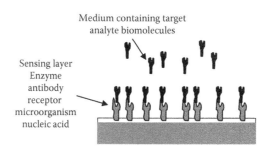

FIGURE 13.16
Typical biosensor surface containing a binding sensing layer, which has specific affinity toward target analyte biomolecules.

[72] in 1994. Since then, extensive experiments have been performed on all kinds of bio-related species to establish a better understanding of binding reactions. Literally, most of the common biomolecules-related life science possessing some form of specific binding would have been tried one way or another. The main reason behind the expansion is that SPR biosensing is a real-time and label-free technique, thus making it well suited for studying reaction kinetics and affinity constants. Affinity-based biosensor systems commonly make use of an immobilized molecule with specific recognition properties to monitor binding events on the sensing surface. A simple illustration is shown in Figure 13.16. Biological-binding reactions such as antibody/antigen, ligand/receptor, and protein/DNA interactions are among the commonest experiments. While the general effort on studying interaction processes between various biomolecular species is ongoing, there is also the trend to push for the detection of low-molecular-weight materials such as DNA. Recently, researchers also use SPR-sensing technology for drug screening and clinical studies [73–76], food and environmental monitoring [44,77–80], and cell membrane mimicry [81–85].

13.5.1 Biomedical Research Applications

Quantitative detection of target biomarkers relies on the recognition element, such as antibody, aptamer, and enzyme, on SPR sensing surface to capture the target biomarkers. Small-sized biomolecule adenosine, which is the basic element to construct RNA, was detected by SPR with aptamer as recognition element. Meanwhile, DNA and protein SPR array have been constructed for high-throughput detection of target DNA and proteins [86–90]. Different protein SPR arrays have been characterized on its sensitivity and specificity intensively these years by sensing interaction between immunoglobulin G (IgG) and its corresponding antibody [88–90]. In pharmaceutical research on drug screening, drug action on cancer response, and drug targeting site have been studied using SPR biosensing. For investigation of drug action on cancer response, cytochrome *c*, an apoptotic biomarker of cancer cell under anti-cancer drug treatment, has been quantified with SPR biosensor functionalized by cytochrome *c* recognizing aptamer [70]. For understanding of drug-targeting site, potential target site such as protein receptor epidermal growth factor (EGFR) functionalized on SPR-sensing surface to screen which compounds could bind with it [91]. Moreover, conventional biochemical assay on drug screening with small molecule promiscuous binders to inhibit target enzyme led to false-positive result by using nonspecific aggregation-type binding to enzymatic sites, which could be overcome by SPR sensing to discriminate various binding mechanisms, that is, specific inhibition on enzyme or aggregation-type binding [92].

13.5.2 Healthcare Applications

Healthcare application is a major aim for bench-to-bedside that benefits most people in the society. Examples include disease diagnosis and regular health monitoring. Rapid detection of disease biomarkers allows early diagnosis for proper treatment before the disease gets worse. Biomarkers of bacterial [93] and viral [94–96] disease have been detected using SPR biosensor. For bacteria, early diagnosis of infection with tuberculosis (TB), the top 10 infectious bacterial killer in developing counties, is done by using SPR with antibody against CFP-10, the tissue-fluid biomarker of TB infection [93]. For virus, protein biomarkers, such as hepatitis B virus surface antigen [94], DNA biomarker such as HIV genomic DNA fragments [95], and microRNA signature such as influenza-induced human microRNA expression quantity [96], have been targeted by using antibodies and oligonucleotide probes, respectively. Meanwhile, cancer diagnosis can be done by detecting DNA or protein biomarker such as mutated DNA sequences or cancer-induced overexpressed proteins [97,98]. An SPR biosensor to detect cancer protein biomarker beta-2 microglobulin is the first one integrated into smartphones for portable use [98]. Moreover, signaling molecules in humans could also be monitored via SPR biosensing [99–101]. For example, human choriogonadotropin (hCG), a signaling molecule that is a sign of pregnancy in female and is also overexpresssed for some types of testicular cancer in male, could be detected with anti-hCG antibody-coated SPR sensing surface to achieve the detection concentration limit of lower than 500 ng/mL [99]. Human interferon gamma (hIFNy), a signal at the site of inflammation, as well as the indicator on diagnosis of rheumatoid arthritis, could be detected with engineered protein derived from albumin-binding domain (ABD) of protein G coated SPR-sensing surface at a concentration as low as 0.2 nM [100]. In addition, allergy also causes inflammation. Therefore, human immunoglobulin E, the antibody responsible for allergy, in serum is screened with SPR technique, with high specificity to discriminate from immunoglobulin isoform human immunoglobulin G, the omnipresent antibody for immunity.

13.5.3 Food Safety Applications

Food safety is essential nowadays since more processed food is available in the market. Food-borne disease causing bacteria, illegal food additives, and food containing toxin are critical issues for food safety [102–104]. Rapid detection of the above three can eliminate most of the food-caused illnesses and casualties. Food-borne disease, bacteria *E. coli* O157:H7 being the most common cause, can be prevented by rapid screening on the food products. A bacterial SPR biosensor to detect *E. coli* O157:H7 directly has been demonstrated [102]. Specificity has been achieved by using cocktail of antibody to construct a sandwich assay to target *E. coli* O157:H7 detection from other bacteria. Onsite detection of melamine, an illegal adulteration in protein products such as milk and infant formula, is becoming necessary these years since its abuse is toxic to humans, especially infants. Portable biosensor SensiQ Discovery system using anti-melamine antibody to target melamine within 15 min with detection limit of 0.02 µg/mL has been developed [103]. Yet, practical detection on food products is still under development. Meanwhile, paralytic shellfish toxins (PSTs) contaminated seafood is conventionally detected using mouse model to assess the toxicity, but the ethical issue and routine use have been challenged. Therefore, an SPR biosensor with optimized assay condition to construct a robust, rapid, and sensitive measurement of PSTs with detection limits well below the regulatory action level that induces toxicity in human has been developed [104].

13.6 Commercial SPR Sensors

SPR sensors have in fact become standard biophysical tools in both academic and industrial research laboratories. Some manufacturers nowadays provide different levels of optical SPR biosensor solutions, from complete to simple miniaturized modules, to suit customers' needs. Table 13.2 provides a list of SPR sensor manufacturers and their websites.

Complete commercial solutions of SPR biosensors consist of three parts: hardware, data analysis program, and sensing chip coated with specific surface chemistry. As far as hardware is concerned, the instrument is required to have high sample measurement speed, full automation, and high sensitivity factor. For the data analysis software, a curve-fitting process is often used to study the biomolecular interaction model such as (A + B = AB), etc., and then calculate the reaction rate constants and the binding coefficients. Software simulations may help user obtain a best reaction model to fit the experimental results. In order to improve measurement consistency, some manufacturers supply surface-treated sensing chips for most common applications. Table 13.3 provides a list of the surfaces available from Biacore and affinity sensors together with their general applications.

From a literature review conducted for commercial optical biosensors in 2007 [47], the majority of reported articles (over 85%) were performed by Biacore systems and the investigation included proteins, antibodies, receptors, peptides, small molecules, oligonucleotides, lipids and self-assembled monolayers, extracellular matrix, carbohydrates, particles and viruses, crude analytes, and other configurations. One rapid developing area is array-based SPR platform, that is, SPRi. It allows real-time detection across the entire sensing surface. Therefore, it offers high throughput analysis of interaction of many targets simultaneously. As for reported cases of biomolecular sensing using SPR, the majority of the experiments performed are on biological binding reactions such as antibody/antigen, ligand/receptor, and protein/DNA interactions. While this general effort on studying interactions is ongoing, there is a growing trend to push for the detection of low-molecular-weight materials for which SPR has been thought to be not a suitable choice due to weak responses. Optimization of detection-sensing surface has been done to suit different applications. One example is the development of different chain length and density of

TABLE 13.2

Information of Commercial SPR Sensing Instruments

Manufacturer	Instrument	Website
GE Healthcare	Biacore 4000, S200, T200, X100…	http://www.biacore.com/
Bio-Rad Laboratories	ProteinOn™ XSPR36	http://www.bio-rad.com/
Affinite Instruments	P4SPR	http://affiniteinstruments.com/
IBIS Technologies	IBIS MX96	http://www.ibis-spr.nl/
Analytical μ-Systems	BIO-SUPLAR 6	http://www.biosuplar.de/
OWLS Sensors	OWLS 200	http://www.owls-sensors.com/
SensiQ Technologies	SensiQ Pioneer	https://www.sensiqtech.com/
Reichert Technologies	Reichert4SPR	http://www.reichertspr.com/
Farfield	IAsys	http://www.farfield-group.com/iasys.asp
GWC Technologies	SPRimager®II	http://www.gwcinstruments.com/
Nicoya	OpenSPR	http://www.nicoyalife.com/

Source: C.L. Baird, D.G. Myszka, *Journal of Molecular Recognition* 14, 2001, 261–268; R.L. Rich, D.G. Myszka, *Journal of Molecular Recognition* 14, 2008, 355–400.

TABLE 13.3

Available SPR-Sensing Surfaces

Chemistry	General Application
Biacore	
CM5—carboxymethyl dextran	Routine analysis
CM7—short chain carboxymethyl dextran	Large particulate analysis
CM7—high capacity carboxymethyl dextran	Small molecule analysis
SA—streptavidin	Biotin conjugation
NTA—nickel chelation	His-tagged conjugation
HPA—hydrophobic monolayer	Create hybrid lipid bilayers
C1—flat carboxymethylated	No dextran
Au—gold surface	User-defined surface
L1—lipophilic dextran	Capture liposomes
Affinity Sensors Surfaces	
CM—carboxymethyl dextran	Routine analysis
Hydrophobic planar	Create lipid monolayer
Amino planar	Alternative coupling chemistry
Carboxylate planar	No dextran
Biotinylated planar	Streptavidin conjugation

Source: R.L. Rich, D.G. Myszka, *Current Opinion in Biotechnology* 11, 2000, 54–61.

carboxymethyl dextran such as CM3 and CM7 chip from Biacore. It allows different levels of immobilization (low or high) to recognition elements to capture target biomolecule of different molecular weights (large or small). As detection instrument improves and better choice of biomolecules has been adopted, real-time label-free detection of low-molecular-weight species is now becoming possible.

Accuracy is an important issue when focusing on biosensing, especially for biomedical applications. It lies on both system stability and correct signal analysis. For system stability, the systematic error should be addressed to each SPR sensor before biosensing applications. Because there is a high variation of target biomolecules in biological samples, large systematic error will lead to large error and variation readout signal to cause false positive or negative mis-interpretations. For correct signal analysis, applying appropriate fitting models is one of the most important step [47]. For example, 1:1 interaction model is applied to kinetic analysis of direct ligand–target binding. Yet, bulk-shift is a common scenario resulting in sudden SPR signal change during biosensing because the SPR running buffer is largely mismatched from biological samples. Interpretation of SPR signal should be carefully done by bulk-shift correction to reveal the true SPR signal before applying fitting curve.

13.7 Conclusion and Future Trends

Biosensor based on SPR phenomenon is continuously growing and improving both in terms of applications and instrumentation development. Sensing applications have already expanded from basic research of biomolecular interaction investigation such as antibody/antigen, ligand/receptor, and protein/polynucleotide interactions to more practical applications such as drug screening, clinical studies, food and environmental monitoring, and cell membrane properties. One important area attracting much attention is integrating SPR with other analytical techniques. For example, the surface-enhanced laser desorption/ionization (SELDI) technology takes advantage of the specific affinity between certain biomolecules so that after immobilization of the target species on an SPR sensor surface, a pulsed laser is then used to ablate the material and the ionized molecules are analyzed by a mass spectrometer. This enables researchers to gain full knowledge of the binding specificity and the mass of the final product [105,106]. Another example is combining SPR with fluorescence studies. The immobilized biomolecules are tagged with a dye and the SPR phenomena will directionally couple out the fluorescence in a narrow range of angles and thus enhance the collection efficiency of the device [107,108]. The future trend of personalized biosensor will be the integration between the SPR-sensing unit with portable detector and analyzer, that is, smart devices such as smart phones and smart watches [109,110]. In addition, discovery of novel data analytical methods [71,111] and sensing surface materials [112,113] that enhance SPR signal will also be the future trend for robust and sensitive biomolecule sensing. Major commercially available sensing surfaces are coated with pure gold. Researches have shown an additional layer of materials, such as silicon oxide layer, graphene sheet, which could increase not only the sensitivity of SPR, but also tolerance to different chemicals from samples to prevent the gold layer from lifting off [111–113]. Moreover, a new trend for future biosensing is the use of localized surface plasmon resonance (LSPR) especially by nanomaterials [114,115]. LSPR provides multiple advantages over current SPR technique. First, temperature controller is no longer needed to stabilize temperature for low fluctuation signal, as LSPR is insensitive to temperature change. Second, nanomaterials such as nanogold coated with recognition element present in solution stage allows simple high-throughput solution-based target detection system, and the signal readout can be obvious color change to allow simple absorbance scanning detection by simple spectrophotometer or camera in smart phone. As the demand from healthcare market grows, it is likely that SPR instruments will go beyond research and academic communities. New designs with multiple analyte-detection capability for industrial, hospital, and home applications such as patient monitoring instruments, clinical diagnostics, food quality control equipment, pollution control devices, and home-use healthcare products may emerge. All these market forces will drive hardware improvement toward low cost, high sensitivity, stable performance, small size, high sample turnover, full automation, user-friendly operation, and minimal sample consumption. A number of publications have reported encouraging results on SPRi sensors [60,89,99,116,117]. 2D arrayed format where multiple biomolecules being measured simultaneously will be common with SPRi. Also, as important is the design of biomolecular interactions, which involve customizing affinity reactions through engineering of biomolecules and optimization of sensor surface chemistry. Nonetheless, the fact that the healthcare market is continuously expanding in light of an aging world population, SPR biosensing technology will continue its current trend of expansion.

References

1. R.W. Wood, On a remarkable case of uneven distribution of light in a diffraction grating spectrum, *Proceedings of the Physical Society of London* 4, 1902, 396–402.

2. R.H. Ritchie, Plasma losses by fast electrons in thin films, *Physical Review* 106, 1957, 874–881.
3. A. Otto, Excitation of surface plasma waves in silver by the method of frustrated total reflection, *Zeitschrift für Physik* 216, 1968, 395–410.
4. E. Kretschmann, H. Raether, Radiative decay of non-radiative surface plasmons excited by light, *Zeitschrift für Naturforsch* 23A, 1968, 2135–2136.
5. B. Liedberg, C. Nylander, I. Lundstrm, Surface plasmons resonance for gas detection and biosensing, *Sensors and Actuators* 4, 1983, 299–304.
6. C. Nylander, B. Liedberg, T. Lind, Gas detection by means of surface plasmons resonance, *Sensors and Actuators* 3, 1982, 79–88.
7. S.E. Harding, B.Z. Chowdhry, *Protein-Ligand Interactions: Hydrodynamics and Calorimetry*, New York: Oxford University Press, 2001.
8. T.P. Vikinge, *Surface Plasmon Resonance for the Detection of Coagulation and Protein Interactions*, Sweden: Linkopings Universitet, 2000.
9. H.C. Hoch, L.W. Jelinski, H.G. Craighead, *Nanofabrication and Biosystems: Integrating Materials Science, Engineering, and Biology*, New York: Cambridge University Press, 1996.
10. J. Homola, S.S. Yee, G. Gauglitz, Surface plasmon resonance sensors: Review, *Sensors and Actuators B* 54, 1999, 3–15.
11. C.L. Baird, D.G. Myszka, Current and emerging commercial optical biosensors, *Journal of Molecular Recognition* 14, 2001, 261–268.
12. R.L. Rich, D.G. Myszka, Advances in surface plasmon resonance biosensor analysis, *Current Opinion in Biotechnology* 11, 2000, 54–61.
13. J.S. Schildkraut, Long-range surface plasmon electrooptic modulator, *Applied Optics* 27, 1988, 4587–4590.
14. G.T. Sincerbox, J.C. Gordon, Small fast large-aperture light modulator using attenuated total reflection, *Applied Optics* 20, 1981, 1491–1494.
15. P.J. Kajenski, Tunable optical filter using long-range surface plasmons, *Optical Engineering* 36, 1997, 1537–1541.
16. Y. Wang, Voltage-induced color-selective absorption with surface plasmons, *Applied Physics Letters* 67, 1995, 2759–2761.
17. G.J. Ashwell, M.P.S. Roberts, Highly selective surface plasmon resonance sensor for NO_2, *Electronics Letters* 32, 1996, 2089–2091.
18. M. Niggemann, A. Katerkamp, M. Pellmann, P. Boismann, J. Reinbold, K. Cammann, Remote sensing of tetrachloroethene with a micro-fibre optical gas sensor based on surface plasmon resonance spectroscopy, *Sensors and Actuators B* 34, 1996, 328–333.
19. C.P. Cahill, K.S. Jahnston, S.S Yee, A surface plasmon resonance sensor probe based on retro-reflection, *Sensors and Actuators B* 45, 1997, 161–166.
20. Y.C. Cheng, W.K. Su, J.H. Liou, Application of a liquid sensor based on surface plasma wave excitation to distinguish methyl alcohol from ethyl alcohol, *Optical Engineering* 39, 2000, 311–314.
21. I. Stemmler, A. Brecht, G. Gauglitz, Compact surface plasmon resonance-transducers with spectral readout for biosensing applications, *Sensors and Actuators B* 54, 1999, 98–105.
22. C.E.H. Berger, J. Greve, Differential SPR immunosensing, *Sensors and Actuators B* 63, 2000, 103–108.
23. T. Akimoto, S. Sasaki, K. Ikebukuro, I. Karube, Refractive-index and thickness sensitivity in surface plasmon resonance spectroscopy, *Applied Optics* 38, 1999, 4058–4064.
24. K.S. Johnston, S.R. Karlsen, C.C. Jung, S.S. Yee, New analytical technique for characterization of thin films using surface plasmon resonance, *Materials Chemistry and Physics*, 42, 1995, 242–246.
25. J. Melendez, R. Carr, D.U. Bartholomew, K. Kukanskis, J. Elkind, S.S. Yee, C. Furlong, R. Woodbury, A commercial solution for surface plasmon sensing, *Sensors and Actuators B* 35–36, 1996, 212–216.
26. W.L. Barnes, A. Dereux, T.W. Ebbesen, Surface plasmon subwavelength optics, *Nature* 424, 2003, 824–830.

72. I. Lundström, Real-time diospecific interaction analysis, *Biosensors and Bioelectronics* 9, 1994, 725–736.
73. R. Karlsson, M. Kullman-Magnusson, M.D. Hämäläinen, A. Remaeus, K. Andersson, P. Borg, E. Gyzander, J. Deinum, Biosensor analysis of drug-target interactions: Direct and competitive binding assays for investigation of interactions between thrombin and thrombin inhibitors, *Analytical Biochemistry* 278, 2000, 1–13.
74. P.O. Markgren, M. Hämäläinen, U.H. Danielson, Kinetic analysis of the interaction between HIV-1 protease and inhibitors using optical biosensor technology, *Analytical Biochemistry* 279, 2000, 71–78.
75. A. Frostell-Karlsson, A. Remaeus, H. Roos, K. Andersson, P. Borg, M. Hämäläinen, R. Karlsson, Biosensor analysis of the interaction between immobilized human serum albumin and drug compounds for prediction of human serum albumin binding levels, *Journal of Medicinal Chemistry* 43, 2000, 1986–1992.
76. E. Danelian, A. Karlén, R. Karlsson, S. Winiwarter, A. Hansson, S. Löfås, H. Lennernäs, M.D. Hämäläinen, SPR biosensor studies of the direct interaction between 27 drugs and a liposome surface: Correlation with fraction absorbed in humans, *Journal of Medicinal Chemistry* 43, 2000, 2083–2086.
77. P. Bjurling, G.A. Baxter, M. Caselunghe, C. Jonson, M. O'Connor, B. Persson, C.T. Elliott, Biosensor assay of sulfadiazine and sulfamethazine residues in pork, *Analyst* 125, 2000, 1771–1774.
78. M. Boström-Caselunghe, J. Lindeberg, Biosensor based determination of folic acid in fortified food, *Food Chemistry* 70, 2000, 523–532.
79. J.D. Wright, J.V. Oliver, R.J.M. Nolte, S.J. Holder, N.A.J.M. Sommerdijk, P.I. Nikitin, The detection of phenols in water using a surface plasmon resonance system with specific receptors, *Sensors and Actuators B* 51, 1998, 305–310.
80. V. Koubová, E. Brynda, L. Karasová, J. Škvor, J. Homola, J. Dostálek, P. Tobiška, J. Rošický, Detection of foodborne pathogens using surface plasmon resonance biosensors, *Sensors and Actuators B* 74, 2001, 100–105.
81. D. Aivazian, L.J. Stern, Phosphorylation of T cell receptor is regulated by a lipid dependent folding transition, *Nature Structural & Molecular Biology* 7, 2000, 1023–1026.
82. B. Bader, K. Kuhn, D.J. Owen, H. Waldmann, A. Wittinghofer, J. Kuhlmann, Biorgainic synthesis of lipid-modified proteins for the study of signal transduction, *Nature* 403, 2000, 223–226.
83. E. Bitto, M. Li, A.M. Tikhonov, M.L. Schlossman, W. Cho, Mechanism of annexim I-mediated membrane aggregation, *Biochemistry* 39, 2000, 13469–13477.
84. R.G. Chapman, E. Ostuni, L. Yan, G.M. Whitesides, Preparation of mixed self-assembled monolayers (SAMs) that resist adsorption of proteins using the reaction of amines with a SAM that presents interchain carboxylic anhydride groups, *Langmuir* 16, 2000, 6927–6936.
85. H.M. Chen, W. Wang, D.K. Smith, Liposome disruption detected by surface plasmon resonance at lower concentrations of a peptide antibiotic, *Langmuir* 16, 2000, 9959–9962.
86. A.R. Halpern, Y. Chen, R.M. Corn, D. Kim, Surface plasmon resonance phase imaging measurements of patterned monolayers and DNA adsorption onto microarrays, *Analytical Chemistry* 83, 2011, 2801–2806.
87. C.L. Wong, H.P. Ho, T.T. Yu, Y.K. Suen, W.W.Y. Chow, S.Y. Wu, W.C. Law, W. Yuan, W.J. Li, S.K. Kong, C. Lin, Two-dimensional biosensor arrays based on surface plasmon resonance phase imaging, *Applied Optics* 46, 2007, 2325–2332.
88. K.-H. Lee, Y.-D. Su, S.-J. Chen, F.-G. Tseng, G.-B, Lee, Microfluidic systems integrated with two-dimensional surface plasmon resonance phase imaging systems for microarray immunoassay, *Biosensors and Bioelectronics*, 23, 2007, 466–472.
89. X. Yu, D. Wang, X. Wei, X. Ding, W. Liao, X. Zhao, A surface plasmon resonance imaging interferometry for protein micro-array detection, *Sensors and Actuators B* 108, 2005, 765–771.
90. Y.-C. Li, Y.-F. Chang, L.-C. Su, C. Chou, Differential-phase surface plasmon resonance biosensor, *Analytical Chemistry*, 80(14), 2008, 5590–5595.

91. Y. Wang, C. Zhang, Y. Zhang, H. Fang, C. Min, S. Zhu, X.-C. Yuan, Investigation of phase SPR biosensor for efficient targeted drug screening with high sensitivity and stability, *Sensors and Actuators B* 209, 2015, 313–322.

92. A.M. Giannetti, B.D. Koch, M.F. Browner, Surface plasmon resonance based assay for the detection and characterization of promiscuous inhibitors, *Journal of Medical Chemistry* 51, 2008, 574–580.

93. S.C. Hong, H. Chen, J. Lee, H.-K. Park, Y.S. Kim, H.-C. Shin, C.-M. Kim, T.J. Park, S.J. Lee, K. Koha, H.-J. Kim, C.L. Chang, J. Lee, Ultrasensitive immunosensing of tuberculosis CFP-10 based on SPR spectroscopy, *Sensors and Actuators B* 156, 2011, 271–275.

94. J.W. Chung, S.D. Kim, R. Bernhardt, J.C. Pyun, Application of SPR biosensor for medical diagnostics of human hepatitis B virus (hHBV), *Sensors and Actuators B* 111–112, 2005, 416–422.

95. N. Bianchi, C. Rutigliano, M. Tomassetti, G. Feriotto, F. Zorzato, R. Gambari, Biosensor technology and surface plasmon resonance for real-time detection of HIV-1 genomic sequences amplified by polymerase chain reaction, *Clinical and Diagnostic Virology* 8, 1997, 199–208.

96. J.F.C. Loo, S.S. Wang, F. Peng, J.A. He, L. He, Y.C. Guo, D.Y. Gu, H.C. Kwok, S.Y. Wu, H.P. Ho, W.D. Xie, Y.H. Shao, S.K. Kong, A non-PCR SPR platform using RNase H to detect microRNA 29a-3p from throat swabs of human subjects with influenza A virus H1N1 infection, *Analyst* 140, 2015, 4566–4575.

97. Y. Shin, A.P. Perera, M.K. Park, Label-free DNA sensor for detection of bladder cancer biomarkers in urine, *Sensors and Actuators B* 178, 2013, 200–206.

98. P. Preechaburana, M.C. Gonzalez, A. Suska, D. Filippini, Surface plasmon resonance chemical sensing on cell phones, *Angewandte Chemie* 124, 2012, 11753–11756.

99. M. Piliarik, H. Vaisocherova, J. Homola, A new surface plasmon resonance sensor for high-throughput screening applications, *Biosensors and Bioelectronics* 20, 2005, 2104–2110.

100. H.S. Ipováa, V.S. Evcu, M. Kuchar, J.N. Ahmad, P. Mikulecky, R. Osickac, P. Maly, J. Homola, Surface plasmon resonance biosensor based on engineered proteins for direct detection of interferon-gamma in diluted blood plasma, *Sensors and Actuators B* 174, 2012, 306–311.

101. Y.H. Kim, J.P. Kim, S.J. Han, S.J. Sim, Aptamer biosensor for lable-free detection of human immunoglobulin E based on surface plasmon resonance, *Sensors and Actuators B* 139, 2009, 471–475.

102. A. Subramanian, J. Irudayaraj, T. Ryan, A mixed self-assembled monolayer-based surface plasmon immunosensor for detection of E. coli O157:H7, *Biosensors and Bioelectronics* 21, 2006, 998–1006.

103. H. Wua, H. Li, F.Z.H. Chua, S.F.Y. Li, Rapid detection of melamine based on immunoassay using portable surface plasmon resonance biosensor, *Sensors and Actuators B* 178, 2013, 541–546.

104. B.J. Yakes, S. Prezioso, S.A. Haughey, K. Campbell, C.T. Elliott, S.L. DeGrasse, An improved immunoassay for detection of saxitoxin by surface plasmon resonance biosensors, *Sensors and Actuators B* 156, 2011, 805–811.

105. J.R. Krone, R.W. Nelson, D. Dogruel, P. Williams, R. Granzow, BIA/MS: Interfacing biomolecular interaction Analysis with mass spectrometry, *Analytical Biochemistry* 244, 1997, 124–132.

106. D. Nedlkov, R.W. Nelson, Surface plasmon resonance mass spectrometry: Recent progress and outlooks, *Trends in Biotechnology* 21, 2003, 301–305.

107. H. Kano, S. Kawata, Two-photon-excited fluorescence enhanced by a surface plasmon, *Optics Letters* 21, 1996, 1848–1850.

108. J.R. Lakowicz, J. Malick, I. Gryczynski, Z. Gryczynski, Directional surface plasmon-coupled: A new method for high sensitivity detection, *Biochemical and Biophysical Research Communication* 307, 2003, 435–439.

109. S.K. Vashist, O. Mudanyali, E.M. Schneider, R. Zengerle, A. Ozcan, Cellphone-based devices for bioanalytical sciences, *Analytical and Bioanalytical Chemistry* 406, 2014, 3267–3277.

110. Q. Wei, H. Qi, W. Luo, D. Tseng, S.J. Ki, Z. Wan, Z. Gorocs, L.A. Bentolila, T.-T. Wu, R. Sun, A. Ozcan, Fluorescent imaging of single nanoparticles and viruses on a smart phone, *ACS Nano* 7, 2013, 9147–9155.

many targets at once. The multiplexed nature of SERS offers significant advantages over other methods, such as fluorescence and chemiluminescence, that exhibit broader emission bands.[8] Large SERS enhancement factors of $10^{12}-10^{15}$ have been reported, inspiring the development of new sensing materials for detection of analytes with highly sensitive detection levels.[7,9–11]

With the advancements in nanotechnology, SERS has opened up many new and exciting possibilities for a large number of biosensing applications.[12–14] However, in order to generate a strong SERS signal and to identify specific target sequences, many existing methods require either multiple incubation/wash steps or target labeling, thus increasing the assay complexity. Over the past several years, we have developed a wide variety of SERS plasmonic platforms for chemical and biological sensing and have been investigating the strength of the SERS effect as part of a nucleic-acid sensing nanoprobe. Particularly, we have demonstrated that highly sensitive, specific, and multiplexed detection of nucleic acids can be achieved by utilizing SERS-active silver nanoparticles[15–17] in the development of a unique SERS-based probe for DNA detection, referred to as the "molecular sentinel" (MS).[16] The MS probe consists of a DNA strand having a Raman label molecule at one end and a metal nanoparticle at the other. The plasmonics nanoprobe uses a hairpin-like stem–loop structure to recognize target DNA sequences. Note that hairpin DNA structures, first developed by Tyagi and Kramer in 1996,[18] have been used in "molecular beacon" systems that are based on optical and electrochemical detection.[19–24] The sensing principle of molecular sentinels, however, is quite different from that of molecular beacons. With MS systems, in the normal configuration (i.e., in the absence of target DNA), the DNA sequence forms a hairpin loop, which maintains the Raman label in close proximity of the metal nanoparticle, thus inducing an intense SERS signal of the Raman label upon laser excitation. Upon hybridization of a complementary target DNA sequence to the nanoprobe hairpin loop, the Raman label molecule is physically separated from the metal nanoparticle, thus leading to a decreased SERS signal. Recently, we have also reported a novel "turn-on" plasmonics-based nanobiosensor, referred to as "inverse molecular sentinel" (iMS) nanoprobes, for the detection of DNA sequences of interest with an "OFF-to-ON" signal switch.[25] The term "inverse" is used to distinguish this new technology from our previously developed "ON-to-OFF" MS nanoprobes.[15,16]

14.2 Development of the iMS Nanoprobe: A New Concept in Label-Free Homogeneous Bioassay

14.2.1 Silver-Coated Gold Nanostars for SERS Detection

The iMS nanobiosensor involves the use of a unique type of SERS-active nanoprobe, that is, silver-coated gold nanostars (AuNS@Ag), as the sensing platform. Nanostars have emerged as one of the best geometries for producing strong SERS in a nonaggregated state due to their multiple sharp branches, each with a strongly enhanced electromagnetic field localized at its tip.[26] AuNS@Ag is a new hybrid bimetallic nanostar platform that exhibits superior resonant SERS (SERRS) properties.[27] We have recently demonstrated that AuNS@Ag provided over an order of magnitude greater signal enhancement compared to uncoated AuNS, rendering it an excellent SERS substrate (Figure 14.1).[27]

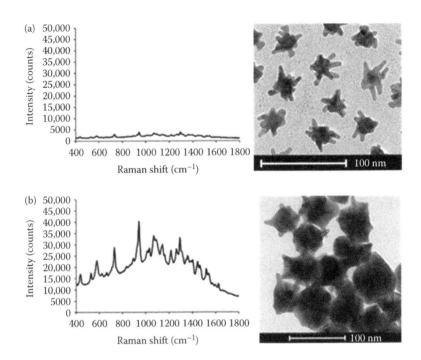

FIGURE 14.1
SERS spectra and TEM images of (a) gold nanostars (AuNS) and (b) silver-coated gold nanostars (AuNS@Ag). (Adapted from Fales, A. M. et al. *J. Phys. Chem.* 2014, 118(7):3708–3715.)

14.2.2 Detection Scheme of the SERS iMS Nanoprobe

The iMS nanobiosensor employs a nonenzymatic DNA strand-displacement process and the conformational change of stem–loop (hairpin) DNA probes for specific target identification and signal switch. As shown in Figure 14.2, a DNA probe, having a Raman label at one end, is immobilized onto a nanostar via a metal–thiol bond. The probe is designed with a sequence that can form a stem–loop, or hairpin, structure upon completion of a strand-displacement process in order to bring the Raman label to the surface of nanostars when the loop is closed, thus producing a strong SERS signal. A single-stranded DNA sequence serves as a placeholder strand and is partially hybridized to the stem–loop probe sequence via the placeholder-binding region. This probe–placeholder duplex keeps the stem–loop structure in an open, or linear, configuration to keep the label farther than 10 nm away from the nanostar surface. In this open configuration, the iMS nanobiosensor exhibits low SERS intensity ("OFF" state) as the SERS enhancement decreases exponentially with increasing separation between the label and the metallic surface. Upon exposure to a target nucleic acid, the placeholder strand leaves the surface following a nonenzymatic strand-displacement process.[28,29] During this displacement process, the target first hybridizes to the overhand region of the probe–placeholder duplex (called the "toehold") and begins displacing the stem–loop probe from the placeholder through a branch migration process, finally releasing the placeholder from the nanobiosensor. This then allows the stem–loop structure to close, thus moving the Raman label in close proximity to the plasmonics-active nanostar surface, yielding a strong SERS signal ("ON" state).[25]

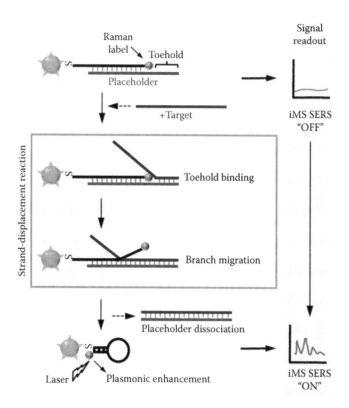

FIGURE 14.2
Detection scheme of the iMS "OFF-to-ON" SERS nanobiosensor. (Adapted from Wang, H. N. et al. *Nanomedicine: NBM* 2015, 11(4):511–520.)

The feasibility of using the iMS SERS nanobiosensor for nucleic acid detection (designed specifically for human radical S-adenosyl methionine domain containing 2 gene, RSAD2) is demonstrated in Figure 14.3a. In the presence of 1 μM target DNA (curve iii), the SERS intensity was significantly increased, indicating that the hybridization between targets and placeholders enabled the formation of the stem–loop structure of the nanobiosensor, thereby moving the SERS dye in close proximity to the nanostar surface and turning the SERS signal "ON." The possibility for quantitative analysis is demonstrated in Figure 14.3b, which illustrates that the SERS intensity at the 557-cm^{-1} Raman peak increases with increasing the amount of target DNA. With over 50 nM of target DNA, the SERS signal is expected to reach a plateau; however, as shown in the inset of Figure 14.3b, a linear trend line was fitted to the data between 0 and 5 nM with a slope (s = 376.47) and a residual standard deviation ($\sigma = 28.87$). The limit of detection (LOD, $3\sigma/s$) for target DNA was determined to be approximately 0.1 nM in the current bioassay system. As a result, the absolute LOD is as low as 200 attomoles. It is noteworthy that this analysis used very small amounts of samples (2 μL).[25]

The placeholder was first designed with a moderate 8-base toehold containing equal numbers of G/C and A/T bases. According to a previous study,[30] the kinetics of strand displacement can be modulated by changing the length or the sequence composition (i.e., G/C content) of toeholds. Thus, the effective reaction time for turning on the SERS signal will depend on the toeholds used, for example, using a strong toehold with a higher G/C content would result in a faster reaction rate.[25]

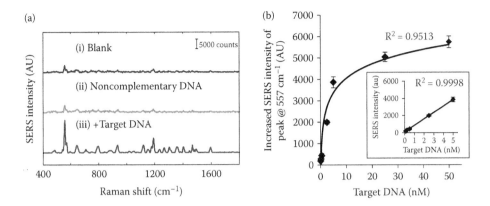

FIGURE 14.3

(a) SERS spectra of the iMS nanobiosensor in the presence or absence of target RSAD2 DNA. (b) Evaluation of the detection sensitivity of the nanobiosensor. (Adapted from Wang, H. N. et al. *Nanomedicine: NBM* 2015, 11(4):511–520.)

The detection specificity was evaluated using the perfectly matched target and sequences having one-, two-, or three-base mismatches. Figure 14.4 shows that the SERS intensity at 557 cm^{-1} decreases with increasing numbers of mismatched bases. In the case of one-base mismatch, the SERS intensity is ~30% less than the perfectly matched target. To further demonstrate the capability of the iMS system to discriminate single-nucleotide differences, the nanobiosensor was tested in solutions with different ionic strength, and it was determined that the SERS signal of the nanobiosensor treated with one-base mismatched sequences decreases with decreasing the NaCl concentration in the reaction buffer. The specificity of detection is critical for many clinical applications, such as single-nucleotide polymorphism (SNP) identification or microRNA detection. Reducing the ionic strength of the media could further improve the detection specificity. This is due to the fact that the mismatched target–placeholder duplex is thermodynamically unstable in low ionic strength solutions compared to the perfectly matched duplex.[25]

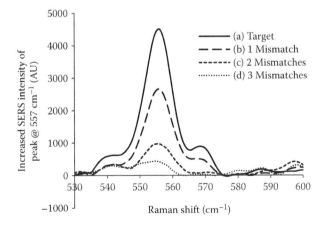

FIGURE 14.4

Evaluation of the detection specificity in the presence of perfectly matched or mismatched sequences. (Adapted from Wang, H. N. et al. *Nanomedicine: NBM* 2015, 11(4):511–520.)

The application of this novel nanostar-based iMS SERS-sensing technology as a useful homogeneous assay for multiplex detection of microRNA (miRNA) has been demonstrated.[31] miRNAs are small, noncoding RNAs of ~20–25 nucleotides in length. This design strategy has the potential for routine clinical use as it overcomes the challenge in current miRNA detection posed by the short length of miRNA sequences. The multiplex capability of the iMS technique was demonstrated using a mixture of the two differently labeled nanoprobes to detect miR-21 and miR-34a miRNAs, which have been recognized as critical biomarkers for breast cancer.[32–35]

An iMS nanoprobe was first designed labeled with Cy5 to detect the mature human miR-21 miRNA. In our previous reports, the physical length of the Raman-labeled stem–loop probes was designed to be greater than 10 nm for a low background SERS signal in their OFF state.[25,36] However, the short sequence lengths of miRNAs would make it difficult to design stem–loop probes with the desired length. To overcome this challenge, a spacer sequence of around 10 nucleotides was added between the loop and the 5′-end stem portion of the sequence (Figure 14.5a). This spacer is used to increase the physical length of the probe in order to keep the Raman label at an appropriate distance away from the nanostar surface while the stem–loop is hybridized to the placeholder strand (open configuration). This "internal spacer," however, does not affect the generation of a strong SERS signal when the probe is in a closed stem–loop configuration (Figure 14.5b).[31]

The iMS nanoprobes have a low SERS background signal in the "OFF" state through the formation of stable probe–placeholder duplexes. Different placeholder strands were investigated for their capability to turn "OFF" the SERS signal of the iMS nanoprobes. A short poly(T) tail (4–6 thymine bases) was added to the 3′-end of the placeholder strand; as such, it can hybridize to a portion of the internal spacer in the probe. The red bars in Figure 14.6a show the SERS background signal (peak height at 558 cm^{-1}) from the iMS nanoprobes in their "OFF" state with the different placeholder strands. Without the poly(T) tail (placeholder-0T), a strong SERS background signal was observed indicating that the placeholder-0T strand did not effectively turn "OFF" the iMS SERS signal. This is due to

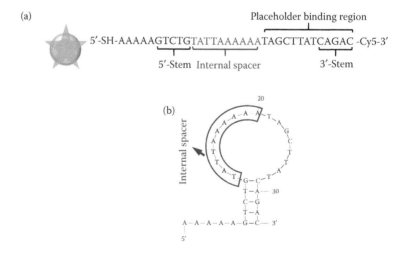

FIGURE 14.5
The design of the iMS nanoprobes for microRNA detection. (a) Sequence structure of the iMS stem–loop probe with an internal spacer for miR-21 detection. (b) Stem–loop configuration of the probe showing the internal spacer is located within the loop region. The stem–loop structure was predicted using the two-state folding tool on the DINAMelt server.[37,38]

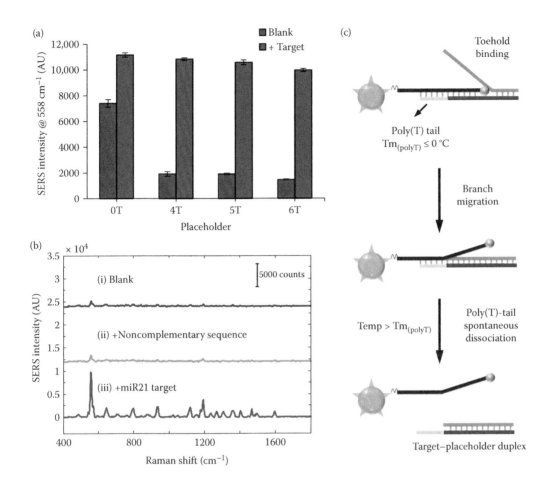

FIGURE 14.6

(a) Investigation of the use of different placeholders with a short poly(T) tail (4–6 thymine bases) in turning off the iMS SERS signal. Red bars represent the peak height SERS background signal at 558 cm^{-1} from the "OFF"-state iMS nanoprobes when using different placeholder strands. Blue bars represent the peak height SERS intensity at 558 cm^{-1} from the "ON"-state iMS nanoprobes after incubation of synthetic DNA targets. Placeholder-0T, -4T, -5T, and -6T contain 0, 4, 5, and 6 thymine bases, respectively. (b) SERS spectra of the Cy5-labeled iMS nanoprobes with placeholder-6T in the presence or absence of miR-21 synthetic target. Spectrum (i): blank. Spectrum (ii): nanobiosensor in the presence of noncomplementary DNA with sequences corresponding to miR-16 miRNA. Spectrum (iii): nanobiosensor in the presence of synthetic miR-21 target DNA. (c) Schematic diagram showing the spontaneous dissociation of the short poly(T) tail of the placeholder from the probe following the branch migration process.

the fact that this probe–placeholder-0T duplex with a melting temperature (Tm) of 38°C is thermodynamically unstable at the reaction temperature of 37°C. In this case, the probe (with the hairpin stem Tm of 44.1°C) has a high tendency to fold into a stem–loop structure, yielding a strong SERS background signal. In contrast, with short poly(T) tails (i.e., placeholder-4T containing four thymine bases, placeholder-5T containing five thymine bases, and placeholder-6T containing six thymine bases) provided low background signals indicating that stable probe–placeholder duplexes were formed. The melting temperatures of these probe–placeholder duplexes are estimated to be 44.7°C, 46.1°C, and 47.4°C for placeholder-4T, -5T, and -6T, respectively. All of the duplexes are higher than the reaction temperature, resulting in greater thermal stability.[31]

To turn "ON" the iMS nanobiosensor, the placeholder strands need to be released from the system upon target binding. It was investigated whether the additional poly(T) tail affects the dissociation of the placeholders from the iMS nanoprobes. The blue bars in Figure 14.6a represent the SERS intensity at the 558 cm^{-1} peak from the iMS nanoprobes after 1-hour incubation of 1 μM of synthetic DNA targets at 37°C. Figure 14.6b shows the SERS spectra for the detection of miR-21 synthetic targets using the placeholder-6T. In the presence of 1 μM miR-21 targets, the SERS intensity was significantly increased (signal "ON") indicating that the hybridization between targets and placeholders enabled the formation of the stem–loop structure by triggering the strand displacement reaction and releasing the placeholder strands. Because of the thermal instability of the short poly(A)–poly(T) duplexes (Tm ≤ poly(T) ≤ 0°C), the additional short poly(T) tail is able to spontaneously dissociate from the probe following the branch migration process at an elevated temperature (Figure 14.6c). However, it was also observed that the intensity decreases slightly with increasing the length of the poly(T) tail in the presence of targets. Thus, the design of the placeholder is critical in ensuring its release from the system following the branch migration process. Nevertheless, the results demonstrated that a poly(T) tail could be added to the placeholder in order to minimize the background signal without affecting the sensor functionality.[31]

The multiplexing capability of SERS is an important feature due to the narrow Raman bandwidths. To demonstrate the multiplexing capability of the iMS technique, a second iMS nanoprobe labeled with a different Raman dye, Cy5.5, was designed to detect miR-34a. Figure 14.7 shows the blank corrected SERS signal of the miR-34a iMS nanoprobes in the presence or absence of target molecules. In the presence of 1-μM miR-34a synthetic RNA targets (spectrum b), the SERS intensity was significantly increased compared with the negative control of 1-μM noncomplementary sequences (spectrum a). The SERS spectrum of the Cy5.5-labeled miR-34a nanoprobes exhibits multiple unique Raman peaks, which are significantly different from that of the Cy5-labeled miR-21 nanoprobes.[31]

Demonstration of the multiplex capability of the iMS technique was first performed with a mixture of the two iMS nanoprobes for simultaneous detection of synthetic miR-21 and

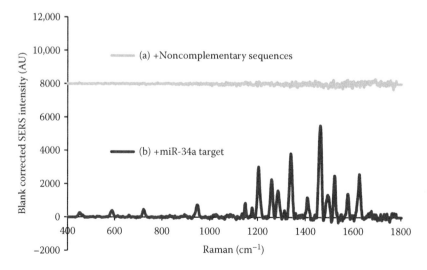

FIGURE 14.7
SERS spectra of the Cy5.5-labeled iMS nanoprobes in the presence or absence of miR-34a synthetic target. Spectrum (a): nanobiosensor in the presence of noncomplementary synthetic RNA with sequences corresponding to SNORD38B small RNA. Spectrum (b): nanobiosensor in the presence of synthetic miR-34a RNA target.

miR-34a miRNAs. The experiments were carried out by mixing different ratios of miR-21 and miR-34a nanoprobes (1:1, 1:2, and 1:4) while keeping the concentration of miR34a nanoprobes constant. Figure 14.8a shows the blank-corrected SERS spectra from the mixtures in the presence of both miR-21 and miR-34a synthetic targets. To identify the SERS signal of each nanoprobe from the mixture, the spectra were analyzed using a spectral decomposition method. As previously described,[39] the spectral decomposition procedure, which was adapted from Lutz et al.,[8] was processed using MATLAB® (R2015b, MathWorks, MA). The decomposition is based on the assumption that the multiplex spectrum comprises the reference spectra and an unknown polynomial. SERS spectra were collected for each probe alone (reference spectra) and for the mixture. The acquired SERS spectra were background subtracted and smoothed using a Savitsky–Golay filter (five-point window and first-order polynomial). The entire spectra ranging from 400 to 1800 cm^{-1}, which contains the distinctive Raman signature of each dye, were loaded into MATLAB's workspace. The decomposition processes utilized MATLAB functions lsqnonneg and fmincon to determine the best fit of the reference spectra to the mixture spectrum. A free-fitting polynomial was introduced to reduce the fitting error.[8] MATLAB analysis subsequently generated the minimally constrained coefficients (i.e., the fractions) for each reference spectrum, which were normalized to 1. As can be seen in Figure 14.8b, the SERS spectra of the mixture can be decomposed into the contributions of the two distinct reporters: Cy5 (miR-21 nanoprobes)

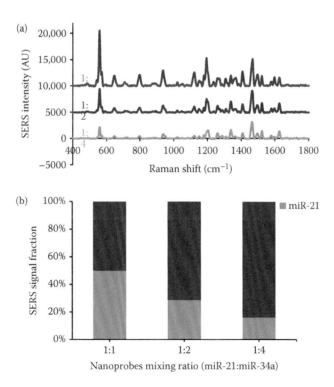

FIGURE 14.8

(a) SERS spectra measured from mixtures of different ratios of Cy5-labeled miR-21 and Cy5.5-labeled miR-34a nanoprobes (1:1, 1:2, and 1:4) in the presence of both miR-21 and miR-34a synthetic targets. The concentration of miR34a nanoprobes was kept constant. (b) SERS signal fractions obtained from the spectral decomposition procedure. The mixture spectra were decomposed into contributions of the two distinct reporters: Cy5 (miR-21 nanoprobes) and Cy5.5 (miR-34a nanoprobes).

and Cy5.5 (miR-34a nanoprobes). The signal fractions were in good agreement with the predetermined ratios of the two nanoprobes.

To demonstrate the detection of the two real miRNA targets in RNA extracts, different amounts (250 ng, 500 ng, and 1 μg) of total small RNA extracted from the MCF-7 breast cancer cell line were added to a 10-μL nanoprobe mixture containing 5 pM of miR-21 nanoprobes and 10 pM of miR-34a nanoprobes. Figure 14.9a shows a typical SERS spectrum of the mixture in the presence of 1 μg of total small RNA sample. The unique SERS peaks of the miR-21 nanoprobes at 558 and 797 cm^{-1}, and of the miR-34a nanoprobes at 1523 and 1626 cm^{-1} were indicated by the arrows. The integrated SERS intensity of these peaks shown in Figures 14.9b and 14.9c were given by the area under the curve over the spectral range of 540–569, 780–810, 1512–1530, and 1615–1640 cm^{-1} for peaks at 558, 797, 1523, and 1626 cm^{-1}, respectively. The SERS intensity of the miR-21 nanoprobes at 558 and 797 cm^{-1} (Figure 14.9b) was found to increase significantly with increasing amounts of small RNA input. For miR-34a SERS signal, a slight increase in the intensity at both 1523 and 1626 cm^{-1} was observed only when 1 μg of the RNA sample was added (Figure 14.9c). Due to the significant difference in the amounts of RNA samples required for a detectable signal change, the abundance of miR-21 in MCF-7 cells was found to be higher than that of miR-34a. This result was consistent with that reported by Fix et al.[40] using microarray. The results of this study demonstrate the feasibility of using the iMS technique for multiplexed detection of

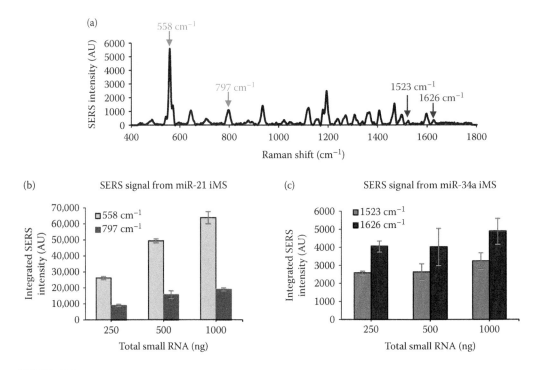

FIGURE 14.9
Multiplexed detection of miR-21 and miR-34a targets in different amounts (250 ng, 500 ng, and 1 μg) of MCF-7 total small RNA extracts using a 10-μL nanoprobe mixture containing 5 pM of miR-21 nanoprobes and 10 pM of miR-34a nanoprobes. (a) SERS spectrum of the mixture in the presence of 1 μg of total small RNA sample. The arrows indicate the unique SERS peaks of the miR-21 nanoprobes at 558 and 797 cm^{-1}, and miR-34a nanoprobes at 1523 and 1626 cm^{-1}. (b) The integrated SERS intensity of the peaks at 558 and 797 cm^{-1} contributed from miR-21 nanoprobes. (c) The integrated SERS intensity of the peaks at 1523 and 1626 cm^{-1} contributed from miR-34a nanoprobes.

miRNAs in real biological samples. Note that the SERS measurements were performed immediately following the incubation of target molecules without any washing steps, which greatly simplifies and accelerates the assay procedure.

In conclusion, an "OFF-to-ON" SERS iMS nanobiosensor has been developed as a homogeneous assay for multiplex detection of short miRNAs in a single sensing platform. As it does not require target labeling and any subsequent washing steps, the positive-readout iMS biosensing platform offers a versatile and powerful tool for a wide variety of applications.

14.3 Applications in Biomedical Diagnostics

The iMS scheme has the great advantage of simple design, specific to any nucleic acid biomarker of choice. In addition to the detection of DNA/RNA biomarkers for infectious disease (dengue virus and RSAD2),[25,36] iMS nanoprobes have been designed for detection of miRNAs for cancer diagnostics.[31] miRNAs are small, noncoding RNAs of approximately 20–25 nucleotides in length that bind to perfect or near-perfect complementary sequences in the untranslated regions (UTRs) of mRNA targets, thereby regulating gene expression at the posttranscriptional level.[41] Dysregulation of miRNAs is often observed in a wide range of cancers. It has been found that miRNAs transcriptionally suppress expression of oncogenes and loss of miRNA expression results in oncogene activation.[42] Recent studies also show that miRNAs regulate not just single oncogenes but entire signaling networks.[43–45] For instance, while miR-21 miRNA is overexpressed and plays an important oncogenic role in triple-negative breast cancer (TNBC), miR-34a is downregulated in TNBC, which has been recognized as the most aggressive subtype of breast cancer.[46,47] miRNAs have been identified that are associated with esophageal adenocarcinoma (EAC) as well as Barrett's esophagus (BE: the premalignant metaplasia associated with EAC), allowing for clinicians to distinguish malignant and pre-malignant samples from squamous mucosa.[48] The dysregulation of miRNAs in both EAC and esophageal squamous cell carcinoma is not limited to tissue, but includes abnormal miRNA expression in serum and plasma.[49] In particular, miR-21 has been described as an esophageal cancer biomarker in tissue and blood; its association with other cancers[50] makes miRNA detection relevant to many clinical areas. miRNAs hold great potential to serve as an important class of biomarkers not only for early diagnosis of cancer, but also for investigation of cancer initiation and progression.[51–54] These small molecules have not been adopted into clinical practice for early diagnostics because of the technical difficulties arising from the intrinsic characteristics of miRNAs, such as the short sequence lengths, low abundance, high sequence similarity, and a wide range of expression levels that could span over four orders of magnitude.[54]

Currently, northern blotting, microarrays, and quantitative reverse transcriptase PCR (qRT-PCR) are often employed as the conventional miRNA detection methods, which involve elaborate, time-consuming, and expensive processes that require special laboratory equipment.[55,56] While northern blotting is the only technique that allows for the quantitative visualization of miRNA, it has low detection efficiency and requires complex methods that can introduce contamination.[57] Microarrays, which allow simultaneous detection of several hundred miRNAs, are only semi-quantitative making them most suitable for comparing relative expression levels of miRNA between different cellular states.[58] Microarrays, therefore, require an additional form of validation, such as qRT-PCR, to quantify expression. qRT-PCR, the gold standard among miRNA detection techniques, allows

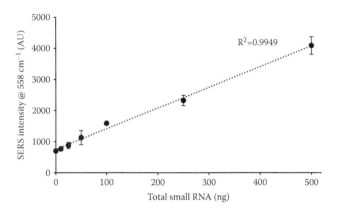

FIGURE 14.11
Evaluation of the dose response of the iMS technique using serial dilutions of total small RNA (10, 25, 50, 100, 250, and 500 ng in a 10-μL assay volume) enriched from breast adenocarcinoma MCF-7 total RNA.

Acknowledgments

This work was sponsored by the National Institutes of Health (R21CA196426) and the Duke Faculty Exploratory Research Fund. This material is based upon work of BMC supported by the National Science Foundation Graduate Research Fellowship under Grant No. 1106401.

References

1. Otto, A., Mrozek, I., Grabhorn, H., Akemann, W. Surface-enhanced Raman-scattering. *J. Phys. Condens. Matter* 1992, 4(5):1143–1212.
2. Vo-Dinh, T. Surface-enhanced Raman spectroscopy using metallic nanostructures. *TrAC, Trends Anal. Chem.* 1998, 17(8–9):557–582.
3. Schatz, G. C. Theoretical-studies of surface enhanced Raman-scattering. *Acc. Chem. Res.* 1984, 17(10):370–376.
4. Campion, A., Kambhampati, P. Surface-enhanced Raman scattering. *Chem. Soc. Rev.* 1998, 27:241–250.
5. Wolkow, R., Moskovits, M. Enhanced photochemistry on silver surfaces. *J. Chem. Phys.* 1987, 87:5858–5869.
6. Kneipp, K., Wang, Y., Kneipp, H., Perelman, L. T., Itzkan, I., Dasari, R. R., Feld, M. S. Single molecule detection using surface-enhanced Raman scattering (SERS). *Phys. Rev. Lett.* 1997, 78:1667–1670.
7. Nie, S., Emory, S. R. Probing single molecules and single nanoparticles by surface-enhanced Raman scattering. *Science* 1997, 275:1102–1106.
8. Lutz, B. R., Dentinger, C. E., Nguyen, L. N., Sun, L., Zhang, J., Allen, A. N., Chan, S., Knudsen, B. S. Spectral analysis of multiplex Raman probe signatures. *ACS Nano* 2008, 2:2306–2314.
9. Kneipp, J., Kneipp, H., Kneipp, K. SERS—A single-molecule and nanoscale tool for bioanalytics. *Chem. Soc. Rev.* 2008, 37(5):1052–1060.

10. Otto, A., Bruckbauer, A., Chen, Y. X. On the chloride activation in SERS and single molecule SERS. *J. Mol. Struct.* 2003, 661:501–514.

11. Xu, H. X., Bjerneld, E. J., Kall, M., Borjesson, L. Spectroscopy of single hemoglobin molecules by surface enhanced Raman scattering. *Phys. Rev. Lett.* 1999, 83(21):4357–4360.

12. Cao, Y. C., Jin, R., Mirkin, C. A. Nanoparticles with Raman spectroscopic fingerprints for DNA and RNA detection. *Science* 2002, 297:1536–1540.

13. Graham, D., Smith, W. E., Linacre, A. M. T., Munro, C. H., Watson, N. D., White, P. C. Selective detection of deoxyribonucleic acid at ultralow concentrations by SERRS. *Anal. Chem.* 1997, 69:4703–4707.

14. Jin, R., Cao, Y. C., Thaxton, C. S., Mirkin, C. A. Glass-bead-based parallel detection of DNA using composite Raman labels. *Small* 2006, 2:375–380.

15. Wabuyele, M. B., Vo-Dinh, T. Detection of human immunodeficiency virus type 1 DNA sequence using plasmonics nanoprobes. *Anal. Chem.* 2005, 77:7810–7815.

16. Wang, H. N., Vo-Dinh, T. Multiplex detection of breast cancer biomarkers using plasmonic molecular sentinel nanoprobes. *Nanotechnology* 2009, 20:065101.

17. Vo-Dinh, T., Liu, Y., Fales, A. M., Ngo, H., Wang, H. N., Register, J. K., Yuan, H., Norton, S. J., Griffin, G. D. SERS nanosensors and nanoreporters: Golden opportunities in biomedical applications. *Wiley Interdiscip. Rev. Nanomed. Nanobiotechnol.* 2015, 7:17–33.

18. Tyagi, S., Kramer, F. R. Molecular beacons: Probes that fluoresce upon hybridization. *Nat. Biotechnol.* 1996, 14:303–308.

19. Zhang, Y., Tang, Z., Wang, J., Wu, H., Maham, A., Lin, Y. Hairpin DNA switch for ultrasensitive spectrophotometric detection of DNA hybridization based on gold nanoparticles and enzyme signal amplification. *Anal. Chem.* 2010, 82:6440–6446.

20. Peng, H. I., Strohsahl, C. M., Leach, K. E., Krauss, T. D., Miller, B. L. Label-free DNA detection on nanostructured Ag surfaces. *ACS Nano* 2009, 3:2265–2273.

21. Dolatabadi, J. E. N., Mashinchian, O., Ayoubi, B., Jamali, A. A., Mobed, A., Losic, D., Omidi, Y., de la Guardia, M. Optical and electrochemical DNA nanobiosensors. *TrAC, Trends Anal. Chem.* 2011, 30:459–472.

22. Bercovici, M., Kaigala, G. V., Mach, K. E., Han, C. M., Liao, J. C., Santiago, J. G. Rapid detection of urinary tract infections using isotachophoresis and molecular beacons. *Anal. Chem.* 2011, 83:4110–4117.

23. Algar, W. R., Massey, M., Krull, U. J. The application of quantum dots, gold nanoparticles and molecular switches to optical nucleic-acid diagnostics. *TrAC, Trends Anal. Chem.* 2009, 28:292–306.

24. Bonanni, A., Pumera, M. Graphene platform for hairpin-DNA-based impedimetric genosensing. *ACS Nano* 2011, 5:2356–2361.

25. Wang, H. N., Fales, A. M., Vo-Dinh, T. Plasmonics-based SERS nanobiosensor for homogeneous nucleic acid detection. *Nanomedicine: NBM* 2015, 11:511–520.

26. Yuan, H., Khoury, C. G., Wang, H. H., Wilson, C. M., Grant, G. A., Vo-Dinh, T. Gold nanostars: Surfactant-free synthesis, 3D modelling, and two-photon photoluminescence imaging. *Nanotechnology* 2012, 23:075102.

27. Fales, A. M., Yuan, H., Vo-Dinh, T. Development of hybrid silver-coated gold nanostars for nonaggregated surface-enhanced Raman scattering. *J. Phys. Chem. C* 2014, 118:3708–3715.

28. Bath, J., Turberfield, A. J. DNA nanomachines. *Nat. Nanotechnol.* 2007, 2:275–284.

29. Zhang, D. Y., Seelig, G. Dynamic DNA nanotechnology using strand-displacement reactions. *Nat. Chem.* 2011, 3:103–113.

30. Zhang, D. Y., Winfree, E. Control of DNA strand displacement kinetics using toehold exchange. *J. Am. Chem. Soc.* 2009, 131(47):17303–17314.

31. Wang, H. N., Crawford, B. M., Fales, A. M., Bowie, M., Seewaldt, V. L., Vo-Dinh, T. Multiplexed detection of microRNA biomarkers using SERS-based inverse molecular sentinel (iMS) nanoprobes. *J. Phys. Chem. C* 2016, 120(37):21047–21055.

32. Selcuklu, S. D., Donoghue, M. T., Spillane, C. miR-21 as a key regulator of oncogenic processes. *Biochem. Soc. Trans.* 2009, 37:918–925.

the absorption continues to increase steeply. This indicates a decrease in quantum efficiency for excitation wavelengths below 350–400 nm.

In addition to the static or CW optical properties illustrated in Figure 15.1, the time-dependence of the PL can have important implications for biophotonic applications. The PL lifetime (representative time for a photon to be emitted following optical excitation) is much longer for silicon nanoparticles than for fluorescent organic dyes, which typically have subnanosecond fluorescence lifetimes, or for nanocrystals of direct band gap semiconductors like CdSe, which typically exhibit PL lifetimes of 20–50 ns. Figure 15.2 shows room temperature PL intensity versus time data for different wavelength ranges. Emission in each wavelength range originates from silicon nanoparticles of a different size, but all within the same sample. The PL lifetimes seen here, of order 10 μs, are comparable to those observed by others and to those observed in porous silicon at these wave lengths. The trend toward decreasing PL lifetime with decreasing particle size can be rationalized in terms of greater spatial overlap of the electron and hole wave functions as they are confined to a smaller volume. For the shorter wavelength emission (400–500 nm) like that seen in the PL spectrum in Figure 15.1, the lifetimes are much shorter, typically of order 10 ns [17]. It is not entirely clear whether the much shorter lifetimes observed for blue emission indicate that it has a different origin than the yellow to red emission, or if they simply reflect an accelerating trend toward shorter lifetimes at smaller particle sizes.

Another important attribute of silicon nanoparticles is their relatively high quantum yield. The earliest reports on free silicon nanoparticles (not incorporated in porous silicon) showed absolute emission quantum yields (ratio of photons emitted to photons absorbed) of about 5% at room temperature, increasing to 50% below 50 K [10]. From the low-temperature behavior in that study, it was concluded that a fraction of the nanoparticles were dark, with a quantum yield of 0, and the remaining bright particles had a quantum yield of unity (100%) at low temperature. Credo et al. [18] came to the same conclusion based on experiments in which they imaged individual nanoparticles on a glass surface by both their topography and their PL. In their sample, only about 3% of the particles that were seen in a topographic scan were photoluminescent. Those that did luminesce had an estimated

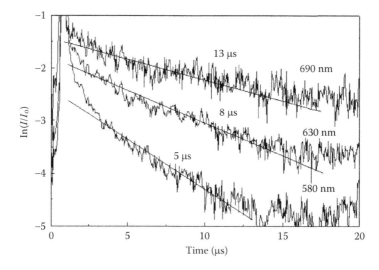

FIGURE 15.2
Time-resolved PL from silicon nanocrystals at different wavelengths. (From Cartwright, A.N. et al., *Proc SPIE: Int Soc Opt Eng* 5222: 134–139. With permission.)

quantum yield near unity (about 88%). Whereas a wide range of quantum yields have been reported in the literature, it seems clear that there is no fundamental limitation on quantum yield from silicon nanoparticles, as would be the case if there were nonradiative recombination pathways that were inherently present in every nanocrystal. Rather, it seems that near-unity quantum yields are possible, provided particles of sufficiently high quality can be prepared, or provided that the bright particles can be separated from the dark ones.

A final important aspect of the PL from silicon nanocrystals is the ease with which this PL can be quenched by interactions with other molecules or entities in solution. This can be considered to be a result of the long PL lifetime of these materials. Because the radiative lifetime is so long, any nonradiative recombination pathways that are introduced by interactions with the environment are likely to be faster than radiative recombination. There have been reports of the quenching of PL from porous silicon or silicon nanoparticles by organic solvents [19], acids and bases [20], amines [21], aromatic nitro compounds [22], anionic and cationic surfactants [23], salts of Cu, Ag, and Au [24], and other semiconductor nanoparticles [25]. Different quenching mechanisms are probably operative in different cases. In porous silicon or agglomerated nanoparticles, there can be quenching by electron transfer between nanocrystals, which is dependent on the dielectric constant of the solvent surrounding the nanocrystals [26,27]. Variations in the effectiveness of different amines in quenching PL from porous silicon have been studied in detail [28], and reduced quenching efficiency of bulkier amines was attributed to steric limitations on their diffusion through the porous silicon network. However, in stable dispersions of individual silicon nanoparticles, neither interparticle electron transfer nor diffusion limitations should be possible. Nevertheless, a recent study of the quenching of PL from free silicon nanoparticles by different amines showed that quenching effectiveness decreased with increasing amine size, from ethylamine to triethylamine [29]. In some cases, PL could be partially restored by addition of a weak organic acid that protonates the amines. Quenching ability could not be simply correlated with the dipole moment, acidity (pKa), or any other single property of the quencher molecules, but appeared to be a more complex function of molecular properties. Strong effects of the solvent on the PL intensity and wavelength have also been observed for these particles [15], but could not be simply correlated with the solvent index of refraction, dipole moment, dielectric constant, or orientational polarizability. In at least some cases, quenching of PL by amines and other compounds can be reduced or eliminated through surface modification [30].

From the point of view of biophotonics, and particularly imaging applications, some of the optical properties of silicon are obviously advantageous, such as their high (or potentially high) quantum yield, their resistance to photobleaching, the large separation between excitation and emission wavelengths, and the possibility of tuning their emission wavelength across the visible spectrum without significantly changing the excitation spectrum. The ability to access emission wavelengths in the near infrared is also an advantage, because tissues are relatively transparent at those wavelengths. However, the need to use blue or UV excitation is probably a disadvantage, because tissues are more absorbing at those wavelengths and both cell and tissue damage and autofluorescence can be problems when short excitation wavelengths are used. A less obvious disadvantage that remains to be fully quantified is simply that the silicon nanoparticles are relatively weak absorbers overall. It may be possible to overcome these disadvantages related to the silicon absorption spectrum and cross section by coupling the particles to another molecule or material. Biteen et al. [31] demonstrated enhancement of not only the absorbance but also the emission rate and quantum efficiency in silicon nanoparticles coupled to a nanoporous layer of gold.

The long PL lifetimes of silicon nanoparticles may be advantageous, in that they allow for time-gated detection strategies where there is a delay between pulsed excitation and detection of emission. This allows emission from the nanoparticles to be easily separated from autofluorescence, which decays much faster (typically with subnanosecond lifetimes). On the other hand, the long PL lifetime limits the overall rate at which a single particle can emit photons, which gives an upper limit for brightness in single particle imaging. In their single-particle studies, Credo et al. [18] measured maximum photon count rates of about 20 kHz (20,000 *detected* photons per second) in a confocal microscope with a collection efficiency of 5%. This was consistent with the measured PL lifetime of 2.2 μs of the particles used in that study, provided that the particles have near-unity quantum yield. For particles with longer PL lifetimes, the saturated photon count rate would be proportionally smaller.

The sensitivity of the PL from silicon nanocrystals to quenching and environmental changes also has both positive and negative aspects. The fact that the PL from bare silicon nanocrystals is readily quenched by amines is an obvious disadvantage for their use in biological systems where molecules with amine groups are omnipresent. As discussed below, it also limits the strategies that can be used for functionalizing the particle surface and making the particles water dispersible, because it may limit the use of amine chemistry. However, it is clear that, at least in some cases, the nanoparticle surface can be coated to dramatically reduce or completely prevent quenching by amines and other potential quenchers. Thus, it seems possible to isolate the particles from their environment such that quenching and environmental effects become unimportant, but some care must be taken to do so. It may be possible to take this one step further and harness the PL quenching to provide specific detection of particular substances for sensing applications. For example, one can imagine attaching amine-containing compounds to the surface of a silicon nanocrystal through a linker molecule that can be selectively cleaved. In this case, the PL would be quenched until the amine groups were removed or protonated. If this protonation of the amines or cleavage of the linker molecule could be carried out selectively, then it would provide a method of sensing the presence of the entity that protonates the amine or cleaves the linker.

15.3 Methods of Preparing Silicon Nanoparticles

A wide variety of methods for producing photoluminescent silicon nanoparticles have been developed over the past 15 years. However, no clearly superior preparation method has emerged. There are no established recipes that can be used to produce high-quality monodisperse samples of photoluminescent silicon nanoparticles like those available for CdSe and some other well studied compound semiconductor quantum dots. Hot colloidal synthesis using surfactants or coordinating solvents, which works well for the II–VI compounds does not work similarly for silicon. Fundamentally, this is because silicon has stronger, more covalent (less ionic) bonding that makes it more difficult to reversibly add and remove atoms at the surface of a growing nanocrystal. Much higher temperature are required to achieve crystallinity in silicon than in compound semiconductors, which is another result of the stronger, covalent bonding present in silicon. In this section, we will review the wide variety of silicon nanoparticle synthesis methods that have been developed and discuss some of the advantages and disadvantages of each method.

15.3.1 Dispersion from Porous Silicon

Soon after the first reports of visible PL from electrochemically etched porous silicon, Heinrich et al. [32] reported the preparation of luminescent colloidal suspensions of silicon particles by ultrasonic treatment of photoluminescent porous silicon. The porous silicon particles that resulted ranged in size from individual nanocrystals (several nanometers in diameter) to much larger porous structures (many microns in size). Bley and coworkers subsequently improved and optimized the anodization (electrochemical etching) conditions, solvent, and sonication time and intensity to produce colloids of more uniform nanoparticles [33]. These colloids contained individual and agglomerated silicon nanocrystals, primarily 2–11 nm in diameter. More recently, Nayfeh and co-workers have further refined this approach [34–36], using hydrogen peroxide to catalyze the anodization reactions, a postetch HF treatment to weaken the connections between nanocrystals in the porous silicon, and centrifugation or gel permeation chromatography to separate the particles into size fractions. They report preparation of discrete particle sizes with diameters of about 1.0, 1.67, 2.15, 2.9, and 3.7 nm. The first four of these sizes had PL emission maxima of 410, 540, 570, and 600 nm, respectively. Dispersion of silicon nanocrystals from electrochemically etched porous silicon is a promising method for producing the small quantities of photoluminescent nanocrystals needed for biophotonic applications. Methods of producing porous silicon have been widely studied, with roughly 10,000 papers related to porous silicon published between 1990 and 2005. With reference to this literature, and particularly the papers cited above in which particles were produced, one can readily establish the experimental capability to produce luminescent particles by this method. Another advantage of this approach is that it starts from high-purity, single-crystalline silicon and is therefore expected to always produce particles with good crystallinity. A disadvantage of this approach is its relatively low yield of individual luminescent nanocrystals, as opposed to larger nanoporous particles, material etched into solution, leftover wafer material, etc. There may be also some disadvantages in terms of the surface state of the particles produced. Their surface termination with hydrogen, oxygen species, or mixtures thereof, is dependent on the composition of the etching bath, which, in turn, is coupled to the etching rate, particle size, etc.

15.3.2 Solution-Phase Methods

Preparation of silicon nanoparticles by liquid-phase solution synthesis was first reported by Heath [37], who reduced a mixture of chlorosilanes with sodium metal in organic solvents at high temperature ($\sim$385°C) and pressure (>100 bar). When a mixture of $SiHCl_3$ and $SiCl_4$ was reduced, the particle sizes ranged from 5 nm to 3 µm, but when octyltrichlorosilane was used instead of trichlorosilane, the particle size range was 2–9 nm. No PL was reported for these particles. Some of the particles were crystalline, but the fraction of the material that was crystalline was low, perhaps due to the relatively low synthesis temperature, relative to the melting point of silicon.

Kauzlarich and coworkers have developed several solution-phase syntheses of silicon nanocrystals. These include the reaction of the Zintl compound KSi with $SiCl_4$ [38,39], reaction of Mg_2Si with $SiCl_4$ followed by addition of a Grignard reagent that provides alkyl groups attached to the surface [40], reaction of Mg_2Si with $SiCl_4$ followed by reaction with $LiAlH_4$, which provides hydrogen termination of the surface [41], reaction of sodium naphthalide with $SiCl_4$ followed by reaction with octanol to yield octanol-capped silicon nanoparticles [42], reaction of sodium naphthalide with $SiCl_4$ followed by reaction with *n*-butyllithium to yield larger, tetrahedral nanocrystals [43], oxidation of Mg_2Si by Br_2

followed by reaction with *n*-butyllithium [44], and reaction of sodium naphthalide with SiCl$_4$ followed by reaction with methanol, then water, then octyltrichlorosilane to give siloxane-coated nanoparticles [45]. Although these particles were, in most cases, photoluminescent, their PL emission was in the blue region of the spectrum rather than the red to near-IR that would be expected for the particle sizes that were observed by transmission electron microscopy (TEM). In the most recent of these publications [45], the particles had an average diameter of 4.5 nm, but had peak PL emission intensity near 400 nm.

Wilcoxon and coworkers prepared silicon nanocrystals by the reduction of silicon halides (SiX$_4$, with X = Cl, Br, or I) with LiAlH$_4$ in inverse micelles [46,47]. They observed PL from these samples at wavelengths spanning the visible spectrum (350–700 nm) but the strongest PL observed was near 400 nm. Longer wavelength PL, near 580 nm, seemed to be independent of particle size. Tilley et al. [48] prepared silicon nanocrystals by the reduction of SiCl$_4$ in reverse micelles with LiAlH$_4$, followed by platinum-catalyzed reaction with 1-heptene that yielded alkyl-terminated particles [48]. These particles were relatively monodisperse, with a mean diameter of 1.8 nm and standard deviation of only 0.2 nm. Their PL emission peaked at 335 nm under 290 nm excitation. This same group prepared water-dispersible silicon nanocrystals by using allylamine instead of 1-heptene as the surface-capping group in the same method [49]. These nanocrystals were 1.4 ± 0.3 nm in diameter, and their PL emission peaked near 480 nm with quantum yields of up to 10%. Despite the widely observed quenching of PL from silicon nanoparticles by amines, at least at longer wavelengths, the direct attachment of an amine to the nanocrystal surface does not seem to have quenched the PL in this case.

Gedanken's group has prepared silicon nanoparticles in solution using a sonochemical approach [50], in which they reduced tetraethyl orthosilicate with colloidal sodium at low temperatures (~200 K) under intense sonication. This seemed to produce agglomerates or a network of porous particles with ethoxy groups on their surface, rather than discrete nanocrystals. These exhibited PL with a peak emission wavelength near 680 nm. Lee et al. [51] combined ultrasonication with the Zintl-salt-based chemistry. They applied high-energy ultrasound to NaSi in glyme for 15–120 min, then cooled the mixture and added SiCl$_4$. After removing excess SiCl$_4$ and adding *n*-butyllithium they obtained butyl-capped nanoparticles with broad PL spectra that peaked in the UV or blue region of the spectrum, but extended as far as 650 nm to give overall PL that appeared blue or white.

Although many of the above-mentioned solution-phase syntheses show promise, none has emerged as a convenient, reproducible method for producing photoluminescent silicon nanoparticles. Potential advantages of solution-phase methods are improved control of particle size distribution and surface termination with surfactants and coordinating solvents. Potential disadvantages of methods developed so far are the need for rigorous air-free reaction conditions, low yields of nanoparticles, and the formation of unwanted reaction by-products. In some cases, crystallinity of the nanoparticles has been questionable. In most cases, solution-prepared nanoparticles have exhibited PL only in the UV and blue regions of the spectrum, and this PL is less easily understood in terms of quantum-confined emission from silicon nanocrystals than the green to red emission observed in porous silicon and silicon nanoparticles prepared by other methods. Blue or UV emission may also be less desirable than longer wavelength emission in bioimaging applications.

15.3.3 Vapor-Phase Methods

Silicon nanoparticles can be prepared in the vapor phase by decomposition of silicon-containing gases or vapors such as silane, disilane, and chlorosilanes, or by thermal, plasma, or laser-driven vaporization (evaporation) of solid silicon followed by nucleation of silicon particles and rapid quenching to prevent their growth and agglomeration. Unwanted silicon particle formation is a common by-product of silicon film deposition by thermal or plasma chemical vapor deposition in the microelectronics industry. Studies of silicon particle formation in the gas phase in that context extend back at least to the 1971 study by Eversteijn [52] that identified the critical temperature and concentration for the onset of particle formation during epitaxial growth of silicon films from silane. There have been many studies of the formation of silicon nanoparticles in the vapor phase since then. Here, we will mainly consider those that resulted in photoluminescent particles and not the many studies of formation of larger silicon nanoparticles. The earliest reports of PL from silicon nanocrystals prepared in the vapor phase coincided with or slightly preceded the discovery of PL from porous silicon. In 1990, Takagi et al. [53] reported production of photoluminescent silicon nanoparticles a few nanometers in diameter by microwave plasma decomposition of silane. These were highly crystalline. The as-produced particles were not photoluminescent, but after heating in humid air to form a passivating oxide layer on their surface, they exhibited room temperature PL at wavelengths of 600–900 nm.

Brus and coworkers produced photoluminescent silicon nanocrystals by thermal decomposition of disilane in a small quartz tube in a pyrolysis oven with short residence time and collected the particles as colloids in ethylene glycol [9,10]. After surface oxidation, these nanocrystals showed size-dependent PL with broad peaks near 660, 770, and 970 nm for particles of different average size. They were able to narrow the PL emission spectra by size-separation of the nanocrystals using size-selective precipitation and size-exclusion chromatography. Their studies on these particles led to major advances in our understanding of PL from silicon nanocrystals. Unfortunately, this is probably not a practical method of making materials for further application, as their reactor produced just a few milligrams of nanoparticles per 24 h day of operation. Coffer's group has used a similar pyrolysis reactor system to prepare erbium-doped silicon nanocrystals that have infrared PL emission near 1550 nm, from the erbium-dopant atoms [54,55] Flagan's group has developed a more sophisticated version of this type of pyrolysis reactor to produce high-quality oxide-capped silicon nanocrystals for use in memory devices [56,57], but in very small quantities, as just a single monolayer of particles is needed in such devices.

From the above studies, it is clear that high-quality nanocrystals can be produced by vapor-phase decomposition of silane or disilane. However, in a conventional heated tube reactor with a residence time of order 1 s, the particle concentration must be below about 10^{12} particles per liter of gas to avoid collision and coagulation of the particles. A 2 nm diameter silicon nanocrystal has a mass of about 10^{-20} g, so this corresponds to a mass concentration of about 10 ng of silicon particles per liter of gas passing through the reactor. The coagulation rate is only weakly dependent on temperature and pressure, so at elevated temperature and reduced pressure, the volume of gas can be increased without increasing the mass flow rate. However, if the pressure is reduced too low, particle losses to the reactor wall will become unacceptably high. So, perhaps 100 ng of silicon particles per standard liter of gas (volume of 1 L at 273 K and 1 bar) can be produced. Thus, about 10 million standard liters of carrier gas would be required to produce 1 g of 2 nm diameter silicon nanoparticles. This corresponds to about 1600 size "A" cylinders of helium. This decreases with the cube of the particle diameter, so just 200 size "A" cylinders would be needed per gram of 4 nm particles. This is consistent with the Brus' group observation that their reactor system, with a somewhat shorter residence time, consumed about 1 size "A" tank of helium per 24 h of operation to produce a few milligrams of silicon nanocrystals [9]. Thus, it is clear that to produce practically useful amounts of silicon nanoparticles in the vapor phase in a reasonable time with reasonable gas consumption and reactor volume, something must be done to either reduce the residence time (by rapidly cooling the gas immediately after particle nucleation) or retard coagulation of the particles. Shorter residence times, or rapid quenching, can be achieved by laser- or plasma-induced heating. Plasma processes can also slow coagulation by producing particles that are all negatively charged and thus repel one another.

One laser-based approach is laser ablation or laser vaporization of solid silicon into a background gas. El-Shall and coworkers have developed a laser vaporization-controlled condensation (LVCC) method in which they focus a frequency-doubled Nd:YAG laser on a silicon target to generate silicon vapor in a background gas [58–61]. The silicon vapor cools very rapidly, nucleating silicon nanocrystals that are collected in the form of web-like agglomerates on a surface above the silicon source. These particles exhibit relatively strong red PL, along with weaker blue PL. Their red PL blueshifted with increased oxidation, consistent with the quantum-confined size-dependent luminescence seen in silicon nanoparticles prepared by other methods. Several other groups have also used laser ablation methods to produce silicon nanocrystals. Among those, the work by Makimura et al. [62] is particularly notable because they observed PL from the nanocrystals within the laser ablation chamber. When H_2 was included in the background gas, they apparently formed hydrogenated silicon nanocrystals that had green PL.

Rather than using laser energy to locally heat and vaporize a solid target, one can use laser energy to dissociate vapor-phase precursor molecules. As early as 1982, Cannon et al. [63,64] used a CO_2 laser to heat silane-containing gases and produce silicon nanoparticles, though these were too much large to exhibit PL. Fojtik et al. [65] used a focused ruby laser to induce breakdown and generate a small volume of plasma in an argon–silane mixture and thereby form silicon nanocrystals. These particles did not initially luminesce, but exhibited red luminescence after etching them with HF and then exposing them to air. Some blue luminescence was also observed. Huisken and coworkers have used a pulsed CO_2 laser to heat silane-containing gas mixtures and generate photoluminescent silicon nanocrystals [66–73]. This produces the nanoparticles as a cluster beam that can be size-separated based on cluster velocity (the smaller particles moving faster than larger ones). These particles do not luminesce immediately after synthesis, but after exposure to air for some time, they exhibit size-dependent PL ranging from orange-yellow to the near IR. The PL continues to blueshift with continued exposure to air, as the surface oxidizes and the crystalline silicon core shrinks. The quantum efficiency of these samples appears to be quite high—30% at minimum, and perhaps much higher [73]. However, this method based on pulsed heating has the obvious disadvantage of a low duty cycle. The pulsed CO_2 laser emits a less than 1 μs pulse, which allows for very rapid heating and cooling of the gases, but when operated at typical repetition rates of a few hertz, results in the formation of very small amounts of material. Botti and coworkers used a continuous CO_2 laser to decompose silane and produce silicon nanocrystals [74–78], similar to the earlier work by Cannon et al [63,64]. By diluting the silane precursor, reducing the total pressure, and other variation of reactor conditions, they were able to reduce the nanoparticle size to the point where some PL was observed.

The final vapor-phase method to be considered here is plasma-based synthesis. This is particularly promising, not only because it can be used to achieve short residence time and selective heating of precursor gases without heating reactor walls, but also because it should initially produce negatively charged silicon nanoparticles that will repel each other and thereby reduce coagulation rates. Two successes in this area have been presented by Mangolini et al. [79] and Sankaran et al [80]. Sankaran et al. [80] used an atmospheric-pressure microdischarge with dimensions of order 1 mm^3 (1 μL) to dissociate a mixture containing 1–5 ppm silane in argon. This produced nanocrystals with relatively narrow size distributions and mean diameters as small as 1–2 nm. These exhibited blue PL with a reported quantum efficiency of 30%. Although a single microdischarge can produce only very small quantities of nanoparticles (tens of micrograms per hour), one can imagine constructing massively parallel arrays of these discharges that could produce macroscopic quantities. Mangolini et al. [79] used a reduced-pressure nonthermal plasma in which the silane-argon gas mixture has a residence time of a few milliseconds. Particles deposited on the quartz tube downstream of this plasma discharge exhibit red to near-infrared PL after a few minutes of exposure to air. The particles had mean diameters of 3–6 nm, controllable by the silane partial pressure and the residence time in the discharge. Most importantly, this method produced luminescent particles at a rate of tens of milligrams per hour, among the highest production rates of all known methods for producing luminescent silicon nanocrystals.

15.3.4 Hybrid Methods

Finally, we will consider some methods that combine aspects of vapor-phase and solution-phase synthesis. The first of these is the thermal decomposition of organosilane precursors in supercritical organic solvents developed by Korgel, Johnston, and coworkers [81,82]. They decomposed diphenylsilane in mixtures of octanol and hexane at 500°C and 345 bar to produce silicon nanocrystals 1.5–4 nm in diameter, capped with organic molecules (presumably octoxy groups). These had PL quantum yields as high as 23%, with shorter PL lifetimes than observed in most other experiments. This method reaches much higher temperatures than conventional solution-phase syntheses, which allows for improved crystallinity like that seen in high-temperature vapor-phase processes. At the same time, it maintains the key advantage of solution-phase synthesis, which is the ability to arrest growth through the adsorption and reaction of ligands on the nanocrystal surface and formation of a stable colloid via the resulting steric stabilization. Thus, this is a promising technique for producing higher quality materials than can be produced by other methods. However, it does not eliminate another problem with solution-phase methods, which is the formation of by-products, which could be organic molecules produced by solvent degradation, polysilanes, organopolysilanes, and so on. If such by-products form, their separation from the desired nanocrystals can be difficult.

In our group, we are applying a method in which we first generate particles in the vapor phase by laser-driven decomposition of silane like that described in the previous section, then etch the particles in solution to reduce their size and passivate their surface such that they become photoluminescent [13,14]. The laser-driven heating of the gas-phase precursors allows us to achieve effective residence times for particle growth of a few milliseconds. This allows us to produce moderately agglomerated particles with primary particle diameters as small as 5 nm at production rates up to ~200 mg/h in a small laboratory-scale reactor. These particles are sufficiently agglomerated and fused that they usually do not show any PL, even after exposure to air. However, etching these particles in a mixture of HF and HNO$_3$ breaks up agglomerates, reduces the particle size, and passivates the surface such that the particles exhibit bright visible PL with a peak wavelength tunable from 500 nm to above 800 nm by varying the etching time and conditions. The yield of particles after etching ranges from a few percent for green-emitting particles to more than 50% for red-emitting particles. The etching can be controlled to produce uniform hydrogen termination of the nanoparticle surface, which is important for subsequent surface functionalization [15]. Recently, Liu et al. [83] have shown that they can produce photoluminescent silicon nanoparticles by simply annealing silicon suboxide (SiO_x with x from 0.4 to 1.8) to nucleate silicon nanoparticles in a matrix of SiO_2, then etching with HF to remove the SiO_2 matrix. These methods involving a second etching step have the disadvantages of being multistep processes and of inevitably wasting material that is etched away. Nonetheless, they have the important advantage of being able to produce macroscopic amounts (hundreds of milligrams per day for a laboratory-scale process) of luminescent material and of being able to control the overall emission color over a broad range.

15.4 Surface Functionalization of Silicon Nanoparticles

If silicon nanoparticles are to be used in a biological environment, then, at minimum, their surface must be modified to make them dispersible in aqueous solutions and to protect them from quenching by compounds present in biological fluids. Surface chemistry is an area where silicon nanoparticles potentially have significant advantages over compound semiconductors. Because silicon forms strong covalent bonds with carbon and oxygen, one can covalently link organic entities to a silicon surface. This contrasts with compound semiconductors where surface functionalization relies on adsorption of bifunctional linker molecules or hydrophobic interactions between surfactants on the nanocrystal surface and a polymer or biomolecule. Covalent attachment not only provides stronger and more robust linkages between the nanoparticle and the molecules attached to it, but also reduces the size of the overall organically capped particle by allowing for shorter linkages between the nanocrystal and the functional groups that make it water dispersible and allow biological functionalization of it.

There are two primary means of attaching organic molecules to a silicon surface: silicon–carbon bonds formed by hydrosilylation reactions, or silicon-oxygen-silicon–carbon linkages formed by silanization reactions. These will be considered separately in Sections 15.4.1 and 15.4.2. Other reactions may also be effective, and will be considered briefly at the end of this section. The surface chemistry of organic molecules on silicon and germanium surfaces has recently been reviewed by both Buriak [84] and Bent [85]. Wayner and Wolkow [86] presented a review that considered only hydrogen-terminated silicon surfaces. Although surface reactions on single-crystalline silicon wafers are often carried out under vacuum, using vapor-phase reagents, this is generally not possible for silicon nanocrystals. Thus, here we consider only solution-phase chemistry. On silicon nanoparticles, it is also not generally possible to carry out reactions that are particular to a certain crystal surface, such as the cycloaddition of alkenes or dienes with silicon dimers on the Si(100) surface. Thus, such reactions are not considered here.

An important fact to keep in mind when considering the attachment of molecules onto freestanding nanocrystals is the effect on the overall size of the structure. Particularly for silicon nanocrystals, which are typically in the 1–5 nm diameter range, the volume of the organic molecules attached to the surface can be comparable to, or even significantly greater than, the volume of the crystalline core. Figure 15.3 shows a model representation of a silicon nanocrystal, roughly 2 nm in diameter (containing 323 silicon atoms) that has been covered only on one hemisphere with $C_{10}H_{20}$ alkyl chains to illustrate the difference in diameter between the coated and the uncoated nanocrystal. The majority of the overall volume of a fully coated nanocrystal like this is made up of the alkyl chains.

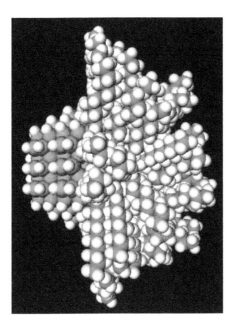

FIGURE 15.3
Model of a ~2 nm silicon nanocrystal with one hemisphere covered with alkyl chains to illustrate the effect of surface functionalization on overall nanoparticle size.

15.4.1 Hydrosilylation Reactions

Hydrosilylation reactions are probably the best-developed solution-phase method for attaching organic molecules to silicon. In this reaction, a silicon–hydrogen bond on a silicon surface reacts with a double or triple bond in an alkene or alkyne to form a direct silicon–carbon bond. Hydrosilylation reactions for attachment of alkenes to a silicon surface was first demonstrated by Linford and coworkers [87,88]. Analogous reactions involving organosilanes in solution are well known. The first application of this reaction to produce stable colloids of free silicon nanoparticles was probably that by Lie et al. [89] Figure 15.4 shows schematically how the hydrosilylation is believed to propagate across the silicon surface. The reaction can be initiated by heating, by visible or UV irradiation, or by a catalyst that generates a free radical site on the surface by removing a hydrogen atom. A double bond reacts with this radical site to form a new silicon–carbon bond and generate a free radical at the neighboring carbon that previously participated in the double bond. That carbon can then abstract a hydrogen atom from a neighboring silicon atom to generate a new radical site, which in turn can react with another alkene and repeat the process. Although some details remain unclear, evidence for this mechanism on both crystalline silicon surfaces and porous silicon has been presented, for example, by de Smet et al. [90] The review by Buriak [84] provides a good overview of hydrosilylation reactions that had been reported in the literature through 2001. The possibility of initiating hydrosilylation reactions using white light, demonstrated for porous silicon by Stewart and Buriak [91] and for H-terminated silicon wafers by Sun et al. [92] may be of particular importance, because it could allow covalent attachment of relatively delicate biomolecules that would not be compatible with high temperature, UV illumination, or some metal-containing catalysts.

Hydrosilylation reactions on hydrogen-terminated crystalline silicon surfaces and porous silicon have been applied to attach a variety of biologically relevant entities to

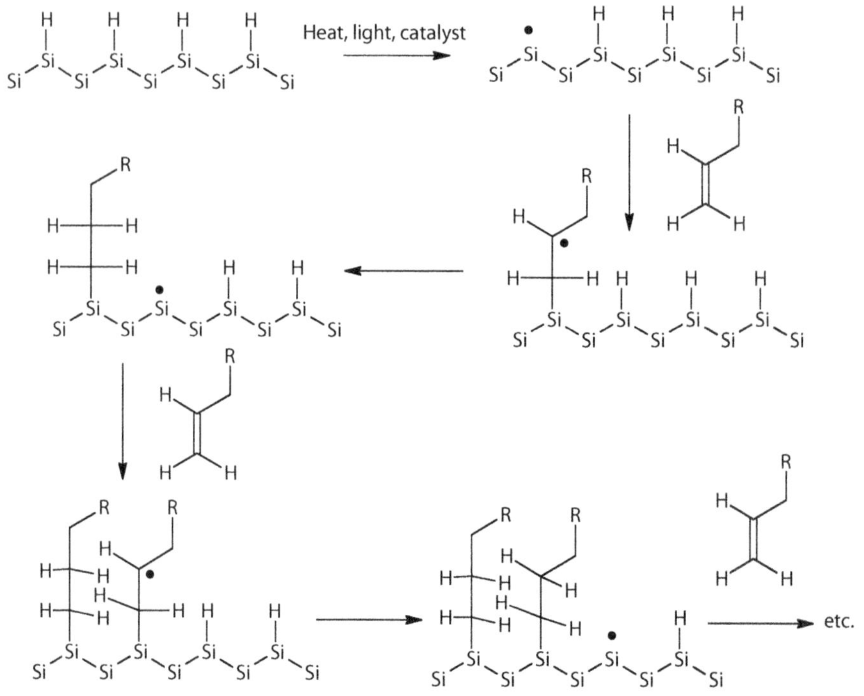

FIGURE 15.4
Schematic mechanism for hydrosilylation of a hydrogen-terminated silicon surface with a terminal alkene.

these surfaces. Presumably, many of the strategies that have been employed will also work for free hydrogen-terminated photoluminescent silicon nanocrystals, though most of them have not been demonstrated. Here, we briefly review early biologically relevant examples of hydrosilylation, in roughly chronological order. In 1997, Wagner et al. [93] reported biofunctionalization of self-assembled monolayers (SAMs) of 1-octene formed by hydrosilylation of an H-terminated Si(111) surface. They used 254 nm UV-initiated hydrosilylation to form a dense monolayer of 1-octene on the wafer surface. They then used two different strategies to create reactive groups for attaching biomolecules to this alkane monolayer. The first was to react it with TDBA-OSu, an aryl-diazirine crosslinker, which created N-hydroxysuccinimidyl groups at the end of about 10% of the octene chains. Amine-terminated molecules can then be attached to these groups, as they demonstrated using amine-functionalized DNA. A problem with this approach was that it left the surface hydrophobic. The second approach was photoinduced chlorosulfonation of the terminal methyl groups of the octene chains, carried out by exposing them to a dilute Cl_2 in SO_2 gas mixture under UV illumination. This created reactive sulfonyl chloride groups that were subsequently reacted with ethylenediamine to form a strong sulfonamide bond and leave amine groups on the surface. This approach involving vapor-phase reagents would be more difficult to apply to free nanocrystals.

In 2000, Strother et al. [94] reported attachment of DNA to silicon wafers by a multistep process.[94] They first attached an ester of undecylenic acid to the H-terminated Si(111) surface by UV-driven hydrosilylation. Then they hydrolyzed the ester using potassium *t*-butoxide in dimethyl sulfoxide (DMSO) to yield a carboxylic acid–covered surface. Then a layer of polylysine was electrostatically bound to the surface through interactions of its terminal amine groups with the carboxylic acid group. To the amine group of this, they attached a heterobifunctional crosslinker molecule with an amine-reactive N-hydroxysuccinimide ester group at one end and a thiol-reactive maleimide group at the other end. Finally, they attached thiol-terminated DNA to the resulting maleimide groups. That same year, Strother et al. [95] reported another strategy for DNA attachment to H-terminated silicon. In this approach, they first synthesized *t*-butoxycarbonyl protected 10-aminodec-1-ene. They attached this to an H-terminated Si(001) surface by UV-initiated hydrosilylation. After removal of the *t*-butoxycarbonyl protecting group, this provided an amine-terminated surface, to which they could attach thiol-terminated DNA with the same heterobifunctional crosslinker mentioned above. Lin et al. [96] carried out a detailed study of this same approach using unprotected and *t*-butoxycarbonyl protected 1-amino-3-cyclopentene, and found that use of the protected amine was essential. In that study, they also compared three different bifunctional linking molecules for attaching thiol-terminated DNA to the amine groups on the silicon surface.

In a series of publications [97], Horrocks and coworkers synthesized DNA directly on planar and porous silicon substrates. They first used thermally driven hydrosilylation to attach 4,4'-dimethoxytrityl-protected ω-undecanol to the silicon surface. Removal of the protecting group then gave a primary alcohol-terminated surface. They used these functionalized flat and porous silicon substrates in a standard DNA synthesizer for solid-state synthesis of oligonucleotides. They also attempted to break up the porous silicon into nanoparticles after the solid-state DNA synthesis to produce DNA-coated nanocrystals. This met with limited success, however, as porous substrates that were sturdy enough to withstand the DNA synthesis were not easily broken up into nanoparticles, and after the DNA attachment, they could not readily be further etched.

de Smet et al. [98] used both thermal and white-light initiated hydrosilylation to attach mixtures of 1-alkenyl saccharides and 1-decene to silicon surfaces in a mixed monolayer. Hart et al. [99] used Lewis acid–catalyzed hydrosilylation to attach hex-5-ynenitrile to the surface of porous silicon. Subsequent treatment with LiAlH$_4$ in ether reduced the nitrile group to a primary amine. Heterobifunctional linker molecules were then used to attach biomolecules of interest, and the PL of the porous silicon was maintained. Cai's group used 254 nm UV-induced hydrosilylation to attach α-oligo(ethylene glycol)-ω-alkenes to H-terminated Si(111) surfaces [100]. The resulting oligo(ethylene glycol) surface was hydrophilic and resistant to protein binding. They were able to pattern this film using conductive atomic force microscopy (AFM) and attach avidin on the patterned spots for subsequent protein binding. Xu et al. [101] attached 4-vinylbenzyl chloride to the Si(111) surface by UV-induced hydrosilylation, then used the Cl-terminated surface to initiate atom transfer radical polymerization of a macromonomer from the surface. They then coupled heparin to the –OH groups of the polyethylene glycol-based polymer layer to produce an anti-thrombogenic surface.

Although many of the above-mentioned investigations used alkenes with a protected amine or carboxyl group at the opposite end, Voicu et al. [102] were able to attach undecylenic acid (10-undecenoic acid) to H-terminated Si(111) selectively at the alkene end, without any apparent reaction of the carboxylic acid group with the Si–H surface. They were then able to convert the carboxylic acid group to the corresponding succinimidyl ester for subsequent linking to primary amines, including amine-terminated DNA.

There are relatively fewer examples of the application of hydrosilylation to free silicon nanoparticles, but the available studies suggest that the chemistry on free nanoparticles is similar to that on porous silicon and silicon wafers. Lie et a. [89] initiated hydrosilylation of silicon nanoparticles thermally, by refluxing porous silicon in a toluene solution of

1-octene, 1-undecene, or other molecules with a terminal alkene group. This yielded stable colloidal dispersions of individual nanocrystals. The hydrosilylation reaction was confirmed by Fourier transform infrared (FTIR) spectroscopy, and the approximate size of the resulting alkylated silicon nanocrystals was determined by time-of-flight mass spectrometry (TOFMS). Under UV excitation, the particles exhibited PL with a peak emission wavelength near 670 nm. In our group we have applied both thermally driven [14,29] and UV-photoinitiated hydrosilylation [15] to photoluminescent silicon particles produced by laser-induced vapor-phase decomposition of silane followed by HF–HNO$_3$ etching. The hydrosilylation reaction was confirmed by FTIR and NMR spectroscopies. The PL of the particles was dramatically stabilized by the attachment of organic molecules to their surfaces. When we attached undecylenic acid to the particles via thermally driven hydrosilylation by refluxing in an ethanol solution, we observed significant oxidation in addition to the desired hydrosilylation reaction [14]. Particles with undecylenic acid or octadecene attached via thermally driven hydrosilylation remained susceptible to PL quenching by amines [29]. However, in more recent work, we have prepared denser monolayers of a variety of alkenoic compounds on nanoparticles with more complete hydrogen termination, and have seen improved resistance to PL quenching [15]. Li and Ruckenstein [103] used UV-driven hydrosilylation to attach acrylic acid to the surface of silicon nanoparticles prepared by this same method and were able to prepare a stable dispersion of them in water that maintained its PL. Warner et al. [49] used platinum-catalyzed hydrosilylation to attach allylamine to blue-emitting silicon quantum dots prepared by the reduction of SiCl$_4$ with LiAlH$_4$ in reverse micelles. They were also able to obtain a stable dispersion in water that maintained its PL. Wang et al. [104] used photoinitiated hydrosilylation to attach 1-octene or 1-hexene to silicon nanocrystals ultrasonically dispersed from porous silicon. They then used TDBA-OSu, an aryl-diazirine crosslinker, which created *N*-hydroxysuccinimidyl groups at the end of some or all of the surface-grafted alkyl chains. This allowed them to attach amine-functionalized DNA to the silicon nanoparticles. The oligonucleotide-conjugated silicon nanoparticles maintained their PL and formed stable dispersions in water. Thus, it appears that the wide range of strategies based on hydrosilylation reactions that have been developed for flat silicon wafer surfaces and porous silicon can, at least in many cases, also be applied to free silicon nanocrystals. This approach provides stable, covalent linkage of biologically relevant molecules to the nanoparticle surface, which should make the resulting nanostructures very robust.

15.4.2 Silanization Reactions

A second general method of attaching organic molecules to silicon is through silanization reactions on hydroxyl-terminated silicon surfaces. This approach uses the wide array of methods and organosilane reagents that have been developed for surface modification of glass. A good review and introduction to the formation of silane monolayers can be found within the comprehensive review of SAMs by Ulman [105]. Ruckenstein and Li [106] have reviewed silane SAM formation in the context of subsequent graft polymerization. As illustrated schematically in Figure 15.5, the silanizing agent (an organosilane) typically has one organic group and three alkoxy or halogen (usually chlorine) groups attached to it. The alkoxy or chlorine groups react with surface hydroxyls to form Si–O–Si linkages. They then condense with each other to form a cross-linked siloxane layer on the surface. In some cases, the silanizing compound has a single-reactive (alkoxysilane or chlorine) group, with methyl groups in place of the other two, in which case a non cross-linked layer can be formed. The chlorosilane reagents can form denser, higher quality monolayers, but are

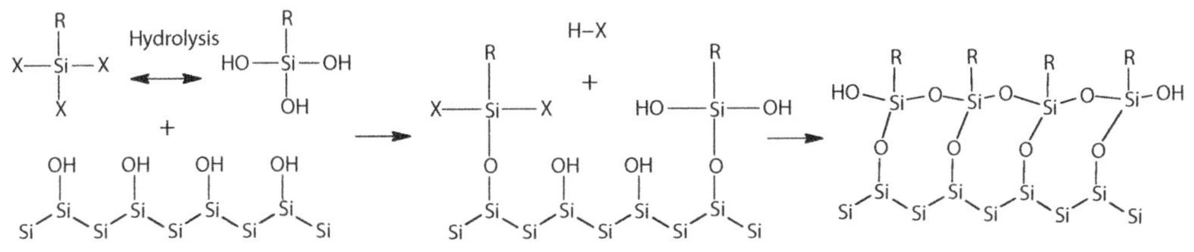

FIGURE 15.5
Schematic mechanism for silanization of a hydroxyl-terminated silicon surface with a chlorosilane or alkoxysilane, where X could be Cl, OCH_3, OCH_2CH_3, etc. The silane can partially or fully hydrolyze in solution before reacting with hydroxyl groups on the surface or afterward. Ultimately, condensation reactions lead to a siloxane layer on the surface, but these condensation reactions can also occur in solution leading to (usually undesirable) polymerization of the silane.

very reactive with water. Trace water is required to prepare a cross-linked layer from them, but in the presence of more than trace amounts of water, they will polymerize in solution. This makes it relatively difficult to achieve reproducible results with them. The alkoxysilanes are much less reactive. They generally will not form dense, high-quality SAMs, but they are much easier to handle and may lead to more reproducible results. Here, we briefly review some examples of this silanization chemistry applied to attach biologically relevant molecules to flat and porous silicon surfaces as well as examples of the application of this approach to silicon nanoparticles.

O'Donnell et al. [107] attached DNA to silicon wafers for subsequent analysis by matrix-assisted laser desorption ionization time-of-flight (MALDI-TOF) mass spectrometry. They first reacted (3-aminopropyl)triethoxysilane with the –OH-terminated silicon surface, prepared by simply washing the wafer with ethanol and flaming it over a Bunsen burner. They then reacted *N*-succinimidyl(4-iodoacetyl)aminobenzoate with the amine groups from the silane to produce an iodoacetamido-terminated surface that could be reacted with thiol-functionalized DNA to attach the DNA to the surface. Sailor's group, in their development of porous silicon-based biosensors, developed multiple silanization-based surface treatment protocols. In one study, they first treated freshly etched H-terminated porous silicon in flowing ozone to create a hydroxylated surface. They then reacted the hydroxylated surface with 2-pyridyldithio(propionamido)dimethylmethoxysilane to produce a noncross-linked layer with pyridyldithio termination on the surface. After cleaving the dithio linkage they reacted the resulting thiol group with the maleimide end of a bifunctional linker molecule having a succinimidyl ester at the other end. Biotinylated bovine serum albumin (BSA) was then bound to the surface by reaction of its amine groups with the succinimidyl ester groups. This provided a biotinylated surface that showed selective reversible binding to appropriate proteins. Their group also reported other similar linking strategies in which (3-aminopropyl)trimethoxysilane was attached to the surface, then glutaraldehyde was used as a linker, or (3-mercaptopropyl)trimethoxysilane was attached to the surface and a maleimide-succinimidyl ester crosslinker was used [108]. In another case, they synthesized a linker molecule (3-bromoacetamidopropyl)trimethoxysilane that could directly provide a bromoacetamido-terminated surface that could be reacted with thiol-functionalized DNA [109].

The above examples, and others not mentioned, clearly demonstrate that silanization chemistry can be used to attach DNA and other biomolecules to silicon wafer surfaces and porous silicon surfaces. This approach also works on free silicon nanoparticles, though there are few published examples. Our group has used nitric acid or sulfuric acid–hydrogen peroxide mixtures (piranha etch) to generate –OH groups on photoluminescent silicon nanoparticles produced by vapor-phase decomposition of silane followed by HF–HNO$_3$ etching [14]. We found that the piranha etch created a much higher density of hydroxyl groups on the surface, compared to nitric acid treatment. Particles treated by both methods were reacted with octadecyltrimethoxysilane to produce alkyl termination. No obvious differences in the quality of this organic layer were observed between the particles that had been surface oxidized by the two methods. In another study, we reacted the piranha etch-treated particles with 3-bromopropyl trichlorosilane, and then used the bromine groups as sites to attach aniline and initiate graft polymerization of polyaniline on the particle surface [30]. Even before aniline substitution and polymerization, the bromopropylsilane monolayer was found to provide substantial protection from chemical degradation and PL quenching of the particles. However, the trichlorosilane chemistry used to prepare the bromopropylsilane monolayer is plagued by sensitivity to trace water, as mentioned above, making consistent reproduction of this protective monolayer difficult.

15.4.3 Other Surface Functionalization Chemistries

In addition to the hydrosilylation and silanization routes described above, other surface functionalization chemistries have been investigated for attaching organic molecules to silicon surfaces. A few of them are briefly discussed here. Direct reaction of alcohols with both porous silicon [110] and silicon nanocrystals dispersed from porous silicon [111] has been reported to form Si–O–R linkages. Surface functionalization of H-terminated silicon with alcohols using iodoform to iodinate the surface in situ was reported to provide much higher coverages of the alcohol on the surface [112]. Light-induced reaction of H-terminated porous silicon with carboxylic acids to produce ester-modified surfaces has been reported by Lee et al. [113,114]. Hydrogen-terminated porous silicon can be reacted electrochemically with organohalides to attach organic molecules to it [115,116], but such electrochemical methods are not practical for free silicon nanoparticles. Lithium reagents (methyllithium, butyllithium, phenyllithium, etc.) can react directly with an H-terminated silicon surfaces to form Si–C bonds [117,118]. However, these highly reactive lithium compounds are much more difficult to handle and work with compared to the simple alkenes or alkynes that can be used in hydrosilylation reactions. Likewise, Grignard reagents have been shown to react directly with hydrogen-terminated porous silicon [119] but may be more difficult to work with and may suffer from limitations on the functional groups that can be included in the Grignard reagents without self-reaction. Attachment of phosponates to the native oxide (nanometer-thick SiO$_2$ layer that forms on silicon wafers in air over time) has been reported and has been used to attach peptides to this surface [120] as well as to form self-assembled alkyl or aryl monolayers [121].

15.5 Applications of Silicon Nanoparticles in Biophotonics

Actual applications of free silicon nanoparticles in biophotonics are still in their infancy. There are two recent reports in which silicon nanoparticles were surface functionalized to make them water dispersible and then used to image cells via fluorescence microscopy. Li and Ruckenstein [103] attached acrylic acid to red-emitting nanoparticles that had been produced by vapor-phase laser pyrolysis followed by solution-phase etching. They incubated fixed Chinese hamster ovary (CHO) cells with an aqueous dispersion of the nanoparticles, and then imaged them, as shown in Figure 15.6. This provides a clear proof-of-concept demonstration of nonspecific cellular imaging, and shows the expected resistance to photobleaching compared to common organic dyes. Similarly, Warner et al. [49] attached allylamine to blue-emitting silicon nanoparticles that had been prepared by solution-phase reduction of $SiCl_4$ in reverse micelles. They incubated HeLa cells with an

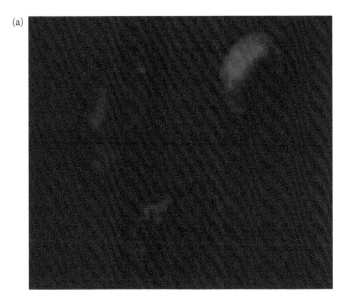

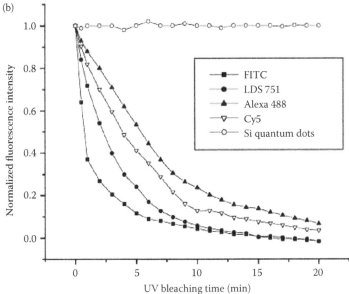

FIGURE 15.6
(a) Fluorescence image of fixed Chinese hamster ovary (CHO) cells stained with acrylic acid-coated luminescent silicon nanoparticles, and (b) bleaching curves for cells stained with these nanoparticles and with several organic dyes commonly used in fluorescence imaging, under continuous illumination from a 100 W mercury lamp using appropriate excitation filters for each wavelength. (Reprinted from Li, Z.F. and E. Ruckenstein, *Nano Lett*, 4, 1463–1467, 2004. © The American Chemical Society. With permission.)

aqueous dispersion of these hydrophilic particles, and then fixed the cells before imaging them in fluorescent microscopy. The particles were clearly visible in the cytosol of the cells, providing a clear demonstration of cellular uptake of the particles. They also demonstrated the resistance of these particles to photobleaching in comparison to organic dyes. Currently, the use of free silicon nanoparticles in biophotonics has not gone beyond these proof-of-principle demonstrations. This is largely due to difficulties in producing high-quality free silicon nanoparticles and making them dispersible in water and in biological fluids. The logical next step beyond these studies is the attachment of molecules to the nanoparticle surface that will allow for specific binding, rather than the nonspecific interaction that has been demonstrated in the above studies. Given that attachment of DNA to silicon nanoparticles has already been reported [104] such demonstrations of specific binding should be forthcoming.

Also of relevance to the application of silicon nanoparticles in biophotonics are their biocompatibility, especially in comparison to the more popular CdSe quantum dots. There have been mixed reports on the biocompatibility of quantum dots of CdSe and other compound semiconductors. Derfus et al. [122] found that mercaptoacetic acid–coated CdSe nanoparticles exhibited acute toxicity toward primary hepatocytes, and they associated this toxicity with the release of Cd^{2+} ions. Surface coating of the CdSe particles with zinc selenide (ZnSe) or BSA substantially reduced their toxicity but did not eliminate it completely. Shiohara et al. [123] observed differences in cytotoxicity for mercapto-undecanoic acid–coated CdSe particles both with respect to the nanoparticle size and the cell type. Lovric et al. [124] studied the cytotoxicity of CdTe quantum dots, and observed greater toxicity for green-emitting particles than for red-emitting particles. Kirchner et al. [125] found that the surface chemistry and propensity of CdSe nanoparticles toward aggregation played an important role in their cytotoxicity, along with their potential to release Cd^{2+}. Selvan et al. encapsulated CdSe and CdSe–ZnS quantum dots within SiO_2 nanoparticles and showed that these particles had dramatically reduced cytotoxicity compared to CdSe quantum dots with other surface treatments. The overall picture that seems to be emerging from these studies is that cytotoxicity of cadmium-containing quantum dots may be controllable, but may remain as a persistent problem.

In contrast to cadmium-containing quantum dots, silicon nanoparticles are expected to be extremely biocompatible. Whereas bulk CdSe is highly toxic, bulk silicon and SiO_2, the product of silicon oxidation, are quite inert. Small SiO_2 particles are a common food additive in products such as frozen orange juice concentrate. Although detailed studies of the cytotoxicity of free, photoluminescent silicon nanoparticles like those cited above for CdSe particles have not yet been presented, the biocompatibility of porous silicon, a closely related material, has been clearly demonstrated. Porous silicon that does not have any protective organic molecules attached to its surface is biodegradable, as shown by Canham in 1995 [126]. Dissolution of porous silicon under physiological conditions releases orthosilicic acid $(Si(OH)_4)$, which is the form in which silicon naturally occurs in blood plasma and other aqueous environments at low concentrations. Surface treatment of porous silicon by attachment of organic molecules can prevent this degradation [127]. The kinetics of this dissolution and their dependence on the porous silicon morphology have been investigated [128]. A variety of mammalian cell types have been cultured on or in the presence of porous silicon, including primary hepatocytes [129], neurons [130–132], and fibroblasts [133]. Silicon nanowires were shown to support the in vitro stability and proliferation of fibroblasts [134]. Nanocomposites of porous silicon with bioerodible polymers such as poly-caprolactone are being investigated as scaffolds for tissue engineering [134], particularly for tissue engineering of bone, because porous silicon has been shown to promote calcification.

Thus, all indications and expectations are that silicon nanoparticles and their potential degradation products (silica and orthosilicic acid) are highly biocompatible and should not pose the difficulties related to cytotoxicity that have arisen for cadmium-containing nanocrystals. Of course, much work remains to be done to demonstrate this. In particular, free silicon nanoparticles can be taken up by cells [49], which may be the intent in imaging studies, whereas it has not been the intent in studies involving porous silicon. This may allow modes of toxicity not possible for porous silicon. It is likely, though, that in experiments where cells have been cultured on porous silicon and some degradation of the porous silicon has occurred, some silicon nanoparticles were released and taken up by the cells with no apparent ill effect. Another important consideration is that molecules attached to the silicon nanoparticle surface, whether to make them water-soluble or to impart specific biological functionality, may induce toxicity. This, of course, is not specific to silicon nanoparticles but would likely occur if the same molecules were attached to nanoparticles of another material. Silicon may have some advantages in this regard as well, as surface-bound molecules can be covalently attached to silicon, rather than linked by a more weakly-bound surfactant. This may make the overall nanoparticles more stable and reduce the release of toxic by-products from it.

15.6 Summary and Conclusions

In this chapter, we have reviewed the optical properties, preparation, surface functionalization, and potential applications of silicon nanoparticles in biophotonics. Compared to compound semiconductor nanocrystals, such as CdSe and CdTe, the biophotonic applications of silicon nanoparticles are much less developed. This is primarily due to the lack of convenient and reliable methods for producing high-quality monodisperse silicon nanoparticles with bright PL and a narrow emission spectrum. The fundamental optical properties of silicon nanocrystals are well-suited for biological imaging and related applications, and silicon nanocrystals can potentially be very efficient light emitters, but there may be some challenges associated with the absorption spectrum of silicon, its relatively long PL lifetime, and the ease with which its PL can be quenched. From about 1990 to 2005, a wide variety of methods was developed for preparing silicon nanocrystals, but none has emerged as the obvious best choice for producing luminescent nanocrystals for biophotonics. These methods continue to improve, and it is now possible to make macroscopic quantities of brightly photoluminescent silicon nanocrystals with peak PL wavelengths spanning the visible spectrum. Silicon has important advantages relative to compound semiconductors in the area of surface functionalization, because silicon forms strong covalent bonds with carbon and oxygen. This allows the attachment of a wide array of organic molecules to silicon with reactions including hydrosilylation and silanization. Strategies for attaching biomolecules to silicon wafers and porous silicon have been developed, and most of these are applicable to free silicon nanocrystals as well. The first demonstrations of the fluorescence imaging of cells using silicon nanocrystals as the fluorophore were published in 2004 and 2005, and attachment of DNA to free silicon nanoparticles was demonstrated around the same time. Thus, imaging studies based on specific attachment of silicon nanocrystals to particular subcellular components should be forthcoming. Whereas the biocompatibility of free silicon nanocrystals has not yet been firmly established, the biocompatibility of nanoporous silicon, silicon nanowires, and the likely degradation products of silicon nanoparticles (silica and orthosilicic acid) are well established. Thus, all indications are that silicon nanocrystals will not present the toxicity concerns that have arisen for cadmium-containing semiconductor nanocrystals. Overall, silicon nanocrystals have tremendous potential for use in biophotonic applications, provided that challenges related to their synthesis can be met. They have important potential advantages over other semiconductor nanocrystals, especially in terms of biocompatibility and flexibility in surface functionalization. In the decade following the first publication of this chapter, many exciting advances in this field were indeed made, including *in vivo* imaging in mice, cytotoxicity testing in mice and monkeys, and integration with other materials to produce multimodal imaging and theranostic agents.

References

1. Bruchez, M., Jr., M. Moronne, P. Gin, S. Weiss, and A.P. Alivisatos. 1998. Semiconductor nanocrystals as fluorescent biological labels, *Science* 281:2013–2015.
2. Chan, W.C.W. and S. Nie. 1998. Quantum dot bioconjugates for ultrasensitive nonisotopic detection, *Science* 281:2016–2018.
3. Michalet, X., F.F. Pinaud, L.A. Bentolila, J.M. Tsay, S. Doose, J.J. Li, G. Sundaresan, A.M. Wu, S.S. Gambhir, and S. Weiss. 2005. Quantum dots for live cells, in vivo imaging, and diagnostics, *Science* 307:538–544.
4. Medintz, I.L., H.T. Uyeda, E.R. Goldman, and H. Mattoussi. 2005. Quantum dot bioconjugates for imaging, labelling and sensing, *Nat Mater* 4:435–446.
5. Smith, A.M. and S. Nie. 2004. Chemical analysis and cellular imaging with quantum dots, *Analyst* 129:672–677.
6. Jaiswal, J.K. and S.M. Simon. 2004. Potentials and pitfalls of fluorescent quantum dots for biological imaging, *Trends Cell Biol* 14:497–504.
7. Canham, L.T. 1990. Silicon quantum wire array fabrication by electrochemical and chemical dissolution of wafers, *Appl Phys Lett* 57:1046–1048.
8. Brus, L. 1994. Luminescence of silicon materials: Chains, sheets, nanocrystals, nanowires, microcrystals, and porous silicon, *J Phys Chem* 98:3575–3581.
9. Littau, K.A., P.J. Szajowski, A.J. Muller, A.R. Kortan, and L. Brus. 1993. *J Phys Chem* 97:1224–1230.
10. Wilson, W.L., P.J. Szajowski, and L. Brus. 1993. A luminescent silicon nanocrystal colloid via a high-temperature aerosol reaction, *Science* 262:1242–1244.
11. Brus, L.E., P.J. Szajowski, W.L. Wilson, T.D. Harris, S. Schuppler, and P.H. Citrin. 1995. Electronic spectroscopy and photophysics of Si nanocrystals: Relationship to bulk c-Si and porous Si, *J Am Chem Soc* 117:2915–2922.
12. Schuppler, S., S.L. Friedman, M.A. Marcus, D.L. Adler, Y.-H. Xie, F.M. Ross, T.D. Harris, W.L. Brown, Y.J. Chabal, L.E. Brus, and P.H. Citrin. 1994. Dimensions of luminescent oxidized and porous silicon structures, *Phys Rev Lett* 72:2648–2650.
13. Li, X., Y. He, S.S. Talukdar, and M.T. Swihart. 2003. Process for preparing macroscopic quantities of brightly photoluminescent silicon nanoparticles with emission spanning the visible spectrum, *Langmuir* 19:8490–8496.
14. Li, X., Y. He, and M.T. Swihart. 2004. Surface functionalization of silicon nanoparticles produced by laser-driven pyrolysis of silane followed by HF– HNO_3 etching, *Langmuir* 20:4720–4727.
15. Hua, F., M.T. Swihart, and E. Ruckenstein. 2005. Efficient surface grafting of luminescent silicon quantum dots by photoinitiated hydrosilylation, *Langmuir* 21:6054–6062.
16. Dash, W.C. and R. Newman. 1955. Intrinsic optical absorption in single-crystal germanium and silicon at 77°K and 300°K, *Phys Rev* 99:1151–1155.

17. Cartwright, A.N., W.D. Kirkey, M.L. Furis, X. Li, Y. He, D. MacRae, Y. Sahoo, M.T. Swihart, and P.N. Prasad. 2003. Ultrafast dynamics in nanostructured materials, *Proc SPIE: Int Soc Opt Eng* 5222:134–139.

18. Credo, G.M., M.D. Mason, and S.K. Buratto. 1999. External quantum efficiency of single porous silicon nanoparticles, *Appl Phys Lett* 74:1978–1980.

19. Lauerhaas, J.M., G.M. Credo, J.L. Heinrich, and M.J. Sailor. 1992. Reversible luminescence quenching of porous silicon by solvents, *J Am Chem Soc* 114:1911–1912.

20. Chun, J.K.M., A.B. Bocarsly, T.R. Cottrell, J.B. Benziger, and J.C. Yee. 1993. Proton gated emission from porous silicon, *J Am Chem Soc* 115:3024–3025.

21. Sweryda-Krawiec, B., R.R. Chandler-Henderson, J.L. Coffer, Y.G. Rho, and R.F. Pinizzotto. 1996. A comparison of porous silicon and silicon nanocrystallite photoluminescence quenching with amines, *J Phys Chem* 100:13776–13780.

22. Germanenko, I.N., S. Li, and M.S. El-Shall. 2001. Decay dynamics and quenching of photoluminescence from silicon nanocrystals by aromatic nitro compounds, *J Phys Chem B* 105:59–66.

23. Canaria, C.A., M. Huang, Y. Cho, J.L. Heinrich, L.I. Lee, M.J. Shane, R.C. Smith, M.J. Sailor, and G.W. Miskelly. 2002. The effect of surfactants on the reactivity and photophysics of luminescent nanocrystalline porous silicon, *Adv Funct Mater* 12:495–500.

24. Andsager, D., J. Hilliard, J.M. Hetrick, L.H. AbuHassain, M. Plisch, and M.H. Nayfeh. 1993. Quenching of porous silicon photoluminescence by deposition of metal adsorbates, *J Appl Phys* 74:4783–4785.

25. Li, S., I.N. Germanenko, and M.S. El-Shall. 1998. Semiconductor nanoparticles in contact: Quenching of the photoluminescence from silicon nanocrystals by WO_3 nanoparticles suspended in solution, *J Phys Chem B* 102:7319–7322.

26. Fellah, S., R.B. Wehrspohn, N. Gabouze, F. Ozanam, and J.-N. Chazalviel. 1999. Photoluminescence quenching of porous silicon in organic solvents: Evidence for dielectric effects, *J Luminescence* 80:109–113.

27. Fellah, S., F. Ozanam, N. Gabouze, and J.-N. Chazalviel. 2000. Porous silicon in solvents: Constant-lifetime PL quenching and confirmation of dielectric effects, *Phys Stat Sol A* 182:367–372.

28. Chandler-Henderson, R.R., B. Sweryda-Krawiec, and J.L. Coffer. 1995. Steric considerations in the amine-induced quenching of luminescent porous silicon, *J Phys Chem* 99:8851–8855.

29. Kirkey, W.D., Y. Sahoo, X. Li, Y. He, M.T. Swihart, A.N. Cartwright, S. Bruckenstein, and P.N. Prasad. 2005. Quasi-reversible photoluminescence quenching of stable dispersions of silicon nanoparticles, *J Mater Chem* 15:2028–2034.

30. Li, Z., M.T. Swihart, and E. Ruckenstein. 2004. Luminescent silicon nanoparticles capped by conductive polyaniline through the self-assembly method, *Langmuir* 20:1963–1971.

31. Biteen, J.S., D. Pacifici, N.S. Lewis, and H.A. Atwater. 2005. Enhanced radiative emission rate and quantum efficiency in coupled silicon nanocrystal-nanostructured gold emitters, *Nano Lett* 5, 1768–1773.

32. Heinrich, J.L., C.L. Curtis, G.M. Credo, M.J. Sailor, and K.L. Kavanagh. 1992. Luminescent colloidal silicon suspensions from porous silicon, *Science* 255:66–68.

33. Bley, R.A., S.M. Kauzlarich, J.E. Davis, and H.W.H. Lee. 1996. Characterization of silicon nanoparticles prepared from porous silicon, *Chem Mater* 8:1881–1888.

34. Yamani, Z., S. Ashhab, A. Nayfeh, W.H. Thompson, and M. Nayfeh. 1998. Red to green rainbow photoluminescence from unoxidized silicon nanocrystallites, *J Appl Phys* 83:3929–3931.

35. Belomoin, G., J. Therrien, A. Smith, S. Rao, R. Twesten, S. Chaieb, M.H. Nayfeh, L. Wagner, and L. Mitas. 2002. Observation of a magic discrete family of ultrabright Si nanoparticles, *Appl Phys Lett* 80:841–843.

36. Belomoin, G., J. Therrien, and M. Nayfeh. 2000. Oxide and hydrogen capped ultrasmall blue luminescent Si nanoparticles, *Appl Phys Lett* 77:779–781.

37. Heath, J.R. 1992. A liquid-solution-phase synthesis of crystalline silicon, *Science* 258:1131–1133.

38. Bley, R.A. and S.M. Kauzlarich. 1996. A low-temperature solution phase route for the synthesis of silicon nanoclusters, *J Am Chem Soc* 118:12461–12462.

39. Mayeri, D., B.L. Phillips, M.P. Augustine, and S.M. Kauzlarich. 2001. NMR study of the synthesis of alkyl-terminated silicon nanoparticles from the reaction of $SiCl_4$ with the Zintl salt, NaSi, *Chem Mater* 13:765–770.

40. Yang, C.-S., R.A. Bley, S.M. Kauzlarich, H.W.H. Lee, and G.R. Delgado. 1999. Synthesis of alkyl-terminated silicon nanoclusters by a solution route, *J Am Chem Soc* 121:5191–5195.

41. Liu, Q. and S.M. Kauzlarich. 2002. A new synthetic route for the synthesis of hydrogen terminated silicon nanoparticles, *Mater Sci Eng B* B96:72–75.

42. Baldwin, R.K., K.A. Pettigrew, E. Ratai, M.P. Augustine, and S.M. Kauzlarich. 2002. Solution reduction synthesis of surface stabilized silicon nanoparticles, *Chem Commun* 17:1822–1823.

43. Baldwin, R.K., K.A. Pettigrew, J.C. Garno, P.P. Power, G.-Y. Liu, and S.M. Kauzlarich. 2002. Room temperature solution synthesis of alkyl-capped tetrahedral shaped silicon nanocrystals, *J Am Chem Soc* 124:1150–1151.

44. Pettigrew, K.A., Q. Liu, P.P. Power, and S.M. Kauzlarich. 2003. Solution synthesis of alkyl- and alkyl/alkoxy-capped silicon nanoparticles via oxidation of Mg_2Si, *Chem Mater* 15:4005–4011.

45. Zou, J., R.K. Baldwin, K.A. Pettigrew, and S.M. Kauzlarich. 2004. Solution synthesis of ultrastable luminescent siloxane-coated silicon nanoparticles, *Nano Lett* 4:1181–1186.

46. Wilcoxon, J.P. and G.A. Samara. 1999. Tailorable, visible light emission from silicon nanocrystals, *Appl Phys Lett* 74:3164–3166.

47. Wilcoxon, J.P., G.A. Samara, and P.N. Provencio. 1999. Optical and electronic properties of Si nanoclusters synthesized in inverse micelles, *Phys Rev B* 60:2704–2714.

48. Tilley, R.D., J.H. Warner, K. Yamamoto, I. Matsui, and H. Fujimori. 2005. Micro-emulsion synthesis of monodisperse surface stabilized silicon nanocrystals, *Chem Commun* 1833–1835.

49. Warner, J.H., A. Hoshino, K. Yamamoto, and R.D. Tilley. 2005. Water-soluble photoluminescent silicon quantum dots, *Angew Chem Int Ed Engl* 44:4550–4554.

50. Dhas, N.A., C.P. Raj, and A. Gedanken. 1998. Preparation of luminescent silicon nanoparticles: A novel sonochemical approach, *Chem Mater* 10:3278–3281.

51. Lee, S., W.J. Cho, C.S. Chin, I.K. Han, W.J. Choi, Y.J. Park, J.D. Son, and J.I. Lee. 2004. Optical properties of silicon nanoparticles by ultrasound-induced solution method, *Jpn J Appl Phys* 43:L784–L786.

52. Eversteijn, F.C. 1971. Gas-phase decomposition of silane in a horizontal epitaxial reactor, *Philips Res Repts* 26:134–144.

53. Takagi, H., H. Ogawa, Y. Yamazaki, A. Ishizaki, and T. Nakagiri. 1990. Quantum size effects on photoluminescence in ultrafine Si particles, *Appl Phys Lett* 56:2379–2380.

54. St. John, J., J.L. Coffer, Y. Chen, and R.F. Pinizzotto. 1999. Synthesis and characterization of discrete luminescent erbium-doped silicon nanocrystals, *J Am Chem Soc* 121:1888–1892.

55. St. John, J., J.L. Coffer, Y. Chen, and R.F. Pinizzotto. 2000. Size control of erbium-doped silicon nanocrystals, *Appl Phys Lett* 77:1635–1637.

56. Ostraat, M.L., J.W. De Blauwe, M.L. Green, L.D. Bell, M.L. Brongersma, J. Casperson, R.C. Flagan, and H.A. Atwater. 2001. Synthesis and characterization of aerosol silicon nanocrystal nonvolatile floating-gate memory devices, *Appl Phys Lett* 79:433–435.

57. Ostraat, M.L., J.W. De Blauwe, M.L. Green, L.D. Bell, H.A. Atwater, and R.C. Flagan. 2001. Ultraclean two-stage aerosol reactor for production of oxide-passivated silicon nanoparticles for novel memory devices, *J Electrochem Soc* 148:G265–G270.

58. Carlisle, J.A., M. Dongol, I.N. Germanenko, Y.B. Pithawalla, and M.S. El-Shall. 2002. Evidence for changes in the electronic and photoluminescence properties of surface-oxidized silicon nanocrystals induced by shrinking the size of the silicon core, *Chem Phys Lett* 326:335–340.

59. Carlisle, J.A., I.N. Germanenko, Y.B. Pithawalla, and M.S. El-Shall. 2001. Morphology, photoluminescence and electronic structure in oxidized silicon nanoclusters, *J Electron Spectrosc* 114–116:229–234.

60. Germanenko, I.N., M. Dongol, Y.B. Pithawalla, M.S. El-Shall, and J.A. Carlisle. 2000. Effect of atmospheric oxidation on the electronic and photoluminescence properties of silicon nanocrystal, *Pure Appl Chem* 72:245–255.

61. Li, S., S.J. Silvers, and M.S. El-Shall. 1997. Surface oxidation and luminescence properties of weblike agglomeration of silicon nanocrystals produced by a laser vaporization—controlled condensation technique, *J Phys Chem B* 101:1794–1802.

62. Makimura, T., T. Mizuta, and K. Murakami. 2002. Laser ablation synthesis of hydrogenated silicon nanoparticles with green photoluminescence in the gas phase, *Jpn J Appl Phys* 41:L144–L146.

63. Cannon, W.R., S.C. Danforth, J.H. Flint, J.S. Haggerty, and R.A. Marra. 1982. Sinterable ceramic powders from laser-driven reactions: I, process description and modeling, *J Am Ceramic Soc* 65:324–330.

64. Cannon, W.R., S.C. Danforth, J.S. Haggerty, and R.A. Marra. 1982. Sinterable ceramic powders from laser-driven reactions: II, powder characteristics and process variables, *J Am Ceramic Soc* 65:330–335.

65. Fojtik, A., M. Giersig, and A. Henglein. 1993. Formation of nanometer-size silicon particles in a laser induced plasma in SiH_4, *Ber Bunsenges Phys Chem* 97:1493–1496.

66. Ehbrecht, M., H. Ferkel, V.V. Smirnov, O.M. Stelmakh, W. Zhang, and F. Huisken. 1995. Laser-driven flow reactor as a cluster beam source, *Rev Sci Instrum* 66:3833–3837.

67. Ehbrecht, M., H. Ferkel, V.V. Smirnov, O. Stelmakh, W. Zhang, and F. Huisken. 1996. Laser-driven synthesis of carbon and silicon clusters from gas-phase reactants, *Surface Surf Rev Lett* 3:807–811.

68. Ehbrecht, M., B. Kohn, F. Huisken, M.A. Laguna, and V. Paillard. 1997. Photoluminescence and resonant Raman spectra of silicon films produced by size-selected cluster beam deposition, *Phys Rev B* 56:6958–6964.

69. Ehbrecht, M. and F. Huisken. 1999. Gas-phase characterization of silicon nanoclusters produced by laser pyrolysis of silane, *Phys Rev B* 59:2975–2985.

70. Huisken, F., B. Kohn, and V. Paillard. 1999. Structured films of light-emitting silicon nanoparticles produced by cluster beam deposition, *Appl Phys Lett* 74:3776.

71. Huisken, F., H. Hofmeister, B. Kohn, M.A. Laguna, and V. Paillard. 2000. Laser production and deposition of light-emitting silicon nanoparticles, *Appl Surf Sci* 154–155:305–313.

72. Ledoux, G., D. Guillois, D. Porterat, C. Reynaud, F. Huisken, B. Kohn, and V. Paillard. 2000. Photoluminescence properties of silicon nanocrystals as a function of their size, *Phys Rev B* 62:15942–15951.

73. Huisken, F., G. Ledoux, O. Guillois, and C. Reynaud. 2002. Light-emitting silicon nanocrystals from laser pyrolysis, *Adv Mater* 14:1861–1865.

74. Borsella, E., M. Falconieri, S. Botti, S. Martelli, F. Bignoli, L. Costa, S. Grandi, L. Sangaletti, B. Allieri, and L. Depero. 2001. Optical and morphological characterization of Si nanocrystals/silica composites prepared by sol–gel processing, *Mater Sci Eng B* B79:55–62.

75. Botti, S., R. Coppola, F. Gourbilleau, and R. Rizk. 2000. Photoluminescence from silicon nanoparticles synthesized by laser-induced decomposition of silane, *J Appl Phys* 88:3396–3401.

76. Botti, S., A. Celeste, and R. Coppola. 1998. Particle size control and optical properties of laser-synthesized silicon nanopowders, *Appl Organometallic Chem* 12:361–365.

77. Borsella, E., S. Botti, M. Cremona, S. Martelli, R.M. Montereali, and A. Nesterenko. 1997. Photoluminescence from oxidised Si nanoparticles produced by CW CO_2 laser synthesis in a continuous-flow reactor, *J Mater Sci Lett* 16:221–223.

78. Borsella, E., S. Botti, S. Martelli, R.M. Montereali, W. Vogel, and E. Carlino. 1997. Optical properties of nanoscale silicon particles obtained by CO_2 laser induced reactions in a flow reactor, *Mater Sci Forum* 235–238:967–972.

79. Mangolini, L., E. Thimsen, and U. Kortshagen. 2005. High-yield plasma synthesis of luminescent silicon nanocrystals, *Nano Lett* 5:655–659.

80. Sankaran, R.M., D. Holunga, R.C. Flagan, and K.P. Giapis. 2005. Synthesis of blue luminescent Si nanoparticles using atmospheric-pressure microdischarges, *Nano Lett* 5:537–541.

81. English, D.S., L.E. Pell, Z.H. Yu, P.F. Barbara, and B.A. Korgel. 2002. Size tunable visible luminescence from individual organic monolayer stabilized silicon nanocrystal quantum dots, *Nano Lett* 2:681–685.

82. Holmes, J.D., K.J. Ziegler, R.C. Doty, L.E. Pell, K.P. Johnston, and B.A. Korgel. 2001. Highly lumi-
    nescent silicon nanocrystals with discrete optical transitions, *J Am Chem Soc* 123:3743–3748.
83. Liu, S.-M., S. Sato, and K. Kimura. 2005. Synthesis of luminescent silicon nanopowders redis-
    persible to various solvents, *Langmuir* 21:6324–6329.
84. Buriak, J.M. 2002. Organometallic chemistry on silicon and germanium surfaces, *Chem Rev*
    102:1271–1308.
85. Bent, S.F. 2002. Organic functionalization of group IV semiconductor surfaces: Principles,
    examples, applications, and prospects, *Surf Sci* 500:879–903.
86. Wayner, D.D.M. and R.A. Wolkow. 2002. Organic modification of hydrogen terminated silicon
    surfaces, *J Chem Soc Perkin Trans* 2:23–34.
87. Linford, M.R. and C.E.D. Chidsey. 1993. Alkyl monolayers covalently bonded to silicon sur-
    faces, *J Am Chem Soc* 115:12631–12632.
88. Linford, M.R., P. Fenter, P.M. Eisenberger, and C.E.D. Chidsey. 1995. Alkyl monolayers on sili-
    con prepared from 1-alkenes and hydrogen-terminated silicon, *J Am Chem Soc* 117:3145–3155.
89. Lie, L.H., M. Deuerdin, E.M. Tuite, A. Houlton, and B.R. Horrocks. 2002. Preparation and char-
    acterisation of luminescent alkylated-silicon quantum dots, *J Electroanal Chem* 538–539:183–190.
90. de Smet, L.C.P.M., H. Zuilhof, E.J.R. Sudholter, L.H. Lie, A. Houlton, and B.R. Horrocks. 2005.
    Mechanism of the hydrosilylation reaction of alkenes at porous silicon: Experimental and
    computational deuterium labeling studies, *J Phys Chem B* 109:12020–12031.
91. Stewart, M.P. and J.M. Buriak. 2001. Exciton-mediated hydrosilylation on photoluminescent
    nanocrystalline silicon, *J Am Chem Soc* 123:7821–7830.
92. Sun, Q.-Y., L.C.P.M. de Smet, B. van Lagen, M. Giesbers, P.C. Thune, J. van Engelenburg, F.A.
    de Wolf, H. Zuilhof, and E.J.R. Sudholter. 2005. Covalently attached monolayers on crystal-
    line hydrogen-terminated silicon: Extremely mild attachment by visible light, *J Am Chem Soc*
    127:2514–2523.
93. Wagner, P., S. Nock, J.A. Spudich, W.D. Volkmuth, S. Chu, R.L. Cicero, C.P. Wade, M.R. Linford,
    and C.E.D. Chidsey. 1997. Bioreactive self-assembled monolayers on hydrogen-passivated
    Si(111) as a new class of atomically flat substrates for biological scanning probe microscopy, *J
    Struct Biol* 119:189–201.
94. Strother, T., W. Cai, X. Zhao, R.J. Hamers, and L.M. Smith. 2000. Synthesis and characterization
    of DNA-modified silicon (111) surfaces, *J Am Chem Soc* 122:1205–1209.
95. Strother, T., R.J. Hamers, and L.M. Smith. 2000. Covalent attachment of oligodeoxyribonucleo-
    tides to amine-modified Si (001) surfaces, *Nucleic Acids Res* 28:3535–3541.
96. Lin, Z., T. Strother, W. Cai, X. Cao, L.M. Smith, and R.J. Hamers. 2002. DNA attachment and
    hybridization at the silicon (100) surface, *Langmuir* 18:788–796.
97. Patole, S.N., A.R. Pike, B.A. Connolly, B.R. Horrocks, and A. Houlton. 2003. STM study of DNA
    films synthesized on Si(111) surfaces, *Langmuir* 19:5457–5463.
98. de Smet, L.C.P.M., G.A. Stork, G.H.F. Hurenkamp, Q.-Y. Sun, H. Topal, P.J.E. Vronen, A.B. Sieval,
    A. Wright, G.M. Visser, H. Zuilhof, and E.J.R. Sudholter. 2003. Covalently attached saccharides
    on silicon surfaces, *J Am Chem Soc* 125:13916–13917.
99. Hart, B.R., S.E. Letant, S.R. Kane, M.Z. Hadi, S.J. Shields, and J.G. Reynolds. 2003. New method
    for attachment of biomolecules to porous silicon, *Chem Commun* 322–323.
100. Yam, C.M., J.M. Lopez-Romero, J. Gu, and C. Cai. 2004. Protein-resistant monolayers prepared
    by hydrosilylation of α-oligo(ethylene glycol)-ω-alkenes on hydrogen-terminated silicon (111)
    surfaces, *Chem Commun* 2510–2511.
101. Xu, F.J., Y.L. Li, E.T. Kang, and K.G. Neoh. 2005. Heparin-coupled poly(poly(ethylene glycol)
    monomethacrylate)-Si(111) hybrids and their blood compatible surfaces, *Biomacromolecules*
    6:1759–1768.
102. Voicu, R., R. Boukherroub, V. Bartzoka, T. Ward, J.T.C. Wojtyk, and D.D.M. Wayner. 2004.
    Formation, characterization, and chemistry of undecanoic acid-terminated silicon surfaces:
    Patterning and immobilization of DNA, *Langmuir* 20:11713–11720.
103. Li, Z.F. and E. Ruckenstein. 2004. Water-soluble poly(acrylic acid) grafted luminescent silicon
    nanoparticles and their use as fluorescent biological staining labels, *Nano Lett* 4:1463–1467.

104. Wang, L., V. Reipa, and J. Blasic. 2004. Silicon nanoparticles as a luminescent label to DNA, *Bioconjugate Chem* 15:409–412.

105. Ulman, A. 1996. Formation and structure of self-assembled monolayers, *Chem Rev* 96:1533–1554.

106. Ruckenstein, E. and Z.F. Li. 2005. Surface modification and functionalization through the self-assembled monolayer and graft polymerization, *Adv Colloid Interface Sci* 113:43–63.

107. O'Donnell, M.J., K. Tang, H. Koster, C.L. Smith, and C.R. Cantor. 1997. High-density, covalent attachment of DNA to silicon wafers for analysis by MALDI-TOF mass spectrometry, *Anal Chem* 69:2438–2443.

108. Tinsley-Brown, A.M., L.T. Canham, M. Hollings, M.H. Anderson, C.L. Reeves, T.I. Cox, S. Nicklin, D.J. Squirrell, E. Perkins, A. Hutchison, M.J. Sailor, and A. Wun. 2000. Tuning the pore size and surface chemistry of porous silicon for immunoassays, *Phys Stat Sol A* 182:547–553.

109. Lin, V.S.-Y., K. Motesharei, K.-P.S. Dancil, M.J. Sailor, and M.R. Ghadiri. 1997. A porous silicon-based optical interferometric biosensor, *Science* 278:840–843.

110. Kim, N.Y. and P.E. Laibinis. 1997. Thermal derivatization of porous silicon with alcohols, *J Am Chem Soc* 119:2297–2298.

111. Swerda-Krawiec, B., T. Cassagneau, and J.H. Fendler. 1999. Surface modification of silicon nanocrystallites by alcohols, *J Phys Chem B* 103:9524–9529.

112. Joy, V.T. and D. Mandler. 2002. Surface functionalization of H-terminated silicon surfaces with alcohols using iodoform as an in situ iodinating agent, *Chem Phys Chem* 11:973–975.

113. Lee, E.J., T.W. Bitner, J.S. Ha, M.J. Shane, and M.J. Sailor. 1996. Light-induced reactions of porous and single-crystal Si surfaces with carboxylic acids, *J Am Chem Soc* 118:5375–5382.

114. Lee, E.J., J.S. Ha, and M.J. Sailor. 1995. Photoderivatization of the surface of luminescent porous silicon with formic acid, *J Am Chem Soc* 117:8295–8296.

115. Gurtner, C., A.W. Wun, and M.J. Sailor. 1999. Surface modification of porous silicon by electrochemical reduction of organo halides, *Angew Chem Int Ed Engl* 38:1966–1968.

116. Lees, I.N., H. Lin, C.A. Canaria, C. Gurtner, M.J. Sailor, and G.M. Miskelly. 2003. Chemical stability of porous silicon surfaces electrochemically modified with functional alkyl species, *Langmuir* 19:9812–9817.

117. Song, J.H. and M.J. Sailor. 1998. Functionalization of nanocrystalline porous silicon surfaces with aryllithium reagents: Formation of silicon–carbon bonds by cleavage of silicon–silicon bonds, *J Am Chem Soc* 120:2376–2381.

118. Song, J.H. and M.J. Sailor. 1999. Reaction of photoluminescent porous silicon surfaces with lithium reagents to form silicon–carbon bound surface species, *Inorg Chem* 38:1498–1503.

119. Kim, N.Y. and P.E. Laibinis. 1998. Derivatization of porous silicon by Grignard reagents at room temperature, *J Am Chem Soc* 120:4516–4517.

120. Midwood, K.S., M.D. Carolus, M.P. Danahy, J.E. Schwarzbauer, and J. Schwartz. 2004. Easy and efficient bonding of biomolecules to an oxide surface of silicon, *Langmuir* 20:5501–5505.

121. Hanson, E.L., J. Schwartz, B. Nickel, N. Koch, and M.F. Danisman. 2003. Bonding self-assembled, compact organophosphonate monolayers to the native oxide surface of silicon, *J Am Chem Soc* 125:16074–16080.

122. Derfus, A.M., W.C.W. Chan, and S. Bhatia. 2004. Probing the cytotoxicity of semiconductor quantum dots, *Nano Lett* 4:11–18.

123. Shiohara, A., A. Hoshino, K.-I. Hanaki, K. Suzuki, and K. Yamamoto. 2004. On the cyto-toxicity caused by quantum dots, *Microbiol Immunol* 48:669–675.

124. Lovric, J., H. Bazzi, Y. Cuie, G.R.A. Fortin, F.M. Winnik, and D. Maysinger. 2005. Differences in subcellular distribution and toxicity of green and red emitting CdTe quantum dots, *J. Mol Med* 83:377–385.

125. Kirchner, C., T. Liedl, S. Kudera, T. Pellegrino, A.M. Javier, H.E. Gaub, S. Stolzle, N. Fertig, and W.J. Parak. 2005. Cytotoxicity of colloidal CdSe and CdSe/ZnS nanoparticles, *Nano Lett* 5:331–338.

126. Canham, L.T. 1995. Bioactive silicon structure fabrication through nanoetching techniques, *Adv Mater* 7:1033–1037.

127. Canham, L.T., C.L. Reeves, J.P. Newey, M.R. Houlton, T.I. Cox, J.M. Buriak, and M.P. Stewart. 1999. Derivatized mesoporous silicon with dramatically improved stability in simulated human blood plasma, *Adv Mater* 11:1505–1507.
128. Anderson, S.H.C., H. Elliot, D.J. Wallis, L.T. Canham, and J.J. Powell. 2003. Dissolution of different forms of partially porous silicon wafers under simulated physiological conditions, *Phys Stat Sol A* 197:331–335.
129. Chin, V., B.E. Collins, M.J. Sailor, and S. Bhatia. 2001. Compatibility of primary hepatocytes with oxidized nanoporous silicon, *Adv Mater* 13:1877–1880.
130. Bayliss, S.C., R. Heald, D.I. Fletcher, and L.D. Buckberry. 1999. The culture of mammalian cells on nanostructured silicon, *Adv Mater* 11:318–321.
131. Ben-Tabou de Leon, S., A. Sa'ar, R. Oren, M.E. Spira, and S. Yitzchaik. 2004. Neurons culturing and biophotonic sensing using porous silicon, *Appl Phys Lett* 84:4361–4363.
132. Bayliss, S.C., L.D. Buckberry, D.I. Fletcher, and M.J. Tobin. 1999. The culture of neurons on silicon, *Sens Actuators A* 74:139–142.
133. Coffer, J.L., M.A. Whitehead, D.K. Nagesha, P. Mukherjee, G. Akkaraju, M. Totolici, R. Saffie, and L.T. Canham. 2005. Porous silicon-based scaffolds for tissue engineering and other biomedical applications, *Phys Stat Sol A* 202:1451–1455.
134. Nagesha, D.K., M.A. Whitehead, and J.L. Coffer. 2005. Biorelevant calcification and non-cytotoxic behavior in silicon nanowires, *Adv Mater* 17:921–924.

Section II

Applications in Biology and Medicine

16

Nanoscale Optical Sensors Based on Surface Plasmon Resonance

Amanda J. Haes, Douglas A. Stuart, and Richard P. Van Duyne

CONTENTS

16.1 Introduction

16.1.1 Importance of Chemical and Biological Sensors

The measurement and detection of molecules and their interactions is the foundation of analytical chemistry as applied to biomedical and environmental sciences. Traditionally, advances in instrumentation and the development of novel detection modalities have resulted in the ability to monitor target species and processes previously inaccessible, generating advances in all realms of science. For example, the development of the portable electrochemical glucometer has improved the routine analysis of blood glucose levels thereby improving the quality of life for millions of diabetics worldwide. Instrumental advances in nuclear magnetic resonance (NMR) spectroscopy and imaging have enabled discoveries in organic chemistry, molecular biology, and cognitive science. Advances in sensor technology are clearly important not only in furthering fundamental biomedical research but also for their direct impact on the general public.

A chemical or biological sensor is a device that responds to varying concentrations of a single analyte or a specific class of chemicals. Fundamentally, a biosensor is derived from the coupling of a ligand–receptor binding reaction [1] to a signal transducer. Much biosensor research has been devoted to the evaluation of the relative merits of various signal transduction methods including optical [2,3], radioactive [4,5], electrochemical [6,7], piezoelectric [8,9], magnetic [10,11], micromechanical [12,13], and mass spectrometric [14,15]. Optical biosensors, in particular, have found a broad base of applications in the detection of a wide range of *biological* molecules such as glucose [16], DNA [17], proteins [18], *Escherichia coli* [19,20], and anthrax [21]. Other optical biosensors present two broad modes of molecular detection: intensity or frequency changes. The first class of optical sensors measure intensity changes at a particular wavelength. For instance, standard ultraviolet–visible (UV–Vis), fluorescence, and infrared absorbance spectroscopies relate the concentration of an analyte to the measured photon throughput. In the other class of optical sensors, such as fluorescence resonance energy transfer (FRET), colorimetry, and surface plasmon resonance (SPR), chemical changes are observed by measuring wavelength shifts.

16.1.2 Optical Sensors: Surface Plasmon Resonance Sensors

Currently, the most widely used optical biosensor is the SPR sensor. This sensor detects changes in the refractive index induced by molecules near the surface of noble metal thin films [22]. Since their original discovery, SPR changes have been used in refractive index-based sensing to detect and monitor a broad range of analyte–surface binding interactions including the adsorption of small molecules [23–25], ligand–receptor binding [26–29], protein adsorption on self-assembled monolayers (SAMs) [30–32], antibody–antigen binding [33], DNA and RNA hybridization [34–37], and protein–DNA interactions [38]. Refractive index sensors have an inherent advantage over other optical biosensors that require a chromophoric group or other label to transduce the binding event. Because all biochemically relevant species have refractive indices greater than air or water, SPR is a universal technique, sensitive to all possible analytes.

Typically, SPR devices utilize one of the three instrumental configurations: angle shift monitored at a fixed input wavelength, wavelength shift measured at a fixed incident angle, and wide-area imaging, which provides multidimensional information. Of all the commercially available SPR instruments, those that detect small angle shifts upon the

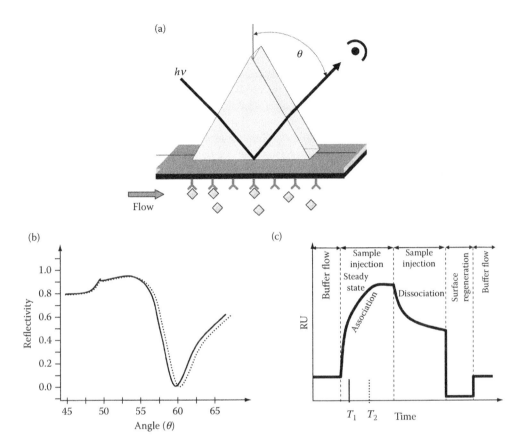

FIGURE 16.1
(a) Experimental setup for an angle shift SPR experiments. Angle shift (and/or wavelength shift) data is recorded as a function of time. This phenomenon occurs when a thin conducting film (such as silver or gold) is placed at the interface between a prism and an external environment. At a specific incident angle, a surface plasmon is formed in the conducting film and is resonant with the light because their frequencies match. As energy is absorbed in this resonance, the intensity of the reflected light minimizes (b). As the dielectric environment at the metal interface increases, the minimum shifts to longer wavelengths. (c) Information about binding kinetics (i.e., association and dissociation rates) is obtained from time-dependent SPR experiments.

absorption of molecules are the most widely implemented. Figure 16.1 depicts the angle shift SPR design. The versatility of this design is manifested by its diverse applications, listed above.

Although SPR spectroscopy is a totally nonselective sensor platform, a high degree of analyte selectivity can be conferred using the specificity of surface-attached ligands and passivation of the sensor surface to nonspecific binding [3,22,39–41]. Chemosensors and biosensors based on SPR spectroscopy possess many desirable characteristics including: (1) a refractive index sensitivity on the order of 1 part in 10^5–10^6 corresponding to an areal mass sensitivity of ~1–10 pg/mm^2 [3,23,24,26]; (2) a long-range sensing length scale determined by the exponential decay of the evanescent electromagnetic field, $L_z \sim 200$ nm [23]; (3) multiple instrumental modes of detection (viz., angle shift, wavelength shift, and imaging) [22]; (4) real-time detection on the 10^{-1}–10^3 s timescale for measurement of binding kinetics [24,25,39,42,43]; (5) lateral spatial resolution on the order of 10 μm enabling multiplexing and miniaturization especially using the SPR imaging mode of detection [22,44];

(6) label-free detection capable of probing complex mixtures, such as clinical material, without prior purification [3,22,40]; and (7) benefits from the availability of commercial instrumentation with advanced microfluidic sample handling [45,46].

The stringent requirements for many advanced applications, such as high-throughput screening, present at least five fundamental challenges to SPR spectroscopy. First, instrument thermostating is required because of the exquisite sensitivity of refractive index changes to temperature. This greatly increases the cost and complexity of the instrument. Second, the SPR angle and wavelength shift detection modes, which have been multiplexed in small arrays, are cumbersome to implement in very large arrays due to the optical complexity of the instrumentation [33,45,47]. Third, while SPR imaging is an important approach to overcoming this problem, it is limited to signal transducer element sizes of a few square micrometers, more typically 10 μm^2 by the excitation wavelength dependent, lateral propagation length, l_d, of the propagating surface plasmon [22]. Fourth, real-time sensing or kinetic measurements using SPR spectroscopy are severely mass transport limited by diffusion to timescales on the order of 10^3–10^4 s for analytes at bulk concentrations, $C_{bulk} < 10^{-6}$–10^{-7} M. Furthermore, as the time required for the analyte surface excess to reach 1/2 saturation coverage scales as the inverse square of C_{bulk} [26], the mass transport problem is greatly exacerbated for C_{bulk} in the low picomolar or high femtomolar domains demanded by many bioassays. Finally, the large size and cost of high-resolution instruments severely limit their application for field portability and low budget projects, respectively.

16.1.3 Motivation for Nanoscale and SPR Integration

The development of nanodevices, including nanosensors that are highly sensitive and selective (give low false positives, low false negatives), has the potential to provide a major improvement over current technologies for disease understanding, treatment, and monitoring. Nanoscale sensors consume less sample volume than conventional instruments because their inherently small size scale in comparison to standard macroscale devices, and permits straightforward integration with microfluidic devices. Additionally, nanoscale systems often exhibit behavior that is markedly different from their macroscale counterparts, thereby providing alternative pathways for obtaining new information.

16.1.4 Localized Surface Plasmon Resonance Sensors

Recently, several research groups have begun to explore alternative strategies for the development of optical biosensors and chemosensors based on the extraordinary optical properties of noble metal nanoparticles. Noble metal nanoparticles exhibit unique extinction spectra (i.e., sum of absorbed and scattered light) that is not observed in their bulk materials, which arises from their localized surface plasmon resonance (LSPR). The LSPR is a collective oscillation of the conduction band electrons at the nanoparticles' surface that develops when incident electromagnetic radiation is of appropriate frequency [48–55]. The LSPR is important not only in phenomena such as surface-enhanced spectroscopies and resonant Rayleigh scattering, but also as a sensitive analytical tool itself. The LSPR of noble metal nanoparticles has been used to detect chemical and biological species because of its sensitivity to refractive index changes near the metal surface.

It was realized that the sensor transduction mechanism of this LSPR-based nanosensor is analogous to that of SPR sensors (Table 16.1). Important differences to appreciate between the SPR and LSPR sensors are the comparative refractive index sensitivities and the characteristic electromagnetic field decay lengths. SPR sensors exhibit large refractive index

TABLE 16.1

Comparison of SPR and LSPR Sensors

Feature/Characteristic	SPR	LSPR
Label-free detection	Yes [25,27,35,40]	Yes [56–59]
Distance dependence	~1000 nm [23]	~30 nm (size tunable) [60,61]
Refractive index sensitivity	2×10^6 nm/RIU [3,23,24,26]	2×10^2 nm/RIU [57,60]
Modes	Angle shift [22] Wavelength shift Imaging	Extinction [56] Scattering [58,62] Imaging [58,62]
Temperature control	Yes	No
Chemical identification	SPR-Raman	LSPR-SERS
Field portability	No	Yes
Commercially available	Yes	No
Cost	$150K–$300K	$5K (multiple particles); $50K (single nanoparticle)
Spatial resolution	~10 μm × 10 μm [22,44]	1 nanoparticle [58,62,63]
Nonspecific binding	Minimal (determined by surface chemistry and rinsing) [3,22,39–41]	Minimal (determined by surface chemistry and rinsing) [56]
Real-time detection	Timescale = 10^{-1}–10^3 s, planar diffusion [24,25,39,42,43]	Timescale = 10^{-1}–10^3 s, radial diffusion [58]
Multiplexed capabilities	Yes [45,46]	Yes (possible)
Small molecule sensitivity	Good [24]	Better [60]
Microfluidics compatibility	Yes	Possible

sensitivities (~2×10^6 nm/RIU) [23]. For this reason, the SPR response is often reported as a change in refractive index units (RIUs). The LSPR nanosensor, on the other hand, has a modest refractive index sensitivity (~2×10^2 nm/RIU) [57]. Given that this number is four orders of magnitude smaller for the LSPR nanosensor in comparison to the SPR sensor, initial assumptions were made that the LSPR nanosensor would be 10,000 times less sensitive than the SPR sensor. This, however, is not the case. In fact, the two sensors are very competitive in their sensitivities. The short (and tunable) characteristic electromagnetic field decay length, l_d, provides the LSPR nanosensor with its enhanced sensitivity [60,61]. Experimental and theoretical results using the LSPR nanosensor indicate that the decay length, l_d, is ~5–15 nm or ~1%–3% of the light's wavelength and depends on the size, shape, and composition of the nanoparticles. This differs greatly from the 200–300 nm decay length or ~15%–25% of the light's wavelength for the SPR sensor [23]. Also, the smallest footprint of the SPR and LSPR sensors differs. In practice, SPR sensors require sufficient area for the establishment of a planar plasmon, at least a 10 μm × 10 μm area for sensing experiments. For LSPR sensing, this spot size can be minimized to a large number of individual sensing elements (1×10^{10} nanoparticles from a 2 mm spot size on samples fabricated using an nanosphere lithography [NSL] mask of 400 nm diameter nanospheres) down to a single nanoparticle (with an in-plane width of ~20 nm) using single nanoparticle techniques [58]. The nanoparticle approach can deliver the same information as the SPR sensor, thereby minimizing the sensor's pixel size to the sub 100 nm regime. Because of the lower refractive index sensitivity, the LSPR nanosensor requires no temperature control whereas the SPR sensor (with a large refractive index sensitivity) requires thermostating. The final and most dramatic difference between the LSPR and SPR sensors is cost. Commercialized SPR instruments can vary between $150K and $300K, whereas the prototype and portable LSPR system costs less than $5K.

There is, however, a unifying relationship between these two seemingly different sensors. Both sensors' overall response can be described using the following equation [23]:

$$\Delta\lambda_{max} = m\Delta n(1 - e^{-2d/l_d})$$ (16.1)

where $\Delta\lambda_{max}$ is the wavelength shift response, m is the refractive index sensitivity, Δn is the change in refractive index induced by an adsorbate, d is the effective adsorbate layer thickness, and l_d is the characteristic electromagnetic field decay length. It is important to note that for planar SPR sensors, this equation quantitatively predicts an adsorbate's effect on the sensor. When applied to the LSPR nanosensor, this exponential equation approximates the response for adsorbate layers but does provide a fully quantitative explanation of its response [60,61]. Like the SPR sensor, the LSPR nanosensor's sensitivity arises from the distance dependence of the average-induced square of the electric fields that extend from the nanoparticles' surfaces.

16.1.5 Nanoparticle Optics: Theory

To more thoroughly understand the LSPR and sensors based thereon, it is necessary to illuminate this phenomenon more thoroughly. Advances in the field of nanoparticle optics have allowed for a deeper understanding of the relationship between material properties such as composition, size, shape, and local dielectric environment and the observed optical properties. An understanding of these properties holds both fundamental and practical significance. Fundamentally, it is important to systematically explore the nanoscale structural and local environmental factors that cause optical property variations, as well as provide access to regimes of predictable behavior. Theoretical insights about the optimal optical response of nanoparticle systems will help to guide future sensor development. Practically, the tunable optical properties of nanostructures can be applied as materials for surface-enhanced spectroscopy [64–68], optical filters [69,70], plasmonic devices [71–74], and sensors [56,59–61,75–85].

The simplest theoretical approach available for modeling the optical properties of nanoparticles is the Mie theory estimation of the extinction of a metallic sphere in the long wavelength, electrostatic dipole limit. In the following equation [86]:

$$E(\lambda) = \frac{24\pi N_A a^3 \varepsilon_m^{3/2}}{\lambda \ln(10)} \left[\frac{\varepsilon_i}{(\varepsilon_r + 2\varepsilon_m)^2 + \varepsilon_i^2} \right]$$ (16.2)

$E(\lambda)$ is the extinction which is, in turn, equal to the sum of absorption and Rayleigh scattering, N_A is the areal density of the nanoparticles, a is the radius of the metallic nanosphere, ε_m is the dielectric constant of the medium surrounding the metallic nanosphere (assumed to be a positive real number and wavelength independent), λ is the wavelength of the absorbing radiation, ε_i and ε_r are the imaginary and real portions of the metallic nanosphere's dielectric function, respectively. The LSPR condition is met when the resonance term in the denominator $((\varepsilon_r + 2\varepsilon_m)^2)$ approaches zero. Even in this most primitive model, it is abundantly clear that the LSPR spectrum of an isolated metallic nanosphere embedded in an external dielectric medium will depend on the nanoparticle radius a, the nanoparticle material (ε_i and ε_r), and the nanoenvironment's dielectric constant (ε_m). As seen in Figure 16.2, as the dielectric of the surrounding environment increases from vacuum, the position of the LSPR spectrum systematically shifts to longer wavelengths. Furthermore, when the nanoparticles

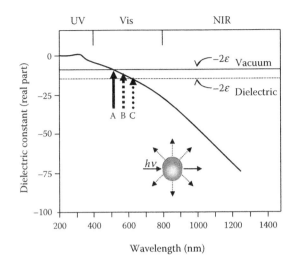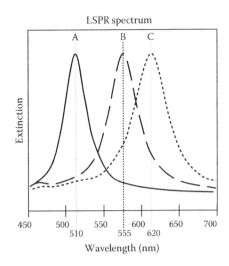

FIGURE 16.2
Dependence of the LSPR spectrum on surrounding dielectric environment. As the medium moves from vacuum (A) to solvents with higher dielectric constants (B–C), the LSPR peak position shifts to longer wavelengths.

are not spherical, as is always the case in real samples, the extinction spectrum will depend on the nanoparticle's in-plane diameter, out-of-plane height, and shape. In this case, the resonance term from the denominator of Equation 16.2 is replaced with

$$(\varepsilon_r + \chi\varepsilon_m)^2 \tag{16.3}$$

where χ, a shape factor term [66], describes the nanoparticle aspect ratio. The values for χ increase from 2 (for a sphere) up to, and beyond, values of 17 for a 5:1 aspect ratio nanoparticle. In addition, many of the samples considered in this chapter contain an ensemble of nanoparticles that are supported on a substrate. Thus, the LSPR will also depend on interparticle spacing and substrate dielectric constant.

Despite the power and popularity of the Mie equations, as deviations from the simple case of a spheroidal particle immersed in a uniform dielectric medium occur, other more complex modeling schemes must be adopted. A number of theories have been advanced and put forward to more accurately describe nanoparticle optics for systems involving coupled nanoparticles, nanoparticles of arbitrary morphology, and multicomponent dielectric environments. Among the more widely used models are the discrete dipole approximation (DDA) and the modified long wavelength approximation (MLWA). Whereas both of models have their individual advantages and disadvantages, the majority of modeling for LSPR nanosensors has been conducted with DDA.

The DDA method [87,88] is a finite element-based approach for solving Maxwell's equations for light interacting with an arbitrary shape and composition nanoparticle. Herein, DDA is used to calculate the plasmon wavelength in the presence or absence of an adsorbate with a wavelength dependent on the refractive index of the layer thickness. For example, to model the optical properties of the nanoparticles used in array-based LSPR sensing, bare silver nanoparticles with a truncated tetrahedral shape were first constructed from cubical elements and one to two layers of the adsorbate were added to the exposed surfaces of the nanoparticle to define the presence of the adsorbate. All calculations refer to silver nanoparticles with a dielectric constant taken from Lynch and Hunter [89].

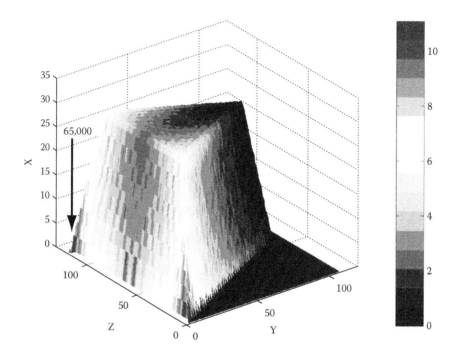

FIGURE 16.3
Local E-field (plotted as contours of $\log|E|^2$) for a Ag nanoparticle ($a = 100$ nm, $b = 30$ nm). (Adapted from Haes, A.J. et al., 2004, *J Phys Chem B* 108(22):6961–6968.)

In this treatment, the dielectric constant is taken to be a local function, as there is no capability for a nonlocal description within the DDA approach. There have been several earlier studies in which the DDA method has been calibrated by comparison with experiment for truncated tetrahedral particles, including studies of external dielectric effects and substrate effects [66,90,91], and based on these results, it is expected that DDA analysis will provide a useful qualitative description of the experimental data. In particular, it has been demonstrated that the plasmon resonance shift caused by molecules located close to the nanoparticle's surface is dominated by hot spots whereas the response from molecules farther away arises from an average of hot and cold regions around the nanoparticle surface [60,61]. By comparing the overall maximum LSPR shifts of the nanoparticles, it has been shown that increasing the aspect ratio of Ag nanotriangles produces larger plasmon resonance shifts and responses dominated by shorter ranged interactions. This property is best displayed in Figure 16.3, which was generated using the DDA method. This geometry leads to fields around the nanoparticles whose magnitudes relative to the incident light are plotted in the figure. The calculations show that the electromagnetic fields are greatly amplified (up to 65,000 times) in the region near the nanoparticle tips, leading to enhanced sensitivity to molecules that might be located there. This information can be used as a basis for constructing nanoparticle structures, and understanding the origins and mechanisms for surface-enhanced spectroscopies.

16.1.6 Outline and Organization

The remainder of this chapter is organized into two sections: (1) LSPR sensing on single nanoparticles and (2) LSPR sensing on nanoparticle arrays. The section on single nanoparticle platforms will be used to illustrate the relative merits of LSPR sensing and experimental apparatus. In the later section, it will be demonstrated that these sensor modalities can be extended to arrays of nanoparticles, and that simple UV–Vis spectroscopy can be used to perform sensing experiments that are comparable to the flat surface SPR sensor technology.

16.2 Single Nanoparticle LSPR Sensing

16.2.1 Introduction to Single Nanoparticle Sensing

It is apparent from Equations 16.1 and 16.2 above that the location of the extinction maximum of noble metal nanoparticles is highly dependent on the dielectric properties of the surrounding environment and that wavelength shifts in the extinction maximum of nanoparticles can be used to detect molecule-induced changes surrounding the nanoparticle. As a result, there are at least four different nanoparticle-based sensing mechanisms that enable the transduction of macromolecular- or chemical-binding events into optical signals based on changes in the LSPR extinction, scattering intensity shifts in LSPR λ_{max}, or both. These mechanisms are: (1) resonant Rayleigh scattering from nanoparticle labels in a manner analogous to fluorescent dye labels [58,62,92–98], (2) nanoparticle aggregation [99–103], (3) charge-transfer interactions at nanoparticle surfaces [57,86,104–107], and (4) local refractive index changes [56,57,59,61,77–79,108–112]. Previous reviews have encompassed particularly the first two mechanisms. Herein, we choose to focus on the final method, as it is not only more readily described by the equations given above, but also more closely analogous to planar SPR experiments.

The key to exploiting single nanoparticles as sensing platforms is to develop a technique, which monitors the LSPR of individual nanoparticles with a reasonable signal-to-noise ratio. UV–Vis absorption spectroscopy does not provide a practical means of accomplishing this task. Even under the most favorable experimental conditions, the absorbance of a single nanoparticle is very close to the shot-noise-governed limit of detection (LOD). Instead, resonant Rayleigh scattering spectroscopy is the most straightforward means of characterizing the optical properties of individual metallic nanoparticles. Similar to fluorescence spectroscopy, the advantage of scattering spectroscopy is that the scattering signal is being detected in the presence of a very low background. The instrumental approach for performing these experiments generally involves using high magnification microscopy coupled with oblique or evanescent illumination of the nanoparticles. Klar et al. [113] utilized a near-field scanning optical microscope coupled to a tunable laser source to measure the scattering spectra of individual gold nanoparticles embedded in a TiO_2 film. Sönnichsen et al. [95] were able to measure the scattering spectra of individual electron beam-fabricated nanoparticles using conventional light microscopy. Their technique involved illuminating the nanoparticles with the evanescent field produced by total internal reflection of light in a glass prism. The light scattered by the nanoparticles was collected with a microscope objective and coupled into a spectrometer for analysis. Matsuo and Sasaki [114] employed differential interference contrast microscopy to perform time-resolved laser scattering spectroscopy of single silver nanoparticles. Mock et al. [115] correlated conventional dark-field microscopy and transmission electron microscopy (TEM) in order to investigate the relationship between the structure of individual metallic nanoparticles and their scattering spectra. McFarland and Van Duyne [58] have also used the same light microscopy techniques to study the response of the scattering spectrum to the particle's local dielectric environment by immersing the nanoparticle in solvents of various refractive indices. Illustrative examples of these and other experiments are presented below.

16.2.2 Single Nanoparticle Experimental Parameters

Colloidal Ag nanoparticles were prepared by reducing silver nitrate with sodium citrate in aqueous solution according to the procedure developed by Lee and Meisel [116]. These nanoparticles were immobilized by drop coating approximately 5 μL of the colloidal solution onto a No. 1 coverslip and allowing the water to evaporate. The coverslip was then inserted into a custom-designed flow cell. Before all experiments, the nanoparticles in the flow cell were repeatedly rinsed with methanol and dried under nitrogen in order to establish equilibrium surface adsorption of solvent molecules and citrate anions. All optical measurements were performed using an inverted microscope (Eclipse TE300, Nikon Instruments) equipped with an imaging spectrograph (SpectroPro 300i, Roper Scientific) and a charge-coupled device (CCD) detector (Spec-10:400B, Roper Scientific). A color video camera was attached to the front port of the microscope to facilitate identification and alignment of the nanoparticles. The experimental apparatus is schematically represented in Figure 16.4. A dark-field condenser (NA = 0.95) was used to illuminate the nanoparticles and a variable aperture 100× oil immersion objective (NA = 0.5–1.3) was used to collect the light scattered by the nanoparticles.

Figure 16.5 illustrates the technique used to acquire the resonant Rayleigh scattering spectrum of single nanoparticles. First, the spectrometer grating was placed in zero order and the spectrometer entrance slit was opened to the maximum setting in order to project a wide-field image onto the CCD. Next, a nanoparticle was placed in the center of the field

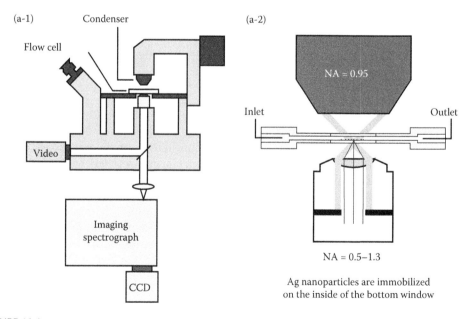

FIGURE 16.4
(a-1) Instrumental diagram used for single nanoparticle spectroscopy. (a-2) Close-up of the flow cell to show illumination and collection geometry. (Adapted from Van Duyne, R.P. et al., 2003, *SPIE* 5223:197–207.)

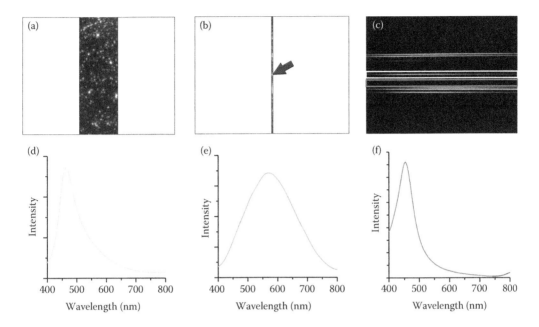

FIGURE 16.5

(a) Wide-field image of immobilized Ag nanoparticles. (b) A nanoparticle is centered and the entrance slit is closed to 50 μm. (c) The grating is rotated into a dispersion configuration and the regions of interest are selected (top box, nanoparticle spectrum; lower box, background). (d) Raw scattering spectrum of a single Ag nanoparticle. (e) Lamp profile used for normalization of the scattering spectrum. (f) Normalized scattering spectrum with the LSPR λ_{max} = 452 nm. (Adapted from Van Duyne, R.P. et al., 2003, *SPIE* 5223:197–207.)

and the entrance slit was closed to 50 μm. Then the spectrometer grating (150 g/mm) was rotated to disperse first-order diffracted light onto the CCD. To ensure that only the scattered light from a single nanoparticle was analyzed, the region of interest was selected using the CCD control software. An adjacent empty region of the CCD with the same dimensions was also collected in order to perform a background subtraction. Integration times varied depending on the lamp intensity and the scattering strength of the nanoparticle, but a typical acquisition comprised of accumulating five exposures, each 15 s in duration. Finally, the raw scattering spectrum was normalized to correct for the lamp spectral profile, spectrometer throughput, and CCD efficiency. This was accomplished by dividing the raw spectrum by the lamp spectrum, which was obtained by increasing the numerical aperture of the objective above 0.95.

16.2.3 Single Nanoparticle Refractive Index Sensitivity

The local refractive index sensitivity of the LSPR of a single Ag nanoparticle was measured by recording the resonant Rayleigh scattering spectrum of the nanoparticle as it was exposed to various solvent environments inside the flow cell. As illustrated in Figure 16.6, the LSPR λ_{max} systematically shifts to longer wavelength as the solvent RIU is increased. Linear regression analysis for this nanoparticle yielded a refractive index sensitivity of 203.1 nm/RIU. The refractive index sensitivity of several individual Ag nanoparticles was measured and typical values were determined to be 170–235 nm/RIU [58]. These are similar to the values obtained from experiments utilizing arrays of NSL-fabricated triangular nanoparticles [57,60].

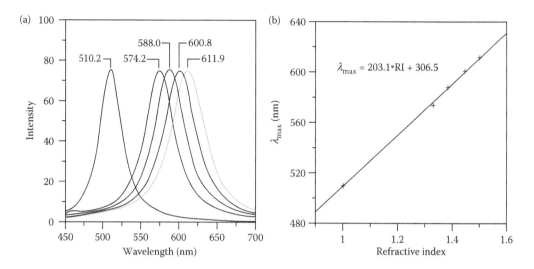

FIGURE 16.6
(a) Single Ag nanoparticle resonant Rayleigh scattering spectrum in various solvent environments (left to right): nitrogen, methanol, 1-propanol, chloroform, and benzene. (b) Plot depicting the linear relationship between the solvent refractive index and the LSPR λ_{max}. (Adapted from McFarland, A.D. and Van Duyne, R.P., 2003, *Nano Lett* 3:1057–1062.)

16.2.4 Streptavidin Sensing with Single Nanoparticles

The biology community would like to reduce the amount of biological sample needed for an assay without amplification. The inherent sensitivity of single nanoparticle systems has the potential to fill this need. To demonstrate that single nanoparticles are useful for the detection of biological molecules rather than immersion in bulk refractive index environments, experiments were conducted using the popular biotin–streptavidin model system (Figure 16.7) [62]. After functionalization with biotin as a capture biomolecule, the LSPR of an individual nanoparticle was measured to be 508.0 nm (Figure 16.7-1). Next, 10 nM streptavidin was injected into the flow cell, and the extinction maximum of the nanoparticle shifted to 520.7 nm (Figure 16.7-2). Based on surface area of the nanoparticle and the footprint of streptavidin, this +12.7 nm shift is estimated to arise from the detection of less than 700 streptavidin molecules. It is hypothesized that as the streptavidin concentration decreases, a fewer number of streptavidin molecules will bind the surface thereby causing smaller wavelength shifts.

16.2.5 Further Developments in Single Nanoparticle LSPR Sensing

The continued development of single nanoparticle LSPR sensors has great potential for miniaturization of refractive index-based biological and chemical sensors. Because of the narrow line widths associated with single nanoparticle scattering spectra, quantitative analysis of molecular species can be more precisely monitored. The size of the nanoparticles allows for straightforward integration of single nanoparticle refractive index sensing in cells without disturbing the chemistry or morphology of the cell. Finally, solutions of nanoparticles with varying sizes or shapes can be easily functionalized with different receptor molecules. These nanoparticles can then be attached to a sensor chip for multiplexed analysis of various species. However, there are some limitations to the immediate implementation of single nanoparticle refractive index sensors. The primary

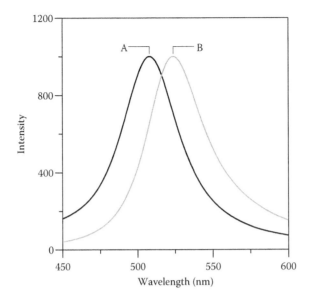

FIGURE 16.7
Individual Ag nanoparticle sensor before and after exposure to 10 nM streptavidin. All measurements were collected in a nitrogen environment. (A) Biotinylated Ag nanoparticle, $\lambda_{max} = 508.0$ nm. (B) After streptavidin incubation, $\lambda_{max} = 520.7$ nm. (Adapted from Van Duyne, R.P. et al., 2003, *SPIE* 5223:197–207.)

difficulty in single nanoparticle sensing is instrumental. The majority of methods employed today use complex and expensive dark field or evanescent wave technology. A more fundamental concern for single nanoparticle LSPR sensing is that the kinetics of reaction for a single particle, either in solution or bound to a surface, are relatively slow. This constraint can partially be relieved by using very small sample volumes, effectively increasing local concentrations, and by forced-flow microfluidics. Also, investigations that involve single nanoparticles inherently remove ensemble averaging from the system. This is beneficial in that unique particle morphologies can be discovered that are particularly sensitive to given optical responses. For example, one can selectively monitor the particles that yield the greatest ΔLSPR λ_{max}, and compare them to Mie theory models.

16.3 Array-Based Nanoparticle Sensing

16.3.1 Introduction to Array-Based Sensing

An approach that combats the difficulties associated with single nanoparticle refractive index sensors is to integrate arrays of independent, homogeneous nanoparticles with a much simpler experimental setup. In order to exploit nanoparticle array sensors, it is necessary to methodically control all parameters that determine the LSPR characteristics. With a sufficiently flexible nanofabrication method, it is possible to manipulate all factors affecting the LSPR, varying one parameter systematically to examine the outcome. For example, Mulvaney wrote a comprehensive review of the tunable optical properties of Au nanoparticles with changing size and dielectric coating [49]. Weimer and Dyer exploited very fine control of substrate temperature, and deposition rate when producing Ag and Au island films in order to systematically tune the LSPR [117]. They were able to empirically create a three-parameter plot whereby knowledge of the chosen metal, substrate temperature and deposition rate allows prediction of the resulting LSPR. Two-dimensional electron beam lithography (EBL) arrays of Ag disks were fabricated by Aussenegg and coworkers in order to probe the LSPR as a function of nanoparticle aspect ratio and the refractive index of the local environment [118]. Both increase in aspect ratio and increase in local refractive index caused systematic shifts of the LSPR to lower energies.

16.3.2 Refractive Index Sensing with Arrays of Chemically Synthesized Nanoparticles

In the most primitive refractive index array sensing methods, chemically synthesized gold or silver nanoparticles are covalently or electrostatically attached in a random fashion to a transparent substrate to detect proteins [83,108]. In this format, signal transduction depends on changes in the nanoparticles' dielectric environment induced by solvent or target molecules (not by nanoparticle coupling). Using this chip-based approach, a solvent refractive index sensitivity of 76.4 nm/RIU has been found and a detection of 16 nM streptavidin can be detected [83,108]. This approach has many advantages including: (1) a simple fabrication technique that can be performed in most laboratories, (2) real-time biomolecule detection using UV–Vis spectroscopy, and (3) a chip-based design that allows for multiplexed analysis. However, the sensitivity of the sensor is greatly limited by nanoparticle coupling.

16.3.3 Nanosphere Lithography: Synthesis and Fabrication

A fabrication platform that combats uncontrolled nanoparticle coupling for the synthesis of these refractive index-based sensors is known as NSL. NSL is a powerful fabrication technique to inexpensively produce nanoparticle arrays with controlled shape, size, and interparticle spacing [119]. The need for monodisperse, reproducible, and materials general nanoparticles has driven the development and refinement of the most basic NSL architecture as well as many new nanostructure derivatives. The NSL fabrication process is shown in Figure 16.8. Every NSL structure begins with the self-assembly of size-monodisperse nanospheres of diameter D to form a two-dimensional colloidal crystal deposition mask. Methods for deposition of a nanosphere solution onto the desired substrate include spin coating [119], drop coating [55], and thermoelectrically cooled angle coating [120]. All of these deposition methods require that the nanospheres be able to freely diffuse across the substrate seeking their lowest energy configuration. This is often achieved by chemically modifying the nanosphere surface with a negatively charged functional group such as carboxylate or sulfate that is electrostatically repelled by the negatively charged surface of a substrate such as mica or glass. As the solvent (water) evaporates, capillary forces draw the nanospheres together, and the nanospheres crystallize in a hexagonally close-packed pattern on the substrate. As in all naturally occurring crystals, nanosphere masks include a variety of defects that arise as a result of nanosphere polydispersity, site randomness, point defects (vacancies), line defects (slip dislocations), and polycrystalline domains. Typical defect-free domain sizes are in the 10–100 μm range. Following self-assembly of the nanosphere mask, a metal or other material is then deposited by thermal evaporation, electron beam deposition, or pulsed laser deposition from a collimated source normal to the substrate through the nanosphere mask to a controlled mass thickness, d_m. After metal deposition, the nanosphere mask is removed, typically by sonicating the entire sample in

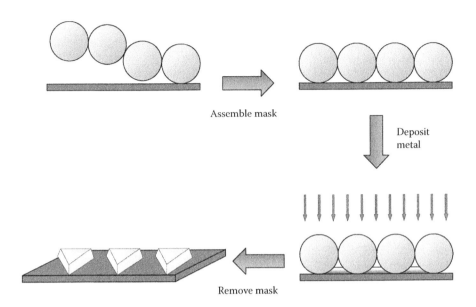

FIGURE 16.8
Schematic illustration of the nanosphere lithography fabrication technique. A small volume of nanosphere solution is drop coated onto the clean substrate. As the solvent evaporates, the nanospheres assemble into a two-dimensional colloidal crystal mask. The desired noble metal is then deposited in a high vacuum thin film vapor deposition system. In the last step of the sample preparation, the lift-off step, the nanospheres are removed by sonication in absolute ethanol.

a solvent, leaving behind the material deposited through the nanosphere mask and onto the substrate.

Using NSL, we have demonstrated that nanoscale chemosensing and biosensing could be realized through shifts in the LSPR extinction maximum (λ_{max}) of these triangular silver nanoparticles [56,57,59,77]. Instead of being caused by the electromagnetic coupling of the nanoparticles, these wavelength shifts are caused by adsorbate-induced local refractive index changes in competition with charge-transfer interactions at the surfaces of nanoparticles. Extensive studies have been completed by Van Duyne and coworkers using NSL-fabricated substrates whereby it is possible to systematically vary nanoparticle aspect ratio [121], shape [122], substrate [123], dielectric environment [124], and effective thickness of a chemisorbed monolayer [57]. In all cases, the experiments revealed systematic shifts in the LSPR: increased aspect ratio shifts the LSPR to lower energies, retraction of sharp tetrahedral tips shifts the LSPR to higher energies, increased refractive index of the substrate or solvent environment shifts the LSPR to lower energies, and increased thickness of chemisorbed molecules shifts the LSPR to lower energies within the limit of the electromagnetic field decay length. It should be noted that the signal transduction mechanism in this nanosensor is a reliably measured wavelength shift rather than an intensity change as in many previously reported nanoparticle-based sensors.

16.3.4 Nanoparticle Array Preparation

For the remainder Ag nanoparticle array studies, NSL was used to fabricate monodisperse, surface-confined triangular Ag nanoparticles. A solution of nanospheres was drop coated onto a clean substrate and allowed to self-assemble into a two-dimensional hexagonally close-packed array that served as a deposition mask. On glass, single layer colloidal crystal nanosphere masks were prepared by drop coating −2 μL of undiluted nanosphere solution on glass. On mica, the nanosphere solution was diluted as a 1:1 solution with Triton X-100 and methanol (1:400 by volume). Approximately 4 μL of this solution was drop coated onto the freshly cleaved mica substrates and allowed to dry, forming a monolayer in a close-packed hexagonal formation, which served as a deposition mask. The samples were then mounted onto a Consolidated Vacuum Corporation vapor deposition chamber system. A Leybold Inficon XTM/2 quartz crystal microbalance (East Syracuse, NY) was used to monitor the thickness of the Ag film deposited on the nanosphere mask. Following Ag vapor deposition, the nanosphere mask was removed by sonicating the samples in ethanol for 3 min.

16.3.5 Ultraviolet–Visible Extinction Spectroscopy for Nanoparticle Arrays

Macroscale UV–Vis extinction measurements were collected using an Ocean Optics spectrometer. All spectra collected are macroscopic measurements performed in standard transmission geometry with unpolarized light. The probe beam diameter was approximately 2–4 mm.

16.3.6 Experimental Setup and Nanoparticle Functionalization for Nanoparticle Arrays

A home built flow cell was used to control the external environment of the Ag nanoparticle substrates (Figure 16.9). Before modification, the Ag nanoparticles were solvent annealed with hexanes and methanol. Dry N_2 gas and solvent were cycled through the flow cell until the λ_{max} of the sample is stabilized.

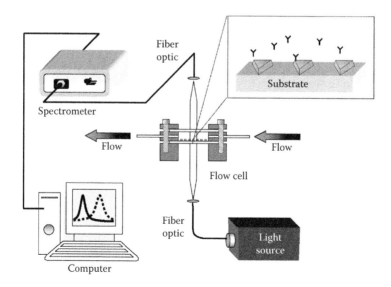

FIGURE 16.9

Instrumental diagram for the array-based LSPR nanosensor experiments. The flow cell is fiber optically coupled to a white light source and miniature spectrometer. The cell is linked directly to either a solvent reservoir or to a syringe containing the desired analyte. (Adapted from Malinsky, M.D. et al., 2001, *J Am Chem Soc* 123(7):1471–1482.)

16.3.7 Tuning the Localized Surface Plasmon Resonance of Ag Nanoparticles

Nanoparticle shape has a significant effect on the λ_{max} of the LSPR. When the standard triangular nanoparticles are thermally annealed at 300°C under vacuum, the nanoparticle shape is modified, increasing in out-of-plane height and becoming ellipsoidal. This structural transition results in a blueshift of $\sim$200 nm in the λ_{max} of the LSPR. It is often difficult to decouple size and shape effects on the LSPR wavelength, and so they are considered together as the nanoparticle aspect ratio (a/b, where a is the in-plane width of the nanoparticle and b is the out-of-plane height of the nanoparticle). Large aspect ratio values represent oblate nanoparticles and aspect ratios with a value of unity represent spheroidal nanoparticles. Figure 16.10 shows a series of extinction spectra collected from Ag NSL-fabricated nanoparticles of varied shapes and aspect ratios on mica substrates. These extinction spectra were recorded in standard transmission geometry. All macroextinction measurements were recorded using unpolarized light with a probe beam size of approximately 2–4 mm². All parameters for Figure 16.10 are listed in Table 16.2. To identify each parameter's effect on the LSPR, one must examine three separate cases. Firstly, extinction peaks labeled F, G, and H all have the same a value (signifying that the same diameter nanosphere mask was used for each sample), but the b value is varied as the shape is held constant. Note that the LSPR λ_{max} shifts to the red as the out-of-plane nanoparticle height is decreased, i.e., the aspect ratio is increased. Secondly, extinction peaks D, E, and H have varying a values, very similar b values, and constant shape. In this case, the LSPR λ_{max} shifts to the red with increased nanosphere diameter (a larger a value). Again, as the aspect ratio increases, the LSPR shifts to longer wavelengths. Finally, extinction peaks C and F were measured from the same sample before and after thermal annealing. Note the slight increase in nanoparticle height (b) as the annealed nanoparticle dewets the mica substrate. The 223 nm blueshift upon annealing is in accordance with the decreasing nanoparticle aspect ratio as the nanoparticles transition from oblate to ellipsoidal geometries.

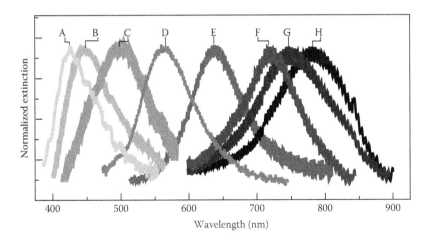

FIGURE 16.10
UV–Vis extinction spectra of Ag nanoparticle arrays on mica substrates. Reported spectra are raw, unfiltered data. The oscillatory signal superimposed on the LSPR spectrum seen in the data is due to interference of the probe beam between the front and back faces of the mica. (Adapted from Jensen, T.R. et al., 2000, *J Phys Chem B* 104:10549–10556.)

TABLE 16.2

Ag Nanoparticle Structural Parameters Corresponding to the UV–Vis Extinction Measurements in Figure 16.10

	A[a]	B[a]	C[a]	D	E	F	G	H
a^b (nm)	120 ± 12	150 ± 15	150 ± 15	90 ± 6	120 ± 6	145 ± 6	145 ± 6	145 ± 6
b (nm)	42 ± 5	70 ± 8	62 ± 8	46 ± 3	46 ± 3	59 ± 4	55 ± 4	50 ± 4
D (nm)	401 ± 7	542 ± 7	542 ± 7	310 ± 9	401 ± 7	542 ± 7	542 ± 7	542 ± 7
d_m (nm)	48	70	62	46	48	60	55	50
Substrate	Mica	Mica	Mica	Mica	Mica	Mica	Mica	Mica
In-plane shape[c,d]	E	E	E	T	T	T	T	T
λ_{max} (nm)	426	446	497	565	638	720	747	782
λ_{max} (cm^{-1})	23,474	22,422	20,121	17,699	15,674	13,889	13,387	12,788
Γ (cm^{-1})	3460	3883	3940	2788	2180	1826	2483	2063
Q	6.78	5.77	5.11	6.35	7.19	7.61	5.39	6.20
Extinction scaling factor	3.7	2.5	2.7	1.6	1.0	1.9	1.3	0.7

Source: Adapted from Jensen, T.R. et al., 2000, *J Phys Chem B* 104:10549–10556.
[a] Annealed at 300°C for 1 hour.
[b] Not corrected for AFM tip convolution.
[c] E, elliptical.
[d] T, triangular.

The range of possible LSPR λ_{max} values extends beyond those shown in Figure 16.10. In fact, λ_{max} can be tuned continuously from $\sim$400 nm to 6000 nm by choosing the appropriate nanoparticle aspect ratio and geometry [99]. Recent experiments exploring the sensitivity of the LSPR λ_{max} to changes in a and b values support the assertion that it is not always possible to decouple the in-plane width and out-of-plane height from one another. Figure 16.10 demonstrates an in-plane width sensitivity of $\Delta\lambda_{max}/\Delta a = 4$ and an out-of-plane height sensitivity of $\Delta\lambda_{max}/\Delta b = 7$. In order to further investigate the out-of-plane height sensitivity, a larger data set was collected in the $\lambda_{max} = 500$–600 nm region using NSL masks made from nanosphere diameter = 310 nm nanospheres. In this case, both the in-plane width and the shape were held constant as the nanoparticle height was varied. From this larger data set, the calculated out-of-plane height sensitivity was $\Delta\lambda_{max}/\Delta b = 2$. The variance in the two $\Delta\lambda_{max}/\Delta b$ values suggests that nanoparticles with smaller in-plane widths ($a = 90 \pm 6$ nm versus $a = 145 \pm 6$ nm) are less sensitive to changes in nanoparticle height. This conclusion supports the relationship between nanoparticle aspect ratio and LSPR shift susceptibility noted above.

16.3.8 Sensitivity of the Localized Surface Plasmon Resonance of Ag Nanoparticles to Its Dielectric Environment

Next, the role of the external dielectric medium on the optical properties of these surface-confined nanoparticles is considered. Just as it is difficult to decouple the effects of size and shape from one another, the dielectric effects of the substrate and external dielectric medium (i.e., bulk solvent) are inextricably coupled because together they describe the entire dielectric environment surrounding the nanoparticles. The nanoparticles are surrounded on one side by the substrate and on the other four sides by a chosen environment. A systematic study of the relationship between the LSPR λ_{max} and

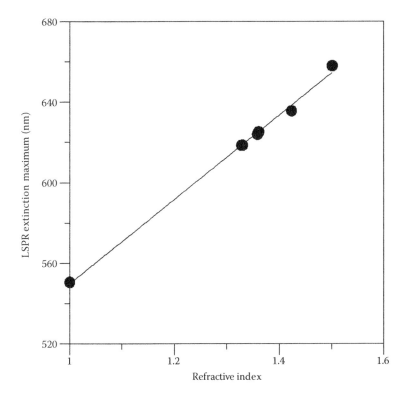

FIGURE 16.11
Solvent sensitivity of unmodified arrays of Ag nanoparticle. Spectral peak shifts were calculated by subtracting the measured extinction maximum, λ_{max}, for the nanoparticles in solvents of $n_{extrernal}$ ranging from 1.33 (methanol) to 1.51 (benzene) from that of a N_2 environment ($n_{external}$) (1.0). Plots display representative measurements from several experiments. The slope of the linear fit shows that the LSPR spectral sensitivity to $n_{external}$ is 191 nm/RIU. (Adapted from Malinsky, M.D. et al., 2001, *J Am Chem Soc* 123(7):1471–1482.)

the external dielectric constant on the four nonsubstrate-bound faces of the nanoparticles was done by immersing a series of varied aspect ratio nanoparticle samples (aspect ratios 3.32, 2.17, and 1.64) in a variety of solvents [101]. These solvents represent a progression of refractive indices: nitrogen (1.0), acetone (1.36), methylene chloride (1.42), cyclohexane (1.43), and pyridine (1.51). Each sample was equilibrated in a N_2 environment before solvent treatment. Extinction measurements were made before, during, and after each solvent cycle. With the exception of pyridine, the LSPR peak always shifted back to the N_2 value when the solvent was purged from the sample cell. The measured LSPR λ_{max} values progress toward longer wavelengths as the solvent refractive index increases (Figure 16.11). A plot of the solvent refractive index versus the LSPR λ_{max} shift is linear for all three aspect ratio samples. The highest aspect ratio nanoparticles (most oblate) demonstrate the greatest sensitivity to external dielectric environment. In fact, the LSPR λ_{max} of the 3.32 aspect ratio nanoparticles shifts 200 nm/RIU.

After completing the aspect ratio and solvent sensitivity experiment, a duplicate of the 2.17 aspect ratio sample was thermally annealed. This annealed sample, with an aspect ratio of 2.14, was then subjected to the same series of solvent treatments and extinction measurements. The resultant solvent refractive index versus LSPR λ_{max} shift still shows a linear trend, but a significantly decreased sensitivity of 150 nm shift per RIU.

16.3.9 Streptavidin Sensing Using LSPR Spectroscopy

The well-studied biotin–streptavidin system with its extremely high binding affinity ($K_a \sim 10^{13}$ M^{-1}) [125] is chosen to illustrate the attributes of these LSPR-based nanoscale affinity biosensors. The biotin–streptavidin system has been studied in great detail by SPR spectroscopy [26,27] and serves as an excellent model system for the LSPR nanosensor [56,126]. Streptavidin, a tetrameric protein, can bind up to four biotinylated molecules (i.e., antibodies, inhibitors, nucleic acids, etc.) with minimal impact on its biological activity [126] and, therefore, will provide a ready pathway for extending the analyte accessibility of the LSPR nanobiosensor.

NSL was used to create surface-confined triangular Ag nanoparticles supported on a glass substrate (Figure 16.12a). The Ag nanotriangles have in-plane widths of ∼100 nm and out-of-plane heights of ∼51 nm as determined by atomic force microscopy (AFM). To prepare the LSPR nanosensor for biosensing events, SAMs were formed on the nanoparticles by incubation in a 3:1 ratio of 1 mM 1-OT:11-MUA in ethanol for 18–36 h. Next, biotin was linked to the surface over a 3 h time period by incubation in a 1:1 ratio of 1 mM EDC:biotin in 10 mM PBS. Samples were then incubated in 100 nM of streptavidin in PBS for 3 h. Samples were rinsed thoroughly with 10 and 20 mM PBS after biotinylation and after detection of streptavidin to ensure removal of nonspecifically bound materials [56].

In this study, the λ_{max} of the Ag nanoparticles was monitored during each surface functionalization step (Figure 16.12b). First, the LSPR λ_{max} of the bare Ag nanoparticles was measured to be 561.4 nm (Figure 16.12b-1). To ensure a well-ordered SAM on the Ag nanoparticles, the sample was incubated in the thiol solution for 24 h. After careful rinsing and thorough drying with N$_2$ gas, the LSPR λ_{max} after modification with the mixed SAM (Figure 16.12b-2) was measured to be 598.6 nm. The LSPR λ_{max} shift corresponding to this surface functionalization step was a 38 nm redshift, hereafter + will signify a redshift and − a blueshift, with respect to bare Ag nanoparticles. Next, biotin was covalently attached by amide bond formation with a two-unit polyethylene glycol linker to carboxylated surface sites. The LSPR λ_{max} after biotin attachment (Figure 16.12b-3) was measured to be 609.6 nm corresponding to an additional +11 nm shift. The LSPR nanosensor has now been prepared for exposure to the target analyte. Exposure to 100 nM streptavidin, resulted in LSPR $\lambda_{max} = 636.6$ nm (Figure 16.12b-4) corresponding to an additional +27 nm shift. It should be noted that the signal transduction mechanism in this nanosensor is a reliably measured wavelength shift rather than an intensity change as in many previously reported nanoparticle-based sensors.

16.3.10 Antibiotin Sensing Using LSPR Spectroscopy

A field of particular interest is the study of the interaction between antigens and antibodies [127]. For these reasons we have chosen to focus the present LSPR nanobiosensor study on the prototypical immunoassay involving biotin and antibiotin, an IgG antibody. In this study, we report the use of Ag nanotriangles synthesized using NSL as an LSPR biosensor that monitors the interaction between a biotinylated surface and free antibiotin in solution [59]. The importance of this study is that it demonstrates the feasibility of LSPR biosensing with a biological couple whose binding affinity is significantly lower ($1.9 \times 10^6 - 4.98 \times 10^8$ M^{-1}) [128,129] than in the biotin–streptavidin model.

NSL was used to create massively parallel arrays of Ag nanotriangles on a mica substrate. A SAM of 1:3 1-MUA:1-OT was formed on the surface by incubation for 24 h. As in the

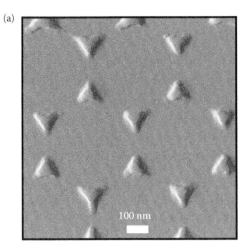

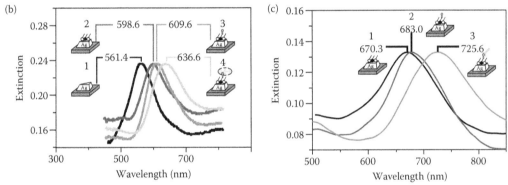

FIGURE 16.12
(a) Tapping mode AFM image of the Ag nanoparticles (nanosphere diameter, $D = 400$ nm; mass thickness, $d_m = 50.0$ nm Ag on a glass substrate). Scan area 1.0 μm². Scan rate between 1 and 2 Hz. After solvent annealing, the resulting nanoparticles have in-plane widths of ~ 100 nm and out-of-plane heights of ~ 51 nm. (b) LSPR spectra of each step in the surface modification of NSL-derived Ag nanoparticles to form a biotinylated Ag nanobiosensor and the specific binding of streptavidin. (1) Ag nanoparticles before chemical modification, $\lambda_{max} = 561.4$ nm. (2) Ag nanoparticles after modification with 1 mM 1:3 11-MUA:1-OT, $\lambda_{max} = 598.6$ nm. (3) Ag nanoparticles after modification with 1 mM biotin, $\lambda_{max} = 609.6$ nm. (4) Ag nanoparticles after modification with 100 nM streptavidin, $\lambda_{max} = 636.6$ nm. All extinction measurements were collected in a N_2 environment. (c) Smoothed LSPR spectra for each step of the preparation of the Ag nanobiosensor, and the specific binding of antibiotin to biotin. (1) Ag nanoparticles after modification with 1 mM 3:1 1-OT/11-MUA, $\lambda_{max} = 670.3$ nm, (2) Ag nanoparticles after modification with 1 mM biotin, $\lambda_{max} = 683.0$ nm, and (3) Ag nanoparticles after modification with 700 nM antibiotin, $\lambda_{max} = 725.6$ nm. All spectra were collected in a N_2 environment. (Adapted from Haes, A.J. and Van Duyne, R.P., 2002, *J Am Chem Soc* 124(35):10596–10604. and Riboh, J.C. et al., 2003, *J Phys Chem B* 107:1772–1780.)

streptavidin experiments, a zero length-coupling agent was then used to covalently link biotin to the carboxylate groups.

Each step of the functionalization of the samples was monitored using UV–Vis spectroscopy, as shown in Figure 16.12c. After a 24 h incubation in SAM, the LSPR extinction wavelength of the Ag nanoparticles was measured to be 670.3 nm (Figure 16.12c-1). Samples were then incubated for 3 h in biotin to ensure that the amide bond between the amine and carboxyl groups had been formed. The LSPR wavelength shift due to this binding event was measured to be +12.7 nm, resulting in an LSPR extinction wavelength of 683.0 nm (Figure 16.12c-2). At this stage, the nanosensor was ready to detect the specific binding of antibiotin. Incubation in 700 nM antibiotin for 3 h resulted in an LSPR wavelength shift of + 42.6 nm, giving a λ_{max} of 725.6 nm (Figure 16.12c-3).

16.3.11 Monitoring the Specific Binding of Streptavidin to Biotin and Antibiotin

The well-studied biotin–streptavidin [56] system with its extremely high binding affinity ($K_a \sim 10^{13}\ M^{-1}$) and the antigen–antibody couple, biotin–antibiotin ($K_a \sim 10^6$–$10^8\ M^{-1}$) [59] have been chosen to illustrate the attributes of these LSPR-based nanoscale affinity biosensors. The LSPR λ_{max} shift, ΔR, versus [analyte] response curve was measured over the concentration range $1 \times 10^{-15}\ M <$ [streptavidin] $< 1 \times 10^{-6}\ M$ and $7 \times 10^{-10}\ M <$ [antibiotin] $< 7 \times 10^{-6}\ M$ (Figure 16.13) [56,59]. Each data point is an average resulted from the analysis of three different samples at identical concentrations. The lines are not a fit to the data. Instead, the line was computed from a response model [59] described by

$$\Delta R = \Delta R_{max}\left(\frac{K_{a,surf}[\text{analyte}]}{1 + K_{a,surf}[\text{analyte}]}\right) \tag{16.4}$$

where ΔR is the nanosensor's response for a given analyte concentration, [analyte], ΔR_{max} is the maximum sensor response for a full monolayer coverage, and $K_{a,surf}$ is the surface-confined binding constant. It was found that the response could be interpreted quantitatively in terms of a model involving: (1) 1:1 binding of a ligand to a multivalent receptor with different sites but invariant affinities and (2) the assumption that only adsorbate-induced local refractive index changes were responsible for the operation of the LSPR nanosensor.

The binding curve provides three important characteristics regarding the system being studied. First, the mass and dimensions of the molecules affect the magnitude of the LSPR shift response. Comparison of the data with theoretical expectations yielded a saturation

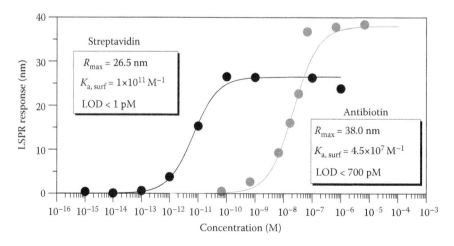

FIGURE 16.13
The specific binding of streptavidin (left) and antibiotin (right) to a biotinylated Ag nanobiosensor is shown in the response curves. All measurements were collected in a N_2 environment. The solid line is the calculated value of the nanosensor's response. (Adapted from Haes, A.J. and Van Duyne, R.P., 2002, *J Am Chem Soc* 124(35):10596–10604. and Riboh, J.C. et al., 2003, *J Phys Chem B* 107:1772–1780.)

response, $\Delta R_{max} = 26.5$ nm for streptavidin, a 60 kDa molecule, and 38.0 nm for antibiotin, a 150 kDa molecule. Clearly, a larger mass density at the surface of the nanoparticle results in a larger LSPR response. Next, the surface-confined thermodynamic binding constant $K_{a,surf}$ can be calculated from the binding curve and is estimated to be 1×10^{11} M^{-1} for streptavidin and 4.5×10^7 M^{-1} for antibiotin. These numbers are directly correlated to the third important characteristic of the system, the LOD. The LOD is less than 1 pM for streptavidin and 100 pM for antibiotin. As predicted, the LOD of the nanobiosensor studied is lower for systems with higher binding affinities such as for the well-studied biotin–streptavidin couple and higher for systems with lower binding affinities as seen in the antibiotin system.

The use of biotin-based model systems provided convenient, well-known, experimental parameters to which LSPR-based nanosensing can be compared and properly evaluated. It is evident that sensing of both streptavidin and antibiotin was successful, thereby establishing the viability of the LSPR technique for biomolecular detection. However, we have also sought to extend the application of the LSPR method to other nonideal and less-studied systems. Therefore, we have conducted experiments using LSPR sensing in two additional circumstances. First, we will discuss the LSPR-based measurement of the concanavalin A (Con A)–carbohydrate interaction. These studies effectively demonstrate how the LSPR sensor might be used in a typical biochemical research setting. The second example is a more complex, biomedical application, that is, the detection of Alzheimer's disease markers.

16.3.12 LSPR Detection of a Carbohydrate-Binding Protein Interactions

Con A is a 104 kDa mannose-specific plant lectin that is comprised of a tetramer with dimensions of 63.2, 86.9, and 89.3 Å [130]. The surface binding constant for Con A to a mannose-functionalized surface was found to be 5.6×10^6 M^{-1} by SPR imaging studies [131]. This section demonstrates the specific binding of Con A to a mannose-functionalized SAM using the LSPR nanosensor [132].

The Ag nanosensor was rinsed with ethanol and placed in a flow cell (see Figure 16.9) after incubation in mannose. The LSPR spectrum of the mannose-functionalized Ag nanosensor in N$_2$ had a λ_{max} of 636.5 nm (Figure 16.14a). Then, 19 µM Con A was injected into the flow cell and the Ag nanosensor was incubated at room temperature for 20 min. The sample was rinsed in buffer solution and dried with nitrogen. The LSPR λ_{max} of the Ag nanotriangles was measured to be 654.3 nm, resulting in a 17.8 nm shift. The LSPR response of Con A binding to mannose-functionalized on the Ag nanosensor was also measured in a buffer environment. Figure 16.14b depicts the 6.7 nm LSPR response in a buffer environment. The 40% reduction of signal in buffer relative to N$_2$ environment is predicted by Equation 16.1 and has been previously observed [59].

16.3.13 Detection of Alzheimer's Disease Markers Using the LSPR Nanosensor Chip

Alzheimer's disease is the leading cause of dementia in people over age 65 and affects an estimated 4 million Americans. Although first characterized almost 100 years ago by Alois Alzheimer, who found brain lesions now called plaques and tangles in the brain of a middle-aged woman who died with dementia in her early 50s [133], the molecular cause of the disease is not understood; and an accurate diagnostic test has yet to be developed. However, two interrelated theories for Alzheimer's disease have emerged that focus on the putative involvement of neurotoxic assemblies of a small 42-amino

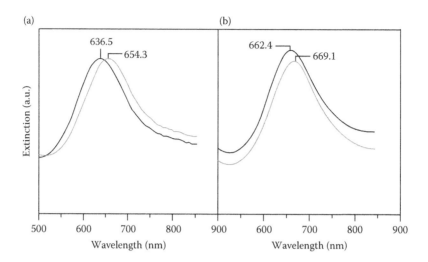

FIGURE 16.14

Detection of Con A with the LSPR biosensor. (a) LSPR spectra of mannose-functionalized Ag nanosensor ($\lambda_{max} = 636.5$ nm) and the specific binding of Con A ($\lambda_{max} = 654.3$ nm) in nitrogen. (b) LSPR spectra of mannose-functionalized Ag nanobiosensor ($\lambda_{max} = 662.4$ nm) and the specific binding of Con A to mannose ($\lambda_{max} = 669.1$) in PBS buffer. (Adapted from Yonzon, C.R. et al., 2003, 2004, *J Am Chem Soc* 126:12669–12676.)

acid peptide known as amyloid beta (Aβ) [134,135]. The widely investigated amyloid cascade hypothesis suggests that the amyloid plaques cause neuronal degeneration and, consequently, memory loss and further progressive dementia. In this theory, the Aβ protein monomers, present in all normal individuals, do not exhibit toxicity until they assemble into amyloid fibrils [136]. The other toxins are known as Aβ-derived diffusible ligands (ADDLs). ADDLs are small, globular, and readily soluble, 3–24 mers of the Aβ monomer [137], and are potent and selective central nervous system neurotoxins, which possibly inhibit mechanisms of synaptic information storage with great rapidity [137]. ADDLs now have been confirmed to be greatly elevated in autopsied brains of Alzheimer's disease subjects [138]. An ultrasensitive method for ADDLs/anti-ADDLs antibody detection potentially could emerge from LSPR nanosensor technology, providing an opportunity to develop the first clinical laboratory diagnostic for Alzheimer's disease. Preliminary results indicate that the LSPR nanosensor can be used to aid in the diagnosis of Alzheimer's disease [139,140].

By functionalizing the Ag nanoparticles with a 3:1 1 mM OT:1 mM 11-MUA SAM layer, the stability of the samples is greatly increased and consistent redshifts are produced upon incubation in given concentrations of both ADDLs and anti-ADDLs. Because it is hypothesized that the presence of anti-ADDLs prohibits the development of Alzheimer's disease, the assay explored here is designed to detect that antibody. Given the size of the ADDL ($4.5 \times N$ kDa, $N = 3$–24) and anti-ADDL (150 kDa) molecules, it was hypothesized that at full coverage, ADDLs may give smaller up to equivalent LSPR shifts in comparison to anti-ADDLs. Additionally, the magnitude of the LSPR shift is larger for molecules closer to the Ag surface rather than farther away from the nanoparticle surface [61]. This effect magnifies the LSPR shift induced by the ADDLs in comparison to the anti-ADDLs.

The first target biomarker analyzed is the anti-ADDL antibody to an ADDL-functionalized nanoparticle surface (Figure 16.15a). After a 24 h incubation in SAM,

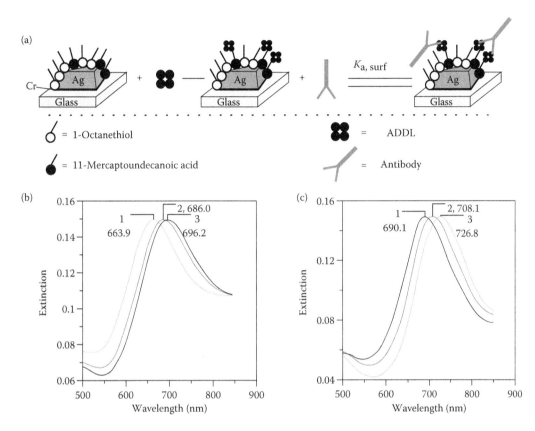

FIGURE 16.15
Design of the LSPR biosensor for anti-ADDL detection. (a) Surface chemistry of the Ag nanoparticle sensor. Surface-confined Ag nanoparticles are synthesized using NSL. Nanoparticle adhesion to the glass substrate is promoted using a 0.4 nm Cr layer. The nanoparticles are incubated in a 3:1 1-OT/11-MUA solution to form a SAM. Next, the samples are incubated in 100 mM EDC/100 nM ADDL solution. Finally, incubating the ADDL-coated nanoparticles to varying concentrations of antibody completes an anti-ADDL immunoassay. (b) LSPR spectra for each step of the preparation of the Ag nanobiosensor at a low concentration of anti-ADDL antibody. Ag nanoparticles after modification with (1) 1 mM 3:1 1-OT/11-MUA, $\lambda_{max} = 663.9$ nm, (2) 100 nM ADDL, $\lambda_{max} = 686.0$ nm, and (3) 50 nM anti-ADDL, $\lambda_{max} = 696.2$ nm. All spectra were collected in a N_2 environment. (c) LSPR spectra for each step of the preparation of the Ag nanobiosensor at a high concentration of anti-ADDL. Ag nanoparticles after modification with (1) 1 mM 3:1 1-OT/11-MUA, $\lambda_{max} = 690.1$ nm, (2) 100 nM ADDL, $\lambda_{max} = 708.1$ nm, and (3) 400 nM anti-ADDL, $\lambda_{max} = 726.8$ nm. All spectra were collected in a N_2 environment. (Adapted from Haes, A.J. et al., 2004, *Nano Lett* 4(6):1029–1034.)

a representative LSPR extinction wavelength of the Ag nanoparticles was measured to be 663.9 nm (Figure 16.15b-1). Samples were then incubated for 1 h in 100 mM EDC/100 nM ADDL to ensure that the amide bond between the amine groups in the ADDLs and the carboxyl groups on the surface had been formed. The LSPR wavelength shift due to this binding event was measured to be +22.1 nm, resulting in an LSPR extinction wavelength of 686.0 nm (Figure 16.15b-2). The Ag nanoparticle biosensor is now ready to detect the specific binding of anti-ADDL. Incubation in 50 nM anti-ADDL for 30 min results in an LSPR wavelength shift of +10.2 nm, giving a λ_{max} of 696.2 nm (Figure 16.15c). The results for this experiment are dramatically different if the concentration of anti-ADDL is increased to 400 nM. At this high concentration, the LSPR wavelength shifts +18.7 nm from 708.1

to 726.8 nm (Figure 16.15c) from the binding of anti-ADDLs to the ADDL-functionalized nanoparticle surface [140].

These preliminary results indicate that the LSPR nanosensor can be used to aid in the diagnosis of Alzheimer's disease by detecting specific antibodies [139,140]. In addition to detecting these antibodies, it is also important to detect the target antigen, in this case, ADDL. In these studies, antibodies that specifically interact with ADDLs were decorated onto the silver nanoparticle surface. The sensor's viability was tested by exposing this surface to model target ADDL molecules. An additional antibody then amplified the LSPR nanosensor's response. A representative assay for the direct detection of ADDLs using a sandwich assay is displayed in Figure 16.16. In this study, anti-ADDL antibody was specifically linked to the SAM-functionalized nanoparticles over a 1 h period. Next, the nanosensor was exposed to a 100 nM ADDL solution for 30 min. During the ADDL incubation, the extinction maximum of the sample shifted from 623.8 to 631.9 nm, an 8.1 nm shift. Next, this shift was amplified by exposing the sample to an additional antibody for 30 min resulting in an additional 10.6 nm wavelength shift or a total LSPR shift of 18.7 nm. It should be noted that the sample was checked for nonspecific binding after steps 1 and 2, and no nonspecific binding was observed.

After the demonstration of the model assay, the target molecules that were sandwiched in the previous experiment were substituted by cerebral spinal fluid from diseased and nondiseased patients [139]. The fluid was centrifuged to remove large pieces of cellular material; otherwise, the samples were not further modified. In preliminary assays, which took less than 2 h to perform, substantial differences were observed based on the analysis of the two types (diseased and nondiseased) of samples.

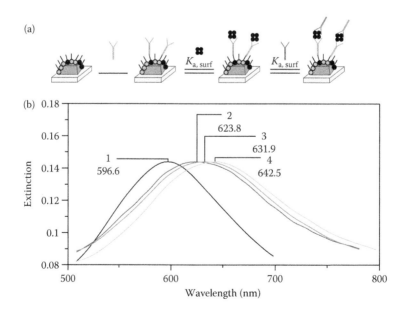

FIGURE 16.16

Sandwich assay for ADDL detection on a mica substrate. (a) Surface chemistry for ADDL detection in an antibody sandwich assay. (b) LSPR spectra for each step in the surface modification for the sandwich assay. Ag nanoparticles after functionalization with (1) 1:3 11-MUA:1-OT (λ_{max} = 596.6 nm), (2) 100 nM anti-ADDL (100 mM EDC) (λ_{max} = 623.8 nm), (3) 100 nM ADDL (λ_{max} = 631.9 nm), and (4) 100 nM anti-ADDL (λ_{max} = 642.5 nm). All measurements were collected in a N_2 environment.

16.4 Summary and Future Outlook

The use of nanoparticles for the highly sensitive and selective detection of biomolecules has been demonstrated with both model systems and complex human samples. Additionally, it has been shown that these assays can be clearly minimized to one nanoparticle with the implementation of dark-field microscopy. The next challenge for sensor development lies in the capability of formatting the sensor into an array format, and integrating the sensor array into a microfluidic chip. By building upon the established (and commercially available) SPR sensing device, parallel, highly sensitive, and affordable nanoparticle array-based systems are foreseeable in the near future. The choice of nanoparticle size and shape will be critical for the optimization of the sensor's response. Future studies will continue to analyze complex biological samples, and if fully realized, will provide vital information regarding disease understanding.

Briefly looking to the future, a reasonable extrapolation of our current data leads us to expect that by optimizing these size- and shape-tunable nanosensor materials, and by using single nanoparticle spectroscopic techniques, it will be possible to (1) reach sensitivities of a few molecules, perhaps even a single molecule, per nanoparticle sensor element; (2) reduce the timescale for real-time detection and the study of protein binding kinetics by 2–3 orders of magnitude because nanoparticle sensor elements will operate in radial rather than planar diffusion mass transport regime; and (3) implement massively parallel bioassays for high-throughput screening applications when maintaining extremely low sample volume requirements. Finally, we point out that LSPR nanosensors can be implemented using extremely simple, small, light, robust, low-cost equipment for unpolarized, UV–Vis extinction spectroscopy in transmission or reflection geometry. The instrumental simplicity of the LSPR nanosensor approach is expected to greatly facilitate field-portable environmental or point-of-service medical diagnostic applications.

Acknowledgment

We acknowledge support of the Nanoscale Science and Engineering Initiative of the National Science Foundation under NSF Award Number EEC-0118025. Any opinions, findings and conclusions, or recommendations expressed in this material are those of the authors and do not necessarily reflect those of the National Science Foundation. We are grateful for useful discussion, technical support, and the expert assistance provided by Dr. Lei Chang, Dr. Christy Haynes, W. Paige Hall, Dr. Eunhee Jeoung, Prof. William Klein, Dr. Michelle Malinsky, Dr. Adam McFarland, Prof. Milan Mrksich, Jonathan Riboh, Prof. George Schatz, Chanda Yonzon, and Dr. Shengli Zou.

References

1. Klotz, I.M. 1997. *Ligand–Receptor Energetics: A Guide for the Perplexed*, 170. New York: Wiley.
2. Lee, H.J, T.T. Goodrich, and R.M. Corn. 2001. SPR imaging measurements of 1-D and 2-D DNA microarrays created from microfluidic channels on gold thin films. *Anal Chem* 73(55):5525–5531.

3. Hall, D. 2001. Use of optical biosensors for the study of mechanistically concerted surface adsorption processes. *Anal Biochem* 288(2):109–125.

4. Wang, J. et al. 1996. DNA electrochemical biosensor for the detection of short DNA sequences related to the human immunodeficiency virus. *Anal Chem* 68(15):2629–2634.

5. Walterbeek, H.T, and A.J. G.M. van der Meer. 1996. A sensitive and quantitative biosensing method for the determination of γ-ray emitting radionuclides in surface water. *J Environ Radioactiv* 33(3):237–254.

6. Thevenot, D.R. et al. 2001. Electrochemical biosensors: Recommended definitions and classification. *Biosens Bioelectron* 16(1–2):121–131.

7. Mascini, M., I. Palchetti, and G. Marrazza. 2001. DNA electrochemical biosensors. *Fresen J Anal Chem* 369(1):15–22.

8. Horacek, J., and P. Skladal. 1997. Improved direct piezoelectric biosensors operating in liquid solution for the competitive label-free immunoassay of 2,4-dichlorophenoxyacetic acid. *Anal Chim Acta* 347(1–2):43–50.

9. Ebersole, R.C. et al. 1990. Spontaneously formed functionally active avidin monolayers on metal surfaces: A strategy for immobilizing biological reagents and design of piezoelectric biosensors. *J Am Chem Soc* 112(8):3239–3241.

10. Miller, M.M. et al. 2001. A DNA array sensor utilizing magnetic microbeads and magnetoelectronic detection. *J Magn Mater* 225(1–2):156–160.

11. Chemla, Y.R. et al. 2000. Ultrasensitive magnetic biosensor for homogeneous immunoassay. *Proc Natl Acad Sci* 97:26.

12. Raiteri, R. et al. 2001. Micromechanical cantilever-based biosensors. *Sensor Actuat B* 79(2–3):115–126.

13. Kasemo, B. 1998. Biological surface science. *Curr Opin Solid State Mater Sci* 3(5):451–459.

14. Natsume, T., H. Nakayama, and T. Isobe. 2001. BIA-MS-MS: Biomolecular interaction analysis for functional proteomics. *Trends Biotechnol* 19(10):S28–S33.

15. Polla, D.L. et al. 2000. Microdevices in medicine. *Annu Rev Biomed Eng* 2:551–576.

16. Iwasaki, Y., T. Horiuchi, and O. Niwa. 2001. Detection of electrochemical enzymatic reactions by surface plasmon resonance measurement. *Anal Chem* 73(7):1595–1598.

17. Maxwell, D.J., J.R. Taylor, and S. Nie. 2002. Self-assembled nanoparticle probes for recognition and detection of biomolecules. *J Am Chem Soc* 124(32):9606–9612.

18. Copeland, R.A., S.P.A. Fodor, and T.G. Spiro. 1984. Surface-enhanced Raman spectra of an active flavoenzyme: Glucose oxidase and riboflavin binding protein on silver particles. *J Am Chem Soc* 106(13):3872–3874.

19. Ivansson, D., K. Bayer, and C.F. Mandenius. 2002. Quantitation of intracellular recombinant human superoxide dismutase using surface plasmon resonance. *Anal Chim Acta* 456(2):193–200.

20. Spangler, B.D. et al. 2001. Comparison of the Spreeta surface plasmon resonance sensor and a quartz crystal microbalance for detection of *Escherichia coli* heat-labile enterotoxin. *Anal Chim Acta* 444:149–161.

21. Ligler, F.S. et al. 1993. Fiber-optic biosensor for the detection of hazardous materials. *Immunol Meth* 3(2):122–127.

22. Brockman, J.M., B.P. Nelson, and R.M. Corn. 2000. Surface plasmon resonance imaging measurements of ultrathin organic films. *Annu Rev Phys Chem* 51:41–63.

23. Jung, L.S. et al. 1998. Quantitative interpretation of the response of surface plasmon resonance sensors to adsorbed films. *Langmuir* 14(19):5636–5648.

24. Jung, L.S., and C.T. Campbell. 2000. Sticking probabilities in adsorption of alkanethiols from liquid ethanol solution onto gold. *J Phys Chem B* 104(47):11168–11178.

25. Jung, L.S., and C.T. Campbell. 2000. Sticking probabilities in adsorption from liquid solutions: Alkylthiols on gold. *Phys Rev Lett* 84(22):5164–5167.

26. Jung, L.S. et al. 2000. Binding and dissociation kinetics of wild-type and mutant streptavidins on mixed biotin-containing alkylthiolate monolayers. *Langmuir* 16(24):9421–9432.

27. Perez-Luna, V.H. et al. 1999. Molecular recognition between genetically engineered streptavidin and surface-bound biotin. *J Am Chem Soc* 121(27):6469–6478.

28. Mann, D.A. et al. 1998. Probing low affinity and multivalent interactions with surface plasmon resonance: Ligands for concanavalin A. *J Am Chem Soc* 120(41):10575–10582.

29. Hendrix, M. et al. 1997. Direct observation of aminoglycoside–RNA interactions by surface plasmon resonance. *J Am Chem Soc* 119(16):3641–3648.

30. Frey, B.L. et al. 1995. Control of the specific adsorption of proteins onto gold surfaces with poly(l-lysine) monolayers. *Anal Chem* 67(24):4452–4457.

31. Mrksick, M., J.R. Grunwell, and G.M. Whitesides. 1995. Biospecific adsorption of carbonic anhydrase to self-assembled monolayers of alkanethiolates that present benzenesulfonamide groups on gold. *J Am Chem Soc* 117(48):12009–12010.

32. Rao, J. et al. 1999. Using surface plasmon resonance to study the binding of vancomycin and its dimer to self-assembled monolayers presenting D-Ala-D-Ala. *J Am Chem Soc* 121(11):2629–2630.

33. Berger, C.E.H. et al. 1998. Surface plasmon resonance multisensing. *Anal Chem* 70(4):703–706.

34. Heaton, R.J., A.W. Peterson, and R.M. Georgiadis. 2001. Electrostatic surface plasmon resonance: Direct electric field-induced hybridization and denaturation in monolayer nucleic acid films and label-free discrimination of base mismatches. *Proc Natl Acad Sci USA* 98(7):3701–3704.

35. Georgiadis, R., K.P. Peterlinz, and A.W. Peterson. 2000. Quantitative measurements and modeling of kinetics in nucleic acid monolayer films using SPR spectroscopy. *J Am Chem Soc* 122(13):3166–3173.

36. Jordan, C.E. et al. 1997. Surface plasmon resonance imaging measurements of DNA hybridization adsorption and streptavidin/DNA multilayer formation at chemically modified gold surfaces. *Anal Chem* 69(24):4939–4947.

37. Nelson, B.P. et al. 2001. Surface plasmon resonance imaging measurements of DNA and RNA hybridization adsorption onto DNA microarrays. *Anal Chem* 73(1):1–7.

38. Brockman, J.M., A.G. Frutos, and R.M. Corn. A multistep chemical modification procedure to create DNA arrays on gold surfaces for the study of protein–DNA interactions with surface plasmon resonance imaging. *J Am Chem Soc* 121(35):8044–8051.

39. Schuck, P. 1997. Use of surface plasmon resonance to probe the equilibrium and dynamic aspects of interactions between biological macromolecules. *Annu Rev Biophys Biomol Struct* 26:541–566.

40. Haake, H.-M., A. Schutz, and G. Gauglitz. 2000. Label-free detection of biomolecular interaction by optical sensors. *Fresen J Anal Chem* 366(6–7):576–585.

41. Garland, P.B. 1996. Optical evanescent wave methods for the study of biomolecular interactions. *Q Rev Biophys* 29(1):91–117.

42. Knoll, W. 1998. Interfaces and thin films as seen by bound electromagnetic waves. *Annu Rev Phys Chem* 49:569–638.

43. Shumaker-Parry, J.S., and C.T. Campbell. 2004. Quantitative methods for spatially resolved adsorption/desorption measurements in real time by surface plasmon resonance microscopy. *Anal Chem* 76(4):907–917.

44. Shumaker-Parry, J.S. et al. 2004. Microspotting streptavidin and double-stranded DNA arrays on gold for high-throughput studies of protein–DNA interactions by surface plasmon resonance microscopy. *Anal Chem* 76(4):918–929.

45. Karlsson, R., and R. Stahlberg. 1995. Surface plasmon resonance detection and multispot sensing for direct monitoring of interactions involving low-molecular weight analytes and for determination of low affinities. *Anal Biochem* 228:274–280.

46. Sjolander, S., and C. Urbaniczky. 1991. Integrated fluid handling-system for biomolecular interaction analysis. *Anal Chem* 63:2338–2345.

47. Zizlsperger, M., and W. Knoll. 1998. Multispot parallel online monitoring of interfacial binding reactions by surface plasmon microscopy. *Prog Coll Pol Sci* 109:244–253.

48. Haynes, C.L., and R.P. Van Duyne. 2001. Nanosphere lithography: A versatile nanofabrication tool for studies of size-dependent nanoparticle optics. *J Phys Chem* 105(24):5599–5611.

49. Mulvaney, P. 2001. Not all that's gold does glitter. *MRS Bulletin* 26(12):1009–1014.

50. El-Sayed, M.A. 2001. Some interesting properties of metals confined in time and nanometer space of different shapes. *Acc Chem Res* 34(4):257–264.

51. Link, S., and M.A. El-Sayed. 1999. Spectral properties and relaxation dynamics of surface plasmon electronic oscillations in gold and silver nano-dots and nano-rods. *J Phys Chem B* 103(40):8410–8426.

52. Kreibig, U. et al. 1998. Optical investigations of surfaces and interfaces of metal clusters. In *Advances in metal and semiconductor clusters*, 345–393, ed. M.A. Duncan. Stamford: JAI Press.

53. Mulvaney, P. 1996. Surface plasmon spectroscopy of nanosized metal particles. *Langmuir* 12(3):788–800.

54. Kreibig, U. 1997. Optics of nanosized metals. In *Handbook of Optical Properties*, 145–190, eds. R.E. Hummel and P. Wissmann. Boca Raton: CRC Press.

55. Hulteen, J.C. et al. 1999. Nanosphere lithography: Size-tunable silver nanoparticle and surface cluster arrays. *J Phys Chem B* 103(19):3854–3863.

56. Haes, A.J., and R.P. Van Duyne. 2002. A nanoscale optical biosensor: Sensitivity and selectivity of an approach based on the localized surface plasmon resonance spectroscopy of triangular silver nanoparticles. *J Am Chem Soc* 124(35):10596–10604.

57. Malinsky, M.D. et al. 2001. Chain length dependence and sensing capabilities of the localized surface plasmon resonance of silver nanoparticles chemically modified with alkanethiol self-assembled monolayers. *J Am Chem Soc* 123(7):1471–1482.

58. McFarland, A.D., and R.P. Van Duyne. 2003. Single silver nanoparticles as real-time optical sensors with zeptomole sensitivity. *Nano Lett* 3:1057–1062.

59. Riboh, J.C. et al. 2003. A nanoscale optical biosensor: Real-time immunoassay in physiological buffer enabled by improved nanoparticle adhesion. *J Phys Chem B* 107:1772–1780.

60. Haes, A.J. et al. 2004. Nanoscale optical biosensor: Short range distance dependence of the localized surface plasmon resonance of noble metal nanoparticles. *J Phys Chem B* 108(22):6961–6968.

61. Haes, A.J. et al. 2004. A nanoscale optical biosensor: The long range distance dependence of the localized surface plasmon resonance of noble metal nanoparticles. *J Phys Chem B* 108(1):109–116.

62. Van Duyne, R.P., A.J. Haes, and A.D. McFarland. 2003. Nanoparticle optics: Sensing with nanoparticle arrays and single nanoparticles. *SPIE* 5223:197–207.

63. Mock, J.J., D.R. Smith, and S. Schultz. 2003. Local refractive index dependence of plasmon resonance spectra from individual nanoparticles. *Nano Lett* 3(4):485–491.

64. Freeman, R.G. et al. 1995. Self-assembled metal colloid monolayers: An approach to SERS substrates. *Science* 267:1629–1632.

65. Kahl, M. et al. 1998. Periodically structured metallic substrates for SERS. *Sensor Actuat B-Chem* 51(1–3):285–291.

66. Schatz, G.C., and R.P. Van Duyne. 2002. Electromagnetic mechanism of surface-enhanced spectroscopy. In *Handbook of Vibrational Spectroscopy*, vol. 1, 759–774, eds. J.M. Chalmers and P.R. Griffiths. New York: Wiley.

67. Haynes, C.L., and R.P. Van Duyne. 2003. Plasmon-sampled surface-enhanced Raman excitation spectroscopy. *J Phys Chem B* 107:7426–7433.

68. Haynes, C.L. et al. 2003. Nanoparticle optics: The importance of radiative dipole coupling in two-dimensional nanoparticle arrays. *J Phys Chem B* 107:7337–7342.

69. Dirix, Y. et al. 1999. Oriented pearl-necklace arrays of metallic nanoparticles in polymers: A new route toward polarization-dependent color filters. *Adv Mater* 11:223–227.

70. Haynes, C.L., and R.P. Van Duyne. 2003. Dichroic optical properties of extended nanostructures fabricated using angle-resolved nanosphere lithography. *Nano Lett* 3(7):939–943.

71. Maier, S.A. et al. 2001. Plasmonics-A route to nanoscale optical devices. *Adv Mater* 13(19):1501–1505.

72. Maier, S.A. et al. 2003. Local detection of electromagnetic energy transport below the diffraction limit in metal nanoparticle plasmon waveguides. *Nat Mater* 2:229–232.

73. Shelby, R.A., D.R. Smith, and S. Schultz. 2001. Experimental verification of a negative index of refraction. *Science* 292(5514):77–78.

74. Andersen, P.C., and K.L. Rowlen. 2002. Brilliant optical properties of nanometric noble metal spheres, rods, and aperture arrays. *Appl Spectrosc* 56(5):124A–135A.

75. Mucic, R.C. et al. 1998. DNA-directed synthesis of binary nanoparticle network materials. *J Am Chem Soc* 120:12674–12675.

76. Hirsch, L.R. et al. 2003. A whole blood immunoassay using gold nanoshells. *Anal Chem* 75(10):2377–2381.

77. Haes, A.J., and R.P. Van Duyne. 2002. A highly sensitive and selective surface-enhanced nanobiosensor. *Mat Res Soc Symp Proc* 723:O3.1.1–O3.1.6.

78. Haes, A.J., and R.P. Van Duyne. 2003. Nanosensors enable portable detectors for environmental and medical applications. *Laser Focus World* 39:153–156.

79. Haes, A.J., and R.P. Van Duyne. 2003. Nanoscale optical biosensors based on localized surface plasmon resonance spectroscopy. *SPIE* 5221:47–58.

80. Fritzsche, W., and T.A. Taton. 2003. Metal nanoparticles as labels for heterogeneous, chip-based DNA detection. *Nanotechnology* 14(12):R63–R73.

81. Aizpurua, J. et al. 2003. Optical properties of gold nanorings. *Phys Rev Lett* 90(5):057401/1–057401/4.

82. Obare, S.O., R.E. Hollowell, and C.J. Murphy. 2002. Sensing strategy for lithium ion based on gold nanoparticles. *Langmuir* 18(26):10407–10410.

83. Nath, N., and A. Chilkoti. 2002. Immobilized gold nanoparticle sensor for label-free optical detection of biomolecular interactions. *Proc SPIE-Int Soc Opt Eng* 4626:441–448.

84. Nam, J.-M., C.S. Thaxton, and C.A. Mirkin. 2003. Nanoparticle-based bio-bar codes for the ultrasensitive detection of proteins. *Science (Washington, DC, United States)* 301(5641):1884–1886.

85. Bailey, R.C. et al. 2003. Real-time multicolor DNA detection with chemoresponsive diffraction gratings and nanoparticle probes. *J Am Chem Soc* 125(44):13541–13547.

86. Kreibig, U., and M. Vollmer. 1995. Cluster materials. *Optical Properties of Metal Clusters*, vol. 25, 532. Heidelberg: Springer-Verlag.

87. Draine, B.T., and P.J. Flatau. 1994. Discrete-dipole approximation for scattering calculations. *J Opt Soc Am A* 11:1491–1499.

88. Kelly, K.L. et al. 2003. The optical properties of metal nanoparticles: The influence of size, shape, and dielectric environment. *J Phys Chem B* 107(3):668–677.

89. Lynch, D.W., and W.R. Hunter. 1985. *Handbook of Optical Constants of Solids*, 350–356, ed. E.D. Palik. New York: Academic Press.

90. Jensen, T.R. et al. 1999. Electrodynamics of noble metal nanoparticles and nanoparticle clusters. *J Clust Sci* 10(2):295–317.

91. Jensen, T.R., G.C. Schatz, and R.P. Van Duyne. 1999. Nanosphere lithography: Surface plasmon resonance spectrum of a periodic array of silver nanoparticles by UV-Vis extinction spectroscopy and electrodynamic modeling. *J Phys Chem B* 103(13):2394–2401.

92. Taton, T.A., G. Lu, and C.A. Mirkin. 2001. Two-color labeling of oligonucleotide arrays via size-selective scattering of nanoparticle probes. *J Am Chem Soc* 123(21):5164–5165.

93. Schultz, S. et al. 2000. Single-target molecule detection with nonbleaching multicolor optical immunolabels. *Proc Natl Acad Sci* 97(3):996–1001.

94. Taton, T.A., C.A. Mirkin, and R.L. Letsinger. 2000. Scanometric DNA array detection with nanoparticle probes. *Science* 289(5485):1757–1760.

95. Sönnichsen, C. et al. 2000. Spectroscopy of single metallic nanoparticles using total internal reflection microscopy. *Appl Phys Lett* 77(19):2949–2951.

96. Sönnichsen, C. et al. 2002. Drastic reduction of plasmon damping in gold nanorods. *Phys Rev Lett* 88:0774021–0774024.

97. Yguerabide, J., and E.E. Yguerabide. 1998. Light-scattering submicroscopic particles as highly fluorescent analogs and their use as tracer labels in clinical and biological applications—II. Experimental characterization. *Anal Biochem* 262:157–176.

98. Bao, P. et al. 2002. High-sensitivity detection of DNA hybridization on microarrays using resonance light scattering. *Anal Chem* 74(8):1792–1797.

99. Connolly, S., S. Cobbe, and D. Fitzmaurice. 2001. Effects of ligand–receptor geometry and stoichiometry on protein-induced aggregation of biotin-modified colloidal gold. *J Phys Chem B* 105(11):2222–2226.

100. Connolly, S., S.N. Rao, and D. Fitzmaurice. 2000. Characterization of protein aggregated gold nanocrystals. *J Phys Chem B* 104(19):4765–4776.
101. Elghanian, R. et al. 1997. Selective colorimetric detection of polynucleotides based on the distance-dependent optical properties of gold nanoparticles. *Science* 227(5329):1078–1080.
102. Mirkin, C.A. et al. 1996. A DNA-based method for rationally assembling nanoparticles into macroscopic materials. *Nature* 382(6592):607–609.
103. Storhoff, J.J. et al. 2000. What controls the optical properties of DNA-linked gold nanoparticle assemblies? *J Am Chem Soc* 122(19):4640–4650.
104. Hilger, A. et al. 2000. Surface and interface effects in the optical properties of silver nanoparticles. *Eur Phys J D* 10(1):115–118.
105. Henglein, A., and D. Meisel. 1998. Spectrophotometric observations of the adsorption of organosulfur compounds on colloidal silver nanoparticles. *J Phys Chem B* 102(43):8364–8366.
106. Linnert, T., P. Mulvaney, and A. Henglein. 1993. Surface chemistry of colloidal silver: Surface plasmon damping by chemisorbed iodide, hydrosulfide (SH-), and phenylthiolate. *J Phys Chem* 97(3):679–682.
107. Kreibig, U., M. Gartz, and A. Hilger. 1997. Mie resonances. Sensors for physical and chemical cluster interface properties. *Ber Bunsen-Ges* 101(11):1593–1604.
108. Nath, N., and A. Chilkoti. 2002. A colorimetric gold nanoparticle sensor to interrogate biomolecular interactions in real time on a surface. *Anal Chem* 74(3):504–509.
109. Eck, D. et al. 2001. Plasmon resonance measurements of the adsorption and adsorption kinetics of a biopolymer onto gold nanocolloids. *Langmuir* 17(4):957–960.
110. Okamoto, T., I. Yamaguchi, and T. Kobayashi. 2000. Local plasmon sensor with gold colloid monolayers deposited upon glass substrates. *Opt Lett* 25(6):372–374.
111. Himmelhaus, M., and H. Takei. 2000. Cap-shaped gold nanoparticles for an optical biosensor. *Sensor Actuat B* 63(1–2):24–30.
112. Takei, H. 1998. Biological sensor based on localized surface plasmon associated with surface-bound Au/polystyrene composite microparticles. *Proc SPIE-Int Soc Opt Eng* 3515:278–283.
113. Klar, T. et al. 1998. Surface-plasmon resonances in single metallic nanoparticles. *Phys Rev Lett* 80:4249–4252.
114. Matsuo, Y., and K. Sasaki. 2001. Time-resolved laser scattering spectroscopy of a single metallic nanoparticle. *Jpn J Appl Phys* 40:6143–6147.
115. Mock, J.J. et al. 2002. Shape effects in plasmon resonance of individual colloidal silver nanoparticles. *J Chem Phys* 116(15):6755–6759.
116. Lee, P.C., and D. Meisel. 1982. Adsorption and surface-enhanced Raman of dyes on silver and gold sols. *J Phys Chem* 86(17):3391–3395.
117. Weimer, W.A., and M.J. Dyer. 2001. Tunable surface plasmon resonance silver films. *Appl Phys Lett* 79(19):3164–3166.
118. Gotschy, W. et al. 1996. Thin films by regular patterns of metal nanoparticles: Tailoring the optical properties by nanodesign. *Appl Phys B* 63:381–384.
119. Hulteen, J.C., and R.P. Van Duyne. 1995. Nanosphere lithography: A materials general fabrication process for periodic particle array surfaces. *J Vac Sci Technol A* 13:1553–1558.
120. Micheletto, R., H. Fukuda, and M. Ohtsu. 1995. A simple method for the production of a two-dimensional, ordered array of small latex particles. *Langmuir* 11:3333–3336.
121. Jensen, T.R. et al. 2000. Surface-enhanced infrared spectroscopy: A comparison of metal island films with discrete and non-discrete surface plasmons. *Appl Spectrosc* 54:371–377.
122. Jensen, T.R. et al. 2000. Nanosphere lithography: Tunable localized surface plasmon resonance spectra of silver nanoparticles. *J Phys Chem B* 104:10549–10556.
123. Duval Malinsky, M. et al.. 2001. Nanosphere lithography: Effect of the substrate on the localized surface plasmon resonance spectrum of silver nanoparticles. *J Phys Chem B* 105:2343–2350.
124. Jensen, T.R. et al. 1999. Nanosphere lithography: Effect of the external dielectric medium on the surface plasmon resonance spectrum of a periodic array of silver nanoparticles. *J Phys Chem B* 103:9846–9853.
125. Green, N.M. 1975. Avidin. *Adv Protein Chem* 29:85–133.

126. Wilchek, M., and E.A. Bayer. 1998. Immobilized biomolecules in analysis. In *Avidin–Biotin Immobilization Systems*, 15–34, eds. T. Cass and F.S. Ligler. Oxford: Oxford University Press.
127. Suzuki, M. et al. 2002. Miniature surface-plasmon resonance immunosensors—Rapid and repetitive procedure. *Anal Bioanal Chem* 372:301–304.
128. Lynch, N.J., R.K. Kilpatrick, and R.G. Carbonell. 1996. Aggregation of ligand-modified liposomes by specific interactions with proteins. II: Biotinylated liposomes and antibiotin antibody. *Biotechnol Bioeng* 50:169–183.
129. Adamczyk, M. et al. 1999. Surface plasmon resonance (SPR) as a tool for antibody conjugate analysis. *Bioconjucate Chem* 10:1032–1037.
130. Hardman, K.D., and C.F. Ainsworth. 1972. Structure of concanavalin A at 2.4-Ang resolution. *Biochemistry* 11:4910–4916.
131. Smith, E.A. et al. 2002. Surface plasmon resonance imaging studies of protein–carbohydrate interactions. *J Am Chem Soc* 125:6140–6148.
132. Yonzon, C.R. et al. 2004. A comparative analysis of localized and propagating surface plasmon resonance sensors: The binding of concanavalin A to a monosaccharide functionalized self-assembled monolayer. *J Am Chem Soc* 126:12669–12676.
133. Alzheimer, A. et al. 1995. An English translation of Alzheimer's 1907 paper, Uber eine eigenartige Erkankung der Hirnrinde. *Clin Anat* 8(6):429–431.
134. Hardy, J.A., and G.A. Higgins. 1992. Alzheimer's disease: The amyloid cascade hypothesis. *Science* 256(5054):184–185.
135. Klein, W.L., G.A. Krafft, and C.E. Finch. 2001. Targeting small A beta oligomers: The solution to an Alzheimer's disease conundrum. *Trends Neurosci* 24(4):219–224.
136. Lorenzo, A., and B.A. Yankner. 1994. Beta-amyloid neurotoxicity requires fibril formation and is inhibited by congo red. *Proc Natl Acad Sci USA* 91:12243–12247.
137. Lambert, M.P. et al. 1998. Diffusible, nonfibrillar ligands derived from A beta(1–42) are potent central nervous system neurotoxins. *Proc Natl Acad Sci USA* 95(11):6448–6453.
138. Gong, Y. et al. 2003. Alzheimer's disease-affected brain: Presence of oligomeric Ab ligands (ADDLs) suggests a molecular basis for reversible memory loss. *Proc Natl Acad Sci USA* 100(18):10417–10422.
139. Haes, A.J. et al. 2004. First steps toward an Alzheimer's disease assay using localized surface plasmon resonance spectroscopy. Northwestern University Invention Disclosure, 24017.
140. Haes, A.J. et al. 2004. A localized surface plasmon resonance biosensor: First steps toward an assay for Alzheimer's disease. *Nano Lett* 4(6):1029–1034.

17

Synthetic Biology: From Gene Circuits to Novel Biological Tools

Nina G. Argibay, Eric M. Vazquez, Cortney E. Wilson,
Travis J.A. Craddock, and Robert P. Smith

CONTENTS

17.1 Introduction

Since its emergence in 2000 (Elowitz and Leibler 2000; Gardner, Cantor, and Collins 2000), synthetic biology has proven to be an incredibly useful approach to engineer cells with novel behaviors. These behaviors are extremely diverse, as are their potential applications. Synthetic systems have been used to gain understanding of naturally occurring cellular networks (Riccione et al. 2012) as well as ecological and evolutionary relationships (Tanouchi, Smith, and You 2012a). Furthermore, they have been used to treat infectious diseases (Ruder, Lu, and Collins 2011), to form the theoretical basis of biological computing (Siuti, Yazbek, and Lu 2013), and in industrial processes (Georgianna and Mayfield 2012).

Interestingly, synthetic biology and nanotechnology have many of the same common broad approaches and goals. Both fields aim to rationally engineer processes that ultimately occur at the molecular level and share common interest in designing systems with broad applications in medicine and industry. Although there are exceptions, what tends to differ between both fields is that synthetic biology uses biological systems to drive molecular interactions that lead to novel behaviors and products. Such molecular interactions are engineered using gene circuits, which are rationally designed, implemented, and optimized using the tools of molecular biology. This process is heavily guided by the creation of mathematical models, which are designed to capture the essential components of the gene circuit (Chandran et al. 2008). Using these models, the behavior can be observed and optimized *in silico*, which serves to focus and refine biological experimentation.

In the following chapter, we begin by covering the basics of gene circuit design and function. We then focus on describing various implementation strategies, with a particular focus on synthetic systems that may have application in medicine. We conclude by reviewing some of the challenges synthetic biology might possibly face in the future.

17.2 Gene Circuit

The basic concept of the gene circuit is built upon the modular fashion in which many genes are arranged in the cell (Jusiak et al. 2014). For example, *Escherichia coli* contains thousands of genes, but not all of these genes are expressed all the time. Conversely, many genes are expressed only when they are needed by the bacterium. This is possible because these genes are separated into discrete units that are activated (or expressed) in response to internal or external signals. Once activated, the gene product will interact with another gene product, itself or the cell, to cause a change in the cell's behavior. It is this very foundation, the independent control of interacting gene products, upon which the concept of the gene circuit has been formed.

Most gene circuits are composed of four main biological parts (Endy 2005), each of which plays a critical role in the function and regulation of the gene circuit. Interestingly, each of these biological parts has demonstrated the ability to be modular. Given such modularity, biological parts from within the same or different organisms can be mixed and matched leading to novel gene circuits. Ideally, the biological parts assembled to create the novel gene circuit should have minimal "cross talk" between each other or existing genes in the cell. Parts with minimal cross talk are termed orthogonal and most often lead to the greatest degree of control (Wang et al. 2011).

While the specific sequence of biological parts that compose the gene circuit can be incredibly diverse, each biological part often has core features that allow for functionality (Figure 17.1). The first part in the gene circuit, the promoter, is a DNA sequence that serves as an input device. The promoter integrates internal or external signals and allows the gene circuit to be either ON or OFF, thus often drawing analogies to a light switch. Signals integrated by promoters can range from sugars and sugar analogs (Lutz and Bujard 1997) to small molecules produced in cell–cell communication (You et al. 2004) to specific wavelengths of light (Olson et al. 2014). At a fundamental level, such signals serve to recruit (activate), or prevent (repress), RNA polymerase binding to the promoter. When bound, RNA polymerase transcribes the information downstream of the promoter and thus "activates" the gene. Ideally, the promoter should be "tunable." Here, RNA polymerase activity increases, or decreases, with increasing signal. To stop RNA polymerase from performing transcription, a terminator is located at the very end of the gene circuit. Note that mechanisms of posttranscriptional regulation (either activating or repressing the gene circuit after transcription has been performed) have also been designed (Isaacs et al. 2004).

Directly downstream of the promoter is the ribosomal binding site (RBS). The RBS is the region where the ribosome binds to the mRNA to initiate protein translation. By varying the DNA sequence of the RBS, one can control the translation initiation rate and thus the level of protein production (Salis, Mirsky, and Voigt 2009). Following the RBS lies the protein coding sequence, which contains the information encoding the protein to be produced. This area is often the most variable and most critical part of the gene circuit as it often directly participates in the desired behavior. Of particular note, some gene circuits

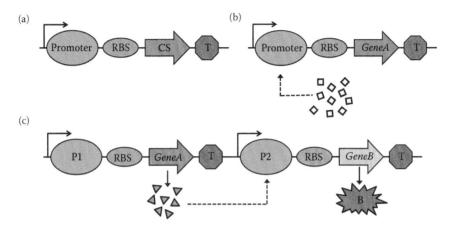

FIGURE 17.1
The basic elements and behaviors of gene circuits. (a) The four main biological parts of a gene circuit. The promoter acts as the ON/OFF switch of the circuit. The ribosomal binding site (RBS) acts to control the rate of translation. The protein coding sequence (CS) specifies the protein to be produced. The terminator (T) stops RNA polymerase from performing transcription. The arrow indicates the direction in which transcription and translation is performed. (b) Activation of a gene circuit in response to an external signal. In the presence of an external signal (yellow squares) the promoter is activated, driving the expression of *geneA*. (c) Assembly of multiple gene circuits that interact together. When promoter "P1" is activated, it expresses *geneA*, which makes an easily diffusible signal (blue triangles). When a sufficient amount of this signal accumulates in the environment, it activates the promoter "P2," which drives the expression of *geneB* and results in the production of protein B. This logic is often observed in synthetic systems that biological parts that allow for cell–cell communication.

will have multiple protein coding sequences downstream of a promoter. This allows parallel production of two or more proteins. In some instances, the first protein will be directly involved in producing the desired behavior. The second protein, often a fluorescent protein, will serve as an output of the promoter, thus verifying the activation of the promoter.

Once the gene circuit is assembled, additional machinery, such as transcriptional and translational proteins and components, is required to fully "express" or activate the gene circuit. The system providing this machinery is often referred to as the "chassis" (Endy 2005). Chassis can be grouped into three general categories: *in vivo*, where a cell drives the gene circuit; *in vitro*, where components extracted from cells drive the gene circuit; or artificial cells, where transcriptional/translational components and the gene circuit are compartmentalized in artificial biomimetic membranes. Each chassis offers its own benefits and drawbacks, which are outlined in greater detail in the sections below.

Once implemented in the chassis of choice, optimizing the behavior of the gene circuit is required. Mathematical modeling can be used to predict parameter spaces that will lead to the desired function. Most often, a mathematical model might simply predict that changing cellular physiology, the rate of protein production, or the environmental conditions will drive the gene circuit toward functionality. Indeed, several manuscripts have demonstrated, both mathematically and experimentally, that simply changing the growth rate of the cells, the translation rate, or the amount of environmental signal will impact the behavior of the circuit and lead to optimization (Ajo-Franklin et al. 2007; Tanouchi et al. 2012b; Payne et al. 2013; Olson et al. 2014). In certain cases, cellular noise can drastically impact the robustness of gene circuit functionality. That is, random fluctuations in cell growth dynamics (Tan, Marguet, and You 2009), external signals, or in transcription/translation rates (Swain, Elowitz, and Siggia 2002; Eldar and Elowitz 2010) can impede the ability of a gene circuit to function correctly. Several approaches to limiting the influence of noise have been developed and include the use of negative feedback motifs (Becskei and Serrano 2000; Dublanche et al. 2006), coherent feed forward loops (Mangan and Alon 2003), or positive feedback loops with bistability (Atkinson et al. 2003) in the gene circuit. Finally, directed evolution remains an alternative mechanism to optimize the behavior of a gene circuit (Cobb, Sun, and Zhao 2013; Currin et al. 2015). Using multiple rounds of mutagenesis and high-throughput screening, the behavior of a gene circuit can be altered and eventually optimized. Indeed, this approach has been used before to convert a nonfunctional gene circuit into one that is fully functional (Yokobayashi, Weiss, and Arnold 2002).

While the examples below are designed primarily to connect to applications in nanotechnology, it should be noted that there are a multitude of behaviors that have been engineered using synthetic biology. These behaviors include, but are not limited to, oscillators (Elowitz and Leibler 2000; Stricker et al. 2008; Tigges et al. 2009; Marguet et al. 2010; Tigges et al. 2010; Chen et al. 2015), switches (Gardner, Cantor, and Collins 2000; Kobayashi et al. 2004; Kramer et al. 2004), transcriptional cascades (Hooshangi, Thiberge, and Weiss 2005), edge detectors (Tabor et al. 2009), and populations of cells that cooperate with each other (Gore, Youk, and van Oudenaarden 2009; Tanouchi et al. 2012a, b; Pai, Tanouchi, and You 2012; Smith et al. 2014).

17.3 Engineering Gene Circuits in Cells

While living organisms are complex and can be unpredictable, they are most often used as the chassis to drive gene circuits (Kelwick et al. 2014). As cells have an active metabolism, they offer sustained expression of a gene circuit over a period of days. However, the inherent complexity of cells can also result in unexpected behavior of the gene circuit, as *in vivo* optimization cannot be fully anticipated ad hoc. Interestingly, such unexpected behavior has driven the field's understanding of engineering principles and has allowed a greater understanding of how natural biological systems operate (Tan, Marguet, and You 2009; Marguet et al. 2010). Nevertheless, given the relative ease at which gene circuits can be synthesized and assembled *in vivo*, cells remain the chassis of choice for the synthetic biologist. While early gene circuits were implemented primarily in *E. coli*, an expanded repertoire of chassis is now available and ranges from additional prokaryotic cells (Prindle et al. 2012), to mammalian cells (Kramer et al. 2004) and yeast (Ajo-Franklin et al. 2007).

Molecular computing has been of interest in synthetic biology as well as nanotechnology (de Silva and Uchiyama 2007). One core component of computing is logic gates (AND, OR, NOR, NOT, NAND, XOR, etc.), which drive multiple processes in computational systems. As such, there has been a significant amount of research devoted to engineering various logic gates. Early manuscripts focused on single input AND, OR, and NOT gates (Mayo et al. 2006; Anderson, Voigt, and Arkin 2007; Rinaudo et al. 2007), the functionality of which was determined by using fluorescent proteins (Gardner, Cantor, and Collins 2000). However, with an increasing understanding of engineering principles and limitations, higher order logic gates have been engineered into cells (Tabor et al. 2009; Tamsir, Tabor, and Voigt 2011; Wang et al. 2011; Moon et al. 2012).

Layered logic gates are of particular interest in molecular computing as they can lead to complex logic functions including EQUAL and XOR logic gates. To create a framework for engineering layered logic gates, Tamsir et al. engineered a series of one function logic gates, such as AND, NOT, and OR gates (Tamsir, Tabor, and Voigt 2011) (Figure 17.2). The majority of the logic gates consisted of two tandem promoters, P_1 and P_2, which are activated using externally supplied chemicals or cell-produced small molecules (Figure 17.2a). As an example, a NOR logic gate, which is only ON when both inputs are OFF, was engineered into *E. coli* using the P_{BAD} and P_{TET} promoters, which are activated by the sugar arabinose and the antibiotic anhydrotetracycline (atc), respectively. Activation of either the P_{BAD} or the P_{TET} promoter results in the expression of the CI repressor that turns off the production of a yellow fluorescent protein (YFP), which serves as the output of the logic gate.

Layering of logic gates was accomplished by engineering different strains of bacteria to each contain a different logic gate, such as the NOR gate described above. However, to "connect" the different strains of bacteria, and thus the logic gates, together, the authors used genes that could produce small diffusible molecules called acyl-homoserine lactones (AHL), which can readily be secreted by bacteria and travel in growth medium (Figure 17.2b). Here, AHL was designed to serve as one of the inputs to the next logic gate in the layer. When cells containing different logic gates were arranged on a nutrient agar plate, they could be connected through AHL, which the authors termed "chemical wires." By arranging these different bacterial strains, and thus logic gates, in a particular spatial order, the authors were able to realize all 16 two-input Boolean logic gates.

Engineering synthetic circuits that operate in the spatial domain (Basu et al. 2005; Sohka et al. 2009; Payne et al. 2013; Liu et al. 2011) is of particular interest within the field, as such systems may have use in industrial processes or may be used to explore the ubiquitous biological phenomenon of pattern formation (Turing 1952). As an example, Payne et al. engineered a gene circuit to create a two-dimensional concentric ring structure (Payne et al. 2013). The gene circuit can be separated into two modules: the activation module and the inhibition module. In the activation module, a T7 RNA polymerase (T7 RNAP) drives a

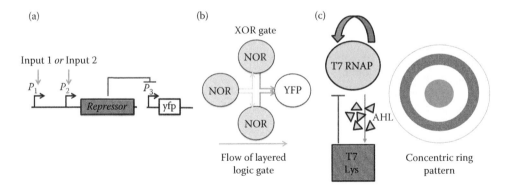

FIGURE 17.2
Logic gates and pattern formation using gene circuits driven by *in vivo* chassis. (a) A gene circuit that leads to a NOR gate. Either Input 1 or Input 2 activates the expression of a repressor protein, which turns off the production of a yellow fluorescent protein (YFP). (b) An XOR gate engineered using a layered logic gate. Each colony of bacteria (circles) contains a different NOR gate. Sequential activation of each NOR gate using chemical wires (green lines, wires realized through the production and secretion of AHL) resulted in an XOR gate with YFP as an output. (c) Spatial pattern formation realized through coupled positive and negative feedback. Left panel: T7 RNA polymerase (RNAP) forms a positive feedback that activates the production of additional T7 RNAP and AHL. AHL activates T7 lysozyme (T7 Lys), which stops production of T7 RNAP and thus serves as the negative feedback. Right panel: This gene circuit results in a concentric ring pattern in *E. coli*, with positive feedback activated in the center (measured using cyan fluorescent protein [CFP], blue) and negative feedback activated on the outside (measured using the mcherry variant of red fluorescent protein, red). The presence of cells without fluorescence is indicated using beige.

positive feedback that drives additional production of T7 RNAP and LuxI, which produces an AHL (Figure 17.2c). As AHL is readily diffusible outside of the cell, it accumulates in the medium, and serves as an input to the inhibition module, which contains a negative feedback gene circuit. This circuit contains a T7 lysozyme (T7 Lys), which, when expressed, inhibits the production of T7 RNAP, and thus inhibits the positive feedback. To visualize expression of the activation and inhibition modules, the authors coupled expression of a cyan fluorescent protein (CFP) to the positive feedback and a modified red fluorescent protein (mcherry) to the negative feedback.

Activation of this circuit resulted in a concentric ring pattern that was tunable. Specifically, activation of the activation module resulted in an intense CFP signal in the interior of the ring. Extending from the center, the cells were colorless until an intense mcherry signal was detected, thus forming a second ring at the edge of the colony. The distance between the CFP center and the mcherry edge, as well as the occurrence of additional mcherry rings, could be tuned using exogenously added AHL. Overall, this study demonstrates that spatial patterns could be rationally engineered into a single bacterial population.

17.4 Driving Gene Circuits Using *In Vitro* Systems

While *in vivo* systems can offer sustained expression of a gene circuit, they are relatively noncustomizable as one cannot directly alter the abundance of components that drive the gene circuit, such as polymerases, ribosomes, and nucleotides. As an alternative to *in vivo* chassis, *in vitro* chassis, which use cell-free extracts (containing, for example, amino acids, ribonucleotides, an ATP regeneration system, tRNA, termination factors, elongation factors, translation initiation factors, ribosomes, and RNA polymerases) to drive the gene circuit, offer a more customizable system. Here, the user can precisely define the concentrations of each component in the system, including those listed above. Furthermore, these systems offer more quantitative measurements of gene circuit dynamics. One drawback of such systems is that the components that drive the gene circuit cannot be replenished by the system, thus limiting the time of gene circuit functionality. Several behaviors have been engineered using *in vitro* chassis (Hodgman and Jewett 2012), including logic gates (Seelig et al. 2006) and oscillators (Kim and Winfree 2011).

Transcriptional cascades are an important part of natural cellular networks and are often observed in genes that govern the development of multicellular organisms (Arnone and Davidson 1997). In transcriptional cascades, the product of one gene circuit directly activates, or represses, a subsequent gene circuit. In nanotechnology, a transcriptional cascade may be analogous to a series of molecular reactions, where the product of one reaction is used directly to activate or repress a second reaction. To determine whether transcriptional cascades could be engineered *in vitro*, Noireaux et al. constructed a two-step and a three-step transcription cascade, which was driven by a cell-free extract from wheat germ (Noireaux, Bar-Ziv, and Libchaber 2003). In both circuits, firefly luciferase, which produces quantifiable light, was used to measure the output (Figure 17.3). In the two-step cascade, a T7 RNAP drives the production of SP6 RNAP, which in turn drives the production of the luciferase gene. To create the three-step cascade, the authors removed the SP6 RNAP driven copy of luciferase and replaced it with the *rpoF* sigma factor. As such, expression of *rpoF* was controlled by SP6 RNAP. When produced, the *rpoF* sigma factor drives the production of luciferase via an *rpoF*-activated promoter. When the output

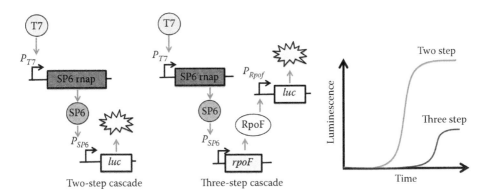

FIGURE 17.3
Driving an engineering transcriptional cascade using an *in vitro* system. Left panel: A two-step transcriptional cascade where T7 RNAP drives the expression of the SP6 RNAP, which drives the expression of firefly luciferase (*luc*). Center panel: A three-step transcriptional cascade. T7 RNAP drives the expression of the SP6 RNAP, which drives the expression of *rpoF*. RpoF subsequently binds to RNA polymerase and drives the expression of *luc*. Right panel: The authors observed that as the number of steps increases in the transcriptional cascade, the time to observe luminescence increased while the total amount of luminescence was reduced.

(amount of luminescence from luciferase) of both gene circuits were compared, the authors observed that the three-step cascade showed a reduction in luminescence and an increase in the time required to observe fluorescence (as compared with the two-step cascade, Figure 17.3). Furthermore, the authors observed that the three-step cascade was more sensitive to the amount of plasmids and additional components added into the system. Here, quantifiable expression could only be observed when the quantity of expression from the first step (T7 RNAP) was low and expression from the second step (SP6 RNAP) was high. This study was of particular interest as it demonstrated that complex behaviors could be engineered using *in vitro* systems. Furthermore, it revealed that the nonintuitive scaling of circuit components is often required to achieve the desired behavior.

17.5 Artificial Cells as an Intermediate Chassis to Drive Gene Circuits

Artificial cells can be used as an "intermediate chassis" to drive a gene circuit. Artificial cells consist of cell-free system, and the gene circuit, which are encapsulated within biomimetic membranes. Such membranes, in contrast to purely *in vitro* expression systems, offer protection and stability, and extend the time for gene expression. In contrast to *in vivo* systems, the user of artificial cells can more precisely define and manipulate the components. Biomimetic membranes that are used to form the artificial cell are generally composed of lipids. Given the correct conditions, bi-lipid layers will spontaneously form, encapsulating the cell-free system and the gene circuit. Environmental factors, including osmotic pressure, pH, and temperature, can affect the stability, size, volume, and yield of artificial cells formed, thus allowing the experimenter to control various facets of the assembly process. A more extensive review of artificial cell properties and fabrication can be found elsewhere (Pohorille and Deamer 2002).

Similar to gene circuits expressed *in vivo*, artificial cells have the potential of providing a greater understanding of the mechanisms occurring in natural cells, and may allow the development of new biological technologies, including the sustained synthesis and delivery of pharmaceuticals. However, expression of a gene circuit encapsulated within a biomimetic membrane is limited by the lack of a constant supply of nutrients and the accumulation of waste products. While the low permeability of biomimetic membranes allows for the spatial restriction of transcription/translation machinery and components, as well as the gene circuit, it also reduces the replenishment of nutrients across the membrane. As such, gene expression and protein synthesis often do not extend for >2 h. To solve this issue, Noireaux and Libchaber extended gene expression by expressing the gene encoding α-hemolysin from *Staphylococcus aureus* in artificial cells (Noireaux and Libchaber 2004). The authors hypothesized that expression of α-hemolysin would create nonspecific pores directly into the biomimetic membrane, facilitating the movement of nutrients and waste into and out of the artificial cell, respectively, thus serving to extend the time allowing for gene expression. Note that addition mechanisms to increase permeability, such as the use of channel proteins, have also been performed (Pohorille and Deamer 2002).

To test their theory, Noireaux and Libchaber used a cell-free system extracted from *E. coli* to drive transcription and translation (Figure 17.4a). A plasmid was used to carry

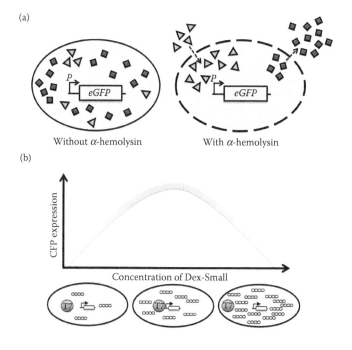

FIGURE 17.4
Artificial cells as an alternative chassis for driving synthetic gene circuits. (a) Expressing α-hemolysin to increase gene circuit activation time and quantity. Left panel: Without the inclusion of α-hemolysin in the expression system, nutrients (green triangles) were quickly depleted while waste (red diamonds) quickly accumulated. This combination served to limit the length and quantity of eGFP expressed. Right panel: In contrast, when α-hemolysin was included in the expression system, eGFP expression time and quantity increased. Here, α-hemolysin served to create pores in the artificial biomimetic membrane, which allowed the diffusion mediated exchange of nutrients and waste with the surrounding environment. *P* and arrow represent promoter. (b) A biphasic relationship between molecular crowding and gene expression in artificial cells. When crowding is low (low Dex-Small concentration, yellow shapes), T7 RNA polymerase (RFP-T7 RNAP, purple circles) finds the promoter, but at a reduced rate. At intermediate crowding, RFP-T7 RNAP is forced to remain on the promoter for extended periods of time, thus increasing CFP expression time. When crowding is high (high Dex-Small concentration), RFP-T7 RNAP is prevented from accessing the promoter, thus decreasing CFP expression.

the DNA coding regions encoding α-hemolysin and enhanced green fluorescent protein (eGFP), which served as a mechanism to measure gene expression. Note that α-hemolysin was chosen for its ability to not completely lyse the biomimetic membrane, but to disrupt membrane structure sufficiently such that nutrients and waste could be exchanged. Encapsulation, and thus the creation of the artificial cells, was performed by placing small quantities of a cell-free system and plasmid (containing the gene circuit) into a tube containing a mixture of oil and dissolved phospholipids. Once isolated, the artificial cells were examined for fluorescence. The authors observed that nonencapsulated cell-free extract was able to sustain expression of eGFP for 5 h. Encapsulation of the cell-free system, without the expression of α-hemolysin, increased total protein yield and expression time to 10 h. Interestingly, the authors observed that the expression of α-hemolysin drastically increased eGFP expression. Here, eGFP production lasted upwards of 100 h. This study provided evidence that expression can be maintained in artificial cells when nutrients, and waste, can be exchanged across the biomimetic membrane.

A clear advantage of artificial cells is reduced complexity as compared with *in vivo* systems. However, it is being increasingly recognized that "less" is not always "more" in synthetic biology. One major difference between *in vivo* systems and artificial cells is the density of molecules that surround the gene circuit. This density may play a critical role in maintaining robust gene expression. To examine how molecular density, or molecular crowding, affects gene expression in artificial cells, Tan et al. used a modified T7 RNAP (fused to a red fluorescent protein, RFP) and a reporter gene (CFP) as an output (Tan et al. 2013). When the RFP-T7 RNAP fusion protein binds to its promoter, P_{T7}, CFP is transcribed. This setup allows the quantification of the binding rate of RFP-T7 RNAP to the promoter (via RFP) and gene expression (via CFP) to be measured independently.

Initially, the authors examined the impact that inert crowding molecules, large and small dextran polymers (Dex-Big and Dex-Small, respectively), had on RFP-T7 RNAP binding to the P_{T7} promoter. In both cases, increasing the density of either Dex-Big or Dex-Small resulted in lower diffusion rates between T7-RNAP and the P_{T7} promoter. Through colocalization experiments, the authors were able to demonstrate that increasing the density of Dex-Big and Dex-Small decreased the disassociation rate (or increased the association rate) between T7-RNAP and the P_{T7} promoter. Dex-Big was observed to increase binding rates (100% increase relative to the control) substantially more as compared with Dex-Small (25% increase relative to the control). As such, the authors demonstrated that molecular crowding could impact the functionality of the transcriptional apparatus.

Building off of their results, the authors encapsulated their gene circuit into artificial cells to examine the influence of molecular crowding on gene expression (Figure 17.4b). Specifically, they encapsulated their gene circuit in a cell-free system containing increasing amounts of Dex-Small. As predicted by a mathematical model, increasing the density of Dex-Small served to initially increase gene expression (as measured through increasing CFP signal). Here, the increasing degree of molecular crowding served to increase RFP-T7 RNAP binding rates, thus serving to increase overall gene expression. However, further increases in the density of Dex-Small resulted in a decrease in gene expression. The reduction in gene expression was hypothesized to be due a significant decrease in the RFP-T7 RNAP diffusion rate, which limited access to the P_{T7} promoter. Overall, this study demonstrates that gene expression could be manipulated based on the molecular crowding of the interior of the artificial cell, a variable that can be challenging to manipulate using *in vivo* systems. Furthermore, it indicates that molecular crowding might have an important role in regulating gene expression in natural systems.

17.6 Applications of Gene Circuits

While many of the aforementioned synthetic systems can be used to develop novel tools with applications in nanotechnology, such as the use of logic gates in developing biologically based computing, this section will detail recent advances in synthetic biology that may have applications in common with nanotechnology.

17.6.1 Drug Delivery

Many studies within the realm of nanotechnology have focused on the development of tools and strategies to deliver pharmaceuticals (Park 2007; Safari and Zarnegar 2014). Similarly, several synthetic biology studies have aimed to engineer systems to deliver pharmaceuticals, including chemotherapeutic drugs. Currently, two general approaches are taken to target cancer cells using synthetic biology. First, a synthetic system is engineered to deliver the chemotherapeutics. In many cases, these systems have the ability to both produce and deliver the chemotherapeutic, thus allowing sustained delivery of the drug. Second, a synthetic system is designed to allow the delivery of genetic material directly into the cancer cell. Here, the genetic material will directly interfere with gene expression in the cancer cell, leading to reduced growth and/or programmed cell death.

Systems designed to treat cancer cells have mostly been engineered using bacteria, which are modified to produce and deliver chemotherapeutics. One particularly challenging feature of engineering such systems is ensuring that the drug is delivered directly into the cancer cell. To address this issue, Anderson et al. engineered a synthetic system consisting of bacteria that express the invasin gene from *Yersinia pseudotuberculosis* (Anderson et al. 2006). Expression of invasin allows bacteria to penetrate into mammalian cells. Once inside, bacterial gene expression can continue by using host-cell provided nutrients. Interestingly, the authors engineered the expression of invasin to only occur in hypoxic (low oxygen) conditions, which are often observed in areas surrounding cancer cells.

Building off of this study, Xiang et al. engineered bacteria (Figure 17.5a) to invade mammalian cells and both synthesize and deliver short hairpin RNAs (shRNa), which would interfere with cell growth (Xiang, Fruehauf, and Li 2006). The shRNA was targeted to interfere with *CTNNB1*, an oncogene that is upregulated in colon and additional cancers. Intravenous injection of these engineered bacteria into mice was able to reduce *CTNNB1* expression in human colon cancer xenografts in mouse intestine. This general strategy of using engineered bacteria to synthesize and deliver shRNAs could be amenable to targeting various genes associated with diverse cancer types, and additional genetic-based diseases.

While not as well developed as *in vivo* delivery systems, nonliving synthetic systems that serve to deliver enzymes directly into cancer cells have been created. In 2006, Link et al. created protein transducing nanoparticles (PTNs), which were designed to target and deliver toxin-producing enzymes to cancer cells (Link et al. 2006). These modified lentiviruses were engineered to not contain viral replication machinery, thus limiting their ability to replicate in cancer cells. Instead, PTNs would fuse with the cell membrane and deliver internalized proteins. As a proof of concept, the authors encapsulated the linamarase protein from cassava (*Manihot esculenta*). This enzyme catalyzes the formation of glucose, acetone, and toxic cyanide from its substrate, linamarin. Upon injection of linamarase via PTNs into rodent and human cell lines, linamarase produced cyanide, which resulted in substantial death of cancer cells. Furthermore, the injection of linamarase containing PTNs into a mouse model of human breast cancer resulted in a marked reduction in tumor

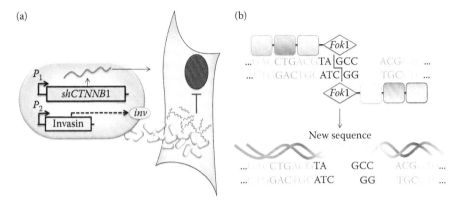

FIGURE 17.5
Applications of synthetic biology align with the interests of nanotechnology. (a) Using cells to deliver pharmaceuticals. In this example, *E. coli* is engineered to express a short hairpin RNA (shRNA) that targets the oncogene *CTNNB1* (purple line), and invasin (*inv*, orange circle), which allows the bacteria to enter into mammalian cells. Upon entry into the mammalian cell via the invasin protein, *shCTNNB1* is exported from the bacteria into the mammalian cell. This shRNA then interferes with the expression of *CTNNB1*, which could reduce the growth of the cell. P_1 and P_2 represent two promoters. (b) Engineering systems for genome editing. One mechanism to edit genomes in mammalian cells is through the use of zinc finger nucleases (ZFN). Zinc fingers (boxes) bind specific DNA sequences. A series of zinc fingers targeting a specific area of the genome is coupled with the nuclease (DNA cutting) domain of the Fok1 restriction endonuclease. Once both ZFNs have found their targeted sequence, the DNA is cut (red line, top panel). While the cell is fixing the cut DNA, a new DNA sequence (yellow area of the helix), which in part matches the genome sequence recognized by the ZFNs (as denoted by the colors in the helix), can be inserted.

growth rates. Overall, this study demonstrated that PTNs might be a viable method to deliver chemotherapeutic proteins directly into cancer cells.

17.6.2 Treatment and Prevention of Infectious Diseases

Infectious diseases remain one of the greatest threats to society. The incidence of infections with antimicrobial resistant bacteria and eukaryotic cells, such as malaria, has seen a drastic increase in the past 20 years (Spellberg et al. 2008). There has been renewed investment in novel strategies to combat infectious disease. Several synthetic systems, which focus on different aspects of the infectious disease process, including transmission and prevention, have been engineered. Synthetic systems that target existing infections (Lu and Collins 2007, 2009) and that act as vaccines (Amidi et al. 2011) have also been engineered.

Mosquito-borne illnesses, such as malaria and dengue, constitute one of the major global sources of mortality and economic burden (Sachs and Malaney 2002). To prevent the spread of these diseases, global efforts have been placed on developing strategies to control mosquitoes. Along this line, Fu et al. engineered male mosquitoes to deliver genetic material to female mosquitoes that would render them flightless (Fu et al. 2010). After an engineered male carrying the gene encoding flightlessness mates with a nonengineered female mosquito, all females from the mating would be flightless and drown. However, the males from this mating, who do not consume blood and thus do not transmit infectious diseases, have the ability to fly and continue to contain the gene encoding flightlessness. These male mosquitoes would pass the flightlessness trait in the mosquito population, thus serving to decrease the quantity of female mosquitoes. To engineer this flightlessness gene, the authors placed a tetracycline transactivator protein (tTA) downstream of an

AeAct-4 promoter. This promoter is predominantly activated in the indirect flight muscles (IFM) of female mosquitoes. High levels of tTA proteins, as well as additional lethal effector genes, are sufficient to interfere with IFM development, thus rendering the females flightless. Interestingly, the release of these engineered mosquitoes into experimental (Wise de Valdez et al. 2011) and natural (Harris et al. 2011) mosquito populations was able to transiently reduce the mosquito populations. Additional methods to control mosquito populations have also been engineered (Windbichler et al. 2011; Galizi et al. 2014).

Recent studies have highlighted the mammalian intestine as a potential location to allow sustained proliferation of engineered bacteria (Ruder, Lu, and Collins 2011; Goh, He, and March 2012; Duan, Liu, and March 2015). In 2010, Duan and March engineered the commensal *E. coli* strain 1917 (Nissle), which can readily grow in the intestines of mice, to prevent colonization of the pathogen *Vibrio cholera* (Duan and March 2010). To successfully infect a host, *V. cholera* relies on the production and secretion of two small molecules, AI-2 and CAI-1, which serve to coordinate the infection process. During the initial stages of infection, the amount of *V. cholera* bacteria is low. As a result, the concentration of AI-2 and CAI-1 in the area surrounding the bacterial infection is also low. Low concentrations of these small molecules result in the expression of toxin and attachment genes, both of which are critical for the initiation of the infection process. As the number of *V. cholera* bacteria increases, so does the concentration of AI-2 and CAI-1. Once a sufficiently high concentration of these two small molecules has been reached, expression of toxin and attachment genes is attenuated. This allows *V. cholera* to escape the host via the intestine. To prevent infection with *V. cholera*, *E. coli* (which already secretes high amounts of CAI-1) was engineered to express high quantities of *cqsA*, which catalyzes the formation of AI-2. The hypothesis was that high concentrations of CAI-1 and AI-2 in the intestine would prevent *V. cholera* from expressing attachment and toxin genes, which would serve to prevent initial colonization. To test this hypothesis, the engineered bacteria were fed to mice, where they became established in the intestine. The mice were then challenged with *V. cholera*. After 48 h, mice that were initially fed the engineered bacteria showed a significant reduction in mortality as compared with mice that were not fed the engineered bacteria. Here, the high amount of AI-2 (and CAI-1) produced by the engineered bacteria served a protective function. As such, this study demonstrated that engineering commensal bacteria might be an effective method to protect against infectious diseases that colonize the intestine.

17.6.3 Genome Editing

Genome editing holds the potential to develop minimalistic organisms "from scratch" as well as the ability to cure genetic diseases (Esvelt and Wang 2013). Historically, genome editing has often lacked specificity and efficiency. However, recent advances in genome editing have unlocked the ability to precisely add, delete, or alter the sequence of genomic DNA *in vivo*.

Two mechanisms for genome editing have come to the forefront, although additional strategies have also been developed (Esvelt and Wang 2013). The first mechanism involves the use of zinc finger nucleases (ZFN) (Urnov et al. 2005) or transcription activator like effector nucleases (TALENS) (Gaj, Gersbach, and Barbas 2013; Joung and Sander 2013). Zinc fingers and transcription activator like effectors recognize and bind to highly specific DNA sequences. These DNA-binding domains are modified to carry the nonspecific nuclease domain of the *Fok1* restriction endonuclease (Figure 17.5b). When two ZFNs or TALENS dimerize to adjacent areas on the DNA, the nuclease activity of *Fok1* is activated causing the

DNA to be cut. The cut DNA can then be repaired by the cell by inserting a user-designed DNA sequence. ZFNs and TALENS have been used in a variety of organisms (Urnov et al. 2010) and have been able to repair DNA associated with disease (Ousterout et al. 2013).

The second mechanism involves the use of the CRISPR (clustered regulator interspaced short palindromic repeat) Cas9 system (Gaj, Gersbach, and Barbas 2013; Jiang et al. 2013). This system, originally found in bacteria and used as a form of protection against viral infections, uses an RNA template, called crRNA, to scan DNA. When an appropriate match to the crRNA is found by CRISPR/Cas9 in DNA, the DNA is cut. In engineered systems, the crRNA is designed to recognize targeted DNA sequences in the genome. Once CRISPR/Cas9 uses the engineered crRNA to locate and cut the intended genomic DNA sequence, a user-designed DNA sequence is then inserted into the cut area using the cell's DNA repair system. Thus far, this system has been shown to be readily amenable for genome editing in a wide range of cells, including mammalian cells (Hwang et al. 2013), stem cells (Ding et al. 2013), and in mouse models (Wang et al. 2013).

17.7 Future Outlook

In recent years, the complexity of gene circuits created using synthetic biology has increased substantially. Coupled with the emergence of functional synthetic systems to treat diseases, the future is bright for synthetic biology as it aims to revolutionize multiple fields. Nanotechnology has also made large strides in similar areas and, as discussed, has many similar goals as synthetic biology. Interestingly, the potential to directly unite aspects of both synthetic biology and nanotechnology remains relatively untapped and thus may serve to drive both fields forward.

Moving forward, however, synthetic biology faces many challenges (Kwok 2010; Fu 2013). Chief among these is achieving a more thorough understanding of cellular complexity and how it might interfere with gene circuit functionality. Although our understanding of engineering principles is increasing, and our appreciation for how normal cellular operations can interfere with the gene circuit (i.e., noise), the field continues to strive for greater mathematically based predictive tools and rules in the design of gene circuits. Toward this end, engineering a minimal cell, containing a minimal genome, is often cited as a large goal of synthetic biology.

In addition to this challenge, the diversity and complexity of biological parts presents a large hurdle to overcome as the field aims to standardize and better define the unique role of each biological part. Furthermore, it is being increasingly recognized that not all biological parts are compatible, thus posing limitations on our ability to engineer increasingly complex behaviors. Despite these challenges, synthetic biology will continue to remain a critical tool in the design of systems to tackle some of the most pressing needs in our society.

References

Ajo-Franklin, C. M., Drubin, D. A., Eskin, J. A. et al. 2007. Rational design of memory in eukaryotic cells. *Genes Dev.* 21(18), 2271–2276.

Noireaux, V., and Libchaber, A. 2004. A vesicle bioreactor as a step toward an artificial cell assembly. *Proc. Natl. Acad. Sci. U.S.A.* 101(51), 17669–17674.

Olson, E. J., Hartsough, L. A., Landry, B. P., Shroff, R., and Tabor, J. J. 2014. Characterizing bacterial gene circuit dynamics with optically programmed gene expression signals. *Nat. Methods* 11(4), 449–455.

Ousterout, D. G., Perez-Pinera, P., Thakore, P. I. et al. 2013. Reading frame correction by targeted genome editing restores dystrophin expression in cells from Duchenne muscular dystrophy patients. *Mol. Ther.* 21(9), 1718–1726.

Pai, A., Tanouchi, Y., and You, L. 2012. Optimality and robustness in quorum sensing (QS)-mediated regulation of a costly public good enzyme. *Proc. Natl. Acad. Sci. U.S.A.* 109(48), 19810–19815.

Park, K. 2007. Nanotechnology: What it can do for drug delivery. *J. Control. Release* 120(1–2), 1–3.

Payne, S., Li, B., Cao, Y., Schaeffer, D., Ryser, M. D., and You, L. 2013. Temporal control of self-organized pattern formation without morphogen gradients in bacteria. *Mol. Syst. Biol.* 9, 697.

Pohorille, A., and Deamer, D. 2002. Artificial cells: Prospects for biotechnology. *Trends Biotechnol.* 20(3), 123–128.

Prindle, A., Selimkhanov, J., Danino, T. et al. 2012. Genetic circuits in *Salmonella typhimurium*. *ACS Synth. Biol.* 1(10), 458–464.

Riccione, K. A., Smith, R. P., Lee, A. J., and You, L. 2012. A synthetic biology approach to understanding cellular information processing. *ACS Synth. Biol.* 1(9), 389–402.

Rinaudo, K., Bleris, L., Maddamsetti, R., Subramanian, S., Weiss, R., and Benenson, Y. 2007. A universal RNAi-based logic evaluator that operates in mammalian cells. *Nat. Biotechnol.* 25(7), 795–801.

Ruder, W. C., Lu, T., and Collins, J. J. 2011. Synthetic biology moving into the clinic. *Science* 333(6047), 1248–1252.

Sachs, J., and Malaney, P. 2002. The economic and social burden of malaria. *Nature* 415(6872), 680–685.

Safari, J., and Zarnegar, Z. 2014. Advanced drug delivery systems: Nanotechnology of health design. *J. Saudi Chem. Soc.* 18(2), 85–99.

Salis, H. M., Mirsky, E. A., and Voigt, C. A. 2009. Automated design of synthetic ribosome binding sites to precisely control protein expression. *Nat. Biotechnol.* 27(10), 946–950.

Seelig, G., Soloveichik, D., Zhang, D. Y., and Winfree, E. 2006. Enzyme-free nucleic acid logic circuits. *Science* 314(5805), 1585–1588.

Siuti, P., Yazbek, J., and Lu, T. K. 2013. Synthetic circuits integrating logic and memory in living cells. *Nat. Biotechnol.* 31(5), 448–452.

Smith, R. P., Tan, C., Srimani, J. K. et al. 2014. Programmed Allee effect results in a tradeoff between population spread and survival. *Proc. Natl. Acad. Sci. U.S.A.* 111(5), 1969–1974.

Sohka, T., Heins, R. A., Phelan, R. M., Greisler, J. M., Townsend, C. A., and Ostermeier, M. 2009. An externally tunable bacterial band-pass filter. *Proc. Natl. Acad. Sci. U.S.A.* 106(25), 10135–10140.

Spellberg, B., Guidos, R., Gilbert, D. et al. 2008. The epidemic of antibiotic-resistant infections: A call to action for the medical community from the Infectious Diseases Society of America. *Clin. Infect. Dis.* 46(2), 155–164.

Stricker, J., Cookson, S., Bennett, M. R., Mather, W. H., Tsimring, L. S., and Hasty, J. 2008. A fast, robust and tunable synthetic gene oscillator. *Nature* 456(7221), 516–519.

Swain, P. S., Elowitz, M. B., and Siggia, E. D. 2002. Intrinsic and extrinsic contributions to stochasticity in gene expression. *Proc. Natl. Acad. Sci. U.S.A.* 99(20), 12795–12800.

Tabor, J. J., Salis, H. M., Simpson, Z. B. et al. 2009. A synthetic genetic edge detection program. *Cell* 137(7), 1272–1281.

Tamsir, A., Tabor, J. J., and Voigt, C. A. 2011. Robust multicellular computing using genetically encoded NOR gates and chemical "wires". *Nature* 469(7329), 212–215.

Tan, C., Marguet, P., and You, L. 2009. Emergent bistability by a growth-modulating positive feedback circuit. *Nat. Chem. Biol.* 5(11), 842–848.

Tan, C., Saurabh, S., Bruchez, M., Schwartz, R., and LeDuc, P. 2013. Molecular crowding shapes gene expression in synthetic cellular nanosystems. *Nat. Nanotechnol.* 8(8), 602–608.

Tanouchi, Y., Pai, A., Buchler, N. E., and You, L. 2012b. Programming stress-induced altruistic death in engineered bacteria. *Mol. Syst. Biol.* 8, 626.

Tanouchi, Y., Smith, R. P., and You, L. 2012a. Engineering microbial systems to explore ecological and evolutionary dynamics. *Curr. Opin. Biotechnol.* 23(5), 791–797.

Tigges, M., Denervaud, N., Greber, D., Stelling, J., and Fussenegger, M. 2010. A synthetic low-frequency mammalian oscillator. *Nucleic Acids Res.* 38(8), 2702–2711.

Tigges, M., Marquez-Lago, T. T., Stelling, J., and Fussenegger, M. 2009. A tunable synthetic mammalian oscillator. *Nature* 457(7227), 309–312.

Turing, A. M. 1952. The chemical basis of morphogenesis. *Philos. Trans. R. Soc. Lond. B Biol. Sci.* 237(641), 37–72.

Urnov, F. D., Miller, J. C., Lee, Y.-L. et al. 2005. Highly efficient endogenous human gene correction using designed zinc-finger nucleases. *Nature* 435(7042), 646–651.

Urnov, F. D., Rebar, E. J., Holmes, M. C., Zhang, H. S., and Gregory, P. D. 2010. Genome editing with engineered zinc finger nucleases. *Nat. Rev. Genet.* 11(9), 636–646.

Wang, B., Kitney, R. I., Joly, N., and Buck, M. 2011. Engineering modular and orthogonal genetic logic gates for robust digital-like synthetic biology. *Nat. Commun.* 2, 508.

Wang, H., Yang, H., Shivalila, C. S. et al. 2013. One-step generation of mice carrying mutations in multiple genes by CRISPR/Cas-mediated genome engineering. *Cell* 153(4), 910–918.

Windbichler, N., Menichelli, M., Papathanos, P. A. et al. 2011. A synthetic homing endonuclease-based gene drive system in the human malaria mosquito. *Nature* 473(7346), 212–215.

Wise de Valdez, M. R., Nimmo D., Betz, J. et al. 2011. Genetic elimination of dengue vector mosquitoes. *Proc. Natl. Acad. Sci. U.S.A.* 108(12), 4772–4775.

Xiang, S., Fruehauf, J., and Li, C. J. 2006. Short hairpin RNA-expressing bacteria elicit RNA interference in mammals. *Nat. Biotechnol.* 24(6), 697–702.

Yokobayashi, Y., Weiss, R., and Arnold, F. H. 2002. Directed evolution of a genetic circuit. *Proc. Natl. Acad. Sci. U.S.A.* 99(26), 16587–16591.

You, L., Cox, R. S., Weiss, R., and Arnold, F. H. 2004. Programmed population control by cell-cell communication and regulated killing. *Nature* 428(6985), 868–871.

18

Recent Trends in Nanomaterials Integration into Simple Biosensing Platforms

Andrzej Chałupniak and Arben Merkoçi

CONTENTS

18.1 Introduction

Since more than five decades when the first biosensor was implemented, research and development in this field has continuously progressed taking advantages of discoveries of new technologies and materials. Currently, due to the strategic importance of nanotechnology in modern science and industry, nanomaterials are increasingly being used in various applications with particular interest in various areas. There is no difference in the case of biosensors, where nanomaterials have found a versatile role as substrates for nanodevices fabrication, as signal transducers and labels.

The development of biosensors runs two ways at the same time. One of them relates to the complex and high-throughput analysis systems for certain compounds. Such biosensing systems can be used in modern analytical laboratories, as well as clinical and industrial applications. The second group consists of biosensing devices suitable for rapid analysis, which do not require the use of sophisticated equipment and trained personnel. This group is particularly promising because it allows the creation of versatile products, available for everyone and easy to use, which are important, for example, in the diagnosis of disease markers, environmental monitoring out of the laboratory (infield), in places with low resources and developing countries, where access to specialized laboratories is limited.

Access to technological innovations always affects the level of advancement of societies. For example, the widespread use of simple biosensors such as glucometers, pregnancy tests, and portable devices can help in further creation of awareness in healthcare prevention. Thus, biosensors can play an important role in education [1].

In this chapter, different biosensing techniques are discussed, from easy paper-based biosensors to more complex lab-on-a-chip (LOC) devices. All of them combine the use of novel nanomaterials.

18.2 Nanomaterials Improve Biosensing

Nanomaterials play a major role in biosensing, due to the versatile functions that they can play in biosensors. Used as labels, they can produce signals that are easy to measure and quantify (i.e., fluorescence emitted by quantum dots [QDs]); as signal amplifiers, they can enhance the response of the sensor (i.e., iridium oxide [IrOX] nanoparticles with their electrocatalytic activity). The unique electrical, chemical, and mechanical properties allow using nanomaterials as substrates for biosensing devices, that is, electrodes covered by graphene oxide (GO). Simultaneously, some nano/microparticles due to their porous structure can be used as carriers for bioreceptors molecules or demonstrate the ability to adsorb selected compounds, providing the ability to use them not only for sensing but also for removal of contaminants. Most of the nanomaterials can be easily conjugated with proteins, enzymes, antibodies, and DNA, what makes possible to obtain specific and sensitive biosensing systems. In this section, some interesting and efficient nanomaterials will be described regarding their application in biosensors.

18.2.1 Nano- and Microparticles

According to IUPAC definition, nanoparticles are particles of any shape with dimensions in the range of 1–100 nm, while microparticles are bigger with size in the range of 100–100 μm [2,3]. Since there are a lot of micro/nanoparticles already reported as useful for biosensors, only some of them, particularly those more promising, will be described.

Gold nanoparticles (AuNP) are one of the most commonly used nanoparticles. Depending on their size, which can be controlled during the synthesis process, they show different parameters both electrochemical and optical. This makes them very versatile and useful for various detection strategies, where AuNPs can work as labels and signal transducers. The conjugation of AuNP with proteins and other biomolecules is very easy and can be obtained without using specific linkers and chemical modification. This feature encourages scientists to use AuNP in biosensing. Due to their intensive color, AuNPs can also be used as labels in colorimetric test like paper-based immunoassay as well as in various electrochemical techniques as explained below.

AuNPs are mainly employed as electrochemical labels in biosensors. In principle, there are two ways to detect them. The first method is differential pulse voltammetry

(DPV), where potential is applied to initiate the reaction of reducing tetrachloroaurate ions to gold metal after (prior oxidation) and measured current used as analytical signal. This approach was reported, that is, for magnetoimmunoassay for *Salmonella* detection [4]. Another strategy for electrochemical detection with AuNP is taking advantage of hydrogen evolution reaction (HER). In this case, AuNP presents catalytic activity (toward hydrogen ions reduction) when they are deposited onto the electrode in the presence of an acid (HCl). After applying a fixed potential, the current signal from the reducing hydrogen is measured and used as analytical signal. This reaction is easy to be followed making this approach very convenient and possible to be employed for various detection strategies like cancer cells [5,6], viruses, DNA, and protein detection. Another interesting example is immobilization of casein peptides to AuNP surface so as to obtain nanocarriers for electrochemical study of the adhesion of casein peptides to pathogen bacteria [7].

For more specific approaches, or detection improvements, AuNPs can be easily modified. For example, when AuNPs are modified with ferrocene derivative (FcD), 1,1'-ferrocenyl bis(methylene lipoic acid ester), such complex can work as an efficient transducing system due to shifting of the ferrocene oxidation potential. This approach was employed by Mars et al. for human IgG detection representing a novel potential shift-based transducing system for sensitive immunosensing [8].

Another interesting example is IrOx nanoparticles, with the size around 12 nm, which are synthesized from potassium hexachloroiridate (IV) (K_2IrCl_6) and sodium hydrogen citrate sesquihydrate ($Na_2C_6H_6O_7 \cdot 1.5H_2O$). Thanks to conductivity and electronic switching properties their possible application in biosensors is varied. IrOx nanoparticles are already reported as a conductivity enhancer for pesticides and phenol detection as well as an electrochemical label for ApoE immunoassay, where the antibodies are labeled with IrOx. They are able to enhance Water Oxidation Reaction (WOR). These nanoparticles electrocatalyze water splitting, which is followed by chronoamperometry by measurements of the current generated during the corresponding reaction [9].

One of the important expectations for the novel materials is to be environmentally friendly and biocompatible. Calcium carbonate microparticles ($CaCO_3MP$) meet these demands. The work of Lopez-Marzo et al. comprehensively describes the possibility of formation of $CaCO_3$-poly(ethyleneimine) (PEI). Such complex shows high potential for biosensing applications due to the possibility to conjugate it with proteins, nucleic acids, and other biomolecules [10]. These particles found applications also in microfluidic systems, working for both sensing and removal of phenol compounds. Biosensing ability was possible due to cross linking of $CaCO_3MPs$ with tyrosinase by glutaraldehyde, while the removal ability is connected with porous structure and capacity of microparticles to adsorb phenolic compounds [11].

18.2.2 Quantum Dots

QDs, fluorescent semiconductors, have recently become increasingly popular tags for molecular imaging and biosensing. Their optical properties like narrow and size-tunable emission spectra (what leads to multicolor detection), various and broad absorption profiles and chemical stability, make them competitive with respect to organic fluorophores, which are commonly used as fluorescent labels [12, 13]. Depending on the type of QDs, their fluorescence lifetime can be 10 times longer comparing to organic fluorophores. However, QDs have some disadvantages like potential toxicity due to the presence of heavy metals, which is not well known yet, but also bigger size (from 2 to 15 nm) than organic fluorophores, what can be a drawback in some approaches [14].

An important advantage is the possibility of QD to be optically and electrochemically detected. Electrochemical method is related to detection of Cd contained in CdS QD (or any other metal depending on the QD). Such detection can be performed on screen-printed electrodes using square-wave anodic stripping voltammetry. While the potential is applied, in the presence of proper buffer, cadmium (Cd^{2+} ion onto the QD surface) is reduced to the working electrode surface and oxidized (electrochemical stripping). The specific peak at the characteristic redox potential of cadmium can be observed [15]. This technique is faster, cheaper, and easier to be performed in comparison with optical detection. Electrochemical detection of QDs can be implemented to high-throughput, automated detection systems. Furthermore, electrochemical detection of QD is really sensitive. In the work reported by Medina-Sánchez et al., QD detection in microfluidic chip was possible in the range between 50 and 8000 ng mL^{-1} with a sensitivity of 0.0009 µA/(ng mL^{-1}) [16]. Since QDs usually contain thiol (–SH) and carboxyl (–COOH) on their surface reactive group, it is possible to easily conjugate them with biomolecules with sensing abilities (i.e., antibodies, aptamers) as well as with proteins like streptavidin and biotin, so as to obtain QDs labels that are able to perform analytical signal enhancement. To make it feasible, it is necessary to use a proper biolinker like primary amines, thioglycolic acid, peptides, or chitosan polymer [17]. QDs were already employed for electrochemical screening of apoptosis of mammalian cells. QDs conjugated with annexin-V can effectively bind to phosphatidylserine, a phospholipid expressed in the outer membrane of apoptotic cells. Electrochemical detection, comparing to optical one, was characterized by shorter time and possibility to perform versatile quantified analysis [18].

In the following parts, application of nanomaterials in biosensing techniques like LOC and microarray will be described.

18.2.3 Carbon Nanomaterials

Carbon is characterized by the presence in various allotropic forms. Physical and chemical properties of a given carbon allotrope depend on hybridization. Carbon materials that are found to be used in biosensing usually have sp^2 hybridization. These are, for example, graphite, carbon nanotubes, and graphene.

Carbon nanomaterials, due to their surface properties, conductivity, and biocompatibility, are good candidates for application in electrochemical biosensors. For example, multiwalled carbon nanotubes (MWCNTs) were reported by Perez Lopez et al. as a material raising the signal intensity of tyrosinase biosensor for catechol detection by 90% [19]. This phenomenon was caused due to improved electronic transference between enzyme and MWCNTs layer. At the same time, surface properties of carbon nanotubes (CNTs) make them useful for enzyme and other biomolecules immobilization. The versatility of this material was also proofed in the work, where MWCNTs was immobilized onto magnetic particles for catechol detection and shown to be very useful and easy for manipulation in ON–OFF response biosensing platform [20].

Carbon nanomaterials also show interesting optical properties. In the work of Morales-Narvaez et al., the Förster resonance energy transfer (FRET) was measured between QD and various carbon nanomaterials: graphite, CNTs, carbon nanofibers (CNFs), and GO. The highest quenching was observed for GO (97 ± 1%), which is what makes this material very useful for future applications employing the phenomena of FRET with interest for biosensing approach (Figure 18.1a) [21]. Moreover, graphene quantum dots (GQD) can be used for small molecules detection (Figure 18.1b) due to the ability of convenient conjugation with

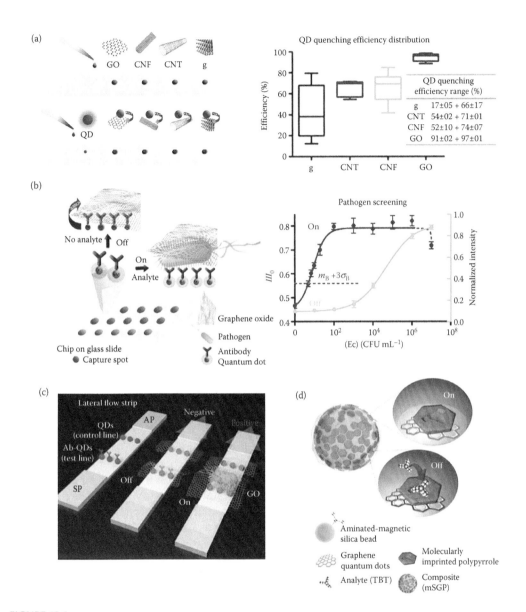

FIGURE 18.1

Carbon nanomaterials in optical biosensing. (a) Comparison of QD quenching caused by various carbon nanomaterials (GO, graphene oxide; CNT, carbon nanotubes; CNF, carbon nanofibers; g, graphite). (Reprinted from *Carbon*, 50(8), Morales-Narvaez, E. et al., Simple Forster resonance energy transfer evidence for the ultrahigh quantum dot quenching efficiency by graphene oxide compared to other carbon structures, 2987–2993, Copyright 2012, with permission from Elsevier.) (b) Pathogen detection in microarray format using GO as a revealing agent. (Modified with permission from, John Wiley & Sons, Copyright 2013, Morales-Narvaez, E. et al. 2013. *Angew. Chem. Int. Ed.* 52(51), 13779–13783.) (c) Pathogen detection in lateral-flow format using GO as a revealing agent. (Reprinted with permission from Morales-Narvaez, E. et al. 2015. Photoluminescent lateral-flow immunoassay revealed by graphene oxide: Highly sensitive paper-based pathogen detection. *Anal. Chem.* 87(16), 8573–8577. Copyright 2015 American Chemical Society.) (d) Small-molecule detection based on molecularly imprinted polymer (MIP) and GO. (Reprinted with permission from Zor, E. et al. 2015. Graphene quantum dots-based photoluminescent sensor: A multifunctional composite for pesticide detection. *ACS Appl. Mater. Interfaces.* Copyright 2015 American Chemical Society.)

magnetic silica beads (used as a carrier) and molecular imprinted polymer (recognition of the analyte) [22].

Carbon nanomaterials can play varied roles in biosensors. For further studies we recommend the following reviews that comprehensively describe them [13, 23]. Some examples of application of carbon materials in optical biosensors are presented in Figure 18.1.

18.2.4 Micro- and Nanomotors

Just as the name suggests, micro- and nanomotors are supposed to perform various mechanical movements like, for example, shuttling, rolling, and rotation on others, due to inducing them by a specific factor. We can distinguish nano/micromotors driven by light, magnetic field, acoustic wave (physical control), as well as by energy resulting from various chemical reactions, where selected compounds act as fuels for motors [26]. Nano/micromotors movement behavior opens a lot of opportunities to take as advantage in biosensing systems. They can support mixing process in microscale, what improves mass transport and increases the intensity of sensors response, for example, in microarray. Acting as carriers, nano/micromotors can be conjugated with biomolecules such as antibodies working as moving, sensing, or capturing probe, which can be applied in microfluidic systems [27]. Nano/micromotors are promising tools not only in biodetection but also in removal. A good example is the work of Guix et al. where alkanethiol-coated Au/Ni/PEDOT/Pt shows ability to capture, transport, and remove oil droplets [28]. Another interesting example is bacteria detection made through concanavalin A (ConA) immobilized onto the nanomotors surface [29].

18.2.5 Micromotors Support Mass Transport

Usability and versatility of different assays performed in liquid phase (such as immunoassays) can be limited by mass transport, which depends on chemical and physical properties of the molecules working as a bioreceptor and target. The induction of any kind of micromixing can be helpful for improving the efficiency of the assay. One of the promising approaches is using self-propelling micromotors. In that work, protein detection (ApoE) based on immunoassay, with fluorophore as a label, was improved 3.5-folds (signal intensity) due to application of polianyline (PANI)-platinum micromotors. Amplification of the signal at the same time can significantly reduce the limit of detection (LOD). The only potential drawback in this study is using hydrogen peroxide as a fuel for micromotors, which in some cases can be a way to degrade the proteins used in a given assay [30].

18.2.6 Nanochannels

Another approach in the application of nanomaterials is using them as filters, which selectively allowed analyte (e.g., protein biomarkers) passing through them. This is the example of biomimetic approach, because in live organisms, cell metabolism is driven by proteins located in cell membranes which are selectively permeable for various ions, nutrients, and other biomolecules [31].

Nanochannels support high sensitivity and selectivity in biosensing. Due to determined size of pores, sample is filtered (interfering big size analytes remain on top of the channel) when it reaches the inner surface of nanochannels where the bioreception occurs. Such behavior reduces the risk of unspecific bonding and introduction of any interference from impurities. Nanochannels can be functionalized inside the membranes, so it is possible to stop the analyte inside them. The nanochannel blocking can be followed by electrochemical techniques [31]. One of the example is voltammetric detection of thrombin, where anodized alumina oxide filter (AAO) membrane was used as a nanochannel, with internal diameter of porous 200 nm. As a bioreceptor, antithrombin IgG labeled with AuNP was used. When the channel is blocked due to immunocomplex formation, diffusion of $[Fe(CN)_6]^{4-}$ is decreased serving such a phenomena (DPV signal decrease) as a source of analytical signal. This system showed very low LOD (LOD 1.8 ng mL^{-1}), even when the analysis was performed in spiked blood samples [32]. Nanochannels can be effectively used for cancer biomarkers sensing from the blood thanks to nanochannel-based filtration of the whole blood. In that case, it was possible to detect 52 U mL^{-1} of CA15-3 breast cancer marker [33]. Another approach for cancer biomarkers was studied using Prussian blue nanoparticles (PBNPs) working as a redox indicator in immunoassay. Application of PBNPs instead of $[Fe(CN)_6]^4$ led to lower detection limit (from 200 µg mL^{-1} to 34 pg human IgG mL^{-1}) (Figure 18.2b and c) [34].

Various materials can be employed for nanochannels preparation. One of the recently published examples is the use of carboxylated polystyrene nanospheres (PS, 500 and 200 nm-sized) coating the working area of flexible screen-printed indium tin oxide/ polyethylene terephthalate (ITO/PET) electrode. Similarly as an example mentioned above, this platform was used for IgG detection as a model protein. Unlike AAO membranes, this approach overcomes many of the limitations in terms of integration and sensitivity, and represents a really disposable biosensing device for a one-step assay (Figure 18.2a) [35].

18.3 Paper Nanobiosensors

The launch of the first pregnancy test in the mid-1970s started the popularity of paper-based biosensors. It was proof that biosensors can be simple devices, possible to be used by everyone, relatively cheap and accessible. Nowadays, when all innovations are also assessed for their impact on the environment, paper-based biosensors positively stand out for their biodegradability and safety for the user and the environment. The portability of paper-based biosensors allows using them in any place on earth, regardless of the availability of specialized laboratories and personnel. This opens the way to improve medical diagnostics and environmental monitoring in developing countries. Paper-based biosensors can also be used in food and pharmaceutical industry to perform preliminary analyses related to quality control, which precede further in-depth test in dedicated laboratories. Furthermore, these biosensors can be both qualitative and quantitative, depending on the needs. Optical detection, possible even with the naked eye, is easy and fast [36, 37].

Paper-based biosensors are already used for sensing proteins, DNA, and other compounds. Like in most biosensors, detection is based on affinity of the specific antibodies or DNA sequences (or aptamers) to target compounds [36]. Although most of the articles present paper-based biosensors for biomarkers detection, these devices also represent an interesting solution for environmental screening. One of the examples is heavy metal detection (cadmium) in paper device reported by López-Marzo et al. [38, 39].

Lateral flow assays (LFAs) are one of the most popular paper-based sensors, and their main advantages are versatility and different detection possibilities (optical, electrochemical).

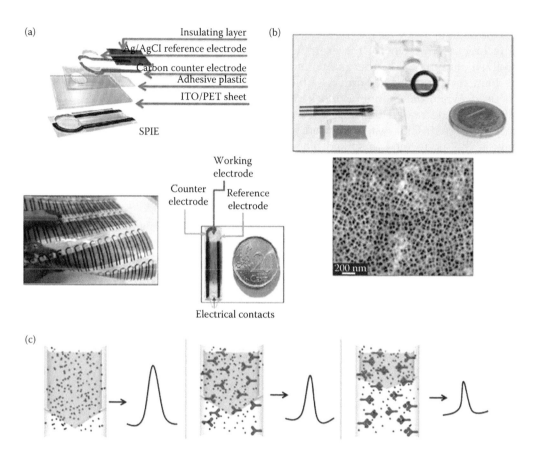

FIGURE 18.2
Application of nanochannels in biosensing. (a) Schematic representation of the different materials and layers, which form the SPIE (screen-printed ITO/PET electrodes). (With kind permission from Springer Science+Business Media: *Nano Res.*, Nanoparticles-based nanochannels assembled on a plastic flexible substrate for label-free immunosensing, 8(4), 2015, 1180–1188, de la Escosura-Muniz, A. et al.) (b) Schematic representation of the process occurring on the nanochannels modified electrode. (Reprinted from *Biosens. Bioelectron.*, 67, Espinoza-Castaneda, M. et al., Nanochannel array device operating through Prussian blue nanoparticles for sensitive label-free immunodetection of a cancer biomarker, 107–114, Copyright 2015, with permission from Elsevier.) (c) Schematic representation of the sensing principle: the voltammetric signal of the PBNPs decrease with the subsequent AAO membranes modification. (Reprinted from *Biosens. Bioelectron.*, 67, Espinoza-Castaneda, M. et al., Nanochannel array device operating through Prussian blue nanoparticles for sensitive label-free immunodetection of a cancer biomarker, 107–114, Copyright 2015, with permission from Elsevier.)

They consist of detection pad, conjugation pad, sample pad, and absorption pad. Detection pad is made of nitrocellulose and its role is capturing the analyte since capturing bioreceptors (i.e., antibodies) are printed onto it. Glass fiber is used for conjugation pad, where conjugate with the color label (e.g., AuNP) is stored. Sample and absorption pads are made of cellulose; those pads support the flow of the liquid and purify the sample [36, 37, 39].

Although paper-based biosensors offer many opportunities, it is still necessary to overcome some drawbacks like improving the limit of detection, sensitivity, and enabling parallel detection of different targets. Application of nanoparticles, through working both as a carrier or as a label, can increase the usability of this type of biosensors.

One of the promising tools in paper-based biosensors is AuNP nanoparticles, due to strong red color, which facilitates optical detection and properties useful for electrochemical sensing. Applications of AuNP for sensing various protein, DNA, and cancer cells in paper biosensors are already reported. To improve detection, different types of particles can be used, that is, QDs or magnetic particles [36].

A very interesting solution to enhance the signal for optical detection in paper-based biosensor is using enzymes able to perform reaction resulting in appearance of color. Such approach was proposed by Parolo et al. where AuNP was conjugated with antibodies previously labeled with HRP. Increase of the color on the strip (by the use of various HRP substrates; in that case TMB, AEC, and DAB) increased the sensitivity of the assay up to 1 order of magnitude in comparison with nonmodified AuNPs [40].

Seeing that paper-based biosensors operation is based on liquid flow, optimization of this parameter can provide interesting results. Changing the geometry characteristics of the pads like width, shape, and length affects flow speed and distribution of the liquid (Figure 18.3b). In the work of Parolo et al., the increase of the sensitivity was up to eightfold due to increasing the size of conjugation and sample pad (Figure 18.3b) [41]. Referring to the idea of liquid flow manipulation to enhance the sensitivity of paper-based device, Rivas et al. proposed the use of delay hydrophobic barriers fabricated by wax printing. For this purpose, wax pillar patterns were printed onto the nitrocellulose membrane to generate delays as well as pseudoturbulence in the microcapillary flow. Biosensing performed for model protein (HigG) showed threefold improvement of the sensitivity (Figure 18.3c) [42].

Another approach increasing sensitivity is triple-line lateral flow sensor reported by Rivas et al. for *Leishmania infantum* DNA detection. Besides control line, two test lines are printed onto the LF strip in order to detect double-labeled (FITC/biotin) amplified *Leishmania* DNA (TL1) and 18S rNA gene (endogenous control) (TL2) [43].

All of the examples mentioned above were focused on colorimetric detection of analytes. However, it is possible also to design photoluminescent LF biosensor, where QD–Ab complex is printed on the substrate and GO is used as a quenching agent. Morales-Narvaez et al. reported photoluminescent detection of *E. coli* and *S. typhimurium* with a LOD of 10 CFU mL^{-1} in standard buffer and 100 CFU mL^{-1} in bottled water and milk [25].

Mechanical strength and chemical neutrality allow fabricating electrochemical paper-based biosensors, where electrodes are printed directly on the various paper material in the same way such as commonly used screen-printed electrodes. An example of this approach was shown by Parolo et al. The nitrocellulose membrane HF240 was used to fabricate screen-printed carbon electrodes (SPCEs), which was tested for AuNP and QD electrochemical detection. Such electrodes, compared with SPCE based on polyester, show significantly higher hydrophilicity of the surface and better electrochemical response. Those properties are linked to 3D structure of the membrane, which enhance the performance of the device [44].

Future development for paper-based devices seems to be so potent because of the need for readily available and inexpensive biosensors. The main challenge is to increase the level of specificity and sensitivity, and improve fabrication process, which would help in attaining better reproducibility of assays performed in paper-based biosensors.

One of the critical points in development of paper-based nanobiosensors is finding the novel type of paper substrates, which are easy to fabricate, biocompatible, and suitable for performing various chemical/biological reactions. One of the promising material is nanopaper (Figure 18.3a). Bacterial cellulose nanopaper is a multifunctional material known for numerous properties such as biocompatibility, optical transparency, sustainability, biodegradability, thermal properties, high mechanical strength, flexibility,

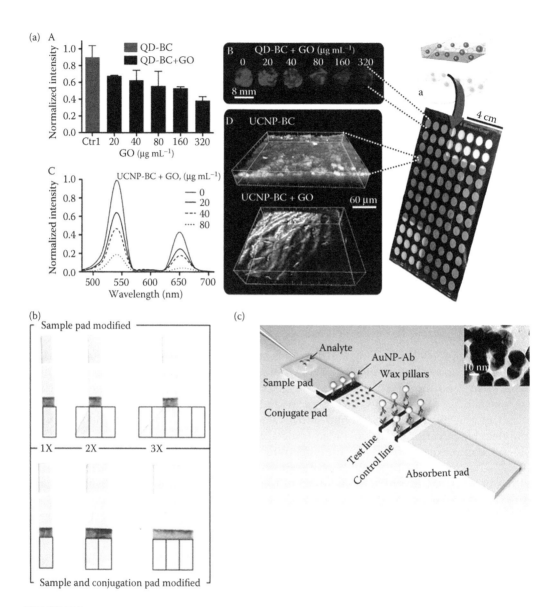

FIGURE 18.3

Novelties in the field of paper biosensors. (a) Nanopaper (BC) multiwell plate for photoluminescent tests. (Reprinted with permission from Morales-Narvaez, E. et al. 2015. Nanopaper as an optical sensing platform. *ACS Nano.* 9(7), 7296–7305. Copyright 2015 American Chemical Society.) (b) Modification of lateral flow test geometry in order to improve sensitivity. (Parolo, C. et al. 2013. Simple paper architecture modifications lead to enhanced sensitivity in nanoparticle based lateral flow immunoassays. *Lab Chip* 13(3), 386–390. Reproduced by permission of The Royal Society of Chemistry.) (c) Application of wax pillars as a hydrophobic barriers in lateral flow test. (Rivas, L. et al. 2014. Improving sensitivity of gold nanoparticle-based lateral flow assays by using wax-printed pillars as delay barriers of microfluidics. *Lab Chip* 14(22), 4406–4414. Reproduced by permission of The Royal Society of Chemistry.)

hydrophilicity, broad chemical-modification capabilities, high porosity, and high surface area. Its interaction with QD, GO, AuNP, and AgNP was already studied by Morales-Narvaez et al. showing a very good suitability of nanopaper for further biosensing applications (Figure 18.2a) [45].

18.4 Microarray Technology

High-throughput technologies are of interest to both scientists and entrepreneurs. Microarrays due to their versatility are currently used for sensing different compounds and molecules such as DNA, RNA, peptides, proteins (including antibodies), and also cells or tissues. Microarray revolutionized genomic studies due to the possibility to measure expressions of various genes in a short time or to perform genotyping. Nowadays, microarrays have gained recognition as a promising tool for biosensing [46]. These platforms are really flexible. As a substrate for spotting, different types of glass (or other materials), with various chemical modifications, that is, silanization, can be used, so as to fit it to the type of spotted molecule. Depending on the willingness of the user/researcher, targets or capture agent can be spotted (i.e., antibodies). In this case, microarray sensing can act as a direct, indirect, or competitive assay—similar to ELISA. In this section, the main lines of development for this biosensing technology including another way to improve their versatility are described [47, 48].

18.4.1 QDs as a Fluorescence Enhancer

Many researchers are looking for new compounds that will be successfully used as fluorescence markers in biodetection systems. QDs can effectively replace organic fluorophores in microarray biosensing. It is worth mentioning that QDs comparing with organic fluorophores have broader and stronger absorption spectra and their emission is characterized by narrow and symmetric spectrum [14]. Depending on the type of QDs, their fluorescence lifetime can be 10 times longer comparing to organic fluorophores. In the work of Morales Narvaez et al., cadmium-selenide/zinc sulfide (CdSe/ZnS) QDs were tested versus fluorescent dye Alexa 647 as reporters in protein immunoassay (ApoE was used as a model protein). Authors observed that QDs are highly effective reporters in microarray, but their properties strongly depend on excitation wavelength. For 532 nm excitation wavelength, using QDs provided five times lower LOD. Those studies open the way for further use of QDs in microarray [14].

18.4.2 GO as a Fluorescence Quenching Factor in Microarray

Due to the ability of GO to quench the fluorescence, described earlier, it is possible to take advantage of this phenomena in biosensing strategy. The properties of fluorescence quenching depend on the distance between donor and acceptor (according to FRET phenomena). Therefore, it can be assumed that increasing the distance between QDs (donor) and GO (acceptor) measurable changes in fluorescence can be observed [12,49]. This approach has been used for bacteria detection (*E. coli* O157:H7). QDs conjugated with antibodies against bacteria were spotted on the glass slide. The slides were incubated with samples with or without (blank) bacteria and subsequently GO was added. The biggest quenching was observed in blanks, while spots with bacteria showed reduced quenching (increase of fluorescence) (Figure 18.1b). However, this response that was found to be as a digital-like test very versatile and efficient for bacteria detection (LOD: 5 CFU mL^{-1}) cannot be used for sensing proteins or small molecules due to their size, which is not sufficient to affect the level of quenching [24, 50].

18.5 Lab-on-a-Chip

Microfluidic LOC devices are characterized by small size of sensing platforms, small volume of reagents (μL, nL, pL, fL) used for performing assay, and low energy consumption. Microfluidic systems can be well automated and used for preconcentration, isolation, or detection of interesting analytes. One of the most interesting applications is microfluidic chips for electrochemical and optical biosensing, which can be applied for detection of various biomarkers, organic compounds, etc. Versatility of LOC devices makes possible to integrate various sensing materials and strategies [51].

LOC microfluidic devices usually consist of microfluidic chip and driving units. Samples and reagents are introduced into the channel. Flow can be driven by different kinds of syringe pumps or dedicated flow control systems. Newest solutions also allow manipulating of liquids by surface acoustic wave and micropumps. Small diameter of microfluidic channel causes relatively low usage of reagents. Thanks to external driving units, parameters of reaction can be well controlled. Environment of microfluidic channel is permanently isolated from outside (unlike most assays performed in microplates, test tubes, etc.), which protects reactions and facilitates reproducible results. Depending on the purpose, different designs of microfluidic chips can be used for mixing, incubation, and preconcentration of the sample. Channels can also work as incubators for cell culturing and carrying out specific chemical reactions. For biosensing applications, chips with incorporated electrodes are particularly interesting to obtain electrochemical sensors. The surface of the channel can be treated by different reagents for further modification by biomolecules (i.e., proteins). In addition, devices equipped with magnets are useful for performing magnetoimmunoassay utilizing magnetic beads (MB) [51,52]. Researchers are focused on miniaturization detection platforms as well as all driving units, which assist them. Recently, LOC devices with all integrated electronic driving units were obtained by using printed circuit board (PCB) as a substrate for microfluidic chip [53, 54].

18.5.1 Unlimited Possibilities of Microfluidic Devices Designs

One of the common techniques for efficient prototyping of microfluidic devices is soft-photolithography, where polydimethylsiloxane (PDMS) chip is fabricated with hollowed microfluidic channel and substrate capable for permanent bonding with PDMS [54]. Apart from glass, one of the most widespread materials used for this purpose is polycarbonate (PC). PC is especially interesting as a substrate in electrochemical microfluidic devices due to the possibility of using electrodes directly printed on its surface. This is possible thanks to well-developed screen-printed technology; however, it can be fabricated using different techniques, for example, ink-jet printing and photolithography. The flexibility of screen-printing technology allows their application in microfluidic platforms obtaining various platforms with customized electrodes. On the other hand, transparency of PDMS enables for visual control of the process either in macro- and microscale (using optical microscopes). More sophisticated microfluidic chips consist of channel layer where all reaction occurs and control layer, also made of PDMS but responsible for controlling the flow of liquid by operating the valves built in the chip and driven by air pressure. This strategy can be used for increasing the sensitivity of assay by recirculation of the compounds that can be detected [16]. Another important feature of PDMS is oxygen permeability; thanks to this, PDMS chips are widely used for cell culturing [51].

18.5.2 Detection and Removal of Hazardous Compounds in Microfluidic Systems

A large number of dangerous compounds such as toxins, pesticides, chemicals, and biocidal compounds require continuous and rapid monitoring of their concentrations. Microfluidic systems due to their automation and small volume of sample needed for analysis can be a promising approach especially for analysis of such kinds of compounds. Microfluidic platforms can combine in one device all analysis steps such as cleaning, preconcentration, detection, and further removal of a given analyte. This also leads to the possibility of creating portable devices, which could work without necessarily using sophisticated professional laboratory equipments. Smart systems for different pollutants detection, where sensing strategy is based on inhibition of enzymatic reactions, are already reported. For example, a simple LOC device is based on biocompatible and biodegradable $CaCO_3$-PEI nanostructured microparticles (MPs) used to detect and remove phenolic wastes. The detection of phenol using a hybrid PDMS/glass chronoimpedimetric microchip and its removal in the same LOC system through the use of an extra $CaCO_3$-PEI MPs microcolumn was achieved (with LOD 4.64 nM) (Figure 18.4d) [11, 55].

Due to the easy fabrication process and flexibility, LOC devices can integrate novel material for electrochemical detection. One of the examples is the work reported by Medina-Sanchez et al. where boron-doped diamond (BDD) electrodes are employed for electrochemical detection and degradation of the pesticide atrazine (Atz). Chronoamperometry revealed a very low LOD of 3.5 pM for Atz (Figure 18.4a) [56].

18.5.3 Effective Sensing of Biomarkers

Common biological interactions usually used for biomarkers detection like antigen–antibody (immunoassays), enzyme–substrate, and aptamer–target can be easily implemented to microfluidic platforms. Thanks to this, it is possible to obtain better LOD, faster response of sensor, and moreover prototyping of multifunctional devices able to analyze a number of biomarkers from one sample in one device. Similarly, just as in microfluidic systems for environment monitoring, it is possible to perform in one chip all necessary steps for sample pretreatment. This is very important taking into account that clinical samples usually consist of blood, serum, etc. In our group, we have already proposed microfluidic devices for different biomarkers (i.e., IgG [57], ApoE) working as electrochemical sensors. Most of obtained solutions are based on sandwich immunoassay, where specific antibodies against given target. Furthermore, secondary antibodies were used and functionalized with QDs, which allowed for sensitive electrochemical detection (Square Wave Anodic Stripping Voltammetry) of interesting biomarkers (Figure 18.4b) [52].

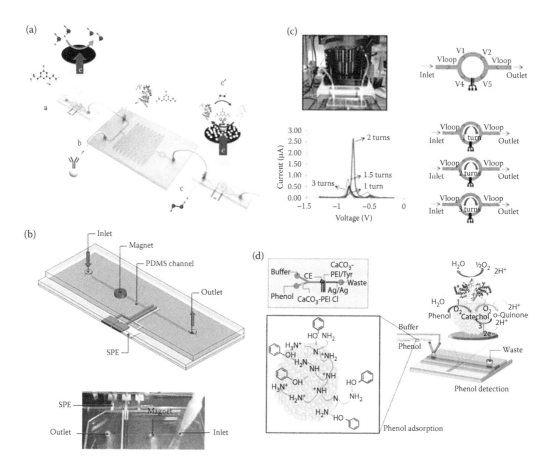

FIGURE 18.4
LOC platforms for biosensing application. (a) Atrazine detection and degradation in LOC device. (Reprinted from *Biosens. Bioelectron.*, Medina-Sánchez et al., Microfluidic platform for environmental contaminants sensing and degradation based on boron-doped diamond electrodes, Copyright 2015, with permission from Elsevier.) (b) Electrochemical detection of ApoE in LOC device using QDs as labels. (Reprinted from *Biosens. Bioelectron.*, 54, Medina-Sanchez, M. et al., On-chip magneto-immunoassay for Alzheimer's biomarker electrochemical detection by using quantum dots as labels, 279–284, Copyright 2014, with permission from Elsevier.) (c) Improved electrochemical QDs detection in recycling flow through mode LOC device. (Medina-Sanchez, M. et al. 2012. On-chip electrochemical detection of CdS quantum dots using normal and multiple recycling flow through modes. *Lab Chip* 12(11), 2000–2005. Reproduced by permission of The Royal Society of Chemistry.) (d) Phenol detection and removal in LOC device with CaCO$_3$ microparticles as an adsorbent. (Reprinted from *Biosens. Bioelectron.*, 55, Mayorga-Martinez, C. C. et al., An integrated phenol "sensoremoval" microfluidic nanostructured platform, 355–359, Copyright 2014, with permission from Elsevier.)

18.6 Conclusions

In this chapter, nanomaterials as flexible building blocks of biosensing devices have been presented. Due to the use of nanomaterials, we can observe progressive change in the field of biosensing devices.

Electrochemical biosensors achieved increased sensitivity and LOD due to the application of various nanoparticles such as AuNP, IrO$_2$, QD, as well as carbon materials. All those materials play a role of effective label, as the chemical reaction they catalyze/perform is related to the quantity of a given analyte. Similarly, optical biosensing became more potent by using QDs in microarray assays compared to commonly known organic fluorophores. Most of the nanomaterials are suitable to be used in various conditions. Because of this property they can be employed for paper-based nanobiosensors as well as LOC devices. Another possible function of nanomaterials is intermediation in (bio)chemical reactions, a good example of which being nanomotors which motion support the mass transport of biomolecules by improving the sensitivity of immunoassays performed in the microarray format. Moreover, nanomaterials can be used as a substrate for biosensing platform and/or environment of reaction. Nanopaper (bacterial cellulose) is a very promising material in various optical-sensing techniques due to its physical and chemical properties. Nanochannels made of different materials can work as effective filters and separate analyte from the solution.

One of the factors facilitating popularization of nanomaterials is its ease of synthesis. Another crucial factor is that even though nanomaterials may be expensive, the quantity used for biosensing assay is sufficiently low (this is especially seen in microfluidic biosensors) to make the final biosensing device relatively cost-effective, which is essential from the point of view of possible commercialization of such novelties.

However, novel nanomaterials present interesting chemical and physical properties and their interaction with humans, animals, and environment is not well known yet. For this reason, implementation of any new nanomaterial in biosensors should be always preceded by a thorough risk analysis.

Acknowledgment

We acknowledge FP7 EU Project "SMS" (contract number 613844). ICN2 acknowledges support from the Severo Ochoa Program (MINECO, Grant SEV-2013-0295) and Secretaria d'Universitats i Recerca del Departament d'Economia i Coneixement de la Generalitat de Catalunya (2014 SGR 260).

References

1. Naik, P. P. et al. 2015. Android integrated urea biosensor for public health awareness. *Sens. Biosens. Res.* 3, 12–17.
2. Aleman, J. et al. 2007. Definitions of terms relating to the structure and processing of sols, gels, networks, and inorganic-organic hybrid materials (IUPAC recommendations 2007). *Pure Appl. Chem.* 79(10), 1801–1827.
3. Vert, M. et al. 2012. Terminology for biorelated polymers and applications (IUPAC recommendations 2012). *Pure Appl. Chem.* 84(2), 377–408.
4. Afonso, A. S. et al. 2013. Electrochemical detection of Salmonella using gold nanoparticles. *Biosens. Bioelectron.* 40(1), 121–126.
5. de la Escosura-Muniz, A. et al. 2009. Rapid identification and quantification of tumor cells using an electrocatalytic method based on gold nanoparticles *Anal. Chem.* 81, 10268.
6. Maltez-da Costa, M. et al. 2012. Detection of circulating cancer cells using electrocatalytic gold nanoparticles. *Small* 8(23), 3605–3612.
7. Espinoza-Castaneda, M. et al. 2013. Casein modified gold nanoparticles for future theranostic applications. *Biosens. Bioelectron.* 40(1), 271–276.

8. Mars, A. et al. 2013. Gold nanoparticles decorated with a ferrocene derivative as a potential shift-based transducing system of interest for sensitive immunosensing. *J. Mat. Chem. B* 1(23), 2951–2955.

9. Mayorga-Martinez, C. C. et al. 2014. Iridium oxide nanoparticle induced dual catalytic/inhibition based detection of phenol and pesticide compounds. *J. Mat. Chem. B* 2(16), 2233–2239.

10. Lopez-Marzo, A., Pons, J., and Merkoçi, A. 2012. Controlled formation of nanostructured CaCO₃-PEI microparticles with high biofunctionalizing capacity. *J. Mat. Chem. B* 22(30), 15326–15335.

11. Mayorga-Martinez, C. C. et al. 2013. Nanostructured CaCO₃-poly(ethyleneimine) microparticles for phenol sensing in fluidic microsystem. *Electrophoresis* 34(14), 2011–2016.

12. Marin, S. et al. 2011. Electrochemical investigation of cellular uptake of quantum dots decorated with a proline-rich cell penetrating peptide. *Bioconjug. Chem.* 22(2), 180–185.

13. Marin, S., and Merkoçi, A. 2012. Nanomaterials based electrochemical sensing applications for safety and security. *Electroanalysis* 24(3), 459–469.

14. Morales-Narvaez, E. et al. 2012. Signal enhancement in antibody microarrays using quantum dots nanocrystals: Application to potential Alzheimer's disease biomarker screening. *Anal. Chem.* 84(15), 6821–6827.

15. Marin, S., and Merkoçi, A. 2009. Direct electrochemical stripping detection of cystic-fibrosis-related DNA linked through cadmium sulfide quantum dots. *Nanotechnology* 20(5), 055101.

16. Medina-Sanchez, M. et al. 2012. On-chip electrochemical detection of CdS quantum dots using normal and multiple recycling flow through modes. *Lab Chip* 12(11), 2000–2005.

17. Mazumder, S. et al. 2009. Review: Biofunctionalized quantum dots in biology and medicine. *J. Nanomater.* 2009, Article ID 815734, 17. doi:10.1155/2009/815734.

18. Monton, H. et al. 2015. Annexin-V/quantum dot probes for multimodal apoptosis monitoring in living cells: Improving bioanalysis using electrochemistry. *Nanoscale* 7(9), 4097–4104.

19. Perez Lopez, B., and Merkoçi A. 2009. Improvement of the electrochemical detection of catechol by the use of a carbon nanotube based biosensor. *Analyst* 134(1), 60–64.

20. Perez-Lopez, B., and Merkoçi, A. 2011. Magnetic nanoparticles modified with carbon nanotubes for electrocatalytic magnetoswitchable biosensing applications. *Adv. Funct. Mater.* 21(2), 255–260.

21. Morales-Narvaez, E. et al. 2012. Simple Forster resonance energy transfer evidence for the ultrahigh quantum dot quenching efficiency by graphene oxide compared to other carbon structures. *Carbon* 50(8), 2987–2993.

22. Zor, E. et al. 2015. Graphene quantum dots-based photoluminescent sensor: A multifunctional composite for pesticide detection. *ACS Appl. Mater. Interfaces.* 7(36), 20272–20279.

23. Perez-Lopez, B., and Merkoçi, A. 2012. Carbon nanotubes and graphene in analytical sciences. *Microchim. Acta* 179(1–2), 1–16.

24. Morales-Narvaez, E., Hassan, A.-R., and Merkoçi, A. 2013. Graphene oxide as a pathogen-revealing agent: Sensing with a digital-like response. *Angew. Chem. Int. Ed.* 52(51), 13779–13783.

25. Morales-Narvaez, E. et al. 2015. Photoluminescent lateral-flow immunoassay revealed by graphene oxide: Highly sensitive paper-based pathogen detection. *Anal. Chem.* 87(16), 8573–8577.

26. Guix, M., Mayorga-Martinez, C. C., and Merkoçi, A. 2014. Nano/micromotors in (bio)chemical science applications. *Chem. Rev.* 114(12), 6285–6322.

27. Garcia, M. et al. 2013. Micromotor-based lab-on-chip immunoassays. *Nanoscale* 5(4), 1325–1331.

28. Guix, M. et al. 2012. Superhydrophobic alkanethiol-coated microsubmarines for effective removal of oil. *ACS Nano* 6(5), 4445–4451.

29. Campuzano, S. et al. 2012. Bacterial isolation by lectin-modified microengines. *Nano Lett.* 12(1), 396–401.

30. Morales-Narvaez, E. et al. 2014. Micromotor enhanced microarray technology for protein detection. *Small* 10(13), 2542–2548.

31. de la Escosura-Muniz, A., and Merkoçi, A. 2012. Nanochannels preparation and application in biosensing. *ACS Nano* 6(9), 7556–7583.

32. de la Escosura-Muniz, A. et al. 2013. Nanochannels for diagnostic of thrombin-related diseases in human blood. *Biosens. Bioelectron.* 40(1), 24–31.

33. de la Escosura-Muniz, A., and Merkoçi, A. 2011. A nanochannel/nanoparticle-based filtering and sensing platform for direct detection of a cancer biomarker in blood. *Small* 7(5), 675–682.

34. Espinoza-Castaneda, M. et al. 2015. Nanochannel array device operating through Prussian blue nanoparticles for sensitive label-free immunodetection of a cancer biomarker. *Biosens. Bioelectron.* 67, 107–114.

35. de la Escosura-Muniz, A. et al. 2015. Nanoparticles-based nanochannels assembled on a plastic flexible substrate for label-free immunosensing. *Nano Res.* 8(4), 1180–1188.

36. Parolo, C., and Merkoçi, A. 2013. Paper-based nanobiosensors for diagnostics. *Chem. Soc. Rev.* 42(2), 450–457.

37. Quesada-Gonzalez, D., and Merkoçi, A. 2015. Nanoparticle-based lateral flow biosensors. *Biosens. Bioelectron.* 73, 47–63.

38. Lopez Marzo, A. M. et al. 2013. All-integrated and highly sensitive paper based device with sample treatment platform for Cd^{2+} immunodetection in drinking/tap waters. *Anal. Chem.* 85(7), 3532–3538.

39. Lopez-Marzo, A. M. et al. 2013. High sensitive gold-nanoparticle based lateral flow immuno-device for Cd^{2+} detection in drinking waters. *Biosens. Bioelectron.* 47, 190–198.

40. Parolo, C., de la Escosura-Muniz, A., and Merkoçi, A. 2013. Enhanced lateral flow immunoassay using gold nanoparticles loaded with enzymes. *Biosens. Bioelectron* 40(1), 412–416.

41. Parolo, C. et al. 2013. Simple paper architecture modifications lead to enhanced sensitivity in nanoparticle based lateral flow immunoassays. *Lab Chip* 13(3), 386–390.

42. Rivas, L. et al. 2014. Improving sensitivity of gold nanoparticle-based lateral flow assays by using wax-printed pillars as delay barriers of microfluidics. *Lab Chip* 14(22), 4406–4414.

43. Rivas, L. et al. 2015. Triple lines gold nanoparticle-based lateral flow assay for enhanced and simultaneous detection of Leishmania DNA and endogenous control. *Nano Res.* 8(11), 3704–3714.

44. Parolo, C. et al. 2013. Paper-based electrodes for nanoparticle detection. *Part. Part. Syst. Charact.* 30(8), 662–666.

45. Morales-Narvaez, E. et al. 2015. Nanopaper as an optical sensing platform. *ACS Nano* 9(7), 7296–7305.

46. Goldsmith, Z. G., and Dhanasekaran, N. 2004. The microrevolution: Applications and impacts of microarray technology on molecular biology and medicine (review). *Int. J. Mol. Med.* 13(4), 483–495.

47. Tu, S. et al. 2014. Protein microarrays for studies of drug mechanisms and biomarker discovery in the era of systems biology. *Curr. Pharm. Des.* 20(1), 49–55.

48. Dufva, M. 2005. Fabrication of high quality microarrays. *Biomol. Eng.* 22(5–6), 173–184.

49. Monton, H. et al. 2012. The use of quantum dots for immunochemistry applications. *Methods Mol. Biol. (Clifton, N. J.)* 906, 185–192.

50. Morales-Narvaez, E., and Merkoçi A. 2012. Graphene oxide as an optical biosensing platform. *Adv. Mater.* 24(25), 3298–3308.

51. Medina-Sanchez, M., Miserere, S., and Merkoçi, A. 2012. Nanomaterials and lab-on-a-chip technologies. *Lab Chip* 12(11), 1932–1943.

52. Medina-Sanchez, M. et al. 2014. On-chip magneto-immunoassay for Alzheimer's biomarker electrochemical detection by using quantum dots as labels. *Biosens. Bioelectron.* 54, 279–284.

53. Chee, P. S. et al. 2013. Micropump pattern replication using printed circuit board (PCB) technology. *Mater. Manuf. Process.* 28(6), 702–706.

54. Duffy, D. C. et al. 1998. Rapid prototyping of microfluidic systems in poly(dimethylsiloxane). *Anal. Chem.* 70(23), 4974–4984.

55. Mayorga-Martinez, C. C. et al. 2014. An integrated phenol "sensoremoval" microfluidic nanostructured platform. *Biosens. Bioelectron.* 55, 355–359.

56. Medina-Sánchez, M. et al. 2015. Microfluidic platform for environmental contaminants sensing and degradation based on boron-doped diamond electrodes. *Biosens. Bioelectron.*75, 365–374

57. Ambrosi, A., Guix, M., and Merkoçi, A. 2011. Magnetic and electrokinetic manipulations on a microchip device for bead-based immunosensing applications. *Electrophoresis* 32(8), 861–869.

19

Nanobiosensors: Carbon Nanotubes in Bioelectrochemistry

Anthony Guiseppi-Elie, Nikhil K. Shukla, and Sean Brahim

CONTENTS

19.1 Introduction

Because of their size and unique material properties, nanomaterials such as semiconductor nanowires (NWs) and carbon nanotubes (CNTs) interact with and influence biological entities in entirely unique ways. Of course, these ways are well known to nature, as many interactions that occur in the natural world arise through hierarchical ordering of molecules and minerals into larger structures through assembles that occur on the nanometer length scale. Nature provides rich examples of elegantly organized functional nanomaterials in biological systems. Among these are bacteria that sense the Earth's magnetic field through the use of nanosized "bar magnets." We are on the cusp of understanding the driving forces, associated chemistries, and assembly strategies for the integration of artificially prepared nanoparticles into functional nanobiosystems. This knowledge has been hampered in the recent past by difficulties associated with the synthesis and fabrication of such materials with defined and reproducible physical and chemical properties. It is this precise and reproducible control of size and chemistry that will enable the "bottom-up" development of engineered devices and systems that exploit the nanoscale.

Accelerated advances in nanotechnology and nanoscience have yielded a host of nanoscale materials possessing enhanced and tailored optical, electrical, magnetic, or catalytic properties [1–4]. The subsequent progression to fabricate functional devices incorporating these nano ensembles has been possible largely due to the array of diversity found in composition, shape, and surface character that such nanomaterials possess [5–7]. Researchers in the life sciences have only recently begun to borrow these nanotools and apply them to a variety of biomedical applications ranging from disease diagnosis to gene

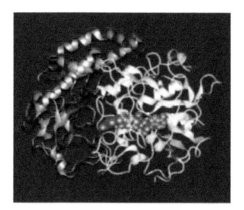

FAD (oxidized)   FADH (radical semiquinone)   FADH₂ (reduced)

FIGURE 19.2
Representation of the oxido-reductase enzyme glucose oxidase showing the flavin adenine dinucleotide (FAD) moiety deeply embedded within a protective glycoprotein shell. (From http://www-biol.paisley.ac.uk/marco/ enzyme_electrode/chapter3/chapter3_page4.htm/. With permission.)

(Figure 19.3). This dramatic current change in the CV is confirmation of the creation of a much larger effective working electrode area of SWNT|GCE compared with GCE alone. Also, the preservation of the general shape of the CV indicates no unusual influence of the SWNT|GCE interface that may be attributable to electron transfer between the nanomaterial and the underlying GCE surface. FAD was also shown to spontaneously adsorb onto and be retained on the SWNT|GCE surface (Figure 19.3). FAD adsorbed onto the SWNT|GCE (FAD|SWNT|GCE) displayed well-defined multiple scan rate CV with a formal potential of −448 mV (all potentials vs. Ag/AgCl) and only small changes in redox peak separation (ΔE_P) and peak widths at half-height of all scan rates investigated. These observations indicate that the adsorbed FAD displayed a quasi reversible one-electron transfer process on SWNT. Furthermore, the linear relationship of the peak current with the scan rate demonstrated that the electron transfer ($k_s = 3.1$ s^{-1}) of the adsorbed FAD was typical for a surface-confined electrode reaction. Having demonstrated the reversible electroactive behavior of the free prosthetic group at the SWNT|GCE system, can a similar direct electroactivity be accomplished with FAD in its natural biological realm, i.e., when buried within the glycoprotein shell of the redox enzyme such as GOx? Figure 19.3c displays the CV of adsorbed GOx onto SWNT|GCE obtained in enzyme-free buffer solution, convincingly showing the electroactivity of the embedded FAD moiety, understandably with suppressed peak currents compared with the solution-borne or surface-adsorbed prosthetic group, but, with coincident electroactive profiles. This suggests that under the present conditions, either the globular protein shell of GOx was still somewhat an obstacle for facile direct electron transfer

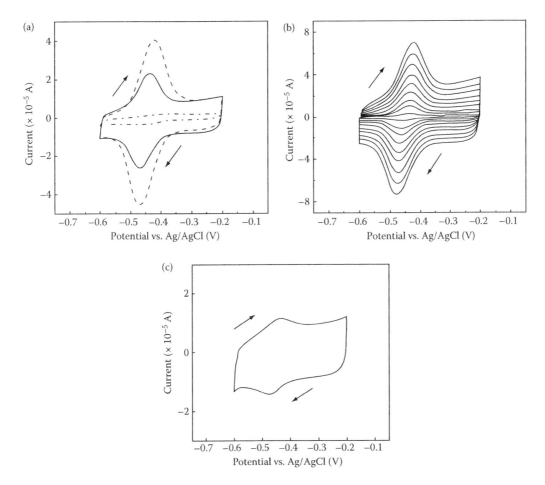

FIGURE 19.3
Cyclic voltammograms of FAD and GOx adsorbed on SWNT|GCE. (a) GCE in pH 7.0 phosphate buffer/0.1 M KCl containing 0.1 mM FAD (dot/dash mixed line); SWNT|GCE in pH 7.0 phosphate buffer/0.1 M KCl containing 0.1 mM FAD (dashed line); FAD|SWNT|GCE in pH 7.0 phosphate buffer/0.1 M KCl (dotted line); FAD|SWNT|GCE in pH 7.0 phosphate buffer/0.1 M KCl after overnight (solid line). Scan rate: 50 mV s^{-1}. (b) FAD|SWNT|GCE in pH 7.0 phosphate buffer/0.1 M KCl at the scan rates (mV s^{-1}) of 10, 25, 35, 50, 65, 80, 100, 120, and 140 (from inner to outer). (c) GOx|SWNT|GCE in pH 7.0 phosphate buffer/0.1 M KCl (solid line); FAD|SWNT|GCE in pH 7.0 phosphate buffer/0.1 M KCl (dotted line). Scan rate: 50 mV s^{-1}. (From Guiseppi-Elie, A. et al. 2002. *Nanotechnology* 13:559. With permission.)

of the enzyme or that the enzyme loading onto such unannealed SWNT|GCE was very limited. The latter possibility was subsequently refuted by AFM analysis [15].

Similarly, FAD and GOx were also adsorbed onto free-standing carbon nanotube paper (SWNTP) in both its unannealed and annealed forms. FAD and GOx adsorbed on unannealed SWNTP displayed similar CVs as those obtained on SWNT|GCE; however, the electrochemical behavior of FAD and GOx on annealed SWNTP was totally and dramatically different (Figure 19.4a and b). Contrary to what was observed with the FAD|SWNT|GCE system, FAD adsorbed on the annealed SWNTP (FAD|SWNTP) displayed an unsymmetrical CV and rather large ΔE_P values with an irreversible transfer process. The electron transfer rate (k_s) of FAD|SWNTP was as low as 0.38 s^{-1} as determined

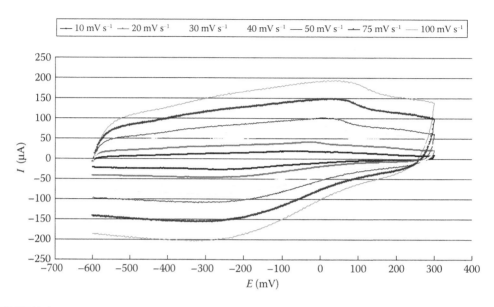

FIGURE 19.6
Cyclic voltammograms of 100 μM pseudoazurin solution at a glassy carbon electrode modified with single wall carbon nanotubes (SWNT|GCE). CVs were obtained in 0.2 M phosphate buffer, pH 6.0 at scan rates of 10, 20, 30, 40, 50, 75, and 100 mV s⁻¹. (From Guiseppi-Elie, A. et al. 2005. Nanotechnology 1:83. With permission.)

k_s value (determined by Laviron method) of 8.9×10^{-2} cm s⁻¹. The ratios of cathodic and anodic peak currents, I_{pc}/I_{pa}, for this system were found to average around 2.8, suggesting the contribution of kinetic or other complications to the electrode process. Interestingly, in the absence of dissolved oxygen, the CV was either not well resolved or there was an absence of the oxidation peak (observed at scan rates >20 mV s⁻¹). The average k_s was now reduced to 1.70×10^{-2} cm s⁻¹. Figure 19.6 shows the MSRCVs of pseudoazurin solution (oxygenated) obtained at a SWNT|GCE electrode. Compared to the bare GCE, there was now increased capacitative charging with about a 1000-fold increase in peak currents (μA range). In sharp contrast to the bare GCE, the I_{pc}/I_{pa} ratio for the SWNT|GCE was approximately 1.0 at all scan rates investigated, indicating a quasi reversible electrochemical reaction for the blue copper protein at the CNT-modified electrode surface. Heterogeneous electron transfer rate constants averaged 5.3×10^{-2} cm s⁻¹. As observed previously with oxidoreductase enzymes, the carbon nanotubes present a modified electrode surface that facilitates enhanced spontaneous adsorption and deposition of pseudoazurin protein compared to the unmodified electrode, evidenced by the dramatic increase in current at all scan rates investigated in comparison with unmodified GCE.

The CVs of adsorbed metalloprotein onto the SWNT|GCE surface in protein-free buffer solution are shown in Figure 19.7. Compared to the voltammograms obtained at the SWNT|GCE in pseudoazurin solution (Figure 19.6), the redox couples at all scan rates investigated occurred at more positive potentials. Although ΔE_p and $E^{o'}$ values were now less than those for pseudoazurin in solution at corresponding scan rates, peak currents were now significantly larger. Of particular significance, k_s was determined to increase to 1.2×10^{-1} cm s⁻¹, a more than two fold increase in k_s compared to the same electrode in psuedoazurin solution. In the absence of dissolved oxygen, peak currents for oxidation and reduction processes were slightly reduced but I_{pc}/I_{pa} ratios still averaged unity, suggesting quasi reversible electron transfer. The calculated heterogeneous electron transfer

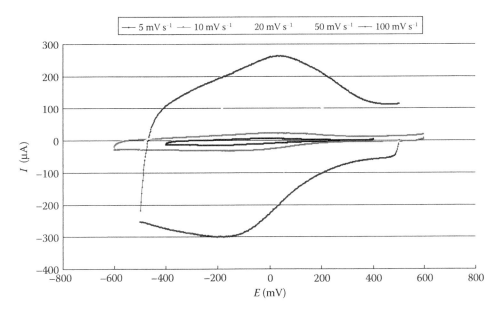

FIGURE 19.7
Cyclic voltammograms of adsorbed pseudoazurin at a glassy carbon electrode surface modified with single wall carbon nanotubes (SWNT|GCE). CVs were obtained in 0.2 M phosphate buffer, pH 6.0 at scan rates of 5, 10, 20, 50, and 100 mV s⁻¹. (From Guiseppi-Elie, A. et al. 2005. Nanotechnology 1:83. With permission.)

rate constant, 1.7×10^{-1} cm s⁻¹, was slightly greater than that obtained in the presence of oxygen and was the largest among all the systems investigated. This finding seems consistent with its presumed role in anaerobic respiration of some bacteria. Table 19.1 summarizes the important electrochemical characteristics obtained for the direct electrochemistry of pseudoazurin at bare glassy carbon electrode and at SWNT-modified glassy carbon electrode, in solution and adsorbed on the nanotube-modified electrode surface.

TABLE 19.1

Comparison of the Electrochemical Characteristics of the Direct Electrochemistry of Pseudoazurin at Glassy Carbon Electrodes

Electrode System	ΔE_p (mV) at 20 mV s⁻¹	$E^{o\prime}$ (mV)[f]	I_{pc}/I_{pa}	D_{appt} (cm² s⁻¹)	k_s (cm s⁻¹)[g]
GCE solution—O_2[a]	166	−38	2.8	1.35×10^{-11}	4.3×10^{-3g} 8.9×10^{-2h}
GCE solution—xO_2[b]	312	90	2.0	7.06×10^{-12}	1.7×10^{-2h}
SWNT | GCE solution—O_2[c]	274	−152	1.1	7.06×10^{-8}	5.3×10^{-2h}
P | SWNT | GCE—O_2[d]	134	−55	1.3	–	1.2×10^{-1h}
P | SWNT | GCE—xO_2[e]	84	−69	1.1	–	1.7×10^{-1h}

a Bare GCE in aerated 100 μM pseudoazurin solution.
b Bare GCE in deaerated 100 μM pseudoazurin solution.
c SWNT-modified GCE in aerated 100 μM pseudoazurin solution.
d Pseudoazurin adsorbed on SWNT-modified GCE in aerated buffer.
e Pseudoazurin adsorbed on SWNT-modified GCE in deaerated buffer.
f Mean formal potential over scan rates 5–100 mV s⁻¹ or at 20 mV s⁻¹.
g k_s determined using the method of Nicholson [50].
h k_s determined using the method of Laviron [22].

CV that shows an almost perfect return to the oxygen-free condition (the dashed lines in Figure 19.9). These results convincingly demonstrate that molecular oxygen was consumed at the electrode interface (glucose $+ O_2 \rightarrow$ gluconolactone acid $+ H_2O_2$) and accordingly confirms that the adsorbed oxidase still maintained its specific enzyme activity.

19.4.2 GOx at Covalently Immobilized Acid-Cut SWNT

As a follow-on of previously documented work, we have investigated the strategic chemical functionalization of SWNTs and the specific covalent immobilization of these to gold electrodes, as well as the conjugation of GOx with such chemically treated and modified SWNTs. Our objective has been to demonstrate the use of chemically modified and bioactive SWNTs in the fabrication of glucose biosensors. The electrodes of this work were gold microdisc electrode arrays (MDEA050-Au, ABTECH Scientific, Richmond, VA). Microdisk electrode arrays are single gold electrodes consisting of an electroactive area that is defined by an array of micron-dimensioned openings (disks) formed through an insulator (silicon nitride). These devices favor hemispeherical diffusion over semi-infinite linear diffusion and so result in broader dynamic range and lower detection limits for amperometric biosensors. The cleaned MDEA050-Au surfaces were functionalized by immersion in 2 mM ethanolic solutions of either cysteamine (CA) or 11-aminoundecanethiol (11-AUT). This treatment resulted in the formation of self-assembled monolayers of the alkanethiols with terminal amine functionalities on the exposed 50 μm diameter gold disks. In parallel, the chemical modification processes of nanotube purification and shortening involved ultra-sonication in a 3:1 ratio of H_2SO_4 and HNO_3 solution, followed by cycles of washing and cross flow filtration [24,25]. This produced open-ended, carboxylic acid-terminated SWNTs (SWNT-COOH). The acid moieties were subsequently activated, and using EDC/NHS coupling chemistry, were covalently attached to the terminal 1° amines of the self-assembled alkane thiol-treated MDEA050-Au. This resulted in devices labeled SWNT|CA|MDEA or SWNT|11-AUT|MDEA.

In a subsequent step, the unreacted acid functionalities of the immobilized SWNT were activated, and again using EDC/NHS coupling chemistry, were covalently attached to the 1° amine functionalities from available lysine residues on GOx. In control experiments, MDEA050-Au devices that were previously functionalized with CA and 11-AUT had GOx adsorbed directly (no SWNT involvement) onto their surfaces to create GOx-CA|MDEA and GOx-11-AUT|MDEA electrodes, respectively. Alternatively, the thiol-functionalized MDEAs that were first coated with SWNTs as described above were followed by covalent coupling of GOx to yield GOx-SWNT|CA|MDEA and GOx-SWNT|11-AUT|MDEA electrodes, respectively. Thus the process resulted in the formation of six different electrode systems. Figure 19.11 shows the scheme for the fabrication of the supramolecular assemblies. Each electrode was in turn made the working electrode in a three-electrode electrochemical cell setup that functioned in chronoamperometric mode. The amperometric response of each biosensor toward various concentrations of glucose was investigated.

The biosensors fabricated without inclusion of nanotubes GOx|MDEA, GOx-CA|MDEA, and GOx-11-AUT|MDEA showed linear dynamic responses to glucose up to 7.5, 20.0, and 15.0 mM, respectively, with sensitivities of 0.0014, 0.0018, and 0.001 μA/mM, respectively. The corresponding SWNT-modified biosensors all exhibited reduced linear dynamic response ranges compared to their unmodified counterparts. However, there were substantial gains in the sensitivities of the SWNT|MDEA electrodes compared to biosensors without nanotubes: 0.1258 μA/mM for GOx-SWNT|MDEA (90-fold increase), 0.2249 μA/mM for GOx-SWNT|CA|MDEA (125-fold increase), and 0.0459 μA/mM for

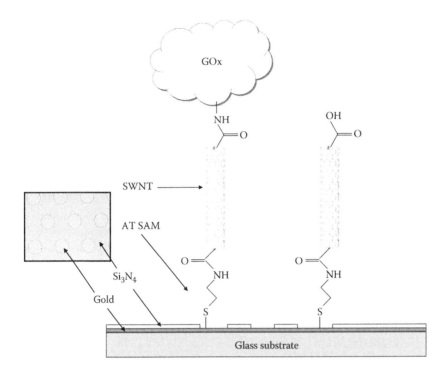

FIGURE 19.11
Schematic illustration of the formation of a supramacromolecular complex of covalently immobilized
GOx-SWNT|SAM|MDEA.

GOx-SWNT|11-AUT|MDEA (46-fold increase). The average response time (t_{95}) of the GOx-
SWNT|CA|MDEA biosensor was 39 s, which was observed to be faster than other biosen-
sors fabricated with or without SWNT.

One of the many advantages of the use of microelectrodes and ultramicroelectrodes in
bioanalytical applications is the promotion of radial diffusion flux patterns of the solution-
borne analyte at the electrode–solution interface [29,30]. The impact of this diffusion geometry on enzyme–substrate kinetic reactions typically manifests itself as an extension of the
observed linear dynamic response range toward the analyte. In contrast, a planar electrode surface facilitates linear diffusion patterns at the electrode–surface interface, usually accompanied by corresponding compromised linear dynamic ranges. In the present
study, the effect of the chemically treated nanotubes was to achieve a high planar electrode
surface area, resulting in much enhanced sensitivities due to greater loading of enzyme
but reduced linear dynamic response ranges due to loss of microdisc electrode geometry.

19.5 Conclusions and Perspectives

The modification of electrodes with carbon nanotubes for biomolecule immobilization is
arguably still in its infancy. In addition to the research presented herein, there have been
several innovative studies on the immobilization of enzymes and other redox proteins

on the ends of aligned nanotube arrays [31], on the walls of nanotubes [32], and inside nanotube bundles [33]. The vast majority of these studies were performed with the goal of realizing functional electrochemical enzyme biosensors. Quite often this surface modification step produces electrodes that are capable of detecting analytes at lower potentials than at other modified electrodes [16,34], an important and highly desirous characteristic for amperometric biosensors.

With respect to the extent of use of nanotube-modified electrodes for investigating direct protein electrochemistry, some exotic and very sophisticated interfaces have been demonstrated that take advantage of the combined excellent electrochemical properties and the "correct" nanoscale dimension of these nanomaterials [35]. In addition to glucose oxidase, pseudoazurin, and other blue copper proteins, nanotube-modified electrodes have been exploited to achieve direct electron transfer to peroxidases [36,37], catalase [38], myogobin [36,39], and cytochrome *c* [40,41]. Chemical modification approaches to introduce purposeful functionalities that "solubilize" and stabilize long and cut SWNTs include arylsulfonation [42], PEGylation [43], and of course, acidification. The use of poly(ethyleneoxide)-based Pluronics [44] and poly(vinyl pyrrolidone) [45] as well as traditional surfactants such as Triton–X [46] have shown promise. Despite these encouraging reports, there still exist numerous challenges related primarily to reproducible fabrication issues of CNT-modified electrodes for biosensing and bioelectrochemistry. Some of these problematic issues include the prevention of nanotube aggregation during electrode modification, the effective separation of semi conducting from conducting tubes (for SWNTs), the separation of nanotubes into uniform lengths (also for SWNTs), and the prevention of non specific adsorption of proteins to the walls of nanotubes. Although some recent advances tackling some of these issues have been demonstrated [47–49], this area of research still demands attention.

Apart from research into the above issues, the next step for extensive investigation using carbon nanotube–modified electrodes is predicted to proceed in three directions: (i) fundamental research into understanding the mechanisms by which CNTs afford the excellent observed electrochemical performances, (ii) research into understanding how to reproducibly fabricate nanostructured electrode surfaces with CNTs for scaled-up operations, and (iii) research into understanding how to successfully integrate these nanomaterials with natural biological systems.

Acknowledgments

This work was supported by the consortium of the Center for Bioelectronics, Biosensors, and Biochips (C3B). The authors thank Prof. R. Baughman, University of Texas at Dallas, for generous donations of single wall carbon nanotubes. N.K.S. acknowledges the Department of Biotechnology, Ministry of Science and Technology, Government of India for the award of the Biotechnology Overseas Associateship.

References

1. Daniel, M.C., and D. Astruc. 2004. *Chem Rev* 104 (1):293.
2. Bruchez, M. et al. 1998. *Science* 281:2013.

3. Hicks, J.F. et al. 2002. *J Am Chem Soc* 124 (44):13322.
4. Mulvaney, P. 1996. *Langmuir* 12 (3):788.
5. Trinidade, T. et al. 2001. *Chem Mater* 13 (11):3843.
6. Schwerdtfeger, P. et al. 2003. *Angew Chem Int Ed* 42 (17):1892.
7. Gangopadhyay, R., and A. De. 2000. *Chem Mater* 12 (3):608.
8. Xiao, Y. et al. 2003. *Science* 299:1877.
9. Elghanian, R. et al. 1997. *Science* 277:1078.
10. Averitt, R.D. et al. 1997. *Phys Rev Lett* 78 (22):4217.
11. Xia, Y. et al. 2003. *Adv Mater* 15 (5):353.
12. Strong, K.L. et al. 2003. *Carbon* 41 (8):1477.
13. Seo, J.W. et al. 2003. *New J Phys* 5:1.
14. Britto, P.J., K.S.V. Santhanam, and P.M. Ajayan. 1996. *Bioelectrochem Bioenerg* 41:121.
15. Guiseppi-Elie, A., C.H. Lei, and R.H. Baughman. 2002. *Nanotechnology* 13:559.
16. Wang, J., M. Musameh, and Y.H. Lin. 2003. *J Am Chem Soc* 125:2408.
17. Gooding, J.J. et al. 2003. *J Am Chem Soc* 125:9006.
18. Moore, R.R., C.E. Banks, and R.G. Compton. *Anal Chem* 76:2677.
19. Frew, J.E., and H.A.O. Hill. 1988. *Eur J Biochem* 172:261.
20. Dryhurst, G., K.M. Kadish, F.W. Scheller, and R. Renneberg. 1982. *Biol. Electrochem.* New York: Academic Press.
21. Baughman, R.H., C. Cui, A.A. Zakhidov, Z. Iqbal, J.N. Barisci, G.M. Spinks, G.G. Wallace, A. Mazzoldi, D. De Rossi, A.G. Rinzler, O. Jaschinski, S. Roth, and M. Kertesz. 1999. *Science* 284:1340.
22. Laviron, E. 1979. *J Electroanal Chem* 101:19 and references therein.
23. Verhoeven, W., and Y. Takeda. 1956. *Inorganic Nitrogen Metabolism*, eds. W.D. McElroy and B. Glass, 156. Baltimore: Johns Hopkins Press.
24. Adman, E.T. 1991. *Adv Protein Chem* 42:145.
25. Adman, E.T. 1986. *Topics in Molecular and Structural Biology*, vol. 6, ed. P.M. Harrison, 1. Weinhein: VCH.
26. Stryer, L. 1988. *Biochemistry*. New York: W.H. Freeman.
27. Kohzuma, T., C. Dennison, W. McFarlane, S. Nakashima, T. Kitagawa, T. Inoue, Y. Kai, N. Nishio, S. Shidara, S. Suzuki, and A.G. Sykes. 1995. *J Biol Chem* 270:25733.
28. Guiseppi-Elie, A., S. Brahim, G.E. Wnek, and R.H. Baughman. 2005. *Nanobiotechnology* 1 (1):083–092.
29. Guiseppi-Elie, A., S. Brahim, G. Slaughter, and K.R. Ward. 2005. *IEEE Sens J* 5 (3):345–355.
30. Bard, A.J., and L.R. Faulkner. 1980. *Electrochemical Methods: Fundamentals and Applications*. John Wiley and Sons.
31. Gao, M., L.M. Dai, and G.G. Wallace. 2003. *Electroanalysis* 15:1089.
32. Xu, J.Z., J.J. Zhu, Q. Wu, Z. Hu, and H.Y. Chen. *Electroanalysis* 15:219.
33. Davis, J.J. et al. 1998. *Inorg Chim Acta* 272:261.
34. Rubianes, M.D., and G.A. Rivas. 2005. *Electrochem Commun* 5:689.
35. Gooding, J.J. 2005. *Electrochim Acta* 50:3049.
36. Yu, X. et al. 2003. *Electrochem Commun* 5:408.
37. Zhao, Y.D., W.D. Zhang, H. Chen, Q.M. Luo, and S.F.Y. Li. 2002. *Sens Actuators B* 87:168.
38. Wang, L., J.X. Wang, and F.M. Zhou. 2004. *Electroanalysis* 16:627.
39. Zhao, G.C., L. Zhang, X.M. Wei, and Z.S. Yang. *Electrochem Commun* 5:825.
40. Davis, J.J., R.J. Coles, and H.A.O. Hill. 1997. *J Electroanal Chem* 440:279.
41. Wang, G., J.J. Xu, and H.Y. Chen. 2002. *Electrochem Commun* 4:506.
42. Pompeo, F., and D.E. Resasco. 2002. *Nano Lett* 2:369.
43. Sun, Y.-P., K. Fu, Y. Lin, and W. Huang. 2002. *Acc Chem Res* 35:1096.
44. Wanka, G., H. Hoffmann, and W. Ulbricht. 1994. *Macromolecules* 27:4145.
45. O'Connell, M.J., P. Boul, L.M. Ericson, C. Huffman, Y. Wang, E. Haroz, C. Kuper, J. Tour, K.D. Ausman, and R.E. Smalley. 2001. *Chem Phys Lett* 342:265.
46. Moore, V.C., M.S. Strano, E.H. Haroz, R.H. Hauge, R.E. Smalley, J. Schmidt, and Y. Talmon. 2003. *Nano Lett* 3:1379.

47. Krupke, R. et al. 2003. *Science* 301:344.
48. Chattopadhyay, D., L. Galeska, and F. Papadimitrakopoulos. 2003. *J Am Chem Soc* 125:3370.
49. Strano, M.S. et al. 2003. *Science* 301:1519.
50. Nicholson, R.S. 1965. *Anal Chem* 37:1351–1355.
51. Hirsch, A. 2002. *Angew Chem Int Ed* 41:1853.
52. http://www-biol.paisley.ac.uk/marco/enzyme_electrode/chapter3/chapter3_page4.htm/.

20

Monitoring Apoptosis and Anticancer Drug Activity in Single Cells Using Nanosensors

Paul M. Kasili and Tuan Vo-Dinh

CONTENTS

20.1 Introduction

20.1.1 Nanotechnology in Biology and Medicine: Single Live Cell Analysis

On December 3, 2003, nearly $4 billion was appropriated for research and development over the following 4 years for nanotechnology. The twenty-first century Nanotechnology Research and Development Act made nanotechnology the highest priority funded science and technology effort since the space race four decades ago [1]. Therefore, the Nanotechnology Age can unequivocally be said to be more significant than any preceding age. Nanotechnology is a collective term that can best be defined as a description of activities at the level of atoms and molecules that have real-world applications in disciplines such as in medicine and biology. The applications of nanotechnology include sensors, robotics, image processing, IT, photovoltaics, instrumentation, new materials, surface coatings, biomaterials, thin films, conducting polymers, displays, photonics, LEDs, liquid crystals, communication, holography, virtual reality, surface engineering, smart materials microelectronics, precision engineering, and metrology. The application of nanotechnology to medicine and biology is no longer an abstract idea but a technologically challenging means that provides us with an extremely novel technological shift from conventional biology and medicine. The last decade or so has seen tremendous growth in nanotechnology, and the profound impact it has had on the development of nanosensors with the major drive being a need for real-time analysis, with low detection limits and high sensitivity. Rapid analysis with the ability to monitor specific biomolecules in specific microenvironments is important for medical applications and we are observing this more in the growing field of personalized medicine, whereby medical decision-making and practices are customized to the individual patient. The focus of this chapter is on the development and application of a unique class of biosensors, that is, optical nanobiosensors, which are an integral part of the current generation of nanotechnology-based devices with applications in biology and medicine. Optical nanobiosensors have and are currently facilitating new ways of approaching basic research in biology and medicine in ways that were unimaginable a few decades ago. One such frontier that is important to biological research and medicine is single living cell analysis. Single cell analysis is a reductionist approach to studying organisms because the cell is the smallest unit of life that can carry out all of life's basic processes. The idea is simple yet powerful, using optical nanobiosensors to study in a minimally invasive manner, single living cells without compromising the integrity of the cell, which is an autonomous system. With this paradigm, the study of single living mammalian cells is of fundamental importance to biology and medicine for a greater understanding of the complex function of biomolecules, subcellular structures and organelles, and biological processes from a holistic perspective.

One of the key challenges in biology and medicine has been molecular analysis of single living cells aimed at providing an understanding of the underlying complex mechanisms in the basic, structural, and functional unit of life. This is important in understanding the normal development of cells, how cells respond to external stimuli, and how cells progress from normal to pathological states. About a decade ago, we envisioned and expressed that the application of nanotechnology to molecular and cell biology and medicine would revolutionize the way in which we study cells, signaling networks, and physiological processes, as well as diseases such as cancer. This has been evident in the development and application of nanotechnology-based tools (e.g., nanofibers, nanoparticles, nanotubes, nanowires, nanomembranes, etc.) for single cell analysis and most recently the push by NIH known as the Single Cell Analysis Program (http://commonfund.nih.gov/Singlecell/index). This is from the understanding that individual cells within a population of cells or in tissue may be different genotypically and these differences can have important consequences phenotypically. Thus, nanotechnology offers tools that give us the capability to address a major challenge that must be addressed in molecular and cellular biology: how do we study and interpret the biochemistry and molecular biology of individual cells and cellular components in the appropriate cellular context using a holistic systems biology approach?

As we have better understood the overall nature of bulk cell assays over time, so too have we achieved a greater understanding of many macroscopic biological processes. This has led us to devise methods and techniques to seek a better understanding of cellular processes at the single cell level and ponder the complexities of the fundamental unit of life. The mammalian cell can be described as a warehouse of nanoscale machines and structures; it is full of proteins that support cellular processes and biochemical pathways, cell replication, and interactions with other cells as well as its environment. The cell in turn regulates the distribution, localization, rate of diffusion, multimerization, and assembly of biomolecules for their specific functioning. Understanding nanoscale machines and structures at the single cell level requires tools capable of minimally invasive analysis of basic molecular processes of the cell.

Since the mid-1940s, biomedical researchers have made enormous progress in identifying and understanding proteins and how they interact in many cellular processes. Much of this research was basic scientific research that sought to discover and unravel how cellular systems work and develop a knowledgebase that other scientists can use in order to achieve practical goals, such as develop treatments and cures for diseases [2,3]. In the 1950s, as scientists began to collaborate, they started to develop the current picture of the cell as a complex and highly organized entity. They found that a typical cell is like a miniature body containing tiny "organs," called organelles. One organelle is the command center, others provide the cell with energy, while still others manufacture proteins and additional molecules that the cell needs to survive and to communicate with its environment. The cell components are enclosed in a plasma membrane that not only keeps the cell intact, but also provides channels that allow communication between the cell and its external environment [2,3].

In recent years, many details of the biochemical mechanisms involved in the cell cycle and cell signaling networks have been discovered using *in vitro* conventional molecular and cell biology techniques that offer a reductionist view of the cell. Using reductionist approaches, scientists have found that cell signaling pathways are regulated by highly complex network of interactions between proteins. However, to unravel and fully understand the mysteries of cellular function and signaling, scientists need to apply both the reductionist and system level approaches to study components of specific biochemical pathways, such as the cell cycle, that involve specific molecules, the individual cells, groups of cells, and whole organisms. The goal, of course, is to be able to put all the parts together to understand normal or abnormal cellular activities, how cells function in normal cellular activities, or how they malfunction in disease states. However, such studies are usually performed in bulk assay format, which is typically blind to the heterogeneity of cells within a population, and the fact that cells in a population may behave asynchronously to external stimuli. A good example is apoptosis; the differences between cells in a population and their ability to respond to external apoptotic stimuli and activate caspases. Thus, to study and understand such molecular mechanisms that underlie differences in how cells respond to apoptotic stimuli, it is necessary to detect and measure caspase activity in intact individual living cells [4–6].

By analyzing caspase activity in single living cells, it should be possible to test hypotheses, and identify and determine the molecular mechanisms that underlie cell-to-cell variability in responsiveness to apoptotic stimuli.

It is noteworthy to point out that there are arguments for the reductionist and systems level approach to single living cell analysis. Both approaches have been applied to monitoring and understanding key cellular events such as apoptosis, and understanding functional biochemical pathways in single living cells, as heterogeneous as they are; however, nanotechnology has made available tools that enable functional analysis and studies at the systems level, mainly because of their scale and ability to achieve spatial and temporal resolution at the single cell level. Therefore, the development of new nanotechnology-based tools for single cell analysis is important because biochemical events and reactions within living cells can be deciphered only if we have access to their cellular distribution and location in a nondestructive and minimally invasive manner. Similarly, as our desire to visualize and decipher biochemical events progresses from bulk cell assays to single living cell studies, high demands are placed on measurement technologies capable of single living cell analysis.

20.2 Apoptosis

The number of cells in tissue of multicellular organisms is tightly regulated by achieving a controlled balance between the rate of cell division and the rate of programmed cell death. Programmed cell death (PCD), also known as apoptosis, is a cellular process that is integral to normal development and is also involved in the development of disease and infection. It is, thus, of importance in biology and medicine. Apoptosis is a morphologically distinct form of cell death that is implemented and preceded by a well-conserved biochemical mechanism involving caspases, a family of cysteine proteases that cleave proteins after aspartic acid residues. They are the main effectors of apoptosis or PCD and their activation leads to characteristic morphological changes of the cell such as shrinkage, chromatin condensation, DNA fragmentation, and plasma membrane blebbing. Induction to commit suicide is required for proper organismal development, to remove cells that pose a threat to the organism (e.g., cell infected with virus, cancer cells), and to remove cells that have damaged DNA. Cells with damaged or mutated DNA that fail to undergo apoptosis become aberrant cells that may be responsible for the development of disease states like cancer. Cells undergoing apoptosis are eventually removed by phagocytosis. Caspases can be monitored to study and detect the effect of various stimuli responsible or capable of inducing apoptosis. Progress in defining the pathways of apoptosis *in vitro* has given new insights into cell dynamics and the role of apoptosis in the dynamics of cell development and disease [7]. An understanding of the pathways of apoptosis, using *in vivo* models, will provide information necessary to bridge the gap between *in vitro* and *in vivo* studies. This is necessary because *in vitro* studies only give us snapshots of a rapidly changing, interactive progression of apoptosis, whereas *in vivo* studies can give us real-time information. However, many questions remain to be answered regarding the real-time link between the diverse proteins involved in the intracellular proteolytic cascade and external stimuli of apoptosis, such as reactive oxygen species (ROS), photodynamic therapy (PDT) drugs, and chemotherapeutic drugs that trigger cellular responses [8].

There are two main pathways by which apoptosis can occur leading to cell death. The first one, the death receptor pathway, involves the interaction of a death receptor, such as the TNF receptor-1 or the Fas receptor with its ligand, and the second one, the mitochondrial pathway, involves physically compromised mitochondria, proapoptotic, and antiapoptotic members of the Bcl-2 family. The end result of both pathways is caspase activation and the cleavage of specific cellular substrates, resulting in the biochemical and morphologic changes associated with the apoptotic phenotype [9]. Keeping in mind that apoptosis is important in the development of disease and disease states, and that there was a limitation in the tools available to analyze the causative relationships in apoptosis at the single cell level, we developed a nanotechnology-based tool, optical nanosensor, and demonstrated, as a proof of principle, its practical application for single living cell analysis. Since apoptosis is a morphological process preceded by biochemical events, the optical nanosenor was used to analyze biochemical events that preceded the morphological process. Human mammary carcinoma (MCF-7) cells were exposed to a PDT drug, δ-aminolevulinic acid (ALA), to trigger the mitochondrial pathway of apoptosis. When activated using the appropriate wavelength, ALA leads to the production of ROS, which compromises mitochondrial structure and function. Upon initiation of the mitochondrial pathway of apoptosis, cytochrome c is released from mitochondria into the cytosol. In the cytosol, cytochrome c interacts with apoptotic protease activating factor-1 (Apaf-1) [10]. The cytochrome c/Apaf-1 complex cleaves the inactive caspase-9 proenzyme to generate active caspase-9 enzyme [11]. Activated caspase-9 exhibits distinct substrate recognition properties and initiates the proteolytic activities of other downstream caspases, which degrade a variety of substrates, resulting in the systematic disintegration of the cell [12–14] and ultimately, cell death. As the sequence of the mitochondrial pathway of apoptosis progresses, the proapoptotic member's cytochrome *c*, caspase-9, and caspase-7 can be detected and identified.

Minimally invasive detection and identification of cytochrome *c*, caspase-9, and caspase-7 at the single cell level requires nanotechnology-based tools capable of performing nondestructive, sensitive, and specific biochemical analysis in single living cells. Optical nanobiosensors due to their nanoscale dimension have the capability for minimally invasive analysis at the single cell level. The ability to perform single living cell analysis without causing cell death is very important for longitudinal studies. In addition, physically disrupting the cell membrane can lead to significant cellular changes in the parameter or biomolecule of interest resulting in an inaccurate view of the actual functional and physiological state of the cell [15,16]. In addition, it is important to understand the dynamic relationships between biomolecules and molecular events in living cells in the context of a cell as an autonomous system rather than a collection of cellular organelles and individual processes as would be the case on *in vitro* assays. The information obtained from such studies can be used to accurately map molecular and cell behavior and function and in the process be applied to personalized medicine. It is also important to analyze cellular components such as proteins in their native physiological environment. Currently used molecular biology protocols for protein analysis often involve exposing proteins to pretreatment conditions, for example, solubilization, denaturation, reduction, etc., which can effectively cause the structural modifications in the protein of interest [16]. Furthermore, it is important to have the ability to monitor and detect slight changes in the quantities of proteins and other biomolecules within the smallest possible detection volumes, such as that of a single cell level, for diagnostic and therapeutic applications. This can be invaluable because it would permit multiplexing that would be otherwise difficult to perform on primary human cells from surgical specimens that are only available in very small and limited quantities.

20.3 Fiber-Optic Nanobiosensors: The Technology

Due to the complexity of the functional and structural unit of life, the cell, and the challenges presented to analyze single living cells, Vo-Dinh and coworkers developed a class of specific and selective antibody-based fiber-optic biosensors. One of the first development and practical application of these class of antibody-based optical fibers to the field of biosensors was reported in 1987 by Vo-Dinh and coworkers [17]. This led to the development of submicron-sized antibody-based optical biosensors [18], which set precedence for the development and application of the first antibody-based optical nanobiosensors capable of minimally invasive analysis at the single cell level. Due to the complex nature of most biological systems, the highly specific and selective binding properties of antibodies make them one of the most powerful bioreceptors to be employed by nanobiosensors [19].

The development of optical nanobiosensors raised tremendous expectations for single living cell analysis while their potential application in the fascinating areas of molecular and cell biology research will help us understand biomolecular interactions involved in cellular pathways such as apoptosis and therefore further our understanding of the dynamics of development, the disease process, and drug development. Optical nanobiosensors can be described as a class of biosensors with dimensions on the order of one billionth of a meter, capable of detecting and monitoring physical stimuli in microenvironments such as that of a single living cell. Optical nanobiosensors incorporate a biological sensing element in close proximity to or integrated with an optical transducer element. Examples of biological sensing elements include nucleic acids (e.g., DNA) or proteins (e.g., enzymes and antibodies). These biological sensing elements can also form the basis of nanobiosensor classification. The sequence of events in nanobioanalysis using optical nanobiosensors is as follows: the nanosensor with chemically immobilized biosensing elements is inserted or introduced into the microenvironment and allowed to equilibriate. The biosensing elements immobilized on the nanobiosensor recognize and bind the targeted biomolecule or if enzymatic reacts with the substrate. The physicochemical perturbation caused by binding or enzymatic reaction can be detected and monitored by a change in fluorescence properties. The fluorescence signal produced and detected can be converted into a measurable signal. Therefore, optical nanobiosensors are capable of providing specific qualitative, quantitative, or semiquantitative analytical information that correlates to a biological or cellular process at the single cell level. The main characteristics of optical nanobiosensors that make them viable tools for single living cell analysis include the following: high sensitivity—the ability to detect small concentration of target molecules or analyte in microenvironments; specificity—the ability to reduce the sensitivity of the nanobiosensor to nontargeted molecules, which are always present in complex mixtures; and bio-stability of the sensing components to maintain functionality [20]. The above characteristics allow for practical applications of optical nanobiosensors, which lie in significant fields such as molecular and cell biology, medicine, and biomedical sciences.

20.4 Sensing Principles of Optical Nanobiosensors

Optical nanobiosensors are capable of performing real-time measurements to monitor changes in small molecules or macromolecules in single living cells by detecting changes in fluorescence signal intensity. They are designed to utilize the exquisite sensitivity of fluorescence that allows qualitatitive, semiquantitative, and quantitative measurements inside living cells. The fluorescence detection mode of optical nanobiosensors utilizes evanescent wave excitation. This is because the dimensions of the nanotip are much smaller than the wavelength of light used. This creates and allows for sensing "hot spots" that permits sensitive and selective detection of the molecule of interest. The fluorescence detected and measured after evanescent wave excitation at the nanotip can be correlated with the presence and concentration of the analyte of interest easily. The limitation, here, is that biological or chemical molecules under investigation should be fluorescent. If they are not fluorescent, they can be chemically labeled or conjugated with fluorescent molecules; or have the ability to interact with fluorescent molecules, causing a variation in the emission of fluorescence that can be detected using evanescent wave excitation [21]. Optical nanobiosensors have been used to selectively monitor and detect biological and chemical components of complex biological systems, and in the process provide unambiguous identification and quantification [22]. Figure 20.1 shows a conceptual illustration of the

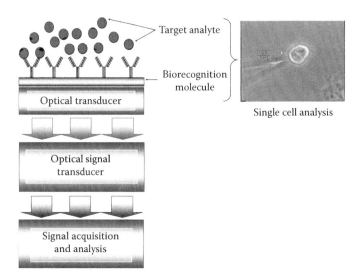

FIGURE 20.1
A schematic representation of the biosensing principle. The interaction between the target analyte and bio-recognition molecule covalently attached to the optical transducer is designed to produce a detectable and mea-surable signal which is captured and recorded for analysis. Inset is an image of an optical nanosensor probing a single living human mammary carcinoma (MCF-7) cell.

biosensing principles of optical nanobiosensors. The interaction between the target ana-lyte and the bioreceptor molecule immobilized on the optical fiber produces an effect that can be detected and measured using a photodetector that converts emitted photons into a measurable signal (e.g., such as an electrical voltage). The observed and measured signal can easily be correlated to the presence or absence of the analyte of interest.

20.5 Optical Nanobiosensor Sensing Formats

One key characteristic in the utility of optical nanobiosensors is their ability to specifically and selectively detect and measure molecules of interest without interference. This speci-ficity and selectivity allows measurements to be made in complex heterogeneous environ-ments such as that of a mammalian cell. Measurements using optical nanobiosensors can be classified based on the type of sensing format, either *direct* or *indirect*. Figure 20.2 shows sensing formats, *direct* and *indirect*, two of which are antibody based, and the third, which incorporates a synthetic peptide conjugated to a fluorophore. The *indirect* sensing format can be further divided into two sensing schemes based on the biological recognition mol-ecule used, in this case an enzyme linked immunosorbent assay (ELISA)-style format that utilizes antibodies as biorecognition molecules and a peptide substrate assay format. The application of antibodies as biorecognition molecules coupled with optical nanobiosensors for measuring caspases activated in the mitochondrial pathway of apoptosis can be done using either the *direct* or the *indirect* immunoassay format. The more straightforward of the two sensing formats is the *direct* sensing format, which is most applicable only when working with molecules that possess intrinsic fluorescent properties. The main advantage of this format is that there is no need to label the interactants. The working principle of the

33. Wang, G.Q. et al., A role for mitochondrial bak in apoptotic response to anticancer drugs. *J Biol Chem*, 2001. 276: 34307–34317.
34. Song, J.M. et al., Detection of cytochrome c in a single cell using an optical nanobiosensor. *Anal Chem*, 2004. 76(9): 2591–2594.
35. Kasili, P.M., Song, J.M., and Vo-Dinh, T., Optical sensor for the detection of caspase-9 activity in a single cell. *J Am Chem Soc*, 2004. 9(126): 2799–2806.
36. Zamzami, N. et al., Reduction in mitochondrial potential constitutes an early irreversible step of programmed lymphocyte death in vivo. *J Exp Med*, 1995. 181(5): 1661–1672.
37. Fletcher, G.C. et al., Death commitment point is advanced by axotomy in sympathetic neurons. *J Cell Biol*, 2000. 150(4): 741–754.
38. Budihardjo, I. et al., Biochemical pathways of caspase activation during apoptosis. *Annu Rev Cell Dev Biol*, 1999. 15: 269–290.

21

Biosensing and Theranostics Applications of Gold Nanostars

Yang Liu, Hsiangkuo Yuan, and Tuan Vo-Dinh

CONTENTS

21.1 Introduction

Cancer has become the second leading death cause in the United States and more than 7 million people are estimated to die due to cancer each year in the world.[1] Early diagnostics and effective therapy of cancer are complex and challenging problems, which require multidisciplinary efforts.[2] Nanotechnology has attracted increasing interest for cancer management because of its strong potential for biomedical applications such as molecular imaging, immunization, theranostics, targeted drug delivery, and therapy.[3–5] Nanoparticles can be easily fabricated to develop contrast agents for various imaging modalities including positron emission tomography (PET), magnetic resonance imaging (MRI), x-ray computer tomography (CT), and high-resolution optical imaging with superior signal-to-noise ratios.[6–8] Furthermore, nanoparticles can also be modified as therapeutic agents, such as drug carriers, radiation therapy enhancers, photothermal transducers, and photodynamic sensitizers.[9–12] The nanoparticles can also be optimized to have desired pharmacokinetics, improved stability, reduced immunogenicity/toxicity, and enhanced accumulation in specific tissues (e.g., tumor).[13–16]

Among different nanoplatforms, gold nanoparticles (AuNPs) are especially suitable for biomedical applications due to their chemical stability, biocompatibility, unique plasmonic properties, and ready multifunctionalization with large surface area.[4,17,18] Plasmonics involves the enhanced electromagnetic (EM) field of metallic nanostructures that can be utilized for sensing and therapy.[6,19] Since plasmonics strongly depends on nanoparticle size and geometry, various kinds of AuNPs, including nanoshells, nanorods, nanocages, nanostars, and hollow nanospheres, have been developed.[2,3,10,20,21] Their plasmonic peak can be tuned into the near-infrared (NIR) region, which is also known as "optical tissue window" with deeper light penetration.[22] Compared with other imaging modalities including PET or MRI, NIR optical imaging offers high spatial resolution and low cost.[8] Various NIR responsive nanomaterials such as polymer and silica nanoparticles loaded with NIR dyes, NIR upconverters, carbon nanotubes, and AuNPs have been developed for *in vivo* imaging.[4,23–25] Among them, AuNPs show outstanding advantages such as high mono dispersity, large surface area for targeting ligands functionalization, as well as intrinsic visibility under multiphoton microscopy without need of extra dye labeling.[26]

Our group has developed a novel method to synthesize star-shaped AuNPs without using toxic surfactant, which is highly suitable for *in vivo* biomedical applications.[26] The synthesized gold nanostars have a tunable plasmonic peak in the NIR region, known as the "tissue optical window," and multiple sharp branches acting as "lighting rods" to significantly enhance the local EM field. With very high EM field enhancement at branch tips and plasmonic tunability in NIR region, they have been successfully applied in various biomedical areas, including surface-enhanced Raman spectroscopy (SERS), photodynamic therapy (PDT), photothermal therapy (PTT), photoacoustic imaging, and biochemical sensing.[27–31] In this chapter, we will discuss the applications of multifunctional GNS nanoprobes in biosensing, cancer imaging, and therapy.

21.2 pH Sensing with SERS

pH has been identified to be a biomarker for cancer detection because tumors have been reported to contain highly acidic region due to high rates of glycolysis metabolism in a hypoxia environment, which generates acidic pyruvate as an end product.[32] A noninvasive spectrochemical sensing method with high sensitivity, specificity, and spatial resolution is of great value for biomedical diagnostics and applications.[33] SERS provides a promising method with high sensitivity and spatial resolution to perform pH sensing for cancer detection as well as other applications.[19] Raman spectroscopy is a kind of noninvasive optical method providing specific chemical identification based on the fingerprint of molecular vibrational spectra, which is directly and specifically related to structural features of the molecule.[34] However, the spontaneous Raman scattering cross section is intrinsically low and molecules with strong Raman scattering have cross sections only on the order of 10^{-29} cm^2 molecule^{-1} sr^{-1} compared with fluorescence cross section usually on the order of 10^{-16} cm^2 molecule^{-1} sr^{-1}.[6] Based on classical EM theory, EM fields can be locally amplified when light encounters metallic nanostructure. The SERS effect involves localized surface plasmons (LSPs) to dramatically amplify Raman scattering when the analytes are on or near metallic nanostructures. The LSPs are associated with oscillating electrons generated when the metallic nanostructure surface is irradiated by an external EM field (e.g., a laser beam). Those LSPs within the conduction band oscillate at the same frequency as that of the incident light and produce a secondary EM field, which is added to the external EM field, resulting in surface plasmon resonance (SPR). The secondary EM field is highly concentrated at points of high curvature on the roughed metallic surface similar to the "lightning rod" effect and "hot spot" effect at narrow regions where multiple nanoparticles are close to each other.[19] The enhancement factor is typically 10^6 to 10^8-fold and has been reported to reach 10^{15}-fold at "hot spots" where the EM field is extremely intense.[35] The high enhancement makes it possible for extremely sensitive detection at single molecule level.

The pH-sensing nanoprobe was developed by functionalizing GNS with para-mercaptobenzoic acid (pMBA), a Raman reporter with a thiol group to be linked to AuNP surface and a carboxylic group, which has a protonated and deprotonated molecular state in acidic and basic environment, respectively.[33] The GNS are synthesized using a nontoxic surfactant-free method previously published by our group.[26] The developed pH-sensing probes were mixed with 20-mM phosphate-buffered saline (PBS) solution (pH in the range between 5 and 9). Raman spectra were collected with a LabRam ARAMIS Raman microscope with 10× objective and a 785-nm laser (30 mW) for excitation. The transmission electron microscope (TEM) image and SERS spectra at pH 5 and 9 of synthesized GNS pH-sensing nanoprobe are shown in Figure 21.1. The observed SERS spectra in the range of 1000–1800 cm^{-1} for the developed pH-sensing GNS nanoprobe in pH 5 and pH 9 buffer solution are different after normalization to the peak intensity at 1078.3 cm^{-1}. The peak position at ~1580 cm^{-1} exhibits a small but noticeable downshift when the pH changes from 5 to 9. The peak intensity at 1010.4, 1136.8, and 1390.1 cm^{-1} increases as the pH changes from 5 to 9. In contrast, the peak intensity at 1700.0 cm^{-1} decreases when the pH changes from 5 to 9. More detailed information about our GNS pH sensing nanoprobe as well as theoretical simulation can be found in Reference 33.

21.3 Multifunctional Gold Nanostar Probe for Cancer Diagnostics

Since each imaging modality has its own advantages and disadvantages, a multifunctional nanoprobe has the potential to provide a complementary platform to perform both macroscopic (PET, MRI, and CT) and high-resolution local scans (optical imaging) for neoplastic lesion assessment.[8] We have developed GNS multifunctional nanoprobes for PET, CT, and optical imaging as well as SERS detection. We have performed dual-energy CT scan on sarcoma-containing mice and the longitudinal imaging shows GNS distribution 30 min (day 1), 24 h (day 2), and 72 h (day 4) after injection (Figure 21.2b).[7] The CT scan at 30 min after IV injection indicates that GNS nanoprobes with both 30 and 60 nm size

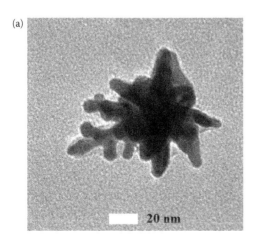
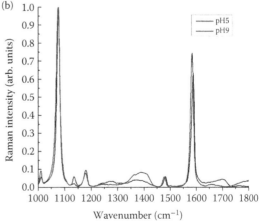

FIGURE 21.1

(a) TEM image of the developed pH-sensing GNS nanoprobe. (b) SERS spectrum of developed pH-sensing gold nanostar probe at pH 5 (black) and pH 9 (red). (Adapted from Liu, Y. et al., *Journal of Raman Spectroscopy* 2013, 44, 980–986.)

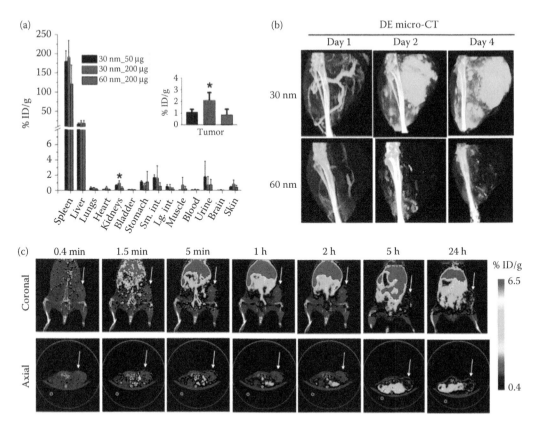

FIGURE 21.2

(a) GNS *in vivo* biodistribution with different injection dose and particle size 2 days after intravenous injection through tail vein. The unit % ID/g is defined as the percentage of total injection dose per gram tissue weight. (b) Maximum intensity projection based on dual-energy (DE) micro-CT data in mice with sarcoma injected with GNS nanoprobes of two different sizes (30 and 60 nm) and imaged at three time points (Day 1 [30 min], Day 2 [24 h], and Day 4 [72 h]). Gold is color coded in green for micro-CT. (c) PET/CT imaging of the tumor-bearing mouse following systemic nanoprobe injection. Images were obtained at 0.4, 1.5, 5 min, 1, 2, 5, and 24 h postinjection, respectively. Tumor is shown under white arrow. Nanoprobes accumulate in tumor gradually within 24 h. (Adapted from Liu, Y. et al., *Sensors* 2015, 15, 3706–3720; Liu, Y. et al., *Theranostics* 2015, 5, 946–960.)

locate mainly in the blood vessels. In addition, it is interesting to find that the majority of blood vessels in the tumor region with GNS nanoprobes perfusion are located outside of the tumor. For the 24 and 72 h time points, a significant amount of GNS nanoprobes accumulate in the tumor tissue and the smaller GNS nanoprobe (30 nm) has higher tumor uptake than its counterpart with larger size (60 nm). We also did a biodistribution study by labeling the GNS nanoprobe with ^{131}I radioisotope (Figure 21.3a). The results show that the GNS nanoprobe with smaller size and higher injection dose has higher tumor uptake in the unit of percentage of total injection dose per gram tissue (% ID/g). Furthermore, we have performed PET/CT scan with the developed multifunctional GNS nanoprobe, which was found to accumulate in the tumor tissue gradually (Figure 21.3c).[6] Those experimental results demonstrate that the developed GNS nanoprobe could be further optimized for future clinical applications including preoperative cancer diagnostics with PET and CT scan.

Figure 21.3a shows GNS distribution inside the tumor after intravenous injection through tail vein as visualized by two-photon microscopy. The smaller GNS nanoprobes (left) can penetrate deeper into tumor interstitial space after leaking through tumor blood vessels due to the enhanced permeation and retention (EPR) effect than those with larger size (right). Gold nanostars, with tip-enhanced plasmonics, have been reported to have two-photon action cross sections (TPACS) more than 1 million GM (Göeppert-Mayer units, equivalent to 10^{-50} cm^4 s photon^{-1}) and much stronger than those of organic fluorophores. Another study also measured two-photon photoluminescence of single AuNP with different shapes and TPACS (define in this sentence—not in the next one) for nanosphere, nanocube, nanotriangle, nanorod, and nanostar was reported to be 83, 500, 1.5×10^3, 4.2×10^4, and 4.0×10^6 GM, respectively.[36] Such high TPACS enable nanostars to be a superior contrast agent for imaging under multiphoton microscopy. GNS nanoprobes also have much stronger SERS enhancement than gold nanospheres (two orders higher) due to their tip-enhanced plasmonics.[37] We have performed *in vivo* SERS measurement with a sarcoma mouse model and results demonstrate that the developed SERS nanoprobes with GNS functionalized by Raman reporter, pMBA, accumulate selectively in tumor. The unique SERS peaks of Raman reporter (pMBA) at

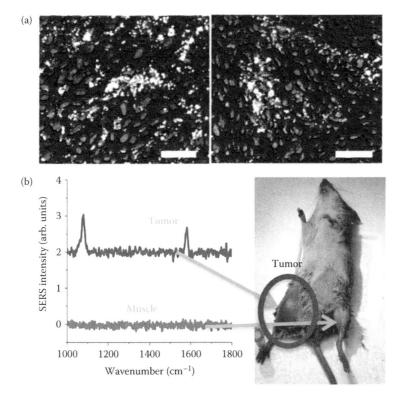

FIGURE 21.3
(a) *Ex vivo* TPL images are shown with GNS (left, 30 nm; right, 60 nm) in white and cell nucleus in blue (DAPI). Blood vessel is stained for CD31 and shown in green. Scale bar is 50 μm. (b) *In vivo* SERS spectrum of GNS nanoprobe in tumor. Mouse with sarcoma was injected with GNS nanoprobe through the tail vein. Three days after injection, a significant amount of GNS nanoprobes accumulate in tumor but not in the normal muscle of the contralateral leg. (Adapted from Liu, Y. et al., *Theranostics* 2015, 5, 946–960.)

1067 and 1588 cm^{-1} can be detected only in tumor tissue but not in the normal muscle of the contralateral leg (Figure 21.3b). Our results indicate that the SERS method using GNS nanoprobe has potential to be used for tumor differentiation in future biomedical applications.

21.4 GNS Nanoprobe for Cancer PTT

PTT, treating cancer through hyperthermia generated by photon energy, is a new promising method with high efficiency and specificity.[12,21,38] Hyperthermia has been used with other cancer therapy methods including radiation therapy and chemotherapy, and many studies have shown that hyperthermia can make some cancer cells more sensitive to radiation therapy and enhance certain anticancer drug effects. Traditionally, hyperthermia was performed using microwave or radiofrequency treatment.[39] Several previous studies have shown tumor size reduction with combined hyperthermia and other treatments. However, these methods can only macroscopically confine energy into the tumor region but are not tumor specific at the cellular level. GNS, with tip-enhanced plasmonics, has been demonstrated to be a superior photon-to-heat transducer (up to 94% conversion efficiency) and has been successfully applied for photothermal ablation to treat tumors with NIR light, which has deep tissue penetration.[7] The small size of GNS (30 nm) is also in favor of long blood circulation time and deep penetration into both tumor interstitial space and tumor cells. We have demonstrated PTT with GNS by both *in vitro* and *in vivo* studies. As shown in Figure 21.4a, the temperature of 0.8 g GNS solution reached 42°C after 10 min 0.8 W/cm^2 980-nm laser irradiation. On the contrary, the pure water temperature only increased to 32°C under the same condition. We have also performed *in vitro* cell studies to demonstrate that the GNS-enhanced PTT can be used to kill cancer cells. Figure 21.4b shows that cell death after laser irradiation with GNS occurs while cells in the nonirradiated area remain alive. Furthermore, we have used a sarcoma mouse model to perform *in vivo* PTT studies with NIR light and GNS photon-to-heat transducer. Figure 21.4c shows NIR imaging of mouse tumor surface temperature during photothermal treatment of primary sarcomas after intravenous injection of GNS or PBS (control). Each image shows the surface temperature plot of the whole mouse. T represents the mouse tumor and H represents the mouse head. The temperature scale bar is between 23°C and 60°C. With a total NIR exposure time of 10 min, the tumor temperature with GNS injection reached >50°C while that with PBS injection reached only around 40°C. After PTT, the sarcoma tumor shrank in size with nanostar injection and laser irradiation (Figure 21.4d). This proof-of-principle experiment demonstrated that the GNS nanoprobe with tip-enhanced plasmonics has the potential to be used for future translation medicine investigations to treat cancer with PTT.

21.5 Conclusion

A multifunctional biocompatible GNS nanoplatform was developed for pH sensing, cancer diangostics, and therapy. A combination of imaging and therapy modalities into a

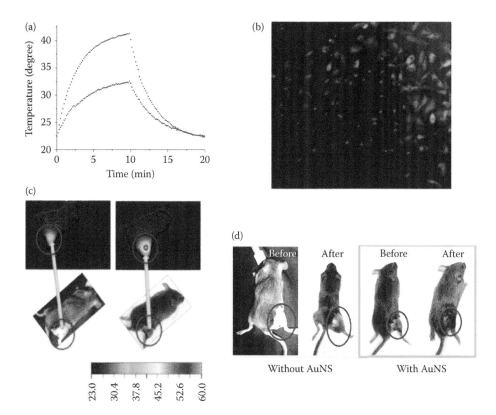

FIGURE 21.4

(a) Temperature profile of 0.2-nM GNS solution (black) and pure water (red) under laser irradiation (0.8 W/cm²). At 10-min time point, the laser irradiation was turned off. (b) *In vitro* PTT on BT 549 cancer cells incubated 24 h with 0.1-nM GNS. Live cells are green while dead cells are red. (c) *In vivo* animal study showing the temperature of tumor is much higher with GNS (right) than that without GNS (left). (d) PTT on mice with sarcoma without and with GNS injection. The mice are shown before PTT and 3 days after PTT. Note that laser irradiation only did not reduce the tumor size when no GNS were injected but it had a clear effect with GNS. The red ellipsoids mark the soft tissue sarcomas. (Adapted from Liu, Y. et al., *Theranostics* 2015, 5, 946–960.)

single nanoprobe makes it possible to perform image-guided therapy with the promise to improve treatment specificity, which makes GNS nanoprobes promising for cancer management. With the aim of future applications in patients, the long-term toxicity of GNS after IV injection should be carefully evaluated.

References

1. Jemal, A.; Siegel, R.; Xu, J.; Ward, E., Cancer statistics, 2010. *CA: A Cancer Journal for Clinicians* 2010, 60, 277–300.
2. Mieszawska, A. J.; Mulder, W. J. M.; Fayad, Z. A.; Cormode, D. P., Multifunctional gold nanoparticles for diagnosis and therapy of disease. *Molecular Pharmaceutics* 2013, 10, 831–847.
3. Kumar, D.; Saini, N.; Jain, N.; Sareen, R.; Pandit, V., Gold nanoparticles: An era in bionanotechnology. *Expert Opinion on Drug Delivery* 2013, 10, 397–409.

4. Wang, L.-S.; Chuang, M.-C.; Ho, J.-A., Nanotheranostics—A review of recent publications. *International Journal of Nanomedicine* 2012, 7, 4679–4695.

5. Rozhkova, E. A., Nanoscale materials for tackling brain cancer: Recent progress and outlook. *Advanced Materials* 2011, 23, H136–H150.

6. Liu, Y.; Yuan, H.; Kersey, F.; Register, J.; Parrott, M.; Vo-Dinh, T., Plasmonic gold nanostars for multi-modality sensing and diagnostics. *Sensors* 2015, 15, 3706–3720.

7. Liu, Y.; Ashton, J. R.; Moding, E. J.; Yuan, H. K.; Register, J. K.; Fales, A. M.; Choi, J. et al., A plasmonic gold nanostar theranostic probe for in vivo tumor imaging and photothermal therapy. *Theranostics* 2015, 5, 946–960.

8. Liu, Y.; Chang, Z.; Yuan, H.; Fales, A. M.; Vo-Dinh, T., Quintuple-modality (SERS-MRI-CT-TPL-PTT) plasmonic nanoprobe for theranostics. *Nanoscale* 2013, 5, 12126–12131.

9. Xing, Y.; Zhao, J.; Conti, P. S.; Chen, K., Radiolabeled nanoparticles for multimodality tumor imaging. *Theranostics* 2014, 4, 290–306.

10. Xia, X.; Xia, Y., Gold nanocages as multifunctional materials for nanomedicine. *Frontiers of Physics* 2014, 9, 378–384.

11. Sun, X.; Huang, X.; Yan, X.; Wang, Y.; Guo, J.; Jacobson, O.; Liu, D. et al., Chelator-free Cu-64-integrated gold nanomaterials for positron emission tomography imaging guided photothermal cancer therapy. *ACS Nano* 2014, 8, 8438–8446.

12. Pekkanen, A. M.; DeWitt, M. R.; Rylander, M. N., Nanoparticle enhanced optical imaging and phototherapy of cancer. *Journal of Biomedical Nanotechnology* 2014, 10, 1677–1712.

13. Ngwa, W.; Kumar, R.; Sridhar, S.; Korideck, H.; Zygmanski, P.; Cormack, R. A.; Berbeco, R.; Makrigiorgos, G. M., Targeted radiotherapy with gold nanoparticles: Current status and future perspectives. *Nanomedicine* 2014, 9, 1063–1082.

14. Black, K. C. L.; Wang, Y.; Luehmann, H. P.; Cai, X.; Xing, W.; Pang, B.; Zhao, Y. et al., Radioactive Au-198-doped nanostructures with different shapes for in vivo analyses of their biodistribution, tumor uptake, and intratumoral distribution. *ACS Nano* 2014, 8, 4385–4394.

15. Ahmadi, A.; Arami, S., Potential applications of nanoshells in biomedical sciences. *Journal of Drug Targeting* 2014, 22, 175–190.

16. Zhang, Y.; Qian, J.; Wang, D.; Wang, Y.; He, S., Multifunctional gold nanorods with ultrahigh stability and tunability for in vivo fluorescence imaging, SERS detection, and photodynamic therapy. *Angewandte Chemie—International Edition* 2013, 52, 1148–1151.

17. Lu, F.; Doane, T. L.; Zhu, J.-J.; Burda, C., Gold nanoparticles for diagnostic sensing and therapy. *Inorganica Chimica Acta* 2012, 393, 142–153.

18. Ahmad, M. Z.; Akhter, S.; Jain, G. K.; Rahman, M.; Pathan, S. A.; Ahmad, F. J.; Khar, R. K., Metallic nanoparticles: Technology overview & drug delivery applications in oncology. *Expert Opinion on Drug Delivery* 2010, 7, 927–942.

19. Vo-Dinh, T.; Fales, A. M.; Griffin, G. D.; Khoury, C. G.; Liu, Y.; Ngo, H.; Norton, S. J.; Register, J. K.; Wang, H.-N.; Yuan, H., Plasmonic nanoprobes: From chemical sensing to medical diagnostics and therapy. *Nanoscale* 2013, 5, 10127–10140.

20. Lee, J.; Chatterjee, D. K.; Lee, M. H.; Krishnan, S., Gold nanoparticles in breast cancer treatment: Promise and potential pitfalls. *Cancer Letters* 2014, 347, 46–53.

21. Zhang, Z.; Wang, J.; Chen, C., Gold nanorods based platforms for light-mediated theranostics. *Theranostics* 2013, 3, 223–238.

22. Weissleder, R., A clearer vision for in vivo imaging. *Nature Biotechnology* 2001, 19, 316–317.

23. Wang, Z.; Niu, G.; Chen, X., Polymeric materials for theranostic applications. *Pharmaceutical Research* 2014, 31, 1358–1376.

24. Wang, L.; Wang, Y.; Li, Z., Nanoparticle-based tumor theranostics with molecular imaging. *Current Pharmaceutical Biotechnology* 2013, 14, 683–692.

25. Sailor, M. J.; Park, J.-H., Hybrid nanoparticles for detection and treatment of cancer. *Advanced Materials* 2012, 24, 3779–3802.

26. Yuan, H.; Khoury, C. G.; Hwang, H.; Wilson, C. M.; Grant, G. A.; Vo-Dinh, T., Gold nanostars: Surfactant-free synthesis, 3D modelling, and two-photon photoluminescence imaging. *Nanotechnology* 2012, 23, 075102.

27. Yuan, H.; Wilson, C. M.; Xia, J.; Doyle, S. L.; Li, S. Q.; Fales, A. M.; Liu, Y. et al., Plasmonics-enhanced and optically modulated delivery of gold nanostars into brain tumor. *Nanoscale* 2014, 6, 4078–4082.

28. Yuan, H.; Liu, Y.; Fales, A. M.; Li, Y. L.; Liu, J.; Vo-Dinh, T., Quantitative surface-enhanced resonant Raman scattering multiplexing of biocompatible gold nanostars for in vitro and ex vivo detection. *Analytical Chemistry* 2013, 85, 208–212.

29. Liu, Y.; Chang, Z.; Yuan, H.; Fales, A. M.; Vo-Dinh, T., Quintuple-modality (SERS-MRI-CT-TPL-PTT) plasmonic nanoprobe for theranostics. *Nanoscale* 2013, 5, 12126–12131.

30. Yuan, H.; Khoury, C. G.; Wilson, C. M.; Grant, G. A.; Bennett, A. J.; Vo-Dinh, T., In vivo particle tracking and photothermal ablation using plasmon-resonant gold nanostars. *Nanomedicine: Nanotechnology Biology and Medicine* 2012, 8, 1355–1363.

31. Yuan, H.; Fales, A. M.; Vo-Dinh, T., TAT peptide-functionalized gold nanostars: Enhanced intracellular delivery and efficient NIR photothermal therapy using ultralow irradiance. *Journal of the American Chemical Society* 2012, 134, 11358–11361.

32. Hashim, A. I.; Zhang, X.; Wojtkowiak, J. W.; Martinez, G. V.; Gillies, R. J., Imaging pH and metastasis. *NMR in Biomedicine* 2011, 24, 582–591.

33. Liu, Y.; Yuan, H.; Fales, A. M.; Vo-Dinh, T., pH-sensing nanostar probe using surface-enhanced Raman scattering (SERS): Theoretical and experimental studies. *Journal of Raman Spectroscopy* 2013, 44, 980–986.

34. Liu, Y.; Yuan, H.; Vo-Dinh, T., Spectroscopic and vibrational analysis of the methoxypsoralen system: A comparative experimental and theoretical study. *Journal of Molecular Structure* 2013, 1035, 13–18.

35. Nie, S. M.; Emery, S. R., Probing single molecules and single nanoparticles by surface-enhanced Raman scattering. *Science* 1997, 275, 1102–1106.

36. Gao, N.; Chen, Y.; Li, L.; Guan, Z.; Zhao, T.; Zhou, N.; Yuan, P.; Yao, S. Q.; Xu, Q.-H., Shape-dependent two-photon photoluminescence of single gold nanoparticles. *Journal of Physical Chemistry C* 2014, 118, 13904–13911.

37. Yuan, H. K.; Fales, A. M.; Khoury, C. G.; Liu, J.; Vo-Dinh, T., Spectral characterization and intracellular detection of surface-enhanced Raman scattering (SERS)-encoded plasmonic gold nanostars. *Journal of Raman Spectroscopy* 2013, 44, 234–239.

38. Ayala-Orozco, C.; Urban, C.; Knight, M. W.; Urban, A. S.; Neumann, O.; Bishnoi, S. W.; Mukherjee, S. et al., Au nanomatryoshkas as efficient near-infrared photothermal transducers for cancer treatment: Benchmarking against nanoshells. *ACS Nano* 2014, 8, 6372–6381.

39. Vogl, T. J.; Farshid, P.; Naguib, N. N. N.; Darvishi, A.; Bazrafshan, B.; Mbalisike, E.; Burkhard, T.; Zangos, S., Thermal ablation of liver metastases from colorectal cancer: Radiofrequency, microwave and laser ablation therapies. *La Radiologia Medica* 2014, 119, 451–461.

22

A Fractal Analysis of Binding and Dissociation Kinetics of Glucose and Related Analytes on Biosensor Surfaces at the Nanoscale Level

Neeti Sadana, Tuan Vo-Dinh, and Ajit Sadana

CONTENTS

A biosensor is "a sensing device with a biological or biologically derived sensing element, which is integrated within or intimately associated with a physical transducer." Biosensors produce discrete or continuous digital electronic signals in proportion to the concentration of an analyte or a group of analytes. The biological component of the sensor can be an enzyme, an antibody, or any biological element that can detect a species.

Biosensors are finding increasing demand in the health-care industry for blood glucose monitoring or for detecting prostate-specific antigen. They are being used extensively in the environmental monitoring of biochemical oxygen demand (BOD), in the food processing industry for food pathogen detection, in military purposes for monitoring biological and chemical warfare, and in the agricultural industry to control crop diseases, plant nutrients, etc.

Detection of analytes related to human health has acquired great importance in recent years. This chapter describes the detection of glucose in diabetic patients. The glucose biosensor has been developed based on carbon nanotube (CNT) nanoelectrode ensembles (NEEs) (Lin et al., 2004). Considerable research has been done in applying CNTs as electrochemical biosensors. Glucose can also be detected using glucose binding protein as the receptor immobilized on a surface plasmon resonance (SPR) biosensor surface (Hsieh et al., 2004). Compounds similar to glucose such as acetaminophen, ascorbic acid, and uric acid can also be detected.

connective tissue, the ratio of the binding and the dissociation rate coefficient (k_1/k_d) is given by

$$(k_1/k_d) = (5.149 \pm 1.912)(D_{f1}/D_{fd})^{1.371 \pm 0.1036} \qquad (22.5)$$

The fit is quite good. Only three data points are available. The availability of more data points would lead to a more reliable fit. Note that the data for glucose and insulin are plotted together. The ratio of the binding and the dissociation rate coefficient (k_1/k_d) exhibits an order of dependence between first and one and one-half order (equal to 1.371) on the ratio of the fractal dimensions (D_{f1}/D_{fd}) that exists on the biosensor surface.

Figure 22.2c and Tables 22.1 and 22.2 show the increase in the ratio of the binding and the dissociation rate coefficient (k_2/k_d) with an increase in the ratio of the fractal dimensions (D_{f2}/D_{fd}). For the data presented in Tables 22.1 and 22.2 and for glucose and insulin present in plasma, and for glucose in interstitial adipose tissue and in interstitial connective tissue, the ratio of the binding and the dissociation rate coefficient (k_2/k_d) is given by

$$(k_2/k_d) = (4.698 \pm 2.022)(D_{f2}/D_{fd})^{1.628 \pm 0.2034} \qquad (22.6)$$

The fit is quite good. Only three data points are available. The availability of more data points would lead to a more reliable fit. Note that the data for glucose and insulin are plotted together. The ratio of the binding and the dissociation rate coefficient (k_2/k_d) exhibits an order of dependence between one and one-half order and second order (equal to 1.628) on the ratio of the fractal dimensions (D_{f2}/D_{fd}) that exists on the biosensor surface. This is slightly more than the order (equal to 1.371) exhibited by the ratio (k_1/k_d) on the ratio of the fractal dimensions (D_{f1}/D_{fd}) that exists on the biosensor surface.

Hsieh et al. (2004) have detected glucose using glucose/galactose-binding protein (GGBP) as the receptor immobilized on an SPR biosensor surface. These authors indicate that the detection of low-molecular-weight analytes such as glucose (180 Da) by an SPR biosensor is difficult since the molecules have insufficient mass to provide a measurable change in the refractive index. These authors have used unlabeled GGBP combined with SPR for the direct detection of glucose.

These authors indicate that GGBP is a bacterial periplasmic binding protein. Upon binding of its ligand proteins, glucose or galactose GGBP exhibits a hinge-twist conformational change (Gerstein et al., 1994; Zou et al., 1993). This conformational change may be used to detect the binding of glucose (Salins et al., 2001) and galactose (Zukin et al., 1977).

Figure 22.3 shows the binding and dissociation of 100 μM glucose in solution to thiol-coupled E149C GGBP (~10,000 RU) immobilized on a CM5 sensor chip. The GGBP was engineered to bind in the physiological range by mutation at additional sites (Hsieh et al., 2004). The E149 is one such mutation site. A dual-fractal analysis is required to adequately describe the binding kinetics. A single-fractal analysis is adequate to describe the dissociation kinetics. The values of (a) the binding rate coefficient, k, and the fractal dimension, D_f, for a single-fractal analysis and (b) the binding rate coefficient, k_1 and k_2, for a dual-fractal analysis and the fractal dimensions, D_{f1} and D_{f2}, for a dual-fractal analysis are given in Tables 22.3 and 22.4. Only one set of data points is available.

It is of interest to note that as the fractal dimension increases by a factor of 1.569 from a value of D_{f1} equal to 1.7992 to D_{f2} equal to 2.824, the binding rate coefficient increases by a factor of 3.31 from a value of k_1 equal to 27.851 to k_2 equal to 92.140.

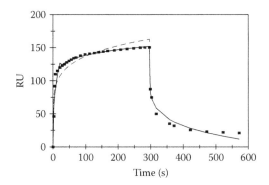

FIGURE 22.3

Binding and dissociation of 100 μM glucose in solution to thiol-coupled E149C GGBP immobilized on a CM5 sensor chip. (Adapted from Hsieh, H.V. et al., 2004. *Biosensors and Bioelectronics*, 9, 653–660.)

The ratio of the binding rate coefficient(s) to the dissociation rate coefficient(s) k_1/k_d is equal to 0.493 and k_2/k_d is equal to 1.632.

Cai et al. (2004) have developed a wireless, remote-query glucose biosensor. The sensor uses a ribbon-like, mass-sensitive magnetoelastic sensor as a transducer. The magnetoelastic sensor is initially coated with a pH-sensitive polymer. A glucose oxidase (GO_x) layer is then coated on the pH-sensitive polymer. These authors indicate that the GO_x-catalyzed oxidation of glucose produces gluconic acid. This induces the pH-responsive polymer to shrink, which decreases the polymer mass. The authors indicate that the magnetoelastic sensor vibrations (characteristic resonance frequency) are inversely dependent on sensor mass loading, and these changes in resonance frequency (from a passive magnetoelastic transducer) may be detected on a remote basis. These authors further emphasize that compounds in clinical samples such as uric acid, acetaminophen, and ascorbic acid may interfere with accurate glucose detection.

Cai et al. (2004) indicate that the normal serum glucose concentration is in the 3.8- to 6.1-mmol/L range under physiological conditions. When the blood glucose concentration is generally higher than 9 mmol/L, then diabetic urine is present. In order that physiological

TABLE 22.3

Binding and Dissociation Rate Coefficients for Glucose in Solution to GGBP Thiol-Immobilized on a CM5 Sensor Chip Surface

k	k_1	k_2	k_d
66.178 ± 9.366	40.325 ± 10.131	93.533 ± 0.668	56.474 ± 3.522

Source: Hsieh, H.V. et al., 2004. *Biosensors and Bioelectronics*, 9, 653–660.

TABLE 22.4

Fractal Dimension Values for the Binding of Glucose in Solution to GGBP Thiol-Immobilized on a CM5 Sensor Chip Surface

D_f	D_{f1}	D_{f2}	D_{fd}
2.6838 ± 0.0414	2.2472 ± 0.2376	2.8297 ± 0.0043	2.6802 ± 0.0268

Source: Hsieh, H.V. et al., 2004. *Biosensors and Bioelectronics*, 9, 653–660.

conditions may be approached with their electrochemical biosensor, Cai et al. (2004) added 0.15 mol/L NaCl, and calibrated their biosensor in the glucose concentration range of 1–15 mmol/L.

Figure 22.4a shows the binding of 1 mmol/L glucose in solution to the sensor that is coated with 0.3 mg of GO$_x$, 0.03 mg of catalase, 0.75 mg of BSA, and 0.3 mg of glutaric aldehyde. A dual-fractal analysis is required to adequately describe the binding kinetics. The values of (a) the binding rate coefficient, k, and the fractal dimension, D$_f$, for a single-fractal analysis and (b) the binding rate coefficients, k$_1$ and k$_2$, and the fractal dimensions, D$_{f1}$ and D$_{f2}$, for a dual-fractal analysis are given in Table 22.5. It is of interest to note that as the fractal dimension increases by a factor of 13.023 from a value of D$_{f1}$ equal to 0.1750 to D$_{f2}$ equal

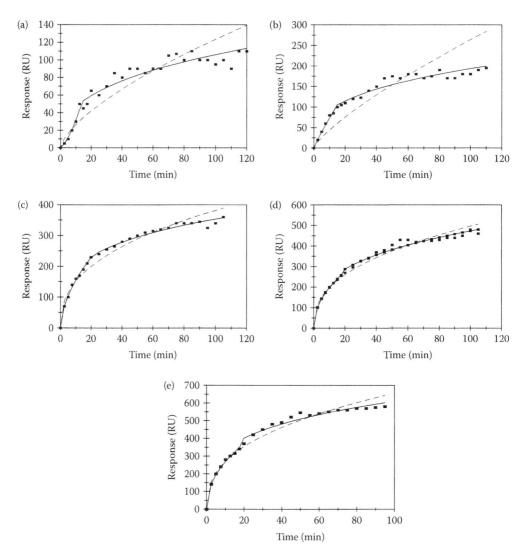

FIGURE 22.4
Binding of different concentrations (in mmol/L) of glucose in solution to the wireless, remote-query glucose biosensor: (a) 1, (b) 4, (c) 7, (d) 10, and (e) 15. (Adapted from Cai, Q. et al., 2004. *Analytical Chemistry*, 76, 4038–4043.)

TABLE 22.5

Rate Coefficients and Fractal Dimensions for the Binding of Different Concentrations of Glucose in Solution (in mmol/L) to 0.3 mg of Catalase, 0.75 mg of BSA (Bovine Serum Albumin), and 0.3 mg of Glutaric Dialdehyde in a Coating on a Magnetoelastic Sensor with a pH-Sensitive Polymer

Analyte (Glucose) Concentration (mmol/L)	k	k_1	k_2	D_f	D_{f1}	D_{f2}
1	5.658 ± 1.879	1.2184 ± 0.198	20.184 ± 2.049	1.661 ± 0.106	0.175 ± 0.238	2.2791 ± 0.058
4	7.461 ± 5.102	9.158 ± 0.666	43.51 ± 3.11	1.451 ± 0.167	1.1914 ± 0.095	2.3522 ± 0.05
7	60.04 ± 5.496	41.852 ± 1.634	106.71 ± 2.91	2.197 ± 0.035	1.8664 ± 0.041	2.4799 ± 0.031
10	78.66 ± 4.912	66.199 ± 1.255	114.71 ± 5.03	2.199 ± 0.024	2.051 ± 0.022	2.3845 ± 0.042
15	111.95 ± 7.64	94.92 ± 2.685	184.213 ± 6.9	2.232 ± 0.028	2.094 ± 0.033	2.4797 ± 0.039

Source: Cai, Q. et al., 2004. *Analytical Chemistry*, 76, 4038–4043.

to 2.2791, the binding rate coefficient increases by a factor of 16.56 from a value of k_1 equal to 1.2184 to k_2 equal to 20.184. An increase in the degree of heterogeneity on the electrochemical biosensor surface leads once again to an increase in the binding rate coefficient.

Figure 22.4b shows the binding of 4 mmol/L glucose in solution to the sensor that is coated with 0.3 mg of GO_x, 0.03 mg of catalase, 0.75 mg of BSA, and 0.3 mg of glutaric aldehyde. A dual-fractal analysis is required to adequately describe the binding kinetics. The values of (a) the binding rate coefficient, k, and the fractal dimension, D_f, for a single-fractal analysis and (b) the binding rate coefficients, k_1 and k_2, and the fractal dimensions, D_{f1} and D_{f2}, for a dual-fractal analysis are given in Table 22.5. It is of interest to note that as the glucose concentration in solution increases by a factor of 4 from 1 to 4 mmol/L, (a) the binding rate coefficient, k_1, increases by a factor of 7.516 from a value of 1.2184 to 9.1577 and (b) the binding rate coefficient, k_2, increases by a factor of 2.155 from a value of 20.184 to 43.056.

Figure 22.4c shows the binding of 7 mmol/L glucose in solution to the sensor that is coated with 0.3 mg of GO_x, 0.03 mg of catalase, 0.75 mg of BSA, and 0.3 mg of glutaric aldehyde. A dual-fractal analysis is required to adequately describe the binding kinetics. The values of (a) the binding rate coefficient, k, and the fractal dimension, D_f, for a single-fractal analysis and (b) the binding rate coefficients, k_1 and k_2, and the fractal dimensions, D_{f1} and D_{f2}, for a dual-fractal analysis are given in Table 22.5. It is of interest to note that as the glucose concentration in solution increases by a factor of 7 from 1 to 7 mmol/L, (a) the binding rate coefficient, k_1, increases by a factor of 34.349 from a value of 1.2184 to 41.852 and (b) the binding rate coefficient, k_2, increases by a factor of 5.286 from a value of 20.184 to 106.709.

Figure 22.4d shows the binding of 10 mmol/L glucose in solution to the sensor that is coated with 0.3 mg of GO_x, 0.03 mg of catalase, 0.75 mg of BSA, and 0.3 mg of glutaric aldehyde. A dual-fractal analysis is required to adequately describe the binding kinetics. The values of (a) the binding rate coefficient, k, and the fractal dimension, D_f, for a single-fractal analysis and (b) the binding rate coefficients, k_1 and k_2, and the fractal dimensions, D_{f1} and D_{f2}, for a dual-fractal analysis are given in Table 22.5. It is of interest to note that as the glucose concentration in solution increases by a factor of 10 from 1 to 10 mmol/L, (a) the binding rate coefficient, k_1, increases by a factor of 54.33 from a value of 1.2184 to 66.199 and (b) the binding rate coefficient, k_2, increases by a factor of 5.683 from a value of 20.184 to 114.714. Once again, and as indicated above, an increase in the glucose concentration in solution leads to an increase in the values of the binding rate coefficients, k_1 and k_2.

Figure 22.4e shows the binding of 15 mmol/L glucose in solution to the sensor that is coated with 0.3 mg of GO$_x$, 0.03 mg of catalase, 0.75 mg of BSA, and 0.3 mg of glutaric aldehyde. A dual-fractal analysis is required to adequately describe the binding kinetics. The values of (a) the binding rate coefficient, k, and the fractal dimension, D$_f$, for a single-fractal analysis and (b) the binding rate coefficients, k$_1$ and k$_2$, and the fractal dimensions, D$_{f1}$ and D$_{f2}$, for a dual-fractal analysis are given in Table 22.5. It is of interest to note that as the glucose concentration in solution increases by a factor of 15 from 1 to 15 mmol/L, (a) the binding rate coefficient, k$_1$, increases by a factor of 77.9 from a value of 1.2184 to 94.918 and (b) the binding rate coefficient, k$_2$, increases by a factor of 9.13 from a value of 20.184 to 184.213.

Figure 22.5a shows the increase in the binding rate coefficient, k$_1$, with an increase in the glucose concentration in solution in the range 1–15 mmol/L. In this concentration range, the binding rate coefficient, k$_1$, is given by

$$k_1 = (1.1764 \pm 0.3813)[glucose]^{1.6910 \pm 0.1335} \tag{22.7}$$

The fit is very good. The binding rate coefficient, k$_1$, exhibits an order of dependence between one and one-half order and second order (equal to 1.6910) on the glucose concentration in solution. The noninteger order of dependence exhibited by the binding rate coefficient, k$_1$, on the glucose concentration in solution lends support to the fractal nature of the system.

Figure 22.5b shows the increase in the binding rate coefficient, k$_2$, with an increase in the glucose concentration in solution in the range 1–15 mmol/L. In this concentration range, the binding rate coefficient, k$_2$, is given by

$$k_2 = (18.283 \pm 4.074)[glucose]^{0.8241 \pm 0.0956} \tag{22.8}$$

The fit is good. The binding rate coefficient, k$_2$, exhibits an order of dependence less than first order (equal to 0.8241) on the glucose concentration in solution. The noninteger order of dependence exhibited by the binding rate coefficient, k$_2$, on the glucose concentration in solution, once again, lends support to the fractal nature of the system.

It is of interest to note that the order of dependence exhibited by k$_2$ (equal to 0.8241) is less than half of that exhibited by k$_1$ (equal to 1.6910) on the glucose concentration in solution. In general, the values of k$_2$ are higher than those of k$_1$ for any particular analyte–receptor reaction occurring on biosensor surfaces, and this is reflected in the constant values for k$_1$ (equal to 1.1764) and k$_2$ (equal to 18.283), respectively.

Figure 22.5c shows the increase in the ratio of the binding rate coefficients, k$_1$/k$_2$, with an increase in the ratio of the fractal dimensions, D$_{f1}$/D$_{f2}$. For the data given in Table 22.5, the ratio of the binding rate coefficients, k$_1$/k$_2$, is given by

$$k_1/k_2 = (0.5371 \pm 0.1469)(D_{f1}/D_{f2})^{0.8812 \pm 0.1175} \tag{22.9}$$

The fit is quite good. There is some scatter in the data, and this is reflected in the error in the constant (0.5371 ± 0.1469). The ratio of the binding rate coefficients, k$_1$/k$_2$, exhibits an order of dependence less than first order (equal to 0.8812) on the ratio of the fractal dimensions, D$_{f1}$/D$_{f2}$.

Lin et al. (2004) have developed a glucose biosensor based on CNT NEEs. These authors indicate that recently the electrochemical properties of CNTs have come into prominence. Considerable research has been done in applying CNTs as electrochemical biosensors

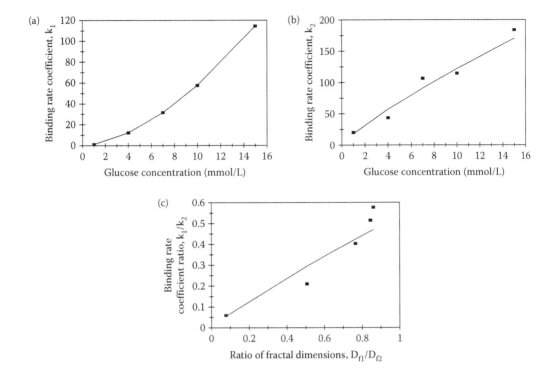

FIGURE 22.5
(a) Increase in the binding rate coefficient, k_1, with an increase in the glucose concentration in solution in the range 1–15 mmol/L. (b) Increase in the binding rate coefficient, k_2, with an increase in the glucose concentration in solution in the range 1–15 mmol/L. (c) Increase in the ratio of the binding rate coefficients, k_1/k_2, with an increase in the ratio of the fractal dimensions, D_{f1}/D_{f2}.

(Azmian et al., 2002; Li et al., 2003; Nguyen et al., 2002; Shim et al., 2002; Sotiropoulou et al., 2003; Yu et al., 2003). Lin et al. (2004) emphasize that their CNT biosensor is able to detect glucose effectively in the presence of common interferents such as acetaminophen, uric acid, and ascorbic acid. These authors emphasize that their biosensor is able to operate without permselective membrane barriers and artificial electron mediators. This simplifies the design and fabrication of the biosensor.

Lin et al. (2004) compared the responses for appropriate physiological levels of glucose, acetaminophen, ascorbic acid, and uric acid at a potential of +0.4 and −0.2 V. These authors noted that at +0.4 V there was significant interference from ascorbic acid, uric acid, and acetaminophen during the detection of glucose. This interference was substantially reduced when operating at −0.2 V. Their glucose biosensor is based on a CNT NEE. The two steps involved include (a) the electrochemical treatment of the CNT NEE for functionalization and (b) the coupling of the glucose oxidase to the functionalized CNT NEE.

Figure 22.6a shows the binding and dissociation of glucose in solution to the CNT NEE biosensor (Lin et al., 2004). A single-fractal analysis is adequate to describe the binding and the dissociation kinetics. The values of the binding rate coefficient, k, and the fractal dimension, D_f, for a single-fractal analysis are given in Tables 22.6 and 22.7.

Figure 22.6b shows the binding and dissociation of ascorbic acid in solution to the CNT NEE glucose biosensor (Lin et al., 2004). A dual-fractal analysis is required to adequately describe the binding kinetics. A single-fractal analysis is adequate to describe the dissociation kinetics. The values of (a) the binding rate coefficient, k, and the fractal dimension,

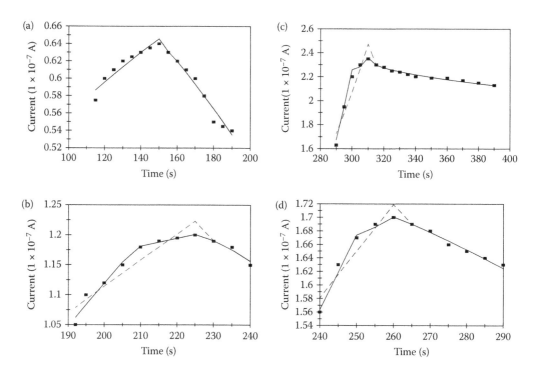

FIGURE 22.6

(a) Binding and dissociation of glucose in solution to the CNT NEE biosensor. (Adapted from Lin, Y. et al., 2004. *Nano Letters*, 4(2), 191–195.) (b) Binding and dissociation of ascorbic acid in solution to the CNT NEE biosensor. (c) Binding and dissociation of acetaminophen in solution to the CNT NEE biosensor. (d) Binding and dissociation of uric acid in solution to the CNT NEE biosensor.

D_f, for a single-fractal analysis, (b) the binding rate coefficients, k_1 and k_2, and the fractal dimensions, D_{f1} and D_{f2}, for a dual-fractal analysis, and (c) the dissociation rate coefficient, k_d, and the fractal dimension, D_{fd}, are given in Tables 22.6 and 22.7. Note that a dual-fractal analysis is required to adequately model the binding kinetics for the interferent ascorbic acid, whereas a single-fractal analysis is adequate to describe the binding kinetics for glucose. This indicates a possible change in the binding mechanism.

Figure 22.6c shows the binding and dissociation of acetaminophen acid in solution to the CNT NEE glucose biosensor (Lin et al., 2004). Once again, a dual-fractal analysis is required to adequately describe the binding kinetics of this interferent. A single-fractal analysis is adequate to describe the dissociation kinetics. The values of (a) the binding rate coefficient, k, and the fractal dimension, D_f, for a single-fractal analysis, (b) the binding rate coefficients, k_1 and k_2, and the fractal dimensions, D_{f1} and D_{f2}, for a dual-fractal analysis, and (c) the dissociation rate coefficient, k_d, and the fractal dimension, D_{fd}, are given in Tables 22.6 and 22.7. It is of interest to note that to describe the binding kinetics of both interferents (for the detection of glucose), a dual-fractal analysis is required to describe the binding kinetics, whereas a single-fractal analysis is adequate to describe the binding kinetics for glucose.

Figure 22.6d shows the binding and dissociation of uric acid in solution to the CNT NEE glucose biosensor (Lin et al., 2004). Once again, a dual-fractal analysis is required to adequately describe the binding kinetics of this interferent. A single-fractal analysis is adequate to describe the dissociation kinetics. The values of (a) the binding rate coefficient, k, and the fractal dimension, D_f, for a single-fractal analysis, (b) the binding rate coefficients, k_1 and k_2,

TABLE 22.6

Binding and Dissociation Rate Coefficients for Glucose, Ascorbic Acid, Acetaminophen, and Uric Acid in Solution to a Nanoelectrode (NEE) Glucose Biosensor

Compound	k	k_1	k_2	k_d
Glucose	0.1057 ± 0.0012	na	na	0.00139 ± 0.00017
Ascorbic acid	0.0169 ± 0.00029	0.00133 ± 0.00002	0.3311 ± 0.0005	0.00095 ± 0.00028
Uric acid	$8.8 \times 10^{-14} \pm 5 \times 10^{-15}$	2.7×10^{-22}	2.3×10^{-5}	0.0229 ± 0.0017
Acetaminophen	0.0052 ± 0.00007	0.000165 ± 0.000001	0.1359 ± 0.0004	0.00166 ± 0.00017

Source: Lin, Y. et al., 2004. *Nano Letters*, 4(2), 191–195.

TABLE 22.7

Fractal Dimensions in the Binding and in the Dissociation Phase for Glucose, Ascorbic Acid, Acetaminophen, and Uric Acid in Solution to a Nanoelectrode (NEE) Glucose Biosensor

Compound	D_f	D_{f1}	D_{f2}	D_{fd}
Glucose	2.2776 ± 0.0928	na	na	0.6544 ± 0.1252
Ascorbic acid	1.4185 ± 0.2278	0.6185 ± 0.3224	2.5244 ± 0.4756	0.1782 ± 0.6622
Uric acid	~0	~0	~0	1.9644 ± 0.0514
Acetaminophen	0.9122 ± 0.4238	$0.0 + 0.5360$	2.0912 ± 0.17164	0.7578 ± 0.131

Source: Lin, Y. et al., 2004. *Nano Letters*, 4(2), 191–195.

and the fractal dimensions, D_{f1} and D_{f2}, for a dual-fractal analysis, and (c) the dissociation rate coefficient, k_d, and the fractal dimension, D_{fd}, are given in Tables 22.6 and 22.7. It is of interest to note, once again, that to describe the binding kinetics of all three interferents (for the detection of glucose), a dual-fractal analysis is required to describe the binding kinetics, whereas a single-fractal analysis is adequate to describe the binding kinetics for glucose.

Tables 22.6 and 22.7 and Figure 22.7a show that the binding rate coefficient, k_2, increases as the fractal dimension, D_{f2}, increases. For the data presented in Figure 22.7a, the binding rate coefficient, k_2, is given by

$$k_2 = (0.0138 \pm 0.0029)D_{f2}^{3.284 \pm 0.1059} \tag{22.10}$$

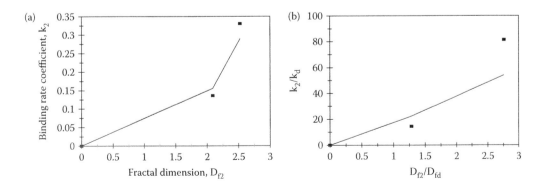

FIGURE 22.7

(a) Increase in the binding rate coefficient, k_2, with an increase in the fractal dimension, D_{f2}. (b) Increase in the ratio of the binding and the dissociation rate coefficient (k_2/k_d) with an increase in the fractal dimension ratio (D_{f2}/D_{fd}).

The fit is quite good. Only three data points are available. Tables 22.6 and 22.7 indicate that one of the data points is actually $D_{f2} \sim 0$. In order that this point may also be used in Figure 22.7a, an arbitrary very small value of $D_{f2} = 1.0 \times 10^{-6}$ was also used. The selection of this very low value point did not make a difference in the fit of the line. The binding rate coefficient, k_2, is sensitive to the degree of heterogeneity that exists on the CNT surface, as noted by the greater than third order dependence on the fractal dimension, D_f. It should be pointed out that the data referred to here are for the interferents, and not for glucose itself.

Tables 22.6 and 22.7 and Figure 22.7b show that the ratio of the binding and the dissociation rate coefficient, k_2/k_d, increases as the fractal dimension, D_{f2}/D_{fd}, increases. For the data presented in Figure 22.7b, the ratio of the binding to the dissociation rate coefficient, k_2/k_d, is given by

$$k_2/k_d = (16.374 \pm 13.723)(D_{f2}/D_{fd})^{1.156 \pm 0.0603} \tag{22.11}$$

The fit is quite good. Only three data points are available. Tables 22.6 and 22.7 indicate that one of the data points is actually $D_{f2} \sim 0$. In order that this point may also be used, an arbitrary very small value of $D_{f2}/D_{fd} = 1 \times 10^{-6}$ was also used. The selection of this very low value point did not make a difference in the fit of the line. The ratio of the binding to the dissociation rate coefficient, k_2/k_d, exhibits an order of dependence slightly higher than first (equal to 1.156) order on the fractal dimension ratio, D_{f2}/D_{fd}. Once again, it should be pointed out that the data referred to here are for interferents, and not for glucose itself.

Cash and Clark (2010) indicate that diabetes is a rapidly growing problem, and it is estimated that 44.1 million people will be affected by 2034. The estimated cost for the that year is $336 billion (in 2007 dollars) (Huang et al., 2009). Diabetes, though relatively easy to manage, leads to lower limb amputations, cardiovascular disease, and blindness.

Diabetics generally take discrete blood samples during the day. By this method, they may miss overlooking hypo- and hyperglycemic excursions between sampling points. Cash and Clark (2010) present continuous monitoring data that permit the more frequent measurements of blood glucose concentration.

Figure not shown. A dual-fractal analysis is required to adequately fit the binding curve. A single-fractal analysis is required to fit the dissociation curve. The values of (a) the binding rate coefficient, k, and the fractal dimension, D_f, for a single-fractal analysis, (b) the binding rate coefficients, k_1 and k_2, and the fractal dimensions, D_{f1} and D_{f2}, for a dual-fractal analysis, and (c) the dissociation rate coefficient, k_d, and the fractal dimension for dissociation, D_{fd}, for a single-fractal analysis are given in Tables 22.8 and 22.9.

Wang et al. (2003) have developed a novel MWNT-based biosensor for glucose detection. Their MWNT-based biosensor exhibits a strong glucose response. They indicate that only recently CNTs have been shown for glucose and DNA detection. CNTs may be considered as a result of folding layers into carbon cylinders. They have presented binding rate for repeat runs. Here, we present data for duplicate runs.

TABLE 22.8

Binding Rate Coefficients for the Continuous Monitoring of Glucose

Analyte	k	k_1	k_2
Glucose	47.09 ± 15.96	48.05 ± 21.23	61.939 ± 2.595

Source: Cash, K.J. and H.A. Clark, 2010. *Trends in Molecular Medicine*, 16(2), 584–593.

TABLE 22.9

Fractal Dimensions for the Binding during the Continuous
Monitoring of Glucose

Analyte	D_f	D_{f1}	D_{f2}
Glucose	$0.0002 + 0.1312$	$0 + 0.608$	0.708 ± 0.278

Source: Cash, K.J. and H.A. Clark, 2010. *Trends in Molecular Medicine*, 16(2), 584–593.

Figure not shown. A dual-fractal analysis is required to adequately describe the binding kinetics. This is Run #1. The values of (a) the binding rate coefficient, k, and the fractal dimension, D_f, for a single-fractal analysis and (b) the binding rate coefficients, k_1 and k_2, and the fractal dimensions, D_{f1} and D_{f2}, are given in Tables 22.10 and 22.11. It is of interest to note that for a dual-fractal analysis as the fractal dimension increases by a factor of 22 from D_{f1} equal to 0.1 to D_{f2} equal to 0.22 the binding rate coefficient increases by a factor of 3.8 from a value of k_1 equal to 0.06 to k_2 equal to 0.228. An increase in the degree of heterogeneity on the MWNT biosensor surface leads to an increase in the binding rate coefficient.

Figure not shown. A dual-fractal analysis is, once again, required to adequately describe the binding kinetics. This is Run #2. The values of (a) the binding rate coefficient, k, and the fractal dimension, D_f, for a single-fractal analysis and (b) the binding rate coefficients, k_1 and k_2, and the fractal dimensions, D_{f1} and D_{f2}, are given in Tables 22.10 and 22.11. It is of interest to note that for a dual-fractal analysis as the fractal dimension increases by a factor

TABLE 22.10

Binding Rate Coefficients (Amperometric Response) of 2 mM Glucose in 0.1 M Phosphate Buffer Solution to an MWNT-Based Biosensor at an Applied Potential of 0.45 versus Ag/AgCl at Room Temperature of 25°C (Influence of Repeat Runs)

Glucose/MWNT Biosensor	k	k_1	k_2
Run #1	0.0909 ± 0.0270	0.06 ± 0	0.2280 ± 0.0104
Run #2	0.1886 ± 0.0748	0.1 ± 0	0.5 ± 0

Source: Wang, G. et al., 2003. *Biochemical and Biophysical Research Communications*, 311, 572–576.

TABLE 22.11

Fractal Dimensions for the Binding Rate Coefficients (Amperometric Response) of 2 mM Glucose in 0.1 M Phosphate Buffer Solution to an MWNT-Based Biosensor at an Applied Potential of 0.45 versus Ag/AgCl at Room Temperature of 25°C (Influence of Repeat Runs)

Glucose/MWNT Biosensor	D_f	D_{f1}	D_{f2}
Run #1	1.8456 ± 0.1452	$0.1 \pm 1.21 \times 10^{-5}$	0.2280 ± 0.0104
Run #2	2.2672 ± 0.1888	$1.0 \pm 2.6 \times 10^{-15}$	3.0 ± 0

Source: Wang, G. et al., 2003. *Biochemical and Biophysical Research Communications*, 311, 572–576.

of 3 from D_{f1} equal to 1.0 to D_{f2} equal to 3.0 the binding rate coefficient increases by a factor of 5 from a value of k_1 equal to 0.1 to k_2 equal to 0.5. This is for Run #2.

22.4 Conclusions

A fractal analysis is used to model the binding and dissociation kinetics of connective tissue interstitial glucose, adipose tissue interstitial glucose, insulin, and other related analytes on biosensor surfaces. The analysis provides insights into diffusion-limited analyte–receptor reactions occurring on heterogeneous biosensor surfaces. The fractal analysis provides a useful lumped parameter(s) analysis for the diffusion-limited reaction occurring on a heterogeneous surface via the fractal dimension and the rate coefficient. It is a convenient means to make the degree of heterogeneity that exists on the surface more quantitative.

Numerical values obtained for the binding and the dissociation rate coefficients are linked to the degree of heterogeneity or roughness (fractal dimension, D_f) present on the biosensor chip surface. The binding and the dissociation rate coefficients are sensitive to the degree of heterogeneity on the surface. The analysis of both binding and dissociation steps describes a clearer picture of the reaction occurring on the surface providing values for ratio of k_1/k_d. Quantitative expressions are developed for (a) k_1 as a function of D_{f1}, (b) k_1 as a function of concentration, (c) k_d as a function of D_{fd}, and (d) k_1/k_d as a function of D_{f1}/D_{fd}.

The values of binding rate coefficient, k, linked with the degree of heterogeneity, D_f, existing on the biosensor surface provide a complete picture of the reaction kinetics occurring on the sensor chip surface. Dual-fractal analysis is used only when the single-fractal analysis did not provide an adequate fit. This was done by regression analysis provided by Quattro Pro 8.0.

It is suggested that roughness on a surface leads to turbulence that enhances mixing and decreases diffusional limitations, leading to an increase in the binding rate coefficients (Martin et al., 1991). The analysis also indicates that along with glucose similar compounds like insulin, acetaminophen, ascorbic acid, and uric acid can also be detected on the same sensor surface whose detection may be useful for other diagnosis such as diabetic urine, etc.

References

American Diabetes Association, 2003. *Diabetic Care*, 26(2), 3359–3360.
Azmian, B.R., J.J. Davis, K.S. Coleman, C.B. Bagshaw, and M.L.H. Green, 2002. *Journal of the American Chemical Society*, 124, 12664–12665.
Butala, H.D., A. Ramakrishnan, and A. Sadana, 2003a. *Biosystems*, 79, 235.
Butala, H.D., A. Ramakrishnan, and A. Sadana, 2003b. *Sensors & Actuators*, 88, 266.
Cai, Q., K. Zeng, C. Ruan, T.A. Sesai, and C.A. Grimes, 2004. *Analytical Chemistry*, 76, 4038–4043.
Cash, K.J. and H.A. Clark, 2010. *Trends in Molecular Medicine*, 16(2), 584–593.
Gerstein, M., A.M. Lesk, and C. Chothia, 1994. *Biochemistry*, 33, 6739–6749.

Harsanyi, G., 2000. *Biomedical Applications: Fundamentals, Technology, and Applications.* Technomic Publications, Lancaster, PA.

Havlin, S., 1989. *The Fractal Approach to Heterogeneous Chemistry: Surface, Colloids, Polymers.* Wiley, New York, NY, pp. 251–269.

Hsieh, H.V., Z.A. Pfeiffer, D.B. Sherman, and J.B. Pitner, 2004. Direct detection of glucose by surface plasmon resonance with bacterial glucose/galactosidase protein. *Biosensors and Bioelectronics,* 19, 653–660.

Huang, E.S., A. Basu, M. O'Grady, and J.C. Capretta, 2009. *Diabetes Care,* 32, 2225–2229.

Lee, C.K. and S.L. Lee, 1995. *Surface Science,* 325, 294.

Leegsma-Vogt, G., M.M. Rhemrev-Boom, R.G. Tiessen, K. Venema, and J. Korf, 2004. *Bio-Medical Materials and Engineering,* 14, 455–464.

Li, J., H.T. Ng, A. Casell, W. Fan, H. Chen, Q. Ye, J. Koehne, J. Han, and M. Meyappan, 2003. *Nano Letters,* 3, 597–602.

Lin, Y., F. Lu, Y. Tu, and Z. Ren, 2004. *Nano Letters,* 4(2), 191–195.

Martin, S.J., V.E. Granstaff, and G.C. Frye, 1991. *Analytical Chemistry,* 65, 2910–2922.

Nguyen, C.V., C. Delzeit, A.M. Cassell, J. Li, J. Han, and M. Meyappan, 2002. *Nano Letters,* 2, 1079–1981.

Pei, J., F. Tian, and T. Thundat, 2004. *Analytical Chemistry,* 76(2), 292.

Rhemberg-Boom, R., 1999. *Biocybernetics and Biomedical Engineering,* 19, 97–104.

Sadana, A. 2003. *Journal of Colloid & Interface Science,* 263, 420.

Salins, L.L.E., R.A. Ware, C.M. Ensor, and S. Daunert, 2001. *Analytical Biochemistry,* 294, 19–26.

Shim, M., N.W.S. Kam, R.J. Chen, Y.M. Li, and H.J. Dai, 2002. *Nano Letters,* 2, 285–288.

Sotiropoulou, S. and N.A. Chaniotakis, 2003. *Analytical and Bioanalytical Chemistry,* 375, 103–105.

Wang, G., Q. Zhang, R. Wang, and S.F. Yoon, 2003. *Biochemical and Biophysical Research Communications,* 311, 572–576.

Yonzon, C.R., C.L. Haynes, X. Zhang, J.T. Walsh, Jr., and R.P. van Duyne, 2004. *Analytical Chemistry,* 76, 78–85.

Yu, X., D. Chattopadhay, I. Galeska, F. Papadimitriakopoulos, and J.F. Rustling, 2003. *Electrochemical Communications,* 5, 408–411.

Zou, J.Y., M.M. Flocco, and S.L. Mowbray, 1993. *Journal of Molecular Biology,* 233, 739–752.

Zukin, R.S., P.R. Hartig, and D.E. Koshland, Jr., 1977. *Proceedings of the National Academy of Sciences of the United States of America,* 74, 1932–1936.

23

Integrated Cantilever-Based Biosensors for the Detection of Chemical and Biological Entities

Elise A. Corbin, Ashkan YekrangSafakar, Olaoluwa Adeniba,
Amit Gupta, Kidong Park, and Rashid Bashir

CONTENTS

23.1 Introduction

Integrated microsystem is an area of intense interest and has proven to be an arena where different fields (physical sciences, engineering, life sciences, medicine) can participate and provide their expertise and unique contributions.[1] Microsystems can be thought of as a single platform having many different functional components integrated together on the microscale, with the goal of miniaturizing laboratory processes, such as concentration and sensing of analytes, that would normally require it to be performed on the macroscale. Figure 23.1 is a schematic diagram showing an example of the conceptual view of an integrated microsystem. The use of microfabrication techniques can result in microfluidic devices and related sensors with very high sensitivity, in addition to reducing the total time to result for chemical and biological analysis. The low cost due to batch fabrication and the reduction in the sample size required for analysis due to miniaturized sensors are also very attractive.

Microscale cantilever beams were first introduced to the nanotechnology field with their use as force sensors in atomic force microscopy (AFM).[2] They have attracted a lot of interest in the MEMS and nanotechnology community in the past few years due to their simple structure and fabrication process flow and versatility in a wide variety of applications.

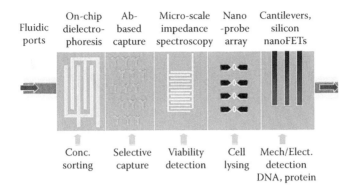

Fluidic ports	On-chip dielectro-phoresis	Ab-based capture	Micro-scale impedance spectroscopy	Nano -probe array	Cantilevers, silicon nanoFETs
	Conc. sorting	Selective capture	Viability detection	Cell lysing	Mech/Elect. detection DNA, protein

FIGURE 23.1
Schematic diagram of an integrated microsystem for the specific detection of biological entities.

These applications range from their use as probes in scanning probe microscopy (SPM)[3] to their use as micromechanical sensors to characterize a wide variety of analytes such as chemical vapors, bacterial cells, viruses, mammalian cells, etc. They are positioned as an ideal candidate to be integrated into microsystems to provide the capability of sensing and actuation.

The purpose of this chapter is to give an overview of various examples of applications of cantilever beams in biochemical sensing that has been reported and some brief future directions in this area of research.

23.2 Background

23.2.1 Motivation for Nanoscale-Thick Cantilever Beams

Decreasing the overall dimensions of the cantilever beam leads to an increase in its mechanical sensitivity to perturbations from the surroundings. This chapter will attempt to lay the groundwork for the basic design rules than can be followed for improving the sensitivity of cantilever-based sensors. There will be different design rules for different types of sensing schemes. But generally speaking, decreasing the dimensions of the cantilever beams can improve the performance of the cantilever beams. The thickness of the cantilever beam is of more interest, as it is the hardest to control during the fabrication process. The thickness is also the dimension that most affects the mechanical sensitivity of cantilever beams.[4] This issue will be discussed in more detail later in the chapter where appropriate.

One issue for cantilever beams with nanoscale thickness has been the problem of low quality factor.[5–8] This issue has been attributed to surface energy loss due to factors such as surface defects and adsorbates,[8] but the mechanism is still not clear. Groups have reported improvements in Q-factors by a factor of around three after annealing either in a N_2 environment[5] or in under ultrahigh vacuum conditions.[8]

23.2.2 Microfabrication Methods of Ultra-Thin Cantilever Beams

Single-crystal materials are preferred materials to make sensor elements due to their high mechanical quality factor.[9] Silicon is usually preferred for fabricating sensor elements due

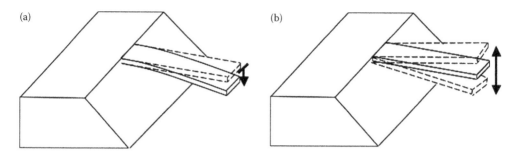

FIGURE 23.2
Schematic diagram representing cantilever beam operating in (a) static mode and (b) dynamic mode.

to advantages such as low stress and controlled material quality, using currently available very-large-scale integration (VLSI) circuit fabrication facilities, miniaturization of devices, high control of dimensions, and the economic advantage of batch fabrication. Various methods for the fabrication of ultrathin cantilever beams have been reported. Among the works that reported fabricating silicon cantilever beams, virtually all of them employ a silicon-on-insulator (SOI) wafer as the starting material.[10–15] Where it has been reported, the SOI wafers used in these processes were obtained from SIMOX (separation by ion implantation of oxygen) process. Work has also been reported on fabricating nanosized cantilever beams using other materials such as silicon nitride,[16] metals,[17] and polymers.[18]

23.2.3 Modes of Operation

The cantilever beam sensor element can be operated in two modes: the static mode (Figure 23.2a) and the dynamic mode (Figure 23.2b). The mode in which the cantilever beam is operated is usually determined by the target entities to be detected. For example, the static mode is usually used when cantilever beam is operated as a surface stress sensor to detect biomolecules in an aqueous environment. The cantilever beam is used in the dynamic mode when the goal is to detect mass change, with the detection being preferably performed in mediums with lower damping.

23.2.4 Deflection Detection Schemes

An important component of any sensor is the detection mechanism that converts the transduction signal into a readout signal that can be recorded and analyzed with high precision in real time. There are various methods to detect the deflection of cantilever beams such as optical reflection, capacitive, piezoelectric, electron tunneling, and piezoresistive.[19,20] Most of these methods can be used in both modes of operation (static or dynamic), as will be discussed below.

23.2.4.1 Optical Detection Methods

Among the optical reflection methods used, the optical lever method is by far the most prevalent.[21] In this method, a laser is focused on the free end of the cantilever beam, and the reflected beam is monitored using a position-sensitive detector (PSD). This method is most commonly used in commercially available atomic force microscope (AFM). Some of the advantages of this method include its simplicity to operate when the system is set up and its sensitivity that is limited by thermomechanical noise, which is able to measure displacements in the order of 0.1 nm. This method allows for the cantilevers to be monitored in the static and dynamic mode. One of the disadvantages is not being able to operate in opaque liquids or in liquids with changing refractive index. This method is also problematic when there is a need for monitoring the deflection of nanosized cantilever beams.[20] A further disadvantage is the low bandwidth, which limits the maximum frequency that can be measured to hundreds of kilohertz.

Another optical method that is commonly used is based on optical interferometry.[22,23] This method involves measuring the interference occurring between the reflected laser signal from the cantilever and a reference laser beam.[19] This technique allows for the monitoring of smaller area cantilever beams (in the size range of 1–4 μm^2) as well as with higher bandwidth in the megahertz range. The system setup is far more complex than the optical lever method. There is also a limit in the maximum displacement that can be measured, and hence, this approach is not conducive to static-mode operation when displacements may be in the micrometer range.

23.2.4.2 Capacitive Detection Method

The capacitive detection method is based on the parallel-plate capacitor principle, with the cantilever being one plate and a conductor on the fixed substrate being the other plate.[24] As the cantilever deforms, the gap between the plates changes, which in turn changes the capacitance, and can hence be monitored electrically. The advantages of this method include its sensitivity, and the possibility of integrating the cantilever beam fabrication with standard microelectronic fabrication processes. The disadvantages of this method is not being able to operate in fluids with changing dielectric constants, such as in liquids with ions, as well not being able to measure large displacements.[19]

23.2.4.3 Piezoelectric Detection Method

Piezoelectric effect is the phenomenon of transient charges being induced due to mechanical strain or deformation. The main advantage of this method is the ability to perform simultaneous sensing of the deformation and actuation of the cantilever.[25] One of the disadvantages of this method is the requirement of depositing thick composite layers on the cantilever beams, which may cause stress buildup leading to curtailment in their proper operation. This method is usually used when the cantilever is operated in the dynamic mode, and is not appropriate to be used in the static mode.

23.2.4.4 Electron Tunneling Detection Method

The measurement of the electron tunneling current between a cantilever beam and a conducting tip, separated by a gap in the nanometer range, has been demonstrated to measure the deflections of cantilever beams.[26] This method was used to monitor the deflection of cantilever beams in the earlier AFM systems.[2] The concept of the system is simple and the method is extremely sensitive to displacements. The method is particularly useful when monitoring the deflections of nanosized cantilever beams. The challenge of this method is that, as it is very sensitive to the material between the gaps, it is usually operated in air or vacuum, and is not conducive to be operated in liquid environments. This method also has the challenge when there is a need monitoring an array of cantilever beams, due to problems of electrically accessing individual cantilever beams.

23.2.4.5 Piezoresistive Detection Method

Piezoresistance is the phenomenon of the variation of conductivity (or resistivity) of a material as a function of applied stress.[27] Piezoresistors are usually fabricated at the fixed end of the cantilever beam where the stress is maximum, with the requirement of a reference resistor near the base of the cantilever, in order to measure the resistance change.[28] One of the advantages of this method is being able to operate the cantilever beam in opaque liquids such as blood as well as in conductive solutions (as long as the electrodes are properly passivated). As in the previous electrical-based methods, this method has the advantage of integrating the cantilever beam fabrication process with standard microelectronic fabrication processes. This method allows for the cantilever beam to be monitored in the static as well as the dynamic mode. One challenge with this method is the problem of using nanoscale-thick cantilever beams due to noise.[11] As piezoresistors are sensitive to temperature change, this method requires proper temperature control.

23.2.5 Applications Reported in Literature

In the last decade, numerous microcantilever sensors have been developed to detect various chemical and biological entities.[29–31] Microcantilever sensors can be largely divided by the operation mode and the measurement environment. The operation mode of microcantilever sensors can be either the static mode based on the cantilever deflection or the dynamic mode based on the resonant frequency. The measurement environment often determines the type of sample entities and limits the sensitivity of the microcantilever sensor. The measurement environment can be in vacuum, in air, or in liquid. In this section, the microcantilever sensors in each category will be reviewed and one exemplary work will be discussed in further detail.

In the static mode, the target entities attach on one side of the cantilever and change the surface stress. Owing to the misbalance of the surface stress on two sides of the cantilever, the cantilever deflects proportionally to the amount of the attached target entities. To maximize the sensitivity of the static-mode cantilever sensor, the probing molecules, which have high selectivity and affinity to the target entities, should be developed. Also, the dimension of the cantilever can be optimized for higher sensitivity. For instance, a thinner cantilever will bend more by the same surface stress. Furthermore, a longer cantilever will produce more displacement at the end of the cantilever. The static-mode cantilever sensors show strong parasitic deflection caused by environmental parameters, such as temperature changes and drift in measurement circuits. To remove such parasitic deflection, a differential measurement mode is often used, where the deflection of the "active" cantilever is subtracted from the deflection of the "passivated reference" cantilever to compensate parasitic deflection.

The static-mode cantilever sensor in liquid shows equally high sensitivity as that in air or vacuum. The detection of numerous biologically important entities in water, including DNA,[32–36] RNA,[37] biotinylated latex beads,[38] protein,[39–41] DNA transcription factors,[42] prostate-specific antigen (PSA),[43–45] *Bacillus subtilis* spore,[46] and glucose[47] have been demonstrated. Also, the measurements of pH[48] and the viability of the bacteria[49] have been demonstrated.

In the work of Fritz et al.,[32] silicon cantilevers with a thickness of 1 μm, width of 100 μm, and length of 500 μm were used to detect DNA hybridization. Synthetic 5′ thio-modified oligonucleotides were covalently attached on the gold-coated cantilever. The bending induced by the hybridization was measured with the optical lever method in a differential mode. The maximum absolute bending was about 100 nm and the relative bending due to the DNA hybridization was around 20 nm at 2000 nM of matching oligonucleotides. The calculated detection limit was 10 nM. The measurement technique showed high selectivity and could detect a single base mismatch in 12-mer oligonucelotides.

A number of studies reported the detection of PSA, which is a well-known biomarker for prostate cancer. In the work of Wu et al.,[43] the top side of a silicon nitride cantilever with a thickness of 0.5 μm, width of 20 μm, and length of 200 μm was coated with gold and the probe molecules; Rabbit Anti-Human-PSA (RAH-PSA) was attached to the gold-coated cantilever through dithiobis-(sulfosuccinimidylpropionate). As the PSA introduced in the flow cell attached to the RAH-PSA on the cantilever, the surface stress was changed to cause the deflection. The deflection of the cantilever was detected with the optical level method. The cantilever was able to detect 0.2 ng/mL to 60 μg/mL of PSA in a background of human serum albumin and human plasminogen.

The static-mode cantilever can be also used in a dry condition to detect a trace of volatile organic vapor,[50–54] hydrogen,[55] and trinitrotoluene[56,57] in air. These cantilevers are coated with a thin layer of absorbent materials to induce the deflection of the cantilever on exposure to target molecules in air. Often, the cantilever sensors are referred to as an "artificial nose," when they are based on multiple cantilevers to distinguish different volatile organic vapors and gases. The deflection of several cantilevers coated with different absorbent material was analyzed via principal component analysis[54] or a neural network[51] to distinguish between different molecules. To detect volatile organic vapor, various polymers[50,51] such as polyvinylpyridine (PVP), polyvinylchloride (PVC), polyurethane (PU), polystyrene (PS), and polymethylmethacrylate (PMMA) can be used. For the detection of water molecules in air, water-soluble polymers such as carboxymethylcellulose (CMC) and polyvinylalcohol (PVA) can be used.

Baller et al. had reported a cantilever array–based artificial nose,[51] which is based on the deflection of cantilevers coated with a polymer layer. A silicon cantilever with a thickness of 1 μm, width of 100 μm, and length of 500 μm was coated with a 5-μm-thick polymer layer. There were eight cantilevers in each array, and the cantilevers were coated with the combination of PVP, PU, PS, and PMMA. The deflection was measured with the optical lever method. The output of the eight cantilevers was analyzed with principal component analysis to distinguish different molecules. Furthermore, the identification of six different natural flavors was demonstrated by analyzing the cantilever array's response with a neural network.

In the dynamic mode,[29,58] the attached target entities increase the total mass of the cantilever and cause the changes in the resonant frequency. In general, the more target entities attaches to the cantilever, the resonant frequency will decrease more. The mass sensitivity of the dynamic cantilever sensor can be defined as the ratio of the resonant frequency shift to the additional mass attached to the cantilever. The mass sensitivity is affected by the location of the target entities on the cantilever and it is proportional to the square of the maximum displacement or the vibration amplitude at each point.[58,59] For instance, the mass sensitivity is maximum at the end of the cantilever and it is minimum at the base of the cantilever. Therefore, to calculate the exact mass of the attached entities, one needs to know where the target entities were attached to. When the number of the target entities on the cantilever is relatively large, one can assume the uniform distribution of the target mass and the average mass sensitivity can be used. The dynamic-mode cantilever sensors can operate in vacuum, in air, and in liquid. However, the amplitude and quality factor of the vibration as well as the mass sensitivity decrease dramatically in liquid, due to the high viscosity of the liquid. Hence, the minimum detectable mass of the cantilever in vacuum or in air is significantly lower than that in water by multiple orders. For this reason, the target entities are often captured in liquid and then the resonant frequency of the cantilever is measured in air or vacuum after drying it.[60–62]

The dynamic-mode cantilever sensors in air or vacuum were used to measure the mass of various biologically relevant entities, including DNA,[63,64] thiol-terminated self-assembled monolayer (SAM),[65,66] virus,[61,67] bacteria,[60,62] and fungal growth.[68] Besides, the dynamic-mode cantilevers were also used to detect mercury vapors,[69] droplet evaporation,[59] and volatile organic vapors[70–73] in air. Owing to the reduced damping in air or vacuum, the cantilever sensors show extremely high mass sensitivity. In vacuum, the mass measurements in the range of zepto-gram[74] and yocto-gram[75] were demonstrated.

Ilic et al. reported the mass measurement of a single DNA molecule using a silicon nitride cantilever[64] with a thickness of 90 nm and lengths of 3.5, 4, and 5 μm. The mass sensitivities were 194, 109, and 54 Hz/ag for the cantilever with the lengths of 3.5, 4, and 5 μm, respectively. To spatially confine the attachment of the DNA, a 40-nm circular gold dot was deposited near the end of the cantilever and the DNA was attached to the gold dot through thiol-group. The cantilever was actuated by optical-thermomechanical motion excitation method, where a laser beam chopped at the resonant frequency of the cantilever was used to induce the cantilever's vibration. The vibration of the cantilever was determined interferometrically. The resonant frequency of the cantilever was measured at $\sim3 \times 10^{-7}$ Torr to achieve the quality factor of 3000–5000. The resonant frequency of the cantilever was measured before and after the DNA attachment. With the 3.5-μm-long cantilever, the binding of single DNA molecules was detected.

The mass measurements of various biological entities in liquid, such as cell,[76–80] bacteria,[81] spore,[82] and micro/nanoparticles,[83,84] were demonstrated. In addition, the concentration of the target molecules, including proteins[85–87] and glucose,[88] was measured with the same method. The vibration of the dynamic-mode cantilever is severely attenuated in liquid due to the high viscosity,[31] and the movement of the liquid surrounding the cantilever causes "added mass effect."[89] These two phenomena severely reduce the sensitivity of the dynamic-mode cantilever sensors. However, most of the biological target requires "wet condition" to maintain their morphologies and metabolism. Hence, various studies have been focused on understanding the behavior and characteristics of resonators immersed in liquid.[90,91] One of the elegant solutions to avoid the degradation of the mass sensitivity in liquid is a suspended microfluidic resonator (SMR).[76,80,83,92] In this approach, a microfluidic channel was integrated into a microcantilever, and the mass of the target particle was measured as the particle flew through the integrated microfluidic channel. As the cantilever itself was in vacuum, the SMR was able to achieve high-quality factor and sensitivity. The SMR was used to measure the buoyant mass of a nanoparticle with an average mass of 10 ag,[83,84] and the mass, density, volume, and deformability of cells.[80,92]

In the work of Campbell et al.,[81] a piezoelectric-excited millimeter-sized cantilever was used to detect Group A *Streptococcus* (GAS). In this work, the cantilever is 127 μm thick, 1–2 mm wide, and 1–2 mm long and made of a composite layer of stainless steel and lead zirconate titannate (PZT). The fabricated cantilever was functionalized with anti-GAS antibody to capture GAS. The vibration of the cantilever was measured with an impedance analyzer. Typical first-mode and second-mode resonant frequencies are 12.8–17.5 and 65.8–72.5 kHz with quality factors of ~30 in PBS. The cantilever was immersed in the suspension of GAS with varying concentration, and the resonant frequency decreased more with higher concentration. Specifically, the resonant frequency of the cantilever decreased by 136, 509, 690, 1130, 1260, and 1782 Hz with the suspension of 7×10^2, 7×10^3, 7×10^5, 7×10^6, 7×10^7, and 7×10^9 cells/mL, respectively.

23.3 Static Mode: Surface Stress Sensor

When operated in the static mode, the cantilever beam is required to be functionalized with receptor molecules on only one side (top or bottom), in order to specifically detect an analyte or an environmental condition. This section will describe the theoretical background of a surface stress sensor, as well as an application in the static mode of a cantilever beam as a pH sensor.

23.3.1 Theoretical Background

For a given change in the surface stress of the cantilever beam, the deflection of the cantilever beam at the free end, Δz, can be given using Stoney's formula,[93]

$$\Delta z = 4\left(\frac{1}{t}\right)^2 \frac{(1-\nu)}{E}(\Delta\sigma_1 - \Delta\sigma_2) \tag{23.1}$$

where $\Delta\sigma_1$ and $\Delta\sigma_2$ are the surface-stress changes in the top and bottom sides of the cantilevers, ν is the Poisson's ratio, E is the Young's modulus of the cantilever material, l is the length, and t is the thickness of the cantilever beam. As can be seen from Equation 23.1, decreasing the thickness and increasing the length can give a larger deflection of the cantilever for the same change in surface stress.

In the static mode, a cantilever beam may be operated as a force sensor as well. This operation will not be described in detail here, but is being mentioned here for completeness. The parameter of interest is the minimum detectable force, which can be given as,[10]

$$F_{min} = \left(\frac{wt^2}{1Q}\right)^{1/2} (E\rho)^{1/4}(k_B TB)^{1/2} \tag{23.2}$$

where w is the width, t is the thickness, l is the length of the cantilever beam, k_B is the Boltzmann's constant, T is the temperature, E is the Young's modulus of the cantilever beam material, ρ is the density of the material, B is the measurement bandwidth, and Q is the mechanical quality factor. Q is defined as the total stored energy in the vibrating structure divided by the total energy loss per unit cycle of vibration. From Equation 23.2, it can be seen that decreasing the thickness and width, and increasing length, along with a high Q can achieve high sensitivity. But it has been shown that decreasing the thickness of the cantilever beam can reduce the Q factor.[5] Hence, this approach of just decreasing the dimensions to improve force sensitivity may not be straightforward, and other factors have to be taken into consideration as well such as the effect of the medium surrounding the cantilever beam, including the temperature, the pressure, etc.

23.3.2 Case Study: Microcantilever Beam–Based pH Microsensor

23.3.2.1 Introduction

A key parameter to measure in most biochemical and biological processes involving microorganisms is the change in pH in very small volumes that are created by release of H^+ ions by transmembrane pumps, by-products of chemical reactions, and other processes related to the functioning of microorganisms. We and our collaborators used intelligent hydrogels

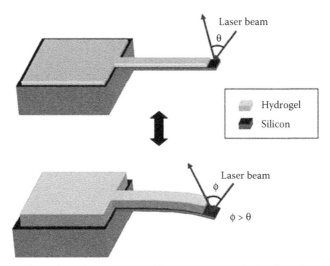

Change in environment, e.g., pH or temperature → hydrogel swells

FIGURE 23.3
Schematic diagram of a cantilever beam–based sensor platform based on a microcantilever patterned with an environmentally responsive hydrogel. (From Hilt, J.Z. et al. *Biomedical Microdevices* 5, 177–184, 2003, with permission.)

as sensing elements that were micropatterned on the surface of cantilevers (see Figure 23.3 for a schematic diagram of the conceptual view) to create highly sensitive microcantilever beam–based pH sensors.[94,95] Hydrogels are mainly hydrophilic polymer networks that swell to a high degree due to an extremely high affinity for water, yet are insoluble because of the incorporation of chemical or physical cross-links. By selecting the functional groups along the polymer chains, hydrogels can be made sensitive to the conditions of the surrounding environment, such as temperature, pH, electric field, or ionic strength.[96] In this work, the polymers used were environmentally responsive hydrogels that are based on ionic networks with a reversible pH-dependent swelling behavior. The hydrogels were based on anionic networks that contain acidic pendant groups,[97,98] which ionize once the pH of the environment is above the acid group's characteristic pK_a. With deprotonation of the acid groups, the network exhibits fixed charges on its chain resulting in an electrostatic repulsion between the chains and, in addition, an increased hydrophilicity of the network (see Figure 23.4). Because of these alterations in the network, water is absorbed into the polymer to a greater degree causing swelling.[99]

23.3.2.2 Materials and Methods

In our work, surface-micromachined cantilevers were fabricated using commercially available SOI wafers with a 2.5-μm silicon layer and 1-μm buried oxide (BOX) layer.[100] A 0.3-μm oxide was grown (Figure 23.5a), and photoresist mask was used to anisotropically etch the oxide, silicon, and the BOX layers consecutively (Figure 23.5b). A 0.1-μm oxide was grown on the sidewall of the SOI layer (Figure 23.5c), and a dry anisotropic etch was used to remove the oxide from the substrate, expose the silicon surface, while leaving it on the sidewalls of the SOI layer. Tetramethylammonium (TMAH) was used to etch the silicon substrate and to release the cantilever/oxide composite structure (Figure 23.5d). The wafers were immersed in buffered hydrofluoric (BHF) acid to etch off all the oxide

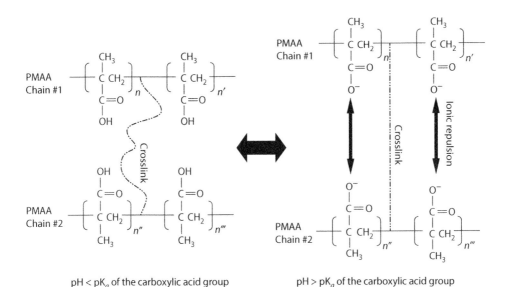

FIGURE 23.4
Schematic of the pH-dependent swelling process of an anionic hydrogel: specifically, a cross-linked poly(methacrylic acid) (PMAA) is illustrated. (From Hilt, J.Z. et al. *Biomedical Microdevices* 5, 177–184, 2003, with permission.)

and release the silicon cantilevers (Figure 23.5e). The cantilevers all had thicknesses of approximately 2.5 μm.

The monomers studied were vacuum-distilled methacrylic acid (MAA) and poly(ethylene glycol) dimethacrylate (designated as PEGnDMA, where n is the average molecular weight of the PEG chain). The initiator used for the UV free-radical polymerization was

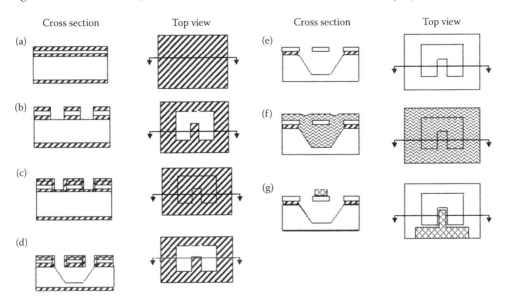

FIGURE 23.5
Fabrication process flow used for the fabrication of the cantilever beams. (From Hilt, J.Z. et al. *Biomedical Microdevices* 5, 177–184, 2003, with permission.)

2,2-dimethoxy-2-phenyl acetophenone (DMPA), and adhesion was gained between the silicon substrate and the polymer using an organosilane coupling agent, γ-methacryloxypropyl trimethoxysilane (γ-MPS). Cross-linked poly(methacrylic acid) (PMAA) networks were prepared by reacting MAA with substantial amounts of PEG200DMA.[101] Photolithography was utilized to pattern these networks onto silicon wafers. The wafers containing cantilevers were soaked in a 10 wt% acetone solution of γ-MPS, which forms a SAM on the native silicon dioxide surface and presents methacrylate pendant groups that react and bond covalently with the polymer film. Next, the hydrogel was defined on the cantilevers. The monomer mixtures were prepared with a mole ratio of 80:20 MAA/PEG200DMA and contained 10 wt% DMPA as the initiator for the UV free-radical polymerization. The monomer mixture was spin-coated at 2000 rpm for 30 seconds onto the silicon samples containing microcantilevers (Figure 23.5f). The sample was exposed to UV light in a Karl Suss MJB3 UV400 mask aligner with an intensity of 23.0 mW/cm² for 2 minutes and then allowed to soak in deionized (DI) distilled water for greater than 24 hours to remove any unreacted monomer (Figure 23.5g). Figure 23.6a shows a scanning electron microscopy (SEM) micrograph of the cantilever/ polymer structure in the dry state, while Figure 23.6b shows an optical micrograph of the dry cantilever/polymer structure. The cantilever deflection was measured using a manually calibrated microscope equipped with a 60× water immersion objective.

23.3.2.3 Results and Discussion

After patterning the hydrogel onto microcantilevers, dynamic and equilibrium bending studies were conducted. The dynamic and equilibrium deflection characteristics of the patterned microcantilevers were examined in various pH solutions at $18 \pm 0.5°C$. The patterned microcantilevers were monitored while exposed to constant ionic strength ($I = 0.5$ M) buffer solutions of varying pH. The buffer solution was composed of a mixture of citric acid, disodiumphosphate, and potassium chloride. The system was shown to have a rapid dynamic response, equilibrating within a few minutes.[100]

The measurement results are shown in Figure 23.7, where the solid line shows the measured deflection of the cantilever versus pH and the dotted line shows the calculated curve from an analytical model. The behavior of the cantilever structure is complicated, but a simplified model can be used if the cantilever structure is examined as a composite beam

(a) (b)

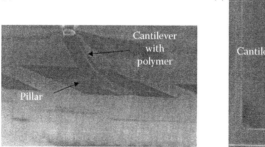
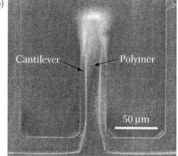

FIGURE 23.6
(a) SEM micrograph taken at an angle of a cantilever/polymer structure in the dry state. The polymer is charging up when viewed in the SEM and hence appears slightly distorted at the top. (b) A top-view optical micrograph of the cantilever/polymer structure in the dry state. The cantilever is bent upwards and hence the tip region is out of focus. The cantilever is 0.8 μm thick and the polymer is 2.5 μm thick. (From Bashir, R. et al. *Applied Physics Letters* 81, 3091–3093, 2002, with permission.)

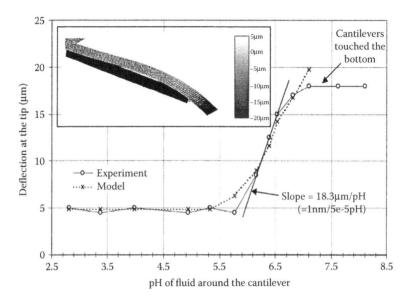

FIGURE 23.7
Equilibrium bending data versus pH (constant ionic strength of 0.5 M) for patterned microcantilever shown in Figure 23.6b. Solid diamonds are for the increasing pH path, while hollow diamonds are for the decreasing pH path (mean ± SD, n = 3). (From Bashir, R. et al. *Applied Physics Letters* 81, 3091–3093, 2002, with permission.)

with no slip at the interface.[102] The inset in Figure 23.7 shows the vertical deflection of the cantilever/polymer structure at pH 7 as obtained by analytical equations. This result demonstrated that the micropatterned hydrogel film was capable of sensing the change in environmental pH, swelling in response and resulting in actuation of the microcantilever. Figure 23.7 shows a linear fit of the deflection response at the region of highest sensitivity, which demonstrated a sensitivity of about 18.3 μm/pH unit. With the detection abilities of optical-based laser detection setups known to easily resolve a 1 nm deflection, this sensitivity can be translated to $5 \times 10^{-5}\ \Delta pH/nm$. This ultrasensitivity is unique to this device and enables for novel applications as a pH sensor in microenvironments. When compared with other microscale techniques such as light-addressable potentiometric sensor (LAPS)[103] or scanning probe potentiometer (SPP),[104] the sensitivity is demonstrated to have increased by at least two orders of magnitude.

23.3.3 Case Study: Nanomechanical Detection of Biomarker Transcripts in Human RNA

23.3.3.1 Introduction

DNA/protein microarray is a widely used technique in studying gene expression across the human genome. It has been more than 15 years that the microarray has been commercialized.[105] Despite its substantial improvement during the past years, it still relies on complicated preconditioning and labeling process.[106] Furthermore, these techniques require target amplification, which can further affect the sample and bias the analysis.[107,108]

Microcantilevers, as a label-free technology, permits parallel detection of multiple proteins and biomolecules, which are essential for diagnosing complex diseases like cancer. In this pioneering work led by Dr. Gerber,[37] the detection of the 1-8U gene's activity induced by interferon-alpha (IFN-α) treatment was demonstrated with microcantilever array. The reports on the therapeutic efficacy of IFN-α in human melanoma[109] and hepatitis C infection[110] have raised the interest on the gene modulation effects of IFN-α. As IFN-α is known to have side effects,[111,112] it is important to evaluate individual patient for the efficacy of IFN-α prior to actual treatments. In this work, a microcantilever array was developed to monitor the expression level of mRNA biomarkers relevant to IFN-α for this purpose.

23.3.3.2 Materials and Methods

23.3.3.2.1 Cantilever Arrays Coated with ssDNA Oligonucleotides

In this work, an array of eight cantilevers has been fabricated from silicon, each with the length of 500 μm, the width of 100 μm, and the thickness of 450 nm. One side of each cantilever has been covered with a thin layer of gold with the thickness of 20 nm. On the gold layers of each cantilever, different sets of single-stranded DNA (ssDNA) were immobilized for various biomarker detection, as shown in Figure 23.8. To ensure full exposure of the short ssDNA (<10 nm) to the solution, the roughness of the gold layer was reduced to ~0.4 nm (Figure 23.8).

23.3.3.2.2 ssDNA Immobilization and Hybridization

Immobilizing a sequence to the gold surface of the cantilever has been done by incubating the cantilevers with an array of microcapillaries for 20 minutes, each filled with a different solution of thiol-modified probe. When an ssDNA fixed on the cantilever combines with a target mRNA from solution, the surface stress of the cantilever changes and it bends accordingly, as shown in Figure 23.9. The bending could be measured with the optical lever method. In readout process, differential deflection measurement with a local reference cantilever is used since the bending of the cantilevers can be affected by the temperature changes, refractive index, and unspecific interactions. In this experiment, cantilevers 1 and 7 are coated with alternating sequences of adenine and cytosine of 12 or 24 bases and they are used as references. For the dehybridization process, the cantilevers have been washed with the solution consisting of 2 mL saline sodium citrate (SSC) with 1 M NaCl, 1 mL SSC with 0.1 M NaCl, and 1 mL 4 M urea. The cleaning step is repeated after performing each experiment.

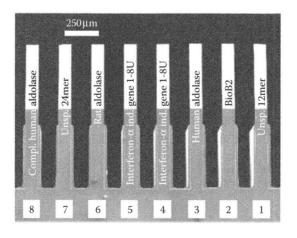

FIGURE 23.8
SEM image of the cantilevers. Each cantilever is functionalized with different set of ssDNA. (From Zhang, J. et al. *Nature Nanotechnology* 1, 214–220, 2006, with permission.)

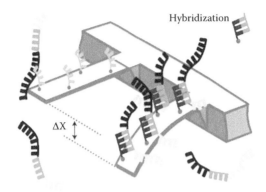

Hybridization

ΔX

FIGURE 23.9
Cantilever array used in the experiment. Differential readout principle, using target-sensitive cantilever and *in situ* reference, was used. (From Zhang, J. et al. *Nature Nanotechnology* 1, 214–220, 2006, with permission.)

23.3.3.2.3 Measurement Procedure

In order to calibrate the device, the cantilever array has been flushed in SSC buffer solution and stabilized at 23°C. Then, a heat pulse of 0.7°C has been applied so that all of the cantilevers bend accordingly. This procedure helps the array to gain the mechanical homogeneity. Subsequently, various concentrations of 12-mer ssDNA BioB2 were injected, and deflection of the cantilever has been observed. The BioB2-sensitized cantilever bends with respect to the BioB2 concentration. The deflection of the cantilever increases with the increase of the concentration. Then, the deflection of cantilevers is restored for the subsequent measurement by injecting the buffer solution. With the cantilever thickness of 450 nm, the sensor showed the minimum detectable concentration of 10 pM, which is sufficient for routine diagnostic tests.

23.3.3.3 Results and Discussion

23.3.3.3.1 Label-Free Gene Fishing in Total RNA

The cantilever array was demonstrated to distinguish similar biomarkers. In order to evaluate the sensitivity of the cantilever to a specific biomarker, rat aldolase messenger RNA (mRNA) has been measured in total complementary RNA (cRNA) from rats. In Figure 23.10a, the black line represents the differential deflection of cantilever coated with rat aldolase with respect to the reference probe. After injection, the deflection increases reaching to 100 ± 10 nm in 20 minutes. The red line indicates the crosstalk between BioB2C-sensitized cantilever and rat aldolase mRNA, which has been previously reported. As shown in Figure 23.10b, the human aldolase cantilever does not respond to rat mRNA and remains at zero, even though these two mRNAs are same except for four bases. This clearly demonstrates that the system is capable of distinguishing between human mRNA and similar but not identical rat mRNA. In Figure 23.10c, the human aldolase mRNA has been detected in total human RNA, using the cantilever coated with complementary human aldolase. The cantilever shows deflection of 55 nm, 30 minutes after the injection.

23.3.3.3.2 Rapid Detection of Gene Expression

The microcantilevers were able to rapidly characterize the expression level of 1-8U gene without amplification and labeling. In the absence of IFN-α treatment, 1-8U gene is inactive

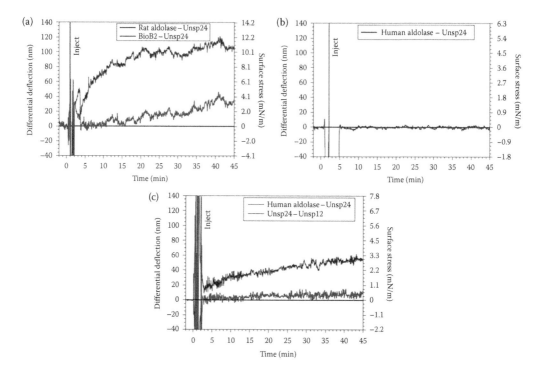

FIGURE 23.10
Differential gene fishing in a complex genomic background. (a) Injection of rat cRNA. The differential deflection of rat aldolase cantilever with respect to reference cantilever is shown in black line. The red line indicates the crosstalk onto prokaryotic sensor. (b) Response of cantilever coated with human aldolase A. (c) Gene fishing in complete genomic background. The black line shows the detection of human aldolase mRNA in total RNA. (From Zhang, J. et al. *Nature Nanotechnology* 1, 214–220, 2006, with permission.)

and does not produce mRNA. In contrast, in the presence of IFN-α treatment, this gene is upregulated. During this experiment, human aldolase A, which can always be found in human cells, has been monitored as a housekeeping gene for internal control. Figure 23.11a represents the cantilever response for a normal I-8U gene that is inactive. The red line is the deflection of the cantilever coated with 1-8U that shows no RNA hybridization as expected. In Figure 23.11b, the cells have been under IFN-α treatment. The 1-8U gene is expressed, and the cantilever sensitized to 1-8U gene can detect the corresponding mRNA readily.

23.4 Dynamic Mode: Resonant Mass Sensor

23.4.1 Theoretical Background

23.4.1.1 Basic Mechanical Parameters of Interest

The spring constant, k, of a rectangular-shaped cantilever beam is given as[113]

$$k = \frac{Et^3w}{4l^3} \qquad (23.3)$$

(a)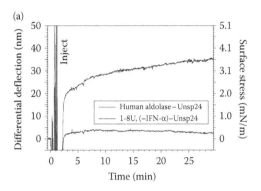

(b)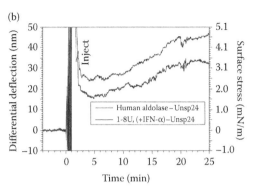

FIGURE 23.11
Nanomechanical measurement of the upregulation of a gene. (a) 1-8U gene shows no response due to the lack of IFN-α treatment (red line). For internal control, the aldolase A gene cRNA is observed (black line). (b) Under IFN-α treatment, the 1-8U gene becomes active and the cantilever coated with that detects this activation. (From Zhang, J. et al. *Nature Nanotechnology* 1, 214–220, 2006, with permission.)

where E is the Young's modulus of the cantilever beam material, l is the length, w is the width, and t is the thickness of the cantilever beam.

The resonant frequency of a cantilever beam in vacuum is given as

$$f_o = \frac{1}{2\pi}\sqrt{\frac{k}{m^*}} \tag{23.4}$$

where for a rectangular-shaped cantilever beam, the effective mass, m^*, is approximately 0.24 times the mass of the cantilever beam. Owing to the low viscosity of air, the resonant frequency in air can be approximated to that of that in vacuum. In liquids, damping occurs which decreases the resonant frequency of the cantilever beam. Assuming the fluid to be inviscid, the resonant frequency of the cantilever immersed in fluid, $f_{0(fluid)}$, can be expressed as a function of the resonant frequency in vacuum, $f_{0(vacuum)}$, as[90]

$$f_{0(fluid)} = \left(1 + \frac{\pi\rho w}{4\rho_c t}\right)^{-0.5} f_{0(vacuum)} \tag{23.5}$$

where ρ_c is the density of the beam and ρ is the density of the fluid.

The change in the resonant frequency of the cantilever beam can be due to change in mass,[114] surface stress,[115,116] or damping due to fluid change.[117] Assuming that the additional mass is uniformly distributed over the cantilever beam and that the spring constant does not change, the mass change as a function of frequency change can be given as[115]

$$\Delta m = \frac{k}{4n\pi^2}\left(\frac{1}{f_1^2} - \frac{1}{f_o^2}\right) \tag{23.6}$$

where n = 0.24 for a rectangular-shaped cantilever beam, f_o is the initial resonant frequency before the addition of the mass, and f_1 is the resonant frequency after the mass addition. The mass sensitivity, S_m, of a resonant frequency–based gravimetric sensor can be given as[118]

$$S_m = \frac{A}{2(m^* + m_d)} \approx \frac{A}{2m^*} \quad (m^* \gg m_d) \tag{23.7}$$

where A is the receptive or active area of the sensor and m_d is the deposited mass. The rightmost expression can be obtained if it is assumed that the mass of the cantilever sensor is much larger than the deposited mass. It should be noted that this would not be the case when the cantilever size is very small (in the nanoscale regime). It can be seen from Equation 23.7 that decreasing the overall mass of the cantilever beam increases the sensitivity to the mass change. Decreasing the dimensions of the cantilever beam can, of course, decrease the overall mass of the cantilever beam.

23.4.1.2 Minimum Detectable Mass

When operated in the dynamic mode, the minimum mass change that can be detected is related to the minimum detectable resonant frequency shift. The main noise source is due to thermomechanical noise.[22] Depending on the mechanical properties, this shift is given as

$$\Delta f_{min} = \frac{1}{\langle \bar{A} \rangle} \sqrt{\frac{f_0 k_B TB}{2\pi kQ}} \tag{23.8}$$

where k_B is Boltzmann's constant, T is temperature in Kelvin, B is the bandwidth measurement of the frequency spectra, Q is the quality factor, and $\langle \bar{A} \rangle$ is the square root of the mean-square amplitude of the vibration.[22,119] From Equation 23.8, it can be seen that if the quality factor Q, a measurement of the sharpness of the frequency spectra peak, and the amplitude of vibration, $\langle \bar{A} \rangle$, are increased, the cantilever will be able to detect smaller frequency shifts and in turn smaller adsorbed masses. The minimum detectable mass change[120,121] can then be obtained by combining Equations 23.6 and 23.8 with $f_1 = (f_0 - \Delta f_{min})$, to obtain

$$\Delta m_{min} = \frac{1}{\langle \bar{A} \rangle} \sqrt{\frac{4 k_B TB}{Q}} \frac{m_{eff}^{5/4}}{k^{3/4}} \tag{23.9}$$

It is also possible to obtain a relation of the quality factor as a function of the dimension of the cantilever beam and the medium in which the cantilever is immersed. From the studies by Sader[90] and Chon et al.,[117] we obtain

$$Q = \frac{(4\mu / \pi \rho w^2) + \Gamma_r(\omega_R)}{\Gamma_i(\omega_R)} \tag{23.10}$$

where μ is the mass per unit length of the beam, ρ is the density of the medium in which the cantilever is immersed, w is the width of the cantilever beam, and $\Gamma_r(\omega_R)$ and $\Gamma_i(\omega_R)$ are, respectively, the real and imaginary parts of the hydrodynamic function $\Gamma(\omega)$.[90]

23.4.2 Case Study 1: Bacterial and Antibody Protein Molecules Mass Detection Using Micromechanical Cantilever Beam–Based Resonators

This section will summarize the use of a surface-micromachined silicon cantilever beam to be used as a resonant biosensor for the detection of mass of bacterial cells and antibodies.[60] Real-time specific detection of bacterial cells has a wide range of applications such as food safety, biodefense, and medical diagnosis. The goal of this particular study was to demonstrate the cantilever beam as a viable candidate as an immunospecific resonant biosensor. Nonspecific binding of bacterial cells on cantilever beams was carried out in order to measure the effective dry mass of the *Listeria innocua* bacteria and to quantify the sensitivity of the sensor to the mass of the bacterial cells. The mass of antibody layer was also measured in order to demonstrate the sensitivity of the cantilever beams to the mass of the protein layers and to demonstrate antigen–antibody interactions of bacterial cells adhering to functionalized surfaces more efficiently than on nonfunctionalized (bare) surfaces.

23.4.2.1 Materials and Methods

23.4.2.1.1 Cantilever Beam Fabrication and Mechanical Characterization

A novel fabrication technique was developed to fabricate thin, low-stress, single-crystal cantilever beam.[122] The process flow involves using merged epitaxial lateral overgrowth (MELO) and chemical mechanical polishing (CMP) of single-crystal silicon. MELO can be regarded as an extension of selective epitaxial growth (SEG) and epitaxial lateral overgrowth (ELO).[122] The fabrication method used in this study has the advantage of fabricating all-silicon structures without any oxide layer being present under the silicon anchor of the cantilever beam. This eliminates any mismatch in material properties between the silicon and silicon dioxide material that exists when using SOI as the starting material. This in turn eliminates, or certainly decreases, the residual stresses in cantilever beams that are a source of vibrational energy loss.[9] The present fabrication method also has the potential of fabricating arrays of cantilever beams with varying length, width, and thickness dimensions on the same substrate. This can allow the fabrication of arrays of cantilever beams with a range of mechanical resonant frequencies and sensitivities.

The cantilever beams were excited with thermal and ambient noise and the corresponding vibration spectra was measured in air using a scanning probe microscope (SPM, Dimension 3100 Series Digital Instruments, Veeco Metrology Group, Santa Barbara, CA) and in certain experiments a laser Doppler vibrometer (LDV, MSV-300 Polytec PI, Auburn, MA). Thermal noise excitation was used since it does not require any power and it does not excite other stiffer, higher mechanical resonance modes such as that of the cantilever holder. The advantage of externally driving the cantilever beam will be a more sensitive mass detection capability by achieving an increase in the amplitude and the quality factor, resulting in the decrease of the minimum detectable frequency shift. The vertical deflection signal of the cantilever beam was extracted from the SPM using a Digital Instrument signal access module (SAM) and then digitized. The power spectral density was then evaluated using MATLAB® software. The thermal vibration spectra data were fit (using the least square method) to the amplitude response of a simple harmonic oscillator (SHO) in order to obtain the resonant frequency and the quality factor. The amplitude response of an SHO is given as[123]

$$A(f) = A_{dc} \frac{f_0^2}{\sqrt{\left[\left(f_0^2 - f^2 \right)^2 + \frac{f_0^2 f^2}{Q^2} \right]}}$$

(23.11)

TABLE 23.1

Planar Dimensions and Measured Values of Unloaded Resonant Frequency, Quality Factor, Spring Constant, and Mass Sensitivity of Representative Cantilever Beams Used to Detect the Mass of Bacterial Cells and Antibody Protein Molecules

Cantilever Designation	Length and Width (μm)	Resonant Frequency (kHz)	Quality Factor (Q)	Spring Constant (N/m)	Mass Sensitivity (Hz/pg)
1	L = 78 W = 23	85.6	56	0.145	65
2	L = 79 W = 24	80.7	54	0.097	90

Source: Gupta, A. et al. *Journal of Vacuum Science & Technology B* 22, 2785–2791, 2004, with permission.

where f is the frequency in Hz, f_0 is the resonant frequency, Q is the quality factor, and A_{dc} is the cantilever amplitude at zero frequency. All the reported values of resonant frequency and quality factor presented and used in this work are those that have been obtained by curve fitting. The cantilever beams were calibrated by measuring their spring constant using the added mass method.[124] Table 23.1 shows the planar dimensions and mechanical characterization results for specifically two cantilever beams, designated cantilever 1 and cantilever 2, whose results have been presented below. Different resonant frequency and spring constant were measured for different cantilever beams with around the same planar dimensions due to difference in thickness of these cantilevers. For the present study, the minimum detectable frequency shift of the cantilever beams was calculated to be around 150–200 Hz using Equation 23.7 with the typical parameters of the cantilevers used in this work.

23.4.2.1.2 Bacterial Growth Conditions and Chemical Reagents Used

L. innocua bacteria were grown in Luria-Bertani (LB) broth at 37°C placed in an incubator. The initial concentration of the bacterial suspension was estimated to be around 5×10^8 cells/mL. The buffer used in all the experiments presented in this section was phosphate buffered saline (PBS) with pH 7.4 (137 mM NaCl, 2.7 mM KCl, and 10 mM phosphate buffer solution). The bacteria were transferred to the PBS buffer for further dilution. Bacterial suspensions in concentration varying from 5×10^6 to 5×10^8 cells/mL were introduced on the cantilever beam surfaces. Goat affinity-purified polyclonal antibody for *L. innocua* was used. BSA (bovine serum albumin) was used as a blocking agent in order to prevent nonspecific binding of bacteria cells in areas not covered by the antibody layer.[125] Tween-20 (0.05% by volume) in PBS was used as a surfactant in order to remove the loosely bound bacteria attached to the surfaces.

23.4.2.1.3 Dry Mass Measurement of Bacterial Cells

Nonspecific binding of bacteria was performed on the cantilever beam surfaces in order to obtain the effective dry mass of *L. innocua* bacteria. All the resonant frequency measurements in the present study were done in air. In order to prevent stiction of the cantilever structures onto the underlying substrate after removal from liquid, the structures were dried using critical point drying (CPD). Following the introduction of the bacterial suspension over the cantilever beam for 30 minutes, the cantilever beams were immersed in ethanol and dried using CPD. The above procedure was repeated on the same cantilever beams with increasing bacteria concentration in order to get frequency shift as a function of cell number bound to the cantilever. The number of bacteria on the cantilever was counted using a dark-field microscope and doubled to account for the bacteria bound at the bottom of the cantilever.

23.4.2.1.4 Antibody Coating of Cantilever Beams

BSA and the antibody to *L. innocua* bacteria were immobilized on the cantilever beam surface using physical adsorption. Both BSA and the antibody solution were introduced on the cantilever beams by dispensing 10–15 μL of the solutions using micropipettes over the cantilever beam locations on the chip.

The cantilever beams were first cleaned using piranha solution ($H_2O_2/H_2SO_4 = 1{:}1$) and then immersed in ethanol before being critical point dried. The resonant frequency was measured in order to obtain the unloaded resonant frequency. The cantilevers were immersed in the *Listeria* antibody, at a concentration of 1 mg/mL, for 15 minutes. The cantilevers were then rinsed for around 30 seconds in DI water and treated with BSA, at a concentration of 2 mg/mL, for 15 minutes. Then the samples were again rinsed in DI for around 30 seconds and then treated in increasing concentrations of methanol in PBS ranging from 1% to 100%. Finally, placing the cantilevers in a 100% methanol solution, they were dried using CPD. The resonant frequency of the cantilever beam was then measured in air in order to get the change in frequency due to the antibody and BSA mass.

The antibody-coated cantilevers were treated in PBS buffer for 15 minutes (in order to rehydrate the antibodies) followed by a short rinse of around 5 seconds in DI. The cantilevers were then treated with a bacterial suspension of *L. innocua*, at an estimated concentration of 5×10^8 cells/mL for 15 minutes. The sample was rinsed in DI for around 30 seconds following which the sample was gently shaken in a solution of 0.05% Tween-20 in PBS for 5 minutes. Following a short rinsing step in DI, the cantilever beams were again treated in increasing concentration of methanol in PBS before being dried using a CPD. The resonant frequency was then measured in order to determine the change in resonant frequency due to the bound cells.

As the resonant frequency of the cantilever beam was measured after both the antibody and BSA were attached to the surface, it was desired to find the separate effects of BSA and antibodies on the resonant frequency due to mass loading. The cantilever beam was initially cleaned to remove all the organics using piranha solution as before and measured to obtain the unloaded resonant frequency. The cantilever was then treated with BSA for around 15 minutes, rinsed in DI for around 30 seconds, and measured to obtain the loaded resonant frequency. The same cantilever was cleaned again in piranha, measured to obtain the unloaded resonant frequency, treated with antibodies, rinsed in DI, and finally, measured to obtain the new loaded resonant frequency.

23.4.2.2 Results and Discussion

23.4.2.2.1 Detection of Bacterial Cell Mass Using Nonfunctionalized Cantilever Beams

The non-specific binding experiment was performed in order to determine the smallest number of bacterial cells that could be detected with the smallest observable shift in resonant frequency of the cantilever, as well as to test whether the shift in resonant frequency was directly proportional to the effective number of bacterial cells, thus proving the validity of the measurements. After the last and highest concentration of bacterial cells was introduced on the cantilever beams and the resonant frequency was measured, cantilever 1 was sputtered with a layer of Au/Pd and SEM micrographs were taken of the cantilever beam. Figure 23.12 shows SEM micrographs depicting unselective binding of bacterial cells on cantilever 1. A uniform distribution of the bacterial cells can be seen over the

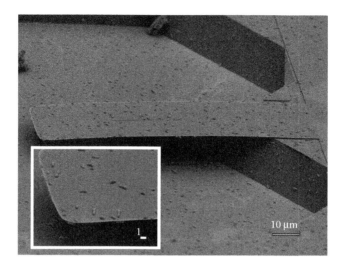

FIGURE 23.12

SEM micrographs showing nonspecific binding of *L. innocua* bacterial cells to cantilever 1 and surrounding area of sample. *Inset*: Higher magnification view showing the individual bacterial cells. (From Gupta, A. et al. *Journal of Vacuum Science & Technology B* 22, 2785–2791, 2004, with permission.)

cantilever surface as well as the surrounding area of the sample. The slight bending of the cantilever beams observed in the micrographs is due to the stress caused by the Au/Pd layer on top of the cantilever beam.

After each bacterial binding and resonant frequency measurement step, the number of bacterial cells were counted (and doubled to account for the top and bottom surface) from the photomicrographs. Equation 23.6 was used to calculate the change in mass with n = 1. The assumption is made that the spring constant does not change after the mass addition. Hence, all the values that were obtained for the mass change assumed that all the mass was concentrated at the free end. In order to get a value for the dry cell mass of a single *L. innocua* bacterium, each of the bacterial cells that were counted on the surface was weighted by a factor of (x/L), where x was the distance from the fixed end and L was the length of the cantilever beam.[126] As the cells attached nonspecifically over the entire cantilever beam and as it was not possible to count the cells on the bottom of the cantilever, it was estimated that the same number of cells attached at the bottom as on the top. Making this assumption and taking an average of the dry cell mass obtained for the three increasing concentrations from different cantilever beams, the dry cell weight was estimated to be around 85 fg. Figure 23.13a shows the frequency shift as a function of effective number of bacterial cells bound on surface of cantilever beams with around the same frequency range. Figure 23.13b shows the vibration spectra measured for cantilever 1 before and after binding of around 180 bacterial cells. The resonant frequency shift was close to 1 kHz.

The typical shape of *Listeria* bacteria is cylindrical with dimensions of length around 0.5–2 μm and width of around 0.4–0.6 μm. Assuming that the density of a bacterial cell is slightly higher than that of water (~1.05 g/cm³) with length of 2 μm and width of 0.4 μm, and that around 70% of the cell mass is due to water, calculations show that the dry cell mass is expected to be around 79 fg. This is certainly close to the measured range of around 85 fg. Ultrasensitive cantilever beams that can detect single cells can be achieved by scaling down the planar dimensions of the cantilever beams,[62,126] with a proportionate decrease in the thickness of the cantilever beams in order to decrease the bandwidth of the

(a)

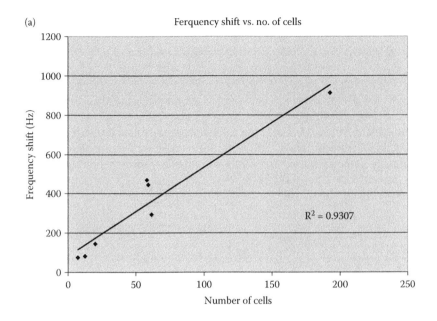

(b)

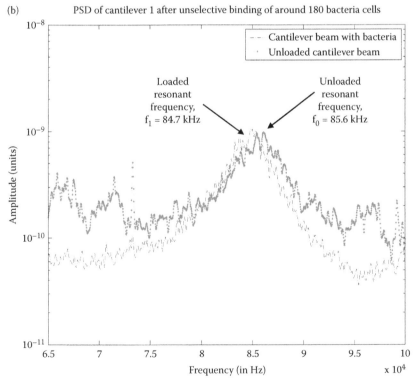

FIGURE 23.13

(a) Measured resonant frequency shift versus effective number of *L. innocua* bacterial cells binding to cantilever 1 (see Table 23.1). (b) Resonant frequency measurement before and after bacterial cell binding on cantilever beam 1. The values of resonant frequencies are extracted from fitting the measured curves to Equation 23.10 (the fitted curves are not shown). (From Gupta, A. et al. *Journal of Vacuum Science & Technology B* 22, 2785–2791, 2004, with permission.)

cantilever beams. The sensitivity of the resonators can also be increased by improving the quality factor (Q) of the cantilever beams, which can be achieved by externally driving the cantilevers, performing the measurement in vacuum and the like.

23.4.2.2.2 Detection of Bacterial Cell Mass Using Antibody-Coated Cantilever Beams

Cantilever 2 was used to measure the mass of the adsorbed antibodies and BSA, followed by the mass of the bacterial cells along with the protein layer. Figure 23.14a shows the change in resonant frequency of cantilever 2 at different stages of the experiment of selectively capturing bacterial cells on the cantilever. The resonant frequency was measured after a piranha clean of the cantilever beams, after the antibody plus BSA immobilization, and finally after the bacterial introduction. It should be pointed out that each of the steps was followed by a CPD step as the resonant frequency needed to be measured in air. As stated before, the

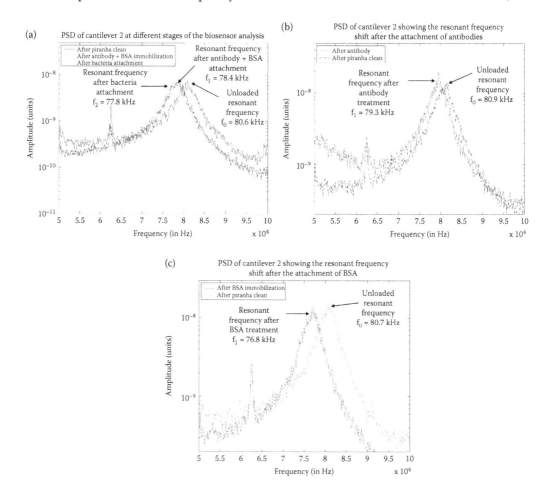

FIGURE 23.14

(a) Resonant frequency measurement showing unloaded cantilever beam, after antibody + BSA immobilization, and after bacterial binding on cantilever 2. The values of resonant frequencies are extracted from fitting the measured curves to Equation 23.1 (the fitted curves are not shown). (b) Resonant frequency shift after the attachment of the antibody to *L. innocua* to cantilever 2. The values of resonant frequencies are extracted from fitting the measured curves to Equation 23.10 (the fitted curves are not shown). (c) Resonant frequency shift after the attachment of BSA to cantilever beam 2. (From Gupta, A. et al. *Journal of Vacuum Science & Technology B* 22, 2785–2791, 2004, with permission.)

CPD was performed to avoid the stiction problem that normally occurs for surface-micromachined structures after being pulled from a liquid medium.

The largest frequency change was measured after the antibody plus BSA immobilization step, which was about 2 kHz. Using Equation 23.6 and assuming that the antibody and BSA form a uniform layer over the cantilever beam surface, the added mass was calculated to be around 93 pg for cantilever 2. The cantilevers were not bent indicating that the adsorption was on both sides of the cantilever. There was a shift in resonant frequency of around 500 Hz (corresponding to a mass change of 5.3 pg) after the attachment of the bacterial cells, as shown in Figure 23.14a. The effective number of bacterial cells that were captured on cantilever 2 was estimated to be around 80 bacterial cells (assuming the mass of each bacterial cell to be around 66 fg).

The mass of the antibody and BSA layer that was adsorbed on the cantilever beam surface was measured to be around 90 pg. In order to estimate whether the values were reasonable, one can make some rough calculations. The molecular weight of an antibody molecule (IgG) is estimated to be around 150 kDa, with an effective area for a single molecule to be around 45 nm². The molecular weight of a BSA molecule is around 66 kDa with an effective area of around 44 nm² (assuming BSA to be an ellipsoid with dimensions of 14 nm by 4 nm). Since the antibody was first attached to the cantilever beam, followed by BSA, it is safe to assume that the BSA covers only those area not covered by the antibody itself and that they do not attach to the antibody layer. It should be reasonable to assume that the antibodies cover the majority of the surface area of the cantilever beam. Assuming total coverage over the entire cantilever surface area (top and bottom of the cantilever) by the antibody, a mass of around 87 pg (mass of antibody layer divided by 0.24) is calculated. Since the antibodies and BSA are nonspecifically adsorbed, they will be randomly attached to the cantilever and could also be attached in multiple layers. The measured values of added mass are, however, in the expected pg range.

In order to better ascertain the effect of mass loading, by BSA and the antibody, on the resonant frequency, they were separately attached to the cantilever beam and the resonant frequency shift was measured (see Section 23.4.2.1.4). The antibody (conc. of 1 mg/mL) gave a frequency shift of around 1.48 kHz, which corresponds to a mass change of 59 pg (see Figure 23.14b). BSA (conc. of 2 mg/mL) gave a frequency change of around 3.94 kHz, which corresponds to a mass change of around 166 pg (see Figure 23.14c). In the case of BSA, theoretical calculations give an expected value of around 40 pg, when making the same assumptions as was made in the case for the antibody. One possible reason for the difference could be that the BSA molecules may be stacking on top of each other forming multiple layers.[127]

23.4.3 Case Study 2: Detection of Spores in Fluids Using Microcantilever Beams

It has become increasingly important to be able to detect biological species in a liquid environment using cantilever beams operated in the dynamic mode. This section will describe work of using resonant cantilever beam–based sensors to detect *B. anthracis* spores in air as well as in a liquid medium.[128]

23.4.3.1 *Materials and Methods*

23.4.3.1.1 *Cantilever Fabrication and Characterization*

P-type (100) 4″ SOI wafers were used as the starting material (see Figure 23.15a). The wafers had an SOI layer of 210 nm thickness and a BOX thickness of around 390 nm.

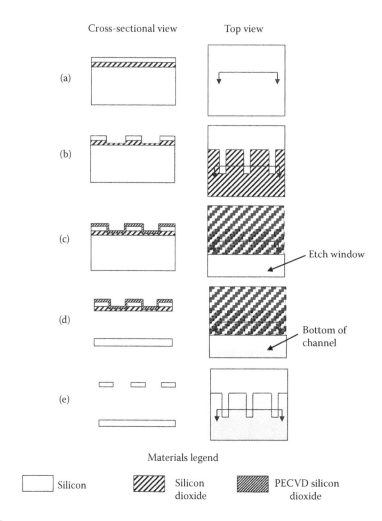

FIGURE 23.15
Process flow used for the fabrication of array of silicon cantilever beams using SOI as the starting material.
(From Gupta, A. et al. *Applied Physics Letters* 84, 1976–1978, 2004, with permission.)

Photolithography followed by reactive ion etching (RIE) using Freon 115 to etch the SOI layer and CHF_3/O_2 in order to thin the BOX layer was performed in order to pattern the cantilever beam shapes (see Figure 23.15b). After depositing a layer of plasma-enhanced chemical vapor deposition (PECVD) oxide as an etch stop layer, an etch window was photolithographically patterned using BHF oxide etch (see Figure 23.15c). In order to etch the underlying exposed silicon and release the cantilever beams, vapor phase etching using xenon difluoride (Xactix, Inc., Pittsburgh, PA) was used (see Figure 23.15d). After the cantilever beams were released, the oxide was etched in BHF, rinsed in DI water, immersed in ethanol, and dried using CPD (see Figure 23.15e).

The measurement of the cantilever resonant frequency was performed using a microscope scanning LDV (MSV-300 from Polytec PI) with a laser beam spot size of around 1–2 μm. The length of cantilever beams varied from 100 to 20 μm, with a width of around 9 μm and thickness of 200 nm.

23.4.3.1.2 Experimental Details for the Resonant Frequency Measurements Performed in Air

The cantilevers were first cleaned using a piranha solution ($H_2O_2/H_2SO_4 = 1:1$ in volume). Afterwards, they were dried with a CPD using ethanol as the exchange liquid, and oxygen plasma etch was performed to remove any other organic compounds that may still have been present. Using the LDV, the thermal noise of the cantilever beam was measured and the unloaded resonant frequency was recorded. The spores were then suspended onto the cantilevers by dropping 20 μL of *B. anthracis* Sterne spores (conc. 10^9 sp/mL) and allowing them to settle for 4 hours on the cantilever beams. Afterward, the cantilevers were dried with CPD, and the measurements of the loaded resonant frequency were performed using the LDV. After measurements, the cantilevers and spores were imaged with SEM for the purpose of counting the number of spores on each cantilever.

23.4.3.1.3 Experiment Details of the Measurements Performed in Liquid

For the liquid measurements, antibodies were used to immobilize the spores onto the cantilever surface. The cantilevers were cleaned and dried as described in Section 23.4.3.1.2 above on air measurements. First 50 μL of sterile DI water was added onto the top of the chip, and the unloaded frequency fluid measurements were performed by measuring the thermal spectra of each cantilever with the LDV. Then, 20 μL of *B. anthracis* spore antibody (conc. 1.15 mg/mL) and 20 μL of BSA (conc. 5 mg/mL) were added to the cantilevers. First, the antibody was added and allowed to bind to the cantilever surface for an hour. The chip was rinsed in PBS for 30 seconds followed by the addition of BSA for 15 minutes. BSA is used to prevent any nonspecific binding of the spores to the cantilever.[125] The cantilevers were then rinsed in PBS Tween (0.05%) for 30 seconds, rinsed in sterile DI water, and then dried with the CPD using methanol as the exchange fluid. Once the cantilevers were dry, 10 μL of PBS was added to the cantilevers to hydrate the antibodies for 1 minute and the PBS was rinsed away with sterile DI water. Afterwards, 20 μL of *B. anthracis* Sterne spore (conc. 10^9 sp/mL) was applied to the cantilevers and allowed to settle for 16 hours. The cantilevers were dried with the CPD using methanol as the exchange fluid. Dark-field photomicrographs were taken before the application of the DI water for measurements. In order to perform loaded fluid measurements, 50 μL of sterile DI water was applied onto the cantilevers and thermal spectra of the resonant frequency were measured using the LDV. The cantilevers were then dried with CPD and imaged again to verify whether or not the spores were displaced before and after fluid measurements. Spores on the cantilever before and after fluid measurement are shown in Figure 23.16, demonstrating that the antibodies were able to immobilize the spores onto the cantilever surface.

23.4.3.2 Results and Discussion

23.4.3.2.1 Measurements Performed in Air

The cantilevers that exhibited a shift in their resonant frequencies were the 50-, 40-, and 25-μm-length cantilevers. The data were analyzed using MATLAB software to extract the resonant frequency and quality factor, and these parameters were obtained by fitting the thermal spectra to the amplitude response of an SHO.[123] The spring constant was calibrated using the Sader method.[129] Resonant frequencies ranged from 150 to 600 kHz for the corresponding lengths mentioned above, with quality factors ranging from 12 to 35. The spores were imaged with a SEM microscope, and the effective number of spores

(a) 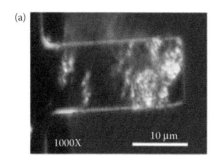   (b)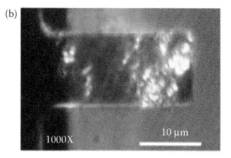

FIGURE 23.16
(a) Cantilever with spores before resonant frequency measurements in liquid medium. (b) Cantilever with spores after liquid measurements. (From Davila, A.P. et al. *Material Research Society, Materials and Devices for Smart Systems*, Fall Meeting, Boston, MA, 2005, with permission.)

was counted for each of the cantilevers. In Section 23.4.2.2.1, it was assumed that there were an equal number of spores on the top and bottom of the cantilever, so the effective number counted on the top was doubled. Typically, a *B. anthracis* spore is 1.5–2 μm long and about 1 μm in diameter. Assuming the shape of the spore to be cylindrical and a density close to that of water (1.0 g/cm^3), the mass of a spore would be approximately between 1 and 2 pg. But, if the spores are dried with CPD, some dehydration in the protein coat is expected due to the immersion in ethanol. Moreover, considering a dehydration of 50%–70%, the resultant mass would be approximately 500–300 fg, respectively. The average mass of a spore according to the mass measurements in air was found to be 367 ± 135 fg.

23.4.3.2.2 Measurements Performed in Liquid

When operating cantilevers in a viscous medium, the thermal spectrum of the cantilever broadens and the resonant frequency shifts to a lower value. In the present study, compared to the values in air, the resonant frequency in liquid was measured to be three to four times less, and the quality factor was found to have decreased by an order of magnitude.[90] As a result, only the two smallest lengths, 40 and 25 μm, were measured for the experiment, to achieve the required sensitivity. For the present study, cantilever beams having resonant frequencies in the tens to hundreds of kHz range were estimated to have the sensitivity to detect at least five spore cells. In order to extract the resonant frequency, quality factor, and spring constant in liquid, the same procedures using MATLAB were carried out as described earlier in Section 23.4.2.1.1. The resonant frequencies, in DI water, of the corresponding lengths mentioned above ranged from 30 to 125 kHz with quality factors ranging from 0.8 to 2.5. The spores were imaged with SEM (after drying them using CPD), and the effective number of spores was counted for each of the cantilevers. SEM images are shown in Figure 23.17. The measured reduction of the resonant frequency from air to liquid is depicted in Figure 23.18.

Again, the same challenge of accurately counting the number of spores on top and bottom was encountered and the same procedure for counting the spores was used as described in the previous section. Consequently, there was a variation in the mass of a spore for each cantilever. The average mass of a spore in liquid was found to be 1.96 ± 1.04 pg and excludes the contribution of the mass by the antibody and the BSA. This mass is in good agreement with the expected mass of 1–2 pg, as described in Section 23.4.3.2.1.

(a)  (b)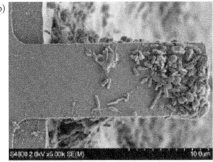

FIGURE 23.17
(a) SEM micrographs of spores on 35-μm-long cantilever after resonant frequency measurements in liquid medium. (b) SEM micrographs of 20-μm-long cantilever after liquid measurements. (From Davila, A.P. et al. *Material Research Society, Materials and Devices for Smart Systems*, Fall Meeting, Boston, MA, 2005, with permission.)

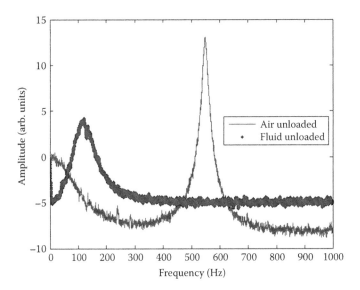

FIGURE 23.18
Decrease in resonant frequency when cantilever beam is immersed in liquid as compared to the measurement in air. (From Davila, A.P. et al. *Material Research Society, Materials and Devices for Smart Systems*, Fall Meeting, Boston, MA, 2005, with permission.)

23.4.4 Case Study 3: Detection of Virus Particles Using Nanoscale-Thick Microresonators

Taking a microcantilever beyond the thermal detection limit in fluid environments would be beneficial to real-time study of biological entities. This section presents a nano-scale thick cantilever, where, under externally driven dynamic conditions, vastly improves the mass detection sensitivity.[130] More specifically, this cantilever will be able to detect the mass of a single vaccinia virus, which is a member of Poxviridae family and forms the basis of the smallpox vaccine.[61]

23.4.4.1 Materials and Methods

23.4.4.1.1 Cantilever Fabrication and Mechanical Characterization

The cantilever fabrication details are similar to that described in Section 23.4.3.1.1, except that wet oxidation, followed by BHF etching, was performed in order to thin the SOI device layer down to 30 nm. The resonant frequencies of typical cantilever beams of length around 5 μm, width around 1.5 μm, and thickness around 30 nm was in the 1- to 2-MHz range with quality factor of around 5–7.

23.4.4.1.2 Nonspecific Virus Capture for Mass Measurement in Air

The vaccinia virus particles were grown and purified according to the established protocols described in Zhu et al.[131] The cantilever beams were first cleaned in a solution of ($H_2O_2/H_2SO_4 = 1:1$), rinsed in DI water, immersed in ethanol, and dried using CPD. The frequency spectra were then measured in order to obtain the "unloaded" resonant frequencies of the cantilever beams. Next, purified vaccinia virus particles at a concentration of *ca.* 10^9 PFU/mL in DI water were introduced over the cantilever beams and allowed to incubate for 30 minutes, following which the cantilever beams were rinsed in ethanol and dried using CPD. The resonant frequencies of the cantilever beams were then measured again in order to obtain the "loaded" resonant frequencies.

23.4.4.2 Results and Discussion

Using the mechanics of a spring-mass system (Equation 23.6), with n = 1, it was possible to determine the added mass for the corresponding change in resonant frequency. The cantilever beams were calibrated by obtaining their spring constant, k, using the unloaded resonant frequency measurement f_0, quality factor Q, and the plan dimensions (length and width) of the cantilever beam (the Sader method).[129] The resonant frequency and the quality factor were obtained by fitting the vibration spectra data to the amplitude response of an SHO (Equation 23.11). The measured spring constant of the cantilever beams was around 0.005–0.01 N/m. The virus particles were counted by observing the cantilever beams and virus particles using a SEM micrograph, as shown in Figure 23.19. The effective mass contribution of the viruses was calculated based on their relative position from the fixed end of the cantilever beams, similar to that done for the bacteria and spore cases, described above. Using the measurements from the various cantilever beams, the resonant frequency shift (decrease) versus the effective number of virus particles that were observed on the cantilever beams was plotted, as shown in Figure 23.20. The relationship was linear, as expected, clearly proving the validity of the measurements. Figure 23.21 shows the resonant frequency shift ($\Delta f = 60$ kHz) after the addition of a single virus particle. This shift is in the range of the theoretical minimum detectable frequency shift ($\Delta f_{min} = 50$ kHz), calculated using Equation 23.8 and the parameters of the cantilever beams used in the study. An average dry mass of 9.5 fg was measured for a single vaccinia virus particle, which is in the range of the expected mass of 5–8 fg.[132] The measured mass sensitivity of the present cantilever beams for a 1-kHz frequency shift is 160 ag added mass (6.3 kHz/fg). Once integrated with on-chip antibody-based recognition and sample concentrators, these nanomechanical resonator devices may prove to be viable candidates for ultrasensitive detection of air-borne virus particles.

23.4.4.3 Improvement in Virus Mass Detection Ability Using Driven Cantilever Beams

Figure 23.22 shows the theoretical minimum detectable mass change of cantilevers with a fixed width and thickness with respect to varying length measurements. The top line

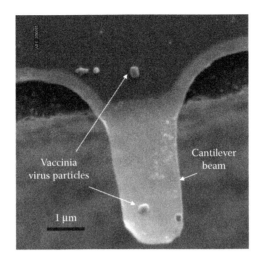

FIGURE 23.19
SEM micrograph showing a cantilever beam with a single vaccinia virus particle. The cantilever beam has plan dimensions of length, L = 4 μm, and width, W = 1.8 μm. (From Gupta, A. et al. *Applied Physics Letters* 84, 1976–1978, 2004, with permission.)

is for cantilevers with a quality factor of 5, while the bottom line is for cantilevers with a quality factor of 500.

The quality factor and vibration amplitude of the cantilever beam can be manipulated (up to a limit), by using a piezoelectric device that augments the natural resonant frequency of the cantilever beam, thereby lowering the limit of detection of the cantilever and improving its sensitivity. In the present study, cantilevers of two different sizes

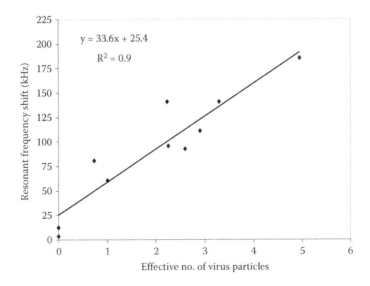

FIGURE 23.20
Plot of measured resonant frequency shift versus the effective number of virus particles on the cantilever beams. A linear fit was performed on the data points. (From Gupta, A. et al. *Applied Physics Letters* 84, 1976–1978, 2004, with permission.)

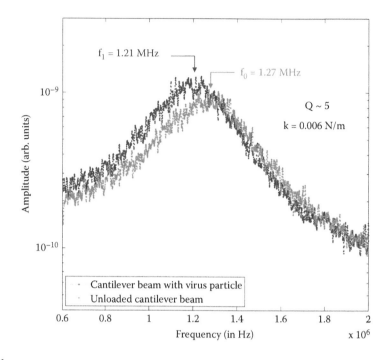

FIGURE 23.21
Plot of resonant frequency shift after loading of a single virus particle. There is a 60-kHz decrease in the resonant frequency of the cantilever beam with plan dimension of L = 3.6 μm and W = 1.7 μm. The unloaded resonant frequency f_0 = 1.27 MHz, quality factor Q = 5, and spring constant k = 0.006 N/m. The resonant frequencies were obtained from a fitting the amplitude response of an SHO to the measured data. (From Gupta, A. et al. *Applied Physics Letters* 84, 1976–1978, 2004, with permission.)

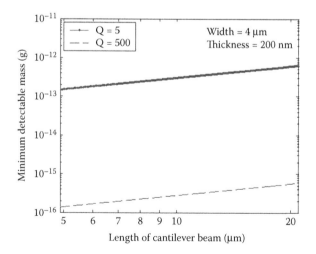

FIGURE 23.22
Plot of the minimum detectable mass versus the length of cantilever beam with fixed width and thickness dimensions. The top line indicates cantilevers with a quality factor of 5, while the bottom line is of cantilevers with a quality factor of 500. Driving the cantilever beam into resonance using an external source such as a piezoelectric oscillator would cause the increase in quality factor. (From Johnson, L. et al. *Sensors and Actuators B: Chemical* 115, 189–197, 2006, with permission.)

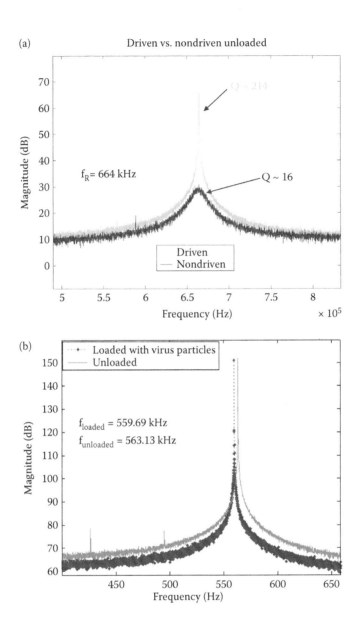

FIGURE 23.23
(a) Frequency spectra of a cantilever driven by thermal noise and by a piezoelectric ceramic. The quality factor of the piezo-driven cantilever is significantly greater than the quality factor of the same cantilever driven by only thermal noise. (b) Frequency spectra of cantilever driven by a piezoelectric ceramic before and after virus loading. Cantilever dimensions: length = 21 μm, width = 9 μm, and thickness = 200 nm. (From Johnson, L. et al. *Sensors and Actuators B: Chemical* 115, 189–197, 2006, with permission.)

were used and the effects of driving them with a piezoelectric ceramic were examined. Figure 23.23a compares the frequency spectra of a cantilever driven by thermal noise and that by a piezoelectric device. Cantilevers driven by the piezoelectric ceramic were shown to be clearly more capable of measuring smaller frequency shifts (see Figure 23.23b). This sensitivity is important when measuring virus particles that have a mass on the order of 10^{-18} g.

23.4.5 Case Study 4: Nanomechanical Effects of Attaching Protein Layers to Nanoscale-Thick Cantilever Beams

23.4.5.1 Introduction

In order to perform selective capture of the target analyte, it becomes necessary to functionalize the sensor surface with the receptor molecules. As was done for the case of the bacterial and spore cells above, the nanoscale cantilever beams described in Section 23.4.4 were coated with antibody molecules to vaccinia virus.[133] Now, the resonant frequency of a cantilever beam is a function of its spring constant and its effective mass. The spring constant of a rectangular-shaped cantilever beam is given by Equation 23.3. As can be seen from Equation 23.3, the thickness of the cantilever beam plays an important role in the overall value of the spring constant. When an attached layer to the cantilever beam is in the same thickness range as the cantilever beam, this should lead to an increase in the overall spring constant value. Protein layers such as BSA and antibodies are in the size range of tens of nanometers, while the cantilever beams used in this work have a thickness of 20–30 nm. The resonant frequency, f, after the attachment of a layer across the entire length of the cantilever beam, as well as on both the top and bottom of the cantilever beam surface, is then given as

$$f = \frac{1}{2\pi} \sqrt{\frac{k + \Delta k}{(m + \Delta m)^*}} \tag{23.12}$$

where m^* is the effective mass (which is 0.24 times the total mass for a rectangular-shaped cantilever beam) and Δk and Δm^* are the changes in the spring constant and effective mass, respectively. If the protein layer is assumed to be attached to the top and the bottom of the cantilever beam, then the overall stress on the cantilever beam can be assumed to be zero due to canceling effects. From Equation 23.12, it can be seen that, on attachment of the protein layer, the change in resonant frequency is due to the change in spring constant as well as the change in mass. The degree of change in resonant frequency (increases or decreases) depends on which of the factors outweighs the other.

23.4.5.2 Materials and Methods

23.4.5.2.1 Cantilever Fabrication and Mechanical Characterization

In the present study, arrays of silicon cantilever beams were fabricated using a combination of wet and dry etching processes, as described in Sections 23.4.3.1.1 and 23.4.4.1.1. The dimensions of the cantilever beams used in this work and the method of their mechanical characterization were similar to those described in Section 23.4.4.1.1.

23.4.5.2.2 Antibody Attachment Scheme

The attachment scheme for the antibody layer to the cantilever beam is shown in Figure 23.24. It has been modified from that used in a previous work.[125] One set of cantilever beams was used to study the selective capture of vaccinia virus particles while another set of cantilever beams in a different chip was used as the control. For the selective capture experiment, cantilevers were cleaned in a standard piranha solution ($H_2O_2/H_2SO_4 = 1:1$ in vol.), rinsed in DI water, immersed in ethanol, dried using a CPD system, and then measured in order to obtain the unloaded resonant frequencies. The cantilevers were then treated with 15 μL of biotinylated BSA (conc. 1.5 mg/mL) for 30 minutes, following by a rinse in

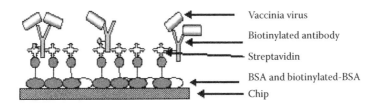

FIGURE 23.24
Scheme for immobilization of antibodies on cantilever beam surface. (From Gupta, A. et al. in *18th IEEE International Conference on Micro Electro Mechanical Systems, 2005, MEMS 2005*, Miami, FL, pp. 746–749, with permission.)

PBS (pH 6.3) for 5 minutes. The cantilevers were then treated with 15 μL of streptavidin (conc. 5 mg/mL) for 15 minutes, rinsed in PBS–Tween (0.05%) for 5 minutes to remove the excess streptavidin, and then treated with 15 μL of biotinylated antibody to vaccinia virus (conc. 5 mg/mL) for 15 minutes. Following a rinse in PBS–Tween for 5 minutes, the sample was then treated with BSA (conc. 5 mg/mL) in PBS (pH 6.3) for 15 minutes and then finally rinsed. The control cantilever beams were treated in similar manner, except that they did not have the antibody layer attached. All chips were then placed in increasing concentration of methanol (25%→50%→75%→100%) for around 1 minute each and then dried using CPD. The cantilever beams were then measured in air in order to obtain their loaded resonant frequency after the antibody/protein stack attachment.

23.4.5.2.3 Selective Capture of Virus Particles from Liquid

Following a PBS (pH 6.3) rinse, a mixture of labeled antigens was then introduced over the cantilever beam (antibody coated as well as the control). The antigens were *L. monocytogenes* V 7 strain (conc. 10^6/mL), *B. anthracis* Sterne strain spores (conc. 10^9/mL), and vaccinia virus Western Reserve strain (conc. 10^{10}/mL). The antigens were allowed to interact with cantilever beams for around 30 minutes. Following this, they were rinsed in PBS–Tween (0.05%) for 15 minutes, in order to detach the nonspecifically bound antigens, and then rinsed in DI water for 30 seconds. They were then immersed in ethanol and dried using CPD. The cantilevers were then measured using the LDV to obtain their loaded resonant frequency after the antigen capture. The cantilevers were imaged using SEM in order to record the specific capture efficiency.

23.4.5.3 Results and Discussion

After the attachment of the antibodies, the resonant frequencies of the cantilevers were found to fall in two categories. One set showed a decrease in resonant frequencies (which was the expected observation) while other set showed an increase in resonant frequencies. However, after the antigen exposure, the resonant frequency decreased (or stayed the same) for all cantilevers, when compared with the resonant frequencies after the antibody attachment. Table 23.2 presents a summary of the results for some of the representative cantilever beams. In Table 23.2, f_1 and f_2 represent the measured resonant frequencies after the antibody attachment and virus capture, respectively.

Figure 23.25a displays a representative plot of the frequency spectra and the fitted data at various stage of the biosensor analysis for the case in which the resonant frequency decreased after the antibody attachment, while Figure 23.25b displays the case when the resonant frequency increased after the antibody attachment. It can be seen in both cases that the resonant frequency decreased after the antigen capture.

TABLE 23.2

Summary of the Mechanical Parameters for Nanoscale-Thick Cantilever Beams (Length, l = 4–5 μm, Width, w = 1–2 μm, and Thickness, t ∼ 30 nm), Used to Attach Protein Layers and Capture Virus Particles

Cantilever No.	Unloaded Resonant Frequency, f_0 (MHz)	Unloaded Spring Constant (N/m)	Unloaded Quality Factor (Q)	Resonant Frequency Change After Antibody Attachment, $\Delta f [f_0 - f_1]$ (MHz)	Spring Constant After Antibody Attachment (N/m)	Resonant Frequency Change After Virus Capture, $\Delta f [f_1 - f_2]$ (MHz)
C1_N1	2.633	0.0173	6.11	0.097	0.0216	0.067
C1_N2	2.751	0.0231	7.70	0.195	0.0318	0.129
C1_N4	2.146	0.0114	4.89	0.233	0.0196	0.066
C1_N6	2.930	0.0226	6.87	0.223	0.0326	0.112
C2_N1	1.278	0.00624	5.92	−0.081	0.0114	0.071
C2_N3	1.264	0.00913	8.23	−0.132	0.0133	0.058
C2_N6	1.308	0.00756	6.57	−0.049	0.0114	0.104
C2_N7	1.300	0.00723	5.91	−0.028	0.0133	0.060

Source: Modified from Gupta, A. et al. in *18th IEEE International Conference on Micro Electro Mechanical Systems,* 2005, MEMS 2005, Miami, FL, pp. 746–749, with permission.

Selective capture efficiency analysis was also performed in a qualitative manner using the SEM micrographs in conjunction with the measured resonant frequency shifts. The increase in mass shown in Table 23.2 corresponded with the count of the effective number of virus particles. For example, no virus particles were observed on cantilever C1_N1 (see Table 23.2) with the change in mass being within the thermomechanical noise limits of ∼5 fg for this work,[22,119] while the mass change in C1_N6 corresponded with a single virus particle observed, as shown in Figure 23.26. Virus particles were observed on the control cantilever beams along with corresponding shifts in the resonant frequencies, but the overall number of such cases was much less than the antibody-coated cantilevers. It was interesting to observe that the number of bacteria and spores captured was negligible in both of the cases.

It can be seen from the experimental results in Table 23.2 that there is a demarcation point at which the resonant frequency increases after the antibody attachment. For the present experiments, the demarcation point at which the two sets of cantilever beams can be differentiated is seen to be around 0.01 N/m, with the spring constants being an order of magnitude higher for one set with respect to the second set. More work is underway in order to explain this phenomenon.

23.4.6 Case Study 5: Diamond Cantilevers for Cellular Studies

23.4.6.1 Introduction

Creating a sensor with an integrated localized heater can be highly desirable for uses in biochemical applications, specifically those utilizing heat-mediated biological reactions. Developing microcantilever sensors as multifaceted bioanalytical tools is advantageous for a wide variety of applications such as controlling and monitoring DNA hybridization, cell lysis, and enzyme reactions. Heated microcantilevers have been used in applications

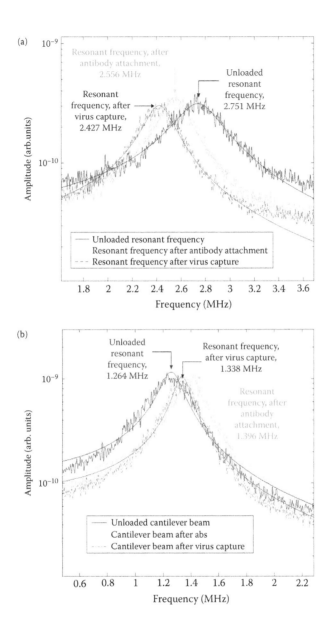

FIGURE 23.25

(a) Vibration spectra of cantilever beam C1_N2 (see Table 23.2) at various stages of the biosensor analysis showing the resonant frequency decreasing after the antibody sandwich attachment and after the antigen capture. (b) Vibration spectra of cantilever beam C2_N3 (see Table 23.2) at various stages of the biosensor analysis showing the resonant frequency increasing after the antibody sandwich attachment and decreasing after the antigen capture. (From Gupta, A. et al. in *18th IEEE International Conference on Micro Electro Mechanical Systems, 2005, MEMS 2005*, Miami, FL, pp. 746–749, with permission.)

of calorimetry,[134] thermogravimetry,[135] and sensing of temperature-dependent chemical reactions.[136,137] The goal of this particular study was to demonstrate the capability and utility of an ultrananocrystalline diamond (UNCD) cantilever with an integrated localized heater. UNCD has been shown to be superior to silicon for cell attachment and biochemical stability,[138–140] and thus is an ideal material for microcantilevers designed for biological

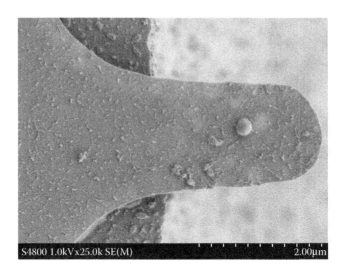

FIGURE 23.26
SEM micrograph of an antibody-coated cantilever beam C1_N6 (see Table 23.2) with virus particle captured right at the free end of the cantilever beam. (From Gupta, A. et al. in *18th IEEE International Conference on Micro Electro Mechanical Systems, 2005, MEMS 2005*, Miami, FL, pp. 746–749, with permission.)

studies. Here, we will discuss a unique silicon–UNCD cantilever design for heat-mediated cell lysis as a simple and effective method for DNA extraction.

23.4.6.2 Materials and Methods

23.4.6.2.1 Material Selection and Properties

Silicon has been the most common material used for microcantilever- and MEMS-based sensors due to well-characterized fabrication methodologies; however, for biological testing, material selection is a crucial component in the design of a sensor. UNCD has shown superior properties to that of silicon in terms of providing an environment for cellular growth. A comparison of PC12 cell growth on UNCD and silicon demonstrated formation of neuronal processes on the UNCD, while closely packed islands of cells formed on the silicon in lieu of neuronal processes that would indicate normal cell growth.[138] A related biocompatibility study showed remarkably improved characteristics with maximum cell attachment, cell spreading, and nuclear area coverage occurring on UNCD as opposed to silicon.[138] The ability of UNCD to promote strong affinity for biological analytes, while minimizing nonspecific binding, makes UNCD an attractive option as a highly translatable material for a range of biosensing applications.[139,141] The chemical inertness, wear resistance, promotion of cell adhesion and proliferation, and Young's modulus make UNCD a stable and selective platform for biomolecular interfaces, including microcantilevers.

23.4.6.2.2 Microcantilever Hotplate Design

Most heated microcantilevers with solid-state heaters take on a U-shape design that forms a continuous electrical path with a heater element at the free end. The heater region on a microcantilever is fabricated through selective doping that creates areas of low electrical conductivity and corresponding interconnects with high electrical conductivity to ensure heating is localized. Current applied across the resistive element—the low doped region—will experience an increase in temperature due to Joule heating. Designing a heated cantilever requires a series of considerations for optimization such as material, geometry, and dopant concentration in order to create the desired cantilever performance.[142] Silicon is a common and versatile material for heated cantilevers because of well-established fabrication techniques and its wide range of operating temperatures. Silicon-heated cantilevers typically operate as high as 500–600°C, but have been shown to go as high as 900°C within a few milliseconds.[143] These devices can also be modified with additional materials to achieve certain surface characteristics for specific applications, such as UNCD for biological applications. By applying a thin layer of UNCD to a silicon cantilever, a high degree of electrical isolation can provide higher fluidic compatibility of silicon-heated cantilever in conductive solutions.

23.4.6.2.3 Silicon–UNCD Cantilever Fabrication

The fabrication of the silicon–UNCD cantilever fabrication is highlighted here.[142] An SOI wafer consisting of a 400-μm silicon handle layer, 1-μm BOX layer, and 2-μm silicon device layer was used as the base material. The geometry of the cantilever devices was patterned and etched using an inductively coupled plasma (ICP) etcher. The silicon cantilever structure was selectively doped twice with phosphorous to form the low doped heater regions at the free end and high doped current paths along the sides of the heater. To ensure uniform heating at the free end of the cantilever, the shape of the doped silicon region was strictly considered and optimized.[144] UNCD was grown over the wafer to produce conformal insulating diamond coating. Using a silicon dioxide mask layer, the UNCD was etched with an oxygen plasma to the same dimensions as the cantilever geometry. The electrical contacts were defined by evaporating aluminum followed by wet etching. Finally, to release the cantilevers, the wafer was etched from the back followed by a short dip into 49% HF to remove any exposed silicon dioxide.

A related design is the microcantilever fabricated exclusively of UNCD in a bilayer format where the bottom layer is a boron-doped UNCD and the top layer is undoped UNCD making it piezoresistive.[145] Understanding the piezoresistive nature and sensitivity of UNCD and the temperature-resistivity dependence is a crucial. There are many other reported MEMS sensors that use polycrystalline diamond, including piezoresistive sensors,[146] chemical sensors,[147] and microcantilevers.[148]

Electrical and thermal characterization was performed to determine the performance of the silicon–UNCD-based microcantilevers. As changes in cantilever geometry, material, and doping concentration influence the resistance, power dissipation, and temperature distribution, it is imperative to experimentally measure these parameters. By placing the cantilever in series with a resistor and stepping through a range of voltages and recording the drop across the series resistor, we can determine the resistance as a function of power dissipated by the device. It is known that there is a strong nonlinear dependence of the resistance of doped silicon with power, as shown in Figure 23.27a. Using Raman spectroscopy, the heater temperature can be determined as a function of the applied power to the cantilever, via the Stokes peak shift. Overall, the resistivity of the cantilever is a nonlinear function of temperature. It was determined that the UNCD layer did not have a significant effect on the temperature variation over the length of the cantilever.[142,149]

23.4.6.2.4 Mammalian Cell and Bacterial Culture and Preparation

NIH 3T3 fibroblast cells were cultured in Eagle's medium with α modification and L-glutamine, supplemented with 10% FBS and 1% penicillin/streptomycin. Cells were maintained at 37°C with 5% CO_2 and 100% humidity. *L. monocytogenes* V7 were incubated

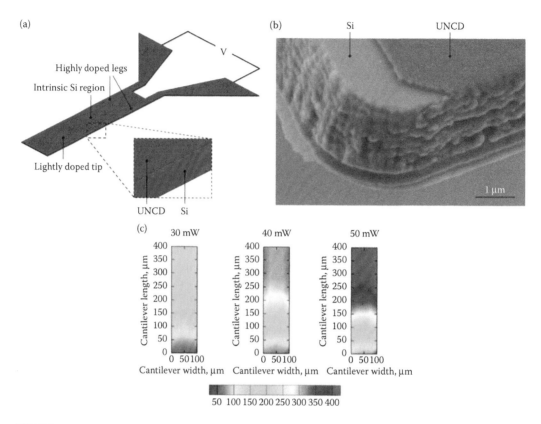

FIGURE 23.27
(a) Schematic of the Si–UNCD cantilever design. (b) SEM image of the free end of the cantilever where the UNCD is grown on top of the silicon. (c) Temperature distribution across the cantilever at three different heating powers. (From Privorotskaya, N. et al. *Lab on a Chip* 10, 1135–1141, 2010, with permission.)

in LB medium at 37°C and harvested at 16 hours after inoculation to guarantee a high percentage of live cell population.

The device was attached and wirebonded to a dual-inline package creating a 20-μL culture well. The entire package was UV sterilized and washed with PBS. For mammalian cell experiments, the devices were functionalized with poly-L-lysine to enhance mammalian cell attachment and adhesion. The 3T3 fibroblast cells were introduced into the well of the device at a final concentration of 10^5 cells/mL. For bacterial experiments, the devices were functionalized with biotinylated BSA followed by streptavidin and finally biotinylated Hsp60 to ensure immobilization of *L. monocytogenes* V7. The *L. monocytogenes* V7 were seeded into the well of the device at a concentration of 10^8 cfu/mL.

23.4.6.3 Results and Discussion

23.4.6.3.1 Mammalian Cell Lysis

NIH 3T3 fibroblasts seeded on the device were stained with three dyes: $DiOC_6(3)$ at 15 μg/mL, Hoechst 33258 at 20 μg/mL, and propidium iodide (PI) at 2 μg/mL.[142] $DiOC_6(3)$ binds to endoplasmic reticulum, vesicle membranes, mitochondria of living mammalian cells, and bacteria's lipid bilayer.[142,150] Hoechst 33258 labels nucleic acid molecules by permeating

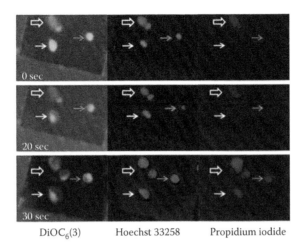

DiOC$_6$(3)   Hoechst 33258   Propidium iodide

FIGURE 23.28
Fluorescent images of NIH 3T3 fibroblast cells grown on the Si–UNCD cantilever hotplate and heat lysed over time. The hollow arrows indicate two cells were dead at the beginning of the experiment; solid arrows show the progression of cell death through heat lysis first over 20 seconds of heating followed by an additional 10 seconds. (From Privorotskaya, N. et al. *Lab on a Chip* 10, 1135–1141, 2010, with permission.)

either into a live or membrane-compromised cell. PI penetrates through the membrane of dead cells and bind to the DNA. The cells were then incubated in the culture conditions previously stated for 2 hours to allow for attachment and spreading. To monitor cell lysis, images were taken under three different conditions: prior to heating, following 20 seconds of heating at 93°C based on pre-calibrated results, and following an additional 10 seconds of heating. Figure 23.28 shows the fluorescent images of the free end of the cantilever of the various dyes over time. Prior to heating, the nonviable cells are indicated by the hollow arrow and the viable cells are indicated by the solid arrows. Heating the cantilever for the first 20 seconds shows the viable cells becoming compromised seen through the PI stain on the right in Figure 23.28; however, the Hoechst 33258 compared to the lipid membrane dye (DiOC$_6$(3)) indicated that the nucleic acids were highly localized within the cell meaning that the nuclear membrane was still intact. After heating for an additional 10 seconds, the PI and Hoechst 33258 dyes covered the entirety of the cell showing that the nuclear membranes were completely compromised.

23.4.6.3.2 Bacterial Lysis

L. monocytogenes V7 were first stained with the same three dyes as with the mammalian cells—DiOC$_6$(3) at 12 µg/mL, Hoechst 33258 at 25 µg/mL, and PI at 2 µg/mL[142]—then added to the functionalized cantilever device and incubated at 37°C for 1 hour. Similar to the mammalian cell experiment, the bacteria lysis was monitored over time under a microscope, where images were taken prior to heating and following 15 seconds of heating at 93°C. Figure 23.29a shows prior to heating that >90% of the cells were viable indicated by the lack of uptake of the PI into the bacteria. Figure 23.29b shows that after 15 seconds of heating that all cells are compromised, indicated by the permeation of PI into the cells. Merged blue and green filter fluorescent images (indicating only DiOC$_6$(3) and Hoechst 33258) are shown in Figure 23.29c on the left prior to and after heating. Also, a merged blue, green, and red filter fluorescent image is shown in Figure 23.29c on the right clearly showing the uptake of PI.

(a)   Before heating

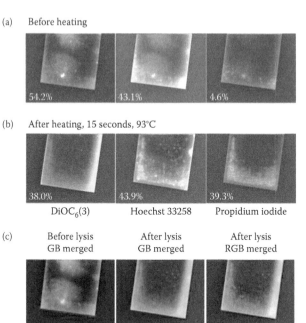

(b)   After heating, 15 seconds, 93°C

                    $DiOC_6(3)$ Hoechst 33258   Propidium iodide

(c)   Before lysis   After lysis   After lysis
           GB merged   GB merged   RGB merged

FIGURE 23.29
Fluorescent images of *L. monocytogenes* V7 grown on the Si–UNCD cantilever hotplate and heat lysed over time.
(a) Before heating, gray scale images were analyzed to determine the viability. (b) After heating at 93°C for 15
seconds, gray images were analyzed for final viability compared to before heat lysis. (c) Before and after
cantilever heating, all fluorescent images combined. (From Privorotskaya, N. et al. *Lab on a Chip* 10, 1135–1141,
2010, with permission.)

23.4.7 Case Study 6: Resonant Mass Sensor for the Mechanical Characterization of Adherent Cells

23.4.7.1 Introduction

The resonance characteristics of cantilever devices have made them ideal for measuring
and observing the mass of biological targets. The resonant frequency of these devices is
very sensitive to the addition of mass at the cantilever tip and thus by sensing the resonant
frequency over time, the growth or mass change rate of cells can be determined. In this
case study, we discuss a cantilever resonant sensor array designed to capture and hold cells
for growth measurements. We also describe a resonant sensor array that extends the functionality of the cantilever devices, which critically have a varying mass sensitivity depending on cell attachment position. By utilizing a pedestal design with four beam springs, as
opposed to the one of traditional cantilevers, this device results in more accurate mass measurements over time. Such a device also allows for the characterization of other biophysical
properties that effect resonant frequency, such as cell viscoelasticity. We introduce the general principles of this phenomenon and provide an overview of reported measurements.

23.4.7.2 Materials and Methods

23.4.7.2.1 "Living Cantilever Array"

The "living cantilever array" platform incorporates a multiplexed iteration of silicon substrates that is encapsulated in a linear PDMS microfluidic channel.[77] The platform is equipped

FIGURE 23.30
FESEM image of the released "living cantilever array." (Adapted from Park, K. et al. *Lab on a Chip* 8, 1034–1041, 2008.)

with positive dielectrophoresis (DEP) capabilities that help to capture cells on a functionalized cantilever surface. The cantilevers are in a parallel format designed for maximum and effective cell capture and provide a means for cells to flow, as shown in Figure 23.30.

The fabrication begins with an SOI wafer with a 240-nm silicon layer, a 400-nm BOX layer, and a 500-μm silicon substrate.[77] The first lithography step entailed defining the pattern of the cantilever. The silicon layer was then etched with RIE to define the cantilever geometry. The second lithography defined the etch windows in order to create an electrical contact to the surface, after which the exposed BOX was etched to produce contact windows on the substrate layer. The third lithography defined the metal electrodes and wirebonding pads through a liftoff process. Subsequently, etch windows for isotropic xenon difluoride (XeF_2) were defined with the fourth lithography step, exposing the BOX layer, which in turn released the cantilevers.

In order to decrease the time to detection, the "living cantilever array" is used as electrodes for DEP to quickly capture the target and detect the physical property. DEP uses nonuniform electric fields between the opposing cantilevers through the use of external sinusoidal sources with a 180° phase shift to draw the cells toward the tip of the cantilever.[77,151] Using a similar setup to that of Section 23.4.2.1.1, an LDV is used to measure the velocity of the vibrating cantilever.

To measure the mass of a cell, three different resonant frequency measurements are required.[77,78] The first measurement is the dry frequency that is taken in air to determine the spring constant of the device. The second measurement is the wet frequency that is taken with growth media to create a reference to determine the mass. Finally, the cells are seeded on the sensors, and the resonant frequency of a cell-filled sensor is measured and compared with the wet reference.

HeLa cells were cultured with Dulbecco's Modified Eagle's Medium/Nutrient Mixture F-12 Ham supplemented with 10% fetal bovine serum (FBS) at 37°C and 5% CO_2, after which the sensor was sterilized with 70% ethanol and functionalized with poly-L-lysine.[77] Prior to the mass experiment, the growth media was exchanged with low conductivity media (8.5% sucrose + 0.3% glucose in DI water), for use with the positive DEP signals applied to the cantilevers. The growth media was replenished every 6–8 hours, and the status of the cells was observed with a microscope. Measurements were carried out after 3 days.

23.4.7.2.2 Pedestal Sensor Array

As demonstrated by this chapter, microcantilevers are an attractive solution to many problems due to their high sensitivity and measurement throughput. However, it is well-known that the cantilever beam structure has a nonuniform mass sensitivity. Through a redesign, the pedestal sensor array was developed as a 9 × 9 array of 81 resonant mass sensors that achieve spatially uniform mass sensitivity.[78] Each sensor within the array comprises a square pedestal suspended by four beam springs over a shallow pit. This unique structure, as shown in Figure 23.31, exhibits a maximum 4% difference in mass sensitivity on any position on the pedestal.[59] The sensor is designed to operate in first resonance mode, where the platform vibrates vertically at approximately 160 kHz in air and 60 kHz in liquid.

The fabrication of the pedestal array starts with an SOI wafer with a 2-μm device layer, a 300-nm BOX layer, and a 500-μm handle.[78] The first lithographic process entailed defining the first metal layer for the sensor area using a liftoff process. The second lithographic step defined the etch mask for silicon etching and ultimately the springs and the platform. A third photolithographic step was used to define the electrodes and interconnects, again using a liftoff process. The final lithographic step defined a release window through RIE etching in the exposed BOX, leaving the silicon handle layer exposed. To suspend the platform and springs, the exposed silicon substrate was isotropically etched using XeF_2. Finally, a passivation layer of SiO_2 was deposited using PECVD. The wafer was diced, and individual chips were attached to a printed circuit board and wire-bonded.

Figure 23.32 shows the experimental setup where there is a uniform magnetic field combined with an actuating current that drives the sensor out of plane, creating a Lorentz force.[78] Similar to previous reports, an LDV system was used to measure the resonant frequency of the vibrating platform by comparing the relative phase between the input signal and the response of the platform. The experiment was fully automated using a motorized

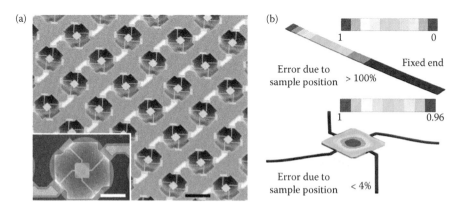

FIGURE 23.31
(a) SEM image showing a sensor array (scale bar: 300 μm); an individual sensor is shown in the inset (scale bar: 100 μm). (b) Simulation of the mass sensitivity over both a simple cantilever geometry and the pedestal sensor. The pedestal design has an error due to cell position less than 4% over the area of the platform. (Adapted from Park, K. et al. *Proceedings of the National Academy of Sciences* 107, 20691–20696, 2010.)

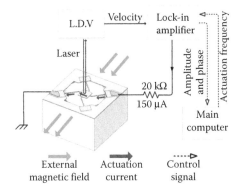

FIGURE 23.32
Overview of the mass measurement setup. (Adapted from Park, K. et al. *Proceedings of the National Academy of Sciences* 107, 20691–20696, 2010.)

stage on a microscope and held at physiological conditions using a fully enclosed temperature-controlled chamber.

A similar protocol for calibration and measurement is used as discussed above with the "living cantilever array." HT29 human colon adenocarcinoma cells were cultured in L-15 medium supplemented with 30% FBS and maintained at 37°C. Collagen type I solution was used to sterilize and functionalize the sensors at 37°C. The cells were seeded into the PDMS culture chamber over the sensor area at 9000 cells per chip. After the cell mass measurements were taken, the cells were fixed with 4% paraformaldehyde for 30 minutes.[78]

Owing to the finite elasticity of the cell, the cell body vibrates out-of-phase with the vibrating sensor skewing the resonant frequency shift created by the target mass.[78] Therefore, the measured apparent mass, which was derived from a resonant frequency shift, should also be a function of the cell stiffness. To more fully understand this relationship between the mass and mechanical properties of the cell, a dynamic model was created, with an idealized cell body and sensor consisting of two mass-spring-dampers as shown in Figure 23.33 and described by its equations of motion in Equation 23.13. The

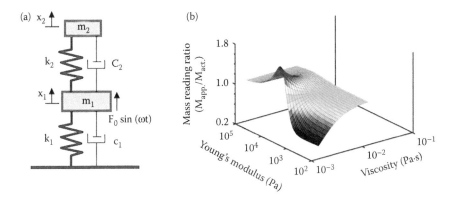

FIGURE 23.33
(a) Schematics of dynamical model to represent the sensor and finite elastic cell. (b) Result of the model, indicating that the viscoelastic properties influence the apparent mass measurement. (Adapted from Park, K. et al. *Proceedings of the National Academy of Sciences* 107, 20691–20696, 2010.)

model shown in Figure 23.33 indicates that the viscoelastic properties of the cell influence the mass measurement. This model shows that the pedestal sensor measurement technique delivers, in addition to the mass of the cell, a combination of complex elastic and viscoelastic dynamic properties of the cell.

$$\begin{bmatrix} m_1 & 0 \\ 0 & m_2 \end{bmatrix} \begin{Bmatrix} \ddot{x}_1 \\ \ddot{x}_2 \end{Bmatrix} + \begin{bmatrix} c_1 + c_2 & -c_2 \\ -c_2 & c_2 \end{bmatrix} \begin{Bmatrix} \dot{x}_1 \\ \dot{x}_2 \end{Bmatrix} + \begin{bmatrix} k_1 + k_2 & -k_2 \\ -k_2 & k_2 \end{bmatrix} \begin{Bmatrix} x_1 \\ x_2 \end{Bmatrix} = \begin{Bmatrix} F_0 \\ 0 \end{Bmatrix} e^{i\omega t} \qquad (23.13)$$

23.4.7.3 Results and Discussion

23.4.7.3.1 "Living Cantilever Array" Cell Mass and Growth

Flowing HeLa human cervical cancer cells through the linear channel of the "living cantilever array," cells were influenced through the use of positive DEP and captured.[77] Once captured, the cells were incubated at 37°C with 100% humidity and 5% CO_2 for 3 days. The masses of two HeLa cells were determined to be 1.01 and 3.57 ng. Following the LDV mass measurement, the cells were stained using $DiOC_6(3)$, a lipophilic fluorescent dye, and confocal measurements of the volumes were acquired to estimate mass through a volume measurement, resulting in masses of 2.48 ng and 4.09 ng. However, owing to the distribution of the cell mass over the length of the cantilever coupled with the nonuniform mass sensitivity along the cantilever beam, the apparent mass LDV measurement is reduced, requiring a correction based on the cell position along the cantilever.

23.4.7.3.2 Pedestal Sensor Cell Mass and Growth

Redesigning a cantilever to have uniform mass sensitivity resulted in the development of the new pedestal sensor.[78] HT29 human colon cancer cells were randomly seeded and captured on the pedestal array and grown on the functionalized surface for 50–60 hours. Figure 23.34 shows the continuous time measurements of single cells grown on the pedestal sensor. Figure 23.34a demonstrates a decrease in mass when a cell is removed from a sensor. Figure 23.34b shows a cell growth curve with a linear increase in mass, whereas Figure 23.34c shows a growth curve with an exponential increase in mass. Finally, Figure 23.34d illustrates a unique feature of the growth curve: a steep dip that is correlated with a cell division event. At division, the cell partially detaches from the surface of the pedestal, thereby affecting the resonant mass measurement. From these growth curves, the growth rate or mass change rate is extrapolated. By binning the data based on the mass, it was concluded that HT29 cells accumulate mass at a rate of 3.25%/h.

For studies on more highly motile cells, such as MCF-7 breast cancer cells, additional techniques are needed to ensure capture and retention throughout the growth measurement. This can be achieved through selective functionalization of the pedestal surface with collagen to promote adhesion, and all other areas with a pluronic to repel cells.[152]

23.4.7.3.3 Pedestal Sensor Cell Stiffness

After the mass measurements, the cells were fixed with 4% paraformaldehyde and the mass was re-measured. Figure 23.35 shows the mass before and after fixation, where the mass after fixation is approximately 1.4 times greater than before. Fixation has been shown to significantly increase the stiffness of tissue.[153,154] This supports our use of a viscoelastic Kelvin–Voigt model to describe the mechanical behavior of the cell on the pedestal sensor, which indicates that the stiffness of the cell influences the mass measurement.[78] Another study using the same pedestal sensors similarly showed that the viscoelastic

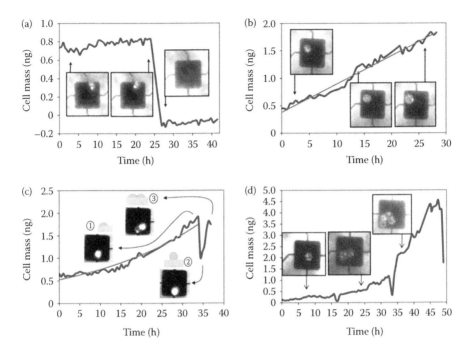

FIGURE 23.34
Mass measurements of HT29 cells over time. (a) Mass drops when a cell is lost from the pedestal sensor. (b) Linear growth profile. (c) Exponential growth profile followed by a division event where the growth profile takes a sharp decrease temporarily. (d) Multiple division events, showing the missing mass events at each division. (Adapted from Park, K. et al. *Proceedings of the National Academy of Sciences* 107, 20691–20696, 2010.)

properties of human breast cancer cells exhibited a change in apparent mass with fixation. Benign (MCF-10A), low malignancy (MCF-7), and high malignancy (MDA-MB-231) breast cancer cells were investigated, and it was shown that after fixation the masses were 1.7, 1.5, and 1.2 times greater than before fixation, respectively (Figure 23.35).[155] A similar dependence of apparent mass on material viscoelasticity was found in a study of tunable

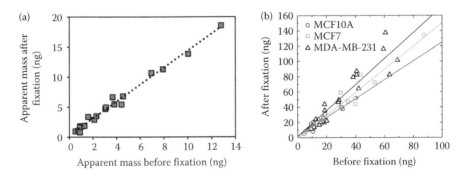

FIGURE 23.35
Comparison before and after fixation. (a) The apparent mass of HT29 cells after fixation is 1.4 times greater than before fixation. (Adapted from Park, K. et al. *Proceedings of the National Academy of Sciences* 107, 20691–20696, 2010.) (b) The apparent mass of MCF-10A, MCF-7, and MDA-MB-231 cells after fixation is 1.2, 1.5, and 1.7 times greater than before fixation. (Adapted from Corbin, E.A. et al. *Lab on a Chip* 15, 839–847, 2015.)

hydrogels where the Kelvin–Voigt model was also inverted in order to estimate stiffness from resonance measurements.[156]

23.4.8 Case Study 7: Suspended Microchannel Resonator for Cell Studies

23.4.8.1 Introduction

Common features of microcantilever resonant sensors are their high sensitivity, scalability, and ability to investigate small samples. In this chapter, however, we have reviewed a series of devices that have shown utility in many areas, but all suffer from sensitivity depletion in a liquid environment, thus hindering detection limits. In this section, we will discuss the suspended microchannel resonator (SMR) as a unique method to overcome the sensitivity degradation by integrating a liquid environment directly on the cantilever device. The SMR device incorporates an integrated microfluidic channel into a cantilever geometry. These devices have been used to weigh single nanoparticles,[84] biomolecules,[157] bacteria,[158] and mammalian cells.[76,92] This method provides excellent sensitivity and it is a suitable platform for the study of single suspended cells, such as bacteria and yeast, or nonadherent mammalian cells, such as lymphocytes.

23.4.8.2 Materials and Methods

23.4.8.2.1 SMR Device Design

Placement of MEMS resonant sensors in a fluid environment severely compromises the quality factor. Creating a biosensor with a high-quality factor for mechanical resonance can be attractive for sensing bio-targets with high sensitivity. The SMR is a standard cantilever geometry with an incorporated microfluidic channel allowing for liquid samples to flow through so that the sensor can be operated in a vacuum environment. It has been shown that dry resonators can achieve quality factors up to 15,000, and that the SMR shows no change in damping characteristics when the microfluidic channel is filled with fluid compared.[157] In addition to the microchannel in the cantilever compartment for mass measurements, the device includes two external channels: one used for storing and growing cells prior to measurement and the other used as a bypass channel to supply fresh medium between measurements.[76] Figure 23.36 shows a schematic of the SMR design and

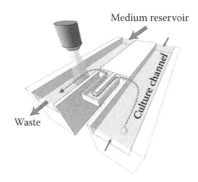

FIGURE 23.36
Schematic of the SMR. Through the microfluidic channel embedded in the cantilever, cells are flowed through while the cantilever is electrostatically driven at resonance. (From Son, S. et al. *Nature Methods* 9, 910–912, 2012, with permission.)

flow during operation.[76] The lack of additional damping typically encountered by cantilevers in liquid environments allows the SMR to measure mass with femtogram precision.

23.4.8.2.2 *Device Operation and Measurement Schemes*

The SMR design was originally based on a flow cytometer configuration.[157] Through many technical innovations, the system now uses an automated external pressure–driven fluid flow to capture and focus a cell at the tip of the cantilever. To measure the resonance of the cantilever, an optical lever method similar to that of AFM is used, where a laser is reflected off the cantilever and recorded by a set of photodiodes. As the cantilever vibrates, the reflected laser changes position on the photodiode sensors, ultimately resulting in the capture of a sinusoidal signal reflective of cantilever vibration. This signal in turn is used as feedback to amplify motion of the cantilever, and the cantilever resonant frequency is measured with a frequency counter.

The measurement of the mass with this device remains similar to other cantilever devices reviewed in this chapter, where an addition of mass is proportional and is measured through a change in the resonant frequency of the device.[159] The critical difference with the SMR is that when an individual mass passes through the microchannel, the frequency shift directly corresponds to the buoyant mass of the target. The buoyant mass is described as the effective mass of an object in fluid, which is the difference of the mass of the target cell and the mass of the displaced fluid, and is described by Equation 23.14:

$$m_B = V_{cell} \cdot (\rho_{cell} - \rho_{fluid}) \tag{23.14}$$

where m_B is the buoyant mass of the cell, ρ_{cell} and ρ_{fluid} are the density of the cell and the fluid, respectively, and V_{cell} is the cell volume. In this scheme, the cell volume is measured through the use of a commercial Coulter counter.[159] The measured buoyant mass from the cantilever is combined with the measured volume from the Coulter counter in Equation 23.14 to extract the cell density. Note that the measured buoyant mass depends on the density of the fluid in the SMR cantilever, which is also included in Equation 23.14. Using the same sample, the data are pooled into histograms of buoyant mass and volume and fitted with a log-normal function to extract the cell density. Alternatively, instead of using a Coulter counter to measure the cell volume and fit the relationship in Equation 23.14, buoyant mass measurements using fluids of two different densities can also be acquired to extract the slope of the line and fit the same relationship.[160] The same authors developed another technological advance for measuring various densities, but instead of using one cantilever they now employ a dual cantilever sensor.[92]

23.4.8.3 **Results and Discussion**

23.4.8.3.1 *Detection of Biomolecules*

Using the same mass and frequency relationship described in Equation 23.14, the SMR was first used to explore the detection of biomolecules.[157] The hollow cantilever channel walls were functionalized by adsorbing NeutrAvidin bound to poly(ethyleneglycol)-biotin grafted poly-L-lysine (PLL-PEG-biotin), and then biotinylated anti-goat IgG antibodies were immobilized onto the biotin. Finally, the binding of goat anti-mouse IgG to the surface was detected. The addition of each functionalization layer mass was quantified by its change in resonance and used as an offset for the baseline of mass measurements. Through changing the concentration of the target analyte to be captured in real time, a limit of detection is reached, as can be seen in Figure 23.37. It was determined, by assuming a monolayer of

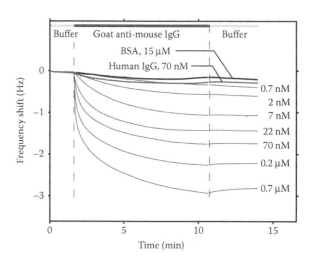

FIGURE 23.37

Goat anti-mouse IgG was injected at concentrations from 0.7 nM to 0.7 mM (blue traces). Between measurements, the surface was regenerated by injecting 200 mM glycine (HCl; pH 2.5), which dissociated the analyte while preserving the activity of the antibodies. Control injections with no IgG (black) or human IgG (red) showed very low levels of nonspecific binding. (From Burg, T.P. et al. *Nature* 446, 1066–1069, 2007, with permission.)

active antibodies in the hollow channel with a dissociation constant of 1 nM, that the lower limit of detection is on the order of 1 pM for a 30-kDa analyte.

23.4.8.3.2 Single-Cell Density

The method of varying the fluid density was found to be extremely useful in identifying *Plasmodium falciparum* malaria–infected erythrocytes,[160] host and donor transfused blood cells,[160] human lung carcinoma cells (H1650), and mouse lymphocytic leukemia cells (L1210).[92] First, using a single SMR configuration, the single-cell volume, mass, and density of human erythrocytes were measured from 16 patients, 9 of whom received blood transfusion prior to analysis. Next, using a dual SMR configuration, H1650 cells and L1210 cells were measured. Figure 23.38 presents the result of these measurements, where the

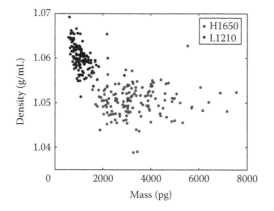

FIGURE 23.38

Density and mass of H1650 and L1210. Density shows much less variation than mass. (From Bryan, A.K. et al. *Lab on a Chip* 14, 569–576, 2014, with permission.)

two cell types were indistinguishable from each other by mass. However, differences in density were detectable by the SMR. A similar result was found with malaria-infected erythrocytes, which had densities distinct from noninfected erythrocytes. Erythrocytes from individuals suspected to have the thalassemia trait, a condition that affects the production of hemoglobin resulting in fewer circulating red blood cells, and comparing with healthy blood cells, were also found to have a different cell density. These individuals also received a blood transfusion prior to donating the sample. The results indicated that erythrocytes of thalassemia patients are offset from the normal sample through a density measurement except for a few erythrocytes that were received during the transfusion.

23.4.8.3.3 Bacterial Growth

The growth of *B. subtilis, Escherichia coli, Saccharomyces cerevisiae,* and mouse lymphoblast cells was investigated.[158] Each individual cell of *B. subtilis* and *E. coli* was trapped for 300 and 500 seconds, respectively, whereas the lymphoblasts and yeast could be trapped for approximately 30 minutes. Growth was observed as an increase in buoyant mass. Figure 23.39 shows that for all four cell types the measured growth rate is size dependent, suggesting that the cells are actively regulating their growth and division. Also, growth rates were compared with the population doubling time of exponential-phase cultures and were found to be consistent. For *B. subtilis,* it was found that an exponential growth model fit best to this data; however, for *E. coli,* identifying the best fit model was inconclusive.

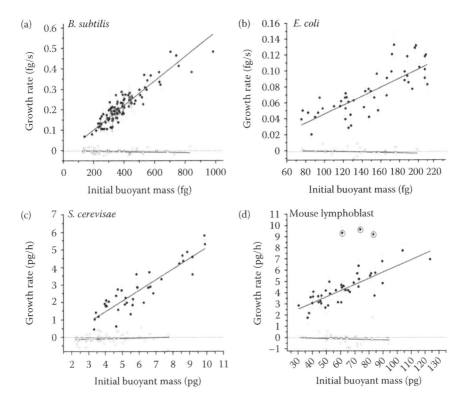

FIGURE 23.39
Growth rate versus initial buoyant mass for (a) *B. subtilis,* (b) *E. coli,* (c) *S. cerevisiae,* and (d) L1210 mouse lymphoblasts. (From Godin, M. et al. *Nature Methods* 7, 387–390, 2010, with permission.)

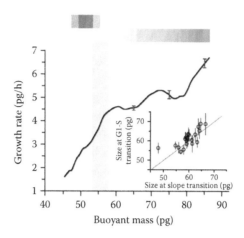

FIGURE 23.40
Growth rate as a function of cell mass, through a measurement of buoyant mass of one cell from the newborn stage through division. (From Son, S. et al. *Nature Methods* 9, 910–912, 2012, with permission.)

23.4.8.3.4 Mammalian Cell Growth and Size Regulation

The cell growth and size regulation of L1210 mouse lymphoblast and pro-B-cell lymphoid cell lines were measured over time.[76] Through the use of fluorescent cell cycle markers (FUCCI), cell cycle events were monitored throughout cell growth simultaneously with mass and size properties. Figure 23.40 shows the growth rate (pg/h) with respect to the buoyant mass, where there is a rapid increase in the growth rate after cytokinesis followed by a slow growth rate period. The yellow strip in Figure 23.40 indicates the G1-S transition; the entry into S phase was strongly correlated with the growth rate transition, suggesting that growth and cell cycle are linked. Upon further investigation of the G1-S transition, by limiting the nutrient supply, the cells were able to maintain their size by slowing cell cycle progression proportional to their decrease in growth rate.

23.5 Future Directions

In earlier decade, research efforts were focused on demonstrating a microcantilever as a highly sensitive sensor and numerous studies were focused on the improvement of its sensitivity. Such research efforts have established the technical background of the static and dynamic microcantilever sensors and produced a powerful platform to detect and characterize biologically relevant entities with unprecedented level of accuracy and precision. On the basis of the developed technologies, the focus of the research in these days has shifted to the applications of the microcantilever sensors to investigate the fundamental aspect of biological process on a new perspective or to develop a rapid, cost-effective, and highly sensitive medical diagnostic tool.

As the microcantilever sensors are being deployed into the real world, new technical challenges are emerging. First, the stochastic nature in capturing the target entities on the microcantilever prevents rapid and highly sensitive biochemical detection on a single molecule level. To detect the target entities with the microcantilever sensors either in static mode or in dynamic mode, the target entities have to approach and attach to the

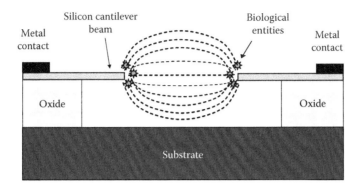

FIGURE 23.41
Cross section along the length of two cantilever beams acting as DEP electrodes showing the field lines as well as the capture of the target analyte.

cantilever's surface. However, in the case of extremely low sample concentration, the target entities may not encounter the miniaturized microcantilever and the microcantilever may fail to detect the target entities. This can be a critical issue, as the microcantilever sensors become sensitive enough to measure a single molecule. We believe the stochastic nature in the capturing process can be addressed by integrating pre-concentration mechanism, such as electrophoresis,[161,162] DEP,[150,163] inertial focusing,[164] and evaporative concentration.[165] Another approach can be a massive array of the microcantilever sensors so that the total sensing area can be significantly increased without losing the sensitivity.

Second, the surface of the microcantilever has to be carefully functionalized so that it is highly specific to the target entity with minimal nonspecific binding. High level of selectivity and specificity in the biochemical detection is essential, when the target entities are in a biological fluid containing undefined proteins and biomolecules. To this aim, the probe molecules that will capture the target entities should have high affinity as well as high selectivity, and the detection assay should be carefully designed. Besides, the surface of the microcantilever has to be passivated with macromolecules such as BSA and PEG, to minimize nonspecific binding. Also, it will be beneficial to add pre-filtration step before the actual assay so that majority of nonrelevant proteins and macromolecules can be removed.

Third, the structure of the microcantilever can be optimized to specific application and various structures can be developed. Earlier, a simple cantilever structure has been widely used in the microcantilever sensor, as it is easy to design and to miniaturize. However, as the microfabrication process evolves, it became feasible to implement different cantilever structures with additional functionalities. For example, a microfluidic channel is integrated into the microcantielver for a very high quality factor.[76,92] A suspended platform structure is developed to achieve uniform mass sensitivity,[59] or a micro-heater is integrated into a cantilever for direct sample processing.[142] Furthermore, electrodes for DEP[163,166] or electrophoresis[161,162] can be integrated on the microcantilever for direct sample concentration (see Figure 23.41).

23.6 Conclusion

The purpose of this chapter was to describe the various works using microcantilevers to detect and characterize biochemical entities, which include mammalian cells, bacteria, spores, viruses, and environmental condition such as pH change. Cantilever beams have immense potential to be used as sensors in an integrated microsystem. The cantilever beam is a very promising device that is being used and will continue to be used as an integral MEMS device for detection and sensing applications.

References

1. Bashir, R. BioMEMS: State-of-the-art in detection, opportunities and prospects. *Advanced Drug Delivery Reviews* 56, 1565–1586, 2004.
2. Binnig, G., Quate, C.F. & Gerber, C. Atomic force microscope. *Physical Review Letters* 56, 930, 1986.
3. Wickramasinghe, H.K. Progress in scanning probe microscopy. *Acta Materialia* 48, 347–358, 2000.
4. Timoshenko, S., Woinowsky-Krieger, S. & Woinowsky-Krieger, S. *Theory of Plates and Shells*, Vol. 2. McGraw-Hill, New York, NY, 1959.
5. Yasumura, K.Y. et al. Quality factors in micron- and submicron-thick cantilevers. *Journal of Microelectromechanical Systems* 9, 117–125, 2000.
6. Yang, J., Ono, T. & Esashi, M. Surface effects and high quality factors in ultrathin single-crystal silicon cantilevers. *Applied Physics Letters* 77, 3860–3862, 2000.
7. Yang, J., Ono, T. & Esashi, M. Energy dissipation in submicrometer thick single-crystal silicon cantilevers. *Journal of Microelectromechanical Systems* 11, 775–783, 2002.
8. Ono, T., Wang, D.F. & Esashi, M. Time dependence of energy dissipation in resonating silicon cantilevers in ultrahigh vacuum. *Applied Physics Letters* 83, 1950, 2003.
9. Stemme, G. Resonant silicon sensors. *Journal of Micromechanics and Microengineering* 1, 113–125, 1991.
10. Stowe, T. et al. Attonewton force detection using ultrathin silicon cantilevers. *Applied Physics Letters* 71, 288–290, 1997.
11. Harley, J. & Kenny, T. High-sensitivity piezoresistive cantilevers under 1000 Å thick. *Applied Physics Letters* 75, 289–291, 1999.
12. Yang, J., Ono, T. & Esashi, M. Mechanical behavior of ultrathin microcantilever. *Sensors and Actuators A: Physical* 82, 102–107, 2000.
13. Saya, D. et al. Fabrication of single-crystal Si cantilever array. *Sensors and Actuators A: Physical* 95, 281–287, 2002.
14. Kawakatsu, H. et al. Towards atomic force microscopy up to 100 MHz. *Review of Scientific Instruments* 73, 2317–2320, 2002.
15. Li, X., Ono, T., Wang, Y. & Esashi, M. Ultrathin single-crystalline-silicon cantilever resonators: Fabrication technology and significant specimen size effect on Young's modulus. *Applied Physics Letters* 83, 3081–3083, 2003.
16. Viani, M.B. et al. Small cantilevers for force spectroscopy of single molecules. *Journal of Applied Physics* 86, 2258–2262, 1999.
17. Chand, A., Viani, M.B., Schaffer, T.E. & Hansma, P.K. Microfabricated small metal cantilevers with silicon tip for atomic force microscopy. *Journal of Microelectromechanical Systems* 9, 112–116, 2000.
18. Genolet, G. et al. Soft, entirely photoplastic probes for scanning force microscopy. *Review of Scientific Instruments* 70, 2398–2401, 1999.
19. Ziegler, C. Cantilever-based biosensors. *Analytical and Bioanalytical Chemistry* 379, 946–959, 2004.
20. Lavrik, N.V., Sepaniak, M.J. & Datskos, P.G. Cantilever transducers as a platform for chemical and biological sensors. *Review of Scientific Instruments* 75, 2229–2253, 2004.

21. Meyer, G. & Amer, N.M. Novel optical approach to atomic force microscopy. *Applied Physics Letters* 53, 1045–1047, 1988.
22. Martin, Y., Williams, C. & Wickramasinghe, H.K. Atomic force microscope—Force mapping and profiling on a sub 100-Å scale. *Journal of Applied Physics* 61, 4723–4729, 1987.
23. Rugar, D., Mamin, H. & Guethner, P. Improved fiber-optic interferometer for atomic force microscopy. *Applied Physics Letters* 55, 2588–2590, 1989.
24. Davis, Z.J. et al. Fabrication and characterization of nanoresonating devices for mass detection. *Journal of Vacuum Science & Technology B* 18, 612–616, 2000.
25. Watanabe, S. & Fujii, T. Micro-fabricated piezoelectric cantilever for atomic force microscopy. *Review of Scientific Instruments* 67, 3898–3903, 1996.
26. Kenny, T. et al. Wide-bandwidth electromechanical actuators for tunneling displacement transducers. *Journal of Microelectromechanical Systems* 3, 97–104, 1994.
27. Kanda, Y. A graphical representation of the piezoresistance coefficients in silicon. *IEEE Transactions on Electron Devices* 29, 64–70, 1982.
28. Tortonese, M., Barrett, R. & Quate, C. Atomic resolution with an atomic force microscope using piezoresistive detection. *Applied Physics Letters* 62, 834–836, 1993.
29. Johnson, B.N. & Mutharasan, R. Biosensing using dynamic-mode cantilever sensors: A review. *Biosensors and Bioelectronics* 32, 1–18, 2012.
30. Arlett, J., Myers, E. & Roukes, M. Comparative advantages of mechanical biosensors. *Nature Nanotechnology* 6, 203–215, 2011.
31. Waggoner, P.S. & Craighead, H.G. Micro- and nanomechanical sensors for environmental, chemical, and biological detection. *Lab on a Chip* 7, 1238–1255, 2007.
32. Fritz, J. et al. Translating biomolecular recognition into nanomechanics. *Science* 288, 316–318, 2000.
33. Hansen, K.M. et al. Cantilever-based optical deflection assay for discrimination of DNA single-nucleotide mismatches. *Analytical Chemistry* 73, 1567–1571, 2001.
34. Mertens, J. et al. Label-free detection of DNA hybridization based on hydration-induced tension in nucleic acid films. *Nature Nanotechnology* 3, 301–307, 2008.
35. Wu, G. et al. Origin of nanomechanical cantilever motion generated from biomolecular interactions. *Proceedings of the National Academy of Sciences* 98, 1560–1564, 2001.
36. McKendry, R. et al. Multiple label-free biodetection and quantitative DNA-binding assays on a nanomechanical cantilever array. *Proceedings of the National Academy of Sciences* 99, 9783–9788, 2002.
37. Zhang, J. et al. Rapid and label-free nanomechanical detection of biomarker transcripts in human RNA. *Nature Nanotechnology* 1, 214–220, 2006.
38. Braun, T. et al. Micromechanical mass sensors for biomolecular detection in a physiological environment. *Physical Review E* 72, 031907, 2005.
39. Savran, C.A., Knudsen, S.M., Ellington, A.D. & Manalis, S.R. Micromechanical detection of proteins using aptamer-based receptor molecules. *Analytical Chemistry* 76, 3194–3198, 2004.
40. Mukhopadhyay, R. et al. Cantilever sensor for nanomechanical detection of specific protein conformations. *Nano Letters* 5, 2385–2388, 2005.
41. Arntz, Y. et al. Label-free protein assay based on a nanomechanical cantilever array. *Nanotechnology* 14, 86, 2003.
42. Huber, F., Hegner, M., Gerber, C., Güntherodt, H.-J. & Lang, H.P. Label free analysis of transcription factors using microcantilever arrays. *Biosensors and Bioelectronics* 21, 1599–1605, 2006.
43. Wu, G. et al. Bioassay of prostate-specific antigen (PSA) using microcantilevers. *Nature Biotechnology* 19, 856–860, 2001.
44. Chou, H.-T. et al. A micro-machined cantilever PSA sensor with digital wireless interface, in *Asia-Pacific Microwave Conference, 2006*, Yokohama, Japan, pp. 1622–1625.
45. Wee, K.W. et al. Novel electrical detection of label-free disease marker proteins using piezoresistive self-sensing micro-cantilevers. *Biosensors and Bioelectronics* 20, 1932–1938, 2005.

95. Peppas, N. & Mikos, A. Preparation methods and structure of hydrogels. *Hydrogels in Medicine and Pharmacy* 1, 1–27, 1986.

96. Peppas, N.A. Physiologically responsive hydrogels. *Journal of Bioactive and Compatible Polymers* 6, 241–246, 1991.

97. Bures, P. & Peppas, N.A. Structural and morphological characteristics of carriers based on poly(acrylic acid). *American Chemical Society*, Polymer Preprints, Division of Polymer Chemistry, 40(1), 345–346, 1999.

98. Scott, R.A. & Peppas, N.A. Compositional effects on network structure of highly cross-linked copolymers of PEG-containing multiacrylates with acrylic acid. *Macromolecules* 32, 6139–6148, 1999.

99. Scott, R., Ward, J. & Peppas, N. Development of acrylate and methacrylate polymer networks for controlled release by photopolymerization technology. *Handbook of Pharmaceutical Controlled Release Technology*, 47–64, 2000.

100. Hilt, J.Z., Gupta, A.K., Bashir, R. & Peppas, N.A. Ultrasensitive bioMEMS sensors based on microcantilevers patterned with environmentally responsive hydrogels. *Biomedical Microdevices* 5, 177–184, 2003.

101. Ward, J., Shahar, A. & Peppas, N. Kinetics of "living" radical polymerizations of multifunctional monomers. *Polymer* 43, 1745–1752, 2002.

102. Young, W.C. & Budynas, R.G. *Roark's Formulas for Stress and Strain*, Vol. 7. McGraw-Hill, New York, NY, 2002.

103. Hafeman, D.G., Parce, J.W. & McConnell, H.M. Light-addressable potentiometric sensor for biochemical systems. *Science* 240, 1182–1185, 1988.

104. Manalis, S. et al. Microvolume field-effect pH sensor for the scanning probe microscope. *Applied Physics Letters* 76, 1072, 2000.

105. Gershon, D. DNA microarrays: More than gene expression. *Nature* 437, 1195–1198, 2005.

106. 't Hoen, P.A.C., de Kort, F., Van Ommen, G. & den Dunnen, J.T. Fluorescent labelling of cRNA for microarray applications. *Nucleic Acids Research* 31, e20, 2003.

107. Larkin, J.E., Frank, B.C., Gavras, H., Sultana, R. & Quackenbush, J. Independence and reproducibility across microarray platforms. *Nature Methods* 2, 337–344, 2005.

108. van Bakel, H. & Holstege, F.C. In control: Systematic assessment of microarray performance. *EMBO Reports* 5, 964–969, 2004.

109. Brem, R., Oraszlan-Szovik, K., Foser, S., Bohrmann, B. & Certa, U. Inhibition of proliferation by 1-8U in interferon-α-responsive and non-responsive cell lines. *Cellular and Molecular Life Sciences* 60, 1235–1248, 2003.

110. Zhu, H., Butera, M., Nelson, D.R. & Liu, C. Novel type I interferon IL-28A suppresses hepatitis C viral RNA replication. *Virology Journal* 2, 80, 2005.

111. Kirkwood, J. Systemic adjuvant treatment of high-risk melanoma: The role of interferon alfa-2b and other immunotherapies. *European Journal of Cancer* 34, 12–17, 1998.

112. Vial, T. & Descotes, J. Clinical toxicity of the interferons. *Drug Safety* 10, 115–150, 1994.

113. Tortonese, M. Cantilevers and tips for atomic force microscopy. *IEEE Engineering in Medicine and Biology Magazine* 16, 28–33, 1997.

114. Thundat, T., Warmack, R.J., Chen, G. & Allison, D. Thermal and ambient-induced deflections of scanning force microscope cantilevers. *Applied Physics Letters* 64, 2894–2896, 1994.

115. Chen, G., Thundat, T., Wachter, E. & Warmack, R. Adsorption-induced surface stress and its effects on resonance frequency of microcantilevers. *Journal of Applied Physics* 77, 3618–3622, 1995.

116. Lee, J.H., Kim, T.S. & Yoon, K.H. Effect of mass and stress on resonant frequency shift of functionalized $Pb(Zr_{0.52}Ti_{0.48})O_3$ thin film microcantilever for the detection of C-reactive protein. *Applied Physics Letters* 84, 3187–3189, 2004.

117. Chon, J.W., Mulvaney, P. & Sader, J.E. Experimental validation of theoretical models for the frequency response of atomic force microscope cantilever beams immersed in fluids. *Journal of Applied Physics* 87, 3978–3988, 2000.

118. Oden, P. Gravimetric sensing of metallic deposits using an end-loaded microfabricated beam structure. *Sensors and Actuators B: Chemical* 53, 191–196, 1998.

119. Albrecht, T., Grütter, P., Horne, D. & Rugar, D. Frequency modulation detection using high-Q cantilevers for enhanced force microscope sensitivity. *Journal of Applied Physics* 69, 668–673, 1991.

120. Ekinci, K., Yang, Y. & Roukes, M. Ultimate limits to inertial mass sensing based upon nano-electromechanical systems. *Journal of Applied Physics* 95, 2682–2689, 2004.

121. Ekinci, K. & Roukes, M. Nanoelectromechanical systems. *Review of Scientific Instruments* 76, 061101, 2005.

122. Gupta, A., Denton, J.P., McNally, H. & Bashir, R. Novel fabrication method for surface micro-machined thin single-crystal silicon cantilever beams. *Journal of Microelectromechanical Systems* 12, 185–192, 2003.

123. Walters, D. et al. Short cantilevers for atomic force microscopy. *Review of Scientific Instruments* 67, 3583–3590, 1996.

124. Cleveland, J., Manne, S., Bocek, D. & Hansma, P. A nondestructive method for determining the spring constant of cantilevers for scanning force microscopy. *Review of Scientific Instruments* 64, 403–405, 1993.

125. Huang, T.T. et al. Composite surface for blocking bacterial adsorption on protein biochips. *Biotechnology and Bioengineering* 81, 618–624, 2003.

126. Ilic, B. et al. Single cell detection with micromechanical oscillators. *Journal of Vacuum Science & Technology B* 19, 2825–2828, 2001.

127. Burghardt, T. & Axelrod, D. Total internal reflection/fluorescence photobleaching recovery study of serum albumin adsorption dynamics. *Biophysical Journal* 33, 455, 1981.

128. Davila, A.P., Gupta, A., Walter, T., Akin, D., Aronson, A., & Bashir, R. Spore detection in air and fluid using micro-cantilever sensors. *Materials Research Society Symposium Proceedings*, Warrendale, PA, USA, vol. 888, pp. 0888–V10, 2005.

129. Sader, J.E., Chon, J.W. & Mulvaney, P. Calibration of rectangular atomic force microscope cantilevers. *Review of Scientific Instruments* 70, 3967–3969, 1999.

130. Johnson, L., Gupta, A.K., Ghafoor, A., Akin, D. & Bashir, R. Characterization of vaccinia virus particles using microscale silicon cantilever resonators and atomic force microscopy. *Sensors and Actuators B: Chemical* 115, 189–197, 2006.

131. Zhu, M., Moore, T. & Broyles, S.S. A cellular protein binds vaccinia virus late promoters and activates transcription in vitro. *Journal of Virology* 72, 3893–3899, 1998.

132. Bahr, G., Foster, W., Peters, D. & Zeitler, E. Variability of dry mass as a fundamental biological property demonstrated for the case of Vaccinia virions. *Biophysical Journal* 29, 305, 1980.

133. Gupta, A., Akin, D., & Bashir, R. Mechanical effects of attaching protein layers on nanoscale-thick cantilever beams for resonant detection of virus particles. *Micro Electro Mechanical Systems, 2005. MEMS 2005. 18th IEEE International Conference*, Fontainebleau Hilton Resort, Miami Beach, FL, USA, 2005.

134. Berger, R., Gerber, C., Gimzewski, J., Meyer, E. & Güntherodt, H. Thermal analysis using a micromechanical calorimeter. *Applied Physics Letters* 69, 40–42, 1996.

135. Berger, R. et al. Micromechanical thermogravimetry. *Chemical Physics Letters* 294, 363–369, 1998.

136. Marie, R., Thaysen, J., Christensen, C.B.V. & Boisen, A. A cantilever-based sensor for thermal cycling in buffer solution. *Microelectronic Engineering* 67, 893–898, 2003.

137. Pinnaduwage, L. et al. Detection of trinitrotoluene via deflagration on a microcantilever. *Journal of Applied Physics* 95, 5871–5875, 2004.

138. Bajaj, P. et al. Ultrananocrystalline diamond film as an optimal cell interface for biomedical applications. *Biomedical Microdevices* 9, 787–794, 2007.

139. Radadia, A.D. et al. Control of nanoscale environment to improve stability of immobilized proteins on diamond surfaces. *Advanced Functional Materials* 21, 1040–1050, 2011.

140. Yang, W. et al. DNA-modified nanocrystalline diamond thin-films as stable, biologically active substrates. *Nature Materials* 1, 253–257, 2002.

141. Stavis, C. et al. Surface functionalization of thin-film diamond for highly stable and selective biological interfaces. *Proceedings of the National Academy of Sciences* 108, 983–988, 2011.

24

Design and Biological Applications of Nanostructured Poly(Ethylene Glycol) Films

Sadhana Sharma, Ketul C. Popat, and Tejal A. Desai

CONTENTS

24.1 Introduction

At first glance, the polymer known as poly(ethylene glycol) (PEG) appears to be a simple molecule. It has the following structure that is characterized by hydroxyl groups at either end of the molecule:

$$HO-(CH_2CH_2O)_nCH_2CH_2-OH$$

It is a linear or branched, neutral polyether available in a variety of molecular weights, and soluble in water and most organic solvents. Despite its apparent simplicity, this molecule is the focus of much interest in the biotechnical and biomedical communities. Primarily, this is because PEG is unusually effective in excluding other polymers from its presence when in an aqueous environment. This property translates into protein rejection, formation of a two-phase system with other polymers, nonimmunogenicity, and nonantigenicity. Also, PEG is nontoxic. The lack of toxicity is reflected in the fact that PEG is one of the few synthetic polymers approved for internal use by the FDA, appearing in food, cosmetics, personal care products, and pharmaceuticals.

The true nature of PEG, however, is revealed by its behavior when dissolved in water. In an aqueous medium, the long chain-like PEG molecule is heavily "hydrated" (meaning water molecules are bound to it) and disordered; measurements using NMR spectroscopy (Breen et al., 1988) and differential thermal analysis indicate that as many as three water molecules are associated with each repeat unit. Gel chromatography experiments show that PEGs are much larger in solution than many other molecules (e.g., proteins) of comparable molecular weight (Hellsing, 1968; Ryle, 1965). Also, the PEG polymer chain is in rapid motion in solution, as demonstrated by relaxation time studies (Nagaoka et al., 1985). This rapid motion leads to the PEG sweeping out a large volume (its "exclusion volume") and prevents the approach of other molecules. In a very real sense, PEG is largely invisible to biological systems and is revealed only as moving bound water molecules. Thus, PEG can be thought of as a "molecular windshield wiper." When this molecular windshield wiper is attached to a molecule or a surface, it prevents the approach of other cells and other molecules. One result of this property is that PEG is nonimmunogenic.

The terminal hydroxyl groups of the PEG molecule provide a ready site for covalent attachment to other molecules and surfaces. Molecules to which PEG is attached usually remain active, demonstrating that bound PEG does not denature or hinder the approach of the other small molecules, and thus, PEG-modified surfaces and PEG-modified proteins are protein rejecting (Harris, 1992). Covalent linkage of PEGs to small-sized molecules increases the size of the molecule to which the PEG is bound, and this property has been utilized to decrease the rate of clearance of molecules through the kidney. Covalent linkage of PEG also alters the electrical nature of the surface, because charges on the surface become buried beneath a viscous hydrated neutral layer. This property has been utilized in capillary electrophoresis to control electroosmotic flow (Harris, 1992). Though PEG polymers have a variety of biological applications, in this chapter we will mainly focus on the design and biological applications of nanostructured PEG films with special reference to silicon-based microelectrical–mechanical systems (MEMS).

24.2 PEG in Action: Mechanism of Biofouling Control

The interaction of a device with a biological environment leads to various challenges that have to be taken into account in order to allow its proper operation. A major issue is biofouling, the strong tendency of proteins and organisms to physically adsorb to synthetic surfaces. The adsorbed protein layer tends to create undesired perturbations to the operation of devices such as pH and glucose sensors. In addition, the protein layer can mediate various biological responses such as cell attachment and activation, which may interfere with the optimal operation of the device by, for example, reducing its life span or increasing its power consumption. It is, therefore, desirable to passivate the surface with another hydrophilic and nonfouling material or polymer. PEG/poly(ethylene oxide) (PEO), a water soluble, nontoxic, and nonimmunogenic polymer, continues to be the favorite of researchers in order to achieve this objective.

There is an abundance of data available in literature that proves the nature of PEG to control biofouling on a variety of surfaces. Nevertheless, the unusual behavior of PEG is still an area of active research and debate. Several theories have been proposed by physicists and chemists but none of them is adequate to explain its protein-resistant behavior under all the conditions. In this section a brief overview of the possible mechanisms of PEG action will be presented.

The unusual efficacy of PEG as an apparent biologically passivating surface film is linked to both the presumed biological inertness of the polymer backbone and also to its solvated configuration. Much of the early theoretical work in this area borrowed the mechanisms that are used to explain the behavior of structureless polymers in isotropic liquids. They treated the proteins as hard spheres and the polymers as random coils (Andrade and Hlady, 1986; Jeon and Andrade, 1991).

Andrade and de Gennes treated the protein resistance of grafted PEG chains theoretically using ideas borrowed from colloid stabilization (Harris, 1992; Andrade and Hlady, 1986; Jeon and Andrade, 1991). According to the mechanism for resistance postulated by Andrade and de Gennes, the water molecules associated with the hydrated PEG chains are compressed out of the PEG layer as the protein approaches the surface. Thermodynamically, the removal of water from the PEG chains is unfavorable, and it gives rise to a steric repulsion that contributes to the inertness of the PEG-terminated surfaces. This theory predicts that the inertness of surfaces will increase with increasing length and density of the PEG chains. In addition, it is unable to explain the high protein resistances offered by PEG thin films made of low molecular weight PEG. Furthermore, the de Gennes–Andrade approach to the rationalization of the properties of inert surfaces based on conformational flexibility and properties of a hydrated polymer–water layer does not provide a general description of inert surfaces. This explanation might contribute to the mechanism of inertness in some cases, but it is clearly irrelevant in others; the ability of the functional groups to interact strongly with water molecules is, however, a common attribute of the inert surfaces in general.

The single-chain mean-field (SCMF) theory proposed by Szleifer for the polymer chains is able to rationalize the inertness of systems with a high density of short $(EG)_nOH$ chains ($n \geq 6$), including that of self assembled monolayers (SAMs) (Szleifer, 1997a; McPherson et al., 1998). The models proposed by Andrade and de Gennes had failed to address such systems. Szleifer's improvements to the model of Andrade and de Gennes do rationalize the inertness of SAMs terminated with $-(EG)_{n<7}OH$, but they do not provide a molecular level explanation of resistance (Satulovsky et al., 2000).

Besseling (1997) suggested that the chemical properties of surfaces might affect their states of hydration and the repulsive or attractive forces that result from the interactions of two such surfaces as they are allowed to interact. Theoretical analysis indicated that the interaction between two surfaces that causes changes in the *orientation* of water molecules (compared to bulk water) is repulsive; such surfaces were identified as having an excess of either proton donors or acceptors.

Wang et al. (1997) suggested that the chain conformation of $-(EG)_nOCH_3$ oligomers at the surface of SAMs seems to be an important determinant of resistance to protein adsorption. The conformation of $-(EG)_nOCH_3$ groups in the SAMs on gold that is inert is the helical conformation (*h*-SAM); when the molecules adopt an all-*trans* conformation on silver (*t*-SAM), the SAM is not inert (Harder et al., 1998). Force measurements on these SAMs suggested the presence of a strong dipole field in the inert *h*-SAM (Feldman et al., 1999). Monte Carlo simulations indicated that the dipole moments of the water molecules at the interface point into the SAM and can orient 3–4 layers of water at the interface (Pertsin and Grunze, 2000; Wang et al., 2000).

Grunze and coworkers have also proposed that the interaction of water with the surface of SAMs is more important than steric stabilization of the terminal $(EG)_nOH$ chains. Theoretical and experimental work from Grunze's group indicates that the conformation and packing of the chains in SAMs affects the penetration of water in the ethylene glycol layer and the inertness of the surface (Perstin and Grunze, 2000; Zolk et al., 2000). Monte Carlo simulations also suggested that *h*-SAMs interact more strongly with water than *t*-SAMs, and sum frequency generation experiments indicate that water penetrates into the (EG)$_n$OH layers of the SAMs and causes them to become amorphous (Zolk et al., 2000). However, it is not clear whether surfaces that are inert induce a particular structure in water molecules near the surface.

Ostuni et al. (2001) studied a variety of SAMs with different functional groups in order to test the theories currently available for explaining the protein-resistant behavior of PEG. They noted that the surface free energy is not a key determinant of inert surfaces. Also, the inertness of the surfaces also does not correlate with hydrophilicity. They suggested that the interaction of the surface with water is a key component of the problem (Besseling, 1997; Feldman et al., 1999; Rau and Parsegian, 1990). Adsorption of proteins, however, consists of two parts. The first and more important part is the formation of an interface between the surface and the protein with the release of water. This interface is generated from two separate interfaces: that between the surface and water, and a corresponding interface between protein and water. The second and probably less important part is reorganization of the protein on adsorption; this reorganization might cause changes in the structure of the protein–water interface. Further studies are however needed to test this hypothesis.

In contrast to the above mentioned studies, Sheth and Leckband (1997) reported direct evidence that PEG is not an inert, simple polymer, but that it can bind proteins. The formation of these attractive interactions is thought to be linked to rearrangements in the polymer configuration. They directly measured the molecular forces between streptavidin and monolayers of grafted M_r 2000 methoxy-terminated PEG. The interactions were investigated as a function of polymer grafting density with chain configurations ranging from isolated "mushrooms" to dense polymer brushes. These measurements provide direct evidence for the formation of relatively strong attractive forces between PEG and protein. At low compressive loads, the forces were repulsive, but they became attractive when the proteins were pressed into the polymer layer at higher loads. The adhesion was sufficiently robust so that separation of the streptavidin and PEG uprooted the anchored polymer from the supporting membrane. These interactions altered the properties of the grafted chains. After the onset of the attraction, the polymer continued to bind protein for several hours. The changes were not due to protein denaturation. These data demonstrate directly that the biological activity of PEG is not due solely to properties of simple polymers such as the excluded volume. It is also coupled to the competitive interactions between solvent and other materials such as proteins for the chain segments and to the ability of this material to adopt higher order intrachain structures.

In essence, in spite of all these investigations, the mechanism of resistance to adsorption of proteins is still a mystery and will continue to attract the attention of researchers in future; what is clear and well established is the proven ability of PEG to control biofouling.

24.3 BioMEMS (Biomedical Microelectrical–Mechanical Systems) and Issues of Biofouling

Microfabrication technology (MEMS or microelectrical–mechanical systems), mainly used in integrated circuits (IC) and the microelectronics industry, has experienced spectacular growth over the past 40 years. The same process technologies used in silicon micro-electronic chip manufacturing are also routinely used in the fabrication of biomedical microelectrical–mechanical systems (BioMEMS). These tiny devices, also referred to as bio-medical microsystems, hold promise for precision surgery with micrometer control, rapid screening of common diseases and genetic predispositions, and autonomous therapeutic management of allergies, pain, and neurodegenerative diseases. The health-care implica-tions predicted by the successful development of this technology are enormous, including early identification of disease and risk conditions, less trauma and shorter recovery times, and more accessible health-care delivery at a lower total cost (Polla et al., 2000). The three rapidly developing areas of microsystem technology are: (*a*) diagnostic BioMEMS, (*b*) sur-gical BioMEMS, and (*c*) therapeutic BioMEMS.

The material–tissue interaction that results from BioMEMS implantation is one of the major obstacles in developing viable, long-term implantable biosystems. The term biofoul-ing refers to the adhesion of proteins or cells onto a foreign material. Biocompatibility is used to describe the formation of encapsulation tissue, typically fibrous, which surrounds nondegradable implants. Encapsulation can occur from days to weeks, whereas biofoul-ing can begin to occur immediately upon implantation (Anderson, 1994). Biofouling and biocompatibility, more specifically fibrous encapsulation, are considered to be the two main reasons for device failure. Both biofouling and fibrous encapsulation have deleteri-ous effects on the performance of a device by retarding access of the sensor to the analyte or release of the drug to the target site, and both effects are functionally intertwined. Here, we will discuss biofouling and biocompatibility issues with special reference to mem-branes for drug delivery, tissue engineering, and biosensor applications.

Membrane biofouling is a process that starts immediately upon contact of the sensor with the body when proteins, cells, and other biological components adhere to the surface, and in some cases, impregnate the pores of the material (et al., 2001). Not only does bio-fouling of the sensor's outer membrane impede analyte diffusion, but it is also believed that the adhering proteins are one of the main factors that modulate the longer term cellu-lar and encapsulation response (Ratner et al., 1996). Electrode fouling, sometimes referred to as electrode passivation, is a completely different process that occurs on the interior of the sensor when substances from the body are able to penetrate the outer membranes and alter the metal electrode surface. Both types of fouling lead to the same sensor outcome—a declining sensor signal, but these are two different phenomena. In vitro protein- and blood-fouling studies and in vivo microdialysis studies (Wisniewski et al., 2001; Ishihara et al., 1998) have clearly shown detrimental effects of membrane biofouling on analyte transport that would lead to a decreased sensor signal. Other researchers have clearly shown that electrode biofouling exists and also causes a decrease in sensor signal.

Biocompatibility is a broad concept for which a variety of definitions exist. The biocom-patibility of a medical device may be defined in terms of the success of that device in ful-filling its intended function. In the context of implantable sensors, tissue engineering, and drug delivery devices, biocompatibility encompasses the body's reaction to the implanted device as well as the device's reaction to the body. Traditionally, an implant is deemed bio-compatible if it invokes a classic foreign body response that concludes in the formation of a thin, avascular, fibrous capsule (FC) around the implant (Sharkawy et al., 1998). However, this may not be the desired effect when communication between blood-borne analytes and the implant is essential, as with, for example, a sensor or immunoisolated cell drug delivery system. In such cases the presence of an FC may impede the transport of molecules dif-fusing from the microvasculature to the implant. This hypothesis is supported by many studies that suggest that after 1 week the response of an implanted sensor constructed with biocompatible inert materials decreases due to a transport barrier created by the FC (Clark et al., 1988; Gilligan et al., 1994; Pfeiffer, 1990; Rebrin et al., 1992). Such findings imply that the tissue response elicited by most inert materials conventionally defined as "biocompatible" may not be suitable for implants that require small molecule concentrations in the surrounding tissue to vary proportionally with those in the blood. It is therefore important to understand the fundamental events and processes that are responsible for implant failure. This knowledge would be useful for devising a strategy to overcome this problem.

The implantation of a biomaterial (without transplanted cells) initiates a sequence of events akin to a foreign body reaction starting with an acute inflammatory response and leading, in some cases, to a chronic inflammatory response or granulation tissue development, a foreign body reaction, and FC development. The duration and intensity of each of these is dependent upon the extent of injury created in the implantation, biomaterial chemical composition, surface free energy, surface charge, porosity, roughness, and implant size and shape. For biodegradable materials, such as those used in many polymer scaffold constructs for cell transplantation, the intensity of these responses may be modulated by the biodegradation process, which may lead to shape, porosity, and surface roughness changes, release of polymeric oligomer and monomer degradation products, and formation of particulates (Zolk et al. 2000). It is the extent and duration of the deviation from the optimal wound-healing conditions that determines the biocompatibility of the material.

The biocompatibility of the biomaterial in soft tissue involves important aspects of protein adsorption, complement activation, and macrophage and leukocyte adhesion and activation with the biomaterial as the agonist as in blood biocompatibility. A central consequence of the inflammatory response to the biomaterial is the activation of macrophages, resulting in the release of cytokines, growth factors, proteolytic enzymes, and reactive oxygen and nitrogen intermediates. The nature of the inflammatory response eventually is related to the degree of fibrosis and vascularization (Babensee et al., 1998) of the tissue reaction. A fibrous tissue reaction surrounding the implanted microcapsule may act as a barrier to nutrient and product diffusion. A thin fibrous tissue reaction may have a negligible diffusion resistance relative to the capsule membrane itself. In contrast, a granular tissue reaction would include vascular structures to facilitate the delivery of nutrients and the absorption of cell-derived therapeutic products. This suggests that a thick, granulous capsule with greater vascularity may be more compatible for implants as compared to thick, avascular, tightly packed repair tissue. A thin tissue of high vascularity can be induced with membranes of particular porosities or architectures or with membranes coated with biocompatible polymers.

Silicon and silicon-based (e.g., silica, glass) materials are the most commonly used materials for MEMS. Excellent micromachinability, enabling high volume fabrication of low-cost microsensors and actuators, and high sensitivity that can be used in a wide range of sensors (e.g., for pressure, motion, and temperature) make it an attractive material for microsystem manufacturing. Nevertheless, silicon itself was not regarded as a biomaterial that could directly interface living tissue. The majority of commercialized electronic implants developed in the beginning like the pacemaker exploited silicon chip technology but completely isolated the CMOS circuitry or sensor chip from the body, usually by a welded titanium package, polymer coating, or ceramic capsule. Due to recent interests in the use of MEMS technology for biomedical applications (BioMEMS), use of silicon as a biomaterial has received considerable attention.

Elemental silicon is nontoxic in nature. Silicon degrades mainly into monomeric silicic acid $(Si(OH)_4)$, which just happens to be the most natural form of silicon in the environment.

In fact, the human body actually needs silicon in this form as an essential trace nutrient (Schwartz and Milne, 1972; Carlisle, 1986). Indeed, silicic acid accounts for 95% of the silicon that is cycled through rivers and oceans, and is present in many foods and drinks. Furthermore, tests using radiolabeled silicic acid drinks given to human volunteers resulted in the concentration of the acid in the bloodstream rising only very briefly above typical values of $\sim$1 mg L^{-1} (Carlisle, 1972). Urine excretion of silicic acid is also highly efficient and expels all the ingested silicon.

Silicon was shown to be extremely bioinert and nontoxic in cortical tissues, suggesting its potential application in implantable microdevice fabrication. Studies on the materials implanted in the cerebral cortex of the animals showed that phosphorous-doped monocrystalline silicon was nonreactive with the absence of any calcification, macrophages, meningeal plasma fibroblasts, and giant cells in the connective tissue capsule (Stenssas and Stenssas, 1978).

Canham (1995) investigated silicon bioactivity with regard to in vivo bonding ability. He observed the growth of hydroxyapatite on porous silicon in simulated body fluids inferring the possible bone implantability of the material. He also found that bulk silicon is relatively bioinert, whereas hydrated microporous silicon coatings were both biocompatible and bioactive with regard to hydroxyapatite nucleation. Furthermore, he reported the manufacturing processes that enhance the biostability of porous silicon (Carlisle, 1986).

Edell et al. (1992) observed gliosis at the tip of the insertable silicon microelectrode arrays in long-term studies, which most probably indicated tissue movement relative to the tips. Although only a single layer of tightly coupled glial cells surrounded the shafts of the arrays, more tissue response was seen at the tip. They demonstrated that tissue trauma could be minimized and biocompatibility could be improved by making appropriate changes in the design of the silicon shafts for the cerebral cortex. Clear silicon dioxide–coated structures showed better long-term biocompatibility in this study. Schmidt et al. (1993) looked into the passive biocompatibility of uncoated and polyamide-coated silicon electrode arrays in feline cortical tissue and observed modest tissue reactions to the implants. Edema and hemorrhage were present around the short-term (24 h) implants, but affected less than 6% of the total area of the tissue covered by the array. With chronic implants (6 months), leukocytes were rarely present and macrophages were found around one-third of the implants. Gliosis was found around all implants, but a fibrotic capsule was not always present and if so, it never exceeded a thickness of 9 μm. The amount of the tissue reaction to the implant suggested that the materials were nontoxic.

It is conceivable that surface properties play a key role in the biocompatibility of silicon. Hence, in order to fully exploit the capabilities of BioMEMS for medical benefits, it is highly desirable to devise strategies to improve the interfacial properties of silicon.

24.4 Nanostructured PEG Films for Silicon-Based BioMEMS

PEG films on surfaces can be prepared by physical adsorption of high molecular-weight PEG or various PEG-containing amphiphilic copolymers on the substrates. This approach may provide a simple, rapid, and effective means of producing PEG surfaces, if the PEG-containing copolymers can be strongly adsorbed onto the surfaces. Nevertheless, the bonding is weak and the immobilized polymers do not permanently remain on the surface.

Covalent grafting of PEG derivatives to polymeric substrates is the most effective way of creating a more stable film on the surface. Several techniques have been used to attach covalently to the surfaces. These include direct chemical coupling of PEG derivatives or through the presence of some silane linker. However, this method is applicable only to the materials that have chemically active functional groups (e.g., –OH, –NH$_2$) at the surface that can react with PEG. In addition, the coupling procedures are usually very complicated and time–consuming. For inert surfaces without any functional groups, PEG attachment can be carried out using UV irradiation, high energy γ irradiation, and plasma glow discharge (Lee et al., 1995). This section will focus on the methods developed by our group to create nanostructured PEG films for silicon-based substrates (e.g., silicon, glass, silicon dioxide, quartz) used for BioMEMS.

Our research group has designed and developed nanostructured PEG films using two different methods: one by coupling PEG–silane in solution phase and another by vapor deposition of ethylene oxide to grow PEG on the surface. Although the solution phase surface modification technique is applicable to a variety of BioMEMS with open channels, it is not appropriate for closed micro- or nanoscale channels. Due to enclosed micro- or nanoscale-size features on the surface, viscosity, and surface tension of solution injected for surface modification may clog the channel, forming lumps and aggregates. In such a case, vapor deposition technique may be more efficient in creating nanostructured, uniform, and conformal PEG films. Our solvent-free vapor deposition technique to modify silicon surfaces involves growing PEG on silicon substrates using ethylene oxide in the presence of a weak Lewis acid as catalyst (Popat et al., 2002).

24.4.1 Solution-Phase Technique for Nanostructured PEG Films

In this method, PEG films on silicon surfaces are created using a covalent coupling technique (Zhang et al., 1998; Sharma et al., 2003). This scheme involves immobilization of PEG to the silicon surface by the functionalization of a PEG precursor through the formation of SiCl$_3$ groups at its chain ends, followed by reaction of surfaces with compounds of the form PEG–OSiCl$_3$ because trichlorosilane derivatives react with surfaces much faster than the other cholorosilane derivatives (Patai and Rappoport, 1989). The hydrolysis of the PEG–OSiCl$_3$ compound by traces of adsorbed water on the silicon surface results in the formation of silanols, which then condense with the silanols at the silicon surfaces to form a network of Si–O–Si bonds, resulting in a silicon surface modified with the PEG chains (Figure 24.1) (Wasserman, 1989; Ulman, 1991). PEG–OSiCl$_3$ (called PEG–silane) is synthesized by reacting PEG with silicon tetrachloride in the presence of triethylamine (catalyst). All the reactions are performed in anhydrous conditions to prevent hydrolysis and other side reactions (Vansant et al., 1995).

PEG concentration (0–50 mM) and time of immobilization (30–120 min) are varied to prepare films of various grafting densities. Figure 24.2 summarizes the thickness of PEG films as measured using ellipsometry. There was no quantitative coupling of PEG at 2 mM concentration. This PEG concentration is probably too low to provide measurable coverage on silicon surface. For the other two concentrations, the thickness of PEG films increases with the concentration of PEG and time of immobilization. This trend is quite significant at 10 mM PEG concentration. However, the variation in the thickness of PEG films with time of immobilization in the case of 5 mM PEG concentration is almost negligible.

PEG films formed by using 10 mM initial PEG concentration (immobilization time = 60 min) for coupling were best suited to our requirement (PEG film thickness = 20 ± 0.93 Å). Nevertheless, we further investigated higher PEG concentrations (upto 50 mM) and

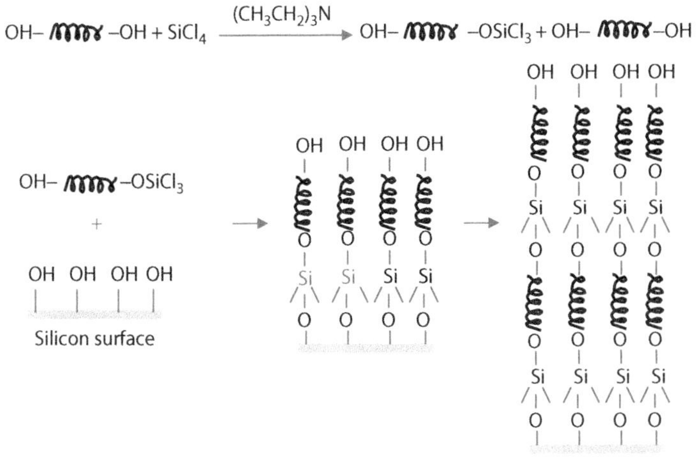

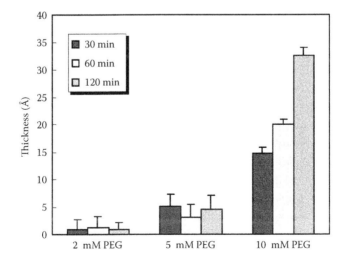

FIGURE 24.1
Reaction scheme illustrating the modification of silicon surface with PEG.

immobilization time (up to 24 h) in order to understand whether the PEG films keep on growing on the surface infinitely or the surface achieves saturation at some point.

The nature (hydrophilic or hydrophobic) of the unmodified and PEG-modified surfaces was assessed by measuring water contact angle. Surfaces with water contact angles in the range of 20–60° are generally considered to be hydrophilic and expected to show minimal protein adsorption. Water contact angle of the bare silicon surface was <6°. PEG-modified surfaces showed water contact angles in the hydrophilic range (20–60°) (Figure 24.3a). Also, PEG concentration (especially in the low concentration, <10 mM, range), rather than the time of immobilization (Figure 24.3b), influences the values of the water contact angles to a greater extent. Lower values of contact angle with higher standard deviations for 5 mM

FIGURE 24.2
Variation in PEG film thickness (measured using ellipsometry) as a function of initial PEG concentration and immobilization time.

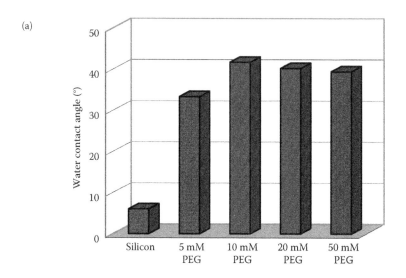

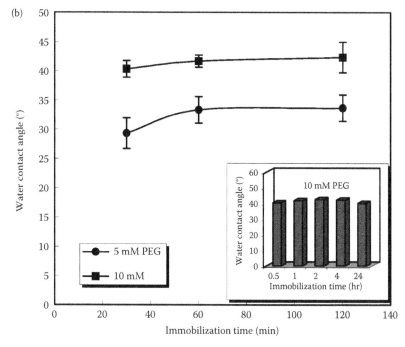

FIGURE 24.3

(a) Variation in water contact angle with PEG concentration and (b) variation in water contact angle with immobilization time.

PEG concentration compared to 10 mM PEG indicate less uniform surfaces at lower PEG concentrations.

X-ray photoelectron spectroscopy (XPS) analysis was performed to ascertain the presence of immobilized PEG on the surface of silicon previously detected by ellipsometry. Survey spectra for an unmodified silicon surface and PEG-modified surface (10 mm PEG, 1 h) illustrated more clearly the change in carbon, silicon, and oxygen composition of the

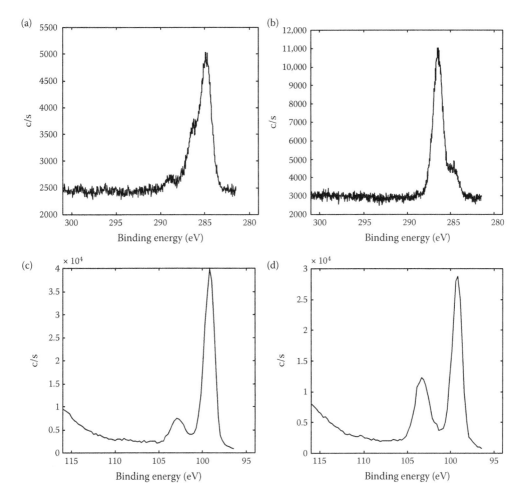

FIGURE 24.4
High-resolution C-1s and Si-2p scans for unmodified bare silicon and PEG-modified (10 mM, 1 h) silicon surfaces. C-1s scans: (a) unmodified silicon, (b) PEG-modified silicon; Si-2p scans: (c) unmodified silicon, (d) PEG-modified silicon. Take off angle: 65°.

silicon surface before and after PEG immobilization. There is a sharp increase in the C-1s (285 eV) and the O-1s (528 eV) peak, and a decrease in the Si-2p (100 eV) peak for the PEG-modified surface compared to bare silicon (not shown).

High-resolution carbon (C-1s) and silicon (Si-2p) scans clearly indicated the existence of PEG moieties on silicon surface (Figure 24.4). We see a distinct increase in C–O (shifted 1.5 eV from C–C peaks) (Figure 24.5b) and Si–O (Figure 24.5d) peaks when PEG is coupled to silicon in comparison to the unmodified silicon surface (Figure 24.5a,c). Since XPS analyzes only the top 50 Å of a surface, these results confirm the presence of PEG moieties on the surface of silicon.

Figure 24.5 highlights the XPS elemental analysis for various PEG concentrations and immobilization times. XPS characterization of PEG-modified silicon surfaces showed an increase in carbon concentration as well as a slight increase in oxygen content compared to unmodified surfaces. In addition, PEG-modified silicon surfaces showed a decrease in silicon concentration compared to unmodified silicon. This trend is followed both with

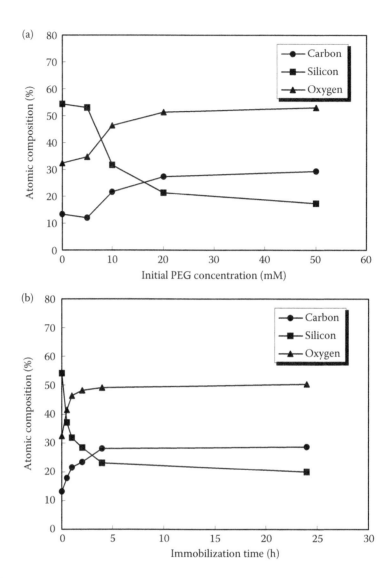

FIGURE 24.5
XPS elemental composition for unmodified and PEG-modified surfaces.

increasing initial PEG concentration (Figure 24.5a) and immobilization time (Figure 24.5b). This indicates building up of the PEG films on the silicon surface.

We did not observe an appreciable difference in surface atomic compositions for 5 mM PEG at various immobilization times (data not shown here). This observation corresponds to the results obtained by ellipsometric measurements where we found almost no variation in film thickness with time of immobilization. The data clearly suggested that 5 mM PEG concentration is too low to provide sufficient coverage to the surface and form a well-defined PEG film. These increases and decreases with time were, however, more significant for 10 mM PEG concentration at different immobilization times and, therefore this concentration was explored for various immobilization times. In this case, we saw a significant increase in carbon and oxygen concentrations; and decrease in silicon concentration with increase in immobilization time (Figure 24.5b). This substantiated previous conclusions drawn from ellipsometric measurements regarding increasing PEG grafting densities with time of immobilization.

High-resolution C-1s scans provide more precise information about PEG grafting as a function of PEG concentration and immobilization time. We saw substantial increase in the intensity of C–O peak with increase in initial PEG concentration and immobilization time (not shown). In order to extract quantitative information from this observation, the curve-fitting software supplied with the instrument was used to calculate the relative contribution of the two peaks (C–O and C–C) to the total carbon concentration. Bare silicon also shows small C–O as well as C-C peaks due to atmospheric impurities. Therefore, these values were subtracted from the [C–O] fraction of the PEG-grafted samples. The fractional area of the [C–O] peak can now be taken as measure of PEG grafting (Figure 24.6).

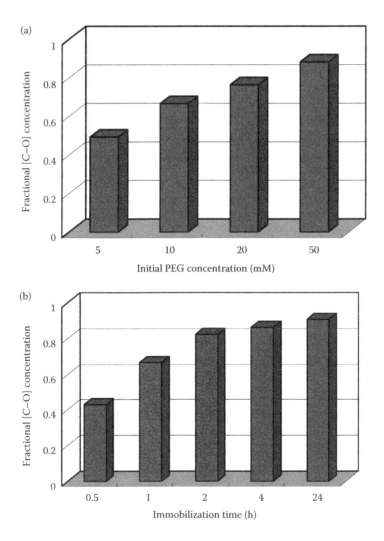

FIGURE 24.6
XPS analysis of the variation in PEG grafting density: (a) effect of initial PEG concentration and (b) effect of immobilization time.

Topography of modified surfaces is extremely important for silicon-based biomedical devices with micro- and nano-sized features on the surface. It directly affects protein adsorption and eventually, the cell adhesion and proliferation on the surface. Atomic force microscopy (AFM) is an extremely useful tool for studying the modified surfaces as it provides real-space film morphology and nanostructure. It also provides detailed topographical information about surface features in terms of roughness parameters of the films. The surface topology and uniformity of the unmodified and PEG-modified silicon surfaces as a function of PEG concentration and immobilization time were characterized using tapping mode AFM.

Figure 24.7 shows 2 μm AFM scans of PEG films developed for various initial PEG concentrations (5, 10, 20, and 50 mM—immobilization time 1h). PEG films developed with 5 mM initial PEG concentration (Figure 24.7b) have more irregularities compared to unmodified silicon (Figure 24.7a). However, the RMS roughness value (R_{rms}) for bare silicon is higher (Figure 24.9). It is important to mention here that RMS roughness (R_{rms}) is a statistical parameter and therefore, averages out the roughness for the entire scan. RMS roughness for 10 and 20 mM initial PEG concentration is almost the same. A comparison of the scans for 10, 20, and 50 mM initial PEG concentrations show flattening or broadening of surface features (Figure 24.7c–e) when immobilization is done at higher concentrations. This indicates more PEG grafting at higher PEG concentrations.

Figure 24.8 shows the AFM pictures of the PEG films developed for different immobilization times (10 mM PEG). The PEG surface density appears to be increasing with immobilization time. For samples immobilized with PEG for 0.5 h (Figure 24.8b), the surface shows close similarity with unmodified silicon (Figure 24.8a) though the RMS roughness is much higher compared to unmodified silicon surface (Figure 24.9). This may be due to less PEG coverage. The surface PEG density and uniformity is significantly improved at higher immobilization times. The RMS roughness parameters increase with the time of immobilization. The RMS roughness values for 1 and 2 h immobilization time do not appear to be significantly different.

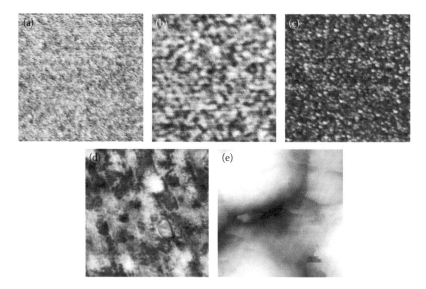

FIGURE 24.7
AFM scans for PEG films formed using different PEG concentrations (immobilization time = 1 h): (a) 0 mM (clean silicon), (b) 5 mM, (c) 10 mM, (d) 20 mM, and (e) 50 mM.

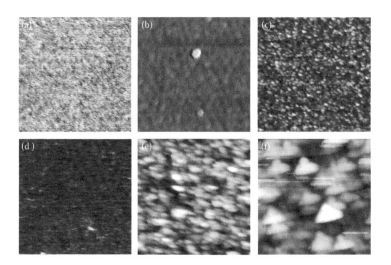

FIGURE 24.8
2-D images of 500 nm scans for PEG films formed at different immobilization times (10 mM PEG concentration): (a) 0 h (clean silicon), (b) 0.5 h, (c) 1 h (d) 2 h, (e) 4 h, and (f) 24 h.

24.4.2 Vapor-Phase Technique for Nanostructured PEG Films

Silanes are often used as precursors or bridges to connect the PEG molecule to a surface. Silane precursors are highly sensitive to moisture. They tend to form aggregates and lumps on the silicon surface in the presence of moisture, which may clog up or mask micro- or nano-size features on devices. The vapor deposition of the silane technique and subsequent PEG coupling on silicon surfaces in a moisture-sfree nitrogen atmosphere (Popat et al., 2002; Wang et al., 1998) allows to overcome this problem.

Silanization of hydrophilized silicon surface with a reactive end group silane like 3-aminotripropyltrimethoxy silane (APTMS) followed by vapor phase ethylene oxide was used to grow PEG films on silicon-based BioMEMS. APTMS is a bifunctional

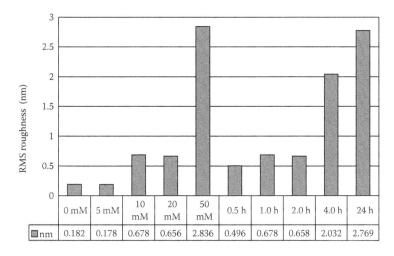

	0 mM	5 mM	10 mM	20 mM	50 mM	0.5 h	1.0 h	2.0 h	4.0 h	24 h
■ nm	0.182	0.178	0.678	0.656	2.836	0.496	0.678	0.658	2.032	2.769

FIGURE 24.9
RMS roughness (R_{rms}) parameters for clean and PEG-coupled silicon surfaces for various PEG concentrations (10 mM PEG concentration) and coupling times (coupling time = 1 h).

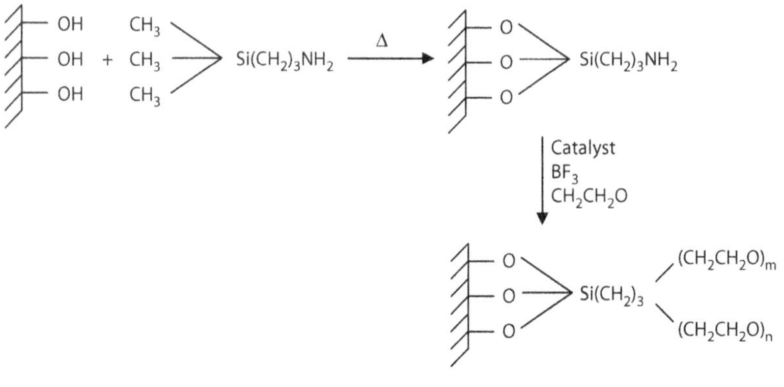

FIGURE 24.10
Proposed chemical reaction on silicon surface using vapor deposition technique.

organosilane possessing a reactive primary amine and a hydrolyzable inorganic trime-thoxy group. It binds chemically to both inorganic materials and organic polymers. It is a short-chained silane with a boiling point of 194°C. It violently reacts with water and tends to polymerize on surfaces forming lumps and aggregates. Boron triflouride was used as a gas catalyst with ethylene oxide because it is a weak Lewis acid. PEG composition could be controlled by the concentration of ethylene oxide and the polymerization reaction time. The reaction could be terminated by flowing inert gas over the surface. Figure 24.10 shows the proposed reaction on the silicon surfaces. Three different concentrations of ethylene oxide were used to create 10, 20, and 40 mmol/cm^2 surface area of silicon. The reaction was allowed to proceed for 1, 2, and 4 h. The ethylene oxide and boron triflouride ratio of 1:2 was maintained in the reaction chamber. An APTMS concentration of 4 mmol/cm^2 surface area of silicon was used.

XPS characterization of PEG-modified silicon surfaces showed an increase in carbon as well as oxygen content. However, there was a sharp decrease in silicon as well as nitrogen concentration (due to silane) compared to unmodified surfaces, with increasing ethylene oxide concentrations and reaction times (Table 24.1). The spectra for PEG-modified silicon surfaces

TABLE 24.1

Atomic Surface Composition for Unmodified and PEG-Modified Silicon Surfaces for Various Ethylene Oxide Concentrations and Reaction Times

	O%	C%	Si%	N%
Clean silicon	28.21	15.84	55.95	0
Silane	9.45	28.95	45.94	15.36
10 mmol/cm^2—1 h	10.73	63.18	13.88	12.21
2 h	12.12	66.76	10.24	10.88
4 h	14.2	68.94	9.8	7.06
20 mmol/cm^2—1 h	10.98	66.21	11.5	11.31
2 h	13.01	70.04	9.71	7.24
4 h	14.35	72.58	7.36	5.71
40 mmol/cm^2—1 h	14.48	69.3	8.16	8.06
2 h	16.06	71.29	7.09	5.56
4 h	16.39	73.54	6.82	3.25

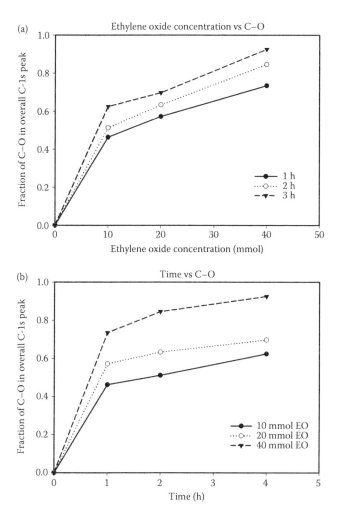

FIGURE 24.11
(a) Variation in fraction of C–O with ethylene oxide concentration for given reaction time and (b) variation in fraction of C–O with reaction time for given ethylene oxide concentration.

for various ethylene oxide concentrations and reaction times show an increase in C-1s (285 eV) and the O-1s (528 eV) peak and a sharp decrease in the Si-2p (100 eV) and N-1s (410 eV) peak for a PEG-modified silicon surface for different reaction times. As the concentration of ethylene oxide increased, more molecules were available on the surface for polymerization, which resulted in more PEG on the surface for a given reaction time. Similarly, for a given ethylene oxide concentration, as the reaction time increased, more PEG is formed on the surface since polymerization is a progressive reaction. This resulted in an increase in carbon and oxygen and a decrease in silicon and nitrogen content on the surface. High-resolution Si-2p, C-1s, O-1s, and N-1s core levels taken on bare silicon and after film deposition showed different line shapes providing evidence for the presence of the polymer film on the surface.

High-resolution C-1s scans confirmed increase in PEG grafting as a function of ethylene oxide concentration and reaction time. We saw substantial increase in the intensity of the C–O peak compared to C–C peak in the total carbon concentration with increase in ethylene oxide concentration (Figure 24.11a) and reaction time (Figure 24.11b).

The intensity of XPS the high-resolution peak for Si-2p was used to determine the thickness of the PEG film on the silicon surface. Figure 24.12 shows the plots for variation in thickness of the film with ethylene oxide concentration for a given reaction time and with reaction times for a given ethylene oxide concentration. The highest thickness of the film is around 38 Å, which is well in the range of applications.

PEG-modified surfaces showed water contact angles in the hydrophilic range (Figure 24.13). As the data indicates, bare silicon is more hydrophilic compared to silane-modified silicon, but PEG surfaces are more hydrophilic as compared to bare silicon that is negatively charged. Also, the contact angle does not depend on the concentration of ethylene oxide, suggesting that the surface is uniformly coated with PEG for all the concentrations tested.

Figure 24.14 shows AFM images of clean and modified silicon surfaces with different concentrations of ethylene oxide for different reaction times. The surface roughness

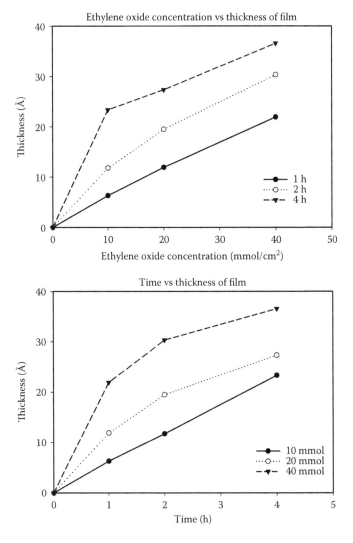

FIGURE 24.12
Thickness of PEG films from the intensities of Si-2p peaks.

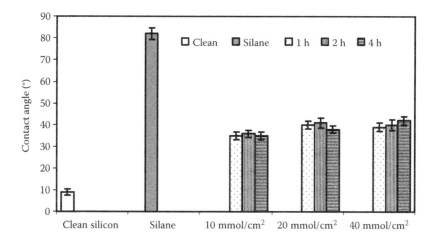

FIGURE 24.13
Contact angle measurements for unmodified and PEG-modified silicon surfaces for various concentrations and reaction times.

increased with increase in the concentration of ethylene oxide and reaction time. For high concentrations, more PEG molecules were available for the reaction and the surface binds more PEG, resulting in higher surface roughness. Similarly, with an increase in reaction time, only the chain length of PEG on the surface increased, resulting in a rougher surface. Figure 24.15 shows the roughness parameters for unmodified and modified silicon surfaces for various reaction conditions. The RMS roughness values for PEG films are extremely low and for all practical purposes can be considered smooth.

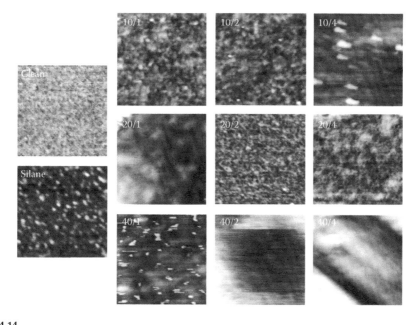

FIGURE 24.14
AFM images for unmodified, silane-modified, and PEG-modified silicon surfaces for various concentrations for ethylene oxide and reaction times.

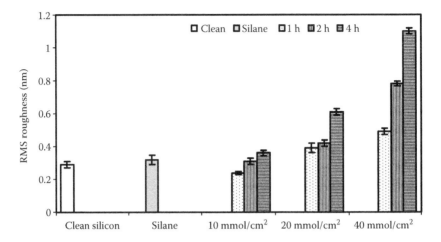

FIGURE 24.15
Surface roughness parameters for unmodified and modified silicon surfaces for various ethylene oxide concentrations and reaction times.

24.5 Biological Applications of Nanostructured PEG Films

24.5.1 Improving Biomolecular Transport through Microfabricated Nanoporous Silicon Membranes

Development of well-controlled, stable, and uniform membranes capable of complete separation of viruses, proteins, or peptides is an important consideration for biofiltration application. The leakage of just one virus or antibody or protein molecule through the membrane

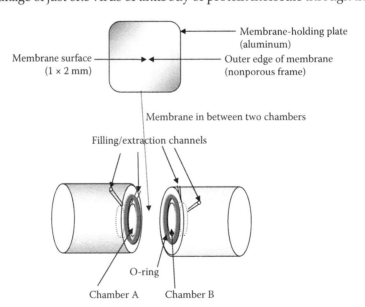

FIGURE 24.16
Diffusion apparatus used for evaluating performance of nanoporous membranes.

will compromise the entire system in such an application. The majority of the membranes currently used for separation of submicron-sized particles in biomedical applications are of the asymmetric or anisotropic variety prepared using polymers such as polysulfones, poly-acrylonitrile, and polyamides. There are several bioincompatibilities associated with these membranes, including sensitization to sterilization, complement activation due to the reactivity of polymeric membranes, and adhesion of various protein and immune components to membranes. Silicon membranes represent an attractive alternative since they are easy to fabricate, chemically and thermally stable, inert, and are capable of postprocessing surface modifications (Sharma et al., 2003, 2004; Popat et al., 2000, 2003). However, the surface of silicon possesses a unique property called "point of zero" charge. Therefore, at physiological pH levels, that is, around pH 7, silicon surface is negatively charged. This may create a streaming potential, resulting in nonspecific adsorption of biomolecules on the surface. A useful strategy to overcome the problem of biofouling is to passivate the charged silicon surface by creating a biocompatible interface (or film) through the coupling of PEG.

Nanoporous silicon membranes of 7 and 19 nm pore sizes were modified with PEG according to the solution-phase covalent coupling procedure described earlier and used for diffusion analysis using glucose (180 D) and lysozyme (14 kD) as model molecules due to the vast differences in their sizes. As stated earlier, the PEG films developed by this method have film thickness of approximately 1.5 nm. It is therefore possible that these PEG films might reduce the effective pore size of the actual membrane and, in turn, affect the diffusion characteristics. Such an effect, if present, would be more pronounced for smaller pore-size membranes and can be evaluated by measuring the diffusion of low molecular-weight solute (e.g., glucose). Figure 24.16 shows diffusion chamber setup used

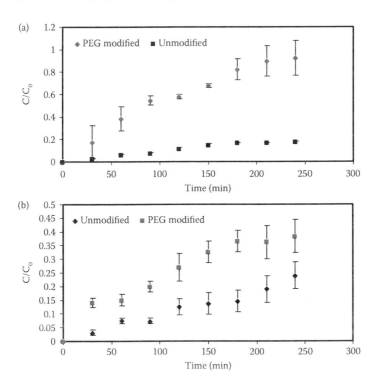

FIGURE 24.17
Diffusion of (a) glucose and (b) lysozyme.

TABLE 24.2

Diffusion Coefficients for Unmodified and PEG-Modified Membranes

Diffusion Coefficient (cm²/s)	Unmodified Membranes	PEG-Modified Membranes
Glucose (7 nm)	6.32×10^{-7}	3.08×10^{-6}
Lysozyme (19 nm)	1.67×10^{-7}	2.18×10^{-7}

in this study. Figure 24.17a shows the glucose diffusion characteristics of nanoporous silicon membranes before and after modification with PEG. We see about a fivefold increase in glucose permeability rather than expected decrease due to anticipated reduced pore size after PEG coupling (Table 24.2). Higher permeability of glucose through the modified membranes can be useful in improving the membrane performance for biofiltration applications. Further, the diffusion of lysozyme through PEG-modified membranes was investigated using 19 nm pore-size membranes. Lysozyme diffusion increased in PEG-modified membranes compared to unmodified membrane as anticipated (Figure 24.17b). This strongly suggests that even though PEG modification may reduce the effective pore size of the membrane due to PEG's nonfouling nature, silicon membrane surface adsorbs less lysozyme compared to unmodified membranes, resulting in higher diffusion rate. Table 24.2 shows the diffusion coefficients for glucose through 7 nm pore-size membrane and lysozyme through 19 nm pore-size membrane.

PEG-modified and unmodified nanoporous membranes were implanted subcutaneously in a Lewis rat and the implanted membranes were retrieved after 17 days. Gross examination of the (subcutaneous) PEG-modified implant shows no scar tissue formation (Figure 24.18). A rich network of blood vessels seems to surround the microfabricated membrane in proximity to the diffusion area. In contrast, the regions where unmodified membranes were implanted show increased fibrosis. This suggests that these nanaostructured PEG

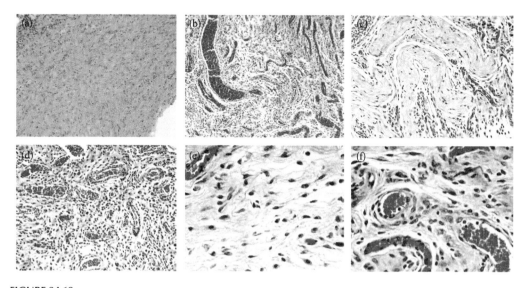

FIGURE 24.18
Histological analysis of tissue surrounding the subcutaneous implants retrieved from Lewis rat after 17 days. Control (unmodified silicon membrane): (a) 10×, (c) 20×, and (e) 50×; PEG-modified membrane: (b) 10×, (d) 20×, and (f) 50×.

films are able to control scar tissue formation in in vivo environments and could be used for implantable BioMEMS.

24.5.2 Nonfouling Microfluidic Systems

One of the important applications in the field of BioMEMS is microfluidic systems that have found many applications in biochemical analysis (Burns et al., 1998; Bernard et al., 2001), chemical reactions (Mitchell et al., 2001), cell-based assays (Fu et al., 1999), and biological analysis (Chiu et al., 2000; Beebe et al., 2002; Thiebaud et al., 2002). The advantages of microfluidic systems are reduced size of operating systems, flexibility in design, reduced use of reagents, reduced production of wastes, decreased requirements for power, increased speed of analyses, and portability. However, as microfluidic technology is rapidly being developed in the laboratory, the effective use of these systems may be improved by developing surfaces that minimally interact with biological solutions. Initial events at the surface include the oriented adsorption of molecules from the surrounding fluid, creating a conditioned interface on which the reagent or sample subsequently adsorbs. The gross morphology, as well as the micro- or nano-topography and chemistry of the surface will determine which molecules can adsorb and in what orientation. Due to the nonspecific surface interactions, the sample gets adsorbed on the surface of the channel, which may result in error in the final analysis. When small quantities of a biological sample are involved, any loss of sample through the system can be critical. Thus, it is useful to focus on fundamental issues related to surface chemistry and topography of microfluidic systems. Silicon-based (e.g., silicon, glass, quartz, silicone) microfluidic systems have become important platforms for diagnostic and therapeutic applications. However, as channel dimensions decrease within these systems, the surface properties of these microchannels become increasingly important. Modifying the inner surface of these channels with PEG can provide a nonfouling interface, which can eliminate the problems associated with microfluidic systems.

For this application, microfluidic channels or capillaries were modified using vapor phase technique as described earlier. FITC-labeled fibrinogen was flowed through unmodified and PEG-modified microchannels or capillaries. The adsorbed surfaces were observed under fluorescence microscope (Figure 24.19) and the fluorescence intensity was directly correlated with the amount of protein adsorbed on the surface. Much lower fluorescence intensity for PEG-modified microchannels or capillaries as compared to unmodified microchannels or capillaries (Figure 24.20) indicated the efficiency of these films in controlling protein interactions with the surface and creating nonfouling interfaces.

24.5.3 Improving the Integrity of Three-Dimensional Vascular Patterns by PEG Conjugation

Damage or loss of an organ or tissue is one of the most frequent and costly problems in health care. Current treatment modalities include transplantation, surgical reconstruction, and mechanical devices such as kidney dialyzers. These therapies have revolutionized medical practice but have limitations. Although transplantation is restricted by ever increasing donor shortage, mechanical devices cannot perform all the functions of a single organ thus providing only temporary benefits. The emerging and interdisciplinary field of tissue engineering offers to solve the organ-transplantation crisis. The creation of functional tissue engineering constructs, however, requires the formation of well-defined biomimetic microenvironments that surround cells and promote controlled cell interactions, maintenance of three-dimensional (3-D) microarchitecture, and proper vascularization.

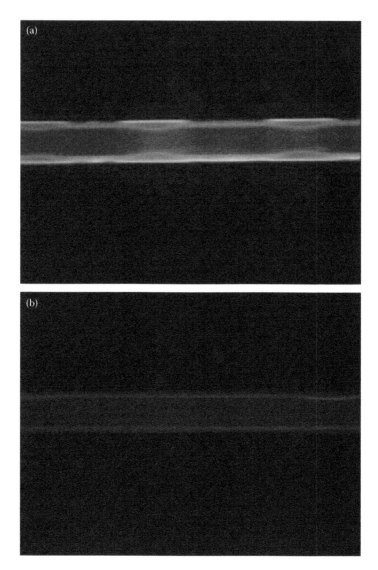

FIGURE 24.19
Fluorescence images for FITC-labeled fibrinogen adsorption in (a) unmodified and (b) PEG-modified microcapillaries.

Vascularization is considered as the main technological barrier for building 3-D human organs, as effective organ perfusion is not possible without an endothelialized vascular tree (Beebe et al., 2002). The extracellular matrix (ECM), serving as a natural scaffold and reservoir of signaling molecules in tissues, may provide 3-D biomimetic environments for proper growth of the cells, and hence vascularization. Chemical modifications of matrix components in vitro may allow ECM scaffolds to be tailored. Patterning of cells within a 3-D matrix provides an approach that allows for shifting from two-dimensional (2-D) patterned cell cultures to 3-D patterned "tissue"-like culture systems (Burns et al., 1998; Bernard et al., 2001).

Microfluidic patterning techniques provide ways to spatially control cells and design appropriate configurations of cells and materials for "engineered" products. Chiu et al.

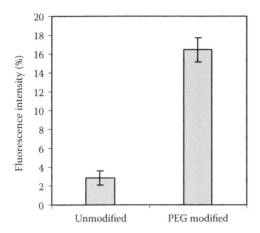

FIGURE 24.20
Percentage florescence for FITC-labeled fibrinogen adsorption in unmodified and PEG-modified microcapillaries.

(2000) fabricated 3-D microfluidic systems and used them to pattern proteins and mammalian cells on a planar substrate. The channel structure, formed by a microstamp in contact with the surface of the substrate, limited the migration and growth of the cells in the channels. Removal of the stamp, however, resulted in the spreading and growth of the different cell types. It was realized that nonselective adhesion and cell migration were the major reasons for the collapse of these patterns over time. PEGs are known for their ability to prevent protein adsorption and cellular adhesion. Therefore, the use of covalently coupled and stable PEG films may be useful for the modification of surfaces for 3-D ECM microfluidic patterning. This may facilitate improved control over cell proliferation and migration and, in turn, pattern maintenance for longer durations.

Pattern integrity is defined here as the ability of cells to stay in the position where they were originally patterned. It is similar to the notion of pattern compliance which has been used elsewhere, except that in this study it refers to a 3-D system. In order to examine the efficacy of PEG in maintaining pattern integrity, first ⟨100⟩ silicon wafers were coated with PEG as described earlier. The change in cellular micropattern width over time was used to evaluate pattern integrity. The width of the cell pattern was determined by drawing two parallel lines as the borders between which most of the cells (>95%) are located, neglecting outliers beyond the border lines. In our previous work, we investigated the influence of the channel size on the compliance and, therefore, defined *compliance multifactor* as the patterned cell area at time t over the original patterned cell area at $t = 0$, divided by channel size in order to eliminate that factor in the final compliance multifactor value. In this study, the *compliance multifactor* was defined as the patterned cell area at time $= t$ over the original patterned cell area at time $= 0$. Smaller values of *compliance multifactor* mean better maintenance of the pattern, that is, greater pattern integrity. Cell number was estimated from micrograph image areas of 1 mm^2.

Figure 24.21 shows the results for the HUVEC-ECM patterns created on unmodified and PEG-modified silicon. The cells patterned on PEG-conjugated substrates displayed greater compliance with the original pattern, with a slower rate of pattern loss in the first 5 days of culture as compared to control silicon. For control silicon substrates, images from two frames were taken to obtain the compliance multifactor after 3 days. Cell migration and

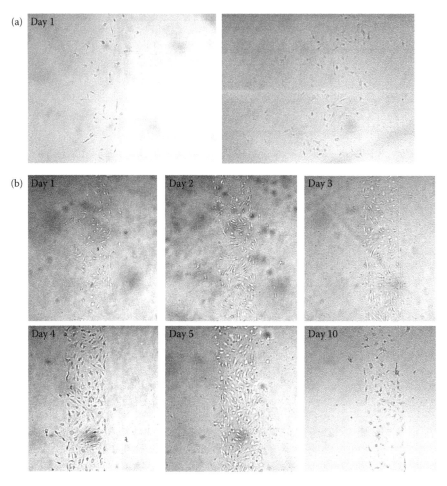

FIGURE 24.21
(a) Control silicon: pictures of cells cultured for 1–2 days. After 3 days, we had to take two frames to add images up to get the compliance multifactor; (b) PEG-conjugated silicon: pictures of cells cultured for 1–5 days and day 10. After 10-day culture, cells apoptosized due to the overconfluency in the channel area.

proliferation contribute to the loss of a cellular pattern over time. Although the compliance multifactor continued to increase for cells patterned on bare silicon, its value begins to plateau for PEG-conjugated silicon. On day 5, the compliance multifactor was about 3.66 ± 0.29 for PEG-conjugated surfaces, compared with 8.23 ± 0.42 for control silicon surfaces (Figure 24.22). Lower values of compliance multifactor for PEG-conjugated surfaces indicate the superior ability of nanostructured PEG films to maintain cell patterns for the period investigated.

24.6 Conclusion

The use of inorganic materials such as silicon is gaining acceptance for use in implantable microdevices. Silicon biomicrodevices are currently being used as implants that can

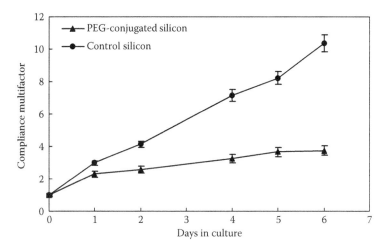

FIGURE 24.22
Compliance multifactor of HUVEC on control silicon and PEG-conjugated silicon wafers.

record from, sense, stimulate, and deliver to biological systems. In this work, we have used two types of nanostructured PEG films to study their interactions with proteins: one by coupling PEG–silane in solution phase and another by vapor deposition of ethylene oxide to grow PEG on surface. We call these PEG films "nanostructured" as these films have thickness in lower nanometer range, and hence suitable for nano- and microdevices. We used PEG–silane coupling procedure for the surface modification of silicon membranes with pores of nanometer dimensions. These membranes are currently being investigated in our laboratory for drug delivery applications and pancreatic islet immunoisolation applications. The PEG films formed by this method were very thin (20 ± 0.93 Å). Such lower thickness values were desired due to the use of these PEG films for nanoporous membranes. Some biosensors involve more complicated patterns such as enclosed microchannels. The solution phase surface modification technique is not appropriate for the micro- or nanoscale channels in these sensors. Due to enclosed micro- or nanoscale-size features on the surface, properties such as viscosity and surface tension of solution injected for surface modification become extremely important. The liquid may clog the channel, forming lumps and aggregates. Thus, a vapor deposition technique may be more efficient in coating closed features since it can more effectively form uniform and conformal films. Therefore, we have developed a solvent-free vapor deposition technique to modify silicon surfaces by growing PEG using ethylene oxide and a weak Lewis acid as catalyst.

   It is believed that the key to many of these processes that are responsible for device rejection is the adsorption of proteins to the surface of the implant, followed by receptor-mediated interactions between cells and the adsorbed proteins. It is known that the silicon surface in water is negatively charged at neutral pH. When exposed to air or water, it develops a native oxide layer with surface silanol groups. These silanol groups are ionizable in water, which results in a negative charge on silicon surface at physiological pH levels. A charged surface will create a streaming potential in the fluid flow and thus may promote protein adsorption, that is, biofouling. This means that by controlling the surface properties of the materials surface it is possible to tailor cell responses rather than allowing uncontrolled and usually unpredictable reactions. A convenient approach is to control protein adsorption by surface modification with a biocompatible material or polymer.

Besides improved biocompatibility, there are several other important factors to be considered for choosing a coating material for silicon. These include ease in surface modification and coupling, availability, reproducibility, patternability, low cost, very little effect on bulk material properties, and compatibility with established silicon-processing steps.

References

J. Anderson. 1994. Inflammation and the foreign body response, In: Klitzman B. (Ed.), *Problems in General Surgery.* J.B. Lippincott, Philadelphia, PA.

J. Andrade and V. Hlady. 1986. *Adv. Polym. Sci.,* 79, 1.

J. Babensee, J. Anderson, L. McIntire, and A. Mikos. 1998. *Adv. Drug Deliv. Rev.,* 33, 111.

D. Beebe, M. Wheeler, H. Zeringue, E. Walters, and S. Raty. 2002. *Theriogenology,* 57, 125.

A. Bernard, D. Fitzli, P. Sonderegger, E. Delamarche, B. Michel, H.R. Bosshard, H. Biebuyck. 2001. *Nat. Biotechnol.,* 19, 866.

N. Besseling. 1997. *Langmuir,* 13, 2113.

J. Breen, D. Huis, J. Bleijser, and J. Leyte, J. 1988. *Chem. Soc. Faraday Trans.,* 84, 293.

M.A. Burns, B.N. Johnson, S.N. Brahmasandra, K. Handique, J.R. Webster, M. Krishnan, T.S. Sammarco, P.M. Man, D. Jones, D. Heldsinger, C.H. Mastrangelo, and D.T. Burke. 1998. *Science,* 282(5388), 484.

L. Canham. 1995. *Adv. Mater.,* 7, 1033.

D. Carlisle. 1972. *Science,* 178, 619.

D. Carlisle. 1986. *Sci. Total Environ.,* 73, 95.

D.T. Chiu, N.L. Jeon, S. Huang, R.S. Kane, C.J. Wargo, I.S. Choi, D.E. Ingber, and G.M. Whitesides. 2000. *Proc. Nat. Acad. Sci. USA,* 97(6), 2408.

L. Clark, R. Spokane, M. Homan, R. Sudan, and M. Miller. 1988. *Am. Soc. Artif. Org. Trans.,* 34, 259.

D. Edell, V. Toi, V. McNeil, and L. Clark. 1992. *IEEE Trans. Biomed. Eng.,* 39(6), 635.

K. Feldman, G. Hahner, N. Spencer, P. Harder, and M. Grunze. 1999. *J. Am. Chem. Soc.,* 121, 10134.

A.Y. Fu, C. Spence, A. Scherer, F.H. Arnold, and S.R. Quake. 1999. *Nat. Biotechnol.,* 17(11), 1109.

B. Gilligan, M. Shults, R. Rhodes, and S. Updike. 1994. *Diabetes Care,* 17, 882.

P. Harder, M. Grunze, R. Dahint, G. Whitesides, and P. Laibinis. 1998. *J. Phys. Chem. B,* 102, 426.

J. Harris (Ed.). 1992. *Poly(ethylene glycol) Chemistry: Biotechnical and Biomedical Applications.* Plenum Press, New York.

K. Hellsing. 1968. *J. Chromatogr.,* 46, 270.

K. Ishihara, N. Nakabayashi, M. Sakakida, N. Kenro, and M. Shichiri. 1998. *American Chemical Society Annual Meeting,* American Chemical Society, Orlando, FL, USA.

S. Jeon and J. Andrade. 1991. *J. Colloid Interface Sci.,* 142, 159.

J. Lee, H. Lee, and J. Andrade. 1995. *Prog. Polym. Sci.,* 20, 1043.

T. McPherson, A. Kidane, I. Szleifer, and K. Park. 1998. *Langmuir,* 14, 176.

M.C. Mitchell, V. Spikmans, A. Manz, and A.J. DeMello. 2001. *J. Chem. Soc. Perkin Trans.,* 1(5), 514.

S. Nagaoka, Y. Mori, H. Takiuchi, K. Yokota, H. Tanzawa, and S. Nishyumi. 1985. In: S. Shalaby, A. Hoffman, B. Ratner, and T. Horbett (Eds.), *Polymers as Biomaterials.* Plenum Press, New York, 361.

E. Ostuni, R. Chapman, R. Holmli, S. Takayama, and G. Whitesides. 2001. *Langmuir,* 17, 5605.

S. Patai and Z. Rappoport (Eds.). 1989. *The Chemistry of Organic Silicon Compounds.* John Wiley & Sons.

A.J. Pertsin and M. Grunze. 2000. *Langmuir,* 16, 8829.

E. Pfeiffer. 1990. *Horm. Metab. Res. Suppl.,* 24, 154.

D. Polla, A. Erdman, W. Robbins, D. Markus, J. Diaz-Diaz, R. Rizq, Y, Nam, H. Brickner, A. Wang, and P. Krulevitch. 2000. *Ann. Rev. Biomed. Eng.,* 2, 551.

K.C. Popat, S. Sharma, and T.A. Desai. 2002. *Langmuir*, 18, 8728.

K.C. Popat, R.W. Johnson, and T.A. Desai. 2002. *Surf. Coat. Technol.*, 154, 253.

K.C. Popat, R.W. Johnson, and T.A. Desai. 2003. *J. Vac. Sci. Tech. B*, 21(2), 645.

B.D. Ratner, A.S. Hoffman, F.J. Schoen, and J.E. Lemons (Eds.). 1996. *Biomaterials Science. An Introduction to Materials in Medicine*. Academic Press, San Diego, CA.

D. Rau and V. Parsegian. 1990. *Science*, 249, 1278.

K. Rebrin, H. Fischer, V. Dorsche, T. Woteke, and P.A. Brunstein. 1992. *J. Biomed. Eng.*, 14, 33.

A.P. Ryle. 1965. *Nature*, 206, 1256.

J. Satulovsky, M. Carignano, and I. Szleifer. 2000. *Proc. Natl. Acad. Sci. USA*, 97, 9037.

S. Schmidt, K. Horch, and R. Normann. 1993. *J. Biomed. Mater. Res.*, 27(11), 1393.

K. Schwartz and D. Milne. 1972. *Nature*, 239, 333.

A.A. Sharkawy, B. Klitzman, G.A. Truskey, and W.M. Reichert. 1998. *J. Biomed. Mater. Res.*, 40, 586.

S. Sharma, R.W. Johnson, and T.A. Desai. 2004. *Langmuir*, 20(2), 348.

S. Sheth and D. Leckband. 1997. *Proc. Natl. Acad. Sci. USA*, 94, 8399.

S. Stenssas and L. Stenssas. 1978. *Acta Neuropath.*, 41, 145.

I. Szleifer. 1997a. *Curr. Opin. Solid State Mater. Sci.*, 2, 337.

I. Szleifer. 1997b. *Physica A*, 244, 370.

P. Thiebaud, L. Lauer, W. Knoll, and A. Offenhausser. 2002. *Biosens. Bioelectron.*, 17, 87.

A. Ulman. 1991. *An Introduction to Ultrathin Organic Films*. Academic Press, Inc., New York.

E. Vansant, P. Voort, and K. Vrancken. 1995. *Characterization and Chemical Modification of the Silica Surface*. Studies in Surface Science and Catalysis (93). Elsevier, Amsterdam, New York, 556.

Y. Wang, M. Ferrari. 1998. In: P.L. Gourlay (Ed.), *SPIE Proceedings of Micro and Nanofabricated Structures and Devices for Biomedical Environmental Applications*, 3258, 20.

R. Wang, H. Kreuzer, and M. Grunze. 1997. *J. Phys. Chem. B*, 101, 9767.

R. Wang, H. Kreuzer, M. Grunze, and A. Pertsin. 2000. *Phys. Chem. Chem. Phys.*, 2, 1721.

S. Wasserman, Y. Tao, and G. Whitesides. 1989. *Langmuir*, 5, 1074.

N. Wisniewski, B. Klitzmann, B. Miller, and W.M. Reichert. 2001. *J. Biomed. Mater. Res.*, 57, 513.

M. Zhang, T. Desai, and M. Ferrari. 1998. *Biomaterials*, 19, 953.

M. Zolk, F. Eisert, J. Pipper, S. Herrwerth, W. Eck, M. Buck, and M. Grunze. 2000. *Langmuir*, 16, 5849.

25

Development of Gold Nanostars for Two-Photon Photoluminescence Imaging and Photothermal Therapy

Hsiangkuo Yuan, Yang Liu, and Tuan Vo-Dinh

CONTENTS

25.1 Introduction

Plasmonic gold nanomaterials offer a unique optical response and biocompatibility for many nanotheranostic applications.[1–5] In the past several decades, scientists have been exploring a variety of nanoplatforms exhibiting breakthrough properties to address current medical needs, especially in the field of cancer therapy. For example, strategies to improve cancer therapy can be achieved by enhancing the targeted delivery, by penetrating the physiological barrier, and by incorporating multiple imaging and therapeutic functionalities into one platform. These issues, which are the fundamental goals in cancer theranostics, have been difficult to solve by conventional medicinal chemistry but have become a strong deliverable by nanotechnology. With greater demands for precision medicine in clinical oncology, nanotechnology-based agent can potentially be an ideal theranostic candidate for improving cancer diagnosis, imaging, and therapy. Readers are advised to read more on cancer nanotheranostics.[6,7] In this chapter, we will discuss one of the most intriguing players in this field, the plasmonic gold nanostars (GNSs).

Although it has yet to be approved for clinical use, many exciting preclinical studies have been published demonstrating their wonderful potentials.

25.2 Rationale for Designing Plasmonic Nanoplatform

An ideal theranostic agent for preclinical research should entail a biocompatible (biodegradable, if possible) nanomaterial with endogenous imaging contrast mechanism and stimuli-responsive therapeutic mechanism. Nanoplatform's physiological stability, contrast intensity, multiplexing capability, *in vivo* trackability, targeting specificity, and controlled localized therapy are of major concerns as well. All of these properties cannot easily be achieved by traditional fluorophore or drug carrier (e.g., liposome). Among many types of nanomaterials, plasmonic gold nanoparticles (AuNPs) not only are biocompatible but also offer strong endogenous optical contrast and therapeutic properties, thus a powerful candidate for nanotheranostics.

Plasmon describes the collective oscillation of surface electron clouds under electromagnetic excitation. When the plasmon is in resonance with the incident optical energy, the local electric field is greatly enhanced, amplifying several interesting properties.[8–10] For example, the plasmon can be tuned into the near-infrared (NIR) tissue optical window by altering the size and geometries (shell, rod, cage, star, etc.).[11–17] When in resonance, the plasmon generates large optical cross sections particularly useful for recently developed optical technologies such as optical coherence tomography (OCT), surface enhance Raman spectroscopy (SERS), photoacoustic tomography (PAT), and multiphoton microscopy (MPM). Such plasmonic resonance can also be exploited for photothermal therapy (PTT) and photodynamic therapy (PDT). The collective advantages inherent to plasmonic nanomaterials have expanded the horizon in the field of cancer theranostics.[7,18]

Among various geometries of AuNP, GNSs consisting of multiple sharp branches on a small core allow for combination of NIR-responsive plasmon resonance and strong lightning rod effect from branch tips rendering GNSs a very powerful plasmonic nanoplatform.[19] Our group has developed plasmon-tunable surfactant-free GNSs, which exhibit strong SERS signal and photoluminescence for multiplexing and cellular imaging,[9,20–25] for multimodal molecular *in vivo* imaging,[4,5,8] and for PTT and PDT.[4,8,20,26,27] In this chapter, we discuss the synthesis of the GNS as well as the application of GNS as a nanotheranostic agent.

25.3 Synthesis and Plasmon Tuning of GNS

Major breakthrough in the past years allowed for fabricating GNS with great quality and reproducibility. Since 2003, numerous GNS synthesis methods have been published. Surfactants, such as poly(N-vinylpyrolidone) (PVP), Triton-100, sodium dodecyl sulfate (SDS), or cetyltrimethylammonium bromide (CTAB), are commonly used to facilitate the process. However, these surfactants limit the use of surfactant-based GNSs due to their toxicity. Surfactant-free GNS synthesis can potentially circumvent the toxicity issue; several surfactant-free methods were thus developed.[21,28]

In 2012, we reported a surfactant-free synthesis of GNS with high monodispersity and plasmon tunability. The process is extremely simple and the reaction is completed in less than 30 seconds, resulting in particles of around 30–60 nm diameters with narrow size distribution (Figure 25.1a). It is worth noting that multiple factors are involved in the synthesis quality of GNS. Our experience suggests that seed-mediated methods are overall better than seed-less method. The seed size and amount determine the GNS size; smaller seed and higher seed concentration make smaller GNS. The reagent's freshness, concentration, and temperature, as well as the mixing sequence of reagents (e.g., $HAuCl_4$, $AgNO_3$, ascorbic acid) also affect the anisotropic branch growth.[21] For example, lower $HAuCl_4$ concentration leads to smaller core and thinner branch growth, hence more red shift and lower scattering. Greater concentration of $AgNO_3$ also leads to more red shift of the plasmon maximum. Lower pH and temperature improve the anisotropic growth quality too. A detailed investigation on GNS size and shape tuning can be found in our previous paper.[21] With all these variables carefully controlled, a reproducible recipe for surfactant-free synthesis of GNS is available. The major advantage of our recipe is the simplicity and absence of surfactant; the GNSs are thus biocompatible and can be conjugated easily with biomolecules for further applications.

Plasmon tunability was achieved by adjusting the Ag^+ concentration. The formation of multiple branches was believed to be related to the underpotential deposition of Ag^+ on certain crystal facets of gold seeds, hence the anisotropic growth on twinned nanoparticles.[29–34] Adding higher concentrations of Ag^+ progressively red-shifted the plasmon band by forming longer, sharper, and more numerous branches with small overall size variation. S5 consists of a few protrusions, while S30 comprises multiple long, sharp branches that appear to

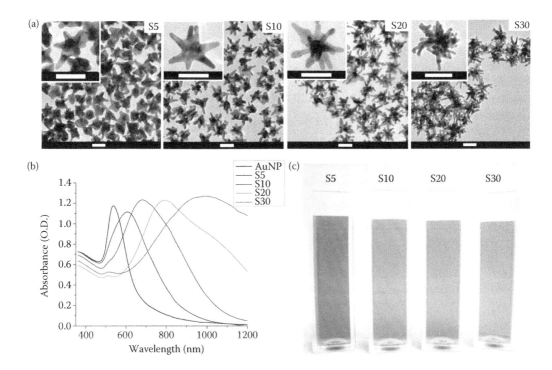

FIGURE 25.1

(a) TEM images of GNS formed under different Ag^+ concentrations (S5: 5 μM, S10: 10 μM, S20: 20 μM, S30: 30 μM). The scale bar is 50 nm. (b) Extinction spectrum of AuNP seeds and S5–S30 solutions (~0.1 nM in DI water). (c) Photograph of the S5–S30 solutions. (Adapted from Yuan H et al. *Nanotechnology.* 2012;23(7):075102.)

branch even further. The nanostars' plasmon peaks are tunable from 600 to 1000 nm by Ag^+ concentration (Figure 25.1b). This is accompanied by a visible change in the solution color from dark blue to dark gray as the plasmon red-shifts and broadens (Figure 25.1c). Both the plasmon peak position and spectral width followed a linear trend with increasing Ag^+ concentration. GNS can therefore be synthesized in a controlled fashion and exploited as an NIR-responsive agent suitable for biomedical applications in the "tissue optical window."

25.4 Analysis of the Plasmon Property of GNSs

We model and compare the optical properties of four GNS solutions (S5–S30). We performed a polarization-averaged simulation using the finite element method (FEM), which yields solutions to the local E-field around 3D metallic nanostructures that are in excellent agreement with the theory.[35,36] Polarization-averaged modeling is more realistic than single polarization methods.

We applied a polarization-average method to different 3D star geometries to model the plasmon shift. In agreement with Hao et al., the plasmon of a nanostar is a result from hybridization of plasmons of the core and the branches.[37] A weak plasmon around 520 nm is attributed to the GNS's core, and a dominant plasmon band at longer wavelengths is supported by the GNS branches. Figure 25.2a depicts that for each of the four nanostars, the local E-field is most greatly enhanced at the tips of those branches that are aligned at least partially parallel to the incident polarization. The enhancement was the greatest when the particle's plasmon matches the incident energy. Also, the E-field along the surface of the branch is enhanced to a value of at least between 1 and 4, suggesting that both plasmon matching and sharp branches contribute to the total E-field enhancement around the nanostar.

Our study suggests that the plasmon shift is controlled mainly by the branch aspect ratio (AR): branch length divided by base width. Figure 25.2b shows the progression of absorption peak position and intensity along with AR. A linear relationship was demonstrated between the absorption peak position and AR, either by adjusting branch length while keeping branch width constant or vice versa. Tip angle, however, does not correlate as linearly as AR to the peak position. Branches with varying tip radius or angle but the same AR result in the same peak position (Figure 25.2c). Consistent with Hao et al.'s finding,[38] the polarization only affects the peak intensity but not the peak position (Figure 25.2d). Also, the peak intensity increases with increasing branch number, branch length, and the core size (Figure 25.2e). It is noteworthy that the core size only contributes to the 520-nm peak, whereas the branch geometry determines the peak position and intensity in the shifted plasmon band. Meanwhile, the geometrical parameters that do not significantly affect the plasmon shift include nanostar core size, branch length (assuming constant AR), and branch number beyond at least two oppositely positioned branches.

25.5 GNSs as an Optical Contrast Agent

The development of ultra-bright optical contrast agent has been an active and exciting field. For many years, fluorescent molecules have been the mainstream product as

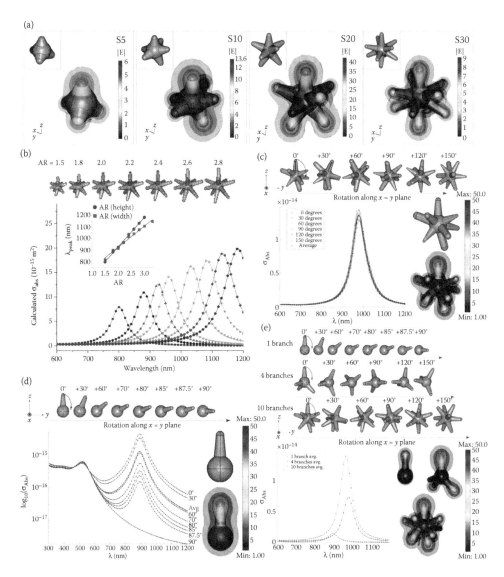

FIGURE 25.2

(a) Simulation of |E| in the vicinity of a nanostar in response to a z-polarized plane wave incident E-field of unit amplitude, propagating in the y-direction, and with a wavelength of 800 nm. E-field enhancement is greatest when plasmon is in resonance with the incident wave. (b) The scatter plots of polarization-averaged absorption against AR. (Inset) The linear relationship between the plasmon peak position and AR, which is tuned by varying branch height (red, $R^2 = 0.997$) or base width (blue, $R^2 = 0.987$) while keeping all other parameters constant. (c) Absorption spectra of 10-branch nanostar with various tip radii of curvature but the same approximate AR. (d) A single-branched nanostar of different orientation to the E-field. (e) The polarization-averaged absorption spectra as a function of number of branches of a nanostar. (Adapted from Yuan H et al. *Nanotechnology.* 2012;23(7):075102.)

molecular probes. However, many organic fluorophores suffer from photobleaching that limit their use for long-term tracking. Quantum dot (QD) offers several advantages over organic fluorophore, such as high quantum yields and low photobleaching,[39,40] but can be cytotoxic due to the release of ions from the cadmium-containing cores.[41] Although QD or fluorophore is commonly used for bioimaging, one of the major disadvantages is the low optical response under NIR irradiation, which is commonly required for deep tissue imaging. Therefore, the development of new NIR-responsive optical probes is highly desirable.

Recent development in NIR-responsive nanoplatform as NIR contrast agent has brought new perspectives to bioimaging. For example, upconverters such as lanthanides can emit visible light under NIR irradiation suitable for *in vivo* detection.[42] Graphene and carbon nanotubes are known to be chemically inert and can also be imaged under NIR and photoacoustic imaging.[43,44] However, functionalizing the surface of both upconverters and carbon-based NPs with biomolecules remains challenging. Other fluorophore-doped polymeric or silica-based NPs, while having been studied extensively in numerous preclinical settings,[45] require the addition of dyes and thus suffering the same problems as organic fluorophores. Plasmonic GNSs, in contrast, feature intrinsic NIR-responsive optical properties, as well as strong emission intensity, multiplexing capability, biocompatibility, and flexibility for surface functionalization. In the following paragraphs, we will discuss a unique GNS contrast mechanism based on two-photon photoluminescence (TPL).

25.6 Characterization of the TPL of GNSs

GNSs exhibit a quadratic dependence of TPL intensity on excitation power, suggesting the existence of an underlying nonlinear two-photon upconverting process on nanostars (Figure 25.3a). Such dependence was not seen on gold or silver nanospheres when examined by the same system (data not shown). However, at higher excitation power, the quadratic dependence shifts toward linear; this may reflect a consequence of the increasing competition between linear decay and upconversion for the depletion of the intermediate excited states.[46] Figure 25.3b illustrates that the TPL excitation spectra of S20 and S30 match their plasmon spectra, indicating that GNSs enhance TPL via plasmon coupling.[47] Interestingly, experimentally measured two-photon action cross section (TPACS) of GNS is more than a million Göeppert–Mayer units (GM). This value is significantly higher than TPACS of QDs (10^5 GM) and organic fluorophores (10^2 GM).[48] The photoluminescence is broad emission, thus appears as a white color on imaging (Figure 25.3c). The broad emission spectrum implies that TPL from nanostars may originate from electron–hole recombination as has been observed on nanorods.[47,49,50] Detailed investigation can be seen in our paper.[21] With such a high TPL emission, we can exploit it for both *in vitro* and *in vivo* imaging.

25.6.1 Investigating Membrane Uptake Mechanism of GNS

We applied both TEM and TPL imaging to assess TAT-GNS's intracellular trafficking pathway (Figure 25.4). BT-549 cells were pretreated with several inhibitors for 30 minutes, incubated with TAT-NS for an hour, then examined under MPM. TAT-NS internalization was inhibited by 4°C (energy blockade), amiloride (AMR; lowering submembraneous pH), cytochalasin D (cytoD; F-actin inhibition), and methyl-β-cyclodextrin (MβCD; lipid raft inhibition), but not chlorpromazine (CPM; clathrin inhibition), genistein (GNT; caveola inhibition), and nocodazole (NCZ; microtubule disruption; data not shown). This highly suggests that the TAT-NS internalization is an energy-dependent, actin-driven, and lipid raft–mediated macropinocytosis.[26]

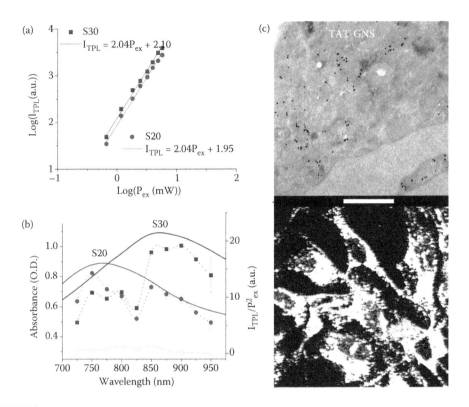

FIGURE 25.3

(a) Quadratic dependence of the I_{TPL} to the P_{ex} from S20 and S30 solutions (0.1 nM). Laser is set at 800 nm with power adjusted between 0.5 and 6 mW. Scatter plots (±1 SD) are displayed with linearly fitted lines. (b) Plasmon spectra (solid lines) and TPL excitation spectra (spline-fitted dashed lines ±1 SD) of S20 (blue) and S30 (red) 0.1 nM in citrate buffer, and rhodamine B (green) 100 nM in MeOH. The spectral dip at 825 nm seen on both nanostars and rhodamine B samples might be a system error from the microscope. (c) TAT-GNS on TEM image (top: N, nucleus; C, cleft; scale bar, 2 μm) correlates to the diffuse white pattern on TPL image (bottom: nucleus stained blue, size 125 × 125 μm²). (Adapted from Yuan H et al. *Nanotechnology*. 2012;23(7):075102; Yuan H et al. *J Am Chem Soc*. 2012;134(28):11358–61.)

25.6.2 High-Resolution Microangiography

Figure 25.5 illustrates an ultra-high-resolution depth-resolved *in vivo* cerebral microangiogram following tail vein injection of PEGylated GNS (PEG-GNS). Through a cranial window, capillaries (5–10 μm in diameter) were visible quickly after injection with minimal tissue autofluorescence background under MPM, demonstrating exceptional microscopic imaging quality and resolution. A vascular tortuosity characteristic of cancerous neovasculature was apparent in the tumor region (Figure 25.5c). In contrast to the commonly used intravascular contrast (e.g., FITC-dextran), which undergoes significant signal decay in less than 30 minutes, the intravascular intensity of PEG-GNS remained stable for hours without significant extravasation, reflecting its intravascular stability inherited from inert gold and surface PEGylation. No apparent behavioral change was noted and the mice appeared healthy for the following week, indicating the likely biocompatibility of the PEG-GNS. Because of GNS's exceptionally intense TPL signal and short fluorescence lifetime (0.2 ns),[21,51] the image can be obtained using low laser energy (e.g., 0.5–1.5 mW at 800 nm wavelength) and fast scanning speed (e.g., 2 μs/pixel).

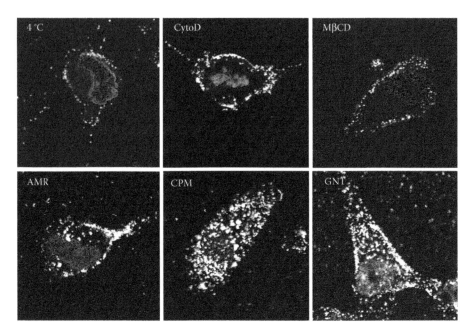

FIGURE 25.4
TPL images of TAT-GNS treated cells under different inhibitors. The cellular uptake of TAT-GNS was inhibited by 4°C, cytochalasin D, methyl-β-cyclodextrin, and amiloride but not chlorpromazine and genistein. TPL images size: $50 \times 50 \ \mu m^2$. Nuclei are stained blue. (Adapted from Yuan H et al. *J Am Chem Soc.* 2012;134(28):11358–61.)

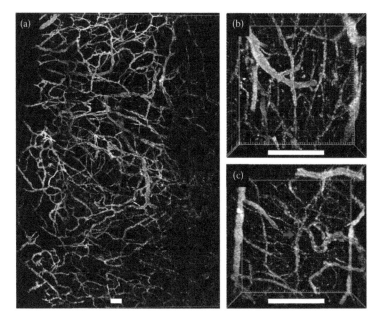

FIGURE 25.5
3D-reconstructed MPM images of brain vasculature through a cranial window chamber on a live mouse within 1 hour after injection of PEG-GNS (1 pmole). (a) Tile images of normal vasculature. (b) Normal vasculature. (c) Tumor vasculature with irregular contour and tortuosity (red arrow). Images were taken under 800 nm, 2 μs/pixel. Scale bar: 200 μm. (Adapted from Yuan H et al. *Nanoscale.* 2014;6(8):4078–82.)

25.6.3 Tissue Imaging of GNS

We examined the distribution of PEG-GNS and TAT-GNS in mice 2 days after systemic administration (i.e., tail injection). *In vivo* TPL imaging after 48-hour incubation shows that, to our surprise, PEG-GNSs accumulate more than TAT-GNSs in tumor parenchyma (e.g., perivascular space, tumor periphery) (Figure 25.6a,f) even though both seem to accumulate inside endothelial cells and perivascular spaces in normal and tumor regions (Figure 25.6b,c,g,h). For PEG-GNS, it is likely that reduced immunoclearance resulted in longer circulatory half-life and hence more passive parenchymal accumulation in the tumor area, particularly in the tumor periphery and perivascular space (Figure 25.6a,c). Although the normal area is well protected by an intact BBB, endothelial cell accumulation and scattered extravasation are still visible (Figure 25.6b). Liver remains the dominant distribution site (Figure 25.6d). For TAT-GNS, in contrast, it seems apparent that a much greater portion

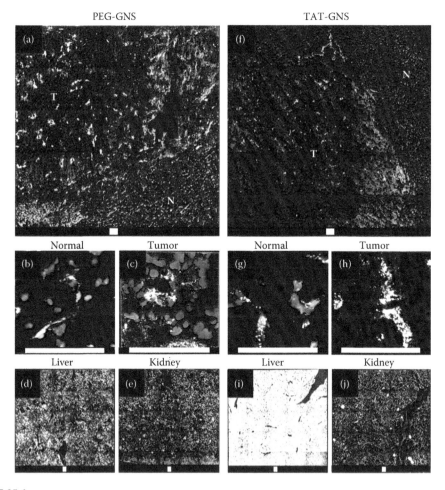

FIGURE 25.6
TPL images of DAPI (blue)- and CD31 (red)-stained cryosectioned specimens from perfused brain excised 48 hours after PEG-GNS (a–e) or TAT-GNS (f–j) injection (5 pmole). Large-area tile images showing distribution of PEG-GNS (a) or TAT-GNS (f) preferentially in the brain tumor. T, tumor; N, normal. Zoom-in images of normal areas (b, g) and tumor areas (c, h). Tile images showing accumulation of PEG-GNS or TAT-GNS in liver (d, i) and kidney (e, j). Scale bar: 100 μm. (Adapted from Yuan H et al. *Nanoscale*. 2014;6(8):4078–82.)

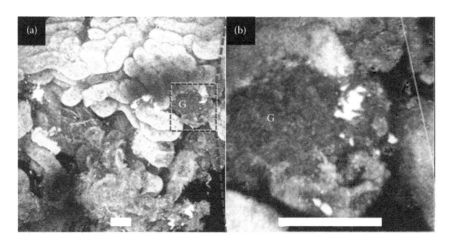

FIGURE 25.7
TPL images from freshly excised half kidney stained with Hoechst 33342 (nuclei; blue). Mice were injected intra-arterially with TAT-GNS-labeled MSCs and sacrificed 5 days afterwards. (a) TPL imaging directly on the half kidney, GNS (white) could be found inside a cell cluster in the JG region. (b) Zoom-in images of a JG region. G, glomerulus. Images were taken under 1000 nm excitation to visualize renal tubules by their autofluorescence. Scale bars are 50 μm. (Adapted from Yuan H et al. *J Biophotonics*. 2016;9(4):406–13.)

of TAT-GNSs are trapped in the liver (Figure 25.6i), which appears bright white on TPL, hence lowering their overall vascular and intratumoral distribution in the brain. Owing to the surface charge, TAT-GNS also accumulated more than PEG-GNS in kidney but not as much as in liver (Figure 25.6e,j).

25.6.4 Stem Cells Tracking

We performed an *in vivo* experiment showing that 5 days after direct arterial injection of TAT-GNS-labeled mesenchymal stem cells (MSCs) to kidney, GNS signals were found in the juxtaglomerular (JG) regions (Figure 25.7). Following the intra-arterial injection, viable MSCs can typically penetrate the glomerular afferent arteriole and migrate to the JG region to become JG cells.[52] PBS injection control showed no distinct GNS signal (data not shown). GNS clusters were visible near some glomeruli (Figure 25.7), reflecting the expected migration of TAT-GNS-labeled MSCs to the JG region.[25]

25.7 GNSs as a Photothermal Therapeutic Agent

Since 2006, plasmonic PTT has been introduced for therapeutic hyperthermia in animal cancer models.[53] By elevating the temperature to more than 42°C, malignant cells are killed through apoptosis or necrosis.[54,55] To date, several PTT studies with plasmon-enhanced local tumor hyperthermia were demonstrated using different classes of NIR-absorbing AuNPs, including silica/gold nanoshells, gold/gold sulfide nanoshells, nanorods, nanocages, hollow nanospheres, nanocubes, and nanostars.[12,14,15,56–64]

The strong and broad plasmon spectrum of GNS matches well with the wavelength of the NIR laser, thus rendering GNS very suitable for *in vivo* photothermal applications.

GNSs, on both simulation and experimental models, predominately absorb rather than scatter the incident photoenergy, particularly at the resonant plasmon wavelength.[27,65] On the basis of our simulation, with small core and thin branches, GNSs scatter less than nanoshell of a similar size, thus a higher C_{abs}/C_{sca} ratio and greater photothermal transduction.[8] The presence of multiple branches on GNS also creates an ensemble of multiple small rods, achieving a comparable absorption but a smaller scattering to nanorod of a similar size, and greater photothermal transduction efficiency.[27] Practically, the photothermal efficiency relies on both the GNS (plasmon position, particle concentration) and photoenergy (wavelength, irradiance, duration).

25.7.1 Ultralow Irradiance for *In Vitro* Photothermolysis

To achieve ultra-efficient photothermolysis, the study was performed on a multiphoton microscope with a pulsed laser (Figure 25.8).[26] The average irradiance (i.e., the power density) was controlled by the acoustic-optic modulator and the scanning area from the microscope's software. While irradiating cells with 0.4 W/cm² on cells treated with PEG-GNS show no laser-induced damage, photothermolysis is clearly present on cells treated with TAT-GNS (Figure 25.8a). At 0.5 mW (6.25 pJ/pulse; irradiance: 0.2 W/cm²), a large portion of cells were damaged (showing red) but still attached on the dish. Such irradiance

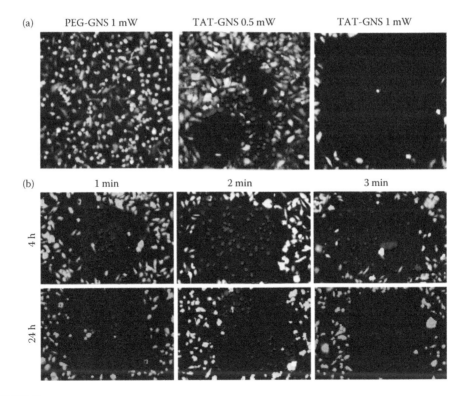

FIGURE 25.8
Live/dead cell imaging after photothermolysis on BT549 cells. (a) Cells were incubated 4 hours with 0.1 nM of PEG-GNS or TAT-GNS followed by 3-minute irradiation (850 nm; 1 mW → 0.4 W/cm², 0.5 mW → 0.2 W/cm²). (b) Comparison of photothermolysis efficiency based on incubation time (4 hours vs. 24 hours) and irradiation time (1–3 minutes; 850 nm, pulse 0.3 W/cm²). (Adapted from Yuan H et al. *J Am Chem Soc.* 2012;134(28):11358–61.)

(0.2 W/cm²) is not only lower than previously reported values using a pulsed laser,[47,66] but also lower than the MPE of skin to laser irradiation (0.4 W/cm² at 850 nm) by ANSI regulation.[67] Also, longer incubation time leads to higher intracellular TAT-NS concentration, and longer irradiation duration leads to greater temperature elevation, hence more prominent photothermolysis (Figure 25.8b). Cells incubated with TAT-NS 24 hours receiving 3 minutes of irradiation have the most prominent photothermal effect where nearly all cells are ablated. Combination of pulsed-laser irradiation and enhanced intracellular delivery of TAT-NS clearly brings forth a very efficient photothermolysis system.

25.7.2 PTT on Animal Tumor Model

A primary sarcoma animal model was used for *in vivo* PTT.[8] Figure 25.9a shows NIR images of surface temperature during the 10-minute PTT process (980 nm, 0.72 W/cm²). Two days prior to the irradiation, mice were treated with 2 mg 30 nm PEG-GNS or PBS systemically (via tail vein injection). Upon irradiation, the tumor surface temperature with PEG-GNS reaches 50°C after only 4 minutes of treatment while the control group maintains at 40°C (Figure 25.9a). After 7 days, there was no sign of tumor recurrence on PTT-treated mouse but not the PBS control. Photographs and x-ray images of these two study conditions are shown in Figure 25.9b. There was mild superficial skin burning directly over the tumor surface, but no other adverse effect was observed in these mice. This proof-of-concept study demonstrated the potential of PEG-GNS for plasmonic-assist PTT. Readers are recommended to read more discussion on practical consideration for *in vivo* PTT.[68,69]

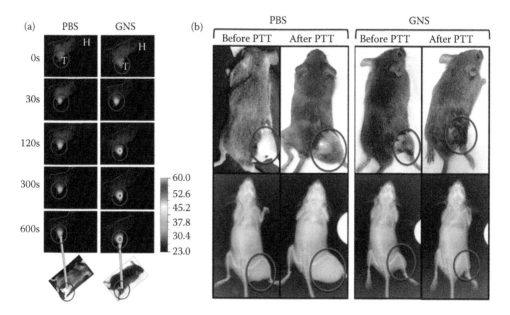

FIGURE 25.9
(a) Near-infrared imaging of mouse surface temperature during photothermal treatment of primary sarcomas after intravenous injection of PBS or PEG-GNS. (b) Photographs (top) and x-ray images (bottom) of mice before and 7-day after PTT with tumors circled in red. X-ray images show a clear decrease in tumor bulk for the mouse with GNS injection, but a significant increase in tumor size for the mouse with PBS injection. Similar results were obtained for the second mouse tested in each group. (Adapted from Liu Y et al. *Theranostics.* 2015;5(9):940–60.)

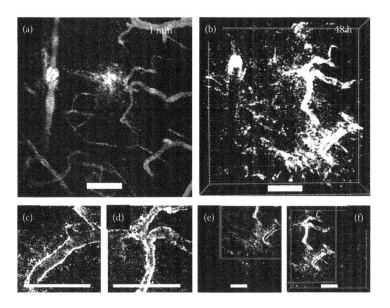

FIGURE 25.10
MPM images of photothermally induced tumor BBB permeabilization through cranial windows on live animals. Tumor vessels were imaged immediately after (a) and 48 hours after (b–f) focal irradiation of 800 nm 20 mW. PEG-GNSs (white) reside near the blood vessels and extravasated deep into the parenchyma (c, d), but not outside the irradiation zone (e, f). Red lines denote the border of irradiation. Scale bar: 100 μm. Image setting: 850 nm, 1% transmission. (Adapted from Yuan H et al. *Nanoscale.* 2014;6(8):4078–82.)

25.7.3 Optically Modulated Blood–Brain Barrier Permeabilization

GNS can also be exploited for photothermally triggered spatially confined microvascular permeabilization.[70] Owing to the high photothermal transduction efficiency of GNS, a focally triggered blood–brain barrier (BBB) permeabilization can be achieved by gentle photothermal process through a cranial window *in vivo* (Figure 25.10). The irradiation (20 mW, 30 seconds) was performed 5 minutes after the PEG-GNS (1 pmole) tail vein injection. At this time point, most PEG-GNSs are intravascular and there is very limited uptake in endothelial cells or RES. After finding the tumor region, the laser irradiation was performed on the same microscope. Immediately following the pulsed-laser irradiation, a small extravasation was visible (Figure 25.10a). After 48 hours, extravasation could be seen confined to tumor vessels in the irradiated area but not the surrounding non-irradiated tumor tissue (Figure 25.10b–f). There was no sign of neurological disability following the laser irradiation over the next 48 hours and the treatment was well tolerated. Following photothermal-triggered BBB permeabilization, an apparent GNS extravasation of 10–20 μm beyond the vessel wall is visible (Figure 25.10c,d); the extravasation depth is much greater than that from merely EPR effect. Such extravasation was not seen in normal brain.

25.8 Conclusion

The surfactant-free plasmon-tunable GNSs have been a game changer in the field of nanotheranostics. It is such a versatile and powerful nanoplatform that pushes the field of plasmonics further into practical biological application, as well as allows us to explore multiple possibilities, which are exemplified by numerous interesting studies in this chapter. Exploiting GNS's high optical contrast intensity and photothermal transduction efficiency, it becomes a powerful tool to investigate many practical challenges inherent to nanoparticles. With more studies, we hope to see more GNS-related clinical translation in the near future.

References

1. Akhter S, Ahmad MZ, Ahmad FJ, Storm G, Kok RJ. Gold nanoparticles in theranostic oncology: Current state-of-the-art. *Expert Opin Drug Deliv.* 2012;9(10):1225–43.
2. Dreaden EC, Alkilany AM, Huang X, Murphy CJ, El-Sayed MA. The golden age: Gold nanoparticles for biomedicine. *Chem Soc Rev.* 2012;41(7):2740–79.
3. Saha K, Agasti SS, Kim C, Li X, Rotello VM. Gold nanoparticles in chemical and biological sensing. *Chem Rev.* 2012;112(5):2739–79.
4. Li J, Huang X, Wang Q, Jing S, Jiang H, Wei Z et al. Pharmacokinetic properties and safety profile of histamine dihydrochloride injection in Chinese healthy volunteers: A phase I, single-center, open-label, randomized study. *Clin Ther.* 2015;37(10):2352–64.
5. Liu Y, Yuan H, Kersey F, Register J, Parrott M, Vo-Dinh T. Plasmonic gold nanostars for multimodality sensing and diagnostics. *Sensors.* 2015;15(2):3706–20.
6. Mura S, Couvreur P. Nanotheranostics for personalized medicine. *Adv Drug Delivery Rev.* 2012;64(13):1394–416.
7. Kunjachan S, Ehling J, Storm G, Kiessling F, Lammers T. Noninvasive imaging of nanomedicines and nanotheranostics: Principles, progress, and prospects. *Chem Rev.* 2015;115(19):10907–37.
8. Liu Y, Ashton JR, Moding EJ, Yuan H, Register JK, Fales AM et al. A plasmonic gold nanostar theranostic probe for in vivo tumor imaging and photothermal therapy. *Theranostics.* 2015;5(9):940–60.
9. Vo-Dinh T, Liu Y, Fales AM, Ngo H, Wang H-N, Register JK et al. SERS nanosensors and nanoreporters: Golden opportunities in biomedical applications. *WIREs Nanomed Nanobiotechnol.* 2015;7(1):17–33.
10. Vo-Dinh T, Fales AM, Griffin GD, Khoury CG, Liu Y, Ngo H et al. Plasmonic nanoprobes: From chemical sensing to medical diagnostics and therapy. *Nanoscale.* 2013;5(21):10127–40.
11. Vo-Dinh T, Dhawan A, Norton SJ, Khoury CG, Wang H-N, Misra V et al. Plasmonic nanoparticles and nanowires: Design, fabrication and application in sensing. *J Phys Chem C.* 2010;114(16):7480–8.
12. Sau TK, Rogach AL, Jäckel F, Klar TA, Feldmann J. Properties and applications of colloidal nonspherical noble metal nanoparticles. *Adv Mater.* 2010;22(16):1805–25.
13. Burda C, Chen X, Narayanan R, El-Sayed M. Chemistry and properties of nanocrystals of different shapes. *Chem Rev.* 2005;105(4):1025–102.
14. Xia Y, Li W, Cobley CM, Chen J, Xia X, Zhang Q et al. Gold nanocages: From synthesis to theranostic applications. *Acc Chem Res.* 2011;44(10):914–24.
15. Zhang J. Biomedical applications of shape-controlled plasmonic nanostructures: A case study of hollow gold nanospheres for photothermal ablation therapy of cancer. *J Phys Chem Lett.* 2010;1(4):686–95.
16. Hahn MA, Singh AK, Sharma P, Brown SC, Moudgil BM. Nanoparticles as contrast agents for in-vivo bioimaging: Current status and future perspectives. *Anal Bioanal Chem.* 2011;399(1):3–27.
17. Weissleder R. A clearer vision for in vivo imaging. *Nat Biotechnol.* 2001;19(4):316–7.
18. Thakor AS, Jokerst J, Zavaleta C, Massoud TF, Gambhir SS. Gold nanoparticles: A revival in precious metal administration to patients. *Nano Lett.* 2011;11(10):4029–36.

19. Nehl CL, Liao H, Hafner JH. Optical properties of star-shaped gold nanoparticles. *Nano Lett.* 2006;6(4):683–8.

20. Fales AM, Yuan H, Vo-Dinh T. Silica-coated gold nanostars for combined surface-enhanced Raman scattering (SERS) detection and singlet-oxygen generation: A potential nanoplatform for theranostics. *Langmuir.* 2011;27(19):12186–90.

21. Yuan H, Khoury CG, Hwang H, Wilson CM, Grant GA, Vo-Dinh T. Gold nanostars: Surfactant-free synthesis, 3D modelling, and two-photon photoluminescence imaging. *Nanotechnology.* 2012;23(7):075102.

22. Yuan H, Fales AM, Khoury CG, Liu J, Vo-Dinh T. Spectral characterization and intracellular detection of surface-enhanced Raman scattering (SERS)-encoded plasmonic gold nanostars. *J Raman Spectrosc.* 2013;44(2):234–9.

23. Yuan H, Liu Y, Fales AM, Li YL, Liu J, Vo-Dinh T. Quantitative surface-enhanced resonant Raman scattering multiplexing of biocompatible gold nanostars for in vitro and ex vivo detection. *Anal Chem.* 2013;85(1):208–12.

24. Fales AM, Yuan H, Vo-Dinh T. Development of hybrid silver-coated gold nanostars for nonaggregated surface-enhanced Raman scattering. *J Phys Chem C.* 2014;118(7):3708–15.

25. Yuan H, Gomez JA, Chien JS, Zhang L, Wilson CM, Li S et al. Tracking mesenchymal stromal cells using an ultra-bright TAT-functionalized plasmonic-active nanoplatform. *J Biophotonics.* 2016;9(4):406–13.

26. Yuan H, Fales AM, Vo-Dinh T. TAT peptide-functionalized gold nanostars: Enhanced intracellular delivery and efficient NIR photothermal therapy using ultralow irradiance. *J Am Chem Soc.* 2012;134(28):11358–61.

27. Yuan H, Khoury CG, Wilson CM, Grant GA, Bennett AJ, Vo-Dinh T. In vivo particle tracking and photothermal ablation using plasmon-resonant gold nanostars. *Nanomedicine: NBM.* 2012;8(8):1355–63.

28. Xie J, Lee JY, Wang DIC. Seedless, surfactantless, high-yield synthesis of branched gold nanocrystals in HEPES buffer solution. *Chem Mater.* 2007;19(11):2823–30.

29. Sau TK, Rogach AL, Döblinger M, Feldmann J. One-step high-yield aqueous synthesis of size-tunable multispiked gold nanoparticles. *Small.* 2011;7(15):2188–94.

30. Guerrero-Martínez A, Barbosa S, Pastoriza-Santos I, Liz-Marzán LM. Nanostars shine bright for you. *Curr Opin Colloid Interface Sci.* 2011;16(2):118–27.

31. Orendorff C, Murphy C. Quantitation of metal content in the silver-assisted growth of gold nanorods. *J Phys Chem B.* 2006;110(9):3990–4.

32. Liu M, Guyot-Sionnest P. Mechanism of silver(I)-assisted growth of gold nanorods and bipyramids. *J Phys Chem B.* 2005;109(47):22192–200.

33. Ahmed W, Kooij ES, van Silfhout A, Poelsema B. Controlling the morphology of multibranched gold nanoparticles. *Nanotechnology.* 2010;21(12):125605–11.

34. Kawamura G, Yang Y, Fukuda K, Nogami M. Shape control synthesis of multi-branched gold nanoparticles. *Mater Chem Phys.* 2009;115(1):229–34.

35. Khoury CG, Norton SJ, Vo-Dinh T. Investigating the plasmonics of a dipole-excited silver nanoshell: Mie theory versus finite element method. *Nanotechnology.* 2010;21(31):315203.

36. Khoury CG, Norton SJ, Vo-Dinh T. Plasmonics of 3-D nanoshell dimers using multipole expansion and finite element method. *ACS Nano.* 2009;3(9):2776–88.

37. Hao E, Bailey R, Schatz G, Hupp J, Li S. Synthesis and optical properties of "branched" gold nanocrystals. *Nano Lett.* 2004;4(2):327–30.

38. Hao F, Nehl CL, Hafner JH, Nordlander P. Plasmon resonances of a gold nanostar. *Nano Lett.* 2007;7(3):729–32.

39. Resch-Genger U, Grabolle M, Cavaliere-Jaricot S, Nitschke R, Nann T. Quantum dots versus organic dyes as fluorescent labels. *Nat Methods.* 2008;5(9):763–75.

40. Smith AM, Duan H, Mohs AM, Nie S. Bioconjugated quantum dots for in vivo molecular and cellular imaging. *Adv Drug Deliv Rev.* 2008;60(11):1226–40.

41. Derfus AM, Chan WCW, Bhatia SN. Probing the cytotoxicity of semiconductor quantum dots. *Nano Lett.* 2004;4(1):11–8.

42. Liu Y, Tu D, Zhu H, Ma E, Chen X. Lanthanide-doped luminescent nano-bioprobes: From fundamentals to biodetection. *Nanoscale*. 2013;5(4):1369–84.

43. Liu Z, Yang K, Lee ST. Single-walled carbon nanotubes in biomedical imaging. *J Mater Chem*. 2011;21(3):586–98.

44. Li B, Cheng Y, Liu J, Yi C, Brown AS, Yuan H et al. Direct optical imaging of graphene in vitro by nonlinear femtosecond laser spectral reshaping. *Nano Lett*. 2012;12(11):5936–40.

45. E Rosen J, Yoffe S, Meerasa A. Nanotechnology and diagnostic imaging: New advances in contrast agent technology. *J Nanomed Nanotechnol*. 2011;02(05).

46. Pollnau M, Gamelin D, Lüthi S, Güdel H, Hehlen M. Power dependence of upconversion luminescence in lanthanide and transition-metal-ion systems. *Phys Rev B*. 2000;61(5):3337–46.

47. Tong L, Wei Q, Wei A, Cheng J-X. Gold nanorods as contrast agents for biological imaging: Optical properties, surface conjugation and photothermal effects. *Photochem Photobiol*. 2009;85(1):21–32.

48. Zipfel WR, Williams RM, Webb WW. Nonlinear magic: Multiphoton microscopy in the biosciences. *Nat Biotechnol*. 2003;21(11):1369–77.

49. Link S, El-Sayed MA. Optical properties and ultrafast dynamics of metallic nanocrystals. *Annu Rev Phys Chem*. 2003;54:331–66.

50. Wang D-S, Hsu F-Y, Lin C-W. Surface plasmon effects on two photon luminescence of gold nanorods. *Opt Express*. 2009;17(14):11350–9.

51. Yuan H, Register JK, Wang H-N, Fales AM, Liu Y, Vo-Dinh T. Plasmonic nanoprobes for intracellular sensing and imaging. *Anal Bioanal Chem*. 2013;405(19):6165–80.

52. Yuan H, Wilson CM, Xia J, Doyle SL, Li S, Fales AM et al. Plasmonics-enhanced and optically modulated delivery of gold nanostars into brain tumor. *Nanoscale*. 2014;6(8):4078–82.

53. Wang H, Gomez JA, Klein S, Zhang Z, Seidler B, Yang Y et al. Adult renal mesenchymal stem cell-like cells contribute to juxtaglomerular cell recruitment. *J Am Soc Nephrol*. 2013;24(8):1263–73.

54. Hirsch LR, Gobin AM, Lowery AR, Tam F, Drezek RA, Halas NJ et al. Metal nanoshells. *Ann Biomed Eng*. 2006;34(1):15–22.

55. Dewhirst MW, Vujaskovic Z, Jones E, Thrall D. Re-setting the biologic rationale for thermal therapy. *Int J Hyperthermia*. 2005;21(8):779–90.

56. Wust P, Hildebrandt B, Sreenivasa G, Rau B, Gellermann J, Riess H et al. Hyperthermia in combined treatment of cancer. *Lancet Oncol*. 2002;3(8):487–97.

57. Van de Broek B, Devoogdt N, D'Hollander A, Gijs H-L, Jans K, Lagae L et al. Specific cell targeting with nanobody conjugated branched gold nanoparticles for photothermal therapy. *ACS Nano*. 2011;5(6):4319–28.

58. Kennedy LC, Bickford LR, Lewinski NA, Coughlin AJ, Hu Y, Day ES et al. A new era for cancer treatment: Gold-nanoparticle-mediated thermal therapies. *Small*. 2011;7(2):169–83.

59. Choi WI, Kim J-Y, Kang C, Byeon CC, Kim YH, Tae G. Tumor regression in vivo by photothermal therapy based on gold-nanorod-loaded, functional nanocarriers. *ACS Nano*. 2011;5(3):1995–2003.

60. Yi DK, Sun IC, Ryu JH, Koo H, Park CW, Youn IC et al. Matrix metalloproteinase sensitive gold nanorod for simultaneous bioimaging and photothermal therapy of cancer. *Bioconjug Chem*. 2010;21(12):2173–7.

61. Wu X, Ming T, Wang X, Wang P, Wang J, Chen J. High-photoluminescence-yield gold nanocubes: For cell imaging and photothermal therapy. *ACS Nano*. 2010;4(1):113–20.

62. Huang X, El-Sayed IH, Qian W, El-Sayed MA. Cancer cell imaging and photothermal therapy in the near-infrared region by using gold nanorods. *J Am Chem Soc*. 2006;128(6):2115–20.

63. Gobin AM, Watkins EM, Quevedo E, Colvin VL, West JL. Near-infrared-resonant gold/gold sulfide nanoparticles as a photothermal cancer therapeutic agent. *Small*. 2010;6(6):745–52.

64. von Maltzahn G, Park J-H, Agrawal A, Bandaru NK, Das SK, Sailor MJ et al. Computationally guided photothermal tumor therapy using long-circulating gold nanorod antennas. *Cancer Res*. 2009;69(9):3892–900.

65. Huff TB, Tong L, Zhao Y, Hansen MN, Cheng J-X, Wei A. Hyperthermic effects of gold nanorods on tumor cells. *Nanomedicine*. 2007;2(1):125–32.

66. Bibikova O, Popov A, Bykov A, Fales A, Yuan H, Skovorodkin I et al. Plasmon-resonant gold nanostars with variable size as contrast agents for imaging applications. *IEEE J Sel Top Quantum Electron*. 2016;22(3):1–8.
67. Au L, Zheng D, Zhou F, Li Z-Y, Li X, Xia Y. A quantitative study on the photothermal effect of immuno gold nanocages targeted to breast cancer cells. *ACS Nano*. 2008;2(8):1645–52.
68. ANSI. *American National Standard for Safe Use of Lasers*. Laser Institute of A, editor. Orlando, FL: Laser Institute of America; 2000.
69. Wilhelm S, Tavares AJ, Dai Q, Ohta S, Audet J, Dvorak HF et al. Analysis of nanoparticle delivery to tumours. *Nat Rev Mat*. 2016;1(5):16014.
70. Abadeer NS, Murphy CJ. Recent progress in cancer thermal therapy using gold nanoparticles. *J Phys Chem C*. 2016;120(9):4691–716.

26

Surface Plasmon–Enhanced Nanohole Arrays for Biosensing

Jean-Francois Masson, Maxime Couture, and Hugo-Pierre Poirier-Richard

CONTENTS

26.1 Introduction

26.1.1 Background on Plasmonic Nanohole Arrays

The field of biosensing has tremendously grown in the past decade for a series of applications ranging from biomedical measurements to environmental monitoring. Biosensors are based on the specific recognition of an analyte by a biomolecular receptor immobilized on a transducer.[1] The capture event of the analyte changes the physical, optical, or electrochemical properties of the transducer, which are thus often based on refractometric measurements, on molecular spectroscopy, including surface-enhanced spectroscopies, or on electrochemistry.

Each type of transducer has advantages and disadvantages, making them suitable for different applications. For example, refractometric measurements such as plasmonic sensing are highly sensitive to biomolecules, but also respond significantly to changes in the refractive index of the bulk solution and to nonspecific adsorption events. Molecular spectroscopies benefit from an added level of selectivity due to the different molecular transitions; however, they typically require a solution exempt of fluorescence and the presence of distinct chromophores, which is not always the case. Electrochemical measurements are selective to electroactive molecules and can be performed in complex fluids, but are not generally applicable to a broad range of analytes. The selection of the transducer therefore depends on the application.

Biosensors have also benefitted from nanotechnology, especially in the field of plasmonics. Specifically, nanoparticles (NPs) and nanostructures have contributed to plasmonics with the advent of several reproducible fabrication methods.[2] Nanostructures have since been implemented in sensing devices for label-free biosensing,[3] and at the present time hundreds of different nanostructures have been reported and applied to different targets. However, there is no consensus on the optimal plasmonic platform, likely due to the relative youth of the field and also due to the different requirements of individual applications. The design of an optimal sensing platform with plasmonic sensors still faces several challenges.[4] For example, pushing the detection limits to lower concentrations, improving the integration of a sensor into devices, designing sensing schemes that work in crude samples, and creating low-cost platforms will facilitate the acceptance of plasmonic devices.

Among all the nanoplasmonic substrates, nanohole arrays stand out due to their high sensitivity and versatility. In brief, nanohole arrays are a thin metallic film in which a series of holes have been incorporated. The nanoholes are plasmonically active with optical properties depending on a series of factors. For example, nanohole arrays may have different geometries, such as ordered arrays[5] (square or hexagonal arrays) or short-range ordered (SRO) arrays (random distribution of nanoholes). The periodicity of nanohole arrays is usually between 100 and 1000 nm; however, larger periodicities have also been reported in the micron range and these are commonly termed microhole arrays. The hole diameter may be smaller than the diffraction limit of light, in which case the transmission is strongly enhanced by the plasmon resonance,[6] while larger nanoholes exhibit strong absorption maxima associated with the plasmon resonance. However, in all cases, nanoholes are plasmonically active and act as excellent refractometric sensors. In addition, nanohole arrays are excellent nanoelectrodes and enhance the intensity of fluorophores and Raman reporters. As detailed in this chapter, their excellent properties in plasmonic sensing, surface-enhanced spectroscopies, and electrochemistry make them ideal candidates for integration into a suite of sensors.

26.1.2 Theory of Light Transmission at Different Length Scales

Soon after the first report on extraordinary optical transmission through nanoholes,[1] the theory of nanohole arrays, with a nanohole diameter smaller than the wavelength of light and a periodicity in the order of 200–600 nm, has been extensively detailed in a series of articles and reviews.[7–10] For these nanohole arrays, light tunnels through the surface plasmon, which is typically excited by momentum matching of the surface plasmon (k_{sp}) and the light wave (k_o):

$$k_{sp} = \frac{\omega}{c} \sqrt{\frac{\varepsilon_m(\lambda)\varepsilon_s(\lambda)}{\varepsilon_m(\lambda) + \varepsilon_s(\lambda)}} \qquad (26.1)$$

$$k_o = \frac{\omega}{c} = \frac{2\pi}{\lambda} \qquad (26.2)$$

where ω is the angular frequency of light, c is the speed of light, λ is the wavelength of light, and ε is the dielectric constant of the metal (m) and of the solution (s). The metal is typically gold or silver, or bimetallic gold and silver,[11] although aluminum nanohole arrays are increasingly reported in the literature.[12–15] The momentum of light, however, never intersects the momentum of the surface plasmon under these conditions. The missing momentum can be gained by the grating structure of nanohole arrays. In this situation, the momentum of light is

$$k_x = \frac{2\pi}{\lambda} \eta_D \sin(\theta) \pm i \frac{2\pi}{P} \pm j \frac{2\pi}{P} \qquad (26.3)$$

where i and j are integers related to the mode excited, θ is the incident angle, η_D is the refractive index of the medium, and P is the periodicity of the array. The factor $2\pi/P$ for square arrays of nanoholes becomes $4\pi/3P$ for hexagonal arrays. In its simplest form, the plasmon resonance wavelengths of hexagonal nanohole arrays are described by Equation 26.4. In this case, the excitation wavelength of the (1,0) plasmon mode is approximately the wavelength of light and thus nanohole arrays of larger periodicities have resonances at longer wavelengths:

$$\lambda = \frac{P}{\sqrt{\frac{4}{3}(i^2 + ij + j^2)}} \sqrt{\frac{\varepsilon_m \varepsilon_d}{\varepsilon_m + \varepsilon_d}} \qquad (26.4)$$

Equation 26.4 is applicable for normal incidence of the light beam; however, the excitation of the plasmon resonance is highly dependent on the relative orientation of the nanohole array with regards to the light beam. An equation describing the plasmon resonance at different incidence angles and rotation angles (ϕ) was derived recently (Equation 26.5).[16]

$$\lambda = \left(\frac{P}{\sqrt{\frac{4}{3}(i^2 + ij + j^2)}} \sqrt{\frac{\varepsilon_m \varepsilon_d}{\varepsilon_m + \varepsilon_d}} \right.$$
$$\left. - \frac{P}{\sqrt{\frac{4}{3}(i^2 + ij + j^2)}} \eta_D \sin\theta (i\cos(\phi) + j\cos(\phi + 60) + i\sin(\phi) + j\sin(\phi + 60)) \right) \qquad (26.5)$$

There are several impacts of the incidence angle dependence on the excitation of the nanohole arrays. Firstly, the nonzero incidence angles causes a shift in the plasmon resonance that is dependent on the mode, the degenerescence of the plasmon mode is lifted, and several plasmon resonances can be excited at different wavelengths[7] (Figure 26.1). Therefore, the plasmon resonance can be accurately tuned to a wavelength. For example, the wavelength can be matched to a Raman laser or to a fluorophore for further enhanced surface-enhanced Raman scattering (SERS)[16] or metal-enhanced fluorescence (MEF).[17] The large blue-shifts of the (1,0) plasmon resonance at higher incident angles can be exploited to increase the bulk sensitivity of sensors. In this situation, the resonance of nanohole arrays with larger periodicities and thus higher sensitivity[18] can be blue-shifted to the Vis–NIR region.[16] The larger incident angle also decreases the field depth, thus increasing the sensitivity to detection events.[16] The decay length can also be decreased in solutions of higher refractive index,[19] also increasing sensitivity to detection events. Hence, several advantages arise from higher incident angles in nanohole array sensors.

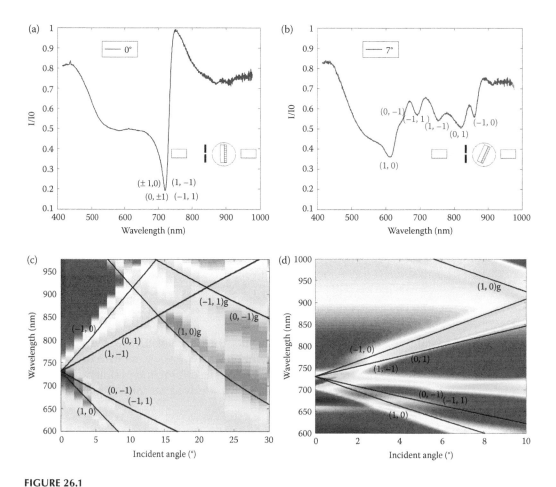

FIGURE 26.1

(a and b) Transmission spectra of nanohole arrays at 0° and 7° incidence angles. The higher incidence angle in (b) resulted in splitting of six plasmon modes at different resonant wavelengths. (c) Experimental and (d) simulated with FDTD plasmon dispersion curve for nanohole arrays. (Couture, M. et al. *Nanoscale* 2013, 5, 12399. Reproduced by permission of The Royal Society of Chemistry.)

26.1.3 Electric Field Location and Surface-Enhanced Spectroscopies

Higher local electric fields arise from the excitation of the plasmon resonance in nanohole arrays. The dispersion properties[20] and the field distribution[21,22] of the plasmon resonance in nanohole arrays have been predicted with FDTD calculations. These simulations revealed that the field is mostly located on the rim of the nanohole (Figure 26.2). Experimental images were acquired to confirm the location of the high electric field using a Raman reporter and Raman imaging.[21,23] The higher local fields make nanohole arrays good substrates for SERS[16,24–27] and MEF.[17,28,29] Tuning the diameter of the nanohole to the period at a ratio of 0.7 increased the SERS intensity of nanohole arrays,[21,27] and while the enhancement factor of nanohole arrays does not rival NP dimers, it can be as high as 10^7 in cavity nanohole arrays.[26]

The MEF fluorescence is a result of the emission of light from a fluorophore located in the proximity of nanohole arrays where the energy of the emitter is coupled to the surface plasmon and radiated in the far-field.[30] For MEF to be optimal, the fluorophore must be located at an optimal distance for minimal quenching and maximal enhancement due to the plasmon resonance. In addition, the plasmon resonance should be tuned to the overlap of the excitation and emission wavelength of the fluorophore. This can be achieved with

2 µm

FIGURE 26.2
(Bottom) FDTD simulation of the electric field distribution of the plasmon on nanohole arrays. (Top) Experimental image of the SERS response of a Raman reporter on the nanohole arrays. The regions of high electric field are co-localized with the high SERS intensity (yellow color). Both the FDTD simulation and experimental images are in good agreement. (Reprinted with permission from Correia-Ledo, D. et al. *Journal of Physical Chemistry C* 2012, 116, 6884. Copyright 2012 American Chemical Society.)

nanohole arrays of different periodicities[28,29] or with different incidence angles.[17] Enhanced FRET was also reported on nanohole arrays.[17] In this case, the plasmon resonance must be tuned to match the overlap of the donor's emission and the acceptor's excitation for optimal fluorescence. Owing to the enhanced fluorescence from the nanoholes, increased sensitivity was reported for PSA with an enhancement in the order of a factor of 3–10 compared to Au mirrors.[31,32] These examples demonstrate the properties of nanohole arrays in surface-enhanced spectroscopies.

26.1.4 Nanohole Arrays as Electrodes

Electrical conductivity is possible in nanohole arrays due to the continuous network of metal. Therefore, nanohole arrays have been used as working electrodes in several experiments.[33–36] For example, cyclic voltammetry was shown for an Os complex mediator on nanohole arrays[34] while nanohole arrays and electrochemistry have also been employed for monitoring DNA hybridization events.[35] The spectroscopic response of nanohole arrays is a function of the potential difference applied to the metal surface.[33] While it was speculated that the change in plasmon resonance might be due to double-layer capacitance, Dahlin et al. observed that the formation of ionic complexes, most likely gold chlorides, explains the change in optical properties of nanohole arrays under applied potential.[33] Electrofocusing of an analyte provides an interesting advantage of nanohole array electrodes as it was shown that applying a potential to the nanohole arrays could increase the diffusion rate of analytes to the sensor and increase the sensitivity of nanohole arrays to a protein.[36] This is a simple method of lowering the limit of detection, which was 2 orders of magnitude better in the reported case.[36] While very few examples of nanohole array electrodes have been reported to date, their potential is undeniable and more studies are likely to be reported in the future.

26.1.5 Sensing with Nanohole Arrays

Nanohole array sensors typically rely on the measurement of the transmission or the reflection spectrum[37] and tracking of the plasmon resonance as a function of time. Changes in plasmon resonance due to binding of analytes are easily tracked via a shift in wavelength and serve for quantification. The bulk sensitivity of nanohole arrays is in the order of 300–1000 nm/RIU[5,38] and can be optimized through tuning of structural parameters. Indeed, increasing the periodicity, tuning the hole diameters or varying the metallic composition (i.e., Ag or Au) is all factors that impact the refractive index sensitivity.[11,21] Alternatively, direct intensity[39] or ratiometric[40] measurements can be implemented for monitoring the plasmon resonance. The refractive index resolution in these configurations is somewhat lower than classical SPR sensors. However, the simpler optical setup for monitoring the plasmon resonance of nanohole arrays compensates for this disadvantage.

To further improve on the resolution of nanohole arrays, several designs and data analysis methods have been evaluated. Data analysis using the integrated response method improved the signal-to-noise ratio up to 90%[41] while phase analysis was also shown to improve the spectral figure of merit of nanohole array sensors.[42] Atomically flat nanohole arrays were fabricated by a template-stripping method and resulted in higher resolution comparable to classical SPR due to the smooth Au film and the implementation of elliptical nanoholes and a Cytop substrate.[43] A recent report showed that sensing can be improved with the distinction of bulk and surface binding events using a ring-hole plasmonic structure[44] as this configuration removes the impact of bulk refractive index changes in sensing experiments. Thus, the performance of nanohole arrays is approaching, albeit still slightly less sensitive, the performance of classical SPR using optimized conditions.

Nanohole arrays hold the advantage of simple integration on-chip for sensing applications.[5,45,46] In these devices, microfluidics are coupled with the nanohole arrays for sample delivery and this is integrated with the optics and detector for monitoring the plasmonic properties of nanohole arrays and for sensing. Highly multiplexed sensing chips were also reported for sensing proteins.[47–49] In these configurations of massively parallel detection with nanohole arrays, intensity-based measurements served to detect the presence of the proteins. Nanohole arrays were also integrated into the tip of optical fibers for SERS[50] or SPR sensing.[51] A detailed review of sensing applications is provided in a following section of this chapter.

26.1.6 Nanohole and Microhole Arrays Excited in Total Internal Reflectance

Nanohole arrays are excellent refractometric sensors in transmission measurements due to their relatively high bulk sensitivities in the order of a few hundreds of nanometers per refractive index unit.[11] This makes them more sensitive than most metallic NPs, albeit slightly less sensitive than SPR in the Kretschmann configuration. The sensitivity of nanohole arrays in transmission measurements was compared to SPR in the Kretschmann configuration with nanohole arrays and with a thin Au film.[38] Nanohole arrays measured in transmission had about 1 order of magnitude less sensitivity than SPR in the classical format with a thin Au film. However, it was shown that nanohole arrays excited in the Kretschmann configuration had a slightly better bulk refractive index sensitivity and had a comparable limit of detection for a protein, although the resolution was 7 times worse due to a broader peak width.[38] If the resonance is investigated at telecommunication wavelengths ($\lambda = 1550$ nm), the sensitivity of nanohole arrays was superior to a thin Au film excited under optimal conditions.[52] Therefore, nanostructuring the gold film in the Kretschmann configuration can improve the analytical performance of plasmonic sensors.

Nanohole arrays have been shown to support both localized and propagating surface plasmons in the Kretschmann configuration of SPR.[53,54] The nanohole array also changes the effective optical constants of the metallic film.[53] The propagating surface plasmon is resonant at a higher energy than the localized surface plasmon, and the interaction of the localized and propagating plasmons results in hybridization of these modes. These localized surface plasmons were also observed for gold particles of near-micron size excited in the Kretschmann configuration,[55] and they had a similar spectral response as reported elsewhere in the literature[53,54] for nanohole arrays supporting localized and propagating surface plasmons (Figure 26.3). The large scattering of the propagating plasmon with nanohole arrays, however, results in a broad plasmon resonance and thus potentially poorer resolution.

Microhole arrays of periodicities exceeding 1 μm have lower scattering and sharper resonances than nanohole arrays excited in the Kretschmannn configuration.[56] They also support localized and propagating surface plasmons.[57] The coexcitation of localized and propagating surface plasmons led to enhanced bulk and surface sensitivity by at least a factor of 3 compared to thin Au films.[57] Microhole arrays also enhance Raman scattering and it was recently reported that they enhance fluorescence[58] and therefore they will be discussed in greater detail later in this chapter. The following sections will detail the fabrication methods for nanohole arrays and their application to sensing.

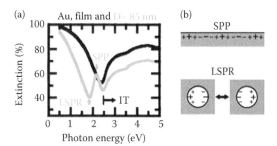

FIGURE 26.3
(a) Spectra of a gold film (black) and nanohole array (blue). The LSPR and SPR bands of the nanohole array is identified in the spectra. (b) Scheme representing the charge distribution of the localized (LSPR) and propagating (SPP) surface plasmons. (Reprinted with permission from Schwind, M.; Kasemo, B.; Zoric, I. *Nano Letters* 2013, 13, 1743. Copyright 2013 American Chemical Society.)

26.2 Fabrication of Nanohole Arrays

Nanopatterned plasmonic structures have been widely studied in the past 10 years for biosensing applications.[59] Several techniques have been proposed over the years to synthesize or fabricate nanostructures[2] and nanohole arrays.[60] Pioneering research on nanohole arrays used FIB, a precise but time-consuming and costly technique, and later on with electron beam lithography (EBL).[4] To achieve wafer scale fabrication of highly ordered nanohole arrays, different methods have been proposed for the efficient manufacturing of these nanostructures. From nanosphere lithography (NSL) to focused ion beam, this section covers the most common fabrication techniques available to produce metallic hole arrays.

26.2.1 Nanosphere Lithography

NSL is one of the most cost-effective methods to produce nanohole and microhole arrays.[61] This process relies on a self-assembled polymer bead array used as a mask for metal deposition. The mask is generally formed by drop coating the polymer beads onto a clean glass substrate. Optimization of the concentration and volume of beads deposited can lead to disperse and low-density colloidal arrays or dense hexagonal packing creating a uniform monolayer. In the case of SRO arrays, a thin metal film is deposited at this stage and removal of the bead mask then leads to the SRO nanohole arrays. For dense and long-range ordered arrays, the substrate covered with dense hexagonal packing of beads is then etched with a plasma cleaner to reduce the diameter of the beads, increasing the spacing between beads previously in close contact. The two defining parameters of a mask are the period and the diameter of the etched beads. The period or center-to-center distance between two nanoholes is set by the original diameter of the polymer beads. The plasma etching time tunes the diameter of the beads and therefore diameter of the nanoholes. Once the mask is etched, the metal film is deposited on top of the beads and the beads are removed via sonication. The whole process is depicted in Figure 26.4.[27]

The main advantages of NSL are simplicity and low cost. By drop coating, samples with a sensing area of a square centimeter are usually obtained. This can also be improved

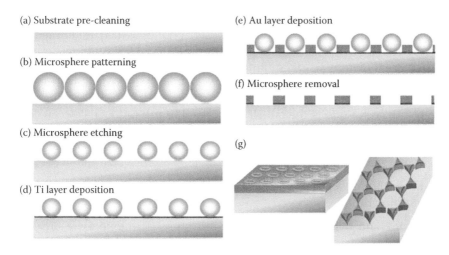

FIGURE 26.4
Steps of nano/microsphere lithography in the process of fabricating hole arrays. (a) Substrate cleaning, typically with strong acids; (b) deposition of a monolayer of polystyrene spheres on the cleaned substrates; (c) polystyrene sphere etching using a reactive ion plasma; (d) adhesion layer deposition; (e) gold layer deposition; (f) polystyrene sphere removal by sonication; (g) resulting formed pattern depending on time of sphere etching. (Adapted from Zheng, P. et al. *Physical Chemistry Chemical Physics* 2015, 17, 21211, with permission from the PCCP Owner Societies. Copyright 2015 PCCP Owner Societies.)

for smaller beads by using other monolayer formation techniques such as the Langmuir–Blodgett film. However, this technique is limited by small areas of perfect ordering, and the many defects of the bead arrays lead to imperfect samples (Figure 26.5).[103] Consequently, it is not ideal for mass production of nanohole arrays and industrial applications.

26.2.2 Focused Ion Beam

Focused ion beam (FIB) is a highly versatile technique, enabling the design of complex 2D and 3D nanostructures with great control over the shape and size. FIB can be used not only for structural design but also for real-time imaging of the sample being processed. The technique relies on a highly energetic beam of ions to mill holes in a metal film (direct milling).[62] The energy of the ions transferred to the surface removes material by sputtering.[63]

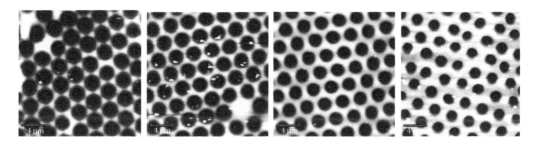

FIGURE 26.5
AFM images of microhole arrays by NSL. Each AFM images corresponds to a $20 \times 20\,\mu m$ scan. (Reprinted with permission from Live, L. S.; Bolduc, O. R.; Masson, J.-F. *Analytical Chemistry* 2010, 82, 3780. Copyright 2010 American Chemical Society.)

The beam can be only tens of nanometers in diameter, providing excellent resolution, an advantage when compared to photolithographic techniques, which are limited to the diffraction limit of light in the UV region. Holes as small as 10 nm in diameter can be produced with a combination of FIB milling and FIB deposition of the same metal.[64] The technique has nearly limitless possibilities for the nanofabrication of plasmonic structures.

FIB milling creates flawless samples due to the high precision of the process.[65,66] The main drawback of this technique is the fabrication time. FIB milling is a slow process, which leads to high fabrication costs and the fabrication of only small areas on a sample. Yet, the resolution and precision of this technique being equal only to EBL, it is often the method of choice if one needs flawless arrays of nanoholes with controlled size and periodicity.

26.2.3 Photolithography

Photolithography is a widely popular microfabrication technique for a broad range of applications. This is due in part to the relatively low cost of the process and to the reliability of the process in fabricating large samples of microhole and nanohole arrays with few defects.[67] The process requires a clean wafer, typically silicon, coated with a photoresist deposited by spin coating. The wafer is then exposed to UV light patterned through a mask to cure regions of the photoresist. Depending on the structure required, two types of photoresists, negative and positive, can be employed in the fabrication process. A positive resist becomes more soluble to the developing solution when exposed to UV light while a negative photoresist crosslinks and becomes insoluble to the developing solution. In the case of fabrication of nanohole and microhole arrays, the mask would be patterned with either disks or the exact replica of the hole array depending on the type of resist used. Once exposed, the wafer is developed with a selective solvent to remove the soluble part of the photoresist and leave an array of pillars. Metal is then deposited on the nanopatterned wafer, and the pillars are removed with a liftoff solution (Figure 26.6).[67] Photolithography could be considered a compromise of the previously described techniques in that it is

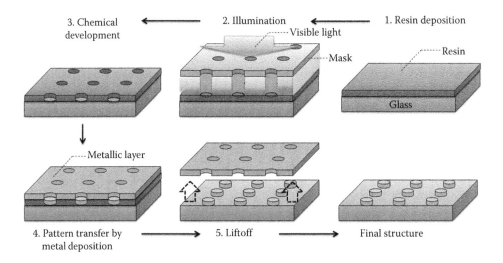

FIGURE 26.6
Schematic of the different steps for the fabrication of an array of cylindrical nanostructures by photolithography with a positive photoresists. (Reproduced from Guillot, N.; de la Chapelle, M. L. *Journal of Nanophotonics* 2012, 6, 064506. Copyright 2012 SPIE publications.)

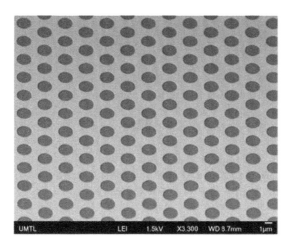

FIGURE 26.7
SEM image of a microhole array prepared by photolithography with a positive photoresist. (Reproduced by permission of the PCCP Owner Societies from Breault-Turcot, J. et al. *Lab on a Chip* 2015, 15, 4433. Copyright 2015 PCCP Owner Societies.)

more affordable than FIB and produces higher quality samples than NSL (Figure 26.7).[58] Photolithography is perfectly suited for the fabrication of hole arrays with a diameter over 100 nm and can produce large-scale samples with a good repeatability.

26.2.4 Laser Interference Lithography

Laser interference lithography (LIL) is a technique that can generate periodic metallic nanostructures on a large uniform area.[68] Fabrication of hole arrays with LIL is similar to conventional photolithography through the utilization of a photoresist, metallization, and liftoff process.[68] However, unlike photolithography, LIL is a mask-free technique, which simplifies the nanofabrication process. The principle of LIL is to expose a photoresist with two coherent beams, which generates an interference pattern through standing waves.[69] A Lloyd's mirror interferometer[70] is usually used to split the two coherent beams from a laser source (Figure 26.8).

The interference between the two plane waves creates a periodic pattern, which can be tuned upon variation of the laser wavelength (λ) and the angle formed between the beams (θ) (see Equation 26.6).

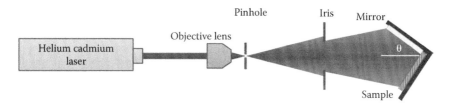

FIGURE 26.8
Lloyd's interferometer setup. A laser beam passed through a lens-pinhole system, which is directed into an iris on the Lloyd's mirror/sample. (Reprinted with permission from Ertorer, E. et al. *Journal of Biomedical Optics* 2013, 18, 8. Copyright 2013 SPIE.)

$$\Lambda = \frac{\lambda}{2\sin\theta} \tag{26.6}$$

In Equation 26.6, Λ corresponds to the fringe period of the interference pattern. A 2D periodic structure is obtained when rotation of the sample (e.g., 90°) is done between two expositions of the fringe pattern. The development of the exposed photoresist will result in an array of pillars, which can generate metallic nanohole arrays in combination with a liftoff process (Figure 26.9).

For example, a 300-mW solid-state laser ($\lambda = 458$ nm) can generate a 2D nanohole array with a periodicity ranging from 450 to 1800 nm by varying θ.[68] Furthermore, the diameter of the holes can be tuned upon variation of the exposition and development time of the photoresist. Nanometric dimensions (<50 nm) can be obtained with LIL by using a UV laser (e.g., ArF at $\lambda = 193$ nm) in combination with deep UV photoresist and immersion techniques.[69] The uniformity of the patterned structure is highly dependent on the stability of the laser source, which is the main expense of LIL. The exposed surface area can be increased by varying the magnification of the objective lens, increasing the distance from the pinhole to the sample, and using a larger Lloyd's mirror.[70] To conclude, LIL is a low-cost fabrication technique that can generate tuneable periodic nanohole arrays on a large scale with high throughput.

26.2.5 Nanoimprinting Lithography

Nanoimprinting lithography (NIL) is a cost-effective technique that can generate nanohole arrays at the wafer scale with great resolution, high uniformity, and high throughput for mass fabrication.[32] Unlike optical lithographic techniques limited by light diffraction, NIL relies on direct mechanical deformation of a resist with a prefabricated mold.[71] Thus, NIL can reach resolution limits below 10 nm facilitating the manufacture of sub-wavelength nanohole arrays.[72] The mold consists of the negative desired pattern, which is then stamped on a substrate coated with either thermoplastic or UV-curable materials.[73] UV nanoimprinting (Figure 26.10) is usually preferred since it can be performed at low pressure and low temperature and is faster than thermal nanoimprinting.[32]

(a)   (b)

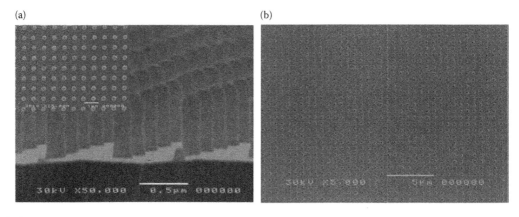

FIGURE 26.9
SEM images of (a) nanopillars array on a glass substrate made by LIL (top view in the inset). (b) Gold nanohole array. (Reprinted with permission from Menezes, J. W. et al. *Advanced Functional Materials* 2010, 20, 3918. Copyright 2010 John Wiley and Sons.)

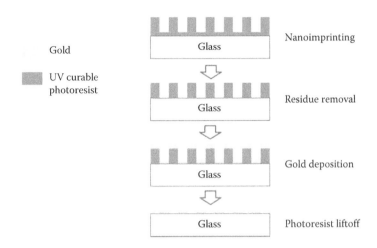

FIGURE 26.10
Principle of the UV NIL process. (Wong, T. I. et al. *Lab on a Chip* 2013, 13, 2405. Reproduced by permission of The Royal Society of Chemistry.)

During the molding step, a pressure (~10 bar) is applied between the mold and the UV resist for few minutes. The substrate, the mold, and the UV resist are then cured by UV light. An anti-adhesive coating on the mold allows for an efficient de-molding of the substrate. Upon separation of the mold and the substrate, residues are removed through oxygen plasma and layers of Cr and Au are deposited. The UV curable photoresist is then lifted off to generate a gold nanohole array structure (Figure 26.11).

The mold is generally made of hard materials that have a high mechanical strength such as silicon, silicon dioxide, and nickel.[32] The manufacture of the mold usually requires EBL and thus remains the main limitation of this technique.[46] However, molds can be used thousands of times without wearing making them cost effective in the long term while generating large areas of nanohole arrays.[46,74]

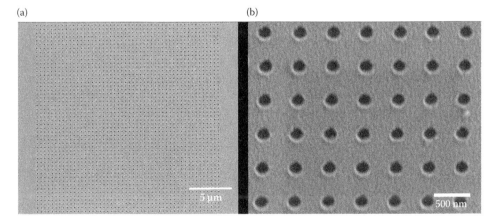

FIGURE 26.11
SEM images at different magnifications (a and b) of a gold nanohole array fabricated with NIL (hole diameter 185 nm, periodicity 450 nm, Ti/Au layer thickness 5/50 nm). (Reprinted from Martinez-Perdiguero, J. et al. *Sensors* 2013, 13, 13960.)

26.3 Biosensing Applications of Nanohole Arrays

Soon after the discovery of the extraordinary optical transmission (EOT) phenomenon,[1] numerous articles reported biosensing applications of nanohole arrays. These early advances in biosensing with nanohole arrays were summarized in reviews published by Gordon et al. and by Genet and Ebbesen.[5,9] As detailed in the introduction of this chapter, the coupling of light with nanohole arrays excites surface plasmon polaritons (SPPs) and is sensitive to refractive index variation. Therefore, the capture of specific analytes on the surface of a nanohole array leads to a small but measureable change in the refractive index, which is observed as a shift in the resonance of the plasmon band. Hence, nanohole arrays have been widely studied for label-free plasmonic biosensing of proteins,[48,75] metal ions,[76] and viruses.[77]

One of the major advantages of using nanohole arrays in biosensing resides in the simplicity of the measurements and optical setup. Sensing with nanohole arrays is typically performed by direct transmission measurements, which require simple and low-cost instrumentation. While commercial UV–Vis instruments are suited for nanohole array analysis, most researchers have designed custom readers. These readers vary in complexity from intensity measurements at a wavelength close to the plasmon resonance to integrated setups for phase analysis using nanohole arrays, and integrated multimodal detection capabilities. The precise control on the position of the SPP allows for optimization of coupling for specific excitation wavelengths,[16] useful in surface-enhanced spectroscopies. Therefore, nanohole arrays have been demonstrated in biosensing experiments using SERS and MEF.[5] The high sensitivity of nanohole arrays in refractive index biosensing as well as in surface-enhanced spectroscopic measurements is an advantage in comparison to the conventional SPR technique.

Additionally, significant efforts have been deployed for the miniaturization of nanohole array devices for higher-throughput biosensing[46,48,78] and for on-chip biosensing applications.[45,46,48] The small footprint of the sensing area of nanohole arrays in the order of square micrometers allows embedding of the sensor into microfluidic devices.[48] The sensor can be excited with a laser or halogen source using microscope objectives. This generates small illumination spot sizes of less than 200 μm^2. A CCD camera usually collects the plasmonic signal, but it has been shown that spectrophotometers equipped with photodiode arrays or single photodiodes can also be used in single channel experiments. The microfabrication process of nanohole array structures generates uniform patterns, which is required for reproducible measurements from chip-to-chip, and indeed device integration using nanohole arrays benefits from the large-scale fabrication processes available for these surfaces, as detailed in the previous section, which will facilitate the bulk manufacture of on-chip sensors. Thus, nanoplasmonic biosensors are expected to be the next generation of plasmonic sensing platforms.

26.3.1 Flow-Over Sensing

A flow-over sensing system consists of the flowing of an analyte solution into a fluidic channel for interaction with the surface of a sensor. With this type of microfluidic device (Figure 26.12), the flow of the analyte is tangential to the surface and the binding kinetics of the analytes are thus mass transport limited.[79] This configuration results in moderately slower binding kinetics; however, it has the advantage of simplicity and robustness.

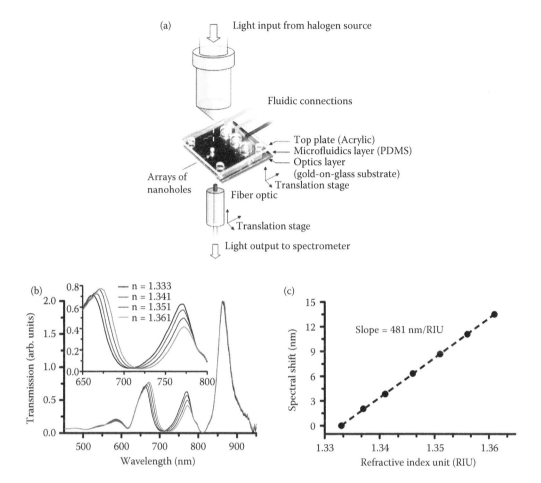

FIGURE 26.12

(a) Transmission setup for microfluidic analysis of nanohole arrays. (b) Plasmonic signal of nanohole arrays in index-calibrated water–glycerol solutions. (c) Calibration curve of sucrose solution. (Reprinted with permission from De Leebeeck, A. et al. *Analytical Chemistry* 2007, 79, 4094. Copyright 2007 American Chemical Society. Reprinted with permission from Im, H. et al. *Analytical Chemistry* 2012, 84, 1941. Copyright 2012 American Chemical Society.)

In this configuration of flow-over analysis and transmission measurements, protein interactions can be detected using nanohole arrays with a sensitivity similar to SPR sensing.[38] The first example of biosensing with nanohole arrays was demonstrated with the well-characterized detection of the streptavidin and biotin interaction.[45] Multiple studies were published soon after to demonstrate the high surface sensitivity of nanohole arrays in a flow-over configuration for the sensing of proteins in the nanomolar range.[4,16]

Arrays of sensors were demonstrated using the flow-over configuration alongside nanohole arrays. The high-throughput performance of imaging systems combined with a six-channel microfluidic chip allowed for parallel biosensing of streptavidin in the nanomolar range with several replicates per channel[47] (Figure 26.13). Furthermore, a chip with 25 sensors was shown to perform an equal number of binding curves simultaneously for the interaction of glutathione *S*-transferase (GST) and anti-GST, again with sensitivity in the nanomolar range.[48] Other reports demonstrated the potential applications of nanohole arrays for multiplexing analysis of antibody–ligand interactions with a 12-channel

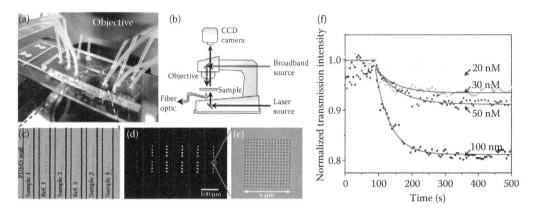

FIGURE 26.13

(a) Image of the microfluidic chip on a microscope stage with Teflon tubes connected to a syringe pump to control the flow of each solution. (b) Schematic of the measurement system for real-time multiplex SPR imaging. (c) A bright-field microscope image of different nanohole array sensors. (d) A transmission mode image. (e) SEM image of the nanohole arrays. (f) Real-time streptavidin–biotin binding kinetics measured for different concentrations of streptavidin. (Reprinted with permission from Lee, S. H. et al. *Langmuir* 2009, 25, 13685. Copyright 2009 American Chemical Society.)

microfluidic chip.[47] More recently, quantification of an ovarian cancer marker (r-PAX8) was demonstrated using nanohole arrays and surface plasmon imaging detection.[80] The microfluidic chips used an integrated six-channel concentration gradient to generate a calibration curve of r-PAX8 with an LOD of 5 nM (Figure 26.14). Each channel contained four identical sensing spots for a total of 28 sensors, which were analyzed simultaneously. A separated channel was used to determine the concentration for an unknown solution of r-PAX8. These examples demonstrate the potential of nanohole arrays in higher-throughput assays.

Different sensing schemes have been applied to a broad variety of biomolecules. For example, studies of cell membrane biorecognition reactions were achieved with nanohole arrays combined with supported lipid bilayers (SLBs).[59] The role of SLB was to mimic cell membrane interactions, which is of great interest for drug development research. Indeed, most approved therapeutic drugs currently available target membrane proteins,[46] and thus it is important to design sensors capable of monitoring drug–receptor interactions. Nanohole arrays in combination with SLB was also demonstrated for biosensing of proteins and DNA.[81] A study of myelin-mimicking lipid bilayers on nanohole arrays kinetically characterized the autoantibodies involved in central nervous system repair.[82] The flow-over sensing methodology, however, can be a limitation to the study of transmembrane proteins due to the underlying substrate.[59] A flow-through sensing approach solved this issue by embedding transmembrane proteins within a lipid membrane into free-standing nanohole arrays, and this sensing methodology will be discussed in greater detail in the following section.[83] The examples herein demonstrate only a subset of the potential applications of nanohole arrays for sensing biomolecules, and it is anticipated that several more applications will be developed in the near future.

26.3.2 Flow-Through Sensing

Mass transport of the analytes to the sensor's surface is an important limitation in the flow-over sensing configuration using nanohole arrays.[59] It has been demonstrated that

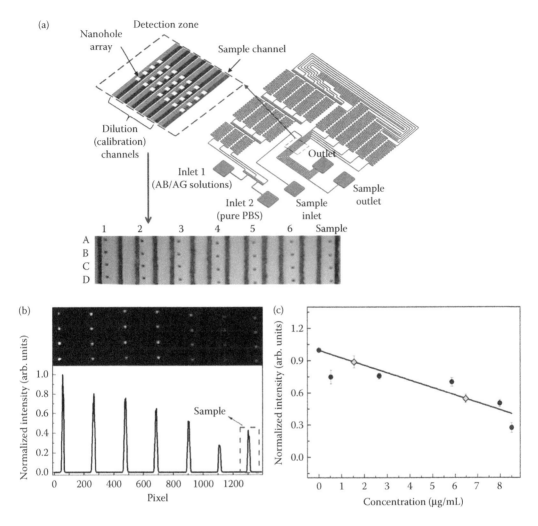

FIGURE 26.14
(a) Scheme of the microfluidic chip. (b) Normalized intensity for detection of various concentrations of r-PAX8. *Insets*: CCD images of the arrays. (c) Average intensities of the sensors as a function of r-PAX8 concentrations ($R^2 = 0.926$). (Escobedo, C. et al. *Analyst* 2013, 138, 1450. Reproduced by permission of The Royal Society of Chemistry.)

molecular binding events occurring on the inner wall of the nanoholes are responsible for most of the sensor's response.[84] However, transport of the analyte into these nanocavities is relatively slow and so in an attempt to increase the transport rate of analytes onto the surface and inside nanoholes, free-standing nanohole array structures were introduced. By using the so-called flow-through sensing approach, a pressure-driven flow is applied to the analyte solution forcing the analyte into the nanoholes, and thus the most sensitive site of the sensor.[79] In this configuration, the nanohole array sensor acts as a membrane and the fluid transport is carried through the nanoholes, which are effectively nanochannels between two fluid reservoirs (Figure 26.15).

The flow-through sensing technique improves the rates of adsorption of analytes to the surface compared to the flow-over sensing approach.[74,79,85] Indeed, it was demonstrated that

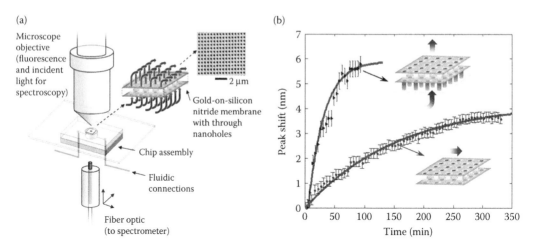

FIGURE 26.15

(a) Optical setup for analysis of flow-through nanohole arrays. (b) Plasmonic response from the adsorption of mercaptoundecanoic acid for flow-through (faster binding) and flow-over (slower binding) sensing. (Reprinted with permission from Eftekhari, F. et al. *Analytical Chemistry* 2009, 81, 4308. Copyright 2009 American Chemical Society.)

a flow-through sensor generates a sixfold enhancement in response time compared to flow-over sensing for the binding of a mercaptoundecanoic acid SAM[85] (Figure 26.15b). On the basis of the same concept, Yanik et al. used a multilayered microfluidic to obtain a 14-fold enhancement in the mass transport rate constant of analytes.[86] Moreover, the flow-through technique improved the plasmonic response to the analyte and decreased the volume required for the analysis. The efficiency of the flow-through sensing approach was demonstrated for label-free biosensing of a cancer biomarker[85] and for viruses.[77] However, while they offer superior analytical performances, flow-through nanohole arrays are more complex to manufacture and require careful handling, as they are more fragile and susceptible to breakage.

26.3.3 96-Well Plasmonic Sensing

Nanohole array technology combined with a multi-well plate reader was recently demonstrated for higher-throughput plasmonic sensing.[78] Gold nanohole arrays can be easily embedded into a 96-well plate for use in combination with transmission measurements. However, a custom built multi-well plate and plate reader was needed to excite nanohole arrays at higher incident angles (Figure 26.16a) so as to benefit from higher surface sensitivity.[16] Such enhancement of the plasmonic response of nanohole arrays allowed for the label-free detection of IgG, prostate-specific antigen (PSA, Figure 26.16b), and methotrexate, an anti-cancer drug, in the nanomolar concentration range. Furthermore, the plasmonic platform was used for the screening of several PSA antibodies for optimal secondary detection of PSA.

26.3.4 Nanohole-Enhanced Spectroscopies

Nanoplasmonic structures hold great promise for biosensing applications with surface-enhanced spectroscopies.[87] As stated before, the electromagnetic (EM) field of a plasmonic structure can enhance the Raman or fluorescence signal of an analyte. For SERS, the plasmonic signal of the nanohole array must be coupled with the excitation wavelength of the

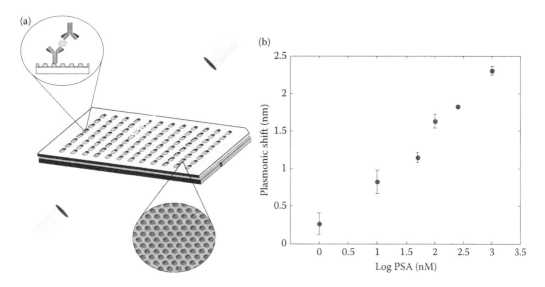

FIGURE 26.16
(a) Scheme of the plasmonic multi-well plate reader. (b) Plasmonic response for the secondary detection of PSA in PBS using the multi-well plate reader. (Reprinted with permission from Couture, M. et al. *ACS Sensors* 2016, 1, 287–294. Copyright 2016 American Chemical Society.)

Raman reporter. Improving the SERS response of nanohole arrays was achieved with the adjustment of the diameter/period ratio of a nanohole array.[21] Adjusting the incident angle of the laser beam to match the plasmon resonance to the laser wavelength also enhanced the SERS response from nanohole arrays.[16]

The plasmonic field of nanohole arrays can be coupled with gold NPs to further enhance the SERS signal. Hence, an example of an SERS biosensing strategy relies on a sandwich assay for the detection of proteins using a functionalized NP (Figure 26.17a).[16] Similarly, a sandwich-structured SERS probes can be used to detect metal ions in human saliva (Figure 26.17b).[76] For this assay, the two ssDNA sequences on the SERS probe hybridize only when Hg^{2+} or Ag^+ ions are present due to nucleotide mismatch. The Ag^+ and Hg^{2+} ions are recognized by DNA via the formation of $C–Ag^+–C$ and $T–Hg^{2+}–T$ base pairs. Upon hybridization of both ssDNA, the NP is brought closer to the sensor surface. The proximity between both plasmonic structures leads to a stronger plasmonic coupling and thereby an amplification of the SERS response from the Raman reporter. The sandwich-structured SERS probes were applied to the detection of Ag(I) and Hg(II) with an LOD in the pico-molar range (Figure 26.17c). The biosensor showed good selectivity and could easily be applied to other targets with different DNA probes. In conclusion, SERS biosensing platforms based on nanohole arrays have great potential for the portable and *in situ* detection of molecules, which could ultimately be detected in bodily fluids.

26.4 Sensing with Microhole Arrays

The field of microhole arrays was initiated by the research from the groups of Coe and Masson. Coe and coworkers used microhole arrays in transmission measurements similar

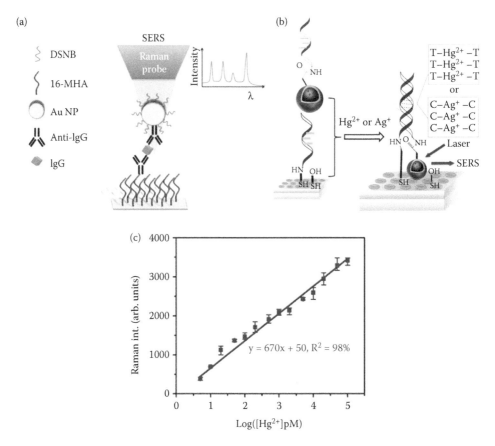

FIGURE 26.17
(a) Scheme of a sandwich SERS bioassay. (b) Principle of the sandwich-structured SERS probes for the detection of Ag(I) or Hg(II). (c) SERS response for detection of Hg(II). (Couture, M. et al. *Nanoscale* 2013, 5, 12399. Reproduced by permission of The Royal Society of Chemistry. Zheng, P. et al. *Nanoscale* 2015, 7, 11005. Reproduced by permission of The Royal Society of Chemistry.)

to the EOT effect described above for nanohole arrays, while Masson and coworkers used the microhole arrays mainly in the prism-coupling configuration of SPR (Kretschmann configuration), analogous to the propagating surface plasmons of thin Au films in classical SPR. The main advantage of the Kretschmann configuration is the ability to do sensing in turbid or absorptive media where transmission measurements can be limited due to the background. The transmission setup makes it easier to multiplex the analysis by running parallel measurements on the same substrate, as described for nanohole arrays in the previous section and shown in Figures 26.12 and 26.13. In the case of microhole arrays, the plasmon can also be excited using grating coupling.[88] This grating coupling makes reflection measurements also possible. The different experimental setup possibilities make microhole arrays a versatile plasmonic substrate for various applications.

26.4.1 Transmission Measurements with Microhole Arrays

In transmission measurements, microhole arrays are plasmonically active in the IR region due to their periodicity, as described by the EOT phenomenon[89,90] and observed from the enhanced transmission in the IR region. The wavelength of the plasmon resonance is

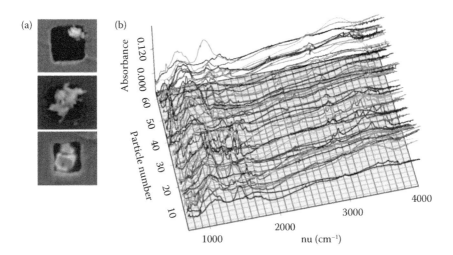

FIGURE 26.18

(a) SEM images of dust particles trapped in the holes of the plasmonic microhole array. (b) Scatter-free IR absorption spectra of 63 individual particles. (Reprinted with permission from Cilwa, K. E. et al. *The Journal of Physical Chemistry C* 2011, 115, 16910. Copyright 2011, American Institute of Physics.)

dependent on the diameter of the holes, on the excitation angle, and on the polarization of the incident light.[91] While the plasmon resonance of microhole arrays is also responsive to refractive index, this feature of microhole arrays has not been exploited as of yet for sensing purposes.

Surface-enhanced spectroscopy is the main advantage of microhole arrays in transmission measurements. Owing to the resonance wavelength in the infrared region of the absorption spectra, the IR spectra of an adsorbate is greatly improved on microhole arrays.[92] The plasmon inside the holes leads to a suppression of the scattering component of the extinction spectrum, leaving only the absorption bands. This was used, for instance, to measure the "scatter-free" absorption spectrum of a yeast cell.[93] The information collected from the absorption bands contributed to the characterization of single cells and could easily be applied to many other species.

In another application, square-shaped microhole arrays were used as a filter to capture micrometric airborne particles. The size range captured by microhole arrays corresponded to the largest particles permeating through the respiratory system to the lungs.[94] Therefore, monitoring these airborne particles represents a target of interest to the analytical sciences. In one instance, 63 different particles were captured with this filter and their individual IR spectra were measured (Figure 26.18).[94] This example demonstrates the possibilities for multiplexed analysis when using microhole arrays in transmission.

26.4.2 Microhole Array Sensing in the Kretschmann Configuration

Microhole arrays support propagating surface plasmons,[56] for which the SPR is similar but slightly larger than for a continuous gold film due to the scattering of the plasmon resonance. The position of the absorption band from both gold films and gold microhole arrays is angle dependent. The SPR bands for both films are co-located for most angles; however, for shorter excitation angles of the microhole arrays, an LSPR band appears around 500 nm (Figure 26.19).[57] This means that microhole arrays support both SPR and LSPR that can be co-excited under specific conditions.

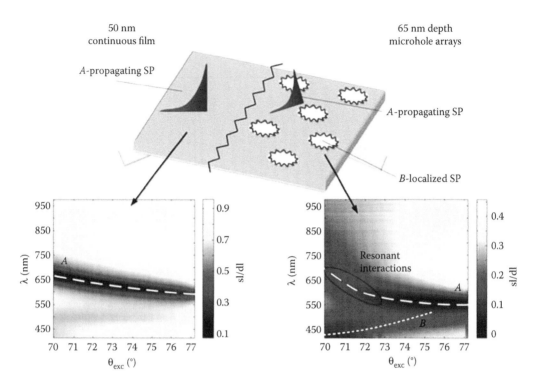

FIGURE 26.19
SPR dispersion in water environment (RI = 1.333 RIU) of continuous films and microhole arrays. (Left) Propagating SP mode (dashed line) is present on continuous films, while 65-nm depth microhole arrays (right) present localized SP (dotted line) and propagating SP modes. (With kind permission from Springer Science + Business Media: *Analytical and Bioanalytical Chemistry*, 404, 2012, 2859, Live, L. S. et al.)

The coupling of localized and propagating surface plasmons in microhole arrays leads to an enhancement in the SPP resonances and higher sensitivity to refractive index.[57,95] Microhole arrays also have greater sensitivity to biomolecules.[57] For example, it was shown that the detection of IgG/anti-IgG was 3–5 times more sensitive with microhole arrays compared to continuous gold films. Thus, a bioassay for PSA, a biomarker for prostate cancer, was constructed using a sandwich assay on microhole arrays. PSA was detected in the low nanomolar range with a detection limit of 0.1 nM.[58]

The plasmon resonance of microhole arrays can be excited by direct illumination due to the grating coupling in reflectance measurements. This configuration is perfectly suited for analysis by surface-enhanced spectroscopies. In this configuration, the difference in plasmonic signal between the holes and the film can be seen in reflectance and SERS (Figure 26.20)[57] where a shift in the plasmon maximum was observed between the holes and terraces. The EM field is stronger at the rim of the holes, where the LSPR is located. This high electric field was observed from the high SERS response located at the rim of the microholes (Figure 26.20b).

Microhole arrays are also suited for MEF where a fluorophore close to the surface could couple with the plasmon, leading to an enhanced fluorescence. It can also be applied to a labeled analyte such as in Figure 26.21[58] where an ELISA-like test was built on a plasmonic sensor. The enhancement factor of microhole arrays (less than 1 order of magnitude) is lower than metallic NPs. However, the SPR sensitivity of microhole arrays is superior to other plasmonic substrates, and thus, microhole arrays are excellent

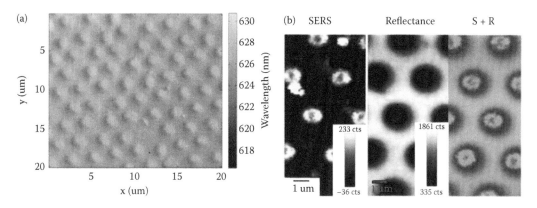

FIGURE 26.20
(a) Image of the minimum wavelength in the reflectance spectrum shows a shift in the plasmon excitation wavelength for the location of the microholes. (b) SERS imaging of the microhole arrays highlighting the enhanced EM field into the holes, a reference image in reflectance of the same area, and finally both images overlaid. (With kind permission from Springer Science + Business Media: *Analytical and Bioanalytical Chemistry*, 404, 2012, 2859, Live, L. S. et al.)

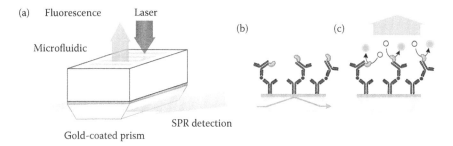

FIGURE 26.21
(a) Concept of the combination of SPR and fluorescence detection on top of a plasmonic sensor through a transparent microfluidic. (b) Schematic of the SPR biodetection and (c) for the fluorescence measurement in a ELISA-type assay. (Reproduced by permission of the PCCP Owner Societies. Copyright 2015 PCCP Owner Societies.)

substrates for combined surface-enhanced spectroscopies and plasmonic sensing on a single substrate.

26.5 Conclusions and Outlook

In less than two decades, nanohole arrays have matured from an optical curiosity to a sensing platform efficient for refractometric sensing, surface-enhanced spectroscopies, and in electrochemical measurements. While there are still several differences in fabrication techniques, geometrical characteristics, and instrumental configurations, it has become clear that nanohole arrays have the potential to rival the sensitivity and resolution of the well-established SPR in the Kretschmann configuration.

Currently, this technique remains in an early stage of development and researchers are still validating different instrumental configurations and integrating them into devices that could serve point-of-care applications. To achieve this goal, the field must also consider the challenges of working in complex fluids and will need to integrate mitigation methods to minimize the impact of nonspecific adsorption as well as the high absorbance and turbidity of clinical samples. Once these developments are complete, we should see an increase in the application of nanohole arrays to detect biomolecules in samples of increased complexity.

The possibility of combining refractometric sensing with surface-enhanced spectroscopies or electrochemistry is another advantage of nanohole arrays. These techniques could improve the sensitivity of nanohole arrays and also provide molecular selectivity to target specific analytes. It has been clearly established that nanohole arrays are good SERS and MEF substrates, but their lower sensitivity than NPs might explain the relatively few articles reporting the use of nanohole arrays in sensing applications with SERS or MEF. However, once the advantages of refractometric and surface-enhanced sensing or electrochemistry are clearly demonstrated, we should likely see an increase in the number of publications in this field. Finally, nanohole arrays are ideal candidates for developing nanoplasmonics sensors due to their versatility and their sensitivity. It is foreseen that devices integrating nanohole and microhole arrays with SPR and SERS or MEF will be packaged in portable devices for a suite of applications.

References

1. Henzie, J.; Lee, J.; Lee, M. H.; Hasan, W.; Odom, T. W., Nanofabrication of plasmonic structures. *In Annual Review of Physical Chemistry* 2009, 60, 147–165.
2. Li, W.; Zhang, L.; Zhou, J.; Wu, H., Well-designed metal nanostructured arrays for label-free plasmonic biosensing. *Journal of Materials Chemistry C* 2015, 3 (25), 6479–6492.
3. Carmen Estevez, M.; Otte, M. A.; Sepulveda, B.; Lechuga, L. M., Trends and challenges of refractometric nanoplasmonic biosensors: A review. *Analytica Chimica Acta* 2014, 806, 55–73.
4. Ebbesen, T. W.; Lezec, H. J.; Ghaemi, H. F.; Thio, T.; Wolff, P. A., Extraordinary optical transmission through sub-wavelength hole arrays. *Nature* 1998, 391 (6668), 667–669.
5. Gordon, R.; Sinton, D.; Kavanagh, K. L.; Brolo, A. G., A new generation of sensors based on extraordinary optical transmission. *Accounts of Chemical Research* 2008, 41 (8), 1049–1057.
6. Garcia-Vidal, F. J.; Martin-Moreno, L.; Ebbesen, T. W.; Kuipers, L., Light passing through sub-wavelength apertures. *Reviews of Modern Physics* 2010, 82 (1), 729–787.
7. Barnes, W. L.; Murray, W. A.; Dintinger, J.; Devaux, E.; Ebbesen, T. W., Surface plasmon polaritons and their role in the enhanced transmission of light through periodic arrays of subwavelength holes in a metal film. *Physical Review Letters* 2004, 92 (10).
8. Gao, H.; Henzie, J.; Odom, T. W., Direct evidence for surface plasmon-mediated enhanced light transmission through metallic nanohole arrays. *Nano Letters* 2006, 6 (9), 2104–2108.
9. Genet, C.; Ebbesen, T. W., Light in tiny holes. *Nature* 2007, 445 (7123), 39–46.
10. Martin-Moreno, L.; Garcia-Vidal, F. J.; Lezec, H. J.; Pellerin, K. M.; Thio, T.; Pendry, J. B.; Ebbesen, T. W., Theory of extraordinary optical transmission through subwavelength hole arrays. *Physical Review Letters* 2001, 86 (6), 1114–1117.
11. Murray-Methot, M.-P.; Ratel, M.; Masson, J.-F., Optical properties of Au, Ag, and bimetallic Au on Ag nanohole arrays. *Journal of Physical Chemistry C* 2010, 114 (18), 8268–8275.
12. Barrios, C. A.; Canalejas-Tejero, V.; Herranz, S.; Moreno-Bondi, M. C.; Avella-Oliver, M.; Puchades, R.; Maquieira, A., Aluminum nanohole arrays fabricated on polycarbonate for compact disc-based label-free optical biosensing. *Plasmonics* 2014, 9 (3), 645–649.

13. Barrios, C. A.; Canalejas-Tejero, V.; Herranz, S.; Urraca, J.; Moreno-Bondi, M. C.; Avella-Oliver, M.; Maquieira, A.; Puchades, R., Aluminum nanoholes for optical biosensing. *Biosensors* 2015, 5 (3), 417–431.

14. Ikenoya, Y.; Susa, M.; Shi, J.; Nakamura, Y.; Dahlin, A. B.; Sannomiya, T., Optical resonances in short-range ordered nanoholes in ultrathin aluminum/aluminum nitride multilayers. *Journal of Physical Chemistry C* 2013, 117 (12), 6373–6382.

15. Schmidt, T. M.; Bochenkov, V. E.; Espinoza, J. D. A.; Smits, E. C. P.; Muzafarov, A. M.; Kononevich, Y. N.; Sutherland, D. S., Plasmonic fluorescence enhancement of DBMBF2 monomers and DBMBF2-toluene exciplexes using Al-hole arrays. *Journal of Physical Chemistry C* 2014, 118 (4), 2138–2145.

16. Couture, M.; Liang, Y.; Richard, H.-P. P.; Faid, R.; Peng, W.; Masson, J.-F., Tuning the 3D plasmon field of nanohole arrays. *Nanoscale* 2013, 5 (24), 12399–12408.

17. Poirier-Richard, H. P.; Couture, M.; Brule, T.; Masson, J. F., Metal-enhanced fluorescence and FRET on nanohole arrays excited at angled incidence. *Analyst* 2015, 140 (14), 4792–4798.

18. Monteiro, J. P.; Carneiro, L. B.; Rahman, M. M.; Brolo, A. G.; Santos, M. J. L.; Ferreira, J.; Girotto, E. M., Effect of periodicity on the performance of surface plasmon resonance sensors based on subwavelength nanohole arrays. *Sensors and Actuators B-Chemical* 2013, 178, 366–370.

19. Mazzotta, F.; Johnson, T. W.; Dahlin, A. B.; Shaver, J.; Oh, S.-H.; Hook, F., Influence of the evanescent field decay length on the sensitivity of plasmonic nanodisks and nanoholes. *Acs Photonics* 2015, 2 (2), 256–262.

20. Odom, T. W.; Gao, H.; McMahon, J. M.; Henzie, J.; Schatz, G. C., Plasmonic superlattices: Hierarchical subwavelength hole arrays. *Chemical Physics Letters* 2009, 483 (4–6), 187–192.

21. Correia-Ledo, D.; Gibson, K. F.; Dhawan, A.; Couture, M.; Vo-Dinh, T.; Graham, D.; Masson, J.-F., Assessing the location of surface plasmons over nanotriangle and nanohole arrays of different size and periodicity. *Journal of Physical Chemistry C* 2012, 116 (12), 6884–6892.

22. Lee, S. H.; Bantz, K. C.; Lindquist, N. C.; Oh, S.-H.; Haynes, C. L., Self-assembled plasmonic nanohole arrays. *Langmuir* 2009, 25 (23), 13685–13693.

23. Gibson, K. F.; Correia-Ledo, D.; Couture, M.; Graham, D.; Masson, J.-F., Correlated AFM and SERS imaging of the transition from nanotriangle to nanohole arrays. *Chemical Communications* 2011, 47 (12), 3404–3406.

24. Brolo, A. G.; Arctander, E.; Gordon, R.; Leathem, B.; Kavanagh, K. L., Nanohole-enhanced Raman scattering. *Nano Letters* 2004, 4 (10), 2015–2018.

25. Chan, C. Y.; Xu, J. B.; Waye, M. Y.; Ong, H. C., Angle resolved surface enhanced Raman scattering (SERS) on two-dimensional metallic arrays with different hole sizes. *Applied Physics Letters* 2010, 96 (3).

26. Tabatabaei, M.; Najiminaini, M.; Davieau, K.; Kaminska, B.; Singh, M. R.; Carson, J. J. L.; Lagugné-Labarthet, F., Tunable 3D plasmonic cavity nanosensors for surface-Enhanced Raman spectroscopy with sub-femtomolar limit of detection. *ACS Photonics* 2015, 2 (6), 752–759.

27. Zheng, P.; Cushing, S. K.; Suri, S.; Wu, N., Tailoring plasmonic properties of gold nanohole arrays for surface-enhanced Raman scattering. *Physical Chemistry Chemical Physics* 2015, 17 (33), 21211–21219.

28. Brolo, A. G.; Kwok, S. C.; Moffitt, M. G.; Gordon, R.; Riordon, J.; Kavanagh, K. L., Enhanced fluorescence from arrays of nanoholes in a gold film. *Journal of the American Chemical Society* 2005, 127 (42), 14936–14941.

29. Guo, P.-F.; Wu, S.; Ren, Q.-J.; Lu, J.; Chen, Z.; Xiao, S.-J.; Zhu, Y.-Y., Fluorescence enhancement by surface plasmon polaritons on metallic nanohole arrays. *Journal of Physical Chemistry Letters* 2010, 1 (1), 315–318.

30. Cao, Z. L.; Ong, H. C., Determination of coupling rate of light emitter to surface plasmon polaritons supported on nanohole array. *Applied Physics Letters* 2013, 102 (24).

31. Wang, Y.; Wu, L.; Zhou, X.; Wong, T. I.; Zhang, J.; Bai, P.; Li, E. P.; Liedberg, B., Incident-angle dependence of fluorescence enhancement and biomarker immunoassay on gold nanohole array. *Sensors and Actuators B-Chemical* 2013, 186, 205–211.

32. Wong, T. I.; Han, S.; Wu, L.; Wang, Y.; Deng, J.; Tan, C. Y. L.; Bai, P.; Loke, Y. C.; Yang, X. D.; Tse, M. S.; Ng, S. H.; Zhou, X., High throughput and high yield nanofabrication of precisely designed gold nanohole arrays for fluorescence enhanced detection of biomarkers. *Lab on a Chip* 2013, 13 (12), 2405–2413.

33. Dahlin, A. B.; Zahn, R.; Voeroes, J., Nanoplasmonic sensing of metal-halide complex formation and the electric double layer capacitor. *Nanoscale* 2012, 4 (7), 2339–2351.

34. Nakamoto, K.; Kurita, R.; Niwa, O., Electrochemical surface plasmon resonance measurement based on gold nanohole array fabricated by nanoimprinting technique. *Analytical Chemistry* 2012, 84 (7), 3187–3191.

35. Patskovsky, S.; Dallaire, A.-M.; Blanchard-Dionne, A.-P.; Vallee-Belisle, A.; Meunier, M., Electrochemical structure-switching sensing using nanoplasmonic devices. *Annalen Der Physik* 2015, 527 (11-12), 806–813.

36. Zhang, J.; Wang, Y.; Wong, T. I.; Liu, X.; Zhou, X.; Liedberg, B., Electrofocusing-enhanced localized surface plasmon resonance biosensors. *Nanoscale* 2015, 7 (41), 17244–17248.

37. Wu, L.; Bai, P.; Zhou, X.; Li, E. P., Reflection and transmission modes in nanohole-array-based plasmonic sensors. *Ieee Photonics Journal* 2012, 4 (1), 26–33.

38. Couture, M.; Live, L. S.; Dhawan, A.; Masson, J.-F., EOT or Kretschmann configuration? Comparative study of the plasmonic modes in gold nanohole arrays. *Analyst* 2012, 137 (18), 4162–4170.

39. Blanchard-Dionne, A. P.; Guyot, L.; Patskovsky, S.; Gordon, R.; Meunier, M., Intensity based surface plasmon resonance sensor using a nanohole rectangular array. *Optics Express* 2011, 19 (16), 15041–15046.

40. Escobedo, C.; Vincent, S.; Choudhury, A. I. K.; Campbell, J.; Brolo, A. G.; Sinton, D.; Gordon, R., Integrated nanohole array surface plasmon resonance sensing device using a dual-wavelength source. *Journal of Micromechanics and Microengineering* 2011, 21 (11), article 115001.

41. Das, M.; Hohertz, D.; Nirwan, R.; Brolo, A. G.; Kavanagh, K. L.; Gordon, R., Improved performance of nanohole surface plasmon resonance sensors by the integrated response method. *IEEE Photonics Journal* 2011, 3 (3), 441–449.

42. Cao, Z. L.; Wong, S. L.; Wu, S. Y.; Ho, H. P.; Ong, H. C., High performing phase-based surface plasmon resonance sensing from metallic nanohole arrays. *Applied Physics Letters* 2014, 104 (17).

43. Tellez, G. A. C.; Hassan, S. A.; Tait, R. N.; Berini, P.; Gordon, R., Atomically flat symmetric elliptical nanohole arrays in a gold film for ultrasensitive refractive index sensing. *Lab on a Chip* 2013, 13 (13), 2541–2546.

44. Zeng, B.; Gao, Y.; Bartoli, F. J., Differentiating surface and bulk interactions in nanoplasmonic interferometric sensor arrays. *Nanoscale* 2015, 7 (1), 166–170.

45. De Leebeeck, A.; Kumar, L. K. S.; De Lange, V.; Sinton, D.; Gordon, R.; Brolo, A. G., On-chip surface-based detection with nanohole arrays. *Analytical Chemistry* 2007, 79 (11), 4094–4100.

46. Escobedo, C., On-chip nanohole array based sensing: A review. *Lab on a Chip* 2013, 13 (13), 2445–2463.

47. Im, H.; Lesuffleur, A.; Lindquist, N. C.; Oh, S.-H., Plasmonic nanoholes in a multichannel microarray format for parallel kinetic assays and differential sensing. *Analytical Chemistry* 2009, 81 (8), 2854–2859.

48. Ji, J.; O'Connell, J. G.; Carter, D. J. D.; Larson, D. N., High-throughput nanohole array based system to monitor multiple binding events in real time. *Analytical Chemistry* 2008, 80 (7), 2491–2498.

49. Wang, Y.; Kar, A.; Paterson, A.; Kourentzi, K.; Le, H.; Ruchhoeft, P.; Wilson, R.; Bao, J., Transmissive nanohole arrays for massively-parallel optical biosensing. *ACS Photonics* 2014, 1 (3), 241–245.

50. Andrade, G. F. S.; Hayashi, J. G.; Rahman, M. M.; Salcedo, W. J.; Cordeiro, C. M. B.; Brolo, A. G., Surface-enhanced resonance Raman scattering (SERRS) using Au nanohole arrays on optical fiber tips. *Plasmonics* 2013, 8 (2), 1113–1121.

51. Jia, P.; Yang, J., Integration of large-area metallic nanohole arrays with multimode optical fibers for surface plasmon resonance sensing. *Applied Physics Letters* 2013, 102 (24), article 243107.

52. Kegel, L. L.; Boyne, D.; Booksh, K. S., Sensing with prism-based near-infrared surface plasmon resonance spectroscopy on nanohole array platforms. *Analytical Chemistry* 2014, 86 (7), 3355–3364.

53. Kekesi, R.; Meneses-Rodriguez, D.; Garcia-Perez, F.; Gonzalez, M. U.; Garcia-Martin, A.; Cebollada, A.; Armelles, G., The effect of holes in the dispersion relation of propagative surface plasmon modes of nanoperforated semitransparent metallic films. *Journal of Applied Physics* 2014, 116 (13).

54. Schwind, M.; Kasemo, B.; Zoric, I., Localized and propagating plasmons in metal films with nanoholes. *Nano Letters* 2013, 13 (4), 1743–1750.

55. Live, L. S.; Murray-Methot, M.-P.; Masson, J.-F., Localized and propagating surface plasmons in gold particles of near-micron size. *Journal of Physical Chemistry C* 2009, 113 (1), 40–44.

56. Live, L. S.; Masson, J.-F., High sensitivity of plasmonic microstructures near the transition from short-range to propagating surface plasmon. *Journal of Physical Chemistry C* 2009, 113 (23), 10052–10060.

57. Live, L. S.; Dhawan, A.; Gibson, K. F.; Poirier-Richard, H.-P.; Graham, D.; Canva, M.; Vo-Dinh, T.; Masson, J.-F., Angle-dependent resonance of localized and propagating surface plasmons in microhole arrays for enhanced biosensing. *Analytical and Bioanalytical Chemistry* 2012, 404 (10), 2859–2868.

58. Breault-Turcot, J.; Poirier-Richard, H. P.; Couture, M.; Pelechacz, D.; Masson, J. F., Single chip SPR and fluorescent ELISA assay of prostate specific antigen. *Lab on a Chip* 2015, 15 (23), 4433–4440.

59. Dahlin, A. B.; Wittenberg, N. J.; Hook, F.; Oh, S. H., Promises and challenges of nanoplasmonic devices for refractometric biosensing. *Nanophotonics* 2013, 2 (2), 83–101.

60. Masson, J.-F.; Murray-Methot, M.-P.; Live, L. S., Nanohole arrays in chemical analysis: Manufacturing methods and applications. *Analyst* 2010, 135 (7), 1483–1489.

61. Cheng, K.; Wang, S.; Cui, Z.; Li, Q.; Dai, S.; Du, Z., Large-scale fabrication of plasmonic gold nanohole arrays for refractive index sensing at visible region. *Applied Physics Letters* 2012, 100 (25), 253101.

62. Jamaludin, F. S.; Mohd Sabri, M. F.; Said, S. M., Controlling parameters of focused ion beam (FIB) on high aspect ratio micro holes milling. *Microsystem Technologies* 2013, 19 (12), 1873–1888.

63. Lugstein, A.; Steiger-Thirsfeld, A.; Basnar, B.; Hyun, Y. J.; Pongratz, P.; Bertagnolli, E., Impact of fluence-rate related effects on the sputtering of silicon at elevated target temperatures. *Journal of Applied Physics* 2009, 105 (4), 044912.

64. Zhou, J.; Yang, G. L., Focused ion-beam based nanohole modeling, simulation, fabrication, and application. *Journal of Manufacturing Science and Engineering-Transactions of the ASME* 2010, 132 (1), article number 011005.

65. Menezes, J. W.; Barea, L. A. M.; Chillcce, E. F.; Frateschi, N.; Cescato, L., Comparison of plasmonic arrays of holes recorded by interference lithography and focused ion beam. *Photonics Journal, IEEE* 2012, 4 (2), 544–551.

66. Si, G.; Danner, A. J.; Teo, S. L.; Teo, E. J.; Teng, J.; Bettiol, A. A., Photonic crystal structures with ultrahigh aspect ratio in lithium niobate fabricated by focused ion beam milling. *Journal of Vacuum Science & Technology B* 2011, 29 (2), 021205.

67. Guillot, N.; de la Chapelle, M. L., Lithographied nanostructures as nanosensors. *NANOP* 2012, 6 (1), 064506-1–064506-28.

68. Menezes, J. W.; Ferreira, J.; Santos, M. J. L.; Cescato, L.; Brolo, A. G., Large-area fabrication of periodic arrays of nanoholes in metal films and their application in biosensing and plasmonic-enhanced photovoltaics. *Advanced Functional Materials* 2010, 20 (22), 3918–3924.

69. Seo, J. H.; Park, J. H.; Kim, S. I.; Park, B. J.; Ma, Z. Q.; Choi, J.; Ju, B. K., Nanopatterning by laser interference lithography: Applications to optical devices. *Journal of Nanoscience and Nanotechnology* 2014, 14 (2), 1521–1532.

70. Ertorer, E.; Vasefi, F.; Keshwah, J.; Najiminaini, M.; Halfpap, C.; Langbein, U.; Carson, J. J. L.; Hamilton, D. W.; Mittler, S., Large area periodic, systematically changing, multishape nanostructures by laser interference lithography and cell response to these topographies. *Journal of Biomedical Optics* 2013, 18 (3), 8.

71. Guo, L. J., Nanoimprint lithography: Methods and material requirements. *Advanced Materials* 2007, 19 (4), 495–513.

72. Song, H. Y.; Wong, T. I.; Sadovoy, A.; Wu, L.; Bai, P.; Deng, J.; Guo, S. F.; Wang, Y.; Knoll, W. G.; Zhou, X. D., Imprinted gold 2D nanoarray for highly sensitive and convenient PSA detection via plasmon excited quantum dots. *Lab on a Chip* 2015, 15 (1), 253–263.

73. Kooy, N.; Mohamed, K.; Pin, L. T.; Guan, O. S., A review of roll-to-roll nanoimprint lithography. *Nanoscale Research Letters* 2014, 9, 13.

74. Kumar, S.; Cherukulappurath, S.; Johnson, T. W.; Oh, S. H., Millimeter-sized suspended plasmonic nanohole arrays for surface-tension-driven flow-through SERS. *Chemistry of Materials* 2014, 26 (22), 6523–6530.

75. Sharpe, J. C.; Mitchell, J. S.; Lin, L.; Sedoglavich, H.; Blaikie, R. J., Gold nanohole array substrates as immunobiosensors. *Anal. Chem.* 2008, 80 (6), 2244–2249.

76. Zheng, P.; Li, M.; Jurevic, R.; Cushing, S. K.; Liu, Y. X.; Wu, N. Q., A gold nanohole array based surface-enhanced Raman scattering biosensor for detection of silver(I) and mercury(II) in human saliva. *Nanoscale* 2015, 7 (25), 11005–11012.

77. Yanik, A. A.; Huang, M.; Kamohara, O.; Artar, A.; Geisbert, T. W.; Connor, J. H.; Altug, H., An optofluidic nanoplasmonic biosensor for direct detection of live viruses from biological media. *Nano Letters* 2010, 10 (12), 4962–4969.

78. Couture, M.; Ray, K. K.; Poirier-Richard, H.-P.; Crofton, A.; Masson, J.-F., 96-Well plasmonic sensing with nanohole arrays. *ACS Sensors* 2016.

79. Escobedo, C.; Brolo, A. G.; Gordon, R.; Sinton, D., Flow-through vs flow-over: Analysis of transport and binding in nanohole array plasmonic biosensors. *Analytical Chemistry* 2010, 82 (24), 10015–10020.

80. Escobedo, C.; Chou, Y. W.; Rahman, M.; Duan, X. B.; Gordon, R.; Sinton, D.; Brolo, A. G.; Ferreira, J., Quantification of ovarian cancer markers with integrated microfluidic concentration gradient and imaging nanohole surface plasmon resonance. *Analyst* 2013, 138 (5), 1450–1458.

81. Dahlin, A.; Zach, M.; Rindzevicius, T.; Kall, M.; Sutherland, D. S.; Hook, F., Localized surface plasmon resonance sensing of lipid-membrane-mediated biorecognition events. *J. Am. Chem. Soc.* 2005, 127 (14), 5043–5048.

82. Wittenberg, N. J.; Im, H.; Xu, X. H.; Wootla, B.; Watzlawik, J.; Warrington, A. E.; Rodriguez, M.; Oh, S. H., High-affinity binding of remyelinating natural autoantibodies to myelin-mimicking lipid bilayers revealed by nanohole surface plasmon resonance. *Analytical Chemistry* 2012, 84 (14), 6031–6039.

83. Im, H.; Wittenberg, N. J.; Lesuffleur, A.; Lindquist, N. C.; Oh, S. H., Membrane protein biosensing with plasmonic nanopore arrays and pore-spanning lipid membranes. *Chemical Science* 2010, 1 (6), 688–696.

84. Ferreira, J.; Santos, M. J. L.; Rahman, M. M.; Brolo, A. G.; Gordon, R.; Sinton, D.; Girotto, E. M., Attomolar protein detection using in-hole surface plasmon resonance. *J. Am. Chem. Soc.* 2009, 131 (2), 436.

85. Eftekhari, F.; Escobedo, C.; Ferreira, J.; Duan, X. B.; Girotto, E. M.; Brolo, A. G.; Gordon, R.; Sinton, D., Nanoholes As nanochannels: Flow-through plasmonic sensing. *Anal. Chem.* 2009, 81 (11), 4308–4311.

86. Yanik, A. A.; Huang, M.; Artar, A.; Chang, T. Y.; Altug, H., Integrated nanoplasmonic-nanofluidic biosensors with targeted delivery of analytes. *Applied Physics Letters* 2010, 96 (2), 3.

87. Couture, M.; Zhao, S. S.; Masson, J. F., Modern surface plasmon resonance for bioanalytics and biophysics. *Physical Chemistry Chemical Physics* 2013, 15 (27), 11190–11216.

88. Homola, J., Surface plasmon resonance sensors for detection of chemical and biological species. *Chemical Reviews* 2008, 108 (2), 462–493.

89. Coe, J. V.; Heer, J. M.; Teeters-Kennedy, S.; Tian, H.; Rodriguez, K. R., Extraordinary transmission of metal films with arrays of subwavelength holes. *Annual Review of Physical Chemistry* 2008, 59 (1), 179–202.

90. Liu, H.; Lalanne, P., Microscopic theory of the extraordinary optical transmission. *Nature* 2008, 452 (7188), 728–731.

91. Cilwa, K.; Teeters-Kennedy, S.; Ramsey, K. A.; Coe, J. V., Infrared plasmonic transmission resonances of gold film with hexagonally ordered hole arrays on ZnSe substrate. *Plasmonics* 2012, 8 (2), 349–355.
92. Heer, J.; Corwin, L.; Cilwa, K.; Malone, M. A.; Coe, J. V., Infrared sensitivity of plasmonic metal films with hole arrays to microspheres in and out of the holes. *The Journal of Physical Chemistry C* 2010, 114 (1), 520–525.
93. Malone, M. A.; Prakash, S.; Heer, J. M.; Corwin, L. D.; Cilwa, K. E.; Coe, J. V., Modifying infrared scattering effects of single yeast cells with plasmonic metal mesh. *The Journal of Chemical Physics* 2010, 133 (18), 185101.
94. Cilwa, K. E.; McCormack, M.; Lew, M.; Robitaille, C.; Corwin, L.; Malone, M. A.; Coe, J. V., Scatter-free IR absorption spectra of individual, 3–5 μm, airborne dust particles using plasmonic metal microarrays: A library of 63 spectra. *The Journal of Physical Chemistry C* 2011, 115 (34), 16910–16919.
95. Live, L. S.; Bolduc, O. R.; Masson, J.-F., Propagating surface plasmon resonance on microhole arrays. *Analytical Chemistry* 2010, 82 (9), 3780–3787.

27

Sensitive DNA Detection and SNP Identification Using Ultrabright SERS Nanorattles and Magnetic Beads for In Vitro Diagnostics

Hoan T. Ngo, Naveen Gandra, Andrew M. Fales, Steve M. Taylor, and Tuan Vo-Dinh

CONTENTS

27.1 Introduction

Early, accurate diagnostics is critical in medicine. Nucleic acid testing (NAT) is an essential class of *in vitro* molecular diagnostics with many advantages including high specificity, high sensitivity, short turnaround time, and mutation identification. *In vitro* molecular diagnostics based on NAT can determine the specific bacteria or virus causing illness. This capability is particularly useful for diagnosing diseases having similar symptoms but being caused by different bacteria or viruses. Not only that, NAT techniques can detect mutations, which is important amid the emergence of drug-resistant bacteria and virus.

Although drug resistance can also be identified using culture techniques, such procedures are slow and usually take several days. In contrast, NAT can provide results within several hours. Furthermore, bacterial or viral load can be quantified, allowing treatment effectiveness to be monitored. Using NAT, host genetic profiles and genomic signatures resulting from host response can be determined. Such information opens opportunities for personalized medicine and allows diagnostics before the development of peak symptoms. With such advantages, NAT has a wide range of applications from screening to diagnostics, therapy selection, monitoring, and risk assessment.

Currently, polymerase chain reaction (PCR) is the gold standard of NAT due to its extremely high sensitivity, with single to a few copies limit of detection. However, it is still quite laborious, time-consuming, and requires skilled workers together with relative expensive and bulky equipment. Until recently, PCR was mainly limited to hospital and laboratories [1]. It is important to develop NAT techniques that can be used in the field, at the point-of-care (POC), and in resource-limited settings. This topic has attracted considerable interest with many techniques reported [1–3]. They can be classified into two main groups: (1) target amplification techniques (e.g., PCR, loop-mediated isothermal amplification, etc.) and (2) signal amplification techniques (e.g., branched DNA, bio-barcode assay, etc.) [4]. While target amplification techniques have been proven to be very sensitive by creating million copies of target sequences through enzymatic reactions, they are susceptible to contamination, inhibitors, and amplification bias, and require extensive sample preparation [5]. Integrating target amplification techniques into miniaturized systems has attracted great interest [6,7]. As an alternative to target amplification, signal amplification techniques directly detect target sequences without enzymatic amplification. Since the copy number of target sequences is usually very low, the main challenge of signal amplification techniques is to achieve clinically relevant limits of detection. To date, several signal amplification techniques showing impressive limits of detection have been reported, including branched DNA [8], bio-barcode [9], assay based on dye-doped silica nanoparticles [10], etc.

SERS is a phenomenon where Raman scattering is enhanced many millionfold [11–14]. SERS sensitivity has been demonstrated by single molecule detection [15–18]. Over the last three decades, our laboratory has been involved in the development and application of various plasmonic nanoplatforms ranging from nanoparticles, nanopost arrays, nanowires, and nanochips for SERS analysis [19–21]. Using these nanoplatforms, we have developed different chemical and biological sensing methods for medical diagnostics and environmental monitoring [22,23]. Compared with fluorescence, SERS has several advantages including being more resistant to photobleaching, more suitable for multiplexing, and having potentially lower limits of detection. Furthermore, SERS can be excited at near-IR wavelengths, thus being less affected by biological matrices.

Many SERS-based DNA-detection methods have been reported. Generally, they can be divided into three main categories. The first involves direct detection of intrinsic SERS signals of DNA targets [24–29]. DNA targets are mixed with SERS nanoparticles or deposited on SERS substrates for SERS measurement. Structural and chemical information of the DNA target can be obtained from SERS spectral fingerprints. DNA target modification is not required and no extrinsic Raman dye is used. Recent works using direct SERS detection showed great potential in characterizing DNA and RNA structure as well as probing their interaction with other analytes [30–34]. Although rapid, acquisition of reproducible SERS fingerprints of large and complex molecules such as DNA is still a challenge [29]. In addition, since intrinsic SERS signals of DNA are very weak compared with that of Raman dyes, direct detection of small copy numbers of DNA, which is often the case in diagnostics applications, is difficult.

The second category involves attachment of Raman dyes to DNA target sequences [35]. SERS signals from the Raman dyes will be used for detection and quantification. The use of Raman dyes allows better sensitivity, multiplex capability, and reproducible signal in complex media. Using this method, multiplex detection of different DNA sequences has been demonstrated [36–38]. However, since modification of DNA target is required, methods using dye-labeled DNA are somewhat cumbersome.

The third category also involves Raman dyes but without modification of DNA target sequences. This confers to this method all the advantages of the second method. Furthermore, since DNA target modification prior to detection is not required, it is more convenient. The Mirkin group used Raman reporter–labeled nanoparticles and silver staining in combination with microarray format chip to distinguish six dissimilar DNA targets, as well as two RNA targets with single nucleotide polymorphisms [39]. Using this method, they could achieve 20 fM limit of detection. The Graham and Faulds group reported the use of streptavidin-coated magnetic beads for capturing followed by lambda-exonuclease digestion and SERS detection for multiplex detection and quantification of DNA from three pathogens with picomolar detection limit [40]. Using this method, they can detect PCR-amplified products of pathogen DNA extracted from clinical samples [41]. The same group also reported the use of Raman reporter–labeled silver and silver-coated magnetic nanoparticles in sandwich assay format for DNA detection. Good discrimination with negative controls was achieved with just 20 fM of target DNA [42]. Using Raman reporter–labeled gold and magnetic nanoparticles, Zhang et al. were able to detect a specific DNA sequence of West Nile virus at 10 pM detection limit using a portable Raman detection platform. In another work, they utilized gold-coated paramagnetic nanoparticles as both a SERS and bioseparation substrate and achieved limit of detection in the range 20–100 nM [43]. Feuillie et al. reported the use of streptavidin-coated magnetic beads, biotin-labeled capture probe, and rhodamine 6G-labeled detection probe in a SERS sandwich-hybridization assay for species detection based on specific DNA detection [44]. Using a portable Raman system, Xu et al. demonstrated multiplex detection of food-borne bacterial pathogens using silica-encapsulated Raman reporter–labeled gold nanoparticles and magnetic microspheres [45]. Particularly, by using a multiple probe approach, they can detect real genomic DNA samples at low femtomolar range. Li et al. developed new nano-SERS-tags composed of (poly(styrene-co-acrylic acid)/(silver nanoparticles)/silica composite nanospheres). By combining these particles with magnetic composite nanospheres, they were able to simultaneously detect up to three different DNA targets in one sample at 10 pM limit of detection [46].

We first reported a new type of DNA gene probe based on SERS to detect DNA biotargets via hybridization to DNA sequences complementary to that probe [47]. We also developed a novel SERS-based DNA detection strategy called the molecular sentinel [48,49]. Since then, we have applied the method on solid SERS Nanowave substrates for singleplex and multiplex DNA detection [50,51]. Compared with other approaches discussed above, the biggest advantage of the molecular sentinel approach is its single-step characteristic. DNA targets were detected by simply depositing sample solutions onto a functionalized SERS Nanowave chip followed by incubation and SERS measurement [50,51]. No posthybridization washing to remove unreacted components is required, making our approach rapid and easy to use. Following our development of the molecular sentinel approach, several other groups have used the approach's sensing mechanism for various applications [52–55]. Recently, we developed another single-step DNA detection strategy having all the merits of the molecular sentinel approach plus signal off-to-on detection strategy upon target hybridization [56,57]. In comparison to the signal-off detection strategy of the original molecular sentinel approach, the new signal-on approach, referred to as inverse molecular sentinel, is more intuitive and less prone to false positives. Details about these two approaches can be found in a recent review published by our group [58]. Besides the aforementioned methods, many other SERS-based DNA detection strategies using Raman dyes without DNA target modification have been reported [59–67].

Herein, we discuss a sandwich hybridization approach using magnetic microspheres and ultrabright SERS nanorattles, referred to as the nanorattle-based method [68]. The novelty of our method lies in the use of the ultrabright SERS nanorattle. Before our work, many hybridization sandwich methods employing magnetic beads and SERS tags have been reported as discussed above. However, those SERS tags' brightness is limited by the number of Raman reporters that can be loaded onto a nanoparticle's surface. In most of the cases, this number is restricted to surface coverage of a self-assembled monolayer or submonolayer [69]. In our work, Raman reporters are loaded into the gap space between the core and the shell. Since the gap is in the nanometer scale, a much higher number of Raman reporters can be loaded into a nanorattle. Furthermore, the plasmonic coupling between the core and the shell creates an intense electromagnetic field (E field) in the gap space. Such a strong E field is another important factor contributing to the extreme SERS brightness of the nanorattle, given fourth power dependence of SERS enhancement on E field enhancement. Experimental results showed that our nanorattle is three orders of magnitude brighter than gold nanoparticle loaded with the same Raman reporter. By employing an ultrabright SERS tag, the nanorattle, we achieved ~100 attomoles limit of detection.

The usefulness of the nanorattle-based method for *in vitro* diagnostics is demonstrated by the detection of a specific DNA sequence of the malaria parasite *Plasmodium falciparum*. Malaria is still a threat in tropical countries, with about 207 million infection cases and over 600,000 deaths in 2013. Mutant malaria parasites, which are resistant to artemisinin drugs, have also been reported. For effective treatment, it is thus important to develop malaria diagnostics methods that are able to not only detect malaria parasites but also distinguish between wild-type (WT) and mutant (Mut) parasites. We demonstrated that the nanorattle-based method can distinguish between WT DNA sequence and mutant DNA sequence with a single nucleotide difference. The use of magnetic beads and the method's simplicity make the nanorattle-based method a promising candidate for integration into lab-on-chip systems for POC *in vitro* diagnostics.

27.2 Material and Methods

27.2.1 Materials

Sodium borohydride ($NaBH_4$), hexadecyltrimethylammonium bromide (CTAB), Au chloride solution 200 mg/dL in deionized water, ascorbic acid (AA), hexadecyltrimethylammonium chloride (CTAC), polyvinylpyrrolidone M_w ~55,000 (PVP), methanol (MetOH), 1,3,3,1′,3′,3′-hexamethyl-2,2′-indotricarbocyanine iodide (HITC), tetradecanol (TD), phosphate buffer saline (PBS), tris-EDTA buffer solution (TE), Tween 20, sodium chloride (NaCl), and hydrochloric acid (HCl) were purchased from Sigma-Aldrich. Sodium citrate dihydrate was purchased from BDH. Methoxy polyethylene glycol thiol M_w 5000 (mPEG-SH) was purchased from Nanocs. Magnetic beads (Dynabeads MyOne Streptavidin C1, 1 μm diameter) were purchased from Life Technologies. All DNA sequences (Table 27.1) were synthesized

TABLE 27.1

DNA Sequences Used in This Work

Name	Sequence
Wild-type *P. falciparum* target	5'-<u>GCT ATG TGT GTT GCT</u> TTT GAT AAT AA**A ATT TAT GTC ATT GGT GGA ACT AAT GG**-3'
Mutant *P. falciparum* target	5'-<u>GCT ATG TAT GTT GCT</u> TTT GAT AAT AA**A ATT TAT GTC ATT GGT GGA ACT AAT GG**-3'
P. falciparum capture probe	5'-Biosg-A_{10}-CCA TTA GTT CCA CCA ATG ACA TAA ATT-3'
Wild-type *P. falciparum* reporter probe	5'-AGC AAC ACA CAT AGC-A_{10}-$(CH_2)_3$-SH-3'
Mutant *P. falciparum* reporter probe	5'-AGC AAC ATA CAT AGC-A_{10}-$(CH_2)_3$-SH-3'

Note: (a) The bolded sequences indicate the regions in target sequences that are complementary to the corresponding capture probes. (b) The underlined sequences indicate the regions in target sequences that are complementary to the corresponding reporter probes. (c) The red letters indicate the mutated position.

by Integrated DNA Technologies (IDT, Coralville, IA). Millipore Synergy ultrapure water (DI) of resistivity = 18.2 MΩ cm was used in all aqueous solutions.

27.2.2 Nanorattle Synthesis

First AuNPs were synthesized using a seed-mediated method. Au seeds were prepared by adding 0.6 mL of 10 mM ice cold $NaBH_4$ to 10 mL of 0.25 mM Au chloride in 0.1 M CTAB solution under vigorous stirring. Upon $NaBH_4$ addition, the solution's color quickly changed from yellow to brown, indicating the formation of small Au seeds. Stirring was stopped after 10 min and the seed solution was aged for 3 h at 27°C to ensure complete decomposition of remaining $NaBH_4$. Growth solution was prepared by adding 600 μL 0.1 M freshly prepared AA to 100 mL of 0.25 mM Au chloride in 0.1 M CTAC (AA final concentration 0.6 mM). Upon AA addition, the Au chloride solution quickly changed color from yellow to colorless, indicating the reduction of Au^{3+} to Au^+. AuNP was synthesized by adding 250 μL Au seeds to the above growth solution under magnetic stirring. Solution color changed from colorless to wine red within a few minutes, indicating the growth of Au seeds into larger AuNP. After the solution color became stable, magnetic stirring was stopped and the solution was left undisturbed overnight in the dark. AuNP's extinction coefficient was measured using FLUOstar Omega microplate reader. Extinction coefficient measurements showed LSPR peak at ~522 nm. The as-prepared AuNPs were centrifuged once and resuspended in DI water before being stored at 4°C. It should be noted that further washing will likely render AuNP aggregation, which can be explained by the decrease in concentration of CTAC stabilizer. For transmission electron microscope (TEM) imaging, 5 μL of the prepared AuNP was dropped on Formvar carbon film-coated 200 mesh copper grids and allowed to dry at room temperature. TEM images were taken using FEI Tecnai G^2 Twin TEM system. TEM image analysis using ImageJ showed that AuNPs have diameter of 19.0 ± 2.7 nm.

27.2.3 Ag Shell Coating and Galvanic Replacement

An Ag shell (~20 nm thick) was coated on AuNP to form Au–Ag core–shell structure (AuNP@Ag). AuNP@Ag was centrifuged once at 8000 rcf for 15 min followed by resuspending in equal volume of 100 mM PVP and of 0.2 M CTAC. The solution was heated to 90°C for 2 min before an Au chloride solution 0.5 mM was slowly added. During Au chloride addition, the solution color changed from orange to dark brown, and finally to purple, indicating that the Ag shells were galvanically replaced by Au^{3+} to turn into porous Au–Ag cages. The solution, containing porous Au–Ag cages with AuNP inside (AuNP@Cage), was further heated for 5 min and then cooled down to room temperature. AuNP@Cage was collected by centrifugation and resuspensded in DI water to remove excess PVP and CTAC. The resuspension was centrifuged once more time and resuspended in MetOH to prepare for dye loading.

27.2.4 Synthesis of AuNP@HITC@Cage by Loading HITC Dye into AuNP@Cage

The AuNP@Cage in MetOH solution was heated to 50°C in a glass bottle followed by addition of 200 mg TD and 20 μL of HITC dye. The solution was heated overnight to completely evaporate MetOH and to allow TD and HITC dyes to go into AuNP@Cage structures. Boiling DI water (3 mL) was added to separate phases, one being TD/dye mixture and the other containing water and a small amount of AuNP@Cage. The majority of AuNP@Cages was, however, stuck on the glass bottle bottom. The solution was kept in an ice bath for 2 min to solidify TD. After TD solidification, sonication was used to redisperse AuNP@Cages stuck at the glass bottle bottom into water, turning the solution color to purple. The mixture was set aside for 5 min to allow TD to form a layer on the solution's surface. The solution, without TD, was carefully transferred into a 15 mL centrifuge tube using a pipette. DI water was added to increase the solution volume to 10 mL. The solution was centrifuged and resuspended in 1 mM CTAC solution several times to remove excess TD.

27.2.5 Final Au Coating

Growth solution was prepared by adding 60 μL of 0.1 M AA (freshly prepared) to Au chloride solution in 0.1 M CTAC. AuNP@HITC@Cage was added to the growth solution under magnetic stirring. Stirring was stopped after 10 min and the solution was left undisturbed for 1 h. The obtained AuNP@HITC@Au was centrifuged and resuspended in DI water once.

27.2.6 Functionalization of Nanorattles with DNA Reporter Probes

Nanorattles were functionalized with DNA reporter probes using a pH-assisted method with slight modifications [70]. First, 50 μL of 100 μM thiolated DNA reporter probe was mixed with 1000 μL of 0.2 nM nanorattle solution. The mixture was placed at room temperature under gentle shaking for 1 h. Then 4.16 μL of citrate-HCl buffer (300 mM trisodium citrate, pH adjusted to 3.1 using 1 M HCl) was added to promote loading of DNA onto nanorattles. One hour later, 5 μL of 1 mM mPEG-SH was added to improve nanorattle's stability. The mixture was allowed to react for 0.5 h. It was then centrifuged (8000 rpm, 5 min) and resuspended in PBS five times to remove unbound DNA reporter probes and thiolated PEG 5 K molecules from DNA-loaded nanorattles. To prevent nanorattles from sticking to the centrifuge tube wall during centrifugation, Tween 20 was added before each centrifugation (final concentration 0.01%). Finally, the as-prepared DNA reporter probe-loaded nanorattle solution was resuspended in TE 1X buffer and stored at 4°C before use.

27.2.7 Functionalization of Magnetic Bead with DNA Capture Probes

Magnetic beads (Dynabeads MyOne Streptavidin C1, 1 μm diameter) were functionalized with DNA capture probes using the manufacturer's protocol. Briefly, 200 μL of 10 mg/mL stock magnetic bead was washed three times using 1 mL washing buffer 1X (5 mM Tris-HCl pH 7.5, 0.5 mM EDTA, 1 M NaCl) and resuspended in 400 µL washing buffer 2X. To load DNA capture probes on magnetic beads, 400 µL of 5 µM biotinylated DNA capture probe was added to the resuspension. The mixture was placed at room temperature under gentle shaking for 0.5 h. Then, DNA capture probe-loaded magnetic beads were washed three times using washing buffer 1X and resuspended in 1600 µL TE 1X (magnetic bead final concentration 1.25 mg/mL). The as-prepared capture probe-loaded magnetic bead solution was stored at 4°C and used within 2–3 months.

27.2.8 DNA Detection

DNA quantification capability of the nanorattle-based method was demonstrated by testing sample solutions containing synthetic DNA sequences of WT malaria parasite *P. falciparum* at different concentrations. Sample solutions were prepared by serial dilution of stock target sequence in hybridization buffer (0.3 M NaCl, 10 mM phosphate-buffered solution at pH 7.4) with Tween 20 0.01%. For each 30 µL sample solution, 2 µL of magnetic beads loaded with capture probes was added. After 1 h of incubation at 37°C, 10 µL of nanorattles loaded with reporter probes (in hybridization buffer) was added. The mixtures were incubated at 37°C for another 1 h before being washed once and pipetted into glass capillary tubes (5–25 µL volume, Sigma-Aldrich). The magnetic bead complexes were concentrated at small spots near the middle of the capillary tubes using a permanent magnet positioned under the tubes.

27.2.9 SERS Measurement

SERS spectra were measured using a lab-built SERS measurement system by focusing the laser beam on the concentrated spots. The lab-built SERS measurement system was composed of a 785 nm laser source (Rigaku Xantus-1), a fiber optic probe (InPhotonics RamanProbe), a spectrometer (Princeton Instruments Acton LS 785), and a CCD camera (Princeton Instruments PIXIS: 100BR_eXcelon). Laser power of the Xantus-1 was set at 300 mW and the CCD camera exposure time was set at 1 s. To confirm the formation of hybridization sandwiches, the magnetic bead complexes were dried at room temperature followed by scanning electron microscope (SEM) imaging using FEI XL30 SEM system.

27.2.10 SNP Discrimination

SNP discrimination capability of the nanorattle-based method was demonstrated by testing samples of *P. falciparum* WT DNA target sequences, *P. falciparum* Mut DNA target sequences with a SNP, nonmalaria DNA sequences, or buffer only (blank) against two different probes: *P. falciparum* WT reporter probe and *P. falciparum* Mut reporter probe. First, 20 µL sample solutions (10 nM in hybridization buffer, Tween 20 0.01%) were mixed in microcentrifuge tubes with 2 µL of magnetic beads loaded with *P. falciparum* capture probes. The mixtures were incubated in a water bath at 40°C for 1 h. Then, 10 µL solutions of nanorattles loaded with either *P. falciparum* WT or Mut reporter probes (in hybridization buffer) were added, and the mixtures were incubated at 40°C for another 1 h. A permanent magnet was used to pull magnetic bead complexes to the wall of the microcentrifuge tubes. Supernatants were discarded and pellets of the magnetic bead complexes were stringently washed by resuspending in a low-salt washing buffer (10 mM NaCl, 10 mM phosphate-buffered solution at pH 7.4) and incubated at 37°C for 5 min. The magnet was applied again and the pellets were washed one more time using low-salt washing buffer with Tween 20 0.01% at room temperature followed by resuspension in low-salt washing buffer. The resuspended solutions were pipetted into glass capillary tubes for SERS measurements, as described above.

27.2.11 Calculation of E Field Enhancement by Nanorattle

E field enhancement of the nanorattles was calculated using the RF module of COMSOL Multiphysics 5.1. First, a 3D model of a nanorattle was built based on TEM images. The model was excited by an incident plane wave propagating along the positive x-axis with E field polarized along the z-axis. Boundary conditions were appropriately applied, and full field was solved. E field enhancement was calculated by dividing total E field by the incident E field.

27.3 Results and Discussion

27.3.1 Detection Scheme

The detection scheme of the nanorattle-based method is illustrated in Figure 27.1. Target DNA is first captured by magnetic beads functionalized with DNA capture probes. Ultrabright SERS nanorattles functionalized with DNA reporter probes are then introduced, inducing the formation of hybridization complexes composing of magnetic beads, target sequences, and ultrabright nanorattles. A permanent magnet is applied to concentrate the hybridization complexes at one spot. Unreacted components are removed and the hybridization complexes are washed several times to remove nonspecific binding before SERS measurement. Herein, streptavidin-coated magnetic beads are used, and biotinylated DNA capture probes are conjugated to the magnetic beads via streptavidin–biotin interaction. Thiolated DNA reporter probes are conjugated to the nanorattles via thiol–gold contact. The nanorattle-based method is a solution-based method, thus possessing faster kinetic than solid chip-based methods. In addition, two signal enhancement

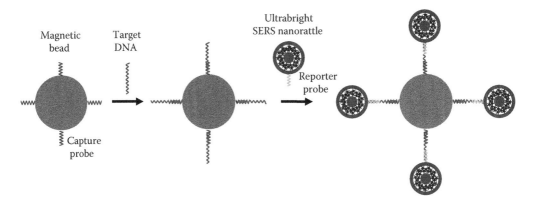

FIGURE 27.1
DNA detection scheme using magnetic beads functionalized with DNA capture probes and ultrabright SERS nanorattles functionalized with DNA reporter probes.

mechanisms are concurrently applied. The first one is the high SERS enhancement of the nanorattles. The second one is the concentration of hybridization complexes at one spot for signal interrogation.

The nanorattle synthesis process is shown in Figure 27.2a. First, ~20 nm AuNPs were synthesized using the seed-mediated growth method. The synthesized AuNPs were coated with Ag shells, followed by galvanic replacement to turn the Ag shells into porous bimetal Au–Ag cages containing AuNPs inside. Raman reporters and TD were loaded into the gap spaces between the AuNPs and the porous bimetal cages. Finally, Au shells were coated on the porous bimetal cages, turning them into complete shells. TD, a phase-change material with melting point of 38–39°C, acts as a gatekeeper, preventing Raman reporters from leaking out during the final coating step. TEM images of nanorattle are shown in Figure 27.2b. The average particle size is ~60 nm. Core–gap–shell structure of nanorattles can be clearly observed. The gap space between the cores and the shells contains Raman reporters and TD. This type of structure has several advantages. First, since the gap is in the nanometer scale, a much higher number of Raman reporters can be loaded into nanorattles than onto nanospheres of same size. Second, plasmonic coupling between the cores and the shells creates an intense E field in the gap spaces. With the large number of Raman reporters located in such an intense E field, nanorattles' SERS intensity is particularly strong. Compared with gold nanoparticles loaded with the same Raman reporter, nanorattles' SERS intensity is three orders of magnitude stronger. Additional advantages of core–gap–shell structure include (1) Raman reporters are protected by complete shells, which prevent signal degradation due to reporter deadsorption, and (2) nanorattle's whole outer surface is available for DNA probe functionalization, allowing more flexibility in tuning DNA probe density for optimal hybridization efficiency without compromising the SERS brightness.

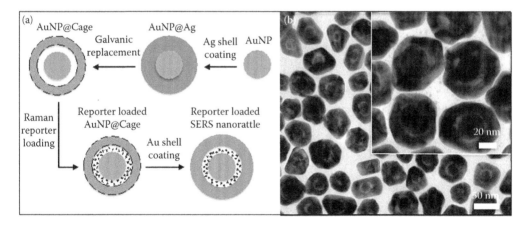

FIGURE 27.2

(a) Ultrabright SERS nanorattles synthesis process starts with coating Au nanoparticle with Ag shell. Galvanic replacement is then used to turn the Ag shell into a porous Au–Ag cage containing Au nanoparticle inside. Raman reporters are loaded into the porous cage and trapped inside with the help of a phase-change material. The porous cage is then coated with a complete Au shell to prevent Raman reporters from leaking out. (b) TEM image of nanorattles shows nanorattles' core–gap–shell structure (inset: higher magnification TEM image). Raman reporters are loaded into the gap spaces between the cores and the shells. (Reprinted from *Biosensors and Bioelectronics*, 81, Ngo, H.T., et al., Sensitive DNA detection and SNP discrimination using ultrabright SERS nanorattles and magnetic beads for malaria diagnostics, 8–14, Copyright (2016). with permission from Elsevier.)

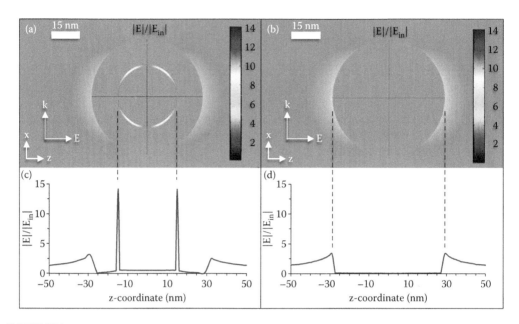

FIGURE 27.3

E field enhancement of nanorattle (a) and Au nanosphere (b). Compared with Au nanosphere, strong E field enhancement in the gap space between nanorattle's core and shell can be observed. E field enhancement along z-axis crossing though center of nanorattle (c) and nanosphere (d). While highest E field enhancement of Au nanosphere is 3.4, nanorattle's is 14.1, which is over four times higher. (Reprinted from *Biosensors and Bioelectronics*, 81, Ngo, H.T., et al., Sensitive DNA detection and SNP discrimination using ultrabright SERS nanorattles and magnetic beads for malaria diagnostics, 8–14, Copyright (2016), with permission from Elsevier.)

E field enhancement by nanorattle and gold nanosphere are calculated using finite element method; as shown in Figure 27.3a, particularly high E field enhancement at the gap space between the core and the shell of nanorattles can be observed. This E field enhancement is estimated to be over four times higher than of gold nanospheres. Since SERS intensity strongly depends on E field intensity with a fourth power dependence [71], such an increase in E field intensity of nanorattle is expected to result in several orders of magnitude increase in SERS intensity. The nanorattle is thus chosen as an ultrabright SERS tag for highly sensitive DNA detection, which will be discussed in following section.

27.3.2 DNA Detection

WT sequence of *P. falciparum* gene PF3D7_1343700 was chosen as a test model to demonstrate performance of the nanorattle-based DNA detection method. In the presence of complementary target sequences, nanorattles were bound onto magnetic beads' surface (Figure 27.4a). In contrast, in the absence of complementary target sequences (i.e., buffer only), almost no nanorattles were found (Figure 27.4b). This result indicated that hybridization sandwiches of (1) magnetic beads loaded with capture probes, (2) target sequence, and (3) nanorattles loaded with reporter probes successfully formed in the presence of complementary target sequences. As the magnetic beads (~1 μm diameter) were much bigger than the nanorattles (~60 nm diameter), tens of nanorattles were found on each magnetic bead.

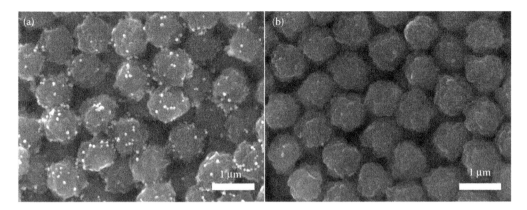

FIGURE 27.4
(a) SEM image of nanorattles (white dots) bound onto magnetic beads' surface in the presence of complementary target sequences. (b) Almost no nanorattle found on magnetic beads in the absence of complementary target sequences. (Reprinted from *Biosensors and Bioelectronics*, 81, Ngo, H.T., et al., Sensitive DNA detection and SNP discrimination using ultrabright SERS nanorattles and magnetic beads for malaria diagnostics, 8–14, Copyright (2016), with permission from Elsevier.)

The quantification capability of the nanorattle-based technique was investigated by testing samples of the WT sequence of *P. falciparum* gene PF3D7_1343700 at different concentrations. Quantification results are shown in Figure 27.5. As the target concentration decreases, SERS intensity also decreases (Figure 27.5a). The decrease in SERS intensity can be explained by a lower number of nanorattles bound on magnetic beads at lower target concentrations. SERS peak intensities at 923 cm^{-1} (normalized against the highest concentration) were used to plot the calibration curve (Figure 27.5b). From the results, the limit of detection was determined to be ~3 picomolar (3×10^{-12} M), which is equivalent to ~100 attomoles (10^{-16} moles) given 30 μL volumes of sample solutions. At 1 nM concentration, the signal detector was saturated due to strong SERS signals. The linear range of our method is expected to be between 10^{-11} and 10^{-10} M target concentrations. For detection

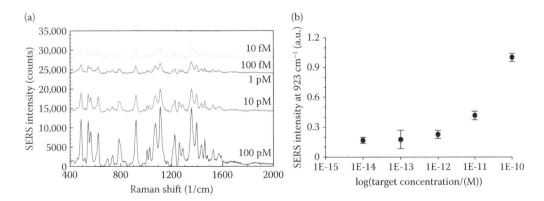

FIGURE 27.5
(a) SERS spectra at different concentrations of wild-type target *P. falciparum* gene *PF3D7_1343700* (vertically shifted for clarity). (b) SERS intensities at 923 cm^{-1} (normalized) versus log(target concentration/(M)). Error bars represent standard deviations (n = 3). (Reprinted from *Biosensors and Bioelectronics*, 81, Ngo, H.T., et al., Sensitive DNA detection and SNP discrimination using ultrabright SERS nanorattles and magnetic beads for malaria diagnostics, 8–14, Copyright (2016), with permission from Elsevier.)

at higher target concentrations without detector saturation, a shorter detector's exposure time or a lower laser power can be used.

27.3.3 SNP Identification

SNP discrimination capability of the nanorattle-based method was demonstrated by distinguishing "WT" and "Mut" DNA sequences of the malaria parasite *P. falciparum* gene with a single nucleotide difference. The Mut sequence has an SNP that confers resistance to Art-R [72,73]. Two reporter probes that are complementary to *P. falciparum* WT and Mut targets and a common capture probe for the two target sequences were designed (Table 27.1). A nonmalaria sequence and blank (buffer only) were used as negative controls. Performance of the two reporter probes are shown in Figure 27.6. Both WT and Mut reporter probe can specifically detect WT and Mut target sequence, respectively. For the WT reporter probe (Figure 27.6a), the SERS intensity was high in the presence of WT target sequences and low in the presence of Mut target sequences. This is due to the perfect matching between the WT reporter probes and the WT target sequences, resulting in stable double strands. More nanorattles were attached to the magnetic beads via sandwich formation, thus resulting in high SERS intensity. In the presence of Mut target sequences, due to the single-base mismatch between the WT reporter probes and the Mut target sequences, double-stranded hybrids are less stable. With the stringent washing step using low-salt washing buffer at 37°C, these unstable double-stranded hybrids will dehybridize. A lower number of nanorattles were attached to the magnetic beads, resulting in low SERS intensity. For nonmalaria sequence and blank, SERS intensities were close to zero (Figure 27.6a). From SERS peak intensity at 923 cm^{-1} of WT probe–WT target and of WT probe–Mut target, SNP discrimination ratio of approximately 8:1 was obtained for the WT probe. Similarly, for the Mut reporter probe (Figure 27.6b), SERS intensity was high in the presence of Mut target sequences and low in the presence of *P. falciparum* WT sequences. SERS intensities were close to zero for both nonmalaria sequences and blank. Using the SERS peak intensity at

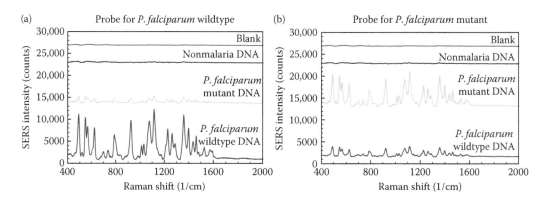

FIGURE 27.6

Detection of wild-type *P. falciparum* and mutant *P. falciparum* with a single nucleotide difference using the nanorattle-based method (vertically shifted for clarity). Two probes, one for wild-type *P. falciparum* (a) and one for mutant *P. falciparum* (b), were designed and tested against *P. falciparum* wild-type DNA, *P. falciparum* mutant DNA, nonmalaria DNA, and blank. The wild-type DNA and the mutant DNA have a single-base difference. (Reprinted from *Biosensors and Bioelectronics*, 81, Ngo, H.T., et al., Sensitive DNA detection and SNP discrimination using ultrabright SERS nanorattles and magnetic beads for malaria diagnostics, 8–14, Copyright (2016), with permission from Elsevier.)

923 cm^{-1} of Mut probe–Mut target and of Mut probe–WT target, the SNP discrimination ratio was determined to be approximately 3.2:1 for the Mut probe. The results clearly demonstrate the SNP discrimination capability of the nanorattle-based method.

It is noteworthy that conditions of the stringent washing step, including temperature and salt concentration, are crucial to the SNP discrimination experiment's success. We investigated the effect of temperature of the stringent washing step by repeating the SNP discrimination experiment except that the stringent washing step was conducted at room temperature as opposed to 37°C. The results show that the discrimination capacity was negatively affected [68]. While the WT probe could still discriminate between WT target and Mut target at lower discrimination ratio 2.5:1, the Mut probe could not discriminate between Mut target and WT target. Selecting temperature for the stringent washing step is thus important to SNP discrimination success. The use of two reporter probes, WT probe and Mut probe, is effective for not only SNP discrimination but also for target composition identification [39].

27.4 Conclusion

We have developed a sensitive DNA-detection method based on sandwich hybridization of magnetic bead, target sequence, and ultrabright SERS nanorattle. The nanorattle has a core–gap–shell structure with Raman reporters loaded into the nanometer gap between the core and the shell. SERS measurements showed that the nanorattles are three orders of magnitude brighter than gold nanospheres loaded with the same Raman reporter. The nanorattles were used as ultrabright SERS tags for DNA detection based on sandwich hybridization of a magnetic bead–target sequence–nanorattle. The current unoptimized detection limit is ~3 pM, equivalent to 100 attomoles, for *P. falciparum* DNA targets. Distinguishing between *P. falciparum* WT and Mut sequence, which has a SNP that confers Art-R resistance, was also demonstrated. The method presented is simple, sensitive, and suitable for automation, making it a promising candidate for integration into portable platforms and lab-on-a-chip systems for *in vitro* diagnostics at the POC.

Conflict of Interest

The authors declare no conflict of interest.

Acknowledgments

This work was sponsored by the National Institutes of Health (R21 AI120981-01) and the Duke Faculty Exploratory Research Fund. Hoan T. Ngo is supported by fellowships from the Vietnam Education Foundation and the Fitzpatrick Scholars Fellowship. Steve M. Taylor is supported by the National Institute of Allergy and Infectious Diseases under award number K08AI100924.

References

1. Niemz, A., T.M. Ferguson, and D.S. Boyle, Point-of-care nucleic acid testing for infectious diseases. *Trends in Biotechnology*, 2011. 29(5): 240–250.
2. Chin, C.D., V. Linder, and S.K. Sia, Commercialization of microfluidic point-of-care diagnostic devices. *Lab on a Chip*, 2012. 12(12): 2118–2134.
3. Choi, S. et al., Microfluidic-based biosensors toward point-of-care detection of nucleic acids and proteins. *Microfluidics and Nanofluidics*, 2011. 10(2): 231–247.
4. Miotke, L., M.C. Barducci, and K. Astakhova, Novel signal-enhancing approaches for optical detection of nucleic acids-going beyond target amplification. *Chemosensors*, 2015. 3(3): 224–240.
5. Zheng, Z., Y.L. Luo, and G.K. McMaster, Sensitive and quantitative measurement of gene expression directly from a small amount of whole blood. *Clinical Chemistry*, 2006. 52(7): 1294–1302.
6. Ahmad, F. and S.A. Hashsham, Miniaturized nucleic acid amplification systems for rapid and point-of-care diagnostics: A review. *Analytica Chimica Acta*, 2012. 733: 1–15.
7. Chang, C.M. et al., Nucleic acid amplification using microfluidic systems. *Lab on a Chip*, 2013. 13(7): 1225–1242.
8. Tsongalis, G.J., Branched DNA technology in molecular diagnostics. *American Journal of Clinical Pathology*, 2006. 126(3): 448–453.
9. Nam, J.M., S.I. Stoeva, and C.A. Mirkin, Bio-bar-code-based DNA detection with PCR-like sensitivity. *Journal of the American Chemical Society*, 2004. 126(19): 5932–5933.
10. Zhao, X.J., R. Tapec-Dytioco, and W.H. Tan, Ultrasensitive DNA detection using highly fluorescent bioconjugated nanoparticles. *Journal of the American Chemical Society*, 2003. 125(38): 11474–11475.
11. Stiles, P.L. et al., Surface-enhanced Raman spectroscopy. *Annual Review of Analytical Chemistry*, 2008. 1(1): 601–626.
12. Schlücker, S., Surface-enhanced Raman spectroscopy: Concepts and chemical applications. *Angewandte Chemie International Edition*, 2014. 53(19): 4756–4795.
13. Lane, L.A., X.M. Qian, and S.M. Nie, SERS nanoparticles in medicine: From label-free detection to spectroscopic tagging. *Chemical Reviews*, 2015. 115(19): 10489–10529.
14. Cialla, D. et al., Surface-enhanced Raman spectroscopy (SERS): Progress and trends. *Analytical and Bioanalytical Chemistry*, 2012. 403(1): 27–54.
15. Kneipp, K. et al., Single molecule detection using surface-enhanced Raman scattering (SERS). *Physical Review Letters*, 1997. 78(9): 1667–1670.
16. Nie, S.M. and S.R. Emery, Probing single molecules and single nanoparticles by surface-enhanced Raman scattering. *Science*, 1997. 275(5303): 1102–1106.
17. Zrimsek, A.B., A.I. Henry, and R.P. Van Duyne, Single molecule surface-enhanced Raman spectroscopy without nanogaps. *Journal of Physical Chemistry Letters*, 2013. 4(19): 3206–3210.
18. Darby, B.L., P.G. Etchegoin, and E.C. Le Ru, Single-molecule surface-enhanced Raman spectroscopy with nanowatt excitation. *Physical Chemistry Chemical Physics*, 2014. 16(43): 23895–23899.
19. Vo-Dinh, T. et al., Surface-enhanced Raman spectrometry for trace organic-analysis. *Analytical Chemistry*, 1984. 56(9): 1667–1670.
20. Vo-Dinh, T., Surface-enhanced Raman spectroscopy using metallic nanostructures. *Trac-Trends in Analytical Chemistry*, 1998. 17(8–9): 557–582.
21. Vo-Dinh, T. et al., Plasmonic nanoparticles and nanowires: Design, fabrication and application in sensing. *Journal of Physical Chemistry C*, 2010. 114(16): 7480–7488.
22. Vo-Dinh, T. et al., Plasmonic nanoprobes: From chemical sensing to medical diagnostics and therapy. *Nanoscale*, 2013. 5(21): 10127–10140.
23. Vo-Dinh, T. et al., SERS nanosensors and nanoreporters: Golden opportunities in biomedical applications. *Wiley Interdisciplinary Reviews-Nanomedicine and Nanobiotechnology*, 2015. 7(1): 17–33.

24. Barhoumi, A. et al., Surface-enhanced Raman spectroscopy of DNA. *Journal of the American Chemical Society*, 2008. 130(16): 5523–5529.

25. Barhoumi, A. and N.J. Halas, Label-free detection of DNA hybridization using surface enhanced Raman spectroscopy. *Journal of the American Chemical Society*, 2010. 132(37): 12792–12793.

26. Papadopoulou, E. and S.E.J. Bell, Label-free detection of single-base mismatches in DNA by surface-enhanced Raman spectroscopy. *Angewandte Chemie-International Edition*, 2011. 50(39): 9058–9061.

27. Panikkanvalappil, S.R., M.A. Mackey, and M.A. El-Sayed, Probing the unique dehydration-induced structural modifications in cancer cell DNA using surface enhanced Raman spectroscopy. *Journal of the American Chemical Society*, 2013. 135(12): 4815–4821.

28. Xu, L.J. et al., Label-free surface-enhanced Raman spectroscopy detection of DNA with single-base sensitivity. *Journal of the American Chemical Society*, 2015. 137(15): 5149–5154.

29. Torres-Nunez, A. et al., Silver colloids as plasmonic substrates for direct label-free surface-enhanced Raman scattering analysis of DNA. *Analyst*, 2016. 141(17): 5170–5180.

30. Morla-Folch, J. et al., Fast optical chemical and structural classification of RNA. *ACS Nano*, 2016. 10(2): 2834–2842.

31. Masetti, M. et al., Revealing DNA interactions with exogenous agents by surface-enhanced Raman scattering. *Journal of the American Chemical Society*, 2015. 137(1): 469–476.

32. Guerrini, L. et al., Direct surface-enhanced Raman scattering analysis of DNA duplexes. *Angewandte Chemie-International Edition*, 2015. 54(4): 1144–1148.

33. Rusciano, G. et al., Label-free probing of G-Quadruplex formation by surface-enhanced Raman scattering. *Analytical Chemistry*, 2011. 83(17): 6849–6855.

34. Morla-Folch, J. et al., Ultrasensitive direct quantification of nucleobase modifications in DNA by surface-enhanced Raman scattering: The case of cytosine. *Angewandte Chemie-International Edition*, 2015. 54(46): 13650–13654.

35. Vo-Dinh, T., L.R. Allain, and D.L. Stokes, Cancer gene detection using surface-enhanced Raman scattering (SERS). *Journal of Raman Spectroscopy*, 2002. 33(7): 511–516.

36. Faulds, K. et al., Multiplexed detection of six labelled oligonucleotides using surface enhanced resonance Raman scattering (SERRS). *Analyst*, 2008. 133(11): 1505–1512.

37. Faulds, K. et al., Quantitative simultaneous multianalyte detection of DNA by dual-wavelength surface-enhanced resonance Raman scattering. *Angewandte Chemie-International Edition*, 2007. 46(11): 1829–1831.

38. Laing, S., K. Gracie, and K. Faulds, Multiplex in vitro detection using SERS. *Chem Soc Rev*, 2016. 45(7): 1901–1918.

39. Cao, Y.W.C., R.C. Jin, and C.A. Mirkin, Nanoparticles with Raman spectroscopic fingerprints for DNA and RNA detection. *Science*, 2002. 297(5586): 1536–1540.

40. Gracie, K. et al., Simultaneous detection and quantification of three bacterial meningitis pathogens by SERS. *Chemical Science*, 2014. 5(3): 1030–1040.

41. Gracie, K. et al., Bacterial meningitis pathogens identified in clinical samples using a SERS DNA detection assay. *Analytical Methods*, 2015. 7(4): 1269–1272.

42. Donnelly, T. et al., Silver and magnetic nanoparticles for sensitive DNA detection by SERS. *Chem Commun (Camb)*, 2014. 50(85): 12907–12910.

43. Zhang, H. et al., Surface-enhanced Raman scattering detection of DNAs derived from virus genomes using Au-coated paramagnetic nanoparticles. *Langmuir*, 2012. 28(8): 4030–4037.

44. Feuillie, C. et al., A novel SERRS sandwich-hybridization assay to detect specific DNA target. *PLoS One*, 2011. 6(5): e17847.

45. Xu, H. et al., Portable SERS sensor for sensitive detection of food-borne pathogens, in *Raman Spectroscopy for Nanomaterials Characterization*, C.S.R. Kumar, Editor. 2012, Springer, Berlin, Heidelberg. 531–551.

46. Li, J.M. et al., Highly sensitive detection of target ssDNA based on SERS liquid chip using suspended magnetic nanospheres as capturing substrates. *Langmuir*, 2013. 29(20): 6147–6155.

47. Vo-Dinh, T., K. Houck, and D.L. Stokes, Surface-enhanced Raman gene probes. *Analytical Chemistry*, 1994. 66(20): 3379–3383.

48. Wabuyele, M.B. and T. Vo-Dinh, Detection of human immunodeficiency virus type 1 DNA sequence using plasmonics nanoprobes. *Analytical Chemistry*, 2005. 77(23): 7810–7815.

49. Wang, H.N. and T. Vo-Dinh, Multiplex detection of breast cancer biomarkers using plasmonic molecular sentinel nanoprobes. *Nanotechnology*, 2009. 20(6): 065101.

50. Ngo, H.T. et al., Label-free DNA biosensor based on SERS molecular sentinel on nanowave chip. *Analytical Chemistry*, 2013. 85(13): 6378–6383.

51. Ngo, H. et al., Multiplex detection of disease biomarkers using SERS molecular sentinel-on-chip. *Analytical and Bioanalytical Chemistry*, 2014. 406(14): 3335–3344.

52. Wei, X.P. et al., A molecular beacon-based signal-off surface-enhanced Raman scattering strategy for highly sensitive, reproducible, and multiplexed DNA detection. *Small*, 2013. 9(15): 2493–2499.

53. Pang, Y.F. et al., SERS molecular sentinel for the RNA genetic marker of PB1-F2 protein in highly pathogenic avian influenza (HPAI) virus. *Biosensors & Bioelectronics*, 2014. 61: 460–465.

54. Wang, H. et al., Hairpin DNA-assisted silicon/silver-based surface-enhanced Raman scattering sensing platform for ultrahighly sensitive and specific discrimination of deafness mutations in a real system. *Analytical Chemistry*, 2014. 86(15): 7368–7376.

55. Qi, J. et al., Label-free, in situ SERS monitoring of individual DNA hybridization in microfluidics. *Nanoscale*, 2014. 6(15): 8521–8526.

56. Ngo, H.T. et al., DNA bioassay-on-chip using SERS detection for dengue diagnosis. *Analyst*, 2014. 139(22): 5655–5659.

57. Wang, H.N., A.M. Fales, and T. Vo-Dinh, Plasmonics-based SERS nanobiosensor for homogeneous nucleic acid detection. *Nanomedicine-Nanotechnology Biology and Medicine*, 2015. 11(4): 811–814.

58. Ngo, H.T. et al., Plasmonic SERS biosensing nanochips for DNA detection. *Anal Bioanal Chem*, 2016. 408(7): 1773–1781.

59. Kang, T. et al., Patterned multiplex pathogen DNA detection by Au particle-on-wire SERS sensor. *Nano Letters*, 2010. 10(4): 1189–1193.

60. Hu, J. et al., Sub-attomolar HIV-1 DNA detection using surface-enhanced Raman spectroscopy. *Analyst*, 2010. 135(5): 1084–1089.

61. van Lierop, D., K. Faulds, and D. Graham, Separation free DNA detection using surface enhanced Raman scattering. *Analytical Chemistry*, 2011. 83(15): 5817–5821.

62. He, S.J. et al., Graphene-based high-efficiency surface-enhanced Raman scattering-active platform for sensitive and multiplex DNA detection. *Analytical Chemistry*, 2012. 84(10): 4622–4627.

63. Jiang, Z.Y. et al., Silicon-based reproducible and active surface-enhanced Raman scattering substrates for sensitive, specific, and multiplex DNA detection. *Applied Physics Letters*, 2012. 100(20): 2031041–2031044.

64. Li, J.M. et al., Multiplexed SERS detection of DNA targets in a sandwich-hybridization assay using SERS-encoded core-shell nanospheres. *Journal of Materials Chemistry*, 2012. 22(24): 12100–12106.

65. Gao, F.L., J.P. Lei, and H.X. Ju, Label-free surface-enhanced Raman spectroscopy for sensitive DNA detection by DNA-mediated silver nanoparticle growth. *Analytical Chemistry*, 2013. 85(24): 11788–11793.

66. Li, M. et al., Plasmonic nanorice antenna on triangle nanoarray for surface-enhanced Raman scattering detection of Hepatitis B virus DNA. *Analytical Chemistry*, 2013. 85(4): 2072–2078.

67. Zhang, J.N. et al., Quantitative SERS-based DNA detection assisted by magnetic microspheres. *Chemical Communications*, 2015. 51(83): 15284–15286.

68. Ngo, H.T. et al., Sensitive DNA detection and SNP discrimination using ultrabright SERS nanorattles and magnetic beads for malaria diagnostics. *Biosensors and Bioelectronics*, 2016. 81: 8–14.

69. Kustner, B. et al., SERS labels for red laser excitation: Silica-encapsulated SAMs on tunable gold/silver nanoshells. *Angewandte Chemie-International Edition*, 2009. 48(11): 1950–1953.

70. Zhang, X., M.R. Servos, and J.W. Liu, Instantaneous and quantitative functionalization of gold nanoparticles with thiolated DNA using a pH-assisted and surfactant-free route. *Journal of the American Chemical Society*, 2012. 134(17): 7266–7269.
71. Schatz, G.C. and R.P. Van Duyne, Electromagnetic mechanism of surface-enhanced spectroscopy, in *Handbook of Vibrational Spectroscopy*. 2006, John Wiley & Sons, Ltd.
72. Ariey, F. et al., A molecular marker of artemisinin-resistant *Plasmodium falciparum* malaria. *Nature*, 2014. 505(7481): 50–55.
73. Mohon, A. et al., Mutations in *Plasmodium falciparum* K13 propeller gene from Bangladesh (2009–2013). *Malaria Journal*, 2014. 13: 431.

28

Gold Nanorods for Diagnostics and Photothermal Therapy of Cancer

Xiaohua Huang and Mostafa A. El-Sayed

CONTENTS

28.1 Introduction

One of the major current interests in nanomedicine is in the use of gold (Au) nanoparticles (NPs) for cancer diagnosis and photothermal therapy (PTT) [1]. The rationale is that Au NPs are nontoxic, photostable, easy to be surface modified, and exhibit strong and tunable localized surface plasmon resonance (LSPR). Due to the unique LSPR, the optical properties of Au NPs are strongly enhanced, orders of magnitude higher in light absorption and scattering efficiencies than those of organic dye molecules [2]. Thus, Au NPs have emerged as novel imaging and detection agents as well as light antenna for PTT by taking advantages of their radiative and nonradiative properties.

By controlling the particle's structure or shape, the LSPR wavelength can be tuned into the near-infrared (NIR) region, an optical window where light has deep penetration in tissue [3]. During the last decade, different types of NIR-absorbing Au NPs have been reported and used for cancer PTT, including Au nanoshells [4–9], nanorods (NRs) [10–21], nanocages [22–27], hollow nanospheres [26,28], nanopopcorns [29], and bellflowers [30]. Particularly, Au NRs have attracted a great deal of attention because of several advantages: (1) small size (around 50 nm in length), (2) ease of preparation, (3) excellent stability (shelf life >1 year), and (4) long blood circulation property, with the half-life >10 h when modified with poly(ethylene) glycol (PEG) [31].

This chapter focuses on the use of Au NRs for light-scattering cancer imaging, spectroscopic detection, and PTT. Optical properties of Au NRs will be introduced at the beginning to understand the principles for diagnostic and therapeutic applications. Spectroscopic detection includes LSPR sensing and surface-enhanced Raman scattering (SERS) detection. PTT includes both *in vitro* and *in vivo* studies followed by combination therapy with chemotherapy and photodynamic therapy (PDT). The chapter will help researchers in cancer nanomedicine to understand fundamental properties of Au NRs and keep up with recent research progress in their use in cancer diagnostics and therapy.

28.2 Optical Properties of Gold NRs

28.2.1 Absorption and Scattering Properties

Different from conventional spherical Au NPs that only have one LSPR band, Au NRs show two bands: a weak band in the visible region corresponding to the electron oscillation along the short axis and a strong band at longer wavelength corresponding to electron oscillations along the long axis (Figure 28.1). They are referred to transverse band and longitudinal band, respectively. This optical behavior has been well understood using

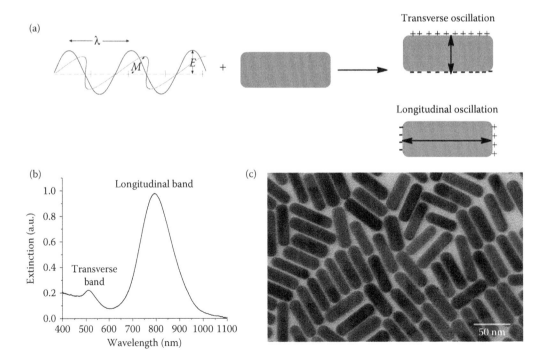

FIGURE 28.1
(a) Schematic illustration of the localized surface plasmon oscillation of gold nanorods. (b) Absorption spectrum of gold nanorods showing the longitudinal and transverse surface plasmon resonance bands. (c) TEM image of gold nanorods with localized surface plasmon resonance absorption around 800 nm.

Gan's theory [32]. Based on Gan's theory, the absorption, scattering, and extinction cross sections are expressed in the following equations [33]:

$$C_{abs} = \frac{2\pi}{3\lambda} \varepsilon_m^{3/2} V \sum_i \frac{\varepsilon_2/(n^{(i)})^2}{\left(\varepsilon_1 + [(1-n^{(i)})/n^{(i)}]\varepsilon_m\right)^2 + \varepsilon_2^2} \tag{28.1}$$

$$C_{sca} = \frac{8\pi^3}{9\lambda^4} \varepsilon_m^2 V^2 \sum_i \frac{\left((\varepsilon_1 - \varepsilon_m)^2 + \varepsilon_2^2\right)/(n^{(i)})^2}{\left(\varepsilon_1 + [(1-n^{(i)})/n^{(i)}]\varepsilon_m\right)^2 + \varepsilon_2^2} \tag{28.2}$$

$$C_{ext} = C_{abs} + C_{sca} \tag{28.3}$$

where λ is the wavelength of the light, V is the volume of the particle, ε_m is the dielectric constant of the medium, ε is the complex dielectric constant of the metal given by $\varepsilon = \varepsilon_1 + i\varepsilon_2$, ε_1 is the real part of the dielectric constant of the metal, ε_2 is the imaginary part of the dielectric constant of the metal, and $n^{(i)}$ is the depolarization factors along the three axes a, b, and c of the NR where $a > b = c$. $n^{(i)}$ is related to aspect ratio R ($R = a/b$) by

$$n^{(a)} = \frac{1}{R^2 - 1}\left(\frac{R}{2\sqrt{R^2-1}} \ln \frac{R+\sqrt{R^2-1}}{R-\sqrt{R^2-1}} - 1\right) \tag{28.4}$$

$$n^{(b)} = n^{(c)} = \frac{1-n^{(a)}}{2} \tag{28.5}$$

The resonance occurs at $\varepsilon_1 = -(1-n^{(i)})\varepsilon_m/n^{(i)}$, where $i = a$ for the longitudinal resonance and $I = b,c$ for the transverse resonance. At such resonance wavelengths, the absorption, scattering, and total extinction are all strongly enhanced, which is the foundation for their application in biomedical detection and treatment.

While the transverse band is insensitive to the aspect ratio of the NR, the longitudinal band is largely dependent on the aspect ratio. Using the linear relationship of the real part of the dielectric constant of gold and light wavelength, the longitudinal LSPR wavelength and the aspect ratio R of Au NRs follow a linear proportional relationship in the formula of [34,35] (Figure 28.2a)

$$\lambda_{max} = 95\,R + 420 \tag{28.6}$$

This provides the possibility to tune the optical properties of gold NRs by simply varying the aspect ratio of the particles (Figure 28.2b). In contrast, the LSPR of Au nanospheres only slightly redshifts with increasing particle size [36].

The LSPR band is also sensitive to the dielectric constant of the surrounding medium. The relationship of LSPR wavelength and ε_m follows the equation of [34]

$$\lambda_{max} = (53.71\,R - 42.29)\varepsilon_m + 495.14 \tag{28.7}$$

Thus, the LSPR wavelength is also linearly proportional to the dielectric constant of the surrounding medium. This provides a way for biosensing based on the changes of the

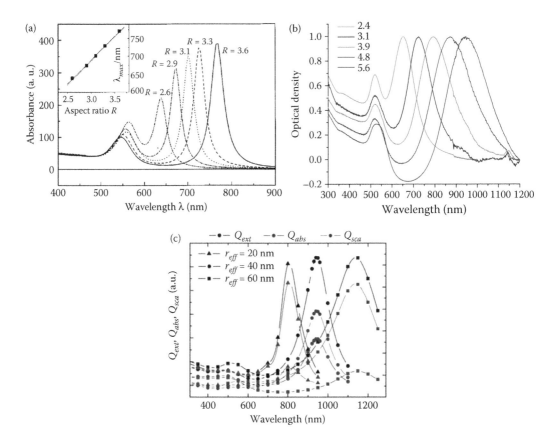

FIGURE 28.2

Optical properties of gold nanorods. (a) Calculated absorption spectra of gold nanorods of different aspect ratios based on Gan's theory. Inset showed the plot of localized surface plasmon resonance wavelength versus the aspect ratio of gold nanorods. (b) Experimental absorption spectra of gold nanorods of different aspect ratios prepared by seed-mediated growth method. (c) Calculated extinction (black), absorption (red), and scattering (blue) spectra of gold nanorods with different effective radius using discrete dipole approximation. (Reprinted with permission from S. Link et al. Simulation of the optical absorption spectra of gold nanorods as a function of their aspect ratio and the effect of the medium dielectric constant. *J. Phys. Chem. B* 1999, 103(16), 3073. Copyright 1999 American Chemical Society. Reprinted with permission from X. Huang et al. Cancer cell imaging and photothermal therapy in the near-infrared region by using gold nanorods. *J. Am. Chem. Soc.* 2006, 128(6), 2115. Copyright 2006 American Chemical Society. Reprinted with permission from K.S. Lee, M.A. El-Sayed. Dependence of the enhanced optical scattering efficiency relative to that of absorption for gold metal nanorods on aspect ratio, size, end-cap shape, and medium refractive Index. *J. Phys. Chem. B* 2005, 109(43), 20331. Copyright 2005 American Chemical Society.)

LSPR wavelength upon the changes of the environment. From the equation, it can be seen that the sensitivity is more prominent for NRs with higher aspect ratios. This relationship has been confirmed with experimental studies by Chen et al. [37].

The optical properties of Au NRs can also be understood using discrete dipole approximation (DDA). DDA is one of the most powerful electrodynamic and numerical methods to calculate optical properties, especially the scattering problem of targets with any arbitrary geometry [38–42]. In this numerical method, the target particle is viewed as a cubic array of point dipoles with polarization α_i. At position r_i, the dipole moment P_i is equal to [43]

$$P_i = \alpha_i E_{\text{loc}}(r_i) \tag{28.8}$$

where E_{loc} is the sum of the incident electric field and the field induced by all other dipoles:

$$E_{loc}(r_i) = E_{inc,i} + E_{other,i} = E_0 \exp(ik \cdot r_i - i\omega t) - \sum_{j \neq i} A_{ij} \cdot P_j \tag{28.9}$$

$$A_{ij} \cdot P_j = \frac{\exp(ikr_{ij})}{r_{ij}^3} \left\{ k^2 r_{ij} \times (r_{ij} \times P_j) + \frac{(1 - ikr_{ij})}{r_{ij}^2} \times [r_{ij}^2 P_j - 3r_{ij}(r_{ij} \cdot P_j)] \right\}, \quad (j \neq i) \tag{28.10}$$

Solving Equation 28.10 by an initial guess of the unknown dipole moment Pj, the extinction and absorption cross sections can be derived from the optical theorem [44]:

$$C_{ext} = \frac{4\pi k}{|E_0|^2} \sum_{j=1}^{N} \{Im\}(E_{inc,j}^* \cdot P_j) \tag{28.11}$$

$$C_{abs} = \frac{4\pi k}{|E_0|^2} \sum_{j=1}^{N} \left\{ \{Im\}\left[P_j\left(\alpha_j^{-1}\right) * P_j^* \right] - \frac{2}{3} k^3 |P_j|^2 \right\} \tag{28.12}$$

Figure 28.2c shows the DDA-calculated optical spectra of Au NRs with different effective radius [43]. As the effective radius increases, the LSPR wavelength is largely redshifted. In addition, the contribution of light absorption (red curves) decreases and that of scattering (blue curves) increases. These results provide the knowledge for the choice of Au NRs for biomedical applications. For imaging applications, larger rods are preferred because of their higher scattering efficiency. For PTT, smaller ones are preferred as light is mainly adsorbed by the particles.

28.2.2 Plasmonic Field Enhancement

The plasmon oscillation of a metal NP can be treated as a photon confined to the small nanoparticle size. This strong confinement of the photon oscillation leads to a large decrease in its wavelength and correspondingly a huge increase in the amplitude of the photon wave. This effect is called plasmonic field enhancement. The field enhancement can be well understood using DDA calculation. As stated in Section 28.2.1, the target is represented by N point dipoles at locations r_j on a cubic lattice according to a specified geometry with material-specific polarizabilities α_j. The polarization (P_j) of a dipole at location r_j is given by

$$P_j = \alpha_j \cdot E_j \tag{28.13}$$

where E_j, the electric field at point r_j, is a sum of the incident electric field, $E_{inc,j}$ (Equation 28.14), plus the electric field effects of all other dipoles, $E_{other,j}$ (Equation 28.15), in the target at locations r_k and A_{jk} in a 3×3 matrix.

$$E_{inc,j} = E_0 \exp(ik \cdot r_j - i\omega t) \tag{28.14}$$

$$E_{other,j} = -\sum_{j \neq k} A_{jk} \cdot P_k \tag{28.15}$$

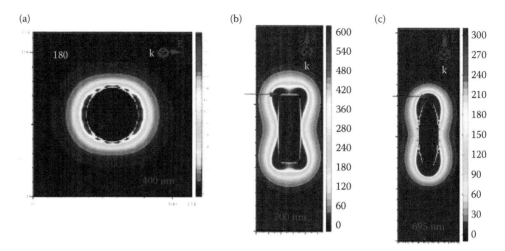

FIGURE 28.3
E fields enhancement contours for a silver nanosphere (a), nanorod (b), and spheroid polarized along their long axis (c). The arrows show where the maximum of E field is. (With kind permission from Springer Science + Business Media: *J. Fluor.*, Synthesis and optical properties of anisotropic metal nanoparticles, 14(4), 2004, 331, E. Hao et al. Reprinted with permission from G. Hao, G.C. Schatz. Electromagnetic fields around silver nanoparticles and dimers. *J. Chem. Phys.* 2004, 120(1), 357. Copyright 2004 American Institute of Physics.)

By solving the resultant set of $3N$ complex linear equations, the DDA can be used to determine the electric field intensity of each point dipole, as well as the unknown polarizabilities P_j, and subsequently, the extinction and absorption cross sections of the target. Using this method, the electric field intensities $|E_j|/|E_0|$ for all dipoles within the NP volume as well as for the dipoles within an extended volume surrounding the target can be calculated. The electric field enhancement factor $<E>$ is calculated by averaging $|E_j|/|E_0|$ for a monolayer of dipoles on the NP surface.

The field enhancement depends on the metal composition and particle size, shape, and structure [45,46] (Figure 28.3). Silver is known to give higher field effects in the visible spectral region due to lower plasmon damping by interband electron transitions. The field increases with increasing particle size, but further size increase decreases the field due to increased radiative damping for larger particles. This explains why 60-nm Au NPs give larger enhancement than those with other sizes [47]. Core–shell NPs give higher field enhancement than solid NPs [45]. The field decays within a distance comparable to the size of the NP [48]. The anisotropic NPs give much higher field enhancement than spherical counterparts due to curvature effects. Prisms, rods, and spheroids with similar size dimension show similar enhancement with $|E|^2/|E^0|^2$ on the scale of $>10^3$, which is significantly higher than that of spheres. The NR shows much higher electric fields at the end of the long axis and weakest fields at the center of the rods.

28.2.3 Photothermal Properties

As mentioned in Section 28.2.1, the total light extinction efficiency is equal to the sum of those of absorbed and scattered light. The scattered light has the same energy as the incident light, referred to as Mie scattering, Rayleigh scattering, or surface plasmon resonance light scattering. The absorbed light is converted into heat by the particle via a series of nonradiative processes [36]. These processes start with electron–hole recombination either within the conduction band (intraband) or between the d band and the conduction band (interband) [49]. These excited electrons cool off rapidly within ~1 ps by exchanging energy with the nanoparticle lattice (electron–phonon interaction) resulting in a hot particle lattice [36]. Subsequently, the lattice cools off by exchanging energy of the phonons with the surrounding medium on the timescale of ~100 ps. Such fast energy conversion and dissipation lead to the heating of surrounding species or environment.

The photothermal conversion efficiency of the metal NPs can be quantitaively measured. This is done by measuring the temperature changes of an NP solution with known optical densitry during laser irradiation and cooling [50,51]. The photothermal conversion efficiency η is calculated using the following equation [50]:

$$\eta = \frac{hS(T_{max} - T_{amb}) - Q_{dis}}{I(1 - 10^{-A_{808}})} \tag{28.16}$$

where h is the heat transfer coefficient, S is the surface area of the container, T_{max} is the maximum equilibrium temperature, T_{amb} is the ambient temperature of the surroundings, Q_{dis} is a parameter expressing the laser-induced heat input by the container, I is the laser power, and A_{808} is the absorbance of the NPs at 808 nm. The same experiment was conducted with water as the control to determine Q_{dis} (mW) by

$$Q_{dis} = \frac{10^3 \cdot m \cdot C \cdot \Delta T}{t} \tag{28.17}$$

where m is the mass of water (g), C is the heat capacity of water (Jg^{-1}K^{-1}), ΔT is the increased temperature (K), and t is the laser exposure time (s). The term hS was calculated based on

$$hS = \frac{\sum_i m_i C_{p,i}}{\tau_s} \tag{28.18}$$

where τ_s is the sample system time constant and i is the system components (NP suspension and sample container). τ_s is related to a dimensionless driving force temperature θ by

$$t = -\tau_s \ln \theta \tag{28.19}$$

where t is the cooling time and θ is given by

$$\theta = \frac{T - T_{amb}}{T_{max} - T_{amb}} \tag{28.20}$$

Figure 28.4a shows the plot of the temperature versus time for an aqueous Au NR sample (OD$_{808}$ = 1.0) during the heating and cooling processes by a diode red laser (λ = 808 nm) [52]. The temperature increased by 50°C when the NP solution was irradiated for 320 s and did not change significantly with further irradiation. The temperature reached a maximum because of the equilibrium between the heat input and output. In contrast, the temperature of the control water increased by only 0.9°C (Figure 28.4b). Figure 28.4c shows the plot of the cooling time t versus the term $-\ln \theta$. The data show a linear relationship, giving a slope of τ_s of 108 s. Based on Equation 28.13, the efficiency of transducing the red laser to heat by Au NRs was calculated to be 75%. This is comparable to Au bellflowers (74%) and nanocages (63%) and much higher than Au nanoshells (13%) [30]. It is worth

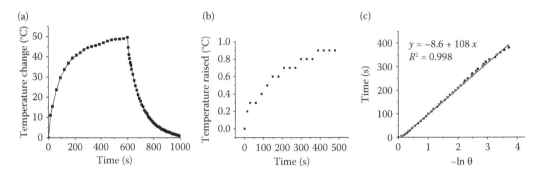

FIGURE 28.4

Photothermal properties of gold nanorods. (a) Plot of the temperature versus time for gold nanorods ($OD_{808} = 1.0$) during laser irradiation (808 nm, 0.55 W/cm²) and cooling (laser off). The temperature increased by 49°C during continuous irradiation for 600 s. (b) Plot of temperature versus time for the water during laser irradiation (808 nm, 0.55 W/cm²). (c) Plot of the cooling time versus $-\ln \theta$. Based on the linear regression analysis, the time constant for heat transfer τ_s was determined to be 108 s. (Reprinted with permission from S. Bhana et al. Huang. Near infrared-absorbing gold nanopopcorns with iron oxide cluster core for magnetically amplified photothermal and photodynamic cancer therapy. *ACS Appl. Mater. Interfaces* 2015, 7, 11637. Copyright 2015 American Chemical Society.)

mentioning that other researchers have reported different photothermal conversion efficiencies of Au NRs (24% [50], 60% [53], and 90% [54]) depending on the dimension of the NRs and the laser source used. For NPs with similar shape and structure, larger NPs have stronger light-scattering contribution in the total extinction efficiency and thus may give lower photothermal conversion efficiencies than smaller NPs. This explains why large gold nanoshells (~120 nm) have lower photothermal conversion efficiency than much smaller Au NRs (50 nm in length and 14 nm in width) and nanocages (35 nm in edge length).

28.3 Gold NRs for Cancer Diagnostics

28.3.1 Light Scattering Cancer Imaging

Au NPs offer an excellent class of contrast agents for imaging by using the strong Mie scattering properties. This light scattering imaging can be easily achieved in dark field using a simple commercial optical microscope (Figure 28.5a) [49]. In the dark field mode, the transmitted light is filtered by a patch stop disc in the dark field condenser. In this way, only scattered light is collected by the objective and thus any particle scattering strongly will lighten up in a dark background. Figure 28.5b and c shows the light-scattering images of normal HaCat cells (human keratinocytes) and HSC3 head and neck cancer cells after inoculation with antibody-conjugated Au NRs [10]. The Au NRs, which had an aspect ratio of 4.0, were synthesized by the seed-mediated growth method by Nikoobakht and El-Sayed [55]. The NRs were conjugated with antiepidermal growth factor receptor (EGFR) antibodies using a polystyrene sulfonate (PSS) linker [10]. The difference between the normal and cancer cells is clearly visualized. While the anti-EGFR-conjugated Au NRs non-specifically adsorbed onto the surface of normal cells, they specifically bound to the cancer cells due to the overexpression of EGFR on the cancer cell surface. The NPs scatter reddish light due to the surface plasmon resonance in the near-infrared region around 800 nm. This provides a simple way to image and detect cancer.

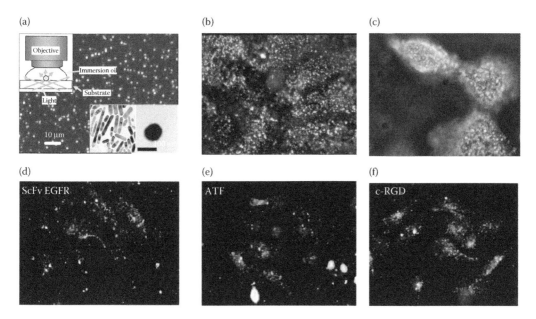

FIGURE 28.5
(a) True color photograph of a mixture of gold nanorods (red) and nanospheres (green) in the dark field illumination (inset). (b) Dark field images of HaCat normal cells after incubation with anti-EGFR-conjugated Au NRs. (c) Dark field images of HSC3 cancer cells after incubation with anti-EGFR-conjugated Au NRs. (d–f) Dark field images of A549 cancer cells after incubation with ScFv EGFR (d), ATF (e), and c-RGD (f) conjugated Au NRs. (Reprinted with permission from C. Sönnichsen et al. Drastic reduction of plasmon damping in gold nanorods. *Phys. Rev. Lett.* 2002, 88(7), 077402. Copyright 2002 American Physical Society. Reprinted with permission from X. Huang et al. *J. Am. Chem. Soc.* 2006, 128(6), 2115. Copyright 2006. Reprinted with permission from X. Huang et al. *ACS Nano.* 2010, 4(10), 5887. Copyright 2010.)

Using the dark field imaging, cancer cells with different marker expression can be detected using Au NRs linked with different targeting ligands. Figure 28.5d through f shows cellular binding of Au NRs on A549 lung cancer cells linked with a single-chain variable fragment (ScFv) peptide, an amino terminal fragment (ATF) peptide, a cyclic RGD peptide, respectively [31]. The ScFv, ATF, and RGD recognize EGFR, urokinase plasminogen activator receptor (uPAR), and the $\alpha_v\beta_3$ integrin receptor on cancer cells, respectively. By labeling the cells with these conjugates and examination with dark field imaging, the presence and the expression level of these receptors can be obtained. The results shown in Figure 28.5d through f demonstrate that A549 lung cancer cells are overexpressed with EGFR, uPAR, and $\alpha_v\beta_3$ integrin receptor. The expression level follows in the order of $\alpha_v\beta_3$ integrin receptor > uPAR > EGFR.

28.3.2 Surface Plasmon Resonance Detection

As described in Section 28.2.1, the LSPR wavelength depends on the dielectric constant of the surrounding medium. Therefore, the LSPR wavelength, either by absorption or scattering, provides a great opportunity to sense the changes of the local environment of the NP [56]. The sensitivity, defined as the plasmon shift per refractive index n_m unit (RIU) change ($d\lambda SPR/dn_m$), depends on the size and shape of the NPs as well as the metal composition [57,58]. Au NRs showed higher sensitivity than spheres. The sensitivity of Au NRs is linearly proportional to the aspect ratio of Au NRs [57]. Au NRs with higher effective radius give higher sensitivity. Using Au NRs of different aspect ratio with different spectroscopic response upon the environment changes, Yu and Irudayaraj have demonstrated a multiplex biosensor assay to detect different antibodies [58]. In this study, human, rabbit, and mouse IgG were conjugated to Au NRs with aspect ratios of 2.3, 3.5, and 5.1 via a 11-mercaptoundecanoic acid (MUA) linker. Binding events of these three molecular probes to their respective complements (anti-IgGs) were monitored and differentiated by the different shifts of the SPR wavelength of the NRs. The limit of detection is found to be in the range of 10^{-9} and 10^{-6} M.

In addition, the LSPR wavelength of Au NRs depends on the proximity of other NPs, like any other metal NPs. NP assembly caused large shift on LSPR wavelength. When Au NRs assemble end to end, the LSPR band is shifted to lower energies. When Au NRs assemble side by side, the LSPR band is shifted to higher energies [59]. An example of using Au NR aggregation for cancer detection is demonstrated by Zhang and Shen who used antibody-conjugated Au NRs to detect cancer antigen 125 (CA125) [60]. After mixing CA125 with anti-CA 125 Au NRs, CA 125 antibody–antigen interactions caused aggregation of Au NRs. This aggregation led to large shift in the scattering plasmon resonance of Au NRs. The method led to a detection limit of 0.4 U mL^{-1}.

28.3.3 SERS Detection

Raman signals are inelastic scattering from molecules upon exposure to the electromagnetic waves. The shift in wavelength (Stokes: red shift; anti-Stokes: blue shift) as compared to the incident light depends on the chemical structure of the molecule and thus provides the most detailed information about the chemical and molecular information of target molecules as compared to other optical spectroscopies such as absorption and Mie scattering. However, only 1 in 10^7 photons is scattered inelastically. This limited Raman technique for sensitive chemical and biomedical detection. The intensity of Raman signals (Stokes signal) I is related to the excitation laser intensity I_0 by

$$I = N\sigma I_0 \tag{28.21}$$

where N is the number of molecules in the probed volume and σ is the Raman cross section. For most molecules, σ is on the order of 10^{-29} cm^2. When a molecule is exposed in the proximity of a metal NP, the Raman signals are strongly enhanced by the particles through chemical and electromagnetic mechanisms [61]. The Raman signal intensity is quantitatively related to the laser intensity I_0 by

$$I = N'\sigma_{ads} \, |A(\nu_L)|^2 |A(\nu_s)|^2 \, I_0 \tag{28.22}$$

where N is the number of molecules involved in the SERS process, σ_{ads} is cross section of the new Raman process of the adsorbed molecule, and $A(\nu_L)$ and $A(\nu_s)$ are the enhancement factors for the laser and the Raman scattered field. $A(\nu)$ is the ratio of the field at the position of the molecules $E_M(\nu)$ and the incoming field $E_0(\nu)$. It shows that the enhancement scales as the fourth power of the local field of the metal nanostructure. As described in Section 28.2.2, Au NRs show enhancement with $|E|^2/|E^0|^2$ on the scale of $>10^3$. This indicates Au NRs may enhance the Raman signals of adsorbed molecules by 10^6. The SERS activity of Au NRs also depends on the aspect ratios of the particle. In general, aspect ratio

1.7 Au NRs give 10–10² greater SERS enhancement than Au NRs of other aspect ratios with an excitation laser of 632.8 nm [62].

SERS-based cancer detection has been performed in two ways: label-free methods and Raman reporter-labeled methods. An example of using label-free method is demonstrated by Huang et al. using anti-EGFR-conjugated Au NRs to differentiate normal and cancer cells [63] (Figure 28.6a through c). When both HaCat and HSC3 cancer cells are treated with anti-EGFR-conjugated Au NRs, Au NRs are assembled onto the cancer cells but not on normal cells due to the overexpression of EGFR on the cancer cells. This assembly strongly enhanced the Raman signals of the CTAB capping materials. Thus, the cancer cells give strong and sharp SERS signals from CTAB, providing a simple and rapid method for cancer detection.

In the Raman reporter-labeled methods, organic dye molecules with highly delocalized free electrons are nonspecifically adsorbed onto Au NPs and stabilized by hydrophilic PEG to serve as SERS NPs [64]. Nie and coauthors demonstrated for the first time SERS-based tumor detection using spherical Au SERS NPs [65]. Au NRs have advantages over Au nanospheres because they provide orders of magnitude higher SERS activities than Au nanospheres [66]. Using SERS Au NRs, von Maltzahn et al. showed that tumor can be facilely detected *in vivo* after direct injection of IR792-coded Au NRs (Figure 28.6d

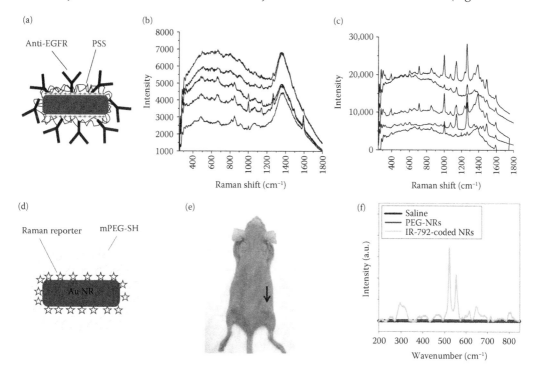

FIGURE 28.6
SERS detection of cancer cells and tumor. (a) Schematic illustration of anti-EGFR-conjugated Au NRs. (b) SERS signals of HaCat normal cells after incubation with anti-EGFR-conjugated Au NRs. (c) SERS signals of HSC3 cancer cells after incubation with anti-EGFR-conjugated Au NRs. (d) Schematic illustration of IR-792-coded Au NR. (e) Photograph of a mouse showing the nanoparticle injection site on the tumor. (f) SERS detection of tumor with IR-792 coded Au NRs. (Reprinted with permission from X. Huang et al. Cancer cells assemble and align gold nanorods conjugated to antibodies to produce highly enhanced, sharp and polarized surface Raman spectra: A potential cancer diagnostic marker. *Nano Lett.* 2007, 7(6), 1591. Copyright 2007 American Chemical Society. G. von Maltzahn et al.: *Adv. Mater.* 1. 2009. Copyright Wiley-VCH Verlag GmbH & Co. KGaA. Reproduced with permission.)

through f) [67]. In contrast, injection of PEGylated Au NRs without Raman reporter or saline solutions does not give any Raman signals. Further, multiplexed multicolor detection is achieved using different Raman reporters (IR792, DTTC765, and DTDC655) without changing the size of Au NRs.

28.4 Gold NRs in PTT

28.4.1 Photothermal Therapy

In 2006, we demonstrated for the first time that Au NRs can be used for PTT of cancer [10]. In this study, EGFR antibodies were adsorbed onto Au NRs through PSS to specifically bind to head and neck cancer cells. PSS helps antibody adsorption by hydrophobic and electrostatic interactions. Au NRs with an aspect ratio of 3.9 were used because their LSPR absorption overlap maximally with the CW NIR laser wavelength (808 nm) from either a commercial portable diode laser or Ti:Sapphire laser (800 nm). Before laser irradiation, Au NR antibody conjugates were incubated with cells on cover slips for 30 min. After washing out free particles, cells were exposed to a focused laser beam (beam size <1 mm) for 4 min. To differentiate dead and live cells, the cells were treated with trypan blue. Dead cells accumulated the dyes making the cells blue while live cells resist the dye molecules and remained colorless. Followed by fixation with 4% paraformaldehyde, cells were examined using a commercial optical microscope. In comparison, normal cells were treated in the same way to examine the capability of Au NRs for selective cancer PTT. Laser intensity was varied to determine the minimal energy for inducing cell death.

Figure 28.7 shows bright field images of normal and cancer cells after treatment with anti-EGFR-conjugated Au NRs and laser irradiation (Ti:Sapphire laser) of HaCat normal cells and HSC 3 and HOC313 head and neck cancer cells. The results showed that both types of cancer cells died under the laser irradiation with a power of 80 mW (10 W/cm^2), whereas normal cells died at 160 mW (20 W/cm^2). This demonstrates that cancer cells can be selectively destroyed at 80 mW without inducing damage to normal cells. This selective destruction was due to the overexpression of EGFR surface proteins on cancer cells, which lead to the bindings of more Au NRs on the cancer cells as compared with the normal cells.

We also first demonstrated the use of Au NRs for PTT of tumors *in vivo* [11]. Au NRs were covalently coated with mPEG-SH (MW5000) to achieve passive tumor targeting. It is well known that tumors' uptake of nanomaterials can be achieved in two ways: passive and active targeting [68]. In passive targeting, NPs are coated with hydrophilic molecules such as PEG to increase blood circulation and biocompatibility. Once the NPs reach tumor cells, they enter tumors, particularly in the interstitial region, by enhanced permeability and retention (EPR) effects [69]. In active targeting, NPs are linked with targeting ligands. Thus, after tumor uptake, the NPs bind to tumor cells and thus enter tumor cells. For PTT, tumor cell uptake is not required since the heating of NPs around tumor cells will increase the temperature of the whole tumor and thus kill tumor cells.

Figure 28.8 shows the results of the changes of tumor volume after laser irradiation of PEGylated Au NR injection in comparison with that in the control groups (without NP injection and laser irradiation) [11]. For the intravenous (I.V.) injection, 100 μL of PEGylated Au NRs (OD$_{\lambda=800}$ = 120) was injected into mouse ($n = 8$) via tail vein. For the direct injection, 15 μL of PEGylated Au NRs (OD$_{\lambda=800}$ = 40) was directly administrated into the tumor

HaCat normal cells HSC cancer cells HOC cancer cells

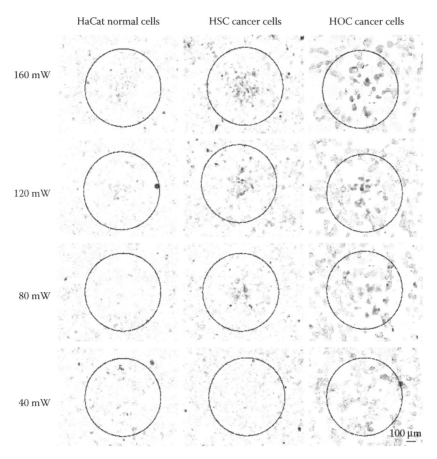

FIGURE 28.7
Selective photothermal therapy of cancer cells with anti-EGFR/Au nanorods. The circles show the laser spots on the samples. (Reprinted with permission from X. Huang et al., Cancer cell imaging and photothermal therapy in the near-infrared region by using gold nanorods. *J. Am. Chem. Soc.* 2006, 128(6), 2115. Copyright 2006 American Chemical Society.)

interstitium. Control tumor sites were injected with 10-mM phosphate buffer solution (PBS). Laser irradiation was performed extracorporally using a CW diode laser at 808 nm (Power Technologies) with a beam size of 1 cm. For the tumors in the tail vein injection group, tumors were exposed to the NIR light with intensity of 1.7–1.9 W/cm^2 after 24-h NP injection. For the tumors in the direct injection group, tumors were exposed to the NI light with intensity of 0.9–11 W/cm^2 after 2 min NP injection. The control tumors were exposed to the laser with intensity of 1.7–1.9 W/cm^2. The micrograph in Figure 28.8a confirmed that Au NRs were indeed accumulated into the tumors (shown as blot spots) under NIR imaging with the same diode laser with increased laser power. Directly injected tumor sites showed about two times higher amount of Au NRs than those subjected to tail vein injection. This is not surprising that usually only <5% of injected Au NRs go to tumors due to the clearance by reticuloendothelial system (RES) [31]. Figure 28.8b shows that the tumors in the direct injection group, recorded over a 13-day period, had >96% decrease in average tumor growth and those in the tail vein injection group has >74% decrease in average tumor growth. Moreover, resorption of >57% of the directly treated tumors and 25% of the intravenously treated tumors was observed over the monitoring period. In contrast, neither

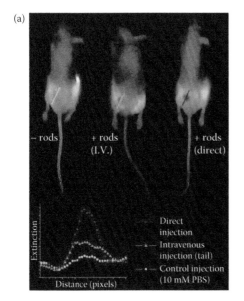 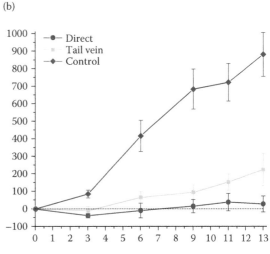

FIGURE 28.8

(a) NIR transmission image of mice prior to PTT treatments and intensity line scans of NIR extinction at tumor sites for control, intravenous, and direct administration of PEGylated gold nanorods. (b) Average change in tumor volumes for HSC3 xenografts following near-infrared PTT treatment by control (blue), intravenous (green), and direct (red) injection of PEGylated gold nanorods. Error bar is standard error of average value. (Reprinted from *Cancer Lett.*, 269(1), E.B. Dickerson et al., Gold nanorod assisted near-infrared plasmonic photo-thermal therapy (PPTT) of squamous cell carcinoma in mice., 57, Copyright 2008, with permission from Elsevier.)

growth suppression nor resorption was observed in any of the control tumors. Direct measurement of the tumor temperature during irradiation using a hypodermic thermocouple showed that laser heating of the Au NRs in tumors led to a temperature rise around 20°C. Such dramatic temperature increase led to the highly efficient ablation of tumor cells.

Following these initial studies, Au NRs have been extensively investigated in cancer PTT [12–21], For example, Tong et al. elucidated the mechanism of photothermal damage by real-time monitoring cell morphology changes under laser irradiation [14]. Using fluorescence staining and imaging, they found that laser heating of Au NRs led to cavitation of plasma membrane. Such membrane perforation resulted in an influx of extracellular Ca^{2+} followed by destruction of the action network and plasma blebbing. This cell death mechanism extends beyond simple hyperthermia. Another study was done by von Maltzahn et al. who conducted Au NR-assisted PTT tumor therapy under computational guidance [16]. Three-dimensional temperature distribution within the NP-injected tumors under laser irradiation was obtained via three-dimensional finite element modeling. The theoretical results helped the selection of experimental parameters to achieve maximally effective tumor ablation. The studies by Xu et al. show that Au NRs can also be used to selectively kill cancer stem cells (CSCs) [20]. This was based on their findings that polyelectrolyte-conjugated Au NRs were internalized by CSCs much faster than noncancerous stem cells (NCSCs). The killing of CSCs was very important as these cells are resistant to many current cancer treatments.

Recently, El-Sayed and coauthors demonstrated for the first time the use of Au NRs for cancer PTT on mammary gland treatment of a cat [70]. In these studies, a large tumor around 14 cm × 12 cm × 10.5 cm (length, width, and depth) was grown on the left caudal

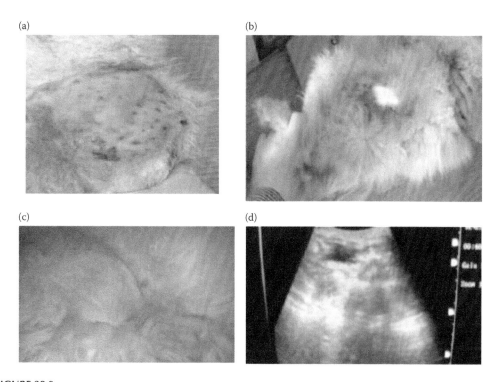

FIGURE 28.9
(a) Photograph showing the mammary gland tumor after Au NR injection. (b) Photograph showing the mammary gland tumor during the laser irradiation. (c) Photograph showing the cat with complete tumor remission. (d) Ultrasound scanning image showing the absence of tumor. (Cited from A.S. Abdoon et al. *J. Nanomed. Nanotechnol.* 2015, 6(5), 1000324. Copyright 2015 Abdoon et al.)

mammary gland on a cat (Figure 28.9a). A 75-μg Au NRs/kg body weight was injected via intratumoral (I.T.) injection, followed by 808-nm laser irradiation for 10 min (Figure 28.9b). The Au NR injection and laser treatment were repeated 15 days later. After 15 days from the first injection, 60% of tumor ablation was achieved. After 17 days from the second treatment, tumors were completely ablated (Figure 28.9b). This complete tumor destruction was further confirmed with ultrasound scanning imaging (Figure 28.9d). Complete blood picture as well as live and kidney functional analyses did not show systemic toxic effects. Most importantly, the studies showed that the cat was able to have normal pregnancy and deliver healthy kittens after the therapy. Breast feeding was also normal from all the nipples including the previously affected nipple. These studies on large animals showed great potential of Au NR-assisted PTT for clinical applications.

28.4.2 Combination Cancer Treatment Involving PTT

Despite the advancement on PTT with light-absorbing plasmonic NPs, complete eradication of tumors remains a challenge with a single irradiation. This is because NPs are mainly confined to the perivascular regions due to the high hydrostatic pressure with tumors and slow diffusion of the large particles, which limits the therapeutic efficacy. This leads to heterogeneous heating of the tumor tissues. Thus, strategies to improve PTT are needed to more effectively kill tumor cells.

One of the most promising methods to enhance cancer therapy is to combine PTT with other treatment modalities. Combination therapy is more effective than individual treatments due to additive or synergistic effects. Significant amount of studies have demonstrated the success of combining Au NR-assisted PTT with chemotherapy [71–81] and Au NR-assisted PTT with PDT [82–84]. When chemotherapy is involved, the chemotherapeutic drugs can be diffused into tumor cells and thus kill tumor cells that cannot be reached by light. When PDT is used, cell death can be induced dually by the localized PDT and PTT. In PDT, photosensitizer absorbs light energy and is excited from ground state to a high-energy state that transfers energy to neighboring oxygen, leading to production of high-energy reactive oxygen species (ROS) mainly singlet oxygen ($^1O_2$) to kill cancer cells [85–87]. Thus, the photosensitizers can also be used to image cancer cell and tumor.

In our recent studies, we developed a simple and highly integrated complex to incorporate PTT and chemotherapy using Au NRs and paclitaxel (PTX) [81] (Figure 28.10a). PTX was loaded to gold NRs with high density (2.0×10^4 PTX per Au NR) via nonspecific adsorption. The PTX–Au NR complex was stabilized with mPEG-SH 5000 that was linked

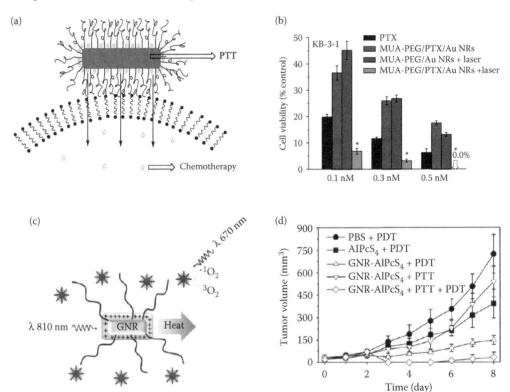

FIGURE 28.10
Combination cancer treatment involving gold nanorod-assisted photothermal therapy. (a) Schematic illustration of the structure of PTX–Au NR complex for photothermal-chemotherapy. (b) Cytotoxicity of PTX, MUA-PEG/Au NRs with laser irradiation, and MUA-PEG/PTX/Au NRs with and without laser irradiation in KB-3-1 cells. The results are mean values ± SD of triplicate experiments. (c) Schematic illustration of PTT and PTD using AlPcS4-loaded Au NRs under excitation with two different laser wavelengths. (d) The tumor volume versus time under different groups of treatment. (Reprinted with permission from F. Ren et al. Gold nanorods carrying paclitaxel for photothermal-chemotherapy of cancer. *Bioconjug. Chem.* 2013, 24, 376. Copyright 2013 American Chemical Society. Reprinted with permission from B. Wang et al. Rose-bengal-conjugated gold nanorods for in vivo photodynamic and photothermal oral cancer therapies. *Biomaterials* 2014, 35, 1954. Copyright 2011 American Chemical Society.)

with MUA. PTX was entrapped in the hydrophobic pocket of the amphiphilic polymeric monolayer. This formulation allows direct cellular delivery of the hydrophobic drugs via the lipophilic plasma membrane. The nanocomplex showed excellent stability under physiological environment, with only 8% PTX release within 6 h. However, when the nanocomplex was exposed to cancer cells, about 64% of PTX was released, indicating the highly efficient drug release process. Figure 28.10b shows the cytotoxicity of PTX, MUA-PEG/Au NRs with laser irradiation, and MUA-PEG/PTX/Au NRs with and without laser irradiation in KB-3-1 cancer cells. The cancer cells treated with 0.1-nM MUA-PEG/PTX/Au NRs and 0.1-nM MUA-PEG/Au NRs plus laser irradiation showed cell viabilities of 36.5% and 45.1%, respectively. However, when these cells were treated with 0.1-nM MUA-PEG/PTX/Au NRs plus laser irradiation (the combination therapy), cell viability remarkably decreased to 6.8%. This combination treatment killed about 30% more cells than chemotherapy and about 40% more cells than the PTT. Similar findings were observed with A549 lung cancer cells. These studies showed that combined PTT and chemotherapy with the PTX-loaded Au NRs were highly effective in killing cancer cells, superior to PTT or chemotherapy alone. By comparing the efficacy of calculated additive results and the experimentally determined combination results, the highly efficient cellular killing was attributed to a synergistic effect between PTT and chemotherapy.

An example to treat cancer using PTT + PDT with Au NRs was demonstrated by Jang et al. who used Al(III) phthalocyanine chloride tetrasulfonic acid (AlPcS4) as the photosensitizer [84] (Figure 28.10c). Short peptides with positively charged arginine were introduced onto Au NRs to help anchor negatively charged AlPcS4 through electrostatic interactions. AlPcS4 is released upon laser heating of the Au NRs and thus restore the fluorescence properties for imaging while generating singlet oxygen for PDT. Figure 28.10d showed that highly effective PTT and PDT were achieved for tumor therapy by using AlPcS4–Au NR complex, superior to PDT and PTT alone.

28.5 Biodistribution of Gold NRs

After I.V. injection, NPs circulate in the bloodstream and then gradually are cleared out by the RES due to the immune response of the body. To ensure sufficient tumor uptake, NPs require a long circulating time. The studies by Von Maltzahn showed that Au NRs coated with mPEG-SH has a blood half-life of 17 h. Niidome compared the biodistribution of as-prepared CTAB-stabilized Au NRs and PEGlyated Au NRs [88]. While most of CTAB-stabilized Au NRs went to major organs right after injection, the PEG-modified Au NRs did not accumulate in major organs except for the liver at least for 72 h. They further investigated the effects of PEG grafting level and injection dose on the biodistribution of Au NRs in tumor-bearing mice [89]. A PEG:gold molar ratio of 1.5 was found sufficient to show both prolonged blood circulation property and EPR effect. At the injection dose of 19.5 μg Au, the liver uptake is saturated. Further increase of Au NRs led to significant increase in the spleen, tumor, and other tissues.

In our previous studies, we found that the targeting ligand had a large effect on the biodistribution of Au NRs using inductively coupled plasmon mass spectrometry (ICP-MS) [31]. We examined three different tumor-targeting ligand, ScFv EGFR, ATF, and c-RGD. The ScFv EGFR-, ATF-, and c-RGD-linked Au NRs showed the blood half-life of 8.3, 6.5, and 9.3 h, respectively (Figure 28.11a). The nontargeted Au NRs coated with a mixture of

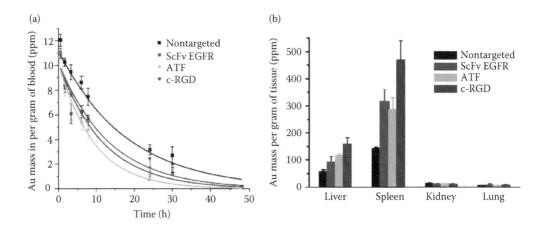

FIGURE 28.11
Blood circulation (a) and organ distribution (b) of gold nanorods linked with different ligands. (Reprinted with permission from X. Huang et al., A reexamination of active and passive tumor targeting by using rod-shaped gold nanocrystals and covalently conjugated peptide ligands. *ACS Nano* 2010, 4(10), 5887. Copyright 2010 American Chemical Society.)

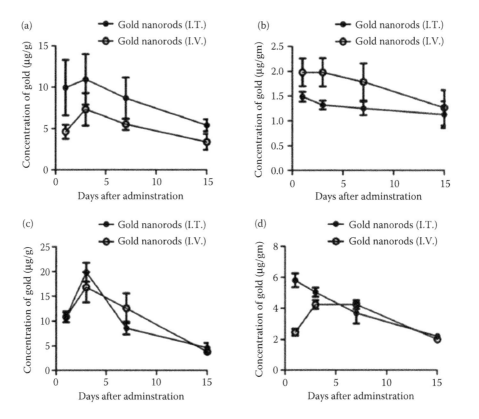

FIGURE 28.12
Tissue pharmacokinetics of Au NRs after intravenous (I.V.) and intratumoral (I.T.) injection on a tumor-bearing mice. (a) Tumor, (b) liver, (c) spleen, and (d) kidney. Au content was checked for each tissue for 2 weeks. Data are presented as mean ±SEM (*n* = 3). (Cited from M.A. El-Sayed et al. *PLoS One* 2013, 8 (10), e76207. Copyright 2013 El-Sayed et al.)

mPEG and HOOC-PEG gave a half-life of 12.5 h. These data indicate that the addition of tumor-targeting ligands accelerate NP blood clearance. All NPs were taken up by the RES organs (liver and spleen), with little accumulation in the kidney or lung (Figure 28.11b). The targeted Au NRs are more efficiently taken up by the RES organs than nontargeted Au NRs. This is not surprising as the Au NRs with exposed peptide ligands are less "stealthy" and thus more readily recognized by the immune system for clearance. These results showed that for photothermal cancer therapy, the preferred route of Au NR administration is I.T. injection instead of I.V. injection.

In 2013, El-Sayed et al. investigated the tissue pharmacokinetics of Au NRs after I.V. and I.T. injection on tumor-bearing mice (Figure 28.12) [90]. The I.T. administrated tumor has higher concentration of Au NRs within tumor tissue than I.V. administrated one for all examined time points within 2 weeks. For both I.V. and I.T. administrated tumors, the concentration of Au NRs increased within 72 h and then decreased, indicating elimination of Au NRs by the tumor (Figure 28.12a). The $t_{1/2}$ of the particle clearance for I.T. and I.V. administrated tumors is 13.8 and 11. 6 days, respectively. These informative results are very helpful in stratifying the time of laser irradiation after particle administration in the treatment. Au NRs in normal organs include liver, spleen, and kidney and has higher concentration for the mouse with I.V. injection than that with I.T. injection (Figure 28.12b through d). Similar to tumor tissues, the concentration of Au NRs reached maximum at 72 h. The $t_{1/2}$ of the particle clearance in different organs ranged from 11.6 to 69.3 days.

Acknowledgment

Huang gratefully acknowledges funding support from the National Institutes of Health/ National Cancer Institutes (grant number R15 CA 195509-01).

References

1. E.C. Dreaden, A.M. Alkilany, X. Huang, C.J. Murphy, M.A. El-Sayed. The golden age: Gold nanoparticles for biomedicine. *Chem. Soc. Rev.* 2012, 41, 2740.
2. P.K. Jain, K.S. Lee, I.H. El-Sayed, M.A. El-Sayed. Calculated absorption and scattering properties of gold nanoparticles of different size, shape, and composition: Applications in biological imaging and biomedicine. *J. Phys. Chem. B* 2006, 110(14), 7238.
3. R. Weissleder. A clearer vision for in vivo imaging. *Nat. Biotechnol.* 2001, 19, 316.
4. L.R. Hirsch, R.J. Stafford, J.A. Bankson, S.R. Sershen, R.E. Price, J.D. Hazle, N.J. Halas, J.L. West. Nanoshell-mediated near infrared thermal therapy of tumors under MR guidance. *Proc. Nat. Acad. Sci. U.S.A.* 2003, 100(23), 13549.
5. D.P. O'Neal, L.R. Hirsch, N.J. Halas, J.D. Payne, J.L. West. Photothermal tumor ablation in mice using near infrared absorbing nanoshells. *Cancer Lett.* 2004, 209(2), 171.
6. C.H. Loo, A. Lin, L.R. Hirsch, M.H. Lee, J. Barton, N.J. Halas, J. West, R.A. Drezek. Nanoshell-enabled photonics-based imaging and therapy of cancer. *Technol. Cancer Res. Treat.* 2004, 3, 33.
7. C. Loo, A. Lowery, N.J. Halas, J.L. West, R. Drezek. Immunotargeted nanoshells for integrated cancer imaging and therapy. *Nano Lett.* 2005, 5(4), 709.
8. A.M. Gobin, M.H. Lee, N.J. Halas, W.D James, R.A. Drezek, J.L. West. Near-infrared resonant nanoshells for combined optical imaging and photothermal cancer therapy. *Nano Lett.* 2007, 7, 1929.

9. P. Diagaradjane, A. Shetty, J.C. Wang, A.M. Elliott, J. Schwartz, S. Shentu, H.C. Park et al. Modulation of in vivo tumor radiation response via gold nanoshell-mediated vascular-focused hyperthermia: Characterizing an integrated antihypoxic and localized vascular disrupting targeting strategy. *Nano Lett.* 2008, 8, 1492.

10. X. Huang, I.H. El-Sayed, W. Qian, M.A. El-Sayed. Cancer cell imaging and photothermal therapy in the near-infrared region by using gold nanorods. *J. Am. Chem. Soc.* 2006, 128(6), 2115.

11. E.B. Dickerson, E.C. Dreaden, X. Huang, I.H. El-Sayed, H. Chu, S. Pushpanketh, J.F. McDonald, M.A. El-Sayed. Gold nanorod assisted near-infrared plasmonic photothermal therapy (PPTT) of squamous cell carcinoma in mice. *Cancer Lett.* 2008, 269(1), 57.

12. H. Takahashi, T. Niidome, A. Nariai, Y. Niidome, S. Yamada. Gold nanorod-sensitized cell death: Microscopic observation of single living cells irradiated by pulsed near-infrared laser light in the presence of gold nanorods. *Chem. Lett.* 2006, 35(5), 500.

13. H. Takahashi, T. Niidome, A. Nariai, Y. Niidome, S. Yamada. Photothermal reshaping of gold nanorods prevents further cell death. *Nanotechnology* 2006, 17, 4431.

14. L. Tong, Y. Zhao, T.B. Huff, M.N. Hansen, A. Wei, J.X. Cheng. Gold nanorods mediate tumor cell death by compromising membrane integrity. *Adv. Mater.* 2007, 19, 3136.

15. T.B. Huff, L. Tong, Y. Zhao, M.N. Hansen, J.X. Cheng, A. Wei. Hyperthermic effects of gold nanorods on tumor cells. *Nanomedicine (Lond.)* 2007, 2(1), 125.

16. G. von Maltzahn, J.H. Park, A. Agrawal, N.K. Bandaru, S.K. Das, M.J. Sailor, S.N. Bhatia. Computationally guided photothermal tumor therapy using long-circulating gold nanorod antennas. *Cancer Res.* 2009, 69, 3892.

17. H.C. Huang, K. Rege, J.J. Heys. Spatiotemporal temperature distribution and cancer cell death in response to extracellular hyperthermia induced by gold nanorods. *ACS Nano* 2010, 4(5), 2892.

18. T. Okuno, S. Kato, Y. Hatakeyama, J. Okajima, S. Maruyama, M. Sakamoto, S. Mori, T. Kodama. Photothermal therapy of tumors in lymph nodes using gold nanorods and near-infrared laser light. *J. Control. Release* 2013, 172, 879.

19. J. Wang, K. Sefah, M.B. Altman, T. Chen, M. You, Z. Zhao, C.Z. Huang, W. Tan. Aptamer-conjugated nanorods for targeted photothermal therapy of prostate cancer stem cells. *Chem. Asian J.* 2013, 8(10), 2417.

20. Y. Xu, J. Wang, X. Li, Y. Liu, L. Dai, X. Wu, C. Chen. Selective inhibition of breast cancer stem cells by gold nanorods mediated plasmonic hyperthermia. *Biomaterials* 2014, 35, 4667.

21. R.K. Kannadorai, G.G.Y. Chiew, K.Q. Luo, Q. Liu. Dual functions of gold nanorods as photothermal agent and autofluorescence enhancer to track cell death during plasmonic photothermal therapy. *Cancer Lett.* 2015, 357, 152.

22. J. Chen, B. Wiley, Z.Y. Li, D. Campbell, F. Saeki, H. Cang, L. Au, J. Lee, X. Li, Y. Xia. Gold nanocages: Engineering their structure for biomedical applications. *Adv. Mater.* 2005, 17, 2255.

23. M. Hu, H. Petrova, J. Chen, J.M. McLellan, A.R. Siekkinen, M. Marquez, X. Li, Y. Xia, G.V. Hartland. Ultrafast laser studies of the photothermal properties of gold nanocage. *J Phys. Chem. B* 2006, 110, 1520.

24. J. Chen, D. Wang, J. Xi, L. Au, A. Siekkinen, A. Warsen, Z.Y. Li, H. Zhang, Y. Xia, X. Li. Immuno gold nanocages with tailored optical properties for targeted photothermal destruction of cancer cells. *Nano Lett.* 2007, 7(5), 1318.

25. S.E. Skrabalak, J. Chen, L. Au, X. Lu, X. Li, Y. Xia. Gold nanocages for biomedical applications. *Adv. Mater.* 2007, 19, 3177.

26. M.P. Melancon, W. Lu, Z. Yang, R. Zhang, Z. Cheng, A.M. Elliot, J. Stafford, T. Olson, J.Z. Zhang, C. Li. In vitro and in vivo targeting of hollow gold nanoshells directed at epidermal growth factor receptor for photothermal ablation therapy. *Mol. Cancer Ther.* 2008, 7, 1730.

27. J. Chen, C. Glaus, R. Laforest, Q. Zhang, M. Yang, M. Gidding, M.J. Welch, Y. Xia. Gold nanocages as photothermal transducers for cancer treatment. *Small* 2010, 6(7), 811.

28. W. Lu, C. Xiong, G. Zhang, Q. Huang, R. Zhang, J.Z. Zhang, C. Li. Targeted photothermal ablation of murine melanomas with melanocyte-stimulating hormone analog conjugated hollow gold nanospheres. *Clin. Cancer Res.* 2009, 15, 876.

29. W. Lu, A.K. Singh, S.A. Khan, D. Senapati, H. Yu, P.C. Ray. Gold nano-popcorn-based targeted diagnosis, nanotherapy treatment, and in situ monitoring of photothermal therapy response of prostate cancer cells using surface-enhanced Raman spectroscopy. *J. Am. Chem. Soc.* 2010, 132, 18103.

30. P. Huang, P. Rong, J. Lin, W. Li, X. Yan, M.G. Zhang, L. Nie et al. Triphase interface synthesis of plasmonic gold bellflowers as near-infrared light mediated acoustic and thermal theranostics. *J. Am. Chem. Soc.* 2014, 136(23), 8307.

31. X. Huang, X. Peng, Y. Wang, X. Wang, D.M. Shin, M.A. El-Sayed, S. Nie. A reexamination of active and passive tumor targeting by using rod-shaped gold nanocrystals and covalently conjugated peptide ligands. *ACS Nano* 2010, 4(10), 5887.

32. R. Gans. Form of ultramicroscopic particles of silver. *Ann. Phys.* 1915, 47, 270.

33. G.C. Papavassiliou. Optical properties of small inorganic and organic metal particles. *Prog. Solid State Chem.* 1980, 12, 185.

34. S. Link, M.B. Mohamed, M.A. El-Sayed. Simulation of the optical absorption spectra of gold nanorods as a function of their aspect ratio and the effect of the medium dielectric constant. *J. Phys. Chem. B* 1999, 103(16), 3073.

35. S. Link, M.A. El-Sayed. Additions and corrections to Simulation of the optical absorption spectra of gold nanorods as a function of their aspect ratio and the effect of the medium dielectric constant. *J. Phys. Chem. B* 2005, 109(20), 10531.

36. S. Link, M.A. El-Sayed. Shape and size dependence of radiative, non-radiative and photothermal properties of gold nanocrystals. *Int. Rev. Phys. Chem.* 2000, 19(3), 409.

37. H. Chen, X. Kou, Z. Yang, W. Ni, J. Wang. Shape- and size-dependent refractive index sensitivity of gold nanoparticles. *Langmuir* 2008, 24, 5233.

38. F.M. Purcell, C.R. Pennypacker. Scattering and absorption of light by nonspherical dielectric grains. *J. Astrophys.* 1973, 186, 705.

39. B.T. Draine. The discrete-dipole approximation and its application to interstellar graphite grains. *Astrophys. J.* 1988, 333, 848.

40. B.T. Draine, J.J. Goodman. Beyond Clausius-Mossotti: Wave propagation on a polarizable point lattice and the discrete dipole approximation. *Astrophys. J.* 1993, 405, 685.

41. B.T. Draine, P.J. Flatau. Discrete-dipole approximation for scattering calculations. *J. Opt. Soc. Am. A* 1994, 11, 1491.

42. B.T. Draine. "The discrete-dipole approximation for light scattering by irregular targets" in *Light Scattering by Nonspherical Particles: Theory, Measurements, and Geophysical Applications.* M.I. Mishchenko, J.W. Hovenier, and L.D. Travis, Editors. New York: Academic Press, 2000, p. 131.

43. K.S. Lee, M.A. El-Sayed. Dependence of the enhanced optical scattering efficiency relative to that of absorption for gold metal nanorods on aspect ratio, size, end-cap shape, and medium refractive Index. *J. Phys. Chem. B* 2005, 109(43), 20331.

44. C.F. Bohren, D.R. Huffman. *Absorption and Scattering of Light by Small Particles.* New York: Wiley, 1983.

45. E. Hao, G. Schatz, J. Hupp. Synthesis and optical properties of anisotropic metal nanoparticles. *J. Fluor.* 2004, 14(4), 331.

46. G. Hao, G.C. Schatz. Electromagnetic fields around silver nanoparticles and dimers. *J. Chem. Phys.* 2004, 120(1), 357.

47. J.T. Krug, G.D. Wang, S.R. Emory, S.M. Nie. Efficient Raman enhancement and intermittent light emission observed in single gold nanocrystals. *J. Am. Chem. Soc.* 1999, 121, 9208.

48. P.K. Jain, W. Huang, M.A. El-Sayed. On the universal scaling behavior of the distance decay of plasmon coupling in metal nanoparticle pairs: A plasmon ruler equation. *Nano Lett.* 2007, 7, 2080.

49. C. Sönnichsen, T. Franzl, T. Wilk, G.V. Plessen, J. Feldmann. Drastic reduction of plasmon damping in gold nanorods. *Phys. Rev. Lett.* 2002, 88(7), 077402.

50. Q. Tian, F. Jiang, R. Zou, Q. Liu, Z. Chen, M. Zhu, S. Yang, J. Wang, J. Wang, J. Hu. Hydrophilic Cu9S5 nanocrystals: A photothermal agent with a 25.7% heat conversion efficiency for photothermal ablation of cancer cells in vivo *ACS Nano.* 2011, 5(12), 9761.

51. D.K. Roper, W. Ahn, M. Hoepfner. Microscale heat transfer transduced by surface plasmon resonant gold nanoparticles. *J. Phys. Chem. C* 2007, 111(9), 3636.
52. S. Bhana, G. Liu, L. Wang, H. Starring, S.R. Mishra, G. Liu, X. Huang. Near infrared-absorbing gold nanopopcorns with iron oxide cluster core for magnetically amplified photothermal and photodynamic cancer therapy. *ACS Appl. Mater. Interfaces* 2015, 7, 11637.
53. J.R. Cole, N.A. Mirin, M.W. Knight, G.P. Goodrich, N.J. Halas. Photothermal efficiencies of nanoshells and nanorods for clinical therapeutic applications. *J. Phys. Chem. C* 2009, 113, 12090.
54. H. Chen, L. Shao, T. Ming, Z. Sun, C. Zhao, B. Yang, J. Wang. Understanding the photothermal conversion efficiency of gold nanocrystals. *Small* 2010, 6, 2272.
55. B. Nikoobakht, M.A. El-Sayed. Preparation and growth mechanism of gold nanorods (NRs) using seed-mediated growth method. *Chem. Mater.* 2003, 15(10), 1957.
56. K. Aslan, J.R. Lakowicz, C.D. Geddes. Plasmon light scattering in biology and medicine: New sensing approaches, visions and perspectives. *Curr. Opin. Chem. Biol.* 2005, 9, 538.
57. S.K. Lee, M.A. El-Sayed. Gold and silver nanoparticles in sensing and imaging: Sensitivity of plasmon response to size, shape, and metal composition. *J. Phys. Chem. B* 2006, 110, 19220.
58. C. Yu, J. Irudayaraj. Quantitative evaluation of sensitivity and selectivity of multiplex nanoSPR biosensor arrays. *Biophy. J.* 2007, 93, 3684.
59. P.K. Jain, S. Eustis, M.A. El-Sayed. Plasmon coupling in nanorod assemblies: Optical absorption, discrete dipole approximation simulation, and exciton-coupling model. *J. Phys. Chem B* 2006, 110, 18243.
60. K. Zhang, X. Shen. Cancer antigen 125 detection using the plasmon resonance scattering properties of gold nanorods. *Analyst* 2013, 138(6), 1828.
61. K. Kneipp, H. Kneipp, I. Itzkan, R.R. Dasari, M.S. Feld. Surface-enhanced Raman scattering and biophysics. *J. Phys. Condens. Matter* 2002, 14(18), R597.
62. C.J. Orendorff, L. Gearheart, N.R. Janaz, C.J. Murphy. Aspect ratio dependence on surface enhanced Raman scattering using silver and gold nanorod substrates. *Phys. Chem. Chem. Phys.* 2006, 8, 165.
63. X. Huang, I.H. El-Sayed, W. Qian, M.A. El-Sayed. Cancer cells assemble and align gold nanorods conjugated to antibodies to produce highly enhanced, sharp and polarized surface Raman spectra: A potential cancer diagnostic marker. *Nano Lett.* 2007, 7(6), 1591.
64. M.Y. Sha, H. Xu, S.G. Penn, R. Cromer. SERS nanoparticles: A new optical detection modality for cancer diagnosis. *Nanomedicine (Lond.)* 2007, 2, 725.
65. X. Qian, X.H. Peng, D.O. Ansari, Q.Y. Goen, G.Z. Chen, D.M. Shin, L. Yang, A.N. Young, M.D. Wang, S. Nie. In vivo tumor targeting and spectroscopic detection with surface-enhanced Raman nanoparticle tags. *Nat. Biotechnol.* 2008, 26, 83.
66. B. Nikoobakht, J. Wang, M.A. El-Sayed. Surface-enhanced Raman scattering of molecules adsorbed on gold nanorods: Off-surface plasmon resonance condition. *Chem. Phys. Lett.* 2002, 366(1), 17.
67. G. von Maltzahn, A. Centrone, J.H. Park, R. Ramanathan, M.J. Sailor, T.A. Hatton, S.N. Bhatia. SERS-coded gold nanorods as a multifunctional platform for densely multiplexed near-infrared imaging and photothermal heating. *Adv. Mater.* 2009, 21, 1.
68. N. Bertrand, J. Wu, X. Xu, N. Kamaly, O.C. Farokhzad. Cancer nanotechnology: The impact of passive and active targeting in the era of modern cancer biology. *Adv. Drug Deliv.* 2014, 66, 2–25.
69. J. Fang, H. Nakamura, H. Maeda. The EPR effect: Unique features of tumor blood vessels for drug delivery, factors involved, and limitations and augmentation of the effect. *Adv. Drug Deliv. Rev.* 2011, 63, 136.
70. A.S. Abdoon, E.A. Al-Ashkar, A. Shabaka, O.M. Kandil, W.H. Eisa, A.M. Shaban, H.M. Khaled et al. Normal pregnancy and lactation in a cat after treatment of mammary gland tumor when using photothermaml therapy with gold nanorods: A case report. *J. Nanomed. Nanotechnol.* 2015, 6(5), 1000324.
71. T.S. Hauck, T.L. Jennings, T. Yatsenko, J.C. Kumaradas, W.C.W. Chan. Enhancing the toxicity of cancer chemotherapeutics with gold nanorod hyperthermia. *Adv. Mater.* 2008, 20(20), 3832.

72. H. Park, J. Yang, J. Lee, S. Haam, I.H. Choi, K.H. Yoo. Multifunctional nanoparticles for combined doxorubicin and photothermal treatments. *ACS Nano* 2009, 3(10), 2919.

73. J. You, G. Zhang, C. Li. Exceptionally high payload of doxorubicin in hollow gold nanospheres for near-infrared light-triggered drug release. *ACS Nano* 2010, 4(2), 1033.

74. J. You, R. Zhang, G. Zhang, M. Zhong, Y. Liu, C.S. Van Pelt, D. Liang, W. Wei, A.K. Sood, C. Li. Photothermal-chemotherapy with doxorubicin-loaded hollow gold nanospheres: A platform for near-infrared light-trigged drug release. *J. Control. Release* 2011, 158, 319–328.

75. Liu, H., D. Chen, L. Li, T. Liu, L. Tan, X. Wu, F. Tang. Multifunctional gold nanoshells on silica nanorattles: A platform for the combination of photothermal therapy and chemotherapy with low systemic toxicity. *Angew. Chem. Int. Ed.* 2011, 50(4), 891.

76. K.C. Hribar, M.H. Lee, D. Lee, J.A. Burdick. Enhanced release of small molecules from near-infrared light responsive polymer-nanorod composites. *ACS Nano* 2011, 5(4), 2948.

77. F.Y. Cheng, C.H. Su, P.C. Wu, C.S. Yeh. Multifunctional polymeric nanoparticles for combined chemotherapeutic and near-infrared photothermal cancer therapy in vitro and in vivo. *Chem. Comm. (Camb.)* 2010, 46(18), 3167–9.

78. J. You, R. Shao, X. Wei, S. Gupta, C. Li. Near-infrared light triggers release of paclitaxel from biodegradable microspheres: Photothermal effect and enhanced antitumor activity. *Small* 2010, 6(9), 1022.

79. T.R. Kuo, V.A. Hovhannisyan, Y.C. Chao, S.L. Chao, S.J. Chiang, S.J. Lin, C.Y. Dong, C.C. Chen. Multiple release kinetics of targeted drug from gold nanorod embedded polyelectrolyte conjugates induced by near-infrared laser irradiation. *J. Am. Chem. Soc.* 2010, 132(40), 14163.

80. W. Wu, J. Shen, P. Banerjee, S. Zhou. Water-dispersible multifunctional hybrid nanogels for combined curcumin and photothermal therapy. *Biomaterials* 2011, 32, 598.

81. F. Ren, S. Bhana, D.D. Norman, J. Johnson, L. Xu, D.L. Baker, A.L. Parrill, X. Huang. Gold nanorods carrying paclitaxel for photothermal-chemotherapy of cancer. *Bioconjug. Chem.* 2013, 24, 376.

82. B. Wang, J.H. Wang, Q. Liu, H. Huang, M. Chen, K. Li, C. Li, X.F. Yu, P.K. Chu. Rose-bengal-conjugated gold nanorods for in vivo photodynamic and photothermal oral cancer therapies. *Biomaterials* 2014, 35, 1954.

83. S.H. Seo, B.M. Kim, A. Joe, H.W. Han, X. Chen, Z. Cheng, E.S. Jang. NIR-light-induced surface-enhanced Raman scattering for detection and photothermal/photodynamic therapy of cancer cells using methylene blue-embedded gold nanorod@SiO$_2$ nanocomposites. *Biomaterials* 2014, 35, 3309.

84. B. Jang, J.Y. Park, C.H. Tung, I.H. Kim, Y. Choi. Gold nanorod–photosensitizer complex for near-infrared fluorescence imaging and photodynamic/photothermal therapy in vivo. *ACS Nano* 2011, 5(2), 1086.

85. T.J. Dougherty, C.J. Gomer, B.W. Henderson. Photodynamic therapy. *J. Natl. Cancer Inst.* 1998, 90(12), 889.

86. C.A. Robertson, D.H. Evans, H. Abrahamse. Photodynamic therapy (PDT): A short review on cellular mechanisms and cancer research applications for PDT. *J. Photochem. Photobiol. B* 2009, 96, 1–8.

87. J.P. Celli, B.Q. Spring, I. Rizvi, C.L. Evans, K.S. Samkoe, S. Verma, B.W. Pogue, T. Hasan. Imaging and photodynamic therapy: Mechanisms, monitoring, and optimization. *Chem. Rev.* 2010, 110, 2795.

88. T. Niidome, M. Yamagata, Y. Okamoto, Y. Akiyama, H. Takahashi, T. Kawano, Y. Katayama, Y. Niidome. PEG-modified gold nanorods with a stealth character for in vivo applications. *J. Control. Release* 2006, 114, 343.

89. Y. Akiyama, T. Mori, Y. Katayama, T. Niidome. The effects of PEG grafting level and injection dose on gold nanorod biodistribution in the tumor-bearing mice. *J. Control. Release* 2009, 139, 81.

90. M.A. El-Sayed, A.A. Shabaka, O.A. El-Shabrawy, N.A. Yassin, S.S. Mahmoud, S.M. El-Shenawy, E. Al-Ashqar et al. Tissue distribution and efficacy of gold nanorods coupled with laser induced photoplasmonic therapy in Ehrlich carcinoma solid tumor model. *PLoS One* 2013, 8(10), e76207.

29

Applications of Nanotechnology in Reproductive Medicine

Celine Jones, Natalia Barkalina, Sarah Francis, Lien Davidson, and Kevin Coward

CONTENTS

29.1 Introduction

Nanoscience is an integrative discipline, which encompasses various aspects of biology, chemistry, physics, computer science, and engineering (Koopmans and Aggeli, 2010; Petros and DeSimone, 2010). The term "nanobiotechnology" refers to the branch of nanoscience that applies specifically to the life sciences and involves the design, synthesis, and deployment of nanomaterials for biological or biochemical applications (Jain, 2010).

The field of nanomedicine is a rapidly developing area of research in which nanoparticles and nanodevices are utilized in diagnostic and therapeutic processes. In this respect, nanomedicine has a clear potential to revolutionize clinical diagnostics and therapy for a range of pathologies. However, the complexity of the human system makes developments in this area extremely challenging. While the first attempt to synthesize nanoparticles dates back to the 1950s (Petros and DeSimone, 2010), it is only over the last two decades or so that nanoparticles have been increasingly deployed with the hope of revolutionizing research tools, diagnostic techniques, and therapeutic approaches in the medical sciences (Salata, 2004; Koopmans and Aggeli, 2010). Among a host of other applications, nanoparticles have been used for the targeted delivery of anticancer therapies for improved pharmacodynamics and pharmacokinetics (Albanese et al., 2012), and also as contrast agents for imaging and diagnosis, with a variety of adaptions tailored to facilitate target specificity, cellular uptake, and cargo release (Solaro, 2008; Gu et al., 2011). This chapter focuses on the use of nanoparticles within reproductive medicine. Further description and evaluation of the use of nanoparticles in other medical science areas can be found elsewhere in this book. Our intention here is to focus predominantly upon reproductive medicine.

29.2 Nanoparticles: Classification and Cellular Uptake

Nanoparticles are small-scale structures ranging in size from 1 to 100 nm (Petros and DeSimone, 2010), which can be synthesized from a variety of precursors including gold, silver, carbon, iron oxide, titanium dioxide, silica, and polystyrene (Albanese et al., 2012). These structures have gained immense popularity due to their ability to act as vehicles for the delivery of biological cargo such as proteins, DNA, and small interfering RNA (siRNA) into target cells and tissues. Numerous types of nanoparticles have been synthesized over the years and classified, depending on the nature of the precursor material (Gu et al., 2011). A more detailed description of this classification system, and the different types of nanomaterials involved, can be found elsewhere in this book.

One particularly valuable feature of nanoparticles is their high surface-area-to-volume ratio, which creates large loading capacities and enables their use as nanocarriers. In addition, by coating nanoparticles with hydrophilic polymers such as polyethylene glycol (PEG), it is possible to prevent the immune system from recognizing the nanoparticles, thus increasing the half-life of the circulating nanoparticle and its cargo. Furthermore, by conjugating cargo with nanoparticles, it is possible to confer a significant amount of protection, thus preventing premature degradation of biological cargo (Solaro, 2008). Recent modifications in surface coating also permit the controlled release of cargo from nanoparticles upon the action of a specific endogenous or exogenous stimulus such as pH, temperature, light, ultrasound, or microwave radiation (Albanese et al., 2012; Rosenholm et al., 2012). Realization of the ability to customize such features led to the evolution of three successive generations of nanoparticles, each with increasing levels of functionality. While toxicity was a pervasive issue for certain nanoparticles, such as those based upon metallic sources, those derived from silica, particularly those of a mesoporous nature, exhibited very favorable biocompatibility (Horie et al., 2012; Tarn et al., 2013; Barkalina et al., 2014a, b).

Several mechanisms have been proposed to account for the cellular uptake of nanoparticles, including phagocytosis, macropinocytosis, caveolar-mediated endocytosis, and clathrin-mediated endocytosis. The exact pathway of uptake is driven by the size of the nanomaterial and the type, and extent, of interaction with the target cell surface (Petros and DeSimone, 2010). The noninvasive nature of the processes involved created a highly attractive alternative to conventional delivery methods such as heat shock or electroporation by minimizing the complexity and risk of these earlier techniques (Petros and DeSimone, 2010). The efficiency of interaction between nanoparticles and the membrane of the target cell, and the potential for subsequent internalization depend directly upon a number of parameters, including the shape, size, composition, and electrostatic charge of the nanomaterial (Chithrani and Chan, 2007; Gratton et al., 2008; Barkalina et al., 2014a; Figure 29.1).

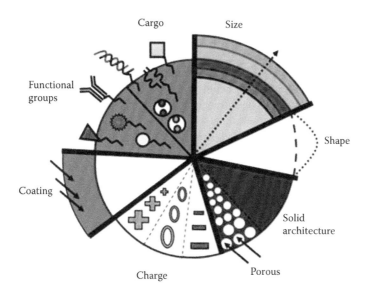

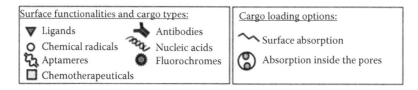

FIGURE 29.1

Nanomaterials represent highly versatile structures with adjustable physical and chemical properties. Changes in the size, shape, architecture, surface charge, and coating of nanomaterials allow manipulate their interaction with cells, thus improving uptake by the target cell population, and minimizing nonspecific interaction. (Reproduced with permission from Barkalina N et al. *Nanomedicine.* 2014a;10:921–938.)

The versatile system created by tailored nanoparticle design and synthesis, together with the significant loading capacity offered by these structures, allows for the storage and targeted delivery of large payload volumes. Furthermore, the biodistribution of cargo adsorbed on a nanoparticle differs dramatically from that of a free molecule, and is determined by the specific profile of the engineered nanoparticle (Pathak and Dhar, 2015). This combination of properties allows for increased payload circulation times, as cargo cannot be released until an appropriate recognition process has been encountered, thus limiting the functional action of the cargo upon a selected population of cells (Albanese et al., 2012). Given their successful use in drug delivery and imaging, nanoparticles have been recognized as powerful candidates for applications in reproductive medicine.

29.3 Current Understanding and Application of Nanoparticles in Reproductive Medicine

As nanoparticles represent highly versatile platforms for delivery of cargo due to their small size, customizable nature, biocompatibility, and the ability to carry large amounts of biological cargo, it is unsurprising that the application of these small-scale vehicles to biomedicine has increasingly captured the focus of researchers and clinicians over the last two decades. Novel possibilities for the use of nanoparticles are coming to light as a result of the wide range of nanomaterials available, and many are already proving advantageous in delivering diagnostic and therapeutic agents in a variety of conditions including cancer, inflammatory diseases, and conditions of the gastrointestinal tract (Ulbrich and Lamprecht 2009; Brakmane et al., 2012; Gmeiner and Ghosh, 2015). Given their growing success in these areas, research is already underway, which would allow the field of reproductive science to benefit from advancements in nanotechnology by the application of nanoparticles to diagnose and treat conditions such as reproductive cancers, endometriosis, uterine fibroids, and sexually transmitted diseases as discussed herein.

29.3.1 Reproductive Cancers

Reproductive cancers are the most commonly diagnosed malignancies and are currently treated by surgery, chemotherapy, radiation, or hormone therapy. However, survival rates remain low and it is therefore crucial that diagnosis occurs during early stages of the disease, as delays can limit treatment options for patients. The use of nanoparticles to deliver chemotherapeutics results in enhanced drug efficiency and also reduces systemic toxicity (Toporkiewicz et al., 2015). Several studies have already demonstrated the potential use of a variety of nanoparticles for reproductive medicine. Derivatives of polylactic acid (PLA), poly(lactic-co-glycolic) acid (PLGA), and magnetic iron have been shown to successfully deliver chemotherapeutic compounds into ovarian, endometrial, and prostate cancer cells (Dhar et al., 2008; Liang et al., 2011; Le Broc-Ryckewaert et al., 2013; Lee et al., 2013; Zhang et al., 2013). Nanoscale gold has also proved to be popular and a series of studies have explored the functionalization of this type of nanomaterial with different targeting ligands to facilitate the delivery of chemotherapeutic agents into ovarian, endometrial, and prostate cancer cells (Stern et al., 2008; Roa et al., 2009; Geng et al., 2011; de Oliveira et al., 2013).

EnGeneIC Delivery Vehicles (EDVs), which are bacterially derived nanospheres, have been shown to exhibit potential in the treatment of cancers involving the reproductive system. These are a novel class of easily synthesized organic nanoparticles with high loading capacities, versatility, and low toxicity. EDVs have been shown to maintain long-term drug stability and are capable of drug release upon endocytosis into target cells. Such engineered bacteria-derived nanospheres, loaded with doxorubicin, have been shown to successfully target epidermal growth factor receptors (EGRFs), thereby promoting the trophoblast-specific delivery of drugs with higher antitumor activity. These nanoparticles demonstrated more pronounced antitumor activity in mouse models of human choriocarcinoma compared to nontargeted doxorubicin-loaded EDVs and the free drug (Kaitu'u-Lino et al., 2013). EDVs are currently undergoing a phase I clinical trial.

Nanoparticles may also allow for multiple types of targeted intracellular payloads to be simultaneously delivered. Qi et al. (2011) used PLGA nanoparticles to simultaneously deliver the proapoptotic human PNAS-4 (hPNAS-4) gene along with cisplatin into mouse ovarian carcinoma cells, with significant antitumor activity observed (Qi et al., 2011). Another report showed the successful delivery of encapsulated (−)-epigallocatechin 3-gallate into PLA-PEG or polysaccharide nanoparticles into prostate cancer cells (Sanna et al., 2011). Long et al. (2013) further utilized a PEGylated liposomal formulation of quercetin to significantly suppress tumor growth, compared to free quercetin, for *in vitro* and *in vivo* models of conventional treatment-resistant human ovarian cancer (Long et al., 2013).

These studies have been able to demonstrate the potential of this technique, which could provide a very effective protocol for combined cancer therapy.

Nanoparticles may also be used to reverse the resistance of tumors to chemotherapy via intracellular delivery and the active targeting and activation of alternative mechanisms of cellular uptake. Metallofullerene nanoparticles encapsulating cisplatin have already been used in mouse models to overcome the resistance of human prostate cancer cell lines and to target ovarian tumor vessels (Liang et al., 2010; Winer et al., 2010). A further report showed that hyaluronic-acid-bound ietrozole polymeric (PLGA-PEG) nanoparticles restored drug sensitivity in letrozole-resistant cells on both *in vitro* and *in vivo* levels (Nair et al., 2014). These findings collectively demonstrate that nanoparticle delivery systems are an effective methodology with which to deliver common drugs for chemotherapy, and are capable of overcoming mechanisms of drug resistance. Indeed, targeted sensitizers represent a new form of antitumor therapy, which can be administered via nanoparticles. These structures are tolerable, effective, and safe, without comprising their antitumor potency. Such sensitizers can be made of magnetic iron oxide or gold and have been deployed during thermal ablation and radiotherapy in patients with prostate cancer (Stern et al., 2008; Johannsen et al., 2005; Nair et al., 2011). In other studies, thio-glucose-bound gold nanoparticles (Glu-GNPs) were shown to exhibit radio-sensitizing effects upon a human ovarian cancer cell line (SK-OV-3). Compared to irradiation alone, there was a 30% increase in the inhibition of cell proliferation in the presence of Glu-GNPs, , primarily, due to elevated production of reactive oxygen species (ROS) (Geng et al., 2011). Similar effects were reported by Zhang et al. (2012) who observed a sensitizing effect, thought to be mediated via an increase in apoptosis, when carbon nanotubes were used to deliver a chemotherapeutic agent (paclitaxel) into the human ovarian cancer cell line OVCAR3 (Zhang et al., 2012).

Nanoparticles are also attracting attention as targeted intracellular gene delivery vectors for gene therapy cancer treatments. The combination of high selectivity, noninvasive delivery, and stable expression of genetic constructs render nanoparticles particularly suitable for intracellular gene therapy. There are several examples of target gene knockdown documented in the literature already, resulting in pronounced antitumor effects (Sun et al., 2011; Steg et al., 2011; Zou et al., 2013). For example, Yang et al. (2013) used chitosan-based human papillomavirus (HPV) nanoparticles to deliver siRNAs against two oncoproteins with therapeutic potential for human cervical cancer, with very promising results (Yang et al., 2013). A further study demonstrated the successful application of magnetic iron oxide (Fe(3)O(4))-dextran-anti-β-human chorionic gonadotropin (hCG) nanoparticles as gene therapy in trophoblastic cancer tissue from an *in vivo* mouse model (Jingting et al., 2011). Therefore, the antitumor effects of nanoparticle-delivered gene therapy offer promising avenues for further development in the treatment of cancers of the reproductive system. Furthermore, the ability of nanoparticles to deliver multiple targeted payloads simultaneously and overcome mechanisms of drug resistance demonstrates their great potential in the treatment of reproductive cancers.

Nanoparticles have been used not only for the targeted delivery of anticancer therapeutic compounds providing improved pharmacodynamics and pharmacokinetics (Albanese et al., 2012) but also as contrast agents for imaging and diagnosis. Such success in the use of nanoparticles in cancer diagnosis and treatment has been achieved due to the ability of nanoparticles to differentiate between malignant and nonmalignant cells. By virtue of the enhanced permeability and retention (EPR) effects, nanoparticle extravasation offers the advantage of being able to target malignant cancer cells (Gmeiner and Ghosh, 2015). Such attributes have allowed the development of nanoparticles for use in the treatment of various cancers, including leukemia, and breast, liver, pancreatic, and non-small cell lung cancers, with many drug-conjugated nanoparticles for these cancers in clinical trials or in the market (Coelho et al., 2015). Extension of these treatments to other areas of medicine such as reproductive cancers is currently under development and appears promising.

29.3.2 Biomarkers

Biomarker screening approaches are being increasingly used to monitor and understand biological processes, and have been successful for some pathologies but not others. Nanoparticles are particularly promising as nanoscale biosensors, recognizing only specific antigens, proteins, nucleic acids, and reactive oxygen and nitrogen species (Medina-Sanchez et al., 2012; Perfezou et al., 2012; Kumar et al., 2013). Owing to their specific optical qualities, nanoparticles provide an excellent tool for such applications compared to traditional techniques. Indeed, superparamagnetic iron oxides and aptamer-conjugated gold nanoparticle formulations have already been utilized extensively in the treatment of prostate and ovarian cancer, in particular for the detection of small and lymph node tumors (Wang et al., 2008).

Nanoparticles therefore present an interesting alternative to common biomarker screening methods, and could have the potential to allow the detection of biomarkers that clinicians have otherwise been unable to find in current biological media such as serum. One such area in reproductive medicine where suitable biomarkers are proving difficult to detect is in the diagnosis of endometriosis.

29.3.3 Endometriosis

Endometriosis is a particularly deleterious condition, affecting 2%–10% of women of reproductive age, and involves the existence of endometrial-like tissue outside the uterine cavity (Dunselman et al., 2014). Currently, there are no suitably sensitive serum biomarkers for endometriosis and this, together with the known resolution limitations of imaging techniques, makes early noninvasive diagnosis of endometriosis highly challenging. Magnetic resonance imaging (MRI) is a commonly used method to detect and localize deep infiltrating endometriosis (Di Paola et al., 2015). Endometriosis is both a diagnostic and a therapeutic challenge as the associated symptoms are usually nonspecific and surgical means are normally required for confirmation of the condition. The need to identify noninvasive biomarkers for endometriosis is therefore apparent. However, in a systematic review assessing the clinical value of serum, plasma, and urine biomarkers spanning 25 years of literature, not a single biomarker or panel of biomarkers for endometriosis were identified as unequivocally clinically useful (May et al., 2010). It is therefore highly important that alternative approaches for the detection and treatment of endometriosis are exploited. Consequently, there is a significant emphasis on research that may reveal novel biomarkers for endometriosis (May et al., 2010). The great diversity of nanomaterials available at present allows for their application with a whole range of biologically and medically significant compounds, including those for antiangiogenesis, anti-inflammation, anticytokine activity, and gene therapy (Laschke et al., 2012; Chaudhury et al., 2013). It therefore seems likely that nanoparticles will be well suited for both the detection (via selective binding and contrast enhancement) and treatment (as a clinical payload delivery system) of endometriosis. Preliminary studies using nanoparticles for the management of endometriosis involved the use of ultrasmall superparamagnetic iron oxide nanoparticles, which were intravenously introduced into a rat model to obtain enhanced MRI contrast of ectopic uterine tissues (EUTs) via the high affinity of nanoparticles toward macrophages (Lee et al., 2012). Later, Zhao et al. (2012) intravenously administered lipid-grafted chitosan micelles, loaded with therapeutic gene encoding pigment epithelium-derived factor (PEDF), a multifunctional protein with antitumor and antiangiogenic properties, to rats with surgically induced endometriosis (Zhao et al., 2012). Compared to placebo, this study reported significant reductions in ectopic lesion volume along with growth inhibition of endometrioid cysts (Zhao et al., 2012). In another study, Chaudhury et al. (2013) reported a reduction in the levels of ROS and angiogenic factors compared to an active control and placebo when cerium oxide nanoparticles (nanoceria) were used in a mouse model with induced endometriosis. Nanoceria also reduced the density of endometrial glands and microvessels, an observation that was indicative of disease regression (Chaudhury et al., 2013). Most recently, Zhang and colleagues detected endometriotic lesions in rat models using magnetic oxide nanoparticles, which had been modified with hyaluronic acid (HA-Fe$_3$O$_4$ NPs) (Zhang et al., 2014), demonstrating the potential uses of nanoparticles in the diagnosis of endometriosis.

The pressing need for the development of novel biomarkers for endometriosis is indicated by the establishment of the Endometriosis Phenome and Biobanking Harmonization Project by the World Endometriosis Research Foundation. This worldwide project aims to identify potential diagnostic biomarkers for endometriosis by standardizing the way in which clinics collect blood and other fluid samples from women both with and without the condition to enable inter-center comparisons and collaborations (Rahmioglu et al., 2014). The current promising research on the use of nanoparticles in endometriosis therefore contributes to the drive to identify the much needed endometriotic biomarkers for both diagnosis and treatment, hopefully leading to quicker diagnoses and the alleviation of suffering in women with the condition.

29.3.4 Uterine Fibroids

Uterine leiomyomas, known more simply as uterine fibroids, are the most common types of pelvic tumor in women of reproductive age. Current treatments require invasive surgery but it is likely that nanoparticles could lead to an improvement in the selectivity and efficacy of antitumor cytokine delivery, a surgically minimal treatment. Nanogold particles conjugated with a tumor necrosis factor alpha (TNF-α) have already been used to evaluate the effect of this cryo-adjuvant during cryosurgery for uterine leiomyoma in a mouse model (Jiang and Bischof, 2010). Furthermore, PLL-PLGA nanoparticles were utilized to deliver an antitumor and antiangiogenic biologically-active metabolite of oestradiol (2-methoxyoestradiol) into a human leiomyoma cell line (huLM) as an alternative to hysterectomy (Ali et al., 2013). Both studies reported an improved antitumor activity compared to the free molecule, indicating the potential use of nanoparticles in the delivery of antitumor agents to uterine fibroids.

29.3.5 Sexually Transmitted Diseases

Sexually transmitted diseases can be extremely resistant to antibacterial treatment and can rapidly lead to tubal infertility. Consequently, it is very important to develop highly effective drug delivery systems in order to improve treatment efficiency. Conjugating antibiotics using neutral PAMAM dendrimer-mediated transfection or gene knockdown in cultures has been shown to specifically target affected cells, thus promoting antimicrobial activity without promoting further infection (Mishra et al., 2011, 2012). Moreover, a principal component of experimental subunit vaccines against chlamydial infection is the membrane protein (MOMP) of *Chlamydia trachomatis*. Biodegradable PLGA nanoparticles have been used to improve the delivery of this protein, thus representing an important step in the search for an effective antichlamydia immunization (Taha et al., 2012; Fairley et al., 2013; Dixit et al., 2014). It is possible therefore that nanoparticles could be successfully used to treat sexually transmitted diseases in the form of nanovaccines, with potential for this to be extended to other areas of reproductive science.

29.4 Nanoparticle-Mediated Delivery to Gametes and Embryos

The continuous discovery, development, and application of engineered nanoparticles in a wide variety of disciplines have understandably raised concerns regarding the potential toxicity of human and environmental exposure. In order for nanotechnology to be successfully applied to reproductive medicine, and indeed, investigative research into the molecular mechanisms underlying human infertility, it is imperative to address the potential impact of nanoparticles upon the viability and function of gametes and embryonic development.

29.4.1 Impact of Nanoparticles upon Oocytes, Embryo Development, and Implantation

The potential impact of nanoparticles upon oocytes, embryo development, and implantation has received only scant attention from researchers thus far. Early work focused upon quantum dots (QDs), semiconductors that are extensively used in medical imaging. Hsieh et al. (2009) demonstrated that short exposure times to cadmium selenide (Cdse)-core QDs exerted detrimental effects upon oocyte maturation, fertilization, and embryogenesis. Using a technique referred to as "*in vitro* maturation," a process in which immature oocytes are aspirated from ovaries and matured *in vitro* for 24 h prior to fertilization. The authors showed that levels of oocyte maturation, blastocyst development, and implantation rate were reduced in the presence of (Cdse)-core QDs. Interestingly, zinc sulfite (ZnS)-coated CdSe QDs had no effect, highlighting the beneficial effect of ZnS, which is able to block the mass transport of cytotoxic Cd^{2+} ions (Hsieh et al., 2009).

Further work examining the effect of nanoparticles on embryos of various species has yielded mixed results. Dubertret et al. (2002) demonstrated an injection of QDs, encapsulated in phospholipid micelles to be nontoxic to *Xenopus* embryos. In contrast, Fynewever et al. (2007) injected polyacrylonitrile nanoparticles into mouse embryos for tagging purposes and showed that this resulted in reduced embryonic development. However, an incubation of mouse embryos with polystyrene-based nanoparticles had no effect (Fynewever et al., 2007), indicating that type of nanoparticle and method of delivery may be key factors in determining toxicity. In a subsequent study, Zielinska et al. (2011) investigated the administration of gold nanoparticles, complexed with heparan-sulfate, to stimulate muscle development in chick embryos, without observing any detrimental effects (Zielinska et al., 2011). Furthermore, Tseng et al. (2013) successfully injected gelatin nanoparticles into chick embryos and monitored the temporal expression of an enhanced green fluorescent protein (*EGFP*) reporter gene for 4 days postinjection, without any adverse consequences upon development (Tseng et al., 2013).

While our present understanding suggests that nanoparticles could be very useful as gene or protein delivery systems for gametes and embryos, there are still many questions left unanswered. One of the main concerns raised by exposing gametes and embryos to nanoparticles is the effect that this exerts upon embryonic development in both the short and the long term. Localization patterns and persistence of nanoparticles in embryos are areas yet to be fully addressed, as well as methods of nanoparticle degradation or exocytosis. While many studies in this area have yielded promising results in the lack of detrimental effects to gametes and embryos, disparity still remains between outcomes, highlighting the requirement for further research in this area using new methods.

29.4.2 Impact of Nanoparticles upon Sperm and Testicular Tissue

Until recently, very little was known with regard to the intentional application of nano-carriers to gametes (sperm and oocytes) and embryos. Makhluf et al. (2008) first exposed bovine sperm to iron oxide nanoparticles conjugated with an antiprotein kinase C (PKC) α antibody. Particles accumulated in the sperm head and antibody activity was maintained, as evidenced by cellular antigen binding. This highlighted the ability of nanoparticles to deliver functional cargo into sperm without adverse effects (Makhluf et al., 2008). In contrast, silver and gold nanoparticles have been shown to compromise sperm DNA integrity and reduce sperm motility (Wiwanitkit et al., 2009; Asare et al., 2012; Taylor et al., 2012). However, recent research in our own laboratory demonstrated that mesoporous silica nanoparticles (MSNPs), functionalized with either polyethileneimine (PEI) or aminoprop-yltriethoxysilane (APTES) surface coatings and loaded with lamin A/C siRNA or mCherry protein, could attach to the surface of boar sperm in a time- and dose-dependent manner (Figure 29.2) (Barkalina et al., 2014a). Most recently, our laboratory also demonstrated that the use of MSNPs associated with an established cell-penetrating peptide (known as C105Y) resulted in increased attraction of the MSNPs toward sperm, thus reducing the time required for association to take place (Barkalina et al., 2015). The studies taking place in our laboratory were the first to use MSNPs in mammalian sperm as a potential delivery method and detected no adverse effects upon sperm motility and viability (Jones et al., 2013; Barkalina et al., 2015b).

In other research, Barkhordari et al. (2013) evaluated the cytotoxicity of zinc oxide nanoparticles upon human sperm viability and demonstrated that cell death was more prominent after 180-min exposure to ZnO nanoparticles and was entirely dose dependent (Barkhordari et al., 2013). In a subsequent report, Yiosungnern et al. (2015) demonstrated that silver nanoparticles (AgNPs) internalized in mouse sperm repressed viability and the acrosome reaction in a dose-dependent manner. Furthermore, these sperm exhibited

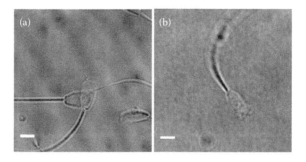

FIGURE 29.2
Association of loaded MSNPs with sperm. (a) Lamin A/C siRNA-loaded MSNPs; (b) mCherry-loaded MSNPs. The density of sperm coating with lamin A/C siRNA-loaded MSNPs was lower, compared to mCherry-loaded MSNPs. Scalebar = 5 μm. (Reproduced with permission from Barkalina N et al. *Nanomedicine.* 2014a; 10:921–938.)

increased abnormalities in both morphology and mitochondrial copy number, resulting in lower rates of fertilization and embryonic development. The proportion of unfertilized oocytes increased following *in vitro* fertilization (IVF) (in which live sperm is simply incubated with oocytes) with sperm exposed to high concentrations of AgNPs, while the rate of blastocyst formation fell.

The spermatogenic cycle takes ~35 days in mice, followed by a 5–10-day period in the epididymis for sperm maturation prior to the acquisition of fertilizing ability. Smith et al. (2015) used the mouse model to investigate the effects of titanium dioxide nanoparticles (ATPNs) upon spermatogenesis (in the testis) and epididymal maturation. Results showed that the incidence of sperm defects was higher after shorter rather than longer exposure times to ATPNs, indicating that such effects were transient. Indeed, sperm showed abnormal flagellae, an excess of residual cytoplasm, and unreacted acrosomes during epididymal maturation, all of which could potentially influence fertility. The observed difference between exposure times in this study was particularly interesting, and suggested that deleterious effects can arise from a variety of factors in unpredictable ways (Smith et al., 2015). For example, earlier experiments in rats showed that single injections of small (20 nm) AgNPs were much more toxic than larger (200 nm) AgNPs and that the smaller particles caused more DNA damage in sperm (Gromadzka-ostrowska et al., 2012). The authors observed that these results were due to the smaller particles being able to enter sperm cells in an easier and swifter manner than the larger particles. Testosterone and dihydrosterone, important testicular hormones, were also abnormally low, indicative of deficiency in testicular function, reduced sperm count, and an increased frequency of abnormal sperm (Gromadzka-ostrowska et al., 2012).

As research into this area is still in its infancy, there appears to be disparity between studies examining the effects of nanoparticles on sperm and testicular tissue. It seems that many factors influence the outcome including the type of nanoparticle used and the length of exposure. However, encouraging results come from studies demonstrating that exposure to nanoparticles produces no detrimental effects in sperm. In particular, the lack of adverse effects upon exposure of sperm to silica-based nanoparticles, which are generally regarded as biocompatible (Tarn et al., 2013), highlights the potential of nanoparticle-mediated delivery systems to sperm and testicular tissue. Nevertheless, extensive research into this area is required to further address questions of safety and efficacy.

29.4.3 Implications for Sperm-Mediated Gene Transfer

Transgenic animals represent a revolutionary research model for the study of development, human diseases, comparative genomics, and xenotransplantation. A range of alternative transfer methods, which utilizes mammalian sperm as gene carriers, have been developed to overcome the numerous limitations associated with conventional transgenic approaches, such as pronuclear microinjection or the injection of transformed embryonic stem cells into the inner cell mass (Coward et al., 2007). These include sperm-mediated gene transfer (SMGT), testis-mediated gene transfer (TMGT), and the use of powerful viruses (Parrington et al., 2011). SMGT whereby the sperm is used as a natural vector to carry exogenous DNA into oocytes has produced mixed results (Figure 29.3). Lavitrano et al. (1989) were the first to show that mouse epididymal sperm incubated with plasmid DNA could take up this DNA and transfer it into an oocyte, producing transgenic offspring. These findings have been repeated in various other species including zebrafish (Khoo 2000) and pig (Lavitrano et al., 2002); however, other groups have failed to find the same SMGT (Brinster et al., 1989). Genetic modification of spermatogenic cells in the testes

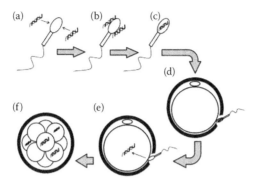

FIGURE 29.3
Sperm-mediated gene and compound transfer. (a–c) Sperm spontaneously bind, internalize, and incorporate exogenous DNA into the genome upon incubation *in vitro*. (d–e) The construct is subsequently delivered into the oocytes at the time of fertilization. (f) Transgenic/mosaic embryos are produced. (Reproduced with permission from Barkalina N et al. *Theriogenology*. 2015c. pii: S0093-691X(15)00289-7.)

using TMGT is an alternative method for introducing transgenes into the sperm. Coward et al. (2006) injected phospholipase C ζ (PLCζ)-EYFP (enhanced yellow fluorescent protein) constructs into the rete testis of mice, using electroporation to allow the DNA to enter the cells. Results demonstrated that not only was the gene expressed in spermatogonia in the testes, but fluorescence was also detected in the head and midpiece of a proportion of mature epididymal sperm, indicating for the first time that *in vivo* gene transfer into the testis could be used to express a fluorescently labeled recombinant sperm-specific protein (Coward et al., 2006). However, while such advancements are exciting, the efficiency of transfer and the level of transgene expression remains disappointingly low (Coward et al., 2007). Nanoparticles have therefore been proposed as alternative, more efficient transgene carriers (Barkalina et al., 2014a) (Table 29.1). However, only scant attention has been paid to this exciting prospect thus far (Table 29.2).

For example, Kim et al. (2010b) demonstrated successful exogenous DNA transfection into boar sperm using magnetic nanoparticles, which resulted in the production of transgenic embryos as detected by the expression of exogenous green fluorescent protein (GFP)

TABLE 29.1

Available Delivery Tools for the Transfer of Molecular Compounds into Gametes

	Electroporation	Viral Vectors	Chemical Reagents	Nanomaterials
Damage to cell membrane	+	−	+	−
Mechanism of uptake	Passage through membrane openings	Receptor-mediated energy dependent	Passage through membrane openings	Mediated/nonmediated energy dependent or independent
Potential for artificial targeting	−	+ (vector engineering)	−	+++ (functionalization)
Risk of integration into the host genome	−	+	−	−
Risk of host infection	−	+	−	−
Preservation of cell function after cargo transfer	Poor	Possible	Poor	Yes

TABLE 29.2

Nanoparticle-Mediated Delivery into Gametes and Intracellular Cell-Labeling *In Vitro*: Experimental Studies in Animal Models

Study	Nanomaterial	Application
Makhluf et al. (2008)	Polyvinylalcohol-coated magnetic iron oxide NPs (Fe_3O_4)	Proof-of-principle transfer of antiprotein kinase C-antibody into sperm
Kim et al. (2010)	Magnetic NPs (commercial agent)	Facilitation of SMGT
Campos et al. (2011a)	Nanopolymer (commercial agent)	Facilitation of SMGT (NanoSMGT)
Campos et al. (2011b)	Nanopolymer (commercial agent) and halloysite clay nanotubes	Facilitation of SMGT (NanoSMGT)
Barchanski et al. (2015)	Nanogold	Proof-of-principle investigation of the potential to label the specific DNA sequences in viable sperm

at the morula stage. More recently, bovine sperm was transfected with exogenous DNA using a nanopolymer and halloysite clay nanotubes as transfectants, in a method known as Nano-SMGT, with no reported effect upon motility or viability (Campos et al., 2011a, b). Furthermore, GFP-conjugated chitosan nanoparticles were successfully used for *in utero* gene delivery into mouse embryos (Yang et al., 2011). One particularly exciting application of such technology is in the field of human fertility treatment. Despite advancements in the field of assisted reproductive technology (ART), a notable proportion of infertility cases are still classified as unexplained, with the underlying causes undetermined. Due to the recognition of nanoparticles as promising multifunctional gene delivery vehicles, there is growing interest in their use in ART as potential diagnostic or therapeutic agents in the clinic, and as a way of investigating, inhibiting, or augmenting the molecular pathways underlying infertile phenotypes.

29.5 Conclusions

Nanoparticles are rapidly proving to be a reliable and efficient platform with which to deliver a range of biological cargoes, including drugs, proteins, and siRNA into target cells and tissues. Consequently, this technology has provided clinicians with a powerful range of exciting and efficient methods for the diagnosis and therapy of numerous pathologies, and has provided scientists with elegant new platforms for investigative research (Richter-Dahlfors and Kjall, 2011). Applying this technology to reproductive biomedicine is likely to assist in the elucidation of mechanisms that underlie unexplained infertile states, and ultimately provide novel options for diagnosis and therapy in the clinic. However, significant research efforts are still a necessity if such goals are to be reached, particularly with regard to specific patterns of internalization and localization, the undocking and functional activation of molecular cargo, the risk of degradation, and the potential effects upon embryo viability and development.

Recently, significant attention has been paid to naturally occurring nanoparticles known as exosomes, which are nanosized vesicles released by a number of cell types into their microenvironment. These exosomes are enclosed by a phospholipid bilayer allowing the particles to carry a wide range of biological material including proteins, cytokines, lipids, and RNAs. They therefore play an essential role in mediating cell communication during fundamental cellular processes such as cell proliferation, apoptosis, tissue repair, and angiogenesis (Barkalina et al. 2015a). Interest in exosomes has led to the realization that they can be purified with relatively straightforward protocols, loaded with molecular cargo, and fluorescently labeled (Nazarenko et al., 2013; Takashi et al., 2013; Tian et al., 2014). Moreover, cells can be engineered to secrete modified exosomes that express specific surface components for targeting, or carry particular cargo (Mizrak et al., 2013; Ohno et al., 2013). As a result of these findings, exosomes have been successfully applied to deliver anticancer therapies (Mizrak et al., 2013; Ohno et al., 2013), and as prototype and experimental therapies for inflammatory and neurodegenerative diseases (Sun et al., 2010; Cooper et al., 2014). A next important step therefore seems to be the application of this technology to reproductive science and medicine. Given that one of the overriding concerns in the application of nanomedicine to reproductive science is the way in which nanoparticles may affect gametes and embryos, it seems logical that naturally derived "nanoplatforms" could be attractive alternatives given their biodegradability and "natural" origins.

Overall, many advancements have been made in nanomedicine over the past few decades with some of these successfully applied to reproductive medicine, and many more with the potential to revolutionize both study and clinical practice in the field. Much research in this area is still required; however, findings thus far indicate a promising future for this application, with many novel and interesting possibilities for the investigation, diagnosis, and treatment of reproductive conditions.

Acknowledgments

The authors wish to thank Miss Charis Charalambous, Dr. Junaid Kashir, and Dr. Helen Townley, for their relative contributions to the nanotechnology research carried out in our laboratory. Our research has been funded by an ESPRC Pathways to Impact Grant and by the Nuffield Department of Obstetrics and Gynaecology (University of Oxford). NB is funded by a Clarendon Scholarship (University of Oxford) and by a Cyril and Phyllis Long Scholarship (Queens College, University of Oxford). LD is also funded by a Clarendon Scholarship. Our research was recently recognized by the Cognizure Publication Corporation in the form of a Nanoscience Research Leadership Award to KC.

References

Albanese A, Tang PS, Chan WCW. The effect of nanoparticle size, shape, and surface chemistry on biological systems. *Annu. Rev. Biomed. Eng.* 2012;14:1–16.

Ali H, Kilic G, Vincent K, Motamedi M, Rytting E. Nanomedicine for uterine leiomyoma therapy. *Ther. Deliv.* 2013;4:161–175.

Asare N, Instanes C, Sandberg WJ, Refsnes M, Schwarze P, Kruszewski M, Brunborg G. Cytotoxic and genotoxic effects of silver nanoparticles in testicular cells. *Toxicology.* 2012;291:65–72.

Barchanski A, Taylor U, Sajti CL, Gamrad L, Kues, WA, Rath D, Barcikowski S. Bioconjugated gold nanoparticles penetrate into spermatozoa depending on plasma membrane status. *J. Biomed. Nanotechnol.* 2015;11:1597–607.

Barkalina N, Charalambous C, Jones C, Coward K. Nanotechnology in reproductive medicine: Emerging applications of nanomaterials. *Nanomedicine.* 2014a;10:921–938.

Barkalina N, Jones C, Coward K. Mesoporous silica nanoparticles: A potential targeted delivery vector for reproductive biology? *Nanomedicine.* 2014b;9:557–560.

Barkalina N, Jones C, Coward K. Nanomedicine and mammalian sperm: Lessons from the porcine model. *Theriogenology.* 2015c; 85:74–82.

Barkalina N, Jones C, Kashir J, Coote S, Huang X, Morrison R, Townley H, Coward K. Effects of mesoporous silica nanoparticles upon the function of mammalian sperm *in vitro. Nanomedicine.* 2014c;10:859–870.

Barkalina N, Jones C, Townley H, Coward K. Functionalization of mesoporous silica nanoparticles with a cell-penetrating peptide to target mammalian sperm *in vitro. Nanomedicine.* 2015b;10:1539–53.

Barkalina N, Jones C, Wood MJ, Coward K. Extracellular vesicle-mediated delivery of molecular compounds into gametes and embryos: Learning from nature. *Hum. Reprod. Update.* 2015a;21:627–39.

Barkhordari A, Hekmatimoghaddam S, Jebali A, Khalili MA, Talebi A, Noorani M. Effect of zinc oxide nanoparticles on viability of human spermatozoa. *Iran. J. Reprod. Med.* 2013;11(9):767–771.

Brakmane G, Winslet M, Seifalian AM. Systematic review: The applications of nanotechnology to gastroenterology. *Aliment. Pharmacol. Ther.* 2012;36(3):213–221.

Brinster RL, Sandgren EP, Behringer RR, Palmiter RD. No simple solution for making transgenic mice. *Cell.* 1989;59:239–241.

Campos VF, Komninou ER, Urtiaga G, de Leon PM, Seixas FK, Dellagostin OA, Deschamps JC, Collares T. NanoSMGT: Transfection of exogenous DNA on sex-sorted bovine sperm using nanopolymer. *Theriogenology.* 2011b;75:1476–1481.

Campos VF, de Leon PMM, Komninou ER, Dellagostin OA, Deschamps JC, Seixas FK, Collares T. NanoSMGT: Transgene transmission into bovine embryos using halloysite clay nanotubes or nanopolymer to improve transfection efficiency. *Theriogenology.* 2011a;76:1552–1560.

Chaudhury K, Babu KN, Singh AK, Das S, Kumar A, Seal S. Mitigation of endometriosis using regenerative cerium oxide nanoparticles. *Nanomedicine.* 2013;9:439–448.

Chithrani BD, Chan WC. Elucidating the mechanism of cellular uptake and removal of protein-coated gold nanoparticles of different sizes and shapes. *Nano Lett.* 2007;7:1542–1550.

Coelho SC, Pereira MC, Juzeniene A, Juzenas P, Ceolho MAN. Supramolecular nanoscale assemblies for cancer diagnosis and therapy. *J. Control. Release.* 2015;213:152–167.

Cooper JM, Wiklander PB, Nordin JZ, Al-Shawi R, Wood MJ, Vithlani M, Schapira AH, Simons JP, El-Andaloussi S, Alvarez-Erviti L. Systemic exosomal siRNA delivery reduced alpha-synuclein aggregates in brains of transgenic mice. *Mov. Disord.* 2014;29:1476–1485.

Coward K, Kubota H, Hibbitt O, McIlhinney J, Kohri K, Parrington J. Expression of a fluorescent recombinant form of sperm protein phospholipase C zeta in mouse epididymal sperm by *in vivo* gene transfer into the tesis. *Fertil. Steril.* 2006;85:1281–1289.

Coward K, Kubota H, Parrington J. In vivo gene transfer into testis and sperm: Developments and future application. *Arch. Androl.* 2007;53:187–197.

Dhar S, Gu FX, Langer R, Farokhzad OC, Lippard SJ. Targeted delivery of cisplatin to prostate cancer cells by aptamer functionalized Pt(IV) prodrug-PLGA-PEG nanoparticles. *Proc. Natl. Acad. Sci. USA.* 2008;105:17356–17361.

Di Paola V, Manfredi R, Castelli F, Negrelli R, Mehrabi S, Pozzi Mucelli R. Detection and localization of deep endometriosis by means of MRI and correlation with the ENZIAN score. *Eur. J. Radiol.* 2015;84(4):568–574.

Dixit S, Singh SR, Yilma AN, Agee RD II, Taha M, Dennis VA. Poly(lactic acid)-poly(ethylene glycol) nanoparticles provide sustained delivery of a *Chlamydia trachomatis* recombinant MOMP peptide and potentiate systemic adaptive immune responses in mice. *Nanomedicine.* 2014;10(6):1311–1321.

Dubertret B, Skourides P, Norris DJ, Noireaux V, Brivanlou AH, Libchaber A. In vivo imaging of quantum dots encapsulated in phospholipid micelles. *Science.* 2002;298(5599):1759–1762.

Dunselman GA, Vermeulen N, Becker C, Calhaz-Jorge C, D'Hooghe T, De Bie B et al. ESHRE guideline: Management of women with endometriosis. *Hum. Reprod.* 2014;29(3):400–412.

Fairley SJ, Singh SR, Yilma AN, Waffo AB, Subbarayan P, Dixit S, Taha MA, Cambridge CD, Dennis VA. *Chlamydia trachomatis* recombinant MOMP encapsulated in PLGA nanoparticles triggers primarily T helper 1 cellular and antibody immune responses in mice: A desirable candidate nanovaccine. *Int. J. Nanomed.* 2013;8:2085–2099.

Fynewever TL, Agcaoili ES, Jacobson JD, Patton WC, Chan PJ. In vitro tagging of embryos with nanoparticles. *J. Assist. Reprod. Genet.* 2007;24(2–3):61–65.

Geng F, Song K, Xing JZ, Yuan C, Yan S, Yang Q et al. Thio-glucose bound gold nanoparticles enhance radio-cytotoxic targeting of ovarian cancer. *Nanotechnology.* 2011;22:285101.

Gmeiner WH, Ghosh S. Nanotechnology for cancer treatment. *Nanotechnol. Rev.* 2015;3(2):111–122.

Gratton SE, Ropp PA, Pohlhaus PD, Luft JC, Madden VJ, Napier ME, DeSimone JM. The effect of particle design on cellular internalization pathways. *Proc. Natl. Acad. Sci. USA.* 2008;105:11613–11618.

Gromadzka-Ostrowska J, Dziendzikowska K, Lankoff A, Dobrzyńska M, Instanes C, Brunborg G, Gajowik A, Radzikowska J, Wojewódzka M, Kruszewski M. Silver nanoparticles effects on epididymal sperm in rats. *Toxicol. Lett.* 2012;214(3):251–258.

Gu Z, Biswas A, Zhao M, Tang Y. Tailoring nanocarriers for intracellular protein delivery. *Chem. Soc. Rev.* 2011;40(7):3638–3655.

Horie M, Kato H, Fujita K, Endoh S, Iwahashi H. *In vitro* evaluation of cellular response induced by manufactured nanoparticles. *Chem. Res. Toxicol.* 2012;25(3):605–619.

Hsieh MS, Shiao NH, Chan WH. Cytotoxic effects of CdSe quantum dots on maturation of mouse oocytes, fertilization, and fetal development. *Int. J. Mol. Sci.* 2009;10(5):2122–2135.

Jain KK. Advances in the field of nanooncology. *BMC Med.* 2010;8:83. doi: 10.1186/1741-7015-8-83.

Jiang J, Bischof J. Effect of timing, dose and interstitial versus nanoparticle delivery of tumor necrosis factor alpha in combinatorial adjuvant cryosurgery treatment of ELT-3 uterine fibroid tumor. *Cryo Lett.* 2010;31:50–62.

Jingting C, Huining L, Yi Z. Preparation and characterization of magnetic nanoparticles containing Fe(3)O(4)-dextran-anti-beta-human chorionic gonadotropin, a new generation choriocarcinoma-specific gene vector. *Int. J. Nanomed.* 2011;6:285–294.

Johannsen M, Gneveckow U, Eckelt L, Feussner A, Waldofner N, Scholz R et al. Clinical hyperthermia of prostate cancer using magnetic nanoparticles: Presentation of a new interstitial technique. *Int. J. Hyperther.* 2005;21:637–647.

Jones S, Lukanowska M, Suhorutsenko J, Oxenham S, Barratt C, Publicover S et al. Intracellular translocation and differential accumulation of cell-penetrating peptides in bovine spermatozoa: Evaluation of efficient delivery vectors that do not compromise human sperm motility. *Hum. Reprod.* 2013;28:1874–1889.

Kaitu'u-Lino TJ, Pattison S, Ye L, Tuohey L, Sluka P, MacDiarmid J et al. Targeted nanoparticle delivery of doxorubicin into placental tissues to treat ectopic pregnancies. *Endocrinology.* 2013;154:911–919.

Khoo HW. Sperm-mediated gene transfer studies on zebrafish in Singapore. *Mol. Reprod. Dev.* 2000;56:278–280.

Kim D, Jeong YY, Jon S. A drug-loaded aptamer-gold nanoparticle bioconjugate for combined CT imaging and therapy of prostate cancer. *ACS Nano.* 2010a;4:3689–3696.

Kim TS, Lee SH, Gang GT, Lee YS, Kim SU, Koo DB et al. Exogenous DNA uptake of boar spermatozoa by a magnetic nanoparticle vector system. *Reprod. Domest. Anim.* 2010b;45:e201–e206.

Koopmans RJ, Aggeli A. Nanobiotechnology—quo vadis? *Curr. Opin. Microbiol.* 2010;13:327–334.

Kumar S, Rhim WK, Lim DK, Nam JM. Glutathione dimerization-based plasmonic nanoswitch for biodetection of reactive oxygen and nitrogen species. *ACS Nano.* 2013;7:2221–2230.

Laschke MW, Menger MD. Anti-angiogenic treatment strategies for the therapy of endometriosis. *Hum. Reprod. Update.* 2012;18:682–702.

Lavitrano M, Bacci ML, Forni M, Lazzereschi D, Di Stefano C, Fioretti D et al. Efficient production by sperm-mediated gene transfer of human decay accelerating factor (hDAF) transgenic pigs for xenotransplantation. *Proc. Natl. Acad. Sci. USA.* 2002;99:14230–14235.

Lavitrano M, Camaioni A, Fazio VM, Dolci S, Farace MG, Spadafora C. Sperm cells as vectors for introducing foreign DNA into eggs: Genetic transformation of mice. *Cell.* 1989;57:717–723.

Le Broc-Ryckewaert D, Carpentier R, Lipka E, Daher S, Vaccher C, Betbeder D, Furman C. Development of innovative paclitaxel-loaded small PLGA nanoparticles: Study of their antiproliferative activity and their molecular interactions on prostatic cancer cells. *Int. J. Pharm.* 2013;454(2):712–719.

Lee HJ, Lee HJ, Lee JM, Chang Y, Woo ST. Distribution and accumulation of Cy5.5-labeled thermally cross-linked superparamagnetic iron oxide nanoparticles in the tissues of ICR mice. *Magn. Reson. Imaging.* 2012;30:860–868.

Lee KJ, An JH, Chun JR, Chung KH, Park WY, Shin JS et al. In vitro analysis of the anti-cancer activity of mitoxantrone loaded on magnetic nanoparticles. *J. Biomed. Nanotechnol.* 2013;9:1071–1075.

Liang C, Yang Y, Ling Y, Huang Y, Li T, Li X. Improved therapeutic effect of folate-decorated PLGA-PEG nanoparticles for endometrial carcinoma. *Bioorg. Med. Chem.* 2011;19:4057–4066.

Liang XJ, Meng H, Wang Y, He H, Meng J, Lu J et al. Metallofullerene nanoparticles circumvent tumor resistance to cisplatin by reactivating endocytosis. *Proc. Natl. Acad. Sci. USA.* 2010;107:7449–4754.

Long QD, Xie Y, Huang YQ, Wu QJ, Zhang HC, Xiong SQ et al. Induction of apoptosis and inhibition of angiogenesis by PEGylated liposomal quercetin in both cisplatin-sensitive and cisplatin-resistant ovarian cancers. *J. Biomed. Nanotechnol.* 2013;9:965–975.

Makhluf SB, Abu-Mukh R, Rubinstein S, Breitbart H, Gedanken A. Modified PVA-Fe3O4 nanoparticles as protein carriers into sperm cells. *Small.* 2008;4(9):1453–1458.

May KE, Condiut-Hulbert SA, Villar J, Kirtley S, Kennedy SH, Becker CM. Peripheral biomarkers of endometriosis: A systematic review. *Hum. Reprod. Update.* 2010;16(6):651–674.

Medina-Sanchez M, Miserere S, Merkoci A. Nanomaterials and lab-on-a-chip technologies. *Lab Chip.* 2012;12:1932–1943.

Mishra MK, Gerard HC, Whittum-Hudson JA, Hudson AP, Kannan RM. Dendrimer-enabled modulation of gene expression in *Chlamydia trachomatis. Mol. Pharm.* 2012;9:413–442.

Mishra MK, Kotta K, Hali M, Wykes S, Gerard HC, Hudson AP et al. PAMAM dendrimer–azithromycin conjugate nanodevices for the treatment of *Chlamydia trachomatis* infections. *Nanomedicine.* 2011;7:935–944.

Mizrak A, Bolukbasi MF, Ozdener GB, Brenner GJ, Madlener S, Erkan EP, Strobel T, Breakfield XO, Saydam O. Genetically engineered microvesicles carrying suicide mRNA/protein inhibit schwannoma tumor growth. *Mol. Ther.* 2013;21:101–108.

Nair HB, Huffman S, Veerapaneni P, Kirma NB, Binkley P, Perla RP et al. Hyaluronic acid-bound letrozole nanoparticles restore sensitivity to letrozole-resistant xenograft tumors in mice. *J. Nanosci. Nanotechnol.* 2011;11:3789–3799.

Nazarenko I, Rupp AK, Altevogt P. Exosomes as a potential tool for a specific delivery of functional molecules. *Methods Mol. Biol.* 2013;1049:495–511.

Ohno S, Takanashi M, Sudo K, Ueda S, Ishikawa A, Matsuyama N et al. Systemically injected exosomes targeted to EGFR deliver antitumour microRNA to breast cancer cells. *Mol. Ther.* 2013;21:185–191.

de Oliveira R, Zhao P, Li N, de Santa Maria LC, Vergnaud J, Ruiz J et al. Synthesis and *in vitro* studies of gold nanoparticles loaded with docetaxel. *Int. J. Pharm.* 2013;454:703–711.

Parrington J, Coward K, Gadea J. Sperm and testis mediated DNA transfer as a means of gene therapy. *Syst. Biol. Reprod. Med.* 2011;57:35–42.

Pathak RK, Dhar S. A nanoparticle cocktail: Temporal release of predefined drug combinations. 2015. *J. Am. Chem. Soc.* [Epub ahead of print].

Perfezou M, Turner A, Merkoci A. Cancer detection using nanoparticle-based sensors. *Chem. Soc. Rev.* 2012;41:2606–2622.

Petros RA, DeSimone JM. Strategies in the design of nanoparticles for therapeutic applications. *Nat. Rev. Drug Discov.* 2010;9:615–627.

Qi X, Song X, Liu P, Yi T, Li S, Xie C et al. Antitumor effects of PLGA nanoparticles encapsulating the human PNAS-4 gene combined with cisplatin in ovarian cancer. *Oncol. Rep.* 2011;26:703–710.

Rahmioglu N, Fassbender A, Vitonis AF, Tworoger SS, Hummelshoj L, D'Hooghe TM, Adamson GD, Giudice LC, Becker CM, Zondervan KT, Missmer SA, WERF EPHect Working Group. World Endometriosis Research Foundation Endometriosis Phenome and Biobanking Harmonization Project: III. Fluid biospecimen collection, processing, and storage in endometriosis research. *Fertil. Steril.* 2014;102(5):1233–1243.

Richter-Dahlfors A., Kjall P. Special issue on Nanotechnologies: Emerging applications in biomedicine. *Biochim. Biophys. Acta.* 2011;(3):237–238.

Roa W, Zhang XJ, Guo LH, Shaw A, Hu XY, Xiong YP et al. Gold nanoparticle sensitize radiotherapy of prostate cancer cells by regulation of the cell cycle. *Nanotechnology.* 2009;20:375101.

Rosenholm JM, Mamaeva V, Sahlgren C, Linden M. Nanoparticles in targeted cancer therapy: Mesoporous silica nanoparticles entering preclinical development stage. *Nanomedicine (Lond.).* 2012;7:111–120.

Salata O. Applications of nanoparticles in biology and medicine. *J. Nanobiotechnol.* 2004;2(3):1–6.

Sanna V, Pintus G, Roggio AM, Punzoni S, Posadino AM, Arca A et al. Targeted biocompatible nanoparticles for the delivery of (−)-epigallocatechin 3-gallate to prostate cancer cells. *J. Med. Chem.* 2011;54:1321–1332.

Smith MA, Michael R, Aravindan RG, Dash S, Shah SI, Galileo DS, Martin-DeLeon PA. Anatase titanium dioxide nanoparticles in mice: Evidence for induced structural and functional sperm defects after short-, but not long-, term exposure. *Asian J. Androl.* 2015;17(2):261–268.

Solaro R. Targeted delivery of proteins by nanosized carriers. *J. Polym. Sci. Part A: Polym. Chem.* 2008;46(1):1–11.

Steg AD, Katre AA, Goodman B, Han HD, Nick AM, Stone RL et al. Targeting the notch ligand JAGGED1 in both tumor cells and stroma in ovarian cancer. *Clin. Cancer. Res.* 2011;17: 5674–5685.

Stern JM, Stanfield J, Kabbani W, Hsieh JT, Cadeddu JRA. Selective prostate cancer thermal ablation with laser activated gold nanoshells. *J. Urol.* 2008;179:748–753.

Sun C, Yi T, Song X, Li S, Qi X, Chen X et al. Efficient inhibition of ovarian cancer by short hairpin RNA targeting claudin-3. *Oncol. Rep.* 2011;26:193–200.

Sun D, Zhuang X, Xiang X, Liu Y, Zhang S, Liu C, Barnes S, Grizzle W, Miller D, Zhang HG. A novel nanoparticle drug delivery system: The anti-inflammatory activity of curcumin is enhanced when encapsulated in exosomes. *Mol. Ther.* 2010;18:1606–1614.

Taha MA, Singh SR, Dennis VA. Biodegradable PLGA85/15 nanoparticles as a delivery vehicle for *Chlamydia trachomatis* recombinant MOMP-187 peptide. *Nanotechnology.* 2012;23(32):325101.

Takahashi Y, Nishikawa M, Shinotsuka H, Matsui Y, Ohara S, Imai T, Takakura Y. Visualization and *in vivo* tracking of the exosomes of murine melanoma B16-BL6 cells in mice after intravenous injection. *J. Biotechnol.* 2013;165:77–84.

Tarn D, Ashley CE, Xue M, Carnes EC, Zink JI, Brinker CJ. Mesoporous silica nanoparticle nanocarriers: Biofunctionality and biocompatibility. *Acc. Chem. Res.* 2013;46(3):792–780.

Taylor U, Barchanski A, Kues W, Barcikowski, S, Rath D. Impact of metal nanoparticles on germ cell viability and functionality. *Reprod. Domest. Anim.* 2012;47(Suppl 4):359–368.

Tian YH, Li SP, Song J, Ji TJ, Zhu MT, Anderson GJ, Wei JY, Nie GJ. A doxorubicin delivery platform using engineered natural membrane vesicle exosomes for targeted tumor therapy. *Biomaterials.* 2014;35:2383–2390.

Toporkiewicz M, Meissner J, Matusewicz L, Czogalla A, Sikorski AF. Toward a magic or imaginary bullet? Ligands for drug targeting to cancer cells: Principles, hopes, and challenges. *Int. J. Nanomed.* 2015;10:1399–1414.

Tseng CL, Peng CL, Huang JY, Chen JC, Lin FH. Gelatin nanoparticles as gene carriers for transgenic chicken applications. *J. Biomater. Appl.* 2013;27(8)1055–1065.

Ulbrich W, Lamprecht A. Targeted drug-delivery approaches by nanoparticulate carriers in the therapy of inflammatory diseases. *J. R Soc. Interface.* 2010;7:S55–66.

Wang AZ, Bagalkot V, Vasilliou CC, Gu F, Alexis F, Zhang L et al. Superparamagnetic iron oxide nanoparticle–aptamer bioconjugates for combined prostate cancer imaging and therapy. *ChemMedChem.* 2008;3:1311–1315.

Winer I, Wang S, Lee YE, Fan W, Gong Y, Burgos-Ojeda D, Spahlinger G, Kopelman R, Buckanovich RJ. F3-targeted cisplatin-hydrogel nanoparticles as an effective therapeutic that targets both murine and human ovarian tumor endothelial cells *in vivo*. *Cancer Res.* 2010;70(21):8674–8683.

Wiwanitkit V, Sereemaspun A, Rojanathanes R. Effect of gold nanoparticles on spermatozoa: The first world report. *Fertil. Steril.* 2009;91:7–8.

Yang J, Li S, Guo F, Zhang W, Wang Y, Pan Y. Induction of apoptosis by chitosan/HPV16 E7 siRNA complexes in cervical cancer cells. *Mol. Med. Rep.* 2013;7:998–1002.

Yoisungnern T, Choi YJ, Woong Han J, Kang MH, Das J, Gurunathan S et al. Internalization of silver nanoparticles into mouse spermatozoa results in poor fertilization and compromised embryo development. *Sci. Rep.* 2015;5:11170.

Zhang HE, Jingchao L, Wenjie S, Yong H, Guofu Z, Mingwu S, Xiangyang S. Hyaluronic acid-modified magnetic iron oxide nanoparticles for MR imaging of surgically induced endometriosis model in rats. *PLoS One.* 2014;9(4):e94718.

Zhang W, Zhang D, Tan J, Cong H. Carbon nanotube exposure sensitize human ovarian cancer cells to paclitaxel. *J. Nanosci. Nanotechnol.* 2012;12:7211–7214.

Zhang X, Chen J, Kang Y, Hong S, Zheng Y, Sun H et al. Targeted paclitaxel nanoparticles modified with follicle-stimulating hormone beta 81-95 peptide show effective antitumor activity against ovarian carcinoma. *Int. J. Pharm.* 2013;453:498–505.

Zhao MD, Sun YM, Fu GF, Du YZ, Chen FY, Yuan H et al. Gene therapy of endometriosis introduced by polymeric micelles with glycolipid-like structure. *Biomaterials.* 2012;33:634–643.

Zielinska M., Sawosz E., Grodzik M., Wierzbicki M., Gromadka M., Hotowy A., Sawosz F., Lozicki A, Chwalibog A. Effect of heparan sulfate and gold nanoparticles on muscle development during embryogenesis. *Int. J. Nanomed.* 2011;6:3163–3172.

Zou L, Song X, Yi T, Li S, Deng H, Chen X et al. Administration of PLGA nanoparticles carrying shRNA against focal adhesion kinase and CD44 results in enhanced antitumor effects against ovarian cancer. *Cancer Gene Ther.* 2013;20:242–250.

30

Theranostic Nanoprobes for SERS Imaging and Photodynamic Therapy

Andrew M. Fales and Tuan Vo-Dinh

CONTENTS

30.1 Introduction

Surface-enhanced Raman scattering (SERS) labeled nanoparticles have shown great promise for molecular imaging applications. Their fingerprint-like spectra, with sharp, narrow peaks, are ideal for multiplex detection.[1,2] SERS tags are also highly photostable, and only require one excitation source to detect multiple labels. This is in contrast to fluorescent probes that have very broad emission, require specific excitation wavelengths, and often suffer from photobleaching. As such, SERS tags have received great interest for use in biolabeling applications.[2–18] When matching the absorption band of the Raman dye with the excitation source, surface-enhanced resonance Raman scattering (SERRS) is achieved, increasing the SERS signal by a couple orders of magnitude.[1] Gold nanoparticles can be used for creating SERS tags, which have been shown to be nontoxic and biocompatible.[19] These nanoparticles are readily modified through gold–thiol bonds, providing an easy way to functionalize the gold surface with targeting moieties. By combining SERS tags with some sort of therapeutic modality, a theranostic nanoparticle construct can be created, allowing for specific detection of a target by SERS detection, followed by treatment using the added form of therapy.

One such therapeutic modality that can be combined with SERS tags is photodynamic therapy (PDT).[20,21] A photosensitizer, excitation light, and oxygen are required for PDT.[22] When light of the appropriate wavelength is applied, the photosensitizer is excited to a higher electronic state. After intersystem crossing to the excited triplet state, the electrons of the photosensitizer can interact with molecular oxygen to produce reactive oxygen species (ROS), mainly singlet oxygen ($^1O_2$). These free radicals can then interact with cellular components, causing injury that leads to cell death.[23] From a practical standpoint, PDT often has limited efficacy due to poor solubility or inactivation of the photosensitizer in a biological environment.[24–27] To overcome this limitation, a drug carrier can be used to protect and deliver the photosensitizer to the desired area. Silica nanoparticles and various core–silica shell nanoparticles have been shown to be effective carriers for a range of different drug molecules.[23,25,28–37] Encapsulation of SERS tags in a silica shell does not impede detection of the SERS signal, while providing the nanoparticles with the drug carrying properties of silica results in a theranostic construct that is capable of SERS detection and PDT.[20,21]

A critical factor involved in the efficacy of nanodrug formulations is sufficient delivery to the target cells. Conjugation of nanoparticles with a cell-penetrating peptide, such as the HIV-1 transactivator of transcription (TAT) peptide, can greatly increase their intracellular accumulation.[38–44] By increasing the number of particles taken up per cell, the overall dose and/or incubation time can be decreased, reducing the risk of side effects.

30.2 Synthesis

While there are a variety of different nanoparticles that could be used for the SERS label, such as nanorods,[45,46] nanorattles,[47–50] and shell embedded dyes,[51–55] we have selected gold nanostars as the core for our theranostic construct. Gold nanostars offer many advantages over other types of nanoparticles for SERS applications. The surface plasmon resonance of the nanostars can easily be tuned in the NIR region, allowing for maximization of electromagnetic (EM) field enhancement by matching the plasmon to the SERS excitation source.[56] Each nanostar also contains multiple "hot-spots" of EM field enhancement at the tips of the branches, which allow for SERS enhancement in a nonaggregated state.[56] By not relying on the random process of aggregation to produce SERS enhancement, reproducibility is greatly improved.[57,58] All of these factors have contributed to the great interest that nanostars have received in recent years for use in SERS studies.

The gold nanostars were synthesized using a surfactant-less, seed-mediated method that was developed in our laboratory.[56] By not requiring any surfactants, we avoid potential issues with toxicity, for example, CTAB on nanorods,[19] and leave the particle surface ready for functionalization. A schematic depiction of the nanoplatform preparation is shown in Figure 30.1. The as-prepared gold nanostars are labeled with a Raman dye, DTDC in this case, prior to silica coating. The photosensitizer, protoporphyrin IX (PpIX), is loaded into the silica shell by simply adding it to the reaction mixture as the silica precursor, tetraethyl orthosilicate (TEOS), undergoes hydrolysis and then condensation onto the nanoparticle

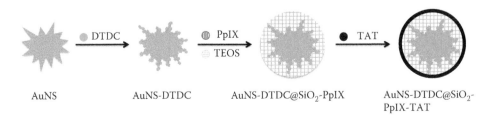

AuNS AuNS-DTDC AuNS-DTDC@SiO$_2$-PpIX AuNS-DTDC@SiO$_2$-PpIX-TAT

FIGURE 30.1
Schematic overview of the theranostic nanoprobe synthesis. Gold nanostars are labeled with a SERS dye, DTDC, prior to silica coating. The silica shell is grown using TEOS as the silica precursor in the presence of the PpIX photosensitizer, which becomes embedded in the silica matrix. The TAT peptide is then electrostatically adsorbed to the silica surface to create the final product.

surface. After the PpIX-embedded silica shell has been formed on the nanostars, the particle surface was modified with the TAT peptide through electrostatic interactions; the silica surface possesses a high negative charge, while the TAT peptide is positively charged.

30.3 Characterization

The synthesized particles were characterized using a variety of methods. Transmission electron microscopy (TEM) was used to verify the silica coating and to inspect the size and shape of the nanostars. A TEM micrograph of the silica-coated nanostars is shown in Figure 30.2a, clearly showing ~10-nm-thick silica shell on nanostars that are ~100 nm in diameter. The extinction spectrum of the gold nanostars exhibited a slight red shift after silica coating, which is to be expected as the refractive index at the particle surface is increased (Figure 30.2b). To verify that PpIX was loaded into the silica shell, the particles were illuminated with light of a wavelength in the excitation band of PpIX and the fluorescence emission spectrum was recorded (Figure 30.2b).

The SERS properties of the nanocomposite were studied on a Raman microscope utilizing a 633-nm HeNe laser for excitation. Figure 30.3 contains the SERS spectra from DTDC-labeled nanostars in solution, the theranostic construct in solution, and the theranostic construct after being taken up by a cell. As shown, there is a significant decrease in the SERS signal after the labeled gold nanostar is coated with silica. This is likely due to displacement of any DTDC molecules that are not bound tightly to the gold nanostar during the silica condensation process. After cellular uptake of the theranostic nanoparticles, the DTDC SERS signal is still readily observed (Figure 30.3).

Two-photon photoluminescence (TPL) imaging was used to verify nanoparticle uptake. We have previously shown that gold nanostars exhibit extremely high two-photon action cross sections with intense, broadband emission under excitation with a femtosecond

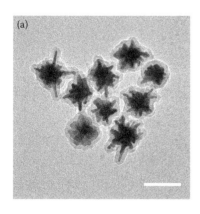
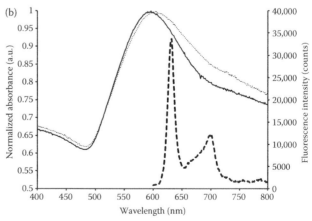

FIGURE 30.2
(a) TEM micrograph of the silica-coated gold nanostars. Scale bar is 100 nm. (b) Extinction spectra of the gold nanostars before (solid) and after (dotted) silica coating, and fluorescence emission, excited at 415 nm, from the PpIX-loaded particles (dashed).

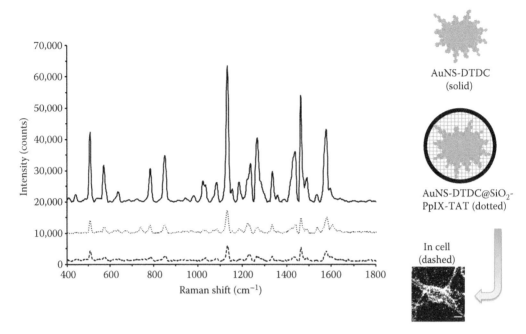

FIGURE 30.3
SERS spectra of DTDC-labeled gold nanostars in solution (solid), the complete theranostic construct in solution (dotted), and the theranostic construct after it has been taken up by a cell (dashed).

laser. This causes the nanostars to show up as white dots on the three-channel composite image from the two-photon microscope. As observed in Figure 30.4, the particles prepared with TAT have much greater accumulation within the cell under the same incubation conditions. Without TAT, there are a few nanoparticles scattered within the cell, whereas the TAT-coated particles are observed throughout the cytoplasm.

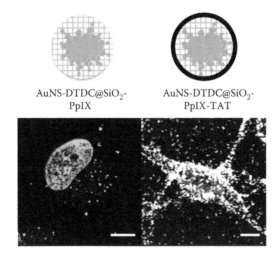

FIGURE 30.4
Two-photon photoluminescence imaging of cells that were incubated with the theranostic construct that was prepared with (right side) or without (left side) TAT. Scale bars are 10 μm.

30.4 SERS Imaging

The application of our theranostic construct for SERS detection was performed on fixed cells after incubation with the particles. SERS maps were created by scanning across the cell in 2 μm steps and recording a SERS spectrum at each point. The integrated intensity of the DTDC SERS peak was then used to create a heat-map that is overlaid on the brightfield image of the cell. Figure 30.5 contains representative SERS images of cells after incubation with the TAT-coated nanoprobes. Without TAT, there was not sufficient uptake of nanoprobes to produce SERS signal from within the cells under the same incubation conditions.

30.5 Photodynamic Therapy

The therapeutic aspect of our nanoprobe was demonstrated with an *in vitro* PDT study. Cells were incubated with particles prior to being exposed to UV light for activation of the photosensitizer. Particle samples prepared without PpIX or without TAT were used as controls. As shown in Figure 30.6, the theranostic nanoprobe was highly effective at destroying cells within the illuminated area. The control sample prepared without PpIX showed some loss of cell density in the irradiated area, possibly due to photothermal heating of the nanoparticles. The control that was prepared without TAT showed no discernable difference between the irradiated and the nonirradiated areas.

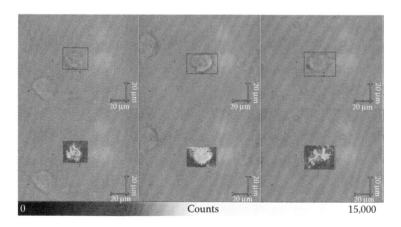

AuNS-DTDC@SiO$_2$-PpIX-TAT

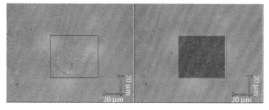

AuNS-DTDC@SiO$_2$-PpIX

FIGURE 30.5

Top: representative SERS maps of cells after incubation with the TAT-coated particles. Bottom: SERS map of a cell after incubation with particles that did not contain TAT, showing little to no SERS signal.

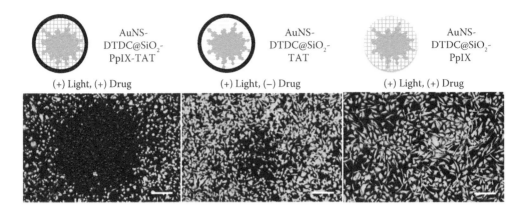

FIGURE 30.6

Live/dead cell staining after photodynamic therapy administration to cells incubated with the theranostic nanoprobe (left) and control particles (center, right). Scale bars are 250 μm.

30.6 Conclusion

In this chapter, we have reviewed the synthesis, characterization, and application of gold nanostar-based theranostic nanoprobes for SERS and PDT. Gold nanostars were selected for their optimal SERS properties and biocompatibility. Encapsulation of the SERS-labeled nanostars in silica provided the particles with drug-carrying capabilities, loading photosensitizer into the silica shell. The drug-loaded, SERS-labeled nanoprobes were shown to be effective for use in SERS imaging and PDT. Future work will investigate the use of these nanoplatforms for targeted, multiplexed SERS detection and photodynamic therapy.

References

1. Yuan, H.; Liu, Y.; Fales, A. M.; Li, Y. L.; Liu, J.; Vo-Dinh, T., Quantitative surface-enhanced resonant Raman scattering multiplexing of biocompatible gold nanostars for in vitro and ex vivo detection. *Anal. Chem.* 2012, 85 (1), 208–212.
2. Zavaleta, C. L.; Smith, B. R.; Walton, I.; Doering, W.; Davis, G.; Shojaei, B.; Natan, M. J.; Gambhir, S. S., Multiplexed imaging of surface enhanced Raman scattering nanotags in living mice using noninvasive Raman spectroscopy. *Proc. Nat. Acad. Sci.* 2009, 106 (32), 13511–13516.
3. Bálint, Š.; Rao, S.; Marro, M.; Miškovský, P.; Petrov, D., Monitoring of local pH in photodynamic therapy-treated live cancer cells using surface-enhanced Raman scattering probes. *J. Raman Spectrosc.* 2011, 42 (6), 1215–1221.
4. Kircher, M. F.; de la Zerda, A.; Jokerst, J. V.; Zavaleta, C. L.; Kempen, P. J.; Mittra, E.; Pitter, K. et al., A brain tumor molecular imaging strategy using a new triple-modality MRI-photoacoustic-Raman nanoparticle. *Nat. Med.* 2012, 18 (5), 829–834.
5. Alvarez-Puebla, R. A.; Liz-Marzán, L. M., SERS-based diagnosis and biodetection. *Small* 2010, 6 (5), 604–610.
6. Kneipp, J.; Kneipp, H.; Wittig, B.; Kneipp, K., Following the dynamics of pH in endosomes of live cells with SERS nanosensors. *J. Phys. Chem. C* 2010, 114 (16), 7421–7426.

7. Kneipp, J.; Kneipp, H.; Rice, W. L.; Kneipp, K., Optical probes for biological applications based on surface-enhanced Raman scattering from indocyanine green on gold nanoparticles. *Anal. Chem.* 2005, 77 (8), 2381–2385.

8. Kneipp, J.; Kneipp, H.; Rajadurai, A.; Redmond, R. W.; Kneipp, K., Optical probing and imaging of live cells using SERS labels. *J. Raman Spectrosc.* 2009, 40 (1), 1–5.

9. Qian, X. M.; Nie, S. M., Single-molecule and single-nanoparticle SERS: From fundamental mechanisms to biomedical applications. *Chemical Society Reviews* 2008, 37 (5), 912–920.

10. Faulds, K.; Smith, W. E.; Graham, D., Evaluation of surface-enhanced resonance Raman scattering for quantitative DNA analysis. *Anal. Chem.* 2003, 76 (2), 412–417.

11. Rodriguez-Lorenzo, L.; Krpetic, Z.; Barbosa, S.; Alvarez-Puebla, R. A.; Liz-Marzan, L. M.; Prior, I. A.; Brust, M., Intracellular mapping with SERS-encoded gold nanostars. *Integrative Biology* 2011, 3 (9), 922–926.

12. Küstner, B.; Gellner, M.; Schütz, M.; Schöppler, F.; Marx, A.; Ströbel, P.; Adam, P.; Schmuck, C.; Schlücker, S., SERS labels for red laser excitation: Silica-encapsulated SAMs on tunable gold/silver nanoshells. *Angewandte Chemie International Edition* 2009, 48 (11), 1950–1953.

13. Cao, Y. C.; Jin, R.; Nam, J.-M.; Thaxton, C. S.; Mirkin, C. A., Raman dye-labeled nanoparticle probes for proteins. *Journal of the American Chemical Society* 2003, 125 (48), 14676–14677.

14. Wang, G.; Park, H.-Y.; Lipert, R. J.; Porter, M. D., Mixed monolayers on gold nanoparticle labels for multiplexed surface-enhanced Raman scattering based immunoassays. *Anal. Chem.* 2009, 81 (23), 9643–9650.

15. Gregas, M. K.; Yan, F.; Scaffidi, J.; Wang, H.-N.; Vo-Dinh, T., Characterization of nanoprobe uptake in single cells: Spatial and temporal tracking via SERS labeling and modulation of surface charge. *Nanomedicine* 2011, 7 (1), 115–122.

16. Gregas, M. K.; Scaffidi, J. P.; Lauly, B.; Vo-Dinh, T., Surface-enhanced Raman scattering detection and tracking of nanoprobes: Enhanced uptake and nuclear targeting in single cells. *Applied Spectroscopy* 2010, 64 (8), 858–866.

17. Keren, S.; Zavaleta, C.; Cheng, Z.; de la Zerda, A.; Gheysens, O.; Gambhir, S. S., Noninvasive molecular imaging of small living subjects using Raman spectroscopy. *Proceedings of the National Academy of Sciences* 2008, 105 (15), 5844–5849.

18. Kim, J.-H.; Kim, J.-S.; Choi, H.; Lee, S.-M.; Jun, B.-H.; Yu, K.-N.; Kuk, E. et al., Nanoparticle probes with surface enhanced Raman spectroscopic tags for cellular cancer targeting. *Anal. Chem.* 2006, 78 (19), 6967–6973.

19. Alkilany, A.; Murphy, C., Toxicity and cellular uptake of gold nanoparticles: What we have learned so far? *Journal of Nanoparticle Research* 2010, 12 (7), 2313–2333.

20. Fales, A. M.; Yuan, H.; Vo-Dinh, T., Silica-coated gold nanostars for combined surface-enhanced Raman scattering (SERS) detection and singlet-oxygen generation: A potential nanoplatform for theranostics. *Langmuir* 2011, 27 (19), 12186–12190.

21. Fales, A. M.; Yuan, H.; Vo-Dinh, T., Cell-penetrating peptide enhanced intracellular Raman imaging and photodynamic therapy. *Mol. Pharm.* 2013, 10 (6), 2291–2298.

22. Lam, M.; Oleinick, N. L.; Nieminen, A.-L., Photodynamic therapy-induced apoptosis in epidermoid carcinoma cells. *Journal of Biological Chemistry* 2001, 276 (50), 47379–47386.

23. Tang, W.; Xu, H.; Kopelman, R.; Philbert, M. A., Photodynamic characterization and in vitro application of methylene blue-containing nanoparticle platforms. *Photochem. Photobiol.* 2005, 81 (2), 242–249.

24. Lee, S. J.; Koo, H.; Lee, D.-E.; Min, S.; Lee, S.; Chen, X.; Choi, Y. et al., Tumor-homing photosensitizer-conjugated glycol chitosan nanoparticles for synchronous photodynamic imaging and therapy based on cellular on/off system. *Biomaterials* 2011, 32 (16), 4021–4029.

25. Rossi, L. M.; Silva, P. R.; Vono, L. L. R.; Fernandes, A. U.; Tada, D. B.; Baptista, M. C. S., Protoporphyrin IX nanoparticle carrier: Preparation, optical properties, and singlet oxygen generation. *Langmuir* 2008, 24 (21), 12534–12538.

26. Orth, K.; Beck, G.; Genze, F.; Rück, A., Methylene blue mediated photodynamic therapy in experimental colorectal tumors in mice. *Journal of Photochemistry and Photobiology B: Biology* 2000, 57 (2–3), 186–192.

27. Konan, Y. N.; Gurny, R.; Allémann, E., State of the art in the delivery of photosensitizers for photodynamic therapy. *Journal of Photochemistry and Photobiology B: Biology* 2002, 66 (2), 89–106.
28. Ohulchanskyy, T. Y.; Roy, I.; Goswami, L. N.; Chen, Y.; Bergey, E. J.; Pandey, R. K.; Oseroff, A. R.; Prasad, P. N., Organically modified silica nanoparticles with covalently incorporated photosensitizer for photodynamic therapy of cancer. *Nano Letters* 2007, 7 (9), 2835–2842.
29. Kim, S.; Ohulchanskyy, T. Y.; Pudavar, H. E.; Pandey, R. K.; Prasad, P. N., Organically modified silica nanoparticles co-encapsulating photosensitizing drug and aggregation-enhanced two-photon absorbing fluorescent dye aggregates for two-photon photodynamic therapy. *Journal of the American Chemical Society* 2007, 129 (9), 2669–2675.
30. Yan, F.; Kopelman, R., The embedding of meta-tetra(Hydroxyphenyl)-chlorin into silica nanoparticle platforms for photodynamic therapy and their singlet oxygen production and pH-dependent optical properties. *Photochemistry and Photobiology* 2003, 78 (6), 587–591.
31. Roy, I.; Ohulchanskyy, T. Y.; Pudavar, H. E.; Bergey, E. J.; Oseroff, A. R.; Morgan, J.; Dougherty, T. J.; Prasad, P. N., Ceramic-based nanoparticles entrapping water-insoluble photosensitizing anticancer drugs: A novel drug–carrier system for photodynamic therapy. *Journal of the American Chemical Society* 2003, 125 (26), 7860–7865.
32. Bechet, D.; Couleaud, P.; Frochot, C.; Viriot, M.-L.; Guillemin, F.; Barberi-Heyob, M., Nanoparticles as vehicles for delivery of photodynamic therapy agents. *Trends in Biotechnology* 2008, 26 (11), 612–621.
33. Lu, J.; Liong, M.; Zink, J. I.; Tamanoi, F., Mesoporous silica nanoparticles as a delivery system for hydrophobic anticancer drugs. *Small* 2007, 3 (8), 1341–1346.
34. Zhao, T.; Wu, H.; Yao, S. Q.; Xu, Q.-H.; Xu, G. Q., Nanocomposites containing gold nanorods and porphyrin-doped mesoporous silica with dual capability of two-photon imaging and photosensitization. *Langmuir* 2010, 26 (18), 14937–14942.
35. Qian, H. S.; Guo, H. C.; Ho, P. C.-L.; Mahendran, R.; Zhang, Y., Mesoporous-silica-coated up-conversion fluorescent nanoparticles for photodynamic therapy. *Small* 2009, 5 (20), 2285–2290.
36. Tada, D. B.; Vono, L. L. R.; Duarte, E. L.; Itri, R.; Kiyohara, P. K.; Baptista, M. S.; Rossi, L. M., Methylene blue-containing silica-coated magnetic particles: A potential magnetic carrier for photodynamic therapy. *Langmuir* 2007, 23 (15), 8194–8199.
37. Zhang, Z.; Wang, L.; Wang, J.; Jiang, X.; Li, X.; Hu, Z.; Ji, Y.; Wu, X.; Chen, C., Mesoporous silica-coated gold nanorods as a light-mediated multifunctional theranostic platform for cancer treatment. *Advanced Materials* 2012, 24 (11), 1418–1423.
38. de la Fuente, J. M.; Berry, C. C., Tat peptide as an efficient molecule to translocate gold nanoparticles into the cell nucleus. *Bioconjugate Chemistry* 2005, 16 (5), 1176–1180.
39. Santra, S.; Yang, H.; Dutta, D.; Stanley, J. T.; Holloway, P. H.; Tan, W.; Moudgil, B. M.; Mericle, R. A., TAT conjugated, FITC doped silica nanoparticles for bioimaging applications. *Chemical Communications* 2004, (024), 2810–2811.
40. Torchilin, V. P., Tat peptide-mediated intracellular delivery of pharmaceutical nanocarriers. *Advanced Drug Delivery Reviews* 2008, 60 (4–5), 548–558.
41. Pan, L.; He, Q.; Liu, J.; Chen, Y.; Ma, M.; Zhang, L.; Shi, J., Nuclear-targeted drug delivery of TAT peptide-conjugated monodisperse mesoporous silica nanoparticles. *Journal of the American Chemical Society* 2012, 134 (13), 5722–5725.
42. Sebbage, V., Cell-penetrating peptides and their therapeutic applications. *Bioscience Horizons* 2009, 2 (1), 64–72.
43. Ruan, G.; Agrawal, A.; Marcus, A. I.; Nie, S., Imaging and tracking of tat peptide-conjugated quantum dots in living cells: New insights into nanoparticle uptake, intracellular transport, and vesicle shedding. *Journal of the American Chemical Society* 2007, 129 (47), 14759–14766.
44. Yuan, H.; Fales, A. M.; Vo-Dinh, T., TAT peptide-functionalized gold nanostars: Enhanced intracellular delivery and efficient NIR photothermal therapy using ultralow irradiance. *Journal of the American Chemical Society* 2012, 134 (28), 11358–11361.
45. Sivapalan, S. T.; DeVetter, B. M.; Yang, T. K.; van Dijk, T.; Schulmerich, M. V.; Carney, P. S.; Bhargava, R.; Murphy, C. J., Off-resonance surface-enhanced Raman spectroscopy from gold nanorod suspensions as a function of aspect ratio: Not what we thought. *ACS Nano* 2013, 7 (3), 2099–2105.

46. Sau, T. K.; Murphy, C. J., Room temperature, high-yield synthesis of multiple shapes of gold nanoparticles in aqueous solution. *Journal of the American Chemical Society* 2004, 126 (28), 8648–8649.

47. Gandra, N.; Portz, C.; Singamaneni, S., Multifunctional plasmonic nanorattles for spectrum-guided locoregional therapy. *Advanced Materials* 2014, 26 (3), 424–429.

48. Liu, K.-K.; Tadepalli, S.; Tian, L.; Singamaneni, S., Size-dependent surface enhanced Raman scattering activity of plasmonic nanorattles. *Chemistry of Materials* 2015, 27 (15), 5261–5270.

49. Jaiswal, A.; Tian, L.; Tadepalli, S.; Liu, K.-k.; Fei, M.; Farrell, M. E.; Pellegrino, P. M.; Singamaneni, S., Plasmonic nanorattles with intrinsic electromagnetic hot-spots for surface enhanced Raman scattering. *Small* 2014, 10 (21), 4287–4292.

50. Pinkhasova, P.; Puccio, B.; Chou, T.; Sukhishvili, S.; Du, H., Noble metal nanostructure both as a SERS nanotag and an analyte probe. *Chemical Communications* 2012, 48 (78), 9750–9752.

51. Gandra, N.; Singamaneni, S., Bilayered Raman-intense gold nanostructures with hidden tags (BRIGHTs) for high-resolution bioimaging. *Advanced Materials* 2013, 25 (7), 1022–1027.

52. Fales, A. M.; Vo-Dinh, T., Silver embedded nanostars for SERS with internal reference (SENSIR). *Journal of Materials Chemistry C* 2015, 3 (28), 7319–7324.

53. Zhou, Y.; Ding, R.; Joshi, P.; Zhang, P., Quantitative surface-enhanced Raman measurements with embedded internal reference. *Analytica Chimica Acta* 2015, 874 (0), 49–53.

54. Zhou, Y.; Lee, C.; Zhang, J.; Zhang, P., Engineering versatile SERS-active nanoparticles by embedding reporters between Au-core/Ag-shell through layer-by-layer deposited polyelectrolytes. *Journal of Materials Chemistry C* 2013, 1 (23), 3695–3699.

55. Lim, D.-K.; Jeon, K.-S.; Hwang, J.-H.; Kim, H.; Kwon, S.; Suh, Y. D.; Nam, J.-M., Highly uniform and reproducible surface-enhanced Raman scattering from DNA-tailorable nanoparticles with 1-nm interior gap. *Nat Nano* 2011, 6 (7), 452–460.

56. Yuan, H.; Khoury, C. G.; Hwang, H.; Wilson, C. M.; Grant, G. A.; Vo-Dinh, T., Gold nanostars: Surfactant-free synthesis, 3D modelling, and two-photon photoluminescence imaging. *Nanotechnology* 2012, 23 (7), 075102.

57. Fales, A. M.; Yuan, H.; Vo-Dinh, T., Development of hybrid silver-coated gold nanostars for nonaggregated surface-enhanced Raman scattering. *J. Phys. Chem. C* 2014, 118 (7), 3708–3715.

58. Zhang, Y.; Walkenfort, B.; Yoon, J. H.; Schlucker, S.; Xie, W., Gold and silver nanoparticle monomers are non-SERS-active: A negative experimental study with silica-encapsulated Raman-reporter-coated metal colloids. *Phys. Chem. Chem. Phys.* 2015, 17 (33), 21261–21267.

31

Virus-Like Particle-Mediated Intracellular Delivery for Nanomedicine

Jadwiga Chroboczek and Inga Szurgot

CONTENTS

31.1 Virus-Like Particles

Virus-like particles (VLPs) are ordered multimeric, occasionally multiprotein nanostructures that are assembled from viral structural proteins and are devoid of any genetic material. These naturally occurring bionanomaterials often emulate the conformation of authentic viruses. VLPs contain repetitive high-density displays of viral surface proteins and as such form a highly adaptable platform for various applications. Importantly, they contain functional viral proteins responsible for cell penetration of the virus, which ensures efficient cell entry.

The foremost application of VLPs is in vaccinology, whereby they provide delivery systems that combine good safety profiles with strong immunogenicity. VLPs structurally resemble live viruses, and thereby are recognized in a similar manner by the immune system, inducing both humoral and cell-mediated immunity. Traditionally, vaccines against viral diseases have been prepared from attenuated or inactivated infectious viral strains but an emerging understanding of how viruses interact with the immune system permits the precise designing of nanoparticles for medical use; VLPs devoid of the viral genome but able to penetrate cells and tissues are a much safer alternative. Moreover, they provide a polyvalent structure that can accommodate multiple copies of antigens, and, finally, they retain the tissue-specific targeting characteristic of the virus of origin.

Like the parental viruses, VLPs can be either nonenveloped or enveloped (formed from capsid proteins enveloped in a lipid membrane derived from the expression host, with glycoproteins embedded in the lipid), and spherical or filamentous (Figure 31.1). Some of them were discovered thanks to electron microscopy studies of the virus life cycle, where

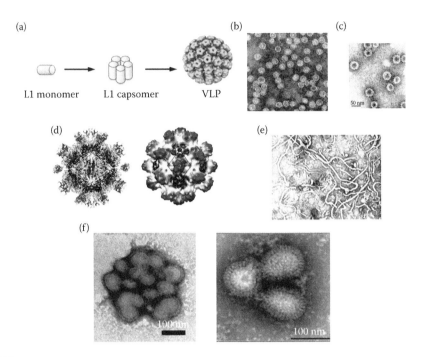

FIGURE 31.1

Virus-like particles. (a) Diagram of assembly of HPV VLP from the L1 capsid protein. Five L1 monomers form spontaneously one pentameric L1 capsomer, 72 of such capsomers self-assemble into a VLP. Adapted from German Cancer Research site, http://www.dkfz.de. (b) Electron microscopy of the VLPs built of the coat protein of coliphage Qβ expressed in *E. coli.* (After Fiedler et al. *Biomacromolecules.* 2012;13:2339–48.) (c) Transmission electron micrographs of polyomavirus-like particles (PyVLPs) produced in insect cells and then purified by sucrose density sedimentation. (After Shin and Folk. *J Virol* 2003;77:11491–8). (d) Structure of the hepatitis E virus-like particle (HEV VLP). Left panel shows the atomic structure of HEV VLP. The three domains, S, P1, and P2 are colored blue, yellow, and red, respectively. Right panel shows cryo-EM reconstruction at 14 Å resolution. (After Guu TS et al. *Proc Natl Acad Sci U S A* 2009;106:12992–7.) (e) Ebola VLP visualized by negative staining. (From https://pathbio.med.upenn.edu/pbr/portal/cell/e30_ebola_med_full.jpg.) (f) Influenza enveloped VLPs. Left panel: EM of negatively stained A/Anhui/1/2013 (H7N9) VLP, assembled from HA and NA proteins expressed from separate baculovirus vectors in insect cells. (After Smith GE et al. *Vaccine* 2013;31:4305–13.) Right panel: influenza VLPs produced in a baculovirus system with the sequences of the genes for HA, NA, and M1 of the 1918 pandemic virus. (After Perrone LA et al. *J Virol* 2009;83:5726–34.).

VLPs appeared together with well-formed viruses. However, in a large majority of cases we do not know the native role of such VLPs. They form spontaneously upon expression in heterologous systems of one or several viral structural proteins. The analysis of published data performed by Zeltins (2013) revealed that at least 110 VLPs have been constructed from viruses belonging to 35 different families. Depending on the complexity of the VLPs, they can be produced from appropriate recombinant vectors in either prokaryotic or eukaryotic expression systems, or assembled in cell-free conditions. Furthermore, they can be formed from proteins derived from a single virus, for example, to obtain immunity to this virus or, alternatively, as a platform derived from a single virus presenting proteins/peptides derived from another microorganism or any cell/tissue (chimeric VLPs).

The yield of *in vitro* VLP production is high and even in eukaryotic systems can approach expression efficiency comparable to that observed for bacterial expression systems. Due to their high molecular weight, VLPs are purified from extracts of expressing cells by sucrose density centrifugation or gel filtration, usually followed by an additional step removing cellular components loosely attached to them.*

31.2 VLP Vaccines

VLPs through mimicking the symmetry of pathogenic viruses offer a ready platform for recognition, uptake, and processing by the immune system. They present both high-density B-cell epitopes for antibody production and intracellular T-cell epitopes, thus inducing, respectively, potent humoral and cellular immune responses (Beyer et al., 2001; Wang and Roden, 2013). These stable and versatile nanoparticles display excellent adjuvant properties capable of inducing innate and cognate immune responses. Indeed, VLPs can enhance the immunogenicity of weakly immunogenic peptides and proteins. They direct antigenic peptides/proteins to immature dendritic cells (DC), activating DC maturation. Mature DC are the key antigen-presenting cells (APC) that efficiently mediate antigen transport to lymphoid tissues for the initiation of T-cell responses and induction of cell-mediated immunity (Zinkernagel, 2014). The uptake of VLPs by APC leads to efficient immune responses and results in the control of pathogenic microorganisms.

VLPs display properties of immune adjuvant and so it was unclear why the adjuvants should be added to VLP-based vaccines. Some empirical answers have been furnished by the following studies. Viral capsomeres (capsid subunits) that could be considered an alternative to VLP-based vaccines can be produced in prokaryotic expression systems. When detailed side-by-side comparison of VLPs and capsomeres was performed for HPV16 L1 protein (a building block of the VLP-based HPV vaccine), VLPs induced consistently higher antibody titers (humoral immunity) than capsomeres; however, the two forms induced similar CD8 T-cell responses (cell-mediated immunity) after subcutaneous, intranasal, and oral immunization. It appeared that at least 20–40 times more L1 in the form of capsomeres than in the form of VLPs was needed to achieve comparable antibody responses, but the lower immunogenicity of capsomeres could be compensated by the use of an adjuvant system containing monophosphoryl lipid A (MPL, Thönes et al., 2008). Accordingly, it has been shown that adjuvanted VLPs are able to elicit a higher titer of total specific IgG compared to VLPs (Visciano et al., 2012). It, thus, seems that the use of adjuvants may be beneficial when improved humoral immunity is sought.

When VLP is used as a platform for the presentation of foreign epitopes, in the majority of cases this is achieved through modification of the VLP gene sequence, so that fused VLP components with foreign epitope are assembled into VLPs during expression, yielding chimeric VLPs. When large protein domains are needed as antigen, such insertions may be incompatible with VLP assembly and in that case chemical conjugation can be a solution. The main problem encountered during VLP manipulations is the proper presentation of epitopes, which must elicit the same type of immune response as the native pathogen.

During the last decade, 2 prophylactic VLP vaccines have received marketing approval for human use, another 12 vaccines have entered clinical development, and many others

---

* This introductory part follows Introduction from our paper "Virus-like particles as vaccine." Chroboczek J, Szurgot I, Szolajska E. *Acta Biochim Pol* 2014;6:531–9.

are in the proof-of-concept stage. The approved VLP-based prophylactic human vaccines currently in use are against hepatitis B virus (HBV) and human papilloma virus (HPV) infections. In July 2015, the European Commission approved a prophylactic VLP human vaccine against malaria.

31.2.1 Hepatitis B Vaccine

Hepatitis B virus infection is one of the most common human diseases. Each year over 1 million people die from HBV-related chronic liver diseases. In 1986, hepatitis B vaccine became the first recombinant protein-based vaccine for humans approved by the Federal Drug Administration. It is based on recombinant HBV surface antigen (HBsAg), which upon production in yeast or mammalian cells forms 22-nm spherical VLPs that are adsorbed onto aluminum hydroxide gel (Greiner et al., 2012). These particles are composed of host cell–derived lipids (30%–50%) and about 70 copies of the ∼25 kDa S protein of 226 amino acid residues, and stabilized by intra- and intermolecular disulfide bonds. Contrary to mammalian cell-derived HBsAg particles, yeast-derived particles contain unglycosylated S protein. The highly hydrophobic S protein in HBsAg VLPs is in tight association with lipids that have been shown to be responsible for the antigenic properties of HBsAg particles stabilizing their structure and protein conformation. Thus, these VLPs exhibit a lipoprotein-like structure with an ordered and rather rigid lipid interface and a more hydrophobic and fluid inner core. The available commercial products include Recombivax HB (Merck), Engerix-B (GlaxoSmithKline Biologicals), and Genevac B (Serum Institute) (Drug master file at http://www.fda.gov/BiologicsBloodVaccines/Vaccines/ApprovedProducts/ucm110102.htm). A course of 2–3 vaccine injections are given intramuscularly, providing protection for at least 25 years in cases of adequate initial response to vaccination, but some guidelines now recommend a single booster at 5 years. The hepatitis B vaccine was found to be generally safe although the Engerix B vaccine appeared to triple the risk of CNS inflammatory demyelination in infant boys (Mikaeloff et al., 2009). A possible culprit is thiomersal, a mercury-containing vaccine preservative that is currently being phased out in many countries.

31.2.2 HPV Vaccine

Approximately 30–40 types of HPV can be transmitted through sexual contact, via the anogenital region. Persistent infection with high-risk HPV types (different from the ones that cause skin warts) may progress to precancerous lesions and invasive cancer. HPV vaccines are VLP assembled from 72 pentamers of HPV major capsid protein L1. Two kinds of prophylactic HPV vaccines exist: a bivalent vaccine (Cervarix) that offers protection against HPV serotypes 16 and 18 and a quadrivalent vaccine (Gardasil) that protects against infection with serotypes 6, 11, 16, and 18. HPV L1 proteins are expressed in yeast, purified and combined to make the vaccine. In addition, the vaccines contain adjuvants such as alumina (aluminum hydroxide) and AS04 with MPL in Cervarix, and amorphous aluminum hydroxyphosphate sulfate in Gardasil. The vaccines are highly efficacious if given before exposure to HPV, that is, to adolescent girls between 9 and 13 years of age in a three-dose schedule. The current duration of protection is 8.4 years with the bivalent vaccine and 5 years with the quadrivalent vaccine. A nine-valent prophylactic HPV vaccine (HPV6/11/16/18/31/33/45/52/58) is currently under development. Research is also directed toward the development of a prophylactic L2 vaccine and therapeutic vaccines (active after infection) (Brotherton and Ogilvie, 2015).

31.2.3 Malaria Vaccine

GlaxoSmithKline scientists worked on the Mosquirix™ (or RTS,S) vaccine for almost 30 years before the vaccine was approved by European drug regulators in July 2015. It should be licensed for use in babies in sub-Saharan Africa at the risk of mosquito-borne disease. The studies were funded by the PATH Malaria Vaccine Initiative and the Bill & Melinda Gates Foundation and assessed by the World Health Organization, which has promised to provide use guidance before the end of this year. The vaccine will be unable to wipe out malaria, as the trial data in 2011 and 2012 showed a reduction of only 27% in the episodes of malaria in babies aged 6–12 weeks, and of around 46% in children aged 5–17 months. The RTS,S vaccine contains the repeat and T-cell epitope of the circumsporozoite protein (CSP) of the *Plasmodium falciparum* malaria parasite on a platform of the hepatitis B virus VLP formed from HBsAg, to which a chemical adjuvant was added (liposome-based AS01) to boost the immune system's response. Malaria infection is prevented thanks to the induction of humoral and cellular immunity that prevent the parasite from infecting the liver (Foquet et al., 2014; Moorthy and Okwo-Bele, 2015).

31.3 Experimental VLP Vaccines

A large number of VLP-based vaccine candidates are undergoing clinical evaluation. The site *clinicaltrials.gov* lists 99 studies for VLP vaccines (10 more than last year), created predominantly against viral infections. Several trials concern new anti-influenza vaccine. A recent trial on a quadrivalent VLP-based influenza vaccine (produced by Novavax in the baculovirus system) that was included in 400 healthy young adults (18–49 years) demonstrated that this VLP vaccine was well tolerated (Novavax communication, 2015). This vaccine consists of VLPs representing four different strains of influenza virus, each expressing strain-specific hemagglutinin and neuraminidase antigens. A trial has been completed on a therapeutic anti-HIV vaccine that investigated the immunity of a VLP made of yeast Ty protein and decorated with HIV p24 protein. However, although this vaccine elicited antibodies against p24 (and Ty), it did not slow the progression of HIV-1 disease (Lindenburg et al., 2002). An interesting approach was used to construct a vaccine against the deadly filoviruses as well as Marburg and Ebola viruses, with no vaccines or therapeutics known. The protein subunit-based and DNA vaccines have only moderate success against these filoviruses, while VLP vaccines have shown promising results when tested in both rodents and nonhuman primates. These VLPs rely on the natural properties of the viral matrix protein (VP) 40 to drive budding of filamentous particles that can also incorporate additional filovirus proteins. Filovirus VLP vaccines have used particles containing two or three viral proteins (GP and VP40, with or without NP) generated in either mammalian or insect cells. It appeared that one or two doses of VLP vaccine could confer protection from lethal filovirus infection (Warfield et al., 2015).

Another group of vaccine trials concerns vaccines based on a bacteriophage VLPs platform prepared in bacteria, against conditions other than viral infections. For example, Cyt003-QbG10 vaccine (Cytos Biotechnology) delivers CpG DNA, an immune modulator, as a potential new treatment for asthma. Several other Qß VLP-based vaccines that are not against viral infections are investigated, such as against type II diabetes mellitus, Alzheimer's, nicotine addiction, etc.

Recently, a Qb VLP conjugated with the nerve growth factor (NGF, vaccine QbNGF) was observed to neutralize NGF and suppress hyperalgesia and weight loss in animal models of chronic pain (Röhn et al., 2011). Another nonviral VLP vaccine, against hypertension, consists of angiotensin II (Ang II), a naturally occurring octapeptide, coupled to the surface of Qβ bacteriophage VLP. This vaccine decreased the blood pressure of Ang II-induced hypertensive mice (Chen et al., 2013).

There are 54 clinical trials on anticancer VLP vaccines, of which the majority are variants of anti-HPV vaccines. Among others, two anti-melanoma vaccines are being studied. The Melan-A VLP and CYT004-MelQbG10 vaccines for melanoma patients consist of the melanocyte differentiation antigen Melan A (also called MART-1) encapsulated in Qβ VLP together with short immunostimulatory oligonucleotides (CpGs) or other adjuvants. Upon administration, these vaccines may activate the immune system to exert a specific cytotoxic T lymphocyte (CTL) response against cancer cells expressing Melan A antigen upregulated in most melanomas. It has been shown that Melan A vaccination resulted in an increase of T-cells at the injection site (Goldinger et al., 2012). However, trial results are not posted yet. In the majority of these trials, apart from investigating the immune response, emphasis is placed on the use of appropriate adjuvants in order to obtain a more pronounced response by immune system.

Our group is investigating the use of the adenoviral dodecahedron VLP that displays strong adjuvant properties. Native adenoviral dodecahedron is a nonenveloped symmetrical VLP built from 12 pentons of human adenovirus serotype 3 (Ad3). The pentons are noncovalent complexes composed of pentameric penton bases and trimeric fiber proteins, both proteins being needed for intracellular penetration of the virus (Figure 31.2). These VLPs, smaller than the virus of origin, are generated during the life cycle of certain adenovirus serotypes, including Ad3, where they conceivably participate in the spread of progeny virus through loosening of tight junctions (Fender et al., 2005, 2012; Lu et al., 2013). These, so-called, dodecahedra-fibers (DFs) of 4.8 MDa and diameter of ~50 nm can also be produced by recombinant expression of Ad3 pentons in the baculovirus system. The formation of these particles is due solely to penton base interactions, as attested, upon expression of penton base protein alone, by the formation of dodecahedra without fibers with molecular weight of 3.6 MDa and diameter of 28 nm (called dodecahedra-base DB, hereafter referred to as Dd, Figure 31.2a). The stability of the dodecameric structure does not depend on disulfide bridges or cations as in other VLPs (Simon et al., 2014). Instead, the major mechanism of stabilization lies in the interlocking of the 60 N-terminal domains derived from 12 pentameric penton bases, which results in the formation of a strong network stabilizing the VLP (Szolajska et al., 2012). Dds retain integrity under different physicochemical conditions, which enables their storage as well as the attachment of therapeutic agents (Zochowska et al., 2009). Importantly, their small size, 28 nm, suggests that the Dds will be able to target not only DC at the injection site, but also lymph node–resident DC. Using confocal microscopy, we observed that human DC readily internalized our vector, with Dd in all DC in the observed field, which demonstrates remarkable vector transduction efficiency. In addition, Dd alone induced maturation of human DC, thus, acting as an adjuvant (Naskalska et al., 2009). Indeed, Dd application resulted in activation of DC maturation markers followed by formation of motile cytoplasmic veils (podosomes), without causing toxicity to DC (Figure 31.3). Furthermore, the antigen attached to Dd was successfully captured and efficiently presented by human DC to antigen-specific CD8+ T lymphocytes (Naskalska et al., 2009; Villegas-Mendez et al., 2010), triggering their responses (Szurgot et al., 2013).

Using the Dd platform built from a unique adenoviral protein, we constructed a novel influenza vaccine carrying the epitopes of influenza virus, with the goal of establishing

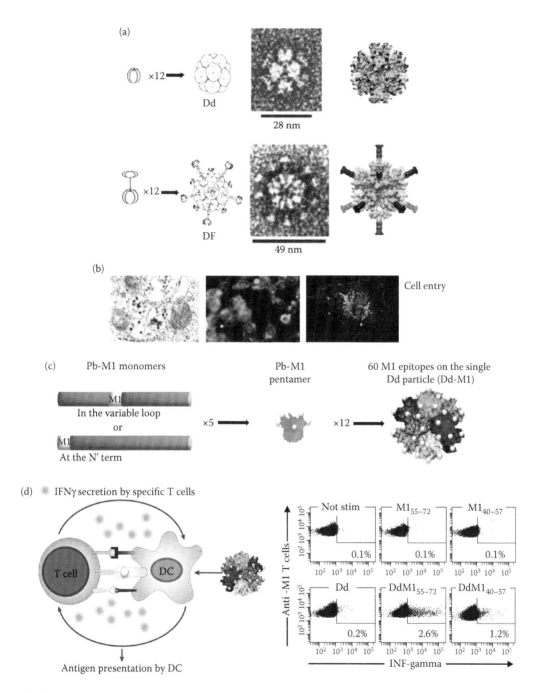

FIGURE 31.2

Adenoviral dodecahedron. (a) Adenoviral dodecahedra. Left panels depict schematically the assembly of Dd (dodecahedron base) and DF (dodecahedron fiber) from adenoviral penton bases or pentons, respectively. Middle panels show cryo-electron microscopy of Dd and DF. (From Fender et al., 1997.) Right panels show Dd and DF structure at 9 Å resolution. (From Fuschiotti et al., 2006, http://pdbj.org/emnavi/emnavi.) **(b)** Cell internalization. Left panel: Dd penetrating HeLa cells for 2 min at 37°C; thin sections were analyzed under electron microscope, without staining. Dds are seen in the form of black spheres in transport vacuoles. (*Continued*)

FIGURE 31.2 (Continued)
Middle and right panels: HeLa cells incubated with Dd-FITC (green signal). The cell nucleus was stained with Hoechst (in blue) or with propidium iodide (in red). (c) Diagram of Dd formation bearing 60 copies of M1 epitopes (Dd-M1). (d) Induction of T cell–mediated immunity by the Dd-M1 vaccine. Left panel: a schematic view of vaccine interaction with DC resulting in epitope presentation to T lymphocytes and induction of secretion of interferon gamma, a marker of cellular immunity. Right panel: presentation of DdM1$_{55-72}$ and DdM1$_{40-57}$ to specific T cells. Secretion of IFN-γ was analyzed by flow cytometry on T cells that were either nonstimulated or stimulated by dendritic cells pulsed with free peptides M1$_{55-72}$ or M1$_{40-57}$, with Dd, or with DdM1$_{55-72}$ or DdM1$_{40-57}$. Neither empty Dd nor free M1 peptides were able to stimulate specific T cells; T lymphocytes did not secrete IFN-γ. But when M1 peptides were presented on the dodecahedral platform, an amplification of T cells was observed, confirming that the Dd-M1 vaccine is able to activate T-cell responses at levels comparable to pandemic influenza vaccine. (Adapted from Schmidt T et al. *Eur J Immunol* 2012;42:1755–66.)

immunity against influenza and not against Ad infection (Szurgot et al., 2013). Influenza virus surface glycoprotein, HA (568 aa), is the key antigen able to induce antibodies neutralizing viral penetration (Knossow and Skehel, 2006). However, due to the high mutation rate, traditional vaccines based on HA confer protection only against the closely related viral strains and are ineffective against strains with serologically distinct HA. In contrast, the internal matrix protein M1 (252 aa) is the most conserved influenza virus protein (Ito et al., 1991), which is capable of stimulating cellular immune responses of the infected hosts and eliciting cytotoxic T-cells (Webster and Hinshaw, 1977; Lee et al., 2008; Garigliany et al., 2010). Therefore, a vaccine containing both antigens should be able to induce both humoral and cell-mediated immunity, possibly providing long-lasting protection against different strains of influenza virus.

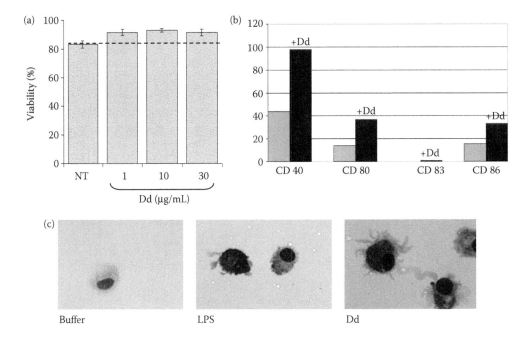

FIGURE 31.3
Adjuvant effect of Dd. (a) Viability after 1-day exposure to Dd analyzed on PBMC from three donors. NT: not treated. (b) Activation of DC maturation markers. (c) Dd adjuvant effect. Left panel: untreated DC. Middle panel: DC treated with LPS (bacterial lipopolysaccharide known to activate DC). Right panel: DC treated with Dd. Note the formation of dendrites upon activation with LPS or Dd.

Instead of attaching the whole M1 protein to Dd, in our vaccine design we decided to use only immunodominant M1 epitopes. Two immunodominant M1 epitopes have been identified through the examination of the virus-specific CD4+ and CD8+ memory T-cell responses to the proteome of two influenza strains (Lee et al., 2008). These highly conserved peptides, $M1_{40-57}$ and $M1_{55-72}$, have been inserted either in the variable loop of Ad3 Pb protein (a dodecahedron building block) or as an extension of the Pb N-terminal domain, without destroying the particles' dodecahedric structure or its cell entry capacity. The candidate vaccine consists of a 1:1 mixture of these two kinds of Dd-M1 particles, each bearing 60 copies of the M1 epitope (Figure 31.2). Dd-M1 strongly induced cell-mediated immunity both *in vitro* and in the animal model. *In vitro* tests showed that Dd carrying M1 epitopes is a potent activator of human myeloid DC (MoDC). M1 peptides were efficiently presented in the context of HLA class II and cross-presented by the HLA class I molecules, activating both CD8+ memory T cells and CD4+ T cells. Importantly, upon chicken vaccination with Dd-M1 immune responses were elicited in the absence of an adjuvant.

The complete vaccine contains both cellular immunity eliciting M1 epitopes, as well as humoral immunity eliciting HA protein that is attached to the vector surface by a fragment of adenovirus fiber protein (Fi), which enables stable interaction with Dd. In this, we have taken advantage of structural data, which show that the N-terminus of a fiber protein nests in the cavity on the penton base surface (Cao et al., 2012). Genes encoding DdM1 and Fi-HA were expressed simultaneously from one recombinant baculovirus. This approach resulted in production of soluble complexes containing DdM1 with attached Fi-HA. DdM1-FiHA caused agglutination of the red blood cells, which suggests that HA in the prepared vaccines is properly folded and biologically active. The complete candidate vaccine will be now tested for its protective properties in an animal model.

Another utilization of VLPs may help to overcome a major difficulty in vaccinology; this involves identifying relevant target epitopes and then presenting them to the immune system in a context that mimics their native conformation. VLP technology is able to display complex libraries of random peptide sequences on surface-exposed loops of a VLP without disruption of protein folding or VLP assembly. It allows the identification of linear epitopes of neutralizing antibodies, and then the same VLP can be used for both affinity selection and immunization. The technique can be used also to identify mutations that improve the immunogenicity of an antigen. For example, a bacteriophage MS2-based VLP display platform was used to develop a vaccine targeting a neutralizing epitope in the minor capsid protein of HPV that provided broad protection from diverse HPV types in a mouse model (Tumban et al., 2012). Similarly, a conserved epitope of *Plasmodium falciparum* antigen AMA1 has been identified using VLP peptide display (Crossey et al., 2015).

31.4 VLP as a Drug Delivery Vector

There is a need in research, biological applications, and medicine for new efficient biocompatible vectors, enabling passage through the plasma membrane. The Achilles heel of many modern therapies is drug delivery; about 40% of newly developed drugs are rejected because of poor bioavailability. Indeed, the toxicity of many drugs makes it difficult to achieve therapeutic concentrations without severe systemic side effects. Efficient nanocarriers are being developed and fully defined self-assembling vectors can be built via approaches based on macromolecular chemistry and physics. However, the limitations of such nanobioconjugates are toxicity, undesired immunogenicity, endosomal entrapment that reduces the amount of biomolecule reaching the cytoplasm, and lack of biodegradability (Sebestik et al., 2011, Van den Berg and Dowdy, 2011; Jian et al., 2012). Even liposomes are known to activate the complement (C) system, which can lead *in vivo* to a hypersensitivity syndrome called C activation–related pseudoallergy (CARPA) (Szebeni and Storm, 2015). In contrast, the proteinaceus VLPs are biocompatible and totally biodegradable polyvalent nanoparticles, able to carry and actively deliver multiple copies of a bioactive molecule. Intracellular delivery, important in both therapeutic and fundamental applications, faces two major challenges: efficient cellular uptake and avoidance of endosomal sequestration. The VLPs show an unsurpassable cell entry capabilities accompanied by efficient liberation in the cytoplasm, features that they inherit from the viruses they are derived from.

We have pioneered the use of VLPs for drug delivery. For this, we took advantage of the extraordinary propensity of adenoviral dodecahedra for cell entry; we estimated that up to 300,000 of these particles could be observed in one cell *in vitro* (Garcel et al., 2006). This remarkable Dd penetration capability results from the recognition of an additional plasma membrane receptor not recognized by the virus-of-origin. Indeed, Dd recognizes two types of receptors; it retains the affinity of the penton bases, its building blocks, for αv integrins (Wickham et al., 1993; Vivès et al., 2004; Fender et al., 2008), but, in addition, it has an affinity for the heparan sulfate, an important constituent of the cell plasma membrane. Heparan sulfate is upregulated in neoplastic blood vessels but not in resting endothelial cells or in most normal organ systems (Pasqualini et al., 1997; Eliceiri and Cheresh, 1999), and Dd recognition suggests its affinity for malignant tumors.

Several years ago, we prepared a chemical conjugate of Dd with the anticancer drug bleomycin (Dd-BLM). In this preparation, each VLP carried on its surface ~60 bleomycin molecules. Upon cell penetration, the Dd-BLM induced cell death. In a similar manner, as free bleomycin, Dd-BLM caused dsDNA breaks but the effective cytotoxic concentration of BLM delivered with Dd was 100 times lower than that of free bleomycin (Zochowska et al., 2009). In the following attempt, we chemically attached a cell-impermeant oncogene inhibitor to Dd (~60 molecules/Dd) for delivery to a rat orthotopic hepatocellular carcinoma (HCC) model. We observed an inhibition of tumor growth, accompanied by near total extinction of two pro-oncogenes, eIF4E and c-myc (Zochowska et al., 2015). These experiments open new vistas for the delivery of nonpermeant labile drugs, conceivably compatible with dose sparing.

Protein-based VLPs could elicit an immune response to the vector, which may significantly restrict the use of such vectors for drug delivery. Dd is composed of only one protein (Ad3 penton base, Pb), one of 11 structural proteins of the adenovirus, so that any eventual immune response after administration of this dodecahedric vector is less likely than the reaction observed after administration of the recombinant adenoviruses used in gene therapy. In addition, Pb seems to be a weakly immunizing protein as the presence of neutralizing Ab directed toward the Pb protein was observed in only approximately half of the 17 cancer patients who had been repeatedly administered the oncolytic adenovirus (Hong et al., 2003). Finally, Dd is derived from Ad serotype 3, which rarely appears in humans. Last but not least, several proteins are used successfully in human therapies (e.g., the anticancer drug Elspar that contains asparaginase).

Like viruses, Dd is able to pass from one cell to another (unpublished results), which might explain the remarkably good intracellular distribution of drugs delivered with Dd seen in an animal model (Zochowska et al., 2015). An additional interesting feature of Dd could be its ability to penetrate drug-resistant malignant cells. Drug resistance (DR) is a major factor in the failure of many forms of chemotherapy. Chemotherapy kills drug-sensitive cells, but leaves behind a large number of drug-resistant cells, which permits the now resistant tumor to grow again. The DR is mainly due to the presence in the cell membrane of the drug efflux protein Pgp that expels membrane-embedded exogenous molecules (Krishna and Mayer, 2000). Multiple strategies have been developed to overcome Pgp-mediated DR, without much success. Since Dd, similar to viruses, enters the cells by rapid endocytosis, it is conceivable that it could be able to bypass this kind of drug resistance. This is confirmed by our preliminary observations, whereby Dd penetrated similar MESSA (uterine sarcoma) and multidrug-resistant (MDR) MESSA-derived cells (unpublished results).

An interesting approach to drug delivery is drug encapsulation inside VLP. All VLPs possess an internal unoccupied cavity of variable diameter. For example, the adenoviral Dd internal cavity is 7 nm in diameter, while *Salmonella typhimurium* bacteriophage P22 VLP with ~64 nm in diameter has a large cavity of 54 nm (Earnshaw et al., 1976). Furthermore, VLPs held together by divalent ions and disulfide bonds may be depolymerized by removal of such ions under reducing conditions and will reassemble on return to initial conditions (Ishizu et al., 2001). Therefore, some VLPs may encapsulate drugs during self-assembly. For example, *Macrobrachium rosenbergii* (giant river prawn) nodavirus (MrNv) capsid protein yields VLPs upon expression in *Escherichia coli*. The disassembly/reassembly of these VLPs is controlled in a calcium-dependent manner permitting encapsulation of DNA molecules (Jariyapong et al., 2014). The VLPs of MS2 bacteriophage self-assemble from 180 copies of a single coat protein (13.7 kDa) in the presence of nucleic acid, enabling their use for delivery of siRNA (Uhlenbeck, 1998). The same VLPs modified with peptide recognizing hepatocellular carcinoma cells were shown to encapsulate various cargo molecules and deliver them to the HCC cells in culture, killing them (Ashley et al., 2011). Interestingly, an analgesic agent ziconotide that is blood–brain barrier (BBB) impermeable and, therefore, can be only used by direct delivery into the spinal fluid, when encapsulated into VLP of phage P22 with attached Tat transducing peptide, was successfully transported in several *in vitro* BBB models of rat and human brain through a recyclable noncytotoxic endocytic pathway (Anand et al., 2015). This is a quite surprising result as at ~54 nm in diameter P22 VLPs are significantly larger than the proteins and quantum dots previously reported in the literature as being translocated to the brain. Clearly, crucial *in vivo* experiments are necessary to determine whether VLP-based drug delivery to the brain can be achieved. However, since viruses are able to enter the brain (e.g., causing viral encephalitis) and may transit human endothelial cells via a transcellular pathway that does not affect the integrity of the BBB (Hasebe et al., 2010), this approach is worth pursuing. An ingenious use of the VLP production process was demonstrated using a gamma-retrovirus, murine leukemia virus (MLV) composed of Gag protein, by the incorporation of nuclear transcription factors during VLP synthesis (Wu and Roth, 2014). The chimeric VLPs released from the cellular expression system were purified and used for the delivery of transcription factors, resulting in the transient delivery of proteins into the target cells. This was possible due to the poor immunogenicity of MLV-based retrovirus vectors.

These examples show an imaginative use of VLPs properties and life cycle, but studies on VLPs as drug delivery vectors are still in their infancy. The greatest hurdle to their medical use is the immune response of these proteinaceous vectors, which might relegate VLPs use to auxiliary therapeutic applications, for example, after the failure of classical chemotherapy. Nevertheless, these biodegradable vectors enabling efficient delivery and significant dose sparing could successfully fit into such a narrow therapeutic window.

Acknowledgments

This work was partially supported by French grants from La Ligue contre le Cancer and Gefluc, and by Polish grants N N302 505738 and UMO2013/09/B/NZ3/02327 of the National Science Centre.

References

Anand P, O'Neil A, Lin E, Douglas T, Holford M. Tailored delivery of analgesic ziconotide across a blood brain barrier model using viral nanocontainers. *Sci Rep*. 2015;5:12497.

Ashley CE, Carnes EC, Phillips GK, Durfee PN, Buley MD, Lino CA et al. Cell-specific delivery of diverse cargos by bacteriophage MS2 virus-like particles. *ACS Nano*. 2011;5:5729–45.

Beyer T, Herrmann M, Reiser C, Bertling W, Hess J. Bacterial carriers and virus-like-particles as antigen delivery devices: Role of dendritic cells in antigen presentation. *Curr Drug Targets Infect Disord*. 2001;1:287–302.

Brotherton JM, Ogilvie GS. Current status of human papillomavirus vaccination. *Curr Opin Oncol*. 2015;27:399–404.

Cao C, Dong X, Wu X, Wen B, Ji G, Cheng L, Liu H. Conserved fiber-penton base interaction revealed by nearly atomic resolution cryo-electron microscopy of the structure of adenovirus provides insight into receptor interaction. *J Virol*. 2012;86:12322–9.

Chen X, Qiu Z, Yang S, Ding D, Chen F, Zhou Y et al. Effectiveness and safety of a therapeutic vaccine against angiotensin II receptor type 1 in hypertensive animals. *Hypertension*. 2013;61:408–16.

Crossey E, Frietze K, Narum DL, Peabody DS, Chackerian B. Identification of an immunogenic mimic of a conserved epitope on the *Plasmodium falciparum* blood stage antigen AMA1 using virus-like particle (VLP) peptide display. *PLoS One*. 2015;10:e0132560.

Earnshaw W, Casjens S, Harrison SC. Assembly of the head of bacteriophage P22: X-ray diffraction from heads, proheads and related structures. *J Mol Biol*. 1976;104:387–410.

Eliceiri BP, Cheresh DA. The role of alpha v integrins during angiogenesis: Insights into potential mechanisms of action and clinical development. *J Clin Invest*. 1999;103:1227–30.

Fender P, Boussaid A, Mezin P, Chroboczek J. Synthesis, cellular localization, and quantification of penton-dodecahedron in serotype 3 adenovirus-infected cells. *Virology*. 2005;340:167–73.

Fender P, Hall K, Schoehn G, Blair GE. Impact of human adenovirus type 3 dodecahedron on host cells and its potential role in viral infection. *J Virol*. 2012;86:5380–5.

Fender P, Schoehn G, Perron-Sierra F, Tucker GC, Lortat-Jacob H. Adenovirus dodecahedron cell attachment and entry are mediated by heparan sulfate and integrins and vary along the cell cycle. *Virology*. 2008;371:155–64.

Fiedler JD, Higginson C, Hovlid ML, Kislukhin AA, Castillejos A, Manzenrieder F et al. Engineered mutations change the structure and stability of a virus-like particle. *Biomacromolecules*. 2012;13:2339–48.

Foquet L, Hermsen CC, van Gemert GJ, Van Braeckel E, Weening KE, Sauerwein R, Meuleman P, Leroux-Roels G. Vaccine-induced monoclonal antibodies targeting circumsporozoite protein prevent *Plasmodium falciparum* infection. *J Clin Invest*. 2014;124:140–4.

Fuschiotti P, Schoehn G, Fender P, Fabry CM, Hewat EA, Chroboczek J, Ruigrok RW, Conway JF. Structure of the dodecahedral penton particle from human adenovirus type 3. *J Mol Biol.* 2006;356:510–20.

Garcel A, Gout E, Timmins J, Chroboczek J, Fender P. Protein transduction into human cells by adenovirus dodecahedron using WW domains as universal adaptors. *J Gene Med.* 2006;8:524–31.

Garigliany MM, Habyarimana A, Lambrecht B, Van de Paar E, Cornet A, van den Berg T, Desmecht D. Influenza A strain-dependent pathogenesis in fatal H1N1 and H5N1 subtype infections of mice. *Emerg Infect Dis.* 2010;16:595–603.

Goldinger SM, Dummer R, Baumgaertner P, Mihic-Probst D, Schwarz K, Hammann-Haenni A et al. Nano-particle vaccination combined with TLR-7 and -9 ligands triggers memory and effector CD8[+] T-cell responses in melanoma patients. *Eur J Immunol.* 2012;42:3049–61.

Greiner VJ, Ronzon F, Larquet E, Desbat B, Estèves C, Bonvin J, Gréco F, Manin C, Klymchenko AS, Mély Y. The structure of HBsAg particles is not modified upon their adsorption on aluminium hydroxide gel. *Vaccine.* 2012;30:5240–5.

Guu TS, Liu Z, Ye Q, Mata DA, Li K, Yin C, Zhang J, Tao YJ. Structure of the hepatitis E virus-like particle suggests mechanisms for virus assembly and receptor binding. *Proc Natl Acad Sci USA.* 2009;106:12992–7.

Hasebe R, Suzuki T, Makino Y, Igarashi M, Yamanouchi S, Maeda A, Horiuchi M, Sawa H, Kimura T. Transcellular transport of West Nile virus-like particles across human endothelial cells depends on residues 156 and 159 of envelope protein. *BMC Microbiol.* 2010;10:165.

Hong SS, Habib NA, Franqueville L, Jensen S, Boulanger PA. Identification of adenovirus (ad) penton base neutralizing epitopes by use of sera from patients who had received conditionally replicative ad (addl1520) for treatment of liver tumors. *J Virol.* 2003;77:10366–75.

Ishizu KI, Watanabe H, Han SI, Kanesashi SN, Hoque M, Yajima H, Kataoka K, Handa H. Roles of disulfide linkage and calcium ion-mediated interactions in assembly and disassembly of virus-like particles composed of simian virus 40 VP1 capsid protein. *J Virol.* 2001;75:61–72.

Ito T, Gorman OT, Kawaoka Y, Bean WJ, Webster RG. Evolutionary analysis of the influenza A virus M gene with comparison of the M1 and M2 proteins. *J Virol.* 1991;65:5491–8.

Jariyapong P, Chotwiwatthanakun C, Somrit M, Jitrapakdee S, Xing L, Cheng HR, Weerachatyanukul W. Encapsulation and delivery of plasmid DNA by virus-like nanoparticles engineered from *Macrobrachium rosenbergii* nodavirus. *Virus Res.* 2014;179:140–6.

Jian F, Zhang Y, Wang J, Ba K, Mao R, Lai W, Lin Y. Toxicity of biodegradable nanoscale preparations. *Curr Drug Metab.* 2012;13:440–6.

Knossow M, Skehel JJ. Variation and infectivity neutralization in influenza. *Immunology.* 2006;119:1–7.

Krishna R, Mayer LD. Multidrug resistance (MDR) in cancer. Mechanisms, reversal using modulators of MDR and the role of MDR modulators in influencing the pharmacokinetics of anticancer drugs. *Eur J Pharm Sci.* 2000;11:265–83.

Lee LY, Ha do LA, Simmons C, de Jong MD, Chau NV, Schumacher R et al. Memory T cells established by seasonal human influenza A infection cross-react with avian influenza A (H5N1) in healthy individuals. *J Clin Invest.* 2008;118:3478–90.

Lindenburg CE, Stolte I, Langendam MW, Miedema F, Williams IG, Colebunders R, Weber JN, Fisher M, Coutinho RA. Long-term follow-up: No effect of therapeutic vaccination with HIV-1 p17/p24:Ty virus-like particles on HIV-1 disease progression. *Vaccine.* 2002;20:2343–7.

Lu ZZ, Wang H, Zhang Y, Cao H, Li Z, Fender P, Lieber A. Penton-dodecahedral particles trigger opening of intercellular junctions and facilitate viral spread during adenovirus serotype 3 infection of epithelial cells. *PLoS Pathog.* 2013;9:e1003718.

Mikaeloff Y, Caridade G, Suissa S, Tardieu M. Hepatitis B vaccine and the risk of CNS inflammatory demyelination in childhood. *Neurology.* 2009;72:873–80.

Moorthy VS, Okwo-Bele JM. Final results from a pivotal phase 3 malaria vaccine trial. *Lancet.* 2015;386:5–7.

Naskalska A, Szolajska E, Chaperot L, Angel J, Plumas J, Chroboczek J. Influenza recombinant vaccine: Matrix protein M1 on the platform of the adenovirus dodecahedron. *Vaccine.* 2009;27:7385–93.

Pasqualini R, Koivunen E, Ruoslahti E. Alpha v integrins as receptors for tumor targeting by circulating ligands. *Nat Biotechnol*. 1997;15:542–6.

Perrone LA, Ahmad A, Veguilla V, Lu X, Smith G, Katz JM, Pushko P, Tumpey TM. Intranasal vaccination with 1918 influenza virus-like particles protects mice and ferrets from lethal 1918 and H5N1 influenza virus challenge. *J Virol*. 2009;83:5726–34.

Röhn TA, Ralvenius WT, Paul J, Borter P, Hernandez M, Witschi R, Grest P, Zeilhofer HU, Bachmann MF, Jennings GT. A virus-like particle-based anti-nerve growth factor vaccine reduces inflammatory hyperalgesia: Potential long-term therapy for chronic pain. *J Immunol*. 2011;186:1769–80.

Schmidt T, Dirks J, Enders M, Gartner BC, Uhlmann-Schiffler H, Sester U et al. CD4+T-cell immunity after pandemic influenza vaccination cross-reacts with seasonal antigens and functionally differs from active influenza infection. *Eur J Immunol*. 2012;42:1755–66.

Sebestik J, Niederhafner P, Jezek J. Peptide and glycopeptide dendrimers and analogous dendrimeric structures and their biomedical applications. *Amino Acids*. 2011;40:301–70.

Shin YC, Folk WR. Formation of polyomavirus-like particles with different VP1 molecules that bind the urokinase plasminogen activator receptor. *J Virol*. 2003;77:11491–8.

Simon C, Klose T, Herbst S, Han BG, Sinz A, Glaeser RM, Stubbs MT, Lilie H. Disulfide linkage and structure of highly stable yeast-derived virus-like particles of murine polyomavirus. *J Biol Chem*. 2014;289:10411–8.

Smith GE, Flyer DC, Raghunandan R, Liu Y, Wei Z, Wu Y, Kpamegan E, Courbron D, Fries LF 3rd, Glenn GM. Development of influenza H7N9 virus like particle (VLP) vaccine: Homologous A/Anhui/1/2013 (H7N9) protection and heterologous A/chicken/Jalisco/CPA1/2012 (H7N3) cross-protection in vaccinated mice challenged with H7N9 virus. *Vaccine*. 2013;31:4305–13.

Spohn G, Jennings GT, Martina BE, Keller I, Beck M, Pumpens P, Osterhaus AD, Bachmann MF. A VLP-based vaccine targeting domain III of the West Nile virus E protein protects from lethal infection in mice. *Virol J*. 2010;7:146.

Szebeni J, Storm G. Complement activation as a bioequivalence issue relevant to generic liposome development. *Biochem Biophys Res Commun*. 2015;468:490–7.

Szolajska E, Burmeister WP, Zochowska M, Nerlo B, Andreev I, Schoehn G et al. The structural basis for the integrity of adenovirus Ad3 dodecahedron. *PLoS One*. 2012;7:e46075.

Szurgot I, Szolajska E, Laurin D, Lambrecht B, Chaperot L, Schoehn G, Chroboczek J. Self-adjuvanting influenza candidate vaccine presenting epitopes for cell-mediated immunity on a proteinaceous multivalent nanoplatform. *Vaccine*. 2013;31:4338–46.

Thönes N, Herreiner A, Schädlich L, Piuko K, Müller M. A direct comparison of human papillomavirus type 16 L1 particles reveals a lower immunogenicity of capsomeres than virus like particles with respect to the induced antibody response. *J Virol*. 2008;82:5472–85.

Tumban E, Peabody J, Tyler M, Peabody DS, Chackerian B. VLPs displaying a single L2 epitope induce broadly cross-neutralizing antibodies against human papillomavirus. *PLoS One*. 2012;7:e49751.

Uhlenbeck O. A coat for all sequences. *Nat Struct Biol*. 1998;5:174–6.

van den Berg A, Dowdy SF. Protein transduction domain delivery of therapeutic macromolecules. *Curr Opin Biotechnol*. 2011;22:888–93.

Villegas-Mendez A, Garin MI, Pineda-Molina E, Veratti E, Bueren JA, Fender P, Lenormand JL. In vivo delivery of antigens by adenovirus dodecahedron induces cellular and humoral immune responses to elicit antitumor immunity. *Mol Ther*. 2010;18:1046–53.

Visciano ML, Tagliamonte M, Tornesello ML, Buonaguro FM, Buonaguro L. Effects of adjuvants on IgG subclasses elicited by virus-like particles. *J Transl Med*. 2012;10:4.

Vivès RR, Lortat-Jacob H, Chroboczek J, Fender P. Heparan sulfate proteoglycan mediates the selective attachment and internalization of serotype 3 human adenovirus dodecahedron. *Virology*. 2004;321:332–40.

Wang JW, Roden RB. Virus-like particles for the prevention of human papillomavirus-associated malignancies. *Expert Rev Vaccines*. 2013;12:129–41.

Warfield KL, Dye JM, Wells JB, Unfer RC, Holtsberg FW, Shulenin S, Vu H, Swenson DL, Bavari S, Aman MJ. Homologous and heterologous protection of nonhuman primates by Ebola and Sudan virus-like particles. *PLoS One*. 2015;10:e0118881.

Webster RG, Hinshaw VS. Matrix protein from influenza A virus and its role in cross-protection in mice. *Infect Immun*. 1977;17:561–6.

Wickham TJ, Mathias P, Cheresh DA, Nemerow GR. Integrins alpha v beta 3 and alpha v beta 5 promote adenovirus internalization but not virus attachment. *Cell*. 1993;73:309–19.

Wu DT, Roth MJ. MLV based viral-like-particles for delivery of toxic proteins and nuclear transcription factors. *Biomaterial*. 2014;35:8416–26.

Zeltins A. Construction and characterization of virus-like particles: A review. *Mol Biotechnol*. 2013;53:92–107.

Zinkernagel RM. On the role of dendritic cells versus other cells in inducing protective CD8+ T cell responses. *Front Immunol*. 2014;5:30.

Zochowska M, Paca A, Schoehn G, Andrieu JP, Chroboczek J, Dublet B, Szolajska E. Adenovirus dodecahedron, as a drug delivery vector. *PLoS One*. 2009;4:e5569.

Zochowska M, Piguet AC, Jemielity J, Kowalska J, Szolajska E, Dufour JF, Chroboczek J. Virus-like particle-mediated intracellular delivery of mRNA cap analog with in vivo activity against hepatocellular carcinoma. *Nanomedicine*. 2015;11:67–76.

32

In Vivo Sensing Using SERS Nanosensors

Janna K. Register, Andrew M. Fales, Hsin-Neng Wang,
Gregory M. Palmer, Bruce Klitzman, and Tuan Vo-Dinh

CONTENTS

32.1 Introduction

Surface-enhanced Raman spectroscopy (SERS) has emerged over the last several decades as an advanced tool for biomedical applications. Since its discovery in the late 1970s, the technique has continued to develop and be reenvisioned in many fields due to its versatility, sensitivity, and selectivity for obtaining molecular information. Prior to the year 2000, the scientific community had recorded just over 2000 publications on the topic related to SERS in various fields, but since then the publications have totaled more than 12,000. The use of SERS as an *in vivo* technique was not as widespread before the year 2000. Most of the biomedical advances for SERS prior to 2000 were related to the fields of bioanalysis, molecular biology, and medicine, with research focused on SERS detection of intracellular interaction [1,2], SERS in living cells as well as drug delivery and distribution in cells [3,4], SERS for the study of cell membrane components [5,6], detection of biomedical extracts [7,8], and the first SERS-based gene probes for *in vitro* biomedical diagnostics [9,10]. The advent of SERS for *in vivo* applications in small animals occurred around 2006 with the initial publications of *in vivo* SERS in small animals [11,12]. Since then, the potential for SERS for *in vivo* biomedical applications has been firmly recognized. From 2006 to 2015, the number of publications per year related to *in vivo* SERS work increased by an order of magnitude, underlining the excitement surrounding this topic. Some of the highlights of the last decade have involved SERS applications for *in vivo* cancer detection [13–15], multiplexed *in vivo* SERS imaging [16], and *in vivo* glucose sensing using SERS [17,18].

32.2 SERS in the Clinic

One of the most difficult hurdles for SERS and its applicability to the clinic has been the biocompatibility of nanoparticles and nanoparticle substrates. Widely believed to be the noble metal exhibiting the largest SERS enhancements, silver is known to be toxic to cells. Researchers have long known that experimenting with the size and shape of the nanostructures brings added enhancement; however, the traditional syntheses of these nanostructures incorporate many chemicals that are not biocompatible and can be difficult to wash out. Our laboratory has pioneered the development of a biocompatible synthesis for gold nanostars, a nanostructure that takes advantage of the electromagnetic enhancement around the sharp spikes as well as the enhancement found in the confined wells between the spikes [19–21]. We have also pushed the envelope of sensitivity by creating nanostars with a silver coating and biocompatible silica shell [22]. We were the first group to show the long-term SERS monitoring in large animals in 2015 [23] and believe that this technology has the capability to be used in a clinical setting.

32.3 SERS in Large Animals

In our large-animal study, the purpose was to create an implanted sensor that could be monitored transdermally for SERS signals. Generally, SERS nanoparticles (NPs) can be delivered into an animal via several ways: injection into the tail vein, direct injection into a tumor, or injection as solution beneath the skin. Another approach for long-term monitoring is to load the SERS-labeled NPs into a biocompatible matrix that would keep the NPs stationary, while inhibiting any foreign body response from interfering with the implant [23].

In our previous study, the SERS nanoparticles were embedded in the microarchitecture of the porous pHEMA hydrogel produced by PROFUSA, Inc. [23]. The interconnected pores promote vascularization throughout the implant, which helps to inhibit fibrous encapsulation of the implant caused by the foreign body response. Two-photon luminescence was used to image the same hydrogel, showing both the porous structure and the incorporation of the SERS nanostars into the pHEMA backbone of the implant. The hydrogels we used were developed for easy implantation and the small, flexible ribbons could fit inside a 12-gauge needle. The needle was loaded with the implant, and, after intradermal insertion, the ribbon was advanced using a trocar-assisted delivery through the inside of the needle, while the needle was removed. This careful placement of the implants in an intradermal location allowed for high-sensitivity SERS signals with minimal disruption to the implant site [23]. Images before (a), during (b), and after (c) insertion of the SERS implant into *ex vivo* human skin are shown in Figure 32.1. The needle is inserted bevel up loaded with the implant (Figure 32.1a). After insertion, the implant is interrogated in a location away from the insertion site particularly for *ex vivo* studies, ensuring the SERS signal is collected through the human skin (Figure 32.1c). The SERS implant was successfully implanted into a live large animal (pig) and monitored over time. Figure 32.2 shows the SERS signal after 1, 2, 3, and 6 days with an image of the implant being monitored. The signal decreased over time but was still detectable after 6 days, which indicates the possibility of long-term SERS measurements *in vivo*. SERS spectra were collected using a 30-mW, 785-nm laser with a fiber probe and 100 ms acquisitions. The dye used for the SERS signal was Cy7.

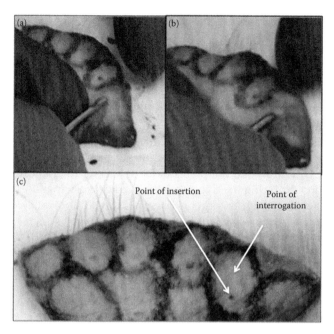

FIGURE 32.1
Intradermal implantation of porous hydrogels loaded with SERS-active nanostars into *ex vivo* human facial skin. (a) The needle placement with bevel up prior to insertion. (b) The needle after insertion. (c) The site with implant pointing out the insertion site where a small hole still exists, and the site of interrogation, which coincides with the location of the implant.

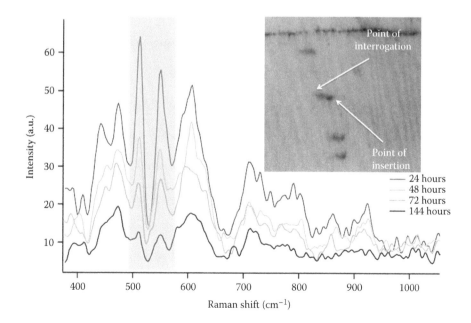

FIGURE 32.2
In vivo detection of SERS implant in a pig after 1, 2, 3, and 6 days. Inset image shows the SERS implant with point of interrogation and point of insertion after 2 days.

32.3.1 Active SERS Nanosensors in Large Animals

One of the long-term goals for *in vivo* SERS nanotechnology is to use advanced nanosensors to detect small molecules in the body over time. Our group has pioneered the development of a DNA-based nanosensor called a molecular sentinel [24]. And recently, we have improved the technology to include a novel "off-to-on" configuration called inverse molecular sentinel (iMS) [25]. In this technology, the SERS dye and plasmonic nanoparticle are separated by a custom-engineered DNA strand and held separate by a placeholder DNA strand that is hybridized to the sensor. In this configuration, the iMS is "off" and produces no SERS signal. However, when the target nucleic acid sequence is available, it will competitively hybridize to the placeholder, releasing the nanosensor to fold into a hairpin configuration, bringing the dye and nanoparticle close together and producing an "on" SERS signal, alerting that the target is present. It was important to show this kind of active SERS nanosensor and its capability of working *in vivo*.

A schematic of the iMS sensor concept is presented in Figure 32.3. Our group has also described the process in more detail elsewhere [25]. The technique has proven to be both sensitive and adaptable. Most recently, we have used the iMS technique for detection on a "Nanowave" chip [25–27] and multiplexed detection of microRNA [25,28]. In order to demonstrate the working capability of this type of sensor *in vivo*, we injected 50 μL of solution containing iMS sensor in the "off position" into the superficial dermis on the dorsum of an anesthetized large male Yorkshire pig (15–20 kg). The SERS signal was measured of the iMS "off" *in vivo* as a reference spectrum shown in Figure 32.4 (left). The injection site was then flooded with an additional 50 μL solution of noncomplementary target sequence (Figure 32.4, center) and the SERS signal was measured. Lastly, an additional injection site of iMS "off" solution was flooded with 50 μL of complementary target sequence and the SERS signal was measured (Figure 32.4, right). Figure 32.5 shows images of the process for nanosensor delivery and functionality testing. It

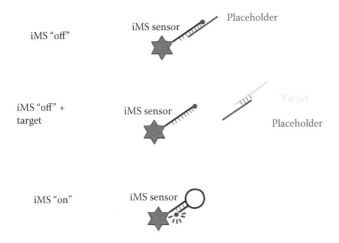

FIGURE 32.3
Schematic representing the process for the iMS "off-to-on" approach. The iMS sensor consists of the gold nanostar with nucleic acid sequence (black) that loosely binds to the placeholder nucleic acid strand (red). When the target sequence (green) is present, it competitively binds to the placeholder, freeing the iMS sensor portion. The iMS sensor then hybridizes to itself to form hairpin structure, which brings the Raman-active dye (blue) close to nanoparticle, which increases the Raman signal, turning the sensor "on."

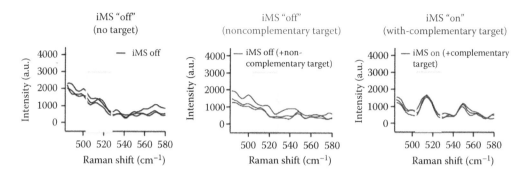

FIGURE 32.4

SERS spectra of the sensor recorded using a 785-nm laser at 28 mW with 10-s acquisition times. The Raman dye on the iMS sensor exhibits a peak at 513 cm⁻¹ when the sensor is turned on (highlighted in yellow box). The black spectra were background reference spectra after injection of 50 µL of iMS "off" sensor. The blue spectra were taken after that injection site of iMS "off" was flooded with 50 µL of noncomplementary target sequence. The red spectra were taken of an additional injection site of iMS "off" sensor that was flooded with 50 µL of complementary target sequence.

was evident that the iMS nanosensors worked correctly and were detectable *in vivo*. The SERS signal for the injection site that was flooded with noncomplementary target sequence shows no significant difference from the reference spectrum of the iMS "off" injection, while the SERS signal for the injection site flooded with the correct target sequence clearly shows the emergence of a SERS peak at 513 cm⁻¹. The SERS signals were measured using a 785-nm laser at 28 mW with 10-s acquisition times. The iMS "on" signal was apparent *in vivo* within 5 min of injection of the complementary target sequence. This proof-of-concept test was followed up by additional testing in our lab using internal reference sensors. Those data show semiquantitatively that the iMS can work *in vivo* to sense different concentrations of complementary target sequences and are published elsewhere [29].

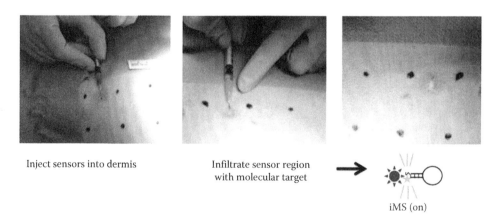

FIGURE 32.5

iMS sensor implanted in male Yorkshire pig. Left image shows injection of iMS sensor "off" while center image shows the additional injection of complementary target sequence to the same site. Right image shows injection site with complementary target and sensor "on."

32.4 Conclusion

We have demonstrated the proof of principle for *in vivo* use of SERS nanosensors. The SERS signals can be monitored *in vivo* through the dermis of a pig, which was used as the medical preclinical model for human skin, over the course of multiple days. Furthermore, we have shown that active SERS nanosensors can function in this environment as well. These results are the stepping stones laying the foundation for multiplexed long-term *in vivo* detection of specific nucleic acid targets, a biosensing capability that would be truly disruptive to the current state of early disease diagnosis with far-reaching implications for the biomedical community at large.

References

1. I. Chourpa, H. Morjani, J. F. Riou, M. Manfait. 1996. *FEBS Lett.* 397(1):61–64. DOI: 10.1016/S0014-5793(96)01141-6
2. I. R. Nabiev, K. V. Sokolov, M. Manfait. 1993. In: *Biomolecular Spectroscopy*, edited by R. J. H. Clark, R. E. Hester, Vol. 21, Chap. 7, Wiley, Chichester, pp. 267–338.
3. I. R. Nabiev, H. Morjani, M. Manfait. 1991. *Eur. Biophys. J.* 19(6):311–316.
4. G. D. Sockalingum, A. Beljebbar, H. Morjani, J. F. Angiboust, M. Manfait. 1998. *Biospectroscopy* 4(5):S71–S78. DOI: 10.1002/(SICI)1520-6343(1998)4:5+<S71::AID-BSPY8>3.0.CO;2-Z
5. K. V. Sokolov, N. E. Byramova, L. V. Mochalova, A. B. Tuzikov, S. D. Shiyan, N. V. Bovin, I. R. Nabiev. 1993. *Appl. Spectrosc.* 47:535.
6. I. R. Nabiev, G. D. Chumanov, R. G. Efremov. 1990. *J. Raman Spectrosc.* 21:49.
7. S. Nie, C. G. Castillo, K. L. Berghauer, J. F. R. Kuck, I. R., Nabiev, N. T. Yu. 1990. *Appl. Spectrosc.* 44:571.
8. K. V. Sokolov, S. V. Lutsenko, I. R. Nabiev, S. Nie, N. T. Yu. 1991. *Appl. Spectrosc.* 45(1):143.
9. T. Vo-Dinh, K. Houck, D. L. Stokes. 1994. *Anal. Chem.* 66(20):3379–3383. DOI: 10.1021/ac00092a014
10. T. Vo-Dinh, D. L. Stokes, G. D. Griffin, M. Volkan, U. J. Kim, M. I. Simon. 1999. *J. Raman Spectrosc.* 30(9):785–793. DOI: 10.1002/(SICI)1097-4555(199909)30:9<785::AID-JRS450>3.0.CO;2-6
11. J. A. Dieringer, A. D. McFarland, N. C. Shah, D. A. Stuart, A. V. Whitney, C. R. Yonzon, M. A. Young, X. Y. Zhang, R. P. Van Duyne. *Faraday Discuss.* 2006. 132:9–26. DOI: 10.1039/b513431p
12. G. R. Souza, C. S. Levin, A. Hajitou, R. Pasqualini, W. Arap, J. H. Miller. 2006. *Anal. Chem.* 78(17):6232–6237. DOI: 10.1021/ac060483a
13. X. M. Qian, X. H. Peng, D. O. Ansari, Q. Yin-Goen, G. Z. Chen, D. M. Shin, L. Yang, A. N. Young, M. D. Wang, S. M. Nie. 2008. *Nat. Biotechnol.* 26(1):83–90. DOI: 10.1038/nbt1377
14. A. Samanta, K. K. Maiti, K. S. Soh, X. J. Liao, M. Vendrell, U. S. Dinish, S. W. Yun, R. Bhuvaneswari, H. Kim, S. Rautela, J. H. Chung, M. Olivo, Y. T. Chang. 2011. *Angew. Chem. Int. Ed.* 50(27):6089–6092. DOI: 10.1002/anie.201007841
15. S. Keren, C. Zavaleta, Z. Cheng, A. de la Zerda, O. Gheysens, S. S. Gambhir. 2008. *Proc. Natl. Acad. Sci. U.S.A.* 105(15):5844–5849. DOI: 10.1073/pnas.0710575105
16. C. L. Zavaleta, B. R. Smith, I. Walton, W. Doering, G. Davis, B. Shojaei, M. J. Natan, S. S. Gambhir. 2009. *Proc. Natl. Acad. Sci. U.S.A.* 106(32):13511–13516. DOI: 10.1073/pnas.0813327106
17. J. M. Yuen, N. C. Shah, J. T. Walsh, M. R. Glucksberg, R. P. Van Duyne. 2010. *Anal. Chem.* 82(20):8382–8385. DOI: 10.1021/ac101951j
18. D. A. Stuart, J. M. Yuen, N. S. O. Lyandres, C. R. Yonzon, M. R. Glucksberg, J. T. Walsh, R. P. Van Duyne. 2006. *Anal. Chem.* 78(20):7211–7215. DOI: 10.1021/ac061238u
19. H. Yuan, C. G. Khoury, H. Hwang, C. M. Wilson, G. A. Grant, T. Vo-Dinh. 2012. *Nanotechnology* 23:075102.

20. H. Yuan, A. M. Fales, C. G. Khoury, J. Liu, T. Vo-Dinh. 2013. *J. Raman. Spectrosc.* 44:234–239.
21. H. Yuan, C. G. Khoury, C. M. Wilson, G. A. Grant, A. J. Bennett, T. Vo-Dinh. 2012. *Nanomedicine* 8:1355–1363.
22. A. M. Fales, H. K. Yuan, T. Vo-Dinh. 2014. *J. Phys. Chem. C* 118:3708–3715.
23. J. K. Register, A. M. Fales, H.-N. Wang, S. J. Norton, E. H. Cho, A. Boico, S. Pradhan, J. Kim, T. Schroeder, N. A. Wisniewski, B. Klitzman, T. Vo-Dinh. 2015. *Anal. Bioanal. Chem.* 407(27):8215–8224.
24. M. B. Wabuyele, T. Vo-Dinh. 2005. *Anal. Chem.* 77(23):7810–7815. DOI: 10.1021/ac0514671
25. H.-N. Wang, B. M. Crawford, A. M. Fales, M. L. Bowie, V. L. Seewaldt, T. Vo-Dinh. 2016. *J. Phys. Chem. C* 120(37):21047–21055. doi: 10.1021/acs.jpcc.6b03299
26. T. Vo-Dinh. 1998. Surface-enhanced Raman spectroscopy using metallic nanostructures. *TrAC, Trends Anal. Chem.* 17:557–582.
27. C. Khoury, T. Vo-Dinh. 2012. Nanowave substrates for SERS: Fabrication and numerical analysis. *J. Phys. Chem. C* 116:7534–7545. DOI: 10.1021/jp2120669
28. H. T. Ngo, H. Wang, A. M. Fales, et al. 2016. *Anal. Bioanal. Chem.* 408:1773. doi:10.1007/s00216-015-9121-4
29. H.-N. Wang, J. K. Register, A. M. Fales, N. Gandra, E. H. Cho, A. Boico, G. M. Palmer, B. Klitzman, T. Vo-Dinh. 2017. SERS nanosensors for *in vivo* detection of nucleic acid targets in a large animal model. *Nat. Biomed. Eng.* Submitted.

Index

A

Printed and bound by CPI Group (UK) Ltd, Croydon, CR0 4YY

01/11/2024

01782604-0018